APPLICATIONS OF GENE-BASED TECHNOLOGIES FOR IMPROVING ANIMAL PRODUCTION AND HEALTH IN DEVELOPING COUNTRIES

Applications of Gene-Based Technologies for Improving Animal Production and Health in Developing Countries

Edited by

HARINDER P.S. MAKKAR

and

GERRIT J. VILJOEN

Animal Production and Health Section,
Joint FAO/IAEA Division of Nuclear Techniques in Food and Agriculture,
International Atomic Energy Agency, Vienna, Austria

A C.I.P. Catalogue record for this book is available from the Library of Congress.

ISBN-10 1-4020-3311-7 (HB)
ISBN-10 1-4020-3312-5 (e-book)
ISBN-13 978-1-4020-3311-7 (HB)
ISBN-13 978-1-4020-3312-4 (e-book)

Published by Springer,
P.O. Box 17, 3300 AA Dordrecht, The Netherlands.

www.springeronline.com

Printed on acid-free paper

Printed in the Netherlands.

PREFACE

Advances in modern biotechnology have provided powerful tools for the identification, mapping, characterization, isolation and manipulation of genes. They provide the keys for understanding and solving many problems associated with animal production and health, and offer exciting opportunities for enhancing agricultural productivity and food security.

Currently, the focus of biotechnological research is on issues and problems of significance for livestock producers and consumers in the developed world. In order to address the problems facing livestock owners in developing countries and to fully realize the benefits from gene-based technologies, there is a need to identify, characterize and apply appropriate gene-based technologies in and for these regions. To this end, an International Symposium was organized by the Joint FAO/IAEA Division of Nuclear Techniques in Food and Agriculture in cooperation with the Animal Production and Health Division of FAO.

The year 2003 marked the 50th anniversary of the discovery of the double helical structure of DNA by Watson and Crick. Given the numerous developments that have taken place in the "-omics" technologies, we thought it opportune to take stock of what has and what could be achieved using these modern scientific tools to maximize the benefits from the so-called "livestock revolution" taking place in many developing countries.

The Symposium comprised a plenary session and four thematic sessions, covering animal breeding and genetics, animal health, animal nutrition and the environment. Other sessions covered ethics, safety and regulatory aspects of gene-based technologies, and there were three thought-provoking panel discussions. One hundred and thirty representatives from 60 Member States of IAEA and FAO attended the Symposium, with a good balance between private and public sector institutions, policy-makers and international assistance agencies. Stimulating debates developed, both from the floor and during more informal sessions. The Symposium received generous support from the private sector, for which the sponsoring Organizations are most grateful.

This publication provides a compilation of peer-reviewed scientific contributions from authoritative researchers attending the Symposium. Valuable information is provided on the role and potential of gene-based technologies for improving animal production and health, the possible applications and constraints in the use of these technologies and the specific research needs and constraints in developing countries. It is hoped that these Proceedings will be of benefit as a reference source for researchers, students and policy-makers, and help bridge the gap between developed and developing countries in the understanding of applications of gene-based technologies in animal agriculture and in exploring future roles of such technologies in the developing world.

James Dargie

Director
Joint FAO/IAEA Division of
Nuclear Techniques in Animal
Production and Health,
IAEA

Samuel Jutzi

Director
Animal Production and Health
Division,
FAO

EXECUTIVE SUMMARY

Research aimed at improving animal production and health has been revolutionized by recent developments in biotechnology, particularly those involving gene-based technologies. These have generated not only new opportunities for knowledge creation but also new options for solving established and emerging problems.

The underlying theme of this Symposium was the exploitation of gene-based technologies for use in the livestock sector in developing countries. In this regard, consideration was also given to inherent problems in developing countries, given their weak research base and lack of training, which in turn limits their ability to effectively utilize gene-based methods. The contrasts between the resources available to developing and developed countries were clearly evident. However, even in technologically advanced countries, there are relatively few examples of novel biotechnological products that can demonstrate key advantages to people and communities. That is not to say that the potential of gene-based technologies were understated, only that tangible proof of practical use was lacking. The main problem seemed to be one of ensuring large-scale, sustained funding of such technologies in the livestock arena. It did, however, become clear that less-developed countries might benefit even more than developed countries from advances in genetic research. Nevertheless, a concerted effort is needed to understand real needs and to make adjustments for the longer developmental time scales needed. It is clear that genetic analyses are already well established, or at least readily achievable in most developing countries, although genomic analysis still rests with a few specialized centres in the developed world.

The scientific information presented at this symposium was of a nature that elicited much excitement regarding the potential applications for gene-based technologies, particularly in those field that have matured more, notably in animal disease diagnostics and therapeutics, rather than in animal production. Developing countries could thus acquire more economic benefits in the short term by focusing research and development work on animal health. In the area of animal production, benefits are more likely to accrue in the medium to long term. The timing of the Symposium was opportune in the face of various recent disease epidemics and the continuing debate on the "livestock revolution" that is expected to address global food requirements and influence trading patterns.

1.1   Animal production

1. The genetic manipulation of grain crops and tropical forages to increase the efficiency of nutrient availability to animals and to decrease phosphorus and nitrogen emission into the environment offers considerable potential in increasing livestock productivity while protecting the environment. This could be achieved through molecular manipulation, transgenesis, and site-directed or conventional mutation breeding.

2. International and national bodies should further promote the use of quantitative methods for polymerase chain reaction (PCR) and oligonucleotide probe-based analysis for reducing methane, countering the effects of antinutritional factors, monitoring gut pathogens associated with food safety, improving utilization of feed resources, and in identifying new pro- and pre-biotics and antibiotic substitutes for addition to feeds in developing countries.

3. The practical applications of genomic studies on rumen microbes are currently unpredictable, but could involve industrial production of key lignocellulytic enzymes for pre-treatment of fibrous residues. Rumen genomic studies might provide insight into the nutrient requirements of these organisms and this knowledge could be employed in practical feeding strategies. Because of present uncertainties and the high cost of genomic research, developing countries should preferably not initiate such studies, but should rather direct resources to enhancing capacity in the area of bio-informatics, since sequence information will become publicly available.

4. In developing countries, the use of genotype information is likely to be more useful in marker-assisted introgression than in selection within breeds. Introgression of a single mutated gene is relatively easy and holds promise for livestock improvement in developing countries and should thus be targeted. Some single-gene mutations are known and information on many more is likely to become available in the near future. A model for introgression is the Booroola gene of the Garole

breed, as demonstrated in India and which could be replicated in other countries.

5. Locally adapted breeds should be characterized for nutritional resilience, behaviour, metabolic rates, energy expenditure, muscle and bone structure, feed conversion efficiencies, female reproductive performance and lactation, protein turnover, mobilization of energy for lactation, etc. In low-input systems in developing countries, complete phenotypic and pedigree information is often not available except in some intensive breeding units. The real value of marker information will thus be under-utilized. A proper recording system must also be in place if the full potential of marker-assisted selection for breed improvement is to be realized.

6. If transgenic animals were to gain acceptance in developing countries, then gene constructs and transgenic technology should be used to modify locally adapted breeds rather than introducing elite transgenic breeds from developed countries. It should be noted that transgenesis is unlikely to be successful for genes encoding traits governed by gene networks, although it is feasible, and should be pursued in the case of single-gene traits.

1.2 Animal health

1. The increasing demand for and trade in animals and animal products is fuelling a livestock revolution in developing countries and posing increased health problems, to both animals and man. Gene-based technologies are playing an increasingly important role in addressing such problems.

2. Major advances have been made in elucidating host-pathogen interactions, diagnosis, molecular epidemiology and predictive epidemiology of veterinary diseases, as well as in vaccine production, through the exploitation of gene-based technologies. Major progress has also been made in the sequencing of animal genomes.

3. Different solutions are required to solve problems of animal health and food safety in developed versus less-developed countries. For the latter, more appropriate exploitation of gene-based technologies could be identified. Developed countries tend to regard developing countries as sources of disease and a threat. There is also a large differential in the attitudes towards various diseases according to the state of economic development. The prospect for increased trade is a major factor in promoting gene-based methods and research in developing countries. More appropriate targeting of research by developed countries is needed

to help produce more relevant vaccines and diagnostic methods for transfer to developing countries.

4. Training is an essential pre-requisite to allow pure and applied research, and, by extension, to produce appropriate reagents, therapeutics and vaccines to deal with problems peculiar to developing countries.

5. Molecular tools such as PCR and micro-arrays have driven developments, particularly within the fields of disease diagnosis and control. Molecular epidemiology based on sequence comparisons is now commonplace, as also are real-time and forensic analyses of diseases and pathogens.

6. Gene-based technologies allow for more direct, non-invasive diagnostic methodologies within a short time frame. An increasing number of faster and more sophisticated molecular diagnostic techniques based on PCR are available for the detection of cellular and viral pathogens. Sensitive and specific techniques will be more readily usable at the site of infection, but the challenge remains of ensuring validation to ascertain true diagnostic sensitivities. Care is needed to ensure that existing serological tests remain in use in tandem with molecular methods.

7. Gene-based technologies have already made a major impact on animal vaccinology. As an illustration, widespread studies are in progress on the use of naked DNA vaccination. For control programmes, it is important that newly designed vaccines include markers to distinguish vaccinated from infected animals. The concept was first tested using DNA viruses whose genomes could easily be manipulated using molecular techniques. Reverse genetics of viral pathogens can provide new immunogenic constructs that act as marker vaccines for many RNA viruses. The development of vaccines against invertebrate disease vectors so as to directly attack the source of the transmission and spread of many important human and veterinary pathogens, offers exciting possibilities. Major constraints on vaccine development occur where there is a high rate of mutation and selection of pathogens, requiring the availability of multivalent vaccines. The physical and environmental stability of such vaccines must also be assured.

1.3   Education, intellectual property rights, ethics, biosafety and related aspects

1. Technologies and expertise should be transferred to developing countries through capacity building programmes whose aims and focus are driven by the countries themselves. Appropriate modalities for achieving this must be determined. Equally important is institutional

strengthening for strategic planning, priority setting and policy development for identification of appropriate gene-based technologies and their applications. Regional and subregional approaches should be used to enhance cooperation, capacity building and sharing of knowledge.

2. Establishing and strengthening the networking of groups in developing countries for research and development work in "modern biotechnology" and application of gene-based technologies will provide the necessary impetus to address the emerging challenges for enhancing animal productivity and health. Increased contact between academics and industry is essential, and is one of the main prerequisites to becoming more competitive in biotechnology. Initiatives to stimulate such interactions should be encouraged and actively supported.

3. There is an urgent need to involve industry in the commercialization of research products, especially in the early stages where venture capital plays an important part in the developed world.

4. Intellectual property rights (IPR) and the use of patented products and procedures are important issues to consider with any application of genomic-based procedures worldwide, including in developing countries. There is a need for the international research community to coordinate their activities efficiently in order to address issues of IPR – issues that have often limited collaboration among public and private institutions.

5. Issues of ethics and biosafety need to be addressed at the institutional and national levels, but harmonization should be at inter-governmental or international level.

6. Gene-based technologies are unlikely to have an impact if applied in isolation. Programmes that work to understand and improve the full socio-economic system of animal production and marketing are more likely to succeed than those that deal with individual components of the system.

7. International funding agencies should assist and support countries to ensure that gene-based technologies are deployed within the framework of national development aspirations.

ACKNOWLEDGEMENTS

The Symposium was organized under the aegis of an Organizing Committee, comprising J.D. Dargie (IAEA), I. Hoffmann (FAO), M.H. Jeggo (CSIRO, Australia), H.P.S. Makkar (IAEA), O. Perera (IAEA), A. Diallo (IAEA), A. Cannavan (IAEA) and J.R. Crowther (IAEA). Their contribution and that of FAO/IAEA support staff before, during and after the Symposium is gratefully acknowledged. Special thanks to Ms Roswitha Schellander and Ms Karen Morrison for their support in organizing the Symposium.

The contributions in this Proceedings have been peer reviewed and we thank all the evaluators.

The Animal Production and Health Section of the Joint FAO/IAEA Division of Nuclear Techniques in Food and Agriculture thanks FAO and IAEA, and the following companies for their support in association with the Symposium:

AMEX Export Import GmbH.
Biacore AB.
Biological Diagnostic Supplies Limited.
Laboratory Supply Company GmbH & Co. KG.
Nunc A/S.
Szabo-Scandic Handels GmbH & Co. KG.
VWR International.

Final editing for consistency of language and style and preparation for publication was by Thorgeir Lawrence, Reykjavík.

DISCLAIMERS

CONTENTS

FAO/IAEA INTERNATIONAL SYMPOSIUM ON *APPLICATIONS OF GENE-BASED TECHNOLOGIES FOR IMPROVING ANIMAL PRODUCTION AND HEALTH IN DEVELOPING COUNTRIES*

Summary, conclusions and recommendations

1.   INTRODUCTION

The symposium was held from 6 to 10 October 2003 in Vienna. One hundred and thirty scientists and decision-makers from 60 Member States participated in the Symposium. A total of 44 oral and 33 poster presentations were made. The programme consisted of opening addresses, an opening session to set the scene and four scientific sessions covering, respectively, animal breeding and genetics; animal health; animal nutrition; and environmental, ethical, safety and regulatory aspects of gene-based technologies. There were also three panel discussions. In the opening address session, three distinguished speakers (Werner Burkart, DDG and Head of the Department of Nuclear Sciences and Applications, IAEA; Samuel Jutzi, Director, Animal Production and Health Division, FAO; and James Dargie, Director, FAO/IAEA Division of Nuclear Applications in Food and Agriculture) presented their views. Mr Burkart stressed the importance of the close relationship between FAO and IAEA for enabling the exploitation and deployment of nuclear technologies in food and agriculture. Mr Jutzi stressed the challenges and opportunities faced by animal agriculture globally, and emphasized the importance and nature of specific and general development policy measures for enhancing the impact

H. P. S. Makkar and G. J. Viljoen (eds.), Applications of Gene-Based Technologies for Improving Animal Production and Health in Developing Countries, 1-21.
© 2005 *IAEA. Printed in the Netherlands.*

of gene-based technologies in animal agriculture in developing countries. Mr Dargie emphasized the need for training, technical support and capacity building in developing countries for enabling the application of gene-based technologies in key areas of the livestock sector.

2. OPENING SESSION: SETTING THE SCENE FOR THE INTERNATIONAL SYMPOSIUM

Samuel Jutzi
Director, Animal Production and Health Division, FAO, Rome

Two keynote addresses were presented: *A vision of gene-based technologies for the livestock industries in the third millennium* by E.P. Cunningham, and *Challenges and opportunities for controlling and preventing animal diseases in developing countries through gene-based technologies* by M.H. Jeggo and J.R. Crowther. The salient points that emerged from this session are summarized below.

2.1   On the technological front

1. The vast expansion of knowledge in the last ten years in the development of new techniques for isolating, amplifying, reading and inserting DNA, the up-scaling and automation of those techniques, and their coupling with information management suggests the continued and accelerated achievement of breakthroughs in the exploration and management of gene function.

2. One of the largest impacts of gene-based technologies on livestock sector so far, given the large economic significance of feed in livestock production, has been through their application in maize and soybean transformation (reduced production costs, reduced mycotoxin levels, higher P- and N-digestion efficiencies, and better balance of amino acids).

3. Non-contentious and fast advancing gene-based technology frontiers are those that relate to genome mapping, exploration of genetic diversity, DNA typing, fingerprinting, genetic tracking, marker technology and functional genomics. Transgenesis is of little concern as long as it is only used for vaccine generation. The need to further strengthen efforts in the production of more efficient vaccines was, however, stressed.

4. Public concerns about animal transgenesis for purposes related to growth and nutritional performance, health management and pharmaceutics or organ supply for human medicine, as well as the application of genetically modified growth hormones, has tended to slow both technological progress and the market introduction of such products and processes.

2.2   In the functional context

Gene-based biotechnology requires an appropriate policy and institutional framework to exploit its full potential. Emphasis should be on developing relevant and effective livestock sector policies, and on development of institutions, organizations and services.

3.   SESSION I: GENE-BASED TECHNOLOGIES APPLIED TO LIVESTOCK GENETICS AND BREEDING

Oswin Perera and Harinder Makkar
Animal Production and Health Section, Joint FAO/IAEA Division,
IAEA, Vienna

Four keynote addresses were presented. These focused on (1) approaches, opportunities and risks of molecular genetics and livestock selection; (2) technologies for germline manipulation; (3) state of the world's animal genetic resources; and (4) integration of gene-based methods and reproductive technologies for genetic improvement of livestock. Molecular genetic and genomic technologies have been applied to gene detection, genetic selection and assessment of genetic diversity and genetic transformation of livestock for the past 20 years. Research has taken place mostly in the developed world and these technologies and their applications are still a young science, but consideration must be given to adapt these technologies to meet the specific needs of developing countries. The paper on the state of world's animal genetic resources revealed a wide gap between developed and developing world in research, development and applications of gene-based technologies. Developing countries should be prepared to receive and fully exploit these technologies.

1. Genetic improvement of livestock is already very effective using conventional technologies, and new technologies, such as use of molecular genetic markers, must compete with these conventional

approaches. Molecular genetic markers are recognized as being useful and cost-effective technologies that can be incorporated into genetic improvement programmes to improve efficiency or rate of genetic gain, or both, but they do not in themselves provide large extra genetic gains and nor do they substitute for existing methods of genetic improvement. There are opportunities for application in the developing world, but the difficulties in terms of needs for advanced training and development of technical capacity should not be underestimated.

2. Marker assisted selection is mainly useful for traits where phenotypic measurement is less valuable because of low heritability, sex-limited expression, non-availability before sexual maturity, or unavailable without sacrificing the animal (e.g. slaughter traits). Traits such as feed intake or disease resistance may also be expensive or difficult to measure, and information on genotype might be useful in selecting for these traits. Genotypic information has extra value in the case of early selection and where within-family variance can be exploited. Reproductive technologies usually lead to early selection and more emphasis on between-family selection. DNA marker and reproductive technologies are therefore highly synergistic and complementary. In the low-input systems existing in developing countries, complete phenotypic and pedigree information is often not available, except in some intensive breeding units. Under these scenarios it would be more difficult to realize the value of the marker information, and it would be harder and more expensive to determine linkage in the case of using linked markers. A proper recording system must be put in place if the full potential of gene-based technologies for breed improvement is to be realized.

3. There are opportunities for application of molecular genetic approaches in the developing world. Rather than exploiting existing quantitative trait loci (QTLs) through within-breed selection, a more likely scenario for developing countries will be that valuable QTLs will be introgressed from one population into another. In developing countries there is a huge variation across breeds, much of it being useful to exploit in genetic improvement programmes. This includes the variation coming from "foreign" breeds in developing countries. Indigenous breeds may contain valuable QTLs, but could benefit from upgrading through crossing with superior exotic breeds. Alternatively, valuable QTLs could be introgressed from exotic breeds. Examples are the Booroola gene in the Garole breed in India, having a moderate and desirable effect on number of lambs weaned, and a number of genes affecting resistance to endemic local diseases. There are many cases of QTLs found in crosses of extreme breeds, and a number of those will be candidates for introgression. In developing countries, use of genotype information is

therefore probably going to be more useful in marker-assisted introgression (MAI) rather than in selection within breeds. Also, in the case of MAI, reproductive technologies will be beneficial because they can help increase the number of animals with the desired genotype.

4. Information is currently available for a number of "direct markers" such as myostatin affecting double muscling in cattle, calipyge doing the same in sheep, and Booroola affecting fecundity. Many of these mutations have a major phenotypic effect. Even if the genetic marker were a direct marker, its effect on phenotype would have to be estimated for the population and the environment in which it is used. The effects of individual genes are likely to be dependent upon the background, which must also be characterized. Phenotypic information may be less used in actual selection in the future, but recording of phenotype will be continuously needed for the purpose of monitoring the QTL effect (retrospectively), and genetic change over time.

5. Genetically modified (GM, transgenic) livestock have not yet entered commercial production. The technology of producing GM livestock remains expensive and difficult, but recent breakthroughs promise substantial reductions in cost and increases in efficiency in the near future. Similarly, increasing knowledge of mammalian and avian genetics and genome function is opening new avenues that could be of economic and social value in both developed and developing countries. Opportunities to create disease resistance seem particularly promising, but applications for livestock improvement in developing countries are likely to be medium to long term rather than short term.

6. The ongoing FAO survey of the State of the World's Animal Genetic Resources has already revealed several consistent demands from developing countries for biotechnology applications. Demands for breed characterization, for cryopreservation of livestock germplasm and for application of embryo transfer technologies were highlighted. There was little expressed demand for advanced genetic technologies, and the gap between the developed and developing worlds in such technologies was clear. Where demand for advanced biotechnologies was expressed, there was little evidence of an underlying strategic plan. These results identify needs for capacity development and technology access in advanced biotechnologies and strategic planning.

7. The country reports within the State of the World's Animal Genetic Resources also confirm observations that changes in status and risks to indigenous livestock genetic resources in West Africa are very rapid, and that several breeds are already extinct and others are under severe threat. Interventions will therefore need to be rapid and comprehensive to prevent permanent loss of a substantial proportion of existing genetic

resources. Global surveys and characterization of livestock diversity, within which molecular diversity is a component, should be completed as an essential starting point for informed actions on conservation and utilization. International support should be provided to accomplish this. An international panel should also be established to review and recommend procedures for characterization and conservation of animal genetic resources. The question of when and how *in vitro* conservation techniques should be applied needs to be defined and specific recommendations provided to users.

8. Gene-based technologies and genetic improvement in general are unlikely to have any impact if applied in isolation. Rather, it is critically important to work with local communities to identify what are the real needs and opportunities for achieving a positive impact on people's lives, and then to work with those communities to test and apply improvements. Programmes that work to understand and improve the full socio-economic system of animal production and marketing are more likely to succeed than programmes that are limited to individual components of the system.

4. SESSION II: GENE-BASED TECHNOLOGIES APPLIED TO PATHOGENS AND HOST-PATHOGEN INTERACTIONS

Paul-Pierre Pastoret,[1] Thomas Barrett[1] and Esteban Domingo[2]
1. *Institute for Animal Health, Compton Laboratory, Compton, Newbury, Berks, UK*
2. *Centro de Investigacion en Sanidad Animal (CISA), 28130 Valdeolmos, Madrid, Spain*

Gene-based technologies applied to animal health must fall in line with the general trends and the needs of animal production and health in developing countries. As indicated in the talk given by Samuel Jutzi, an increasing demand for animal products is foreseen in developing countries over the next 20 years – the so-called "livestock revolution". This increased demand will no doubt be accompanied by an increase in the exchange of animals and animal products, with consequent greater risks of transboundary spread of disease. There are therefore likely to be major problems concerning food security in developing countries, not forgetting the many problems associated with food safety.

As exemplified by the recent outbreaks of foot-and-mouth disease and classical swine fever in Europe, the control of OIE List A diseases in developed countries is still achieved through the implementation of sanitary policies that include mass slaughtering of infected and in-contact animals. One cannot expect such measures to be adopted in developing countries where food security has not yet been achieved. In any event, public opinion in developed countries is increasingly concerned with animal welfare and the ethical justification of a mass slaughter policy; this has led to the new concept of "vaccination for life". Additional biological facts relating to developing countries must be taken into account, such as the high level of animal biodiversity in some of them and the existence of wildlife reservoirs of infections and infestations, which can be a source of new emerging diseases in the human and domestic animal populations. Compared with wildlife species, the existence of domestic livestock is a recent event in evolutionary terms, and therefore the domestic animal populations have had little time to adapt to certain pathogens. Coupled with this is the fact that the evolutionary rate of viruses is of another order of magnitude compared with that of their host species, and leads to the establishment of quasi-species populations with multitudes of variant genomes. To date, a total of about 3600 viral species and their variants have been described.

Animal health is an important component of sustainable agriculture; however, biosecurity – the need to protect the environment and maintain biodiversity – must also be taken into account. By the end of the last century, molecular biology had made outstanding progress. Molecular tools – PCR, micro-array technology, etc. – had all made a substantial impact in the field of disease diagnosis and control. The complete genome sequences of the majority of relevant viruses, as well as those of numerous bacteria, are now available, and the completion of the human genome sequence has been achieved. Genome sequence information will soon be available for most of the important livestock species, including poultry, pigs and cattle. These advances will also affect the use of gene-based technologies in animal health in developing countries, and this will need to be taken into account in future studies on host–pathogen interactions (interactome).

4.1 Impact of gene-based technologies on animal health

The impact of gene-based technologies on animal health has primarily been on infectious or parasitic diseases, and these have made important contributions in four main areas:

- host–pathogen interactions;
- diagnosis;
- molecular epidemiology and predictive epidemiology; and

- vaccines.

Nowadays, the three disciplines of epidemiology, diagnosis and vaccinology are more closely linked than ever, since one of the main advances in veterinary vaccinology has been the development of marker vaccines, obtained by recombinant DNA technology. These marker vaccines, used for "vaccination for life", must, however, be linked with suitable companion diagnostic tests. Epidemiological studies will then indicate if vaccination campaigns are necessary and, if so, whether or not they are successful.

It is worth keeping in mind that these three disciplines, when translated into practice, have no negative impact on animal welfare, animal rights or animal integrity; in fact, the use of vaccines should improve animal welfare.

4.1.1   Host-pathogen interactions

New gene-based technologies such as micro-arrays and knowledge of genomic sequences of both the pathogen and the host have allowed more refined pathogenesis studies to be carried out, i.e. the "interactome" between gene products of the pathogen and those of the host.

It is becoming obvious that many viruses, microbes and parasites are very "clever" at evasion, they have co-evolved with their hosts and therefore it can sometimes be very difficult to find the best way to fight them. This brings to mind a well-known paper by Rolf Zinkemagel entitled *Immunology taught by viruses* and no doubt they will also "teach" us much about pathogenesis.

4.1.2   Diagnosis

Serological, indirect methods of diagnosis of infectious diseases are still very useful but have some drawbacks, such as the lag period between the initial infection and the appearance of detectable amounts of antibodies. Gene-based technologies are allowing us to increasingly use more direct non-invasive diagnostic methodologies in a more rapid time frame.

There are an increasing number of faster and more sophisticated molecular diagnostic techniques based on PCR available for the detection of cellular and viral pathogens. These are suited to surveillance of pathogens in environmental samples (water, soil, etc.), in food products and in biological specimens from humans, animals and plants. Nucleic acid fingerprinting and sequencing can be applied to the identification of pathogens (species, types, subtypes, variants, etc.). Portable instruments enabling this work to be carried out in the field are also becoming available.

For diagnostic applications in developing countries, the main recommendations are to:

- adopt technologies that use fluorescent probes and accessible bio-informatics programmes;
- develop diagnostic tools that can be used in the field (pen-side technologies);
- train personnel in the techniques to ensure proper interpretation of positive and negative results; and
- establish networks of epidemio-surveillance laboratories that function in an integrated manner.

4.1.3 Epidemiology

Molecular epidemiology is one of the most powerful applications of gene-based technology in animal health. Bearing in mind that viruses, especially RNA viruses, are quasi-species, that is to say highly variable, it is very important to be able to rapidly identify and characterize them to enable effective control measures to be implemented without undue delay. A further development could be the new science of predictive epidemiology using sequence data, and also micro-arrays, to search for new pathogens. As exemplified by several recent examples (Hendra, Nipah, SARS, etc.), wildlife can act as reservoirs for numerous potentially emerging, often zoonotic, infections. It would now be feasible to check for potential disease risks in wildlife using highly conserved regions of nucleic acid sequence in known viral genomes.

For application to epidemiological surveillance in developing countries, two main recommendations emerge:

- It is essential to implement surveillance of the genotypic or antigenic types and variant forms of the pathogens to be controlled (see also under *Diagnosis*). Phylogenetic methods to identify lineages and to trace the origin of pathogens and quantitative relatedness among different forms of the same pathogen should be used systematically, notably for predictive epidemiological purposes.
- Studies on wildlife should be undertaken, especially in regions with very high animal biodiversity

4.1.4 Vaccinology

Compared with other technologies, vaccines offer the single most cost-effective measure to control or even eradicate an infectious disease, as exemplified by the eradication of smallpox, the near eradication of rinderpest and the foreseen elimination of wildlife rabies from mainland Europe. There are no broad spectrum antivirals yet available and so, for most viral diseases, vaccines are the only cost-effective solution to disease control, apart from some cases where no vaccine is available and slaughtering may be the only control option (e.g. African swine fever), especially when confronted with zoonotic pathogens, as was the case with the Hendra and Nipah virus infections in domesticated species.

Whatever the technology used in their production, vaccines fall into two categories: they are either attenuated or inactivated products. The immunological response in the vaccinated animal depends on whether the vaccine is attenuated or inactivated, with attenuated vaccines generally giving more effective and long-lasting immunity.

Gene-based technologies have already had a major impact in animal vaccinology, the most spectacular example being the demonstration that it is now possible to vaccinate using naked DNA. For the first time this makes it possible to vaccinate animals against viruses that cannot be cultured *in vitro*. Nevertheless, vaccines are not mere scientific concepts, they must work effectively to protect the animal against the clinical signs of the infection or, better still, against infection itself and, clearly, their efficacy does not depend on the level of sophistication used in their production. New vaccines produced using gene-based technology must either show an improvement in comparison with vaccines made using the available technologies, or at least be equivalent in effectiveness to the classical products. The history of vaccinology shows us that each new technology finds its own application niche.

One of the presentations related to the production of a vaccine against East Coast Fever using genomic and post-genomic technologies. Until now, the only available vaccination protocol for this disease was infection with a fully virulent *Theileria parva* followed by treatment to reduce the clinical disease. This is reminiscent of the early days of variolation to protect against smallpox. The availability of safe vaccines as a result of gene-based technologies would be a major improvement in such cases.

To facilitate control programmes, and for economic reasons, it is important that newly designed vaccines used for vaccination campaigns include markers in order to distinguish vaccinated from infected animals. The concept was first tested using DNA viruses whose genomes could easily be manipulated using molecular techniques to produce marker vaccines. Reverse genetics of viral pathogens can now provide new immunogenic constructs that, if properly used, can act as effective marker vaccines for many viruses with RNA genomes. Another important development is the possibility we now have to devise vaccines against invertebrate vectors (ticks, insects, etc.) that are responsible for the transmission and spread of many important human and veterinary pathogens.

Due to the highly dynamic nature of the genomes of many of the pathogens that need to be controlled, and also the genetic polymorphism of vector species where these are involved in transmission, new vaccines may have their limitations.

An important feature to be considered in vaccine design is environmental stability. Several of the newly engineered vaccines (e.g. the vaccinia-rabies recombinant) are sufficiently stable to be used in developing countries, which may have only modest storage and distribution infrastructure.

Finally, intellectual property rights (IPR) and the use of products and procedures already patented is an important issue to consider with any application of genomic-based procedures worldwide, including in developing countries.

For application to vaccine design and production in developing countries, the main recommendations are:

- Vaccines should be multivalent in order to elicit ample protection against a range of related forms of pathogen, wherever antigenic diversity has been documented. For specific geographical areas, it may be necessary to tailor a vaccine according to the antigenic strains encountered in a defined area.

- Special attention should be paid to the stability of the vaccine under adverse storage and delivery conditions.

- IPR issues should be discussed at a specific meeting involving not only scientists but also representatives of industry, patent lawyers and specialists in international trade.

5. SESSION III: GENE-BASED TECHNOLOGIES APPLIED TO PLANTS, RUMEN MICROBES, AND SYSTEMS BIOLOGY

C.S. McSweeney[1] and Harinder Makkar[2]
1. CSIRO Livestock Industries, Queensland Biosciences Precinct, Carmody Rd., St. Lucia, Brisbane, Australia
2. Animal Production and Health Section, Joint FAO/IAEA Division, IAEA, Vienna

This session focused on the role of gene-based technologies in improving nutrient supply and utilization by the animal. The session dealt with three main components: (1) forage nutrient supply, (2) efficiency of digestion by gut micro-organisms, and (3) nutrient-tissue interactions in the animal. Each of these areas was addressed by a series of plenary papers. Research undertaken in each of these areas should be considered in terms of the potential impact on the animal. An integrated approach combining these elements is required to achieve an improvement in nutritional status of animals. For example, attempts to modify the nutritive quality of forage should take into account the effect on efficiency of rumen fermentation and the flow of nutrients for absorption and utilization by the animal. Similarly, optimizing nutrient–gene interactions for production traits must take into account which forages are likely to produce these benefits after they have been digested by gut micro-organisms, and nutrients absorbed by the animal.

5.1   Forage nutrient supply

The paper in this component demonstrated that gene-based technologies are well advanced in programmes to develop transgenic forages or use molecular genetics tools to breed and select for forages with specific nutritive qualities. Molecular assisted breeding and transgenics could currently be used to:

- Reduce the concentration of compounds that retard digestion, especially:
    - lignin (in grasses);
    - tannins (in legumes); and
    - toxins (in legumes).
- Optimize the concentration of desirable compounds, including:
    - proteins resistant to rumen degradation;
    - sulphur-containing amino acids; and
    - soluble carbohydrates (e.g. fructose).
- Enhance resistance to diseases and drought.

It was also apparent that applications of gene-based technologies towards these objectives are well advanced compared with animal approaches, which are technically more difficult, have much longer generation times and are affected by more complex ethical issues.

In the forage plant area, current work focuses mainly on temperate species, with little effort in tropical plants, but, comparatively, the effort in forage plants is miniscule compared with grain-based crops for human consumption. One major reason for this bias is that return on investment from grain crops is huge compared with that from forage plants. Also, altering structural characteristic, such as lignin composition, and protective chemicals, like tannins, may compromise survival mechanisms of tropically adapted plants. Substantial progress could be made in improving nutritive value of tropical forages through conventional breeding and selection. However, gene-based technologies would be particularly useful in identifying the genes or QTL associated with desirable nutritive traits. An example of improving nutritive value in a plant used for forage could be molecular marker-assisted breeding for desirable digestibility characteristics in fibrous residues from the major grain crops in the tropics, i.e. rice, sorghum, maize and millet.

5.2   Efficiency of digestion by gut micro-organisms

A paper on genomics of rumen bacteria outlined the exciting and rapid progress that is being made in sequencing the entire genomes of representatives of the predominant rumen bacteria. This work is partially completed, and will soon result in publicly available databases containing the annotated sequences from these organisms. However, research will need to continue to assign activity to genes of unknown function, which could represent a large proportion of these genomes. Already it has been demonstrated that the cloned cellulase genes from some of the cellulolytic bacteria only represent a small fraction of the genes encoding for this function based on the genome sequence analysis. During discussion on this paper, several points were emphasized.

Genomics of rumen micro-organisms – especially fibre-degraders – is of critical importance to developing countries that rely on low-quality roughages and crop residues as basal feeds.

Genomic studies of fibrolytic bacteria in the short to medium term will be predominantly in developed countries, since the expertise and equipment required will not be available to developing countries in the time frame considered.

Developing countries will be able to engage in this gene-based technology through bio-informatics studies, since sequence information will be publicly available and accessible to developing country scientists. However, scientists from developing countries currently are not skilled in this rapidly expanding area of biology.

Practical application of information from such research is currently unclear but could involve industrial production of key lignocellulytic enzymes for pre-treatment of fibrous residues. Also, additional information could be generated from genomics studies that provide insight into the nutrient requirements of rumen and gut organisms that in turn could be employed in practical feeding strategies.Several papers, including a presentation in another session, identified the greatest current demand and potential application of gene-based technologies in developing countries as being the need for molecular-based tools for studies of gut microbial ecology. During the discussions, it was clear that quantitative methods for PCR and oligonucleotide probing based on analysis of the small sub-unit ribosomal genes are in demand, and already being employed by some scientists from developing countries. These methods are affordable and

within the current capacity of molecular biology laboratories in such regions. There was unequivocal support for the application of molecular ecological tools as they will be essential for developing strategies to:
–   reduce methane;
–   counter the effects of anti-nutritional factors;
–   predict rumen microbial responses to bio-active compounds;
–   improve the utilization of varying feed resources through a better understanding of the responses of major functional groups to these nutrients; and
–   monitor gut pathogens associated with issues of food safety. These tools will need to be employed if nutritional research in the animal sciences is going to make progress in the future.

A paper was presented on the development of a genetically modified pig, which expressed a transgenic phytase into the digestive tract via the saliva. This was an elegant demonstration of the progress that is being achieved in animal transgenesis. These animals were shown to be healthy and could digest and absorb more phosphorus; in addition the gene insert (or trait) was shown to be stable over several generations. The biological success of this project and its potential use in developing countries was overshadowed by the reluctance of several countries to engage in the development or use of transgenic animals. It was also recommended that if transgenic animals gained acceptance in developing countries, then the gene constructs and transgenesis technology should be used to modify locally adapted breeds rather than introduce elite transgenic breeds from developed countries.

The main conclusions and recommendations on transgenics for improved digestion were that:
•   limited current applications exist due to consumer and government resistance;
•   the genetic constructs should be used in indigenous breeds rather than introducing unadapted high producing exotic genotypes; and
•   it was unlikely that there would be any success when using traits governed by gene networks, but the use of single gene effects were feasible (e.g. phytase).

5.3   Nutrient-tissue interactions in the animal

One of the papers in this session described in detail the potential impact of nutrient supply on host-tissue metabolism, production performance and on gene expression. This is an emerging and complex area of animal science, but the first experiments are now being conducted on the influence of nutrients on gene expression; on animal tissue function and muscle development; and on the programming of gene expression by the nutritional environment in the developing foetus. The conclusions were that gene-based technologies (e.g. marker-assisted selection) used in combination with conventional breeding and selection programmes could be used to make rapid progress in developing superior phenotypes for production traits. The priority area identified for developing countries was determining the characteristics associated with survival and superior productivity during seasonal nutrient deprivation, which is experienced in many developing countries, especially in the tropics. Issues related to product quality (tenderness, marbling etc.) were not considered currently important for poverty alleviation.

The following strategy was outlined as an approach for developing ruminant animals that show superior resilience during nutrient deprivation:

- characterize locally adapted breeds for "nutritional resilience" using phenotypic attributes, e.g. adaptive behaviour, metabolic rate, muscle and bone structure;
- identify the genes (using quantitative trait loci (QTLs) or single nucleotide polymorphisms (SNPs)) that account for significant variance in weight loss and compensatory growth during periods of nutrient deprivation followed by improved nutrition; and
- consider the possible positive and negative implications of selecting for "nutritional resilience", e.g. toughness, feed conversion efficiency, and female reproductive performance and lactation.

The need to obtain insight into nutritional influences on growth rate, immune competence, stress responsiveness, reproduction and *in utero* effects was stressed. The potential implications of this rapidly emerging field of research within the agricultural context specifically for livestock production systems, are that the management of the breeding herd, especially during pregnancy, could have long-term impacts on health, growth rate, reproductive efficiency, milk and wool yield of the offspring. All these in turn could have a significant economic impact.

Developing countries would benefit greatly from the development of basic reagents and tools, including protocols for detection, expression and transfer of genes; training; and improvements in laboratory infrastructure to undertake research based on gene-based technologies. An exponential growth in molecular tools in the last decade should not discourage scientists in developing countries from using them to solve problems relevant to the developing world. Promoting the networking of groups in developing countries for research and development work in "modern biotechnology" will bring the necessary impetus for addressing the emerging challenges of providing adequate, safe and good nutrition to livestock, while conserving the environment and utilizing available resources in a sustainable manner.

Table 1. Status of gene-based technologies and factors influencing their applicability.

	Relative cost	Research capacity and progress		Feasibility	Public and government acceptance	Atractive-ness
		Developing country	Developed country			
Forage plants						
• Molecular selection and breeding	Moderate	Low	High	High	High	Moderate
• Transgenics	High	Low	High	Low	Low	Moderate
Rumen and gut microbes						
• Ecology	Low	Moderate	High	High	High	High
• Genomics	High	Low	Moderate	High	Moderate	Moderate
Animal nutrition–genetic interaction						
• Marker and gene assisted selection	Moderate	Low	High	High	High	High
• Transgenics	High	Low	Moderate	Low	Low	Moderate
Forage plants	Moderate	Low	High	High	High	Low

6. PANEL DISCUSSION I: WHICH GENE-BASED TECHNOLOGIES ARE MOST LIKELY TO SUCCEED IN ENHANCING ANIMAL PRODUCTIVITY IN DEVELOPING COUNTRIES?

John Donelson
Department of Biochemistry, University of Iowa, Iowa City, Iowa 52242, USA

The panel focused primarily on three gene-based technologies. Dr Rod Mackie (University of Illinois, USA) described the analyses of specific genes in the microbial flora that populate gastrointestinal tracts and discussed how the expression of these genes can affect the productivity and health of livestock. Dr John Gibson (ILRI, Kenya) discussed the identification and analysis of various genetic parameters of biodiversity and how these parameters can potentially be used to maintain and exploit diversity in livestock breeding programmes. Dr John Egerton (University of Sydney, Australia) described his experience in developing a recombinant vaccine for footrot in sheep and goats, and some lessons learned from its application in the developing world. The conclusions and recommendations from the discussion session following these three presentations are given below.

1. Microbial populations of the gut exhibit a great diversity of organisms, ecology, evolution and function. Many of these gut microbes have yet to be identified and characterized because of difficulties in culturing them *in vitro*. Further understanding of their inter-relationships in the gut and their functions for their livestock hosts is required for improving the nutritional status of livestock in developing countries. Since much of the analysis conducted to date has been on gut microbes of livestock in developed countries, it is recommended that gut microbes of livestock in developing countries be given a high priority for future studies.

2. The breadth of biodiversity amongst livestock breeds existing in the world today is very large and should be taken into account in any long-term breeding programme. In many cases, the inclusion of existing breeds of livestock into a breeding programme will provide greater benefits than breeding for specific traits within a breed. It is recommended that this large diversity of existing traits amongst breeds be considered whenever possible in breeding programmes for livestock in developing countries.

3. New and better vaccines, some of which will come from the application of gene technology, are needed to improve the health and productivity of livestock in the developing world. Finding the correct antigens for these vaccines requires knowledge of the pathogenesis of a wide spectrum of diseases, and so the research involved in identifying these antigens will be long term and costly. Adoption of improved vaccines by poor farmers will depend on containment of vaccine cost. Also, it is essential that animal health infrastructure exists at the local level to ensure accurate disease diagnosis and to disseminate information about the correct use of vaccines in the control of disease. Despite these obstacles, vaccines are a proven way to successfully combat infectious diseases and it is recommended that their development against livestock diseases of the developing world be both continued and accelerated.

4. The gap between "a practical problem in search of a solution" and "a technique in search of an application" can only be closed by those individuals who clearly understand both the problems and the techniques. Hence, meetings such as this symposium, which are attended both by individuals familiar with livestock problems in the field and by scientists actively using gene-based technologies, are essential means of communication. It is recommended that similar symposia be held in future years and that developing countries around the world continue to be well represented at the meetings. Equally important is the identification of funding opportunities to hold regional

meetings (e.g. in West Africa or in East and Central Africa) that bring together livestock scientists of developing countries who share common research objectives.

5. Gene-based technologies are not a "silver bullet" for solving all livestock issues of developing (or developed) countries. They are, however, one of many different approaches to be considered and utilized. The first gene was cloned about 25 years ago and many of the associated technologies are still in their infancy. Ultimately, these technologies have the potential to be of major benefit to owners of livestock in the developing world. Thus, it is recommended that gene-based technologies continue to be considered within the context of the broad spectrum of possible solutions to problems faced by livestock owners in developing countries.

6. Priorities should be determined by the needs of livestock owners in the many different regions and environments where livestock are kept. In each country, policies and procedures should be established for setting such priorities. The challenge and the recommendation of this Panel are to establish mechanisms that ensure gene-based technologies are widely available and properly understood so that they may be incorporated into these policies and procedures when appropriate.

7.   SESSION IV: ROLE OF INTERNATIONAL ORGANIZATIONS AND FUNDING AGENCIES IN PROMOTING GENE-BASED TECHNOLOGIES IN DEVELOPING COUNTRIES
AND
PANEL DISCUSSION 2: GENE-BASED TECHNOLOGIES IN ENVIRONMENT, FOOD SAFETY AND ANIMAL INDUSTRY, AND RELATED ETHICAL AND INTELLECTUAL PROPERTY RIGHT ISSUES

John Hodges
Lofererfeld 16, A-5730 Mittersill, Austria

Actions should follow the procedure of identifying a need and then looking for a solution, not the reverse: i.e. not "We have a gene-based technique – how can we use it?" This principle of starting with the need is evident in the attitude to using gene-based technologies for biodiversity. If we start by thinking about the technology we are led to conclude that it can be used for genome mapping at the molecular level and thus identify diversity between breeds of livestock. Enormous investment of funds and time would be needed to analyse so many genomes or to select those thought to be critically endangered or similar and to put the results into a database.

The right way is to identify the need, i.e. conservation of livestock biodiversity. Gene-based technology enables us to meet that need without genome analysis. Simply collect and store, in secure locations and by using long-term methods, the cells of each breed, which contain the record of molecular diversity. When the need arises in the future, samples can be sequenced for the differences in bases to be revealed. In this way the objective of conservation is achieved quickly and relatively cheaply.

7.1 Identification of needs in developing countries

Parameters that should be considered include:
- poverty;
- reduction of human labour in caring for livestock;
- local energy; and
- improved local food production.

The local socio-economic and cultural values must be included, and address:
- – stakeholder involvement in identifying needs and in decision-making;
- – cultural sensitivity to understand local values; and
- – ethical standards, applicable in both donor and recipient countries.

This action probably needs to be addressed on the basis of groupings of countries or livestock farming systems, or both, as needs will vary.

7.2 Design and search for gene-based solutions applicable to the livestock sector

This will require characterization of the broad categories of existing gene-based options for livestock.
- Characterization must recognize the contrast in needs between farmer families and intensive units near urban centres.
- There must be awareness of which technologies are contentious and which are non-contentious.
- Classify existing gene-based technologies according to:
  - – opportunities – matching the needs for which they may offer solutions;
  - – capabilities – defining the extent to which each technology actually can address an identified need;

- dangers – recognizing dangers, and whether they are science-based, social-based, economics-based, ethics-based or insensitive to local values;
- costs, time frames and scale of benefits; and
- whether local, regional or global in scope.

A prerequisite for the design and search for appropriate gene-based solutions is to identify needs of developing countries, as mentioned above.

The broad categories of existing gene-based options to be characterized include:

- animal health: diagnosis, protection and treatment;
- ruminant and non-ruminant nutrition and metabolism;
- reproduction enhancement;
- transgenic livestock;
- germline manipulation;
- gene-based trait selection;
- molecular analysis of genetic diversity; and
- animal identification and traceability.

This is not a comprehensive list, and each item could be subdivided.

7.3 New gene-based technologies suited to identified needs

Novel and currently unavailable gene-based technologies, that might be designed to address some of the identified needs, should be sought. Emphasis should be placed upon simpler gene-based technologies, such as those that are small-scale, cost-effective and use local resources. For example, to provide energy to relieve the heavy labour undertaken mainly by women in caring for livestock.

7.4 Special study of genomic impacts upon the environment, food chain and human genome

- Understanding the genetic linkages between species that are involved in the animal food chain when innovative gene-technologies are either introduced or contemplated.
- Clear recognition of the difference in dangers between "substantial equivalence applying to DNA" and the outflows from later-generation gene-based technologies in terms of proteomics and functional genetics when metabolic and physiological changes are the target.
- Documentation of cases in which there are interactions between a transgene and the recipient animal. Several important examples were

cited in the Symposium, such as the response of the Boran breed to trypanotolerant genes from N'Dama cattle.

7.5   New approach to risk assessment

- We need to move on from the bland scientific statement that "There is no evidence that …" – implying that therefore there is no risk.
- Recognition of uncertainty and unknown risks in gene-based technologies must be given a new foundation based upon anticipation and testing, not solely using probabilities.

Probabilities are not good enough when a catastrophe may happen. Fears of unknown and unexpected risks of a science-based nature have been recognized recently by the National Research Council of the National Academies of the United States of America (*Animal Biotechnology: Science Based Concerns.* See: http://www.nap.edu/books/0309084393/html/) and a similar recent report from the UK makes the same points (Food Ethics Council, UK, 2003, *Engineering Nutrition: GM crops for global justice.* See: http://www.foodethicscouncil.org/library/reportspdf/gmnutritionfull.pdf).

7.6   IPR and TRIPS, and safety regulations for gene-based technologies with animals

Developing countries need assistance in several fields:
- better understanding the present and developing situation regarding IPR and TRIPS for livestock using gene-based technologies;
- obtaining resources and advice on the design of national IPR legislation; and
- designing national legislation for safety, testing and approval of new food products from animals.

7.7   Education and training

Training and education in gene-based technologies should focus on younger scientists through network links.

OPENING ADDRESS

W. Burkart
Deputy Director General, Department of Nuclear Sciences and Applications, IAEA.

Distinguished Delegates, Ladies and Gentlemen,

On behalf of the Directors General of the International Atomic Energy Agency and the Food and Agriculture Organization of the United Nations, and on my own behalf, I have great pleasure in welcoming you to this International Symposium on *Applications of Gene-Based Technologies for Improving Animal Production and Health in Developing Countries*.

At the outset, let me briefly highlight the mandate and objectives of the International Atomic Energy Agency, and in particular, the mission of my Department, the Department of Nuclear Sciences and Applications.

The Agency serves as the world's foremost intergovernmental organization for scientific and technical cooperation in the peaceful uses of nuclear technology, safety and verification. The Agency was established as an autonomous organization of the United Nations in 1957, with a statutory mandate "to accelerate and enlarge the contribution of atomic energy to peace, health and prosperity throughout the world", as well as to improve nuclear safety and safeguard the non-proliferation of nuclear weapons. Thus, the Agency assists its 137 Member States in the use of nuclear technology, promotes radiological and nuclear safety; and verifies, to the extent possible, that nuclear materials are not diverted away from legitimate peaceful uses to military purposes.

The Agency works to foster the role of nuclear science and technology in support of sustainable human development. In the field of Nuclear Sciences and Applications, this involves both advancing knowledge through research and disseminating this knowledge to Member States through technical cooperation to tackle pressing worldwide challenges for mankind – hunger

23

H. P. S. Makkar and G. J. Viljoen (eds.), Applications of Gene-Based Technologies for Improving Animal Production and Health in Developing Countries, 23-25.
© 2005 *IAEA. Printed in the Netherlands.*

and malnutrition, disease, degradation of natural resources, and climate change.

The Agency implements research through its Research Contract Programme. It supports and coordinates research through international networks of scientists from developing countries and senior advisors from both developed and developing countries. Where appropriate, the Agency facilitates the transfer of nuclear technology to Member States through its Technical Cooperation Programme. Interregional, regional and national projects in food and agriculture, human health and nutrition, industry, environmental studies and other applications are implemented in Member States. Many of these projects contribute directly or indirectly to the goals of sustainable development and protection of the environment, as set out in Agenda 21 of the 1992 UN Conference on Environment and Development.

The research and technical cooperation activities are supported by two laboratories, located in Seibersdorf, Austria, and in Monaco. These laboratories provide scientific and analytical services to research projects, and training and quality assurance services in the area of technical cooperation.

This Symposium, which I have the privilege to open today, is convened by the Animal Production and Health Sub-programme of the Joint FAO/IAEA Division of Nuclear Techniques in Food and Agriculture. Livestock have played a pivotal role in human development since time immemorial. In the twenty-first century, the role of livestock has been further enhanced. A "Livestock Revolution" is now taking place and it is projected that, by 2020, developing countries will consume 100 million tonnes more meat and 223 million tonnes more milk than they did in 1993. Considerable evidence exists that the rural poor and landless, especially women, get a higher share of their income from livestock than better-off rural people. In addition, livestock provide the poor with draught power and manure, along with the opportunity to utilize resources that do not compete with human food, and to diversify income. The Livestock Revolution could well become a key means of alleviating poverty in the next 20 years. Currently, around 1.2 billion people live on less than US$ 1 a day, and 800 million people are malnourished and go to bed hungry.

The FAO report on *World agriculture: towards 2015/2030* and the discussions in the Scientific Forum at the World Food Summit in Johannesburg clearly highlight the role and importance of modern biotechnology in addressing food security problems. This Symposium could not have been convened at a better time, since it marks the 50th anniversary of the discovery by Watson and Crick of the double helical structure of DNA. Numerous exciting developments have taken place since then in the fields of genomics and transgenesis. This Symposium will provide an interactive environment to:

- discuss the role and future potential of gene-based technologies for improving animal production and health,
- identify constraints in the use of these technologies in developing countries and how to use them in simple practical ways appropriate for developing countries,
- identify and prioritize specific research needs,
- explore the possibility of international coordination in the area of biotechnology in animal agriculture, and
- examine ethical, technological, policy and environmental issues and the role of nuclear techniques in the further development and application of genetic manipulation with respect to livestock.

This Symposium is unique in the sense that it focuses on the relevance of, and the need for, gene-based technologies in developing countries.

I am sure that you will take advantage of this important gathering for the exchange of information and ideas in an interdisciplinary setting. I am confident that this meeting, during which 37 oral contributions and some 33 posters will be presented and three exciting Panel Discussions will be conducted, will help bridge the current gap in the knowledge and application of gene-based technologies between developed and developing countries.

The challenges are significant, but I am confident that you will succeed in your objectives. I wish you a very pleasant stay in Vienna, and a productive, scientifically rewarding and successful meeting.

I declare this Symposium open.

OPENING ADDRESS

Samuel C. Jutzi
Director, Animal Production and Health Division, FAO, Rome, Italy

1.   INTRODUCTION

I am very pleased to have the opportunity to address this important animal biotechnology symposium on behalf of FAO's Division for Animal Production and Health. I am most grateful to the Joint FAO/IAEA Division of Nuclear Techniques in Food and Agriculture for having called this Symposium. We at FAO in Rome look to our sister Division in Vienna when it comes to advice and support for biotechnology applications in agriculture and their transfer into practical use. This involves a leading role for this Division in directing FAO's biotechnology work; my colleague, Jim Dargie, Director of the Joint Division, is the accepted leader of the corporate FAO Working Group on Biotechnology, and I wish to thank him here for his efforts in this function, and also for helping to maintain the high profile of the animal biotechnology agenda in FAO.

2.   FOCUS ON GENE-BASED TECHNOLOGIES

A deliberate effort was made to focus this Symposium on more recent *Gene-based Technologies for Improving Animal Production and Health in Developing Countries*. More conventional biotechnology areas, some of which have had certain commercial success in developing countries, are therefore left aside.

H. P. S. Makkar and G. J. Viljoen (eds.), Applications of Gene-Based Technologies for Improving Animal Production and Health in Developing Countries, 27-32.
© 2005 *IAEA. Printed in the Netherlands.*

3. DIFFERENTIATION BETWEEN BIOTECHNOLOGIES IN THE PLANT AND ANIMAL DOMAINS

Such focus is well justified in the attempt to advance the technical discussion. This focus also serves to differentiate the animal biotechnology work from the crops world, which is strongly dominated by transformation technology – to an extent that will possibly never be the case in the animal sector. Animal cells generally lack the totipotency of plant cells. This precludes the use of very low frequency transformation methods such as the ones used in plant transformation. In addition, reproduction technologies are less developed than in the plant biotechnology sector. Such characteristics restrict the potential for rapid and large-scale market penetration by GM genotypes as has occurred for crops. However, the way the world's animal agriculture is evolving and expanding, there are not only unprecedented challenges to deal with, but there are also unprecedented opportunities to exploit, including in the biotechnology sphere.

4. DEVELOPMENT OF THE LIVESTOCK SECTOR

Globally, livestock production currently accounts for about 43 percent of the gross value of agricultural production. In developed countries this share is more than half, while in developing countries it accounts for one-third of agricultural production. This latter share, however, is rising quickly following rapid increases in livestock production as a result of population growth, urbanization, changes in life styles and dietary habits, and increasing disposable incomes. The total demand for animal products in the developing countries is expected to more than double by 2030. As a result, the share of livestock production in total agricultural output is growing rapidly and is expected to have passed the 50 percent threshold by 2020. Important to note for the assessment of the market potential of new technologies in the animal sector is that the annual incremental output of the livestock sector is almost three times that of the crop sector.

Satisfying the increasing and changing demands for animal food products, while at the same time sustaining the natural resource base, is one of the major challenges facing world agriculture today. Significant among these challenges are:
– a geographic shift of livestock production from temperate and dry areas to warmer, more humid and disease-prone environments,

- a decreasing importance of ruminant vis-à-vis monogastric livestock species,
- a substantial rise in the use of grain-based feed,
- a change in livestock production practices from a local multi-purpose activity into a market-oriented and increasingly integrated process, and
- more large-scale, industrial production located close to urban centres, with associated environmental and public health risks.

Facing and managing these challenges raises a number of substantive global and national public policy issues that will have to be addressed. Broadly, these encompass issues associated with equity and poverty alleviation, the environment and natural resource management, and public health and food safety.

5. BIOTECHNOLOGY OPPORTUNITIES IN ANIMAL AGRICULTURE

Agricultural biotechnology has long been a source of innovation in production and processing profoundly affecting the livestock sector; however, this impact has been and continues to be primarily notable in animal agriculture in developed countries, while the adoption of even early-generation biotechnology in the livestock sector of developing countries tends to be far less. However, with the market demand for food of animal origin growing dynamically in many developing countries, there will be a commercial, often industrial, livestock subsector emerging in many of these countries. This subsector is likely to more readily pick up modern biotechnology options than the traditional small-scale subsector.

In order to arrive at some meaningful conclusions with respect to possible policy guidance for the deployment of animal biotechnologies in support of the overall development objectives, it may be useful to go a little further into the analysis of the forces driving design and use of such technologies.

Investments in animal biotechnology research are determined largely by market demand or market size, by technology opportunities, and by the ability of a research establishment to capture the economic benefits from research, the so-called "appropriability". This appropriability is determined by the use of patents, trademarks and trade secrets, and by exerting market power, all mechanisms for protecting intellectual property. Private, but increasingly also public, research tends to be biased toward those production systems, commodities or technologies that favour such appropriation opportunities. In international livestock development these are the intensive, increasingly industrialized, poultry and swine production systems that offer huge returns to investments because of their amenability to such profit appropriation. High growth rates and reproductive efficiencies are factors that give poultry and swine a decisive edge when it comes to investment decisions. As much of the expansion of the world's poultry and swine production is located in or even transferred to developing countries due to the easy opportunity to externalize environmental costs, profit opportunities are further enhanced in the frequent absence of respective policy and its enforcement.

The anticipated expanding application of molecular technological products and processes will contribute to significantly enhanced livestock productivity and production in the commercial, fast expanding, capital-intensive subsector in developing countries. The traditional livestock subsector, and particularly livestock-dependent poor farmers, are likely to benefit much less from the output of such investments. The current dichotomy between the modern and the traditional subsectors is therefore likely to be exacerbated.

6.   OPPORTUNITIES TO STEER DEVELOPMENTS IN FAVOUR OF SMALL-SCALE AND POOR FARMERS

The challenge, therefore, is how to make sure that modern biotechnology, applied to livestock, can help enhance agricultural productivity in developing countries in a way that reduces poverty, improves food security and nutrition, and respects sustainable use of natural resources while promoting rural development. We know that most of biotechnology R&D activities are conducted by large private companies for commercial exploitation and are designed to meet the requirements of developed markets; we also know that the gap between the industrialized and developing countries in technical expertise and relevant capacities is widening.

Two areas of policy design and enforcement in support of the desired outcomes are singled out: (1) biotechnology specific, and (2) livestock sector development oriented policies.

6.1   Biotechnology specific

There are perhaps only two ways to expand biotechnology R&D for the benefit of the poor: First, allocate additional public resources to biotechnology research that promises large social benefits, i.e. with a focus on "commercially orphaned" products, traits and processes. Second, expand private-sector research for the poor by converting some of the social benefits of research for the private sector, in the form of innovative public-private partnerships. The public sector can entice the private sector to develop technologies for the poor by offering up-front to buy the exclusive rights to newly developed technology and make it available either for free or for a nominal charge to small-scale farmers [similar to J. Sachs' proposal for developing vaccines for tropical diseases]. A variant thereof might also be the philanthropy-oriented donation of private sector resources for such research and development work.

Such biotechnology-specific interventions need to be enhanced by the design and implementation of coherent national strategies and capacity building plans for biotechnology development, and by the enforcement of enabling trade policies with respect to biotechnology products and processes.

6.2   Livestock sector development policy

Agricultural problems are multidisciplinary in nature and biotechnology in isolation is unlikely to solve them. Agricultural biotechnology is only one tool in addressing poverty and food security. Ultimately, the reduction of poverty and related malnutrition and hunger requires political solutions. Technology applications by developing country farmers are often hampered by limited access to appropriate capacity building, delivery systems, extension services, productive resources and markets, as well as by poorly developed rural infrastructure. Biotechnology is no "quick fix" for the infrastructural, political and institutional constraints and policies that are required to facilitate the incorporation of smallholders into commercial production. This implies fundamental revision of policies that tend to favour large-scale, industrial livestock production through artificial economies of scale that are enabled by the externalization of negative environmental impacts. Linking small-scale producers vertically with larger-scale marketers and processors would combine the environmental and poverty-alleviation benefits of small-scale livestock production with the economies of scale that derive from larger-scale processing. Regulatory systems are required that are compatible with international best practices to ensure compliance with agreed biosafety and food safety standards for consumer protection.

And, finally, there are clear public concerns expressed over potential negative consequences of biotechnology in general, and of genetic engineering products and processes in particular. These concerns are for human and animal health and welfare, the socio-economic and the bio-physical environment, procedures for risk assessment, management and communication – all these need to be observed insofar as they exist or need to be elaborated and negotiated where incremental guidance is required. With respect to the assessment of products and processes of gene technology, FAO supports a science-based evaluation that would objectively determine benefits and risks of each product or process. To quote FAO's Statement on Biotechnology: "This calls for a cautious case-by-case approach to address legitimate concerns for the biosafety of each product or process prior to its release".

I am convinced that this conference will enhance our collective knowledge on how gene-based technologies can best help improve animal production and health in developing countries, and I would like to thank all those who will contribute to this effort.

OPENING ADDRESS

James Dargie

Director, Joint FAO/IAEA Division of Nuclear Techniques in Food and Agriculture, Vienna, Austria

Ladies and Gentlemen,

Like the speakers before me, I would like to welcome you all to this Symposium, and to say how pleased I am that we have been able to attract such a high quality field of livestock scientists and managers from all over the world. I believe we share a common vision – namely of ensuring that new scientific and technological developments are channelled in appropriate directions and with sufficient intensity to grasp the opportunities available through the livestock sector for eliminating the hunger and malnutrition that still afflict so many of our societies. I also wish to record my appreciation of the support given to this Symposium by the private sector, which has enabled us to channel much more resources than would otherwise have been possible into ensuring that so many researchers and decision-makers from developing countries could be here this week.

The presence of Dr Jutzi and others from FAO headquarters at this Symposium is not simply symbolic of the existence of a Division operated jointly between FAO and the IAEA that deals with many different aspects of food and agriculture. In fact, it is yet another example of two Divisions with staff that have a particularly close and effective track record of working together to help countries to analyse and solve the challenges they face in improving the performance of their livestock sectors. So, as we set about our work this week, bear in mind that although this meeting has been largely organised by the Joint Division and is taking place in Vienna, the lessons and recommendations from it will also be duly carried to FAO, and most importantly, I hope to your own countries and institutions.

Dr Jutzi has given us an excellent entrée to the topic of this Symposium. He has painted the big picture in terms of global trends within the livestock

H. P. S. Makkar and G. J. Viljoen (eds.), Applications of Gene-Based Technologies for Improving Animal Production and Health in Developing Countries, 33-40.
© 2005 *IAEA. Printed in the Netherlands.*

sector, the opportunities it presents and the challenges it faces if it is going to play its full part in meeting people's expectations for diets that are not simply nutritionally better, but offer variety, quality and safety. These are expectations – indeed rights – common to essentially everyone whether they live in a modern city environment, or in a village in the Sahara or high Andes, and no power in the world can change it.

The problem is that, despite many positive trends, over a billion people are living on less than $ 1 a day, and around 850 million are malnourished, and that over the next 30 years there will be an additional 2 billion people to feed. Most of these people are and will be in South Asia and sub-Saharan Africa – the very parts of the world where the International Livestock Research Institute and FAO have recently shown from their poverty-livestock mapping studies that livestock intersect most forcefully with the poor. While this certainly helps to focus policy, institutional and research agendas, it raises two questions: first, what tools do we have now, or will be needed in the future, to intervene at the points of highest relevance to the people involved in the production and marketing systems existing in these regions; and, second, how do they gain access to them?

In attempting to answer these questions, we must bear in mind that not only do the majority of the poor engage in integrated crop-livestock or rangeland systems, but also the reality that their assets are often not simply in cattle – the main focus of so much research in the past. They are also in crops, grasslands, sheep, goats, pigs and poultry, and in varieties, landraces, breeds or types that are indigenous to the regions, countries or localities concerned. We also cannot neglect the clear trend towards expansion of peri-urban and landless systems to cater for the demands of increasingly urbanized populations, and the opportunities and risks that these bring in terms of production, markets and people's health and livelihoods. And while the issues to be tackled and the options for intervening to bring real and lasting benefits to people involved in these enterprises may differ, we must recognize the close inter-dependence of these systems, both nationally and internationally. Outbreaks of foot-and-mouth disease, for instance, in Europe and elsewhere, and crop failures in Africa, have surely taught us that by now.

Notwithstanding these and other considerations, the crux of the matter is that the parts of the world where poverty and livestock husbandry intersect most are the very parts of the world where the productivity of livestock and the supply of products to consumers through formal and informal markets are simply insufficient in quantity or quality to meet or create demand nationally, let alone compete in the international marketplace. In fact, a closer look at FAO's statistics on the livestock revolution referred to by Dr Jutzi reveals the stark reality that when countries like China, Brazil, Thailand

and some others are removed from the equation, people in more than 40 developing countries do not have access to any more meat or milk than they had 10 years ago; and very few countries are trading internationally in these and other livestock products.

So, in a nutshell, many – indeed, most – developing countries are simply not meeting or creating the demand for livestock or their products. There are of course very many reasons for this: internal and external; political and institutional; social and economic; environmental and technical – they vary widely between and within countries, and they are certainly both complex and evolving.

Dr Jutzi has mentioned the great, possibly even overriding, importance of sound livestock development policies within the framework of overall agricultural and national development plans. In effect, the need is for governments to take stock, consider long-term goals and options for achieving these, and then not only to formulate policies but also to devise ways of ensuring that the domestic and international institutional and other barriers to their adoption are progressively broken down. I could not agree more, and would only add that this process must start by encouraging policy- and decision-makers to engage and stay engaged with the people for whom livestock matter – and increasingly that also means consumers – to understand their needs and concerns, and then, in addressing these, to exploit the complementary roles of regulatory, economic and technical measures.

Putting all this into practice, and in particular ensuring poverty-focused development, is indeed the major challenge facing very many Member States of FAO and IAEA, and which they, and the international community as a whole, are committed to achieving through the Millennium Development Goals. While many initiatives are taking place in this area, the contribution being made by Dr Jutzi's Division in collaboration with ILRI and the United Kingdom in promoting a livelihoods orientation to policy development in the livestock sector is particularly noteworthy. In essence, it involves gathering and interpreting information from national case studies about policies and strategies and communicating the results to all stakeholders with a view to improving policy transparency among farmers, traders and governments; in effect, getting everyone to "buy in" to the processes of identifying their problems, assessing their options and making sensible selections and management decisions. There will clearly be a critical role for social scientists in the mix of expertise needed to do this effectively, something that for too long has been neglected in policy and technical decision-making.

Within that overarching framework, the policy objectives that governments set for their countries on science and technology, and the strategies they follow to achieve these objectives, will be critical for the future well-being of the livestock sector. Horizontal expansion is no longer a viable option for most countries if they are to respond to effective demand for livestock commodities. The goal must be intensification – higher output of meat, milk, eggs and other products per animal, coupled with less wastage and improved penetration into national and global markets. At the same time it will be essential to ensure responsible stewardship of natural resources, including animal genetic resources, and the issue of animal welfare must not be neglected.

Formidable as the challenges may be as the move from extensive to more intensive yet sustainable production evolves – access to new technologies and to the knowledge to use or adapt them locally will be a vital element. Technology has progressively transformed the way livestock products are produced and processed, and helped deliver both more and a wider variety of higher quality products to consumers. The fundamental policy challenge faced by governments becomes: How to bring about the changes needed? The changes are needed to empower communities to capture the economic and social benefits from the food and product chains derived from livestock that are available now or will come along in the future, from both traditional and advanced science and technology, while managing the risks involved in doing so.

If we take this line of argument further, and accept that the key factors enabling such changes are public, private and international investments in education, research and extension, appropriately channelled, of course, and supported by the right local, national and international incentives and infrastructure, then we must accept also that the public sector, supported by the international development community, must continue to play a major role in the training of animal scientists, veterinarians and others. This can be fully justified by the public good component of knowledge, of human capital development and of information – the main products of research, and which collectively, by increasing the supply of technical opportunities and research resources, provide incentives to private investment and public-private sector partnerships for the benefit of local producers.

A major problem here is that research is in difficulty: not only public sector livestock research and extension in most developing countries, but also international research, are in trouble, with budgets stagnating or tightening, while private sector investments are growing in all too few of them. Since demand for technology is driven by consumer demand for the final products and the market size, and because biological cycles are much longer than in crops, R&D investments in livestock-related technology are seen as higher risk and attract substantially less investment than they deserve. This Catch-22 situation can only be resolved by a much greater commitment at national, regional and international levels to reshaping institutional structures and market signals. Within that, the commitment to capture the increasing returns to scale on the supply side and creating critical masses of scientific creativity by establishing or strengthening R&D consortia and partnerships, including with private sector entities, would seem particularly important.

Dr Jutzi has described some options for enhancing public-private sector partnerships in this era of privatization and intellectual property rights (IPR). Certainly the public and international livestock research sectors will need to adjust to the realities of life, and carefully define and continuously refine their areas of comparative advantage, and we might debate what these should be, using examples that you yourselves may have or experiences derived from the crop sector.

Unquestionably, however, as a rule of thumb, the largest social returns will come from focusing on research directed at carefully identified problem areas, and with clear public good components. This means putting much greater emphasis on people and systems than on livestock themselves. I believe also that society benefits when the public sector has freedom to operate, and when it maintains public access to genetic resources and research tools subject to IPR protection. At the same time, I accept that considerably more work needs to be done to improve our understanding of the influence of IPRs on both public and private sector investments in livestock research, and the public sector's freedom both to operate and collaborate with the private sector.

While many factors have combined to drive change over the last 20–30 years, foremost amongst them was the rapid strengthening of the impulse of biological science and the technology that developed with it, from a phenotypic to a genotypic orientation. Over this period, we have witnessed an increasingly rapid succession of advances leading to the development of recombinant DNA and molecular marker technology and of other techniques, allowing genome mapping and the function and regulation of genes and gene combinations to be studied. Terms like proteomics, metabolomics, transcriptomics, micro-arrays and others, never before heard in the English language, are now commonplace in research circles – although still, I suspect, not part of the vocabulary of veterinarians treating calf scours on the farm!

In any case, these techniques, which are now included under the heading of modern biotechnology, are currently being applied to unravel the structure and function of the bovine and other genomes of livestock; the plants that they eat and the micro-organisms that they use to digest their feed; as well as many of the pathogens and vectors of disease that affect them. A large part of what this represents today is new knowledge and better understanding of the mechanics of cell biology, hereditary and immune processes and the like, and, I suspect, a mass of data sitting in computers awaiting analysis using sophisticated software algorithms to handle gene prediction in different genomes.

There are, nevertheless, high expectations that out of all this will sooner or later come the knowledge and technologies to develop better livestock; better ways of feeding them, including reducing pollution from intensive systems; better ways of controlling pests and diseases through vaccines and therapeutic agents; better ways of characterizing and preserving genetic resources; and more.

Even allowing for the tendency of scientists and companies to over-emphasize the importance of their work, and the reality that advances in knowledge in gene structure and function will involve increasingly complex and expensive technology and data handling, and levels of multi-disciplinarity and interactions between academia and industry not hitherto seen, there has already been a steady stream of new information and technology resulting in farm-level application, and more will surely come.

This is most obvious in the crop sector, with the appearance in the shops and supermarkets of some countries over the last decade of genetically modified crops and of foods derived from these, and including all types of livestock products. Recombinant DNA technology has also been successfully used to produce a number of livestock vaccines and some of these are commercially available and being used in both developed and developing countries; and although we have yet to see GM livestock and products from them on farms and in the market, the pace of progress in this field is rapid, and expectations of commercial use in agriculture are increasing.

Yet, at the same time, we have seen the rather negative reactions of some societies in different parts of the world to this kind of science and the products it has produced, and we can only conclude, in hindsight, that if we are to allow science to progress and society to benefit, we must do better in preparing our societies – not simply by proclaiming the benefits, but by fully addressing the potential risks and hazards to food safety and the environment, and to the welfare of animals.

As Dr Jutzi has pointed out, this Symposium is not about artificial insemination, embryo transfer, cloning, large-scale culture of micro-organisms to produce vaccines and feed additives, or any of the other techniques or products applied to living organisms. It is about reviewing and dissecting advances in genomics and molecular genetic approaches for improving livestock productivity, and about discussing future directions of R&D, with a focus on developing countries.

After the two opening review papers, the Symposium will cover gene-based applications in the areas of livestock genetics and breeding; pathogens and host-pathogen interactions; plants, rumen microbes and systems biology;

and food and environmental safety, ethics and IPRs. So the programme is comprehensive, and because of the transcending nature of the approaches and techniques being used, and the wider issues involved, it does, I believe, provide an opportunity for you all to stay engaged throughout, whatever your specific field of interest.

You will also see that a number of Panel Discussion Sessions have been arranged during the week. One is to consider which gene-based technologies are most relevant and likely to succeed in improving the sustainability of livestock production systems in developing countries. In doing this, I would like you also to identify the constraints and to come up with recommendations on how these technologies can be used in a simple and practical way and for what purpose, but always keeping in mind the need above all to secure and enhance the assets of poor livestock-keeping communities. And I would like you to ask yourself two questions – difficult or uncomfortable as they may be to answer. The first is: given that artificial insemination (AI) has been around for 50 years, that a vaccine has been available for rinderpest for about the same time and that contagious bovine pleuropneumonia (CBPP) was eradicated from some African countries many years ago but is now back with a vengeance, what makes you think that the new gene-based technologies will be any better or more widely used for societies' benefit than these and other existing technologies? Surely, this shows that sound policies coupled with economics are often more important than technology. My second question is this: about 35 years ago, I worked at the Glasgow Veterinary School, where we succeeded in developing and commercializing with the private sector a radiation attenuated vaccine against cattle lungworm, one which is still widely used. Since then, tremendous efforts have gone into developing vaccines against other parasites, including through molecular approaches. But none has so far been delivered. So the question is: what timescale are we really looking at before these and other new products are available?

The second Panel Discussion is about the role of international organizations and funding agencies in furthering the use of gene-based technologies for livestock production in developing countries. Without pre-empting your deliberations on this, and before you set about it, I would invite you to have a look at the FAO Web pages on biotechnology to see what has been done so far. Certainly both FAO and IAEA believe that sharing knowledge and experiences between countries through meetings like this, and disseminating the information that comes from them through publishing the proceedings, are among their strongest comparative advantages. But there are other possibilities as well, such as our roles in promoting international coordination of effort and steering resources towards the problems of developing countries. In this regard, I would particularly highlight the IAEA Research Contracts scheme, which, through support for Coordinated Research Projects, offers significant opportunities for developing and developed countries to work together in planning and carrying out activities involving nuclear techniques. Since isotopic markers are extensively used in molecular biology, and IAEA will soon be planning its Programme and Budget for the next biennium, it would be helpful if you could identify and prioritize potential areas for support through this mechanism. In short, I would like you to develop a plan for translating the recommendations from this Symposium into actions that can be considered by both IAEA and FAO.

Ladies and Gentlemen,

The social and economic disparities between the industrialized and developing countries are substantial, and in many cases growing even wider. Genomics research requires high capital investment and there are fears that the forces underpinning the livestock and genomics revolutions will combine to widen that gap further by diverting scarce funds and human resources in developing countries away from more traditional livestock systems and R&D activities, and through the much greater strength of scientific endeavour and appropriation of benefits from it by industrialized countries. These fears are well-founded, but, while all countries will have to consider the opportunities and risks of these technologies in relation to the interest of their people, perhaps the greatest risk for the developing countries is not getting involved at all, and to let the global markets for agricultural products and science and technology run their present course. This, I believe, is not an option for anyone who believes in social equity. The livestock sector – indeed, the whole agricultural sector – is so important to most developing countries for securing their social and economic progress that surely they must be given the chance to decide for themselves the paths to take in using science and technology for their future development. Surely, in the spirit of solidarity, indeed self interest, the industrialized countries and international community must use all the means at their disposal to support these aspirations.

In concluding, I would like to think that during this week we will have laid the grounds for contributing to that process.

Thank you.

GENE-BASED TECHNOLOGIES FOR LIVESTOCK INDUSTRIES IN THE 3RD MILLENNIUM

E.P. Cunningham
Trinity College, Dublin, Ireland

Abstract: The first complete genome sequence of an organism was for yeast, in 1996. Since then, the much larger task of doing a complete human sequence has been completed. Those of major domestic animals are following rapidly. It will always be impossible to foresee the full potential of such an explosion in knowledge, but aspects of gene-based technologies are already beginning to have an impact in the livestock sector.

The first and most obvious area of impact concerns feed supply, which constitutes 50–75 percent of total costs in many livestock systems. Production costs for maize and soybean are being reduced by genetic modification of the crop for herbicide and insect resistance. Maize has been modified to reduce phosphorous and nitrogen excretion in swine and poultry, and also to provide a more valuable amino acid balance.

Genetic modification of the animal is also possible. Most dramatically, the insertion of a growth hormone in the DNA of fish accelerates growth. However, in this and all other cases, the genetic modification (GM) of animals has produced profound physiological disturbances. At the same time, the administration of GM-produced growth hormone to dairy cows is now routine in the United States of America and several other countries. This is not permitted in Europe, where the attitude to all GM technologies has been much more cautious.

Conventional selection programmes continue to deliver steady genetic improvement in all animal populations. New molecular methods offer the prospect of enhancing genetic gains, particularly for traits that are difficult or expensive to measure, or which have low heritability.

41

H. P. S. Makkar and G. J. Viljoen (eds.), Applications of Gene-Based Technologies for Improving Animal Production and Health in Developing Countries, 41-51.
© 2005 *IAEA. Printed in the Netherlands.*

Gene technologies have much to contribute to the control of disease in animals. As pressure to reduce antibiotic and drug use increases, genetically modified vaccines with proven specificity and distinguishable from natural infections are already in use. DNA typing is helping with rapid and precise diagnosis. In addition, the interaction of some pathogens (e.g. scrapie) with the genotype of the animal calls for the application of DNA technologies.

Following the BSE epidemic in Europe, safety of livestock-derived foods is high on research and regulatory agendas. DNA techniques are already in use for tracking of sources of *Salmonella enterica* and *Escherichia coli* outbreaks, as well as for traceability of product in the food chain.

Finally, gene-based technologies can facilitate the measurement and conservation of genetic diversity in animal populations.

1.   INTRODUCTION

The last fifty years have seen dramatic increases in output and efficiency in the livestock sector in developed countries. Indicators such as milk yield per cow and rate and efficiency of meat production in pigs and poultry have more than doubled. Driven by economic pressures and opportunities, and made possible by technical innovation in genetics, nutrition and management, these changes have made it possible for the livestock industries to deliver their products to the market at a price that has been dropping relentlessly at about 3 percent per annum.

Across the range of developing countries, technical development has been less rapid. The considerable expansion of output has generally been through increased animal numbers. Inhibiting factors have included more challenging climatic and disease constraints, as well as issues of land tenure, social structure, market structure, education and access to technology.

While market demand in the developed world is fully met by production, growing populations and rising incomes in developing countries will create a demand for a doubling of output (Delgado *et al.*, 1999). This demand cannot be met simply by a lateral expansion of existing practices. Greater efficiency in resource use is required, not just for economic reasons, but also to improve human and animal welfare; to conserve depleting energy resources and living biological resources; and to halt and reverse environmental degradation brought on by wasteful production practices.

Much of this agenda can be served only by the development and deployment of appropriate technologies. Advances in our knowledge of genetic mechanisms have opened a world of new possibilities for addressing these objectives. While knowledge of DNA structure is now fifty years old, technologies for intervening in genetic structure effectively date back little more than a decade. From small beginnings with restriction enzymes, we have in the last ten years seen great advances in techniques for isolating, amplifying, reading and inserting DNA segments. These techniques have been scaled up and automated to make possible mass genotyping. Coupled with developments in information management, this wealth of data facilitates prediction of genetic structure and function, and thus accelerates the pace of knowledge accumulation. This explosion in knowledge has been compared in significance to the development of the periodic table one hundred years ago, and to the expansion of horizons to encompass the whole globe in the sixteenth century (Lander, 1999; Browne and Botstein, 1999).

2.   GM FEED

The study of Delgado *et al.* (1999) estimated that global meat consumption would grow at nearly 2 percent per annum for the period to 2020. Most of this growth would take place in developing countries (with a growth rate of 2.8 percent per annum as against 0.6 percent in developed countries). This would lead to a doubling of demand by 2020, driven by a combination of expanding populations and rising incomes. China alone will account for a high proportion of the total growth, with meat consumption there increasing at a steady kilogram per capita per annum.

Most of this additional meat will come from modern systems of pig and poultry production. These in turn are almost totally reliant on feed grains from outside the system. Meat demand therefore translates largely into demand for increased feed grains, estimated at an additional 300 million tonnes per annum by 2020 in developing countries.

The most significant applications of molecular genetics in agriculture to date have been in the development of genetically modified (GM) crops. About 70 million hectares of GM crops were grown in 2002. Approximately 90 percent of these were grown in two countries, the United States of America and Argentina. Furthermore, 75 percent of the genetic modification was for one purpose, herbicide tolerance, with insect resistance making up most of the remainder. Finally, 80 percent of the acreage was for the two main internationally traded feed crops: maize and soybean.

These applications of GM technology are aimed at reducing production cost. Though the use of GM crop technology has been resisted strongly in Europe and some other countries, its widespread and growing use in the main feed exporting countries is likely to maintain downward pressure on world feed prices. This will have the effect of both encouraging production and holding down consumer meat prices, adding impetus to the demand in developing countries.

Additional applications of GM technology in feed crops include:

- Improving nutritional value, through higher digestibility and better amino acid balance.
- Lower mycotoxin contamination in maize through reduction in maize borer damage.
- Reduced manure volume through higher net energy value.
- Reduced pollution through better availability to the animal of nitrogen and phosphorous.

A good example of this last application is given by Etherton *et al.* (2003). They reported that maize and soybean genetically modified to have reduced phytate levels led to increased availability of phosphorous to the animal and reduced phosphorous excretion rates. With low-phytate maize, phosphorous bio-availability increased from 22 percent to 77 percent in pigs and from 10 percent to 52 percent in poultry, leading to reductions in phosphorous excretion by 30 to 40 percent. Similar results were obtained with soybean, and feeding rations of low-phytate maize and soybean led to reductions in phosphorous excretion of 50–60 percent.

3.   GM ANIMALS

The introduction of extraneous genetic material into animal genotypes has been pursued most energetically in mice, and more than 1000 lines of GM mice have been produced for research purposes. Transgenic pigs, sheep, goats and cattle have also been produced. This has mostly been for pharmaceutical purposes, seeking to use the animals as bioreactors ("biopharming"). The use of animals for this purpose is an expensive business, because of high animal maintenance costs, low success rates, and long generation intervals.

The incentive to use transgenic livestock in food production systems is considerably less. The same high development costs are involved. Furthermore, it has proven difficult to control the rate, timing and tissues involved in the expression of transgenic products in the animal, resulting in often severe physiological disturbance. A further disincentive has been the fear of public reaction to the idea of transgenic food animals. For these reasons, it is difficult to envisage transgenic livestock being part of production systems in the medium term.

If applications are developed, they are likely to be in the area of improved animal health and welfare. A good example is the search for transgenic dairy cows resistant to mastitis. This disease is estimated to cost in excess of US\$ 2 billion per annum in the United States of America. Progress has been made on enhancing mastitis resistance by incorporating into the cow's genome the ability to produce antibacterial enzymes in the udder (Kerr *et al.*, 2002). Applications in swine are reviewed by Prather *et al.* (2003).

4.   MARKER-ASSISTED SELECTION

The range of possibilities for using molecular methods to contribute to genetic improvement programmes has been the subject of many reviews (e.g. Bulfield, 2000; Georges, 2001; Niemann *et al.*, 2003; Dekkers and Hospital, 2002). Some traits of interest are controlled by one or very few loci. In such cases – such as presence or absence of horns in cattle, coat colour in many animals, naked neck in poultry – it was possible even before the advent of molecular methods to determine the genotype of an individual and to use this knowledge in a breeding programme. Molecular methods have made it possible to characterize individuals for additional traits: double muscling in cattle, hyperfertility in sheep, and meat quality in pigs.

A separate application of gene technology to assist selection programmes can be found in cases where traits need to be recorded in related animals, but the cost or inconvenience of individual animal identification has made the establishment of the necessary relationships difficult. Such has been the case, for example, in beef cattle breeding, where it would be desirable to recover carcase data from the offspring of bulls, but animals are not individually identified, or identity is lost because they move several times during their lifetime. In salmon breeding, it is possible, though expensive, to rear family groups together. In addition to cost, this can also create non-genetic differences between families, which can reduce the effectiveness of selection. Norris, Bradley and Cunningham (2000) have developed a system for allocating individual fish to their parental groups by using a panel of microsatellite loci. The fish can then be reared in commercial conditions, and at harvest time tissue samples are analysed for DNA, and the fish are allocated to their parental groups until a sufficient number have been accumulated for each family. Genetic assessment of sires and dams can then be carried out for traits like meat colour and fat content, based on the carcase characteristics of their offspring. In this way, genetic technologies can make selection possible for traits that formerly could not be included in the breeding objective.

Most traits of interest are controlled by an unknown number of loci whose relative importance may vary, depending on interactions with genes for other loci, or with environmental factors. Loci involved in such traits are called quantitative trait loci (QTLs).

The focus of most work has been on the search for QTLs through close linkage to neutral markers. The neutral markers used are generally the microsatellite markers of published marker maps of each species. The object is to identify markers closely linked to a functional gene or QTL. Many promising results have been published, such as for markers linked to QTLs for trypanotolerance in mice. Markers identified in this way are being used by private companies in pig and poultry breeding, and are being marketed as aids to improving tenderness and intramuscular fat in beef cattle. However, the overall impact on breeding programmes has been modest.

5.   GENE TECHNOLOGIES IN ANIMAL HEALTH

The prospects for using gene technologies in the area of animal health are reviewed in a recent issue of *Genetics Selection Evolution* (No.1, 2003).

While few applications have been developed to the point of general use, genetic management strategies are under development for a number of diseases. Much work is being done to identify functional candidate genes for mastitis resistance in cattle. Concerted work is also focused on exploiting genetic mechanisms involved in bovine trypanotolerance.

One area where genetic tools are already in application is that of scrapie in sheep. This is a prion disease, with long incubation period and uncertain modes of transmission. Prevalence varies between countries (some countries being free of the disease) and breeds. Susceptibility to the disease (or perhaps variation in the incubation period) has been shown to be associated with a particular locus, for which DNA test methods have been developed. Though the disease is not known to present risks to humans, there are concerns because of its similarity to BSE. Throughout the EU, an eradication programme based on genetic screening will therefore be launched in 2005.

Much attention is being given to the genetic control of immune responses. Because of similarities across species, comparative genomics is particularly useful, and the use of micro-array technologies to track the flow of gene effects looks particularly promising.

In addition to manipulating or characterizing the genome of the animal, diseases can be addressed by improving the efficiency of diagnostics, and the specificity and effectiveness of vaccines (Yilma *et al.*, 2003).

6. IMPROVING FOOD SAFETY

The BSE epidemic in Europe, which began in 1986, has transformed public attitudes to food safety (Cunningham, 2003). Reinforced by other food scares, it has led to significant changes in consumer behaviour, and to a greatly increased demand for better controls throughout the food chain.

Because of totally separate developments in food production, distribution and consumption, this challenge has in some respects become greater. The larger scale of processing and retailing companies, the increasingly open and free market at all stages of food production through to delivery to the consumer, and the increasing complexity of many foods mean that identity and origin of foods and their components are more difficult to preserve. At the same time, traceability of product is an essential requirement for any food safety system.

As a result of these concerns, the EU and each of its member countries have now established Food Safety Authorities and developed a legislative framework that requires identification and registration of animals, and traceablity of their products. A recent report (EU, 2003) concluded that while the live animal traceability systems put in place in recent years have worked reasonably well, traceability from point of slaughter to the consumer continues to present many problems.

For several decades, parentage verification of herd book animals has been carried out using blood groups. Within the last ten years, this technology has largely been replaced by use of DNA profiles. This DNA technology can now be adapted to link meat products reliably to the animal from which they are derived. The principal challenge has been to develop operational systems that can be applied on a large scale, and at a cost that can accommodate modern livestock processing and distribution. Such systems have now been developed (Cunningham and Meghen, 2001).

The advantages of DNA traceability are considerable. It offers a degree of accuracy that is unequalled. It does not require a parallel paper trail. Minute biological samples can be taken at the point of slaughter, or at any point earlier in the animal's life. Such earlier sampling can be facilitated by devices that harvest a bio-sample at the same time as they insert a conventional ear tag. The bio-samples, whether taken from the live animal, or from the carcase, are archived. Analyses are then carried out on samples taken at any subsequent stage in product processing, distribution or sale, and resulting profiles can be matched to archived routine samples from the carcases or live animals. Because the system requires no change in normal work practices, it can be implemented at reasonable cost.

Such a system provides multiple benefits throughout the food chain.
– Customer assurance concerning integrity of origin of products.
– Verification of origin, which is independent of all interested parties in the chain.
– Facilitation of rapid source identification and product recall.
– Compliance, at reasonable cost, with national and international requirements.
– Feedback of information on aspects of product quality to producers and seed stock breeders.

As operational procedures are refined and reduced in cost, DNA-based traceability is likely to become universal in beef, and probably also in swine.

7.   UNDERSTANDING AND MANAGING GENETIC DIVERSITY

The discovery of restriction enzymes, which cut DNA strands at specific sites, led to the first practical measure of genetic diversity at the molecular level, namely restriction fragment length polymorphisms (RFLPs). The length of the DNA fragments created by specific RFLPs could be measured using gel electrophoresis. The pattern of fragment lengths then was an indicator of the pattern of presence or absence of specific nucleotides at specific sites.

In 1989 in the animal genetics laboratory at Trinity College Dublin, we set up a project to use this technique to measure genetic divergence between Indian, African and European cattle. The intention was to provide some measure of the genetic distance between breeds to help in the interpretation, and perhaps prediction, of dairy cattle cross-breeding results.

About the same time, the measurement of DNA sequence was becoming feasible, and we began comparing mitochondrial sequence data from the same breeds. In addition, a further technique, based on what are now known as microsatellites, became available at about that time. The existence of short repeated sequences of DNA throughout the non-coding regions of the genome had been discovered. The number of such repeats at a particular locus proved to be highly polymorphic, and therefore to be a good measure of genetic diversity. As with RFLPs, they could be measured readily by electrophoresis, because they gave rise to variation in the length of DNA fragments. We used the twenty microsatellites that had been characterized in cattle to open a parallel line of measurement of genetic differences between breeds.

In the years since then, measurement of genetic diversity both within and between populations has expanded rapidly, and productively, with most of the results coming from the use of sequence and microsatellite data. The results of our own research are indicative of the kind of insights that these techniques can bring.

The first results (Loftus *et al.*, 1994) produced some dramatic new understanding of the domestication origins of the world's cattle populations. Mitochondrial sequence comparisons showed that the genetic distance between European and Indian cattle breeds was so great (more than 300 000 years) that they could not have originated from a common domestication event. This evidence of (at least) a dual domestication in cattle has been of great value not only in understanding the reasons for the amount of heterosis observed in crosses between *Bos indicus* and *Bos taurus* (Rutledge, 2001), but also in the fields of anthropology and archaeology. Subsequent work has shown similar dual domestications in sheep and goats, but not in horses (Hill *et al.*, 2002).

This work gave a peculiar result in African cattle. All African breeds studied, including several with a *Bos indicus* phenotype, had *Bos taurus* or European-type mitochondria. Parallel work with microsatellites (MacHugh *et al.*, 1994) confirmed the hybrid nature of most African breeds, and showed a clear gradation in the balance of Indicus and Taurus genes, which largely reflected the degree of challenge from trypanosomiasis. Further parallel work on Y chromosome haplotypes confirmed that African cattle populations are largely derived from a *Bos taurus* maternal base, with varying degrees of *Bos indicus* male mediated intrusion (Bradley *et al.*, 1996).

Further work (Troy *et al.*, 2001) came to the conclusion that the cattle populations of Europe were not derived by domestication of native wild cattle (Aurochs) but from Near-Eastern genetic sources and were spread through Europe with the Neolithic transition in the period 10 000 to 5 000 B.C. Molecular characterization of genetic diversity in many animal populations is now being undertaken using standardized panels of microstaellites (FAO, 2004). These data can be used to assist in the planning of conservation programmes. For example, we have recently shown (Cunningham *et al.*, 2001) in studies in thoroughbred horses how a combination of molecular methods and pedigree analysis can be used to estimate the degree of genetic diversity in founder populations.

REFERENCES

Bradley, D.G., MacHugh, D.E., Cunningham, E.P. & Loftus, R.T. 1996. Mitochondrial diversity and the origins of African and European cattle. *Proceedings of the National Academy of Sciences, USA,* **93**: 5131–5135

Brown, P.O. & Botstein, D. 1999. Exploring the new world of the genome with DNA microarrays. *Nature Genetics,* **21**(1 Suppl): 33–37.

Bulfield, G. 2000. Farm animal biotechnology. *Trends in Biotechnology,* **18**: 10–13.

Cunningham, E.P. 1999. The application of biotechnologies to enhance animal production in different farming systems. *Livestock Production Science,* **58**: 1–24.

Cunningham, E.P. (ed). 2003. *After BSE – A future for the European livestock sector?* EAAP Publication, No. 108.

Cunningham, E.P. & Meghen, C.M. 2001. Biological identification systems: genetic markers. *OIE Review of Science and Technology,* **20**: 491–499.

Cunningham, E.P., Dooley, J.J., Splan, R.K. & Bradley, D.G. 2001. Microsatellite diversity, pedigree relatedness and the contributions of founder lineages to thoroughbred horses. *Animal Genetics,* **32**: 360–364.

Dekkers, J.C.M. & Hospital, F. 2002. The use of molecular genetics in the improvement of agricultural populations. *Nature Reviews Genetics,* **3**: 22–32.

Delgado, C., Rosengrant, M., Steinfeld, H., Ehui, S., & Courbois, C. 1999. Livestock in 2020. The next food revolution. IFPRI/FAO/ILRI.

Etherton, T.D. and a team of nine scientists. 2003. Biotechnology in animal agriculture: An overview. *CAST Issue Paper,* No. 23.

EU [European Union]. 2003. Overview report of a series of missions carried out in all member states during 2002 in order to evaluate the operation of controls over the traceability and labelling of beef and minced beef. DG Sanco Directorate F, Report 9505/2003.

FAO [Food and Agriculture Organization of the United Nations]. 2004. Domestic Animal Diversity Information System (DAD-IS). See: www.fao.org/dad-is.

Georges, M. 2001. Recent progress in livestock genomics and potential impact on breeding programs. *Theriogenology,* **55**: 15–21.

Hill, E.W., Bradley, D.G., El-Barody, M.A., Ertugrul, O., Splan, R.K., Zakharov, I. & Cunningham, E.P. 2002. History and integrity of thoroughbred dam lines revealed in equine mtDNA variation. *Animal Genetics,* **33**: 287–294.

Kerr, D.E., Wellnitz, O., Mitra, A. & Wall, R.J. 2002. Potential transgenic animals for agriculture. *In:* Proceedings of the 53rd EAAP Meeting. Cairo, September 2002.

Lander, E.S. 1999. Array of hope. *Nature Genetics,* **21**(1 Suppl): 3–4.

Loftus, R.T., MacHugh, D.E., Bradley, D.G., Sharp, P.M. & Cunningham, E.P. 1994. Evidence for 2 independent domestications of cattle. *Proceedings of the National Academy of Sciences, USA,* **91**: 2757–2761.

MacHugh, D.E., Loftus, R.T., Bradley, D.G., Sharp, P.M. & Cunningham, E.P. 1994. Microsatellite DNA variation within and among European cattle breeds. *Proceedings of the Royal Society of London, Series B, Biology,* **256** (1345): 25–31.

MacHugh, D.E., Shriver, M.D., Loftus, R.T., Cunningham, E.P. & Bradley, D.G. 1997. Microsatellite DNA variation and the evolution, domestication and phylogeography of taurine and Zebu cattle (*Bos taurus* and *Bos indicus*). *Genetics,* **146**: 1071–1086.

Niemann, H., Rath, D. & Wrenzycki, C. 2003. Advances in biotechnology: new tools in future pig production for agriculture and biomedicine. *Reproduction in Domestic Animals,* 38: 82–89.

Norris, A.T., Bradley, D.G. & Cunningham, E.P. 2000. Parentage and relatedness determination in farmed Atlantic salmon (*Salmo salar*) using microsatellite markers. *Aquaculture,* **182**: 73–83.

O'Brien, S.J., Menotti-Raymond, M., Murphy, W.J., Nash, W.G., Wienberg, J., Stanyon, R., Copeland, N.G., Jenkins, N.A., Womack, J.E. & Graves, J.A.M. 1999. The promise of comparative genomics in mammals. *Science,* **286**(5439): 458–462, 479–481.

Prather, R.S., Hawley, R.J., Carter, D.B., Lai, L. & Greenstein, J.L. 2003. Transgenic swine for biomedicine and agriculture. *Theriogenology,* **59**(1): 115–123.

Rutledge, J.J. 2001. Greek temples, tropical kine and recombinant load. *Livestock Production Science.* **68**: 171–179.

Troy, C.S., MacHugh, D.E., Bailey, J.F., Magee, D.A., Loftus, R.T., Cunningham, E.P., Chamberlain, A.T., Sykes, B.C., Bradley, D.G. 2001. Genetic evidence for Near-Eastern origins of European cattle. *Nature,* **410**(6832): 108891.

Yilma, T., Ahmad, S., Jones, L., Ngotho, R., Wamwai, H., Berhanu, B., Mebratu, G., Berhe, G., Diop, M., Sarr, J., Aziz, F. & Verardi, P. 2003. Inexpensive vaccines and rapid diagnostic kits tailor-made for the global eradication of rinderpest. pp. 99–111, *in:* F. Brown and J.A. Roth (eds). *Vaccines for OIE List A and Emerging Animal Diseases.* Proceedings of the International Symposium, Ames, Iowa, 16-18 September 2002. [Developments in Biologicals, No. 114]. Basle, Switzerland: Karger.

CHALLENGES AND OPPORTUNITIES FOR CONTROLLING AND PREVENTING ANIMAL DISEASES IN DEVELOPING COUNTRIES THROUGH GENE-BASED TECHNOLOGIES

J.R. Crowther[1] and M.H. Jeggo[2]

1. *Joint FAO/IAEA Division, Animal Production and Health Section, Wagramer Strasse 5, A1400 Vienna, Austria.*
2. *Director, Australian Animal Health Laboratory, Geelong, Private Bag 24, Geelong, 3220 Victoria, Australia.*

Abstract: Polymerase Chain Reaction (PCR) technology allows scientist to amplify, copy, identify, characterize and manipulate genes in a relatively simple way. Exploitation of the technology to devise new products and translate these to the commercial sector has been remarkable. Molecular technologies are not difficult to establish and use, and can appear to offer developing countries many opportunities. However, developing countries should look in a different way at the apparent advantages offered. Whilst molecular biological science appears to offer solutions to many problems, there are a number of drawbacks. This desire to adopt the latest technology often overrides any considerations of the use of more conventional technologies to address needs. The conventional, and often more practical, methods already provide many specific tools in the disease control area. Changing the technology can also deflect critical resources into the molecular field in terms of laboratory funding and training. This may cause redundancy of staff, limit further development in conventional techniques, and polarize scientists into the older (less glossy) and newer (molecular) camps.

Animal disease diagnosis still primarily utilizes conventional techniques such as Enzyme Linked Immunosorbent Assay (ELISA). This will not change drastically in developing countries, but developments will combine such methods with more discriminatory molecular techniques, and a balanced and parallel development is needed. An understanding of the use and possible advantages of the various technologies is required by both scientists and policy-makers in developing nations.

53

H. P. S. Makkar and G. J. Viljoen (eds.), Applications of Gene-Based Technologies for Improving Animal Production and Health in Developing Countries, 53-71.
© 2005 *IAEA. Printed in the Netherlands.*

Vaccines based on molecular science could have a real impact in developing countries, but "vaccinology" needs to examine both the animal (immunology of target species) and the disease agent itself. This is a research-based science and, as such, is expensive, with no surety of success. Developing countries should exploit links with developed countries to provide the "field" genetic resource (endemic disease situation) in order to devise and test vaccines developed through molecular studies.

Development of technologies cannot be divorced from an understanding of the epidemiology of the diseases found in developing countries. This is frequently not undertaken due to the many competing demands on the scarce resources available. However, increased livestock trade possibilities may provide the focus and catalyst needed to ensure that animal health science is applied appropriately and usefully for the benefit of developing countries.

1.   INTRODUCTION

There is plenty of evidence that the livestock revolution is providing a real opportunity for the rural poor in developing countries to escape the poverty trap. Beyond the market opportunities created by the increased demand for livestock product, there will be an improvement in quality and safety of food, highlighting another area where relevant gene technologies might be advantageous in the medium to long term. However, focusing on animal disease, the developmental state of any country will be crucial in catalysing the national commitment to control and eventually eradicate. The possibilities of increasing trade, restricted by a country's contamination with disease agents, are probably the major stimulus in facilitating better planning and resources for control. The relative importance of disease in terms of time and geography is also very varied. Control requires a complete package comprising good veterinary infrastructure, reporting systems, sample submission, laboratories with trained staff, validated methods, epidemiological units able to plan surveys, vaccination teams, and so forth.

Constraints to improving livestock productivity should be considered under the three headings of improving reproduction; nutrition; and disease. There is seldom a holistic approach to these disciplines; however, there are two major areas, which if addressed, would have the greatest short-term impact on productivity in developing countries, greatly overshadowing all foreseeable technological exploitation. The first is good management of existing resources at the farm level. The second is unrestricted or poorly understood movement of animals and animal products, whether legal or illegal.

This second area is usually ignored because of the apparent enormity of the problem, but, without proper attention, most interventions, including those considered in this paper, are likely to have little impact. This is highlighted in the section dealing with foot-and-mouth disease (FMD). It is relevant to compare control measures aided by vaccines and diagnostics, with those involving identification and effective manipulation of gene(s) for increasing resistance and tolerance factors against disease, since developing solutions will require completely different time scales. Appreciating this is crucial in deciding where efforts should be made by individual countries in terms of resources and short, medium and long term planning. The costs and benefits of one, or another, or a combination of approaches, should always be borne in mind, and, to allow best advice, calculations should be made independent of political considerations. Basically, this involves risk assessment of all parameters.

Countries must cooperate since many of the limiting diseases are trans-boundary in nature and activities in one country can severely affect another's efforts. Regional projects should be one of the main tools, with additional advantages in understanding, planning and managing disease control. Such control must include considerations of veterinary infrastructural requirements, disease reporting systems, contingency planning, epidemiological units, systems for data retrieval and analysis, vaccine campaign advice, sero-monitoring and surveillance systems, provision of and validation of kits, laboratory guidelines, accreditation guidance, training and expertise. It is difficult to achieve this complete package; in reality, most support is a reaction to a disease emergency rather than a planned preventive programme. In order to achieve better success, better planning and awareness of the importance of livestock disease is in general needed.

FAO has a programme specifically dedicated to the development of systems for transboundary animal and plant pests and diseases control, called EMPRES, which includes emergency prevention systems. FAO plays a catalytic role in monitoring and coordinating regional and international efforts and providing technical advice and other forms of support to its Members.

Animal disease management strategies in any country involve consideration of the genetic nature of the causative agent, as well as that of the host. The major strategy for micro-organisms is to exploit their inherent high rate of replication, so that mutations can overcome host defence mechanisms, and it might be concluded that the simpler the agent of disease in terms of genetic material, the more risk there is to livestock, e.g. virus diseases, which are the majority of the Office International des Epizooties (OIE) List A pathogens! This selection process vastly outweighs any of the possible genetic attributes of the far more slowly reproducing hosts.

Approaches looking at genetic advantage in the host are therefore inherently disadvantaged. There are examples of such gene-advantaged animals, but the challenge data is limited and focused on single animals and no account is taken of the multiplicity of agents available to infect an animal independent of the gene(s) introduced.

There is a critical lack of fundamental knowledge on immune mechanisms in livestock, but ongoing gene mapping projects in chickens, cattle, pigs, sheep and goats may accelerate our understanding in this vital area. Training cannot be underestimated. This must be continuous and should emphasize the relationship between serological and molecular techniques and the advantages and drawbacks inherent in the use of particular methods. The technology gap between scientists is most obvious in countries that have developed some potential through institutions to exploit biotechnology. Such specialized institutions require considerable funding, too often at the expense of established institutes performing valuable routine diagnostics This sets up a divide, centred on resource management, between the more active field workers and the more showcase-research oriented few. The split between the two must be avoided through a better understanding and planning of problem-oriented tasks, rather than adopting technology *per se* as a solution in itself.

Diagnostics involve the use of defined reagents in the laboratory or in the field. This is usually extended to the use of diagnostic kits and this whole area is beset with problems, including supply, validation and use. Measuring diagnostic sensitivity and specificity is extremely demanding, and compounded by the poor use of good reagents by untrained staff. The production of kits of good quality with robust reagents is difficult enough, but the distribution and sustainability of supply is practically impossible. There are few candidates for profit in the diagnostic area to encourage the commercial sector to produce and distribute diagnostic kits, and this applies to both serological-based and molecular-based approaches.

The situation is compounded, since national, regional and international reference laboratories for conventional or newer technologies simply do not adequately underpin routine diagnostic laboratories. Although international organizations identify and appoint reference laboratories and collaborating centres, there is little evidence of their effective support to efforts in this area. Support to reference laboratories must be adequate and sustained. The exact role of any reference laboratory has to be defined from the start, and national and regional acceptance is vital. Given that, it is also inappropriate to nominate a national laboratory for such a role with the underlying premise that this will then enable the laboratory to seek funding to fulfil the role! It must from the outset have demonstrable expertise, facilities and resources.

Vaccine development involves both conventional approaches and all the possibilities of genetic manipulation. It requires extensive knowledge of animal immunology, gene manipulation, adjuvant use, delivery systems, large-scale production, testing and certification (quality and safety). Any serious national production usually has to be commercial. Unfortunately, commercial companies assessing vaccine as a product for sale identify only areas where there might be a profit. So their estimation of markets may vary considerably, depending on the needs of any particular region. Commercial success is difficult to assure in the livestock area, and even current, highly purified, FMD vaccines are barely profitable. As with the diagnostics, the vaccine market is highly fragmented. Development of vaccines for some important agents for use in developing countries is non-existent, and coupled with the poor research training and facilities in such countries, ensures that no successful tool is likely to be developed to aid control. The high cost of registering vaccines is also a limiting factor, and this may well exacerbate bad practice through the use of "national" vaccines. Some vaccines are more applicable to exploitation than others. Scaling up "genetically" associated vaccines may be very difficult, whereas the more conventional (attenuated vaccines, ideally) are relatively easy to prepare (hence the success of the rinderpest (RP) campaign and the development of the attenuated peste des petits ruminants (PPR) vaccines). Figure 1 attempts to condense the factors involved in livestock disease control.

In Figure 1, it is assumed that there is a livestock revolution to allow the world to be fed safely with livestock products. Production elements (central grey box) involve good management of reproductive, nutritional and diseases factors. The importance of all these has to be recognized. Probably the most impact can be made on any situation by improving management, even with existing facilities and resources. The need for quantity of food is giving way to strong need to improve quality (safety). The impetus for management comes from both local and more international needs (trade pressure). The balance is debatable of who pays for improvements and what benefits there are directly to the poor or the commercial sector (social economics). Other technologies, such as information communication technologies (ICT), are maybe a more important development need in the short term. Gene technologies can be exploited directly or indirectly over a longer term. Use of existing technologies should be maximized and is most important in the shorter term compared with gene technologies. Training is highlighted for newer and existing methods. All developments have to be viewed within the context of national and cross-border needs. This can lead to conflict when cross-border control is lacking. Often the consideration of who pays dominates, although outside influences, including funding agencies, affect that equation.

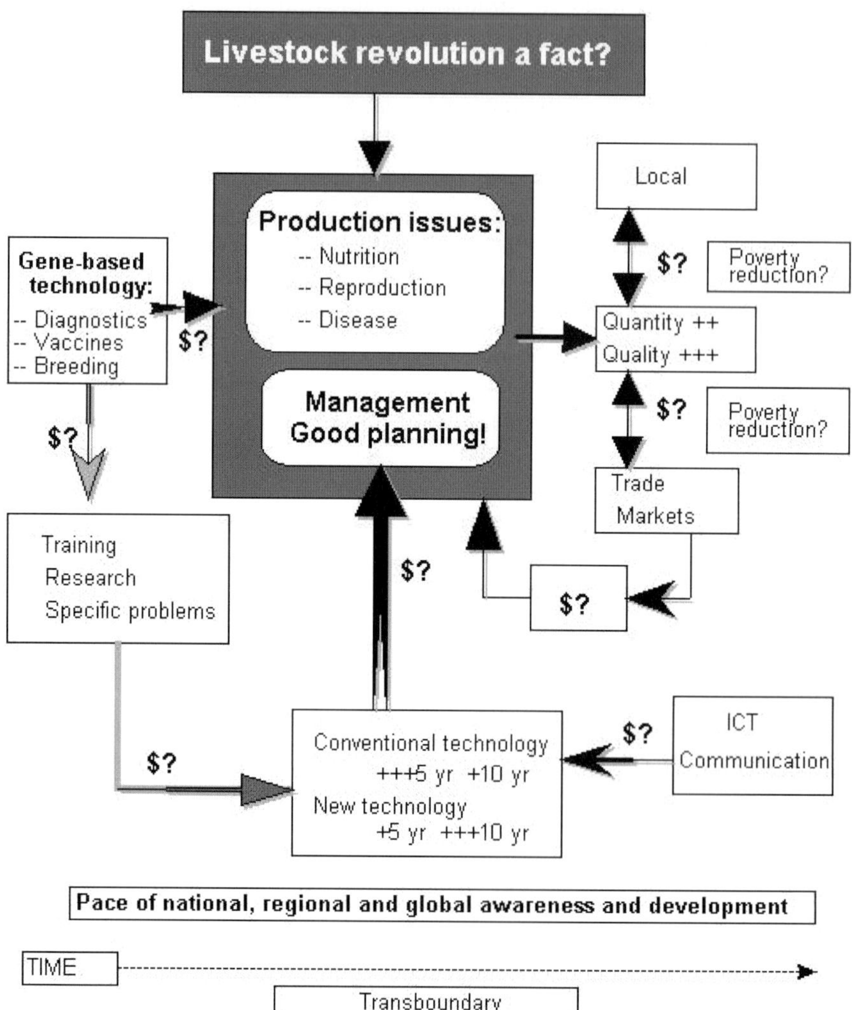

Figure 1. Issues in gene technology transfer

2.   OPPORTUNITIES

There are two spheres for exploitation of gene-based technologies: in research and in applied areas. It is clear that "advancement" of research is more and more dependant on such techniques. The aim of research in this area would be to develop better diagnostics and vaccines. This must be planned in the context of what is already available and what is, ultimately, specifically required by a country (or better a group of countries). It is essential to apply science in areas that are most relevant to country's needs, and although molecular techniques offer very rapid manipulative methods to create new products, they may not be appropriate for many countries. Planning of research within the context of overall developmental needs is critical to ensure the maximum use of the usually scarce resources, but such planning is too often conspicuous by its absence. Research is costly, and it is essential to determine if the approach being applied is the most effective way to solve the particular problem.

2.1   Diagnostics

Ideally diagnostics employ methods that allow:
- identification of disease agents or parts of agents (confirmation of clinical diagnosis);
- differentiation of agents (differential diagnosis);
- assessing the epidemiology of disease agents (surveys); and
- monitoring of control interventions (e.g. measuring efficacy of vaccines, drugs).

There is a wide range of techniques currently available, providing an arsenal of "conventional" tests:
- Tissue culture to allow isolation of viruses etc.
- Cultivation of bacteria on solid or in liquid media.
- Neutralization testing.
- Complement fixation.
- ELISA.
- Haemagglutination (HI) and inhibition tests (HAI).
- Immunohistochemistry.
- Microscopy (including fluorescence).
- Electron microscopy.
- Polyacrylamide gel electrophoresis (PAGE).
- Nucleic acid hybridization.
- Immunoblotting.
- Restriction endonuclease mapping.

2.2   Diagnostic ideals

The needs and location of where assays are carried out is related to a range of national issues, including infrastructure, lines of communication, reporting systems and taking and transportation of samples. Thus, the merits of technologies have to be judged according to needs. Certain advancements, e.g. pen-side or field tests, offer tremendous potential for more remote on-site testing, whereas others have to be regarded as simply complementing existing techniques. It should be noted that communications are vital to improving disease reporting and control; indeed, it might be that the priority in most remote areas is technical solutions to communications. Table 1 indicates the appropriateness of tests for different situations.

2.3   Existing diagnostics

Often conventional technologies are ignored or made redundant through a lack of equipment maintenance. This is most obvious in microscopy (a large number of microscopes have been provided worldwide for fluorescence techniques, but are useless for want of new objectives!). Over the years, "new" technologies have been introduced requiring different equipment and the associated training.

Table 1. Situations and the diagnostics that can be used.

Situation	Test and ancillary requirements
In the field situation	Pen-side tests and biosensors, requiring low expertise levels and training.
In small, "local" laboratories	ELISA, agar gel tests and pen-side test strips needing limited equipment, and requiring better expertise and some training. Communication with Reference Centres.
In Regional Reference Laboratories	Good equipment needed to carry out all conventional tests, including tissue culture, as well as equipment and reagents for molecular-based tests (e.g. PCR). Well-trained and experienced staff needed. Constant communication with other Reference Centre(s) and World Reference Centre(s). Epidemiology units needed for design and analysis of work. Links to veterinary managers for disease control.
World Reference Laboratories	Fully equipped laboratory needed, requiring expert staff. Activities involve a research component, reference status, standards elaboration, and solving problems using samples from the world collection. Constant communication and data retrieval from other Reference Centres and Collaborating Centres (all well equipped and properly staffed laboratories with the support to act as a reference source).

Perhaps the most potent force in diagnostics in the past 30 years has been the advent of ELISA. This remains the most useful assay for measuring and comparing antibodies and antigens and is likely to remain a major technique, since it fulfils most testing criteria. Many kits are available, and ELISA is the basis for most prescribed tests for the OIE List A diseases. As such, countries will continue to use this technology. Unfortunately the extent of training remains inadequate for countries to fully exploit ELISA's potential.

2.4   Newer technologies in diagnostics

Advances in molecular biological techniques and related instrumentation have great potential for improving disease diagnosis. This includes genetic engineering hybridoma technology and large-scale (industrial production) tissue culture; and molecular modelling for designing active anti-disease agents (vaccines and pharmaceuticals). Some of these improvements are through a better understanding of the immunology of the host, as well as through unravelling the genetic character of the pathogen. Advances in other technologies are relevant, in particular the performance of computers in collecting, analysing and storing data, and developments in instantaneous and continuous communication. The massive leap in the ability to manipulate genomic information is linked absolutely to the development of the core polymerase chain reaction (PCR) technology. Table 2 summarizes "modern" approaches to aiding diagnosis.

Great emphasis is now being placed on the use of PCR. Laboratories able to perform this assay should nevertheless also be able to perform basic serological techniques. Screening tests followed by more time consuming and costly confirmatory tests is a good working practice, and laboratories should not shy away from using more specialized institutes to undertake the confirmatory testing. Tests must also be considered in the context of whether vaccines are being used and the responses this will engender in the animal.

2.5   Diagnostics technology transfer in practice

The Joint FAO/IAEA Division in Vienna has had a long history in technology transfer of diagnostics, including the ELISA and PCR technologies. The transfer goes beyond the mere provision of reagents, and training was found to be the most vital element. Technology is best transferred within the context of defined scientific projects: the technologies providing the tool to obtain data on which decisions could be made. As such, the whole package, from sample taking, to analysis and processing of data, to action, has to be considered. Effective use of technologies only comes through such a holistic approach.

Table 2. Technologies to improve diagnostics

Technology	Some applications and comments
DNA manipulation	Expression-specific proteins for use as diagnostic reagents. Expression systems: *E. coli*, yeasts, mammalian cells, Baculovirus. Gene deletion. Linking diagnosis to vaccine used. Differentiation of vaccinated and infected animals, e.g. Pseudorabies vaccine.
DNA	Hybridization reactions. *In situ* hybridization in diagnosis. Restriction endonuclease mapping. Comparison of strains.
Polymerase Chain Reaction (PCR)	Amplification of genes. Rescue and amplification from samples (RNA and DNA viruses). Detection and differentiation of genes with specific primers. Multiplex systems to allow rapid differentiation of disease agents. Rapid sequencing and comparison of products using differential diagnosis and confirmation. Molecular epidemiology. Portable PCR machines.
Real-time PCR	Direct detection of products. Massive expansion of technology. Fastest growing research area. Multiplex systems. Robotics (genome projects) for automation for testing large numbers of samples. Expensive start-up costs (but getting cheaper). Closed systems (reducing cross-contamination). Allows easier development of multiplex systems. Very potent research tool.
Synthetic proteins	Peptides identified and produced as reagents for diagnosis. Epitope characterization. Pepscan, phage libraries.
Hybridoma technology	Large supply of monoclonal antibodies (MAbs) from tissue culture. Production of defined product for use in assays to detect antigens and antibodies (for ELISA).
Monoclonal antibodies (MAbs)	Improved specificity and sensitivity for polyclonal serum-based assays (standards easier). Panels of MAbs for qualitative comparison of strains. Rapid differentiation between and within closely related strains. Production of MAb escape mutants to allow characterization of antigens. Characterization of epitopes at molecular level. Paratope profiling (determination of antibody spectrum).
Biosensors.	"Instant" measurement in a single instrument. Pen-side possibilities.
Pen-side tests	Strong developments for diagnosis and environmental monitoring. Dip-stick technologies. Rinderpest, PPR, FMDV antigen detection. FMD DNA chip available!
Instruments	Rapid measurement of various signals in immunoassays. Fluorescence polarization, ELISA, bioluminometry, chemoluminescence.
Availability of commercial reagents and equipment	Restriction enzymes, DNA polymerases, reverse transcriptases, labelled bases, conjugated antibodies (enzymes, gold particles, fluorescent markers), dig-labelling, cell culture, affinity purification, cytokines, MAbs, microtitre equipment, thermocyclers, kits, PCR. PCR/ELISA allows sequencing, labelling, oligonucleotide primers, antibody production (including MAbs). Primers. Many enzyme immunoassays (viral diseases of swine, sheep, cattle, poultry, fish, dogs, cats).
Services	Sequence databanks, host and agent. Sequencing. Out-sourced testing. Training. Comparative data accessible to all.
Computers	Data collection, analysis, storage, communication of results. Databanks of sequences. Essential in sequencing and comparative studies relating large amounts of information.

3.   ELISA

This technology has for some 15 years been the focus of some of the programmes run by the Joint FAO/IAEA Division, and the transfer of this technology to developing countries has involved a complex array of activities and programmes. The need for ELISA technology was identified as a result of control programmes for specific diseases, e.g. RP, FMD, brucellosis and trypanosomiasis. Here, specific ELISA kits were developed for sero-monitoring or surveillance. The transfer has been successful, with networks of laboratories in most continents able to perform good assays. Such networks can also serve to further validate assays. The instrumentation to read ELISA has proved remarkably robust, with few servicing problems. The peripheral equipment is also durable (multi channel pipettes, tips washers, incubators). Training in all aspects of the project is invariably necessary and has to continue through the life of the project or programme as new people come on board and as new initiatives (e.g. Internal Quality Control (IQC) and External Quality Assurance (EQA)) are introduced. This approach crucially allows the development of local expertise, with the trainees becoming the trainers.

The cost of setting up an ELISA laboratory varies depending on the extent to which it is to be used and the facilities currently available. Table 3 indicates the inherent costs. The extent of samples examined, the real price for kits and their exact make up (do they contain plates, etc?) would depend on the activities envisaged. Research activities will increase miscellaneous reagent costs, and no allowance has been made for costs of experiments.

Table 3. Costs associated with establishing an ELISA capability.

Item	Approximate cost (US$)
ELISA reader	6,000–8,000
Computer	2,000
Calibration plate	700
Tips (per 1000). Need depends on samples run	15
Microtitre plates, 10 and 80 samples per plate (depending on kit supply)	3–8 per plate
Miscellaneous reagents	1,000
Pipettes (multi- and single-channel)	1,500
Storage for samples, racks and containers	1,000 per 2,500 samples
Freezers	1,000–3,000
Kits (variable)	0.5–1.5 max. per sample examined
Incubators/shaker	3,000
Distilled water apparatus (or supply of good water)	2,000
Washer	1,500
Training (3 work-months)	6,000–10,000
Books	200
Training expert visit, short-term home training course	1,000

3.1   Kits

Critical to the use of ELISA technology is the basic need for kits. These comprise sets of stable, quality controlled reagents and materials that allow a specific test to be standardized and effectively validated. This area is beset with difficulties. Although many kits are available, the validity of their use, in many cases, is dubious. Validation *per se* is the subject of OIE guidelines. It is inherently difficult to define the diagnostic sensitivity and specificity of kits. Validation is primarily concerned with establishing specificity with regard to samples from non-infected countries since they are produced in developed countries. There is an international move to focus validation on "fitness for purpose", linked to a process for registration. The main concern will be "Who will pay?" for this, but without it there will be a continuation of the practice of kit supply with no quality control elements (IQC or EQA), and lack of validation. An important component is support for sustainable kit production and distribution in developing countries. The Joint FAO/IAEA Division has been instrumental in helping Senegal assemble and produce quality-controlled kits for African Swine Fever. This is coupled to EQA rounds to assure quality. An associated factor is the need for laboratories to gain accreditation. Standards have now been devised by IAEA, through OIE, to accredit laboratories. It is vital that there be regional cooperation to enable this process, particularly in the light of developing trading requirements.

4.   DIAGNOSTICS AND PCR

The fundamental advantage of PCR is that minute amounts of genetic material can be amplified millions of times in a short time, allowing detection from samples of a single copy of a genome or part genome. PCR products can then be identified precisely through sequencing. The ability to amplify genomes allows genetic manipulation of genes, which is the basis of the gene revolution. The very fact of the ultimate sensitivity of the PCR technique produces some of the problems in use of the method in routine terms.

PCR amplification has many advantages, including:
- rapid diagnosis (within hours);
- high sensitivity and specificity;
- decreasing costs for each PCR assay;
- automation;
- less training and experience required compared with virus isolation;

- detection of viral nucleic acids in samples that are unsuitable for virus isolation, such as toxic, mummified foetuses; semen; or organs carrying only viral DNA copies, e.g. during latent herpes viral infections;
- use of "general" or "universal" PCR primers, which can amplify any one of the members of an entire virus family;
- relative ease of standardization of the PCR assays;
- quantitative analysis possibilities; and
- multiplex PCR assays that can simultaneously provide diagnosis for a whole disease complex, like respiratory disease, which can involve a wide range of possible agents – viruses, bacteria and parasites.

Besides these and other positive features, PCR assays also have some limitations:
- PCR cannot discriminate between viable and non-viable agents;
- it is still relatively expensive;
- due to its extreme sensitivity, PCR can produce false positives; and
- false negatives may also occur, due to enzyme-inhibitory substances in the sample or failure to add all of the components essential to the reaction.

4.1   Transfer of PCR

Initially it was thought that the PCR technology would be difficult to transfer and use in the conditions found in developing countries, but this has proven wrong. Intrinsically, PCR can be transferred provided that:
1. The design of the laboratory in which PCR is used is correct. This is vital, since contamination by minute amounts of genes can totally destroy the effectiveness of the laboratory to diagnose anything.
2. Training is given before any work commences on PCR.
3. Equipment is suitable and complete.
4. Strict laboratory practices are adhered to and rigidly enforced.

4.2   Cost involved in setting up a basic PCR facility

Setting up a basic laboratory is essential in developing more expansive methods in PCR using the fast developing PCR technologies. Table 4 indicates the basic requirements and their costs.

Table 4. Basic requirements and costs for a PCR laboratory

Item	Approximate cost (US$)
Laboratory refurbishment, 3–4 small laboratory stations or areas needed	Local costs (1,000–10,000)
Thermocycler	3,500–8,000
Pipettes 3 sets × 3 Only for PCR station	1,500
Special tips (aerosol resistant) 10,000	3,000
Enzymes for PCR	2,000
Electrophoresis (power packs, visualization of gels, recording film)	6,000
Primers for agents	1,000
Miscellaneous chemicals	2,500
Protective clothing (gloves, lab coats)	1,000
Work stations (hoods) (×3)	6,000
Refrigerators and freezers	2,000
Labelled reagents (Note supply problems)	2,000
PCR tubes	1,000
Micro-centrifuge	3,000
UV lamps for decontamination	1,500–2,000
Special tube holders to avoid contamination	2,000
Training (3–6 person months per person for at least 2 persons)	12,000–24,000

5. VACCINES

Vaccines are still fundamental to many disease control and eradication programmes. They are usually best supplied through commercial sources in terms of sustainable quality, and should be regarded as expensive items in terms of delivery and maintenance of product quality, and so their use should be planned carefully. This requires all the necessary veterinary infrastructural factors to be in place, including facilities to test the efficacy of vaccines and an understanding of local strain variation that might affect vaccine performance.

Conventional vaccines typically fall into four categories:
1. Attenuated strains of live agents.
2. Crude preparations of killed agents.
3. Pure killed agents, in association with adjuvants.
4. Parts of agents or the products they produce, e.g. toxins.

The ability to manipulate genes using PCR has allowed the development by various means of new types of vaccines:
- Removing pathogenic genes from agents so that they immunize without causing disease.
- Selecting genes that produce certain immunogenic (protection inducing) proteins for insertion into genomes of other agents (vectors).
- Identifying immunogenic proteins and selecting specific genes for expression in other systems to produce large quantities of immunizing

protein (subunits, polypeptide, peptide) or for chemical synthesis of peptides from an identified known structure.

– Using DNA alone as a vaccinating agent.
– Preparing agents with protein but no, or a defective, genome.
– Expressing proteins on the surface of vectors as a chimera protein.
– Producing marker vaccines through insertion of genes using recombinant technologies to produce a protein associated with the replication of the vaccinating agent to distinguish vaccinated from naturally infected livestock. Can also be used with subunit vaccines.
– Producing RNA virus vaccines by reverse genetics.
– Expressing in plants.

Research into new-generation vaccines will invariably require gene manipulation (PCR, etc.). Thus any vaccine development work will require a full-scale molecular biology laboratory. Any decision to produce such vaccines locally should take fully into account the costs and technology needs. It must also be appreciated that producing the candidate vaccine is just the beginning, and that registration, scaling up to production capacity, and packaging and distribution are other critical and costly areas. Care should be taken in determining the real need for new vaccines, and whether current or conventional vaccines could satisfy requirements. Another critical factor is the issue of Intellectual Property Rights (IPR) concerning genes and methods involved. This must be carefully addressed before any research is undertaken. General features, which must be considered for vaccines, include:

1. The host species' immunology and protective mechanisms (humoral or cellular).
2. The agent structure and function (antigenicity, pathogenicity, variation).
3. The vaccine formulation in terms of agent (whole attenuated agent, whole inactivated agent, large mixture of antigens, polypeptides, peptides).
4. Delivery systems for vaccines (injection, oral, water, aerosol, particulate, etc.).
5. Physical stability of vaccines (heat stability affecting efficacy).
6. Need (quantity of vaccine required, scaling up, industrialization).
7. Safety to animals and humans (reversion of attenuated trains, sterility, residues due to adjuvants).
8. Animal experimentation facilities to assess vaccine usefulness and safety.

Gene-based technologies may certainly help in some of these areas when compared with more conventional approaches, and offer distinct advantages for developing countries.

5.1   Conclusion on vaccines

Genetic technologies offer novel routes to constructing new vaccines. This development potential may require a relatively small outlay for equipment and reagents, but a high investment in training. There is also no guarantee that such approaches will be successful.

Conventional approaches should always be looked at first where there is an immediate need for control. The bad use of good vaccines must be avoided through good planning and coordination between field, diagnostic and epidemiological workers involved in disease control. If there is a need for a new-generation vaccine, a more effective approach may be the utilization of products developed in other laboratories (in developed countries). Care has to be taken where such exploitation is made, with regard to patenting and IPR.

6.   FORCES OF CHANGE

An examination of attempts to control two major livestock diseases – RP and FMD – illustrates the forces involved in change, and good and bad practice, as well as highlighting the gap that exists between the needs and influences on activities in general for developed and developing nations. Overall, trade-related issues dominate over the risk of the disease itself. The examination also demonstrates that high-technology solutions are not always necessary to control and eradicate disease. Good veterinary infrastructure linked to effective planning and implementation is the critical element. Where technology can play a part is in limiting the negative impact of ineffective management and resource allocation.

The problems associated with RP and FMD are summed up in Table 5. From the factors listed there, it is understandable how RP is now close to being eradicated and why FMD is spreading. The obvious factors are that RP is a killer disease, it is not particularly contagious, there is a vaccine giving life-long immunity, the vaccine protects against all types of the virus, there is a restricted host range, and, at least in the past 30 years, the disease has had a somewhat restricted global occurrence. FMD in contrast, is highly infectious, consists of seven serotypes with numerous sub-types, occurs globally, affects many species, and vaccines give poor and limited immunity. Thus we assume that rinderpest was easy to control and FMD was not. Rinderpest was important and FMD was not. Today, FMD remains the biggest disease threat to trade, both national and international, and is now considered the single most important disease to be dealt with in terms of alleviating poverty in rural communities. Why, therefore, has this not resulted in a similar effort and success story as RP? If there was an FMD vaccine that gave the same level of immunity as that for RP, would we still be seeing outbreaks of FMD? The answer is almost certainly that, with an effective vaccine, FMD would be a disease of the past. The fact remains that we do not have vaccines for FMD that will work in the developing country situation, with all the limitations that this implies. Whilst this is a simplistic approach, and ignores a range of other issues that make the global eradication of FMD a somewhat different proposition to that of RP, effective vaccines will be a prerequisite. The critical question then becomes "Can the use of gene-based technologies deliver such a vaccine?"

Table 5. Factors involved in the control of rinderpest and of foot-and-mouth disease.

Parameter	Rinderpest	Foot-and mouth disease
Disease character	High mortality.	High morbidity and highly contagious.
	Not so infectious unless very close contact.	Infectious by close contact, and by aerosol over long distances.
	Mild to very severe (death).	Mild (sub clinical to severe). Seldom death except for young and abortion.
	Low virus excretion.	Very high virus excretion.
	Kills	Affects draught power (debilitating)
Species infected	Cattle, buffalo and some wildlife.	All even-toed ungulates (cattle, buffalo, sheep, goats, swine, wild animals).
Distribution	Limited now to possible foci in Somalia and NE Kenya. Was a major problem in Africa and Asia.	Very widespread and increasing in S. America, Africa, Mid- and Far East, S.E. Asia, Ex- CIS republics.
Vaccine	Live attenuated, safe (non reversion).	Purified, inactivated whole virion.
	Easy to produce in tissue culture	Hard to produce
	Life-long immunity after one dose.	7 serotype vaccines needed (and some strain specific).
	Single vaccine protects against all strains.	Immunity only while antibody lasts (3-6 months).
	Blanket vaccination and then vaccination of young animals possible.	Continuous vaccination policy needed. Blanket vaccination not feasible.
	Cessation of vaccination possible when herd immunity above certain level.	US$ 0.5 to 1.0 per serotype per dose. Different species need different formulations.
	US$ 0.3 per dose.	
	Single species.	
Diagnostics	Antigen ELISA (capture)	Antigen ELISA (7 serotypes needed).
	Antibody for sero-monitoring vaccine efficacy and surveillance (competitive, indirect).	Antibody ELISA (competitive for each of 7 serotypes).
	Dip-stick for antigen.	Antibody ELISA to distinguish vaccinated and infected.
	Monoclonal antibodies in tests.	Monoclonal antibodies for research and characterization.

Parameter	Rinderpest	Foot-and mouth disease
Molecular	PCR diagnosis. PCR and sequencing (extensive molecular biology – 4 lineages). Virus rescue. Marker vaccines. Genes in vaccinia recombinant vaccine potential.	PCR diagnosis (multiple primers needed plus RT-PCR). Extensive sequence database of strains worldwide. Attempts at chimeric vaccines, DNA vaccines, peptide vaccines, vectored vaccines, antigen genes plus interferon approaches, adjuvant trials, drug trials.
Immunity	Single vaccination adequate. Humoral antibodies for life. No obvious carrier state problems.	Multiple life-long need to vaccinate. Humoral antibodies are protective. Mucosal immunity important (no solution). Carrier animals in wild and after infection (with risk that they may infect non-immune animals) – major problem for trading nations.
Control	Large campaign in Africa (US$ 250 million). Vaccination ceased. Constant surveillance (clinical and serological). Almost eradicated after 10-year campaign.	No large campaigns. Vaccination sporadic. Mainly commercial. Diagnosis varied. Spreading. Very poor or **no** animal movement control is a major constraint (legal and illegal). Products are infectious. Culling (stamping out) effective but many environmental and socio-economic problems.

6.1　Technology needs and opportunities

Although this exemplifies the situation with regard to FMD, it is generally applicable to many livestock diseases, where the emergence of new strains, and increase in challenge from new and emerging viruses in developing countries, threatens the developed world. The threat also has to be considered in the context of bio-terrorism. Together, these threats may act as a positive catalyst for funding research using gene-based technologies to develop effective vaccines against such agents.

7.　CONCLUSIONS

The desire to utilize the latest technology in the belief that this will provide a solution must be tempered with the need to carefully consider current well-proven and available technologies and products. There is often a drive for funding organizations to invest in technology transfer without a full study of whether this is the best approach to solving a particular problem. The desire to create research capacity to locally solve a problem is admirable, but often fails to deliver the required solution.

- Can gene-based technologies make a difference?
  There are papers in the symposium that clearly identify applications:
  - Understanding the disease process and epidemiology of the infection (Domingo).
  - Understanding the basis of genetic resistance and then creating this state (Mackie and Cann).
  - Understanding the causative agent and how it causes disease (Vanderplasschen).
  - Creating a new generation of vaccines and diagnostics based on genetic manipulation of the causative agents (Parida *et al.*).
  - Using gene expression systems for vaccines, diagnostics and therapeutics (Viljoen, Romito and Kara).
- Issues relating to the development and use of these technologies
  - The acceptance of genetically-modified organisms (GMOs) (animals, vaccines, therapeutics).
  - The cost of the science.
  - The ability to conduct the research.
  - The focus on developing country problems.
  - The research drivers, e.g. cost recovery.
  - A new research paradigm – global solutions to global problems.
  - Separation of research solutions from research capacity building.
- The opportunities
  - The livestock revolution is a genuine opportunity.
  - Livestock disease will limit this opportunity.
  - Technologies can assist disease risk management.
  - Gene-based technologies offer a quantum leap in both understanding the disease process and developing solutions.The challenge
  - Accepting GMOs.
  - Funding research for solutions.
  - Focusing on developing country problems.
  - Creating national infrastructure for technology application.
  - Creating a genuine "level playing field" for trade, through OIE and WTO.

MOLECULAR GENETICS AND LIVESTOCK SELECTION

Approaches, opportunities and risks

John L. Williams

Department of Genomics and Bioinformatics, Roslin Institute (Edinburgh), Roslin, Midlothian, Scotland EH25 9PS, UK.

Abstract: Following domestication, livestock were selected both naturally through adaptation to their environments and by man so that they would fulfil a particular use. As selection methods have become more sophisticated, rapid progress has been made in improving those traits that are easily measured. However, selection has also resulted in decreased diversity. In some cases, improved breeds have replaced local breeds, risking the loss of important survival traits. The advent of molecular genetics provides the opportunity to identify the genes that control particular traits by a gene mapping approach. However, as with selection, the early mapping studies focused on traits that are easy to measure.   Where molecular genetics can play a valuable role in livestock production is by providing the means to select effectively for traits that are difficult to measure.   Identifying the genes underpinning particular traits requires a population in which these traits are segregating. Fortunately, several experimental populations have been created that have allowed a wide range of traits to be studied. Gene mapping work in these populations has shown that the role of particular genes in controlling variation in a given trait can depend on the genetic background. A second finding is that the most favourable alleles for a trait may in fact be present in animals that perform poorly for the trait. In the long term, knowledge of the genes controlling particular traits, and the way they interact with the genetic background, will allow introgression between breeds and the assembly of genotypes that are best suited to particular environments, producing animals with the desired characteristics. If used wisely, this approach will maintain genetic diversity while improving performance over a wide range of desired traits.

H. P. S. Makkar and G. J. Viljoen (eds.), Applications of Gene-Based Technologies for Improving Animal Production and Health in Developing Countries, 73-88.

1.   INTRODUCTION

Cattle were first domesticated about 12,000 years ago (Grigson, 1989; Loftus *et al.,* 1994; Bradley *et al.*, 1998), and now there are over 1,200 million cattle worldwide that provide a source of food, motive power and clothing. In order to thrive, cattle naturally became adapted to local environments. Improvement of the local populations was then achieved by selective breeding, which was initially carried out at a local level using a limited number of shared bulls. This selective breeding led to the development of animals with characteristic phenotypes suited to local uses and environments, which became identified as breeds. In 1993 there were 783 cattle breeds recognized worldwide (FAO, 1993). The wide diversity of phenotypes displayed by the various breeds provides the opportunity for the selection of animals with improved production, suited to local needs and uses.

1.1   Improvement versus genetic diversity

Breed improvement has been enhanced over recent years by the use of artificial insemination and the development of statistical methods to maximize genetic gain. Selection programmes, coupled with improvements in management, have resulted in dramatic improvement in simple production traits that can be readily measured. Consequently, where the economic environment supports high-input agriculture, there has been a dramatic increase in milk yield and meat produced from the improved stock. The unfortunate consequence has been the reduction of genetic diversity, both within the improved breeds, as the superior individuals are preferentially used as breeding stock, and also through the displacement of traditional "unimproved" breeds. The latter is the greater cause for concern in less developed and environmentally less favoured areas. Traditional breeds are often adapted to thrive in their local environments, e.g. with increased tolerance of extremes in temperature, or in the face of particular disease or parasite challenge. By replacing the local breeds with improved stock, it is hoped that production will be increased, when in reality attempts to introduce improved dairy breeds into some areas have met with disastrous consequences. The replacement of local stock that are adapted for survival in the face of disease challenge, with disease-sensitive stock may mean the irretrievable loss of the genetically controlled basis of disease resistance. Natural genetic resistance allows stock to survive without the requirement for expensive protective measures. The very areas where this resistance is of greatest importance are those that do not have the resources to provide the extensive veterinary care required to maintain the "improved", but disease-sensitive, stock. Genetic diversity is being lost along with specific alleles that may confer specific survival traits. In 1993, 112 of the 783 cattle breeds worldwide were at risk (FAO, 1993)

1.2   Molecular genetics

There are opportunities for using molecular biology to identify genes that are involved in variety of traits. Armed with this information, it should be possible to select improved stock on the basis of their genetic make-up. In theory the genes controlling all genetically controlled traits could be identified, but the risks are that attention will focus on traits that are important for production in high-input, expensive agricultural systems, rather than on survival in diverse environments, for example. If applied with care, the use of molecular information in selection programmes has the potential to increase productivity, enhance environmental adaptation and maintain genetic diversity. The first task is to understand the genetic control of the traits of interest, and then to identify the genes involved. The approaches to identify genes controlling important traits currently focus on genetic mapping, which initially localizes the genes to chromosomal regions, before attempting to identify the genes themselves, starting from knowledge of their chromosomal position. The approaches for gene mapping are described below, along with some examples, and the way the information can be applied is discussed.

2.   WHAT IS REQUIRED FROM SELECTION?

The requirement from livestock varies depending on the environment and circumstances. In developed countries the focus has traditionally been on increasing the quantity of production, whether that is meat, milk, wool or other saleable produce. However, consumer pressure is now demanding high quality product, and that animals are farmed in sustainable, environmentally and welfare friendly, systems. Selection criteria are therefore likely to shift to quality, efficiency and health traits. In less-developed and disadvantaged areas, the requirements are more simple: survival in the face of disease challenge, without veterinary intervention, and in the face of adverse environmental conditions.

3.   SELECTION FOR IMPROVED PRODUCTION

Selection for improved milk production has been achieved through progeny testing. This approach uses the production records from daughters of test sires to calculate their genetic merit. Sires with above average predicted genetic merit for traits such as milk production are then used as "elite" sires to produce both more daughters and also additional sires. These sires are themselves progeny tested for selection of the next generation of elite sires. This approach has been used widely by breeding companies and has resulted in the milk yields of the Holstein breed nearly doubling over the past 40 years, to as much as 12,000 litres per 305-day lactation in the top Holstein cows today. In developed countries, the Holstein represents the most numerous cattle breed and accounts for the majority of milk produced.

3.1   Potential problems

The consequence of intensive selection has been the narrowing of the genetic base of the cattle population through the widespread use of a relatively small number of "elite" sires, many of which are closely related. As a result, the dairy industry has very large half-sib families in commercial herds, sired by a limited number of sires. Alongside the dramatic increase in milk yield, there has been an associated decrease in protein content of the milk, decreased fertility and increased incidence of lameness. The unfortunate consequence of focusing selection on a single trait is that detrimental genes may inadvertently be selected for in other traits. This intensive selection from a small number of elite sires inevitably leads to inbreeding and the potential concentration of deleterious recessive defects. As the level of inbreeding increases – defined by the inheritance of the same ancestral genes from both parents – the likelihood that an individual will inherit the same deleterious mutation from both parents also increases. This has been manifested in the elite Holstein dairy population by the appearance of genetic diseases such as DUMPS (Deficiency of Uridine Mono-Phosphate Synthase) (Schwenger, Schober and Simon, 1993), BLAD (Bovine Leukocyte Adhesion Deficiency) (Shuster and Kehrli, 1993), and, most recently, a bulldog-calf syndrome.

3.2   How could molecular genetics help?

Until now, selection programmes have been based on phenotypic selection, where traits are measured directly and animals with superior performance in the trait are used as breeding stock. Where the trait is sex limited, such as milk production, progeny test schemes have allowed the genetic merit of the sex not displaying the trait to be estimated. There are several problems associated with phenotypic selection: first, narrowing of the genetic base of a population, as discussed above; second, the approach can only be applied to traits that are easily measured; and, third, high costs. In traits that are displayed only in the adult, which includes most production traits now being selected for, it is necessary to raise a large number of individuals for which the trait is recorded, so that a few can be chosen for breeding. In the case of progeny testing for milk production, the costs are very high, as the test sires have to be raised and then the daughters themselves raised and bred from before the trait can be measured and the elite sires selected.

In slow growing or late maturing species, one way to speed up selection and reduce costs is to identify juvenile predictors of adult performance (Meuwissen, 1998). Such predictors would allow earlier selection of breeding stock, before incurring many of the rearing costs. However, the reliability of juvenile predictors is often low. The use of molecular markers potentially offers a way to select breeding animals at an early age – indeed as embryos; to select for a wide range of traits; and to enhance reliability in predicting the mature phenotype of the individual.

3.3 Application of genetic markers

The identification of the genes controlling a particular trait is a two-step process. Initially the location of the gene is identified, and then this information is used to find the gene itself. Most traits involved in livestock production are not under the control of a single gene, but are controlled by several genes, which have an additive effect, so that the trait has a continuous distribution. Such traits are referred to as quantitative traits and a locus controlling them is termed a quantitative trait locus (QTL).

If the chromosomal location of a QTL is known, DNA markers close to that QTL can be used to enhance selection programmes by identifying animals that carry the favourable allele at the QTL. This process is called marker-assisted selection (MAS) (Kashi, Hallerman and Soller, 1990). However, as the markers are likely to be at some distance from the gene, there is the possibility of recombination occurring between the marker and the trait gene. Thus, to use MAS, it is first necessary to determine the phase of the markers, i.e. which marker alleles predict the favourable or unfavourable alleles at the trait gene. This phase determination has to be done within a family, as the phase could be different in different families. In addition, the phase of flanking markers can change through recombination over generations. In contrast once the trait gene is known, and the allelic variations in the gene characterized, the information can be used directly in the population, without having to reconfirm the phase in each family or for each generation.

4. LOCALIZING GENES FOR SPECIFIC TRAITS

Identifying the genes that control a trait can be approached in a number of ways.

One approach is to use information on the physiology of the trait to identify the biochemical pathways involved in the trait, and hence suggest the genes involved. This information can be coupled with patterns of expression among tissues to facilitate cloning of the gene(s) likely to affect the trait. These "candidate" genes are then studied in the context of the trait to identify if they play a role in controlling the observed variation. This approach clearly requires a good *a priori* knowledge of the trait and its physiology. However, even with good knowledge of the trait, important genes are likely to be missed, as many may not be obviously involved in the known physiology.

A second approach is to start with no prior assumptions, and to use a genome mapping approach. A mapping approach is often necessary, particularly with complex traits, where the various gene interactions cannot be predicted. Mapping the genes controlling particular traits requires families, which have to be segregating for the trait, in which to track the inheritance of chromosomal regions, and information quantifying the trait. The mapping process involves tracking inheritance of chromosomal regions, using DNA markers, and then correlating inheritance of the markers with inheritance of characters associated with the trait. Therefore, to map trait genes, there are two requirements: families and markers.

Once the chromosomal location of the trait gene(s) is known, it is then necessary to identify the gene itself. In practice identification of the "trait genes" is achieved using a combination of the genetic mapping and candidate-gene approaches.

4.1 Genome maps

Over the past decade, genetic and physical maps have been developed for the genomes of all the major domestic species. Two types of genome maps exist: genetic and physical maps. Genetic maps are based on linkage between markers determined from their inheritance in families (Barendse *et al.*, 1997; Georges *et al.*, 1995; Bishop *et al.*, 1994). Genetic maps were predominantly composed of anonymous markers, but more recently genes have been added.

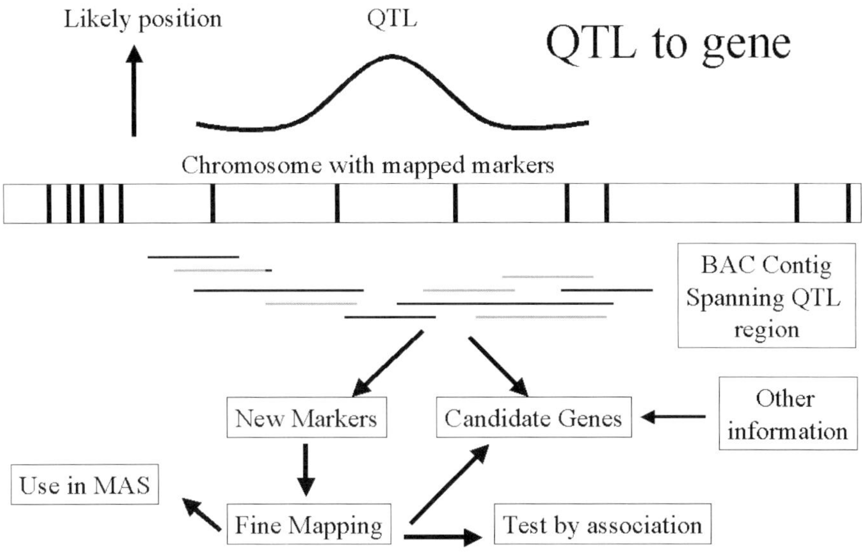

Figure 1. Strategy for identifying the trait gene starting from a QTL location. Identification of the gene controlling a trait from a QTL location is not easy. The first step is to construct a physical map of the region from bacterial artificial chromosome (BAC) clones. This physical map can then be used to identify new markers to fine-map the region. In some cases, a good candidate gene that can be tested may be found within the QTL region; otherwise it may be necessary to sequence across the region, using the BAC clones as the starting point.

These maps have been used to select markers for mapping production related traits (e.g. Kühn *et al.*, 1999; Stone *et al.*, 1999). Two notable successes have used genetic mapping to identify major genes involved in increased muscling (Grobert *et al.*, 1997; McPheron and Lee, 1997), and in milk production (Grisart *et al.*, 2002).

The first physical maps used *in situ* hybridization to localize DNA probes on chromosomes fixed on microscope slides. This approach, when used with probes derived from individual chromosomes from other species, demonstrated that the genes within particular chromosomal regions are conserved between species (Solinas-Toldo, Lengauer and Fries, 1995). This "conservation of synteny" across species means that the regions flanking a QTL that has been mapped in one species can be examined in another, usually a species with extensive genomic information, primarily man. The cross-species comparison may identify putative candidate genes for the traits, based on knowledge of function and position.

4.2   Populations for QTL mapping

Most of the data available on QTLs involved in dairy-associated traits has come from studies on the commercial dairy population, where there are large families produced as a result of the extensive use of artificial insemination (AI). This population structure is well suited to genetic mapping studies. The use of data collected on commercial herds to map QTLs was pioneered by Georges *et al.* (1995), who used the USA Holstein population to map QTLs involved in milk yield and quality. The study used 1518 progeny-tested bulls with production data from over 150,000 cows to derive their performance information. The sires were genotyped with 159 microsatellite loci, and analysis of the phenotypic and DNA marker information identified five QTLs for milk yield, fat and protein content.

In order to localize a QTL controlling a trait, it is necessary to have performance recorded for the trait. For the dairy industry, a few traits are routinely recorded, including milk production traits, somatic cell scores (which are associated with mastitis), and conformation. The situation in the beef industry is worse, where the use of AI is more limited and so few large families exist. A more efficient way of mapping QTLs is to breed experimental populations specifically for investigating particular traits.

5.   DOUBLE MUSCLING

The genome mapping approach has identified many QTLs for production traits in livestock, but up to now few trait genes have been identified. One gene that has been identified is a gene involved in double muscling in cattle. Selection for beef cattle in Europe has produced animals that are highly efficient and develop large muscles. Several breeds now carry a trait known as double muscling, which is associated with economically beneficial production traits, including muscular hypertrophy, and reduced intra-muscular fat and connective tissue (Ménissier, 1982). However, double muscling is also associated with dystocia, and calving difficulties, necessitating a high rate of caesarean sections. The most extreme form of double muscling is seen in the Belgian Blue breed, where the trait behaves as if it is controlled by a single major gene, with the extreme form of the trait expressed in homozygous individuals.

5.1   Mapping the double-muscling gene

The gene responsible for the double-muscled phenotype was mapped using a research population of cattle in which double-muscled Belgian Blue cattle were crossed to a non-carrier breed. The F_1 cross-bred animals were backcrossed to the double-muscled Belgian Blue to produce a second-generation cross, some of which showed double muscling (homozygotes for the gene), and some of which did not (heterozygotes). The animals in the study were genotyped with DNA markers from a panel that covered the whole genome and the genotypes correlated with the phenotypes to localize the gene responsible. This approach identified a region on bovine chromosome 2 as the most likely location of the gene responsible for the trait (Charlier *et al.*, 1995).

5.2   Finding the gene

Identifying the trait gene from the chromosomal location can be achieved by a number of strategies. Currently, the initial, and easiest strategy is to identify and test positional candidate genes, i.e. genes whose physiological function is likely to affect the trait and that map to the correct chromosomal location. Candidate genes within the region on chromosome 2 where the double-muscling gene had been localized did not prove to be to be the trait genes. Therefore a fine mapping and positional cloning strategy was initiated by constructing a contiguous, or overlapping, set of large-fragment DNA clones spanning the region (see Figure 1).

As this work was being done, an apparently unconnected piece of work in mouse identified a new member of the transforming growth factor (TGFβ) family of genes. The new member, initially called growth differentiation factor 8 (GDF-8), was expressed mainly in skeletal muscle. Mice where expression of the gene was knocked out developed hyper-muscularity, similar to the double-muscling phenotype in cattle. The GDF-8 gene product was found to be a negative regulator of muscle growth and was therefore called myostatin. Subsequent studies in Belgian Blue cattle showed that double-muscled animals carried an 11 bp deletion within the coding region of the myostatin gene that destroyed the functional protein (Grobert *et al.*, 1997). Work on other double-muscled breeds of cattle showed that they all carried a mutation in the coding region of their myostatin gene, lending support to this being the gene controlling the double-muscled phenotype.

5.3   Effects of genetic background

The 11 bp deletion in the myostatin gene is consistently associated with the double-muscled phenotype in the Belgian Blue breed. However, the same 11 bp deletion allele was also found in the South Devon breed in the UK (Smith *et al.*, 2000), where the phenotype associated with the mutation was found to be variable: some homozygous animals having an almost double-muscled phenotype, whereas others had poor conformation. Although the phenotype is not as extreme in South Devon compared with Belgian Blue cattle, the 11 bp deletion is associated with significantly increased muscle and decreased back fat in this breed (see Table 1). A similar observation of a variable phenotype associated with homozygosity of the 11 bp deletion has been made in the Spanish Casina breed (S. Dunner, pers. comm.).

Although myostatin clearly plays a role in controlling muscle development, there must be additional gene products that interact with myostatin and modify the phenotype. This interaction of myostatin and other modifier genes explains the variable phenotype. This example shows that if a major gene for a trait is mapped in one population, extrapolation of the likely effects into other populations should be done with care.

Table 1. Effect of the 11 bp deleted allele in South Devon cattle (Wiener *et al.*, 2002).

Trait	Significant effects	
Muscle score	+0.81	additive
Fat depth	-4.0 mm	additive
Muscle depth	+1.4 mm	additive
Calving difficulty score	+0.28	recessive

6. WHERE TO FIND THE BENEFICIAL GENES?

As a rule, selection programmes for a particular trait start with animals that display superior characteristics in that trait. Over successive generations, the best animals for the trait are then used for breeding. There are two inherent problems with this strategy: first, it supposes that the individuals that display the superior characteristics for a trait possess the best alleles, at least at the majority of loci. In fact there is no reason why animals with poor characteristics for the trait should not possess better alleles at a few loci. The second inherent problem, which extends the first, is that subsequent selection for a related trait will also start with the already "improved" animals, even though the prior selection was for another trait. An example of this is selection for protein production in dairy cattle that have been selected for milk yield.

6.1 Genes controlling fatness in pigs

Food preferences vary across the world. In Europe, consumers require meat that is low in fat, whereas meat with a high fat content is sought in

Japan. Selection of stock suited to particular markets has produced breeds with widely divergent characteristics. Cattle breeds, such as the Belgian Blue, discussed above, and Charolais, Limousin, etc., are widely used in Europe because they are fast growing and produce lean carcases. In contrast, the Wagu cattle breed is used to produce beef for the Japanese market because it has exceptionally high levels of intra-muscular fat. A similar situation is found in pig production, where Large White pigs are used extensively in Europe because of their lean growth, whereas the Meishan, from China, lays down large quantities of fat.

An experimental cross between Large White and Meishan pigs has been used to explore the genetic control of lean vs fat growth and fertility. A two-generation cross-bred population was established and the F_2 population slaughtered and fat content determined. The F_2 animals were also genotyped with markers covering the whole genome. A QTL analysis correlating production data with the marker information revealed several QTLs associated with carcase fat, but one QTL, on chromosome 7, had a particularly large effect, accounting for about 30% difference in back fat thickness (see Table 2).

The surprising finding, however, was that the "thin allele" originated from the Mieshan breed, which phenotypically is very fat. This QTL has now been verified in other populations.

Table 2. The allelic effect of the QTL on pig chromosome 7 on back fat thickness. (Numbers in parentheses are standard errors of the estimated allelic effect).

Position	Additive effect (mm)
Shoulder	2.5 (0.5)
Mid-Back	3.4 (0.4)
Loin	3.4 (0.4)

SOURCE: Drs A. Archibald and C. Haley, pers. comm.

6.2 Trypanosomiasis

In many countries the major concern for livestock breeding is not increasing productivity *per se*, but ensuring, firstly, that stock survive, and, secondly, that they thrive. In many areas of the world, diseases that are potentially fatal to livestock are endemic. Tropical theileriosis, a potentially fatal disease of cattle, is caused by the tick-borne *Theileria annulata* and is present in a world-spanning belt from Morocco to China. East Coast Fever, caused by the related parasite, *T. parva*, is a major obstacle to cattle production in Kenya and neighbouring eastern African countries. In West Africa, sleeping sickness in man, caused by *Trypanosoma brucei rhodesiense* and *T. gambiense*, has a related disease in cattle caused by *T. congolense* and *T. vivax*. Susceptible cattle become anaemic, lose weight and will die if not treated. Although most cattle are highly susceptible to this disease, breeds are known that show resistance, e.g. the N'Dama cattle found in West Africa, where the disease is endemic.

In order to identify the genetic control of resistance to *T. congolense* infection, an experimental population of cattle was established at the International Laboratory for Research on Animal Diseases (ILRAD; now part of the International Livestock Research Institute, ILRI) in Kenya, by crossing N'Dama with Boran cattle. The N'Dama cattle from West Africa can be infected with the parasite, but they are able to rapidly control the infection and survive without treatment. Boran cattle, however, are susceptible to trypanosomiasis and lose condition rapidly, and invariably die unless treated. The F_1 Boran $\times$ N'Dama cross-bred animals were inter-crossed to produce an F_2 population of 177 animals. Response to disease was tested by challenging the F_2 with *T. congolense*, and various measures of resistance to the disease recorded, including parasitaemia, weight loss and red blood cell volume. These traits were correlated with genotypes for 477 markers covering all 29 chromosomes to localize QTLs for resistance to the disease (Hanote *et al.*, 2003).

The study identified QTLs on 18 chromosomes that were associated with resistance traits. Interestingly, for about eight of these, the resistance allele originated from the susceptible Boran breed. There is some trypanosomiasis in areas of Sudan, where the Boran breed originated, which may explain why this breed harbours some tolerance genes. Nevertheless, the "improved" Boran individuals that were the founders for the experimental population were tested and found to be highly susceptible to the disease.

7.   USING GENOMIC DATA

Genetic improvement of livestock has, up to now, focused on those traits that are easy to measure and, in general, on those that affect profitability of agricultural enterprises. Traditionally, this meant increased output. However, using modern methods, selection for a broader spectrum of traits can be considered and other traits – such as those affecting welfare, health, disease resistance, fertility, efficiency or product quality – could be included in selection objectives. The first outputs from the gene mapping studies are information on the chromosomal locations of genes and QTLs controlling traits. This information identifies marker loci linked to the genes. These markers could be used to aid traditional selection programmes in making choices among individuals. Because markers are separated from the trait gene, recombination will occur between them and change the "phase"; thus, as discussed above, information has to be derived and used within families.

Progeny testing is often carried out using several sons of elite sires, which sometimes include full-sibs produced by embryo transfer. The estimated genetic merit for these individuals is the same, therefore the choice of which to use in a testing scheme is no better than random. Marker information on currently identified QTLs, e.g. for milk yield or composition, could be used to select the individuals carrying the highest proportion of beneficial alleles for those traits. Progeny testing remains necessary, but this pre-selection should increase the proportion of better-than-average individuals tested. The testing programmes in themselves will then confirm and strengthen the information on the QTLs.

Information on the genes controlling these traits is much more powerful than using linked markers, as the phase does not have to be determined and selection can be made directly on genotype. Gene-based selection has already been carried out in cattle; until now, this has mainly focused on selection against heritable diseases, e.g. bovine leukocyte adhesion deficiency (BLAD). However, genes are being identified that control phenotypes of production traits, e.g. information on myostatin genotypes is being used in different breeds to select both for and against double muscling.

7.1 Future perspectives

Most production traits are quantitative and hence are controlled by several genes. Knowledge of these genes will allow selection for individuals carrying the most beneficial gene at all of the loci controlling the trait, whereas phenotypic selection generally only identifies individuals with beneficial alleles at some of the loci. In theory, once sufficient knowledge is available, individuals carrying beneficial genes for several traits could be identified using DNA information. Animals could then be selected on a variety of trait combinations suited to particular environments or production goals, and matings could also be established to produce progeny with the desired characteristics. Whilst we are a long way from this level of knowledge, work is underway to identify the genes involved in a wide range of traits.

As the examples discussed above demonstrate, the behaviour of even major genes controlling a trait is dependent on genetic background. Thus information is required not only on the genes involved, but also on their interactions with other genes. Further information on gene interactions will come from gene expression studies. Micro-arrays have now been produced for the majority of livestock species, allowing the expression patterns of many thousands of genes to be assayed simultaneously. These arrays have produced valuable data when used to examine pure-cell populations, and are now being tested for the study of more complex traits.

The most beneficial alleles for a trait may not occur in the population that is apparently the most favourable for the trait. Knowledge of the genes underpinning a trait will allow a search for novel alleles across different populations, with the potential to set up breeding schemes to create and test novel combinations of alleles at different loci. Some of these novel allele combinations may allow further progress in improving traits beyond that which would be possible by selecting within breeds. Thus it is important to maintain a diverse range of genetic backgrounds to provide sources of variation.

Perhaps an application of DNA-aided selection that is closer to implementation is the introgression of genes between breeds. Introgression has been detected retrospectively for the myostatin gene, where we now know that double muscling in several different breeds is controlled by the same allele, which must have been transferred between breeds by crossing. In the future, using DNA tests, it will be much easier to transfer a particular allele of a gene, like the myostatin double-muscling allele, between breeds. Back-crossing and selection, using DNA information to track the introgressed allele, will allow recovery of the majority of the original genetic composition of the recipient breed whilst retaining the new allele. The application of this technology will be important in exploiting information on disease response, where resistance alleles could be bred into more productive breeds to allow increased productivity in areas where disease is endemic.

8.   CONCLUSIONS

Knowledge of the loci controlling individual traits will allow direct selection for favourable alleles at these loci. In the first instance, this can be done by marker-assisted selection using markers linked to the gene involved in the trait. However, ultimately, knowledge of the allelic variation within that gene will allow more efficient selection. There are several advantages of using markers in selection programmes, rather than relying on phenotype-based selection. In using markers it will be possible to introgress favourable alleles for particular traits from one breed into another, taking advantage of specialized characteristics of different breeds, e.g. to maintain disease resistance while increasing production. By using information on the markers spanning the genome, as well as the genes being selected for, it will also be possible to maintain the widest possible genetic diversity within a breed. Thus a considered and well-managed use of molecular information will help preserve the genetic diversity of livestock populations while improving production and quality of life.

REFERENCES

Barendse, W., Vaiman, D., Kemp, S., Sugimoto, Y., Armitage, S., Williams. J.L. *et al.* 1997. A medium density genetic linkage map of the bovine genome. *Mammalian Genome*, **8**: 21–28.

Bishop M.D., Kappes, S.M., Keele, J.W., Stone, R.T., Sunden, S.L.F., Hawkins, G.A., Toldo, S.S., Fries, R., Grosz, M.D., Yoo, J.Y. & Beattie, C.W. 1994. A genetic linkage map for cattle. *Genetics*, **136**: 619–639.

Bradley, D.G., Loftus, R.T., Cunningham, P. & McHugh, D.E. 1998. Genetics and domestic cattle origins. *Evolutionary Anthropology,* **6**: 79–86.

Charlier, C., Coppieters, W., Farnir, F., Grobet, L., Leroy, P.L., Michaux, C., Mni, M., Schwers, A., Vanmanshoven, P., Hanset, R. & Georges, M. 1995. The mh gene causing double-muscling in cattle maps to bovine Chromosome 2. *Mammalian Genome*, **6**: 788–792.

FAO [Food and Agriculture Organization of the United Nations]. 1993. *World Watch List for Domestic Animal Diversity.* Edited by R. Loftus and B. Scherf. FAO, Rome.

Georges, M., Nielsen, D., Mackinnon, M., Mishra, A., Okimoto, R., Pasquino, A.T., Sargeant, L.S., Sorensen, A., Steele, M.R., Zhao, X., Womack, J.E. & Hoeschele, I. 1995. Mapping quantitative trait loci controlling milk production in dairy cattle by exploiting progeny testing. *Genetics*, **139**: 907–920.

Grigson, C. 1989. Size and sex: the evidence for domestication of cattle in the near east. pp. 77–109, *in:* A. Milles, D. Williams and G. Gardener (eds). *The Beginnings of Agriculture. BAR [British Archaeological Reports] International Series*, No. 496.

Grisart, B., Coppieters, W., Farnir, F., Karim, L., Ford, C., Berzi, P., Cambisano, N., Mni, M., Ried, S., Simon, P., Spelman, R., Georges, M. & Snell, R. 2002. Positional candidate cloning of a QTL in dairy cattle: identification of a missense mutation in the bovine DGAT1 gene with major effect on milk yield and composition. *Genome Research*, **12**: 222–231.

Grobet, L., Martin, L.J.R., Poncelet, D., Pirottin, D., Brouwers, B., Riquet, J., Schoeberlein, A., Dunner, S., Ménissier, F., Massabanda, J., Fries, R., Hanset, R. & Georges, M. 1997. A deletion in the bovine myostatin gene causes the double-muscled phenotype in cattle. *Nature Genetics*, **17**: 71–74.

Hanotte, O., Ronin, Y., Agaba, M., Nilsson, P., Gelhaus, A., Horstmann, R., Sugimoto, Y., Kemp, S., Gibson, J., Korol, A., Soller, M. & Teale, A. 2003. Mapping of quantitative trait loci controlling trypanotolerance in a cross of tolerant West African N'Dama and susceptible East African Boran cattle. *Proceedings of the National Academy of Sciences, USA,* **100**: 7443–7448.

Kashi, Y., Hallerman, E. & Soller, M. 1990. Marker-assisted selection of candidate bulls for progeny testing programs. *Animal Production,* **51**: 63–74.

Kühn, C.H., Freyer, G., Weikard, R., Goldammer, T. & Schwerin, M. 1999. Detection of QTL for milk production traits in cattle by application of a specifically developed marker map of BTA6. *Animal Genetics*, **30**: 333–340.

Loftus, R.T., MacHugh, D.E., Bradley, D.G., Sharp, P.M. & Cunningham, E.P. 1994. Evidence for two independent domestications of cattle. *Proceedings of the National Academy of Sciences, USA,* **91**: 2757–2761.

McPheron, A.C. & Lee, S.-J. 1997. Double muscling in cattle due to mutations in the myostatin gene. *Proceedings of the National Academy of Sciences, USA,* **94**: 12457–12461.

Ménissier, F. 1982. General survey of the effect of double muscling on cattle performance. pp. 437–449, *in:* J.W.B. King and F. Ménissier (eds). *Muscle Hypertrophy of Genetic Origin and its Use to Improve Beef Production.* London: Martinus Nijhoff Publishers.

Meuwissen, T.H.E. 1998. Optimizing pure line breeding strategies utilizing reproductive technologies. *Journal of Dairy Science,* **81** (Suppl. 2): 47–54.

Schwenger, B., Schober, S. & Simon, D. 1993. DUMPS cattle carry a point mutation in the uridine monophosphate synthase gene. *Genomics,* **16**: 241–244.

Shuster, D. & Kehrli, M. 1993. Bovine leukocyte adhesion deficiency (BLAD) among Holstein cattle – molecular biology, immunology and clinical disease. *Journal of Immunology,* **150**: a167–a167 part 2.

Smith, J.A., Lewis, A.M., Wiener, P. & Williams, J.L. 2000. Genetic variation in the bovine myostatin gene in UK beef cattle: Allele frequencies and haplotype analysis in the South Devon. *Animal Genetics,* **31**: 306–309.

Solinas-Toldo, S., Lengauer, C. & Fries, R. 1995. Comparative genome map of human and cattle. *Genomics,* **27**: 489–496

Stone, R.T., Keele, J.W., Shackelford, S.D., Kappes, S.M. & Koohmaraie, M. 1999. A primary screen of the bovine genome for quantitative trait loci affecting carcass and growth traits. *Journal of Animal Science,* **77**: 1379–1384.

Walling, G.A., Archibald, A.L., Cattermole, J.A., Downing, A.C., Finlayson, H.A., Nicholson, D., Visscher, P.M., Walker, C.A. & Haley, C.S. 1998. Mapping of quantitative trait loci on porcine chromosome 4. *Animal Genetics,* **29**: 415–424.

Wiener, P., Smith, J.A., Lewis, A.M., Woolliams, J.A. & Williams, J.L. 2002. Muscle-related traits in cattle: The role of the myostatin gene in the South Devon breed. *Genetic Selection and Evolution,* **34**: 221–232.

FIRST REPORT ON THE STATE OF THE WORLD'S ANIMAL GENETIC RESOURCES
Views on biotechnology as expressed in country reports

R. Cardellino, I. Hoffmann and K.A. Tempelman
Animal Production and Health Division, FAO, 00100 Rome, Italy.

Abstract: As part of the country-driven strategy for the management of farm animal genetic resources, FAO invited 188 countries to participate in the First Report on the State of the World's Animal Genetic Resources, with 145 consenting. Their reports are an important source of information on the use of biotechnology, particularly biotechnical products and processes. This paper analyses information from country reports so far submitted, and is therefore preliminary. There is clearly a big gap in biotechnology applications between developed and developing countries, with artificial insemination the most common technology used in developing countries, although not everywhere. More complex techniques, such as embryo transfer (ET) and molecular tools, are even less frequent in developing countries. Most developing countries wish to expand ET and establish gene banks and cryoconservation techniques. There are very few examples in developing countries of livestock breeding programmes capable of incorporating molecular biotechnologies in livestock genetic improvement programmes.

1.   INTRODUCTION

The Food and Agriculture Organization of the United Nations (FAO) has been requested by its member countries to develop and implement the Global Strategy for the Management of Farm Animal Genetic Resources. The global livestock sector is faced with the challenge of the fast increasing demand for animal products in developing countries. FAO has estimated that demand for meat will double by 2030 (2000 basis) and demand for milk will more than double in this 30-year period.

89

H. P. S. Makkar and G. J. Viljoen (eds.), Applications of Gene-Based Technologies for Improving Animal Production and Health in Developing Countries, 89-98.

Animal genetic resources, however, are disappearing worldwide at a fast rate. Over the past 15 years, 300 out of 6000 breeds identified by FAO have become extinct (FAO, 1993, 1999). There are few successful genetic improvement programmes in adapted indigenous breeds. In many developing countries, considerable efforts have been made in training professionals in animal genetics, but breeding programmes applied to livestock for low-input farming systems have largely failed.

As part of the country-driven strategy for the management of farm animal genetic resources, FAO has invited 188 countries to participate in the First Report on the State of the World's Animal Genetic Resources, to be completed before 2006. To date, 145 countries have agreed to submit country reports. Drafting of country reports has been under way in most of these countries, with the end of 2003 as a target date for submission. Country reports will hopefully reflect problems, needs and opportunities for the management of farm animal genetic resources at country and regional levels. Country reports are basically strategic policy documents approved by FAO's governing bodies and designed to answer three questions regarding animal genetic resources: where the country is now, where it needs to be, and how to get there (FAO, 2001).

Each country report has five main elements. Part 1 reports on the state of genetic resources in the farm animal sector, covering both *in situ* and *ex situ* conservation, as well as the state of the art of techniques being used in the context of both local production systems and socio-economic conditions. Part 2 describes the changing demands on the farm animal sector and the implications for future national policies in conservation and utilization of animal genetic resources. Part 3 is a review of the state of national capacities related to farm animal genetic resources and an overall assessment of capacity building requirements. Part 4 contains identification of national priorities covering diverse fields of activity, animal species and breeds as well as short- and long-term needs for institutional building, research, information systems, policy, legislation and regulations. Part 5 deals with recommendations for international cooperation, indicating the areas and modes of cooperation that the country wishes to follow, as well as proposed contributions and requirements.

FAO expected that countries would include views on biotechnology in relation to farm animal genetic resources, within the context of the recommended country report structure, particularly in parts 1 and 4. The country reports therefore represent a prime source of information on the state of the use of biotechnology. This paper focuses on the extraction from the reports of information on the use of biotechnical products and processes. It should be stressed that this is only a preliminary report based on an analysis of a still-expanding database, therefore the conclusions are not final. It also focuses on animal genetics and breeding, not covering uses of biotechnology for animal health.

2. MATERIAL AND METHOD

By October 2003, 47 country reports had officially been submitted to FAO, of which 41 – representing all regions – were analysed for this paper. Information on biotechnology used in animal breeding and reproduction, conservation of animal genetic resources and commercial purposes was examined. In addition, 32 draft country reports and information from the regional facilitators involved in supporting countries in report preparation were included in the analysis. Information also came from discussions held during regional workshops in Latin and Central America, covering ten countries each. The results are presented descriptively. No statistical analysis is possible because the information available is not the result of a formal sampling procedure. No individual countries are mentioned in the present paper, but the regions and sub-regions where the information originated are identified. The number of countries within each region in which specific events occur are reported. Where useful, comparisons between developed and developing countries are presented.

3. RESULTS

3.1 West Africa

West Africa is represented by 15 country reports. With some exceptions, artificial insemination (AI) is or has been used to a limited extent, mostly in cattle in the vicinity of AI centres. In general, infrastructure for the use of AI is described as almost non-existent, very poor or disappearing. Eight countries perform AI on local cattle using semen of imported exotic breeds with the purpose of improving dairy production in the favourable areas of the country or in the context of peri-urban dairy production. Two of these countries have cooperation programmes with universities from developed countries, and, in these projects, AI is the main component as a tool for cross-breeding. In two other countries of the region, AI is used to upgrade trypanotolerant cattle with breeds of larger body size, either exotic or local, but more susceptible to trypanosomiasis. In three countries, AI is utilized for experimental purposes only. One country noted that AI is seen as a cause of erosion of local animal genetic resources when practiced indiscriminately for cross-breeding in cattle and horses with commercial objectives. AI is mainly used in public, government-financed programmes, with no clear plans or genetic improvement objectives. No policy or legislation issues regarding the use of AI were mentioned, with the exception of one country planning to declare AI and embryo transfer (ET) a state-controlled activity. Three countries regard AI as essential for the development of *in situ* conservation programmes.

ET is only used in a few specific cases, such as the introduction of a South American zebu breed via ET into one country of the region. In another country, embryos of a local breed are conserved at two locations. Two countries specifically indicate the absence of cryoconservation of semen, embryos or oocytes, and the absence of cryoconservation can be assumed for practically all countries. Many countries express the need for the development and use of ET in the future, but without stating specific intentions for its use. Limited molecular characterization has been carried out in some countries, mainly as part of international development projects and research studies. These were, however, of limited scope and duration.

Capacity building and training in AI and ET in the context of performance and genetic evaluations of livestock, and molecular techniques aimed at characterization of local animal genetic resource, were reported as priorities. Other countries expressed a need for AI/ET and molecular techniques, but without specifying whether for characterization only or other purposes. Major constraints are financial resources and the lack of skilled human resources to undertake in-country training in the respective technologies.

3.2   Eastern Europe

Eastern Europe was represented by ten country reports from three regions: Caucasus, Baltic and the Eastern countries of Europe. In all countries of the Caucasus region, infrastructure for carrying out AI has practically disappeared after the collapse of the former Soviet Union, and even basic activities such as animal identification and registration no longer exist. In two countries, AI is used in cattle in a very limited fashion. One country reports use of ET in cattle at an experimental level. All countries mention capacity building in AI for both breeding and conservation purposes as priorities, and the creation of the necessary infrastructure. There are no animal gene banks in these countries.

In the Baltic countries, AI is widely used in cattle, pigs and horses. These activities are often connected to national AI programmes and breeding programmes through breeders' associations. ET is used experimentally in a limited number of cases, but two countries also report some commercial use.

AI and ET are mentioned as tools to be utilized in implementing national plans for conservation and management of animal genetic resources. All countries have legal instruments to regulate AI and ET. Some research on molecular characterization of local breeds using microsatellite markers is being carried out, and DNA is sampled in cattle, sheep and pigs. All countries have gene banks in place, all still requiring further development. The setting up of gene banks, building technical expertise in AI and ET, and cryoconservation are reported as priorities. The major limiting factor mentioned is the availability of financial resources for carrying out activities of conservation and management of animal genetic resources based on biotechnical techniques.

In the Eastern European region, one country reports very limited use of AI in cattle as the only biotechnique currently practiced. AI is used in cattle in the other three countries, and also in pigs in two of these three. Only one country reports the existence of a semen bank, without specifying if it involves commercial or conservation activities. In all countries, ET is used experimentally and in a limited number of cases also commercially. One country mentions use of ET for conservation of animal genetic resources. Research projects using microsatellites and DNA sampling in pigs are carried out in two countries, while one country also conducts DNA sampling in cattle and sheep.

3.3   Latin America and the Caribbean

As a result of two regional meetings on country report preparation, and identification of national and regional priorities, ten South American and ten countries in Central America and the Caribbean identified AI as the most effective means of diffusion of genetic improvement. ET is used commercially in a limited way because of high costs, but it was considered a biotechnology of interest for the future. Because of the different approach – analysis of regional workshops instead of country reports – the results reflect regional rather than national priorities.

In both meetings, AI and ET capacity building were considered a regional need, with some countries being able to provide consultants and services in these areas to other countries in the region. Emphasis was placed on the need to conduct projects on molecular characterization of local breeds, especially criollo cattle, sheep and goats, with regional coordination for increased efficiency. All countries identified the need for national programmes in conservation and utilization of animal genetic resources, including development of biotechnology and updated legislation to deal with exchange of genetic material and intellectual property (IP) issues related to animal genetic resources and to products modified or obtained as a result of the applications of biotechnology. *Ex situ* conservation was seen as important at the regional level, and research and capacity building in cryoconservation were identified as priorities. A model suggests establishing a gene bank in each country for conservation and distribution of material, with specific objectives. Collaboration between these gene banks would subsequently function as a regional germplasm bank. Regulated access and material transfer agreements (MTAs) should be jointly developed. FAO was requested to support, coordinate and help find financial means for all these regional activities.

3.4   Central Asia

Six Central Asian countries reported very limited infrastructure as a constraint on the application of biotechnology, in addition to obstacles to identification and registration of livestock. Although in some countries AI centres were formerly active and mostly state-run, AI currently has very limited use and only in cattle. Two countries mention some use of ET in an experimental context. Training of experts in AI and the establishment of gene banks are considered a priority. One country proposes the establishment of laboratories to produce cattle, sheep and goat embryos, and to expand to bactrian camels and yak. This country reports storing of frozen semen of highly productive breeds and breeds at risk of extinction.

3.5   West and east Asia

Three countries in west and east Asia reported that AI is commonly used. In one country, private rural households are entitled to AI services, whereas in the other two countries AI is organized by provincial or district governments, mainly in nucleus herds and small-scale commercial farm units under intensive or semi-intensive management. The use of ET is limited to an experimental scale, mainly due to costs of commercial applications and to few existing facilities. Priorities noted include additional training in applying AI in particular, and expanding the national gene pool by the creation of gene banks for cryoconservation, in general. Updating existing regulations for conservation of animal genetic resources is also a priority. One of these countries reported the use of microsatellite DNA technology for research on molecular genetic diversity. No specific priorities were mentioned, but the country is implementing plans for a centre for animal germplasm.

3.6 Western and southeastern Europe

Reports from 15 Western and southeastern European countries represented the more developed countries. All countries have national AI programmes in place and AI is used widely throughout the farming sector in private commercial enterprises. AI is mainly used in dairy cattle, but also in beef cattle, pig, sheep, horses and poultry. In southeastern European countries (3), most AI activities depend on governmental support and ET is not used except for one state-run project on the application of ET in cattle breeding for multiple ovulation and embryo transfer (MOET) schemes. All countries of southeastern Europe report the need for capacity building in AI, and eventually ET.

In countries of Western Europe (12), commercial use of ET is common in cattle. There are national and private gene banks for both commercial and conservation purposes in all countries. Several countries have a dual strategy of conserving material of rare breeds and backing-up commercial breeds in the form of semen, embryos and oocytes. Some countries carry out research in the production of transgenic animals for pharmaceutical and production purposes, in cloning and other advanced biotechnologies, all of which are well documented in the available scientific literature. The commercial use of transgenics was not reported. Several countries report activities in the use of molecular markers for animal breeding, and one country sets as priority the integration of new biotechnologies with traditional animal breeding expertise. Activities are carried out in DNA-typing to confirm pedigree information and its eventual use in traceability. Research in sperm and embryo sexing is mentioned by a few countries. Priorities are expressed in cryoconservation of genetic material, molecular characterization of endemic animal genetic resources, expansion of gene bank activities, and changes required in national and international policies on the use, conservation and exchange of genetic material.

3.7 Other regions

The Near East was represented by one country report. AI is mostly used in cattle, no ET is used, and work is in progress regarding regulation on use of genetically-modified organisms (GMOs). Priorities were expressed for the creation of ET facilities, training in new biotechnological methods and establishment of gene banks. Major constraints are funding and the lack of skilled human resources. One report from a country in the Arab Peninsula indicates that AI is only used experimentally, although it is seen as a way to expand breeding programmes, and ET is not used.

Two countries in East Africa indicate the use of AI to improve dairy production in the favourable areas and in peri-urban systems, through crossbreeding with exotic genetic material. One country has stored small samples of germplasm of two indigenous breeds. The other country representing the region has a characterization research project, with a molecular component, funded by a donor government. Capacity building in basic biotechnology is considered a priority.

4.   CONCLUSION

The various sources of information used for this paper provide a coherent picture. The gap in biotechnology application between developed and developing countries is reconfirmed by the country reports, and this gap is not expected to change as the analysis of additional country reports progresses.

In developing countries, AI is the most common technology used, but it is not used in all countries. This is due to organizational, logistical and farmer training constraints. Farm-level problems occur in heat detection of females, which is difficult to achieve in extensive pasture-based systems. Even in developed countries, AI is less frequent in beef cattle than in dairy cattle, where close, daily contact with humans facilitates heat detection and the handling or fixation of the cow during the AI procedure. Even if farmers are trained to detect heat, poor communication networks or road infrastructure may hinder the timely insemination of the cow. The liquid nitrogen supply needed to deep-freeze the semen is also a constraint in some countries. In such environments, pregnancy rates after AI can be very low and the use of natural service is preferred since genetic gains may be outweighed by higher fertility.

The more complex and demanding techniques, such as ET or molecular techniques, are applied even less frequently in developing countries. This is more pronounced for GMOs, which are only mentioned to express the lack of proper regulations and guidelines for their eventual production, use and exchange.

Most countries expressed a desire for the introduction or expansion of ET and, to a lesser extent, molecular technologies. The majority of these requests do not indicate clear objectives for the use of these techniques or their use as a tool in a targeted process of performance recording and genetic improvement. Therefore, this seems more a wish of not being left behind in technology development than a part of the national livestock development strategy.

Increased awareness of the value of local animal genetic resources is evident from the fact that country reports often mention AI as being introduced without proper planning, and could be seen as a potential threat to the conservation of local breeds. Breed characterization is perceived as becoming important, and many countries that have expressed a wish for the introduction and development of molecular techniques regard them as a complement to phenotypic breed characterization. Cryoconservation was identified as a priority by most countries and gene banks were recommended, but at the same time funding and local technical capacity remain major constraints.

The following policy recommendations can be drawn from this very preliminary analysis of country reports:

- To promote a strong public sector involved in biotechnology issues dealing with conventional research and development. Together with strategic policy setting, training and extension, and all related institutional and regulatory matters, it would offer farmers a range of technology options.
- To involve farmers, their associations and NGOs in priority setting to ensure that decision-making is a demand-driven process.
- To promote animal identification and performance recording as prerequisites for efficient breeding programmes.

While the use of biotechnology has expanded in developed countries, there has been limited public investment in animal biotechnology in most developing countries, and only modest support for more conventional livestock research and development to improve productivity, nutrition and the health of farm animals. It should be emphasized that biotechnology, as with all other technologies, is merely a tool in achieving the goal of improved livestock production for improving the livelihoods of farmers, particularly of poor farmers in developing countries.

Few livestock breeding programmes exist that are capable of incorporating molecular biotechnology in livestock genetic improvement activities in developing countries. Marker-assisted selection, for example, could be used in programmes for improving livestock breeds. This situation is unlikely to change without significant investment in the public and private sectors. It should be noted that such programmes would need linkages to strong conventional animal breeding programmes, since the interpretation of the genomics data requires information on observed production traits.

FAO will continue its efforts to implement the Global Strategy for the Management of Farm Animal Genetic Resources and support priority actions in the utilization, development and conservation of animal genetic resources as developed by the countries.

REFERENCES

FAO [Food and Agriculture Organization of the United Nations]. 1993. *World Watch List for Domestic Animal Diversity.* 1st edition. Edited by R. Loftus and B. Scherf. Rome: FAO.

FAO. 1999. *World Watch List for Domestic Animal Diversity.* 3rd edition. Edited by B.D. Scherf. Rome: FAO.

FAO. 2001. *Animal Genetic Resources Information*, 30. See: www.fao.org/DAD-IS.

DEVELOPMENT OF GERMLINE MANIPULATION TECHNOLOGIES IN LIVESTOCK

C. Bruce A. Whitelaw
Department of Gene Expression and Development, Roslin Institute, Roslin, Midlothian EH25 9PS, UK

Abstract: Genetic improvement by conventional breeding is restricted to those genetic loci present in the parental breeding individuals. Gene addition through transgenic technology offers a route to overcome this restriction. The transgene can be introduced into the germ cells or the fertilized zygote, using viral vectors, by simple co-culture or direct micro-injection. Alternatively, the transgene can be incorporated into a somatic cell, which is then incorporated into a developing embryo. This latter approach allows gene-targeting strategies to be employed. Using pronuclear injection methods, transgenic livestock have been generated with the aim of enhancing breeding traits of agricultural importance, or for biomedical applications. Neither has been taken beyond the development phase. Before they are, in addition to issues of commercial development, basic technological issues addressing inefficiency and complexity of the methodology need to be overcome, and appropriate gene targets identified. At the moment, perhaps the most encouraging development involves the use of viral vectors that offer increased simplicity and efficiency. By combining this new technology with transgenes that evoke the powerful intracellular machinery involved in RNA interference, pioneering applications to generate animals that are less susceptible to infectious disease may be possible.

1.   INTRODUCTION – GENETIC MODIFICATION

Man has been genetically modifying livestock (and companion animals) for centuries. The traditional approach involves breeding two animals and

H. P. S. Makkar and G. J. Viljoen (eds.), Applications of Gene-Based Technologies for Improving Animal Production and Health in Developing Countries, 99-109.

choosing the desired offspring for further breeding. The breeder specifically selects (unidentified) allelic variations of a gene or genes for the resulting phenotype. More recently, the genomic revolution has enabled selection to be performed on genotype, increasing the sophistication of this technology. Alternatively, genetic modification involves the introduction of genetic material by laboratory and surgical methods, with the generated animal being termed transgenic.

This presentation introduces the range of nuclear technologies that are available for germline manipulation of livestock, focusing on the generation of transgenic livestock. The limitations of this technology are discussed while identifying what has been achieved, concluding with a view of what may be the next revolution in the exciting opportunities evident in the application of this technology.

2.   HOW TO INTRODUCE THE TRANSGENE

Although robust and successful, conventional breeding is limited in that animals produced by mating two selected individuals are a genetic mixture of the two parents. Thus, unknown or undesirable traits can be co-selected inadvertently. Even more fundamental, the genetic improvement desired in the offspring is restricted to those genetic loci present in the parental breeding individuals. Gene addition through transgenic technology offers a route to overcome these limitations.

A transgenic animal is one that carries a piece of DNA – the transgene – inserted into its germline, such that the transgene is inherited in a Mendelian manner. The transgene can be a copy of a gene already present in the host animal or be derived from a different species. For example, there are transgenic mice that glow green because they carry a transgene encoding the green fluorescent protein from a jellyfish. Alternatively, the transgene can be a hybrid gene, i.e. a composite of two or more different genes. Transgenic livestock therefore overcome some of the limitations of classical animal breeding regimes, where importation of genes by cross-breeding is limited to those traits already present within a given species (Table 1).

Table 1. Milestones in the development of transgenic livestock technology.

Date	Event	Reference
1981	First transgenic mice	Gordon and Ruddle, 1981
1985	First transgenic livestock	Hammer *et al.*, 1985
1996	Nuclear transfer technology developed	Campbell *et al.*, 1996
2000	First gene targeting in livestock (knock-in)	McCreath *et al.*, 2000

The technical step is how to incorporate the transgene into the germline. There are two approaches to this challenge. The first approach is to introduce the transgene into the germ cells or fertilized zygote. This can be accomplished by the use of viral vectors, by simple co-culture or by direct micro-injection. The second approach is to incorporate the transgene into a cell that is then incorporated into a developing embryo.

3.   PRONUCLEAR INJECTION

In the early 1980s, it quickly became apparent that technology enabling the transfer of genes into the mouse genome (Gordon and Ruddle, 1981) would revolutionize how biology would be investigated. It was plainly obvious that this same technology offered the potential for completely new and exciting opportunities if it could be applied to livestock. These first studies utilized the pronuclear injection technology. This is a conceptually simple method that has proven robust since the first generation of transgenic livestock was reported (Hammer *et al.*, 1985). Nevertheless it has been limited in its application. The reason for this is twofold. First, it is inefficient, with at best 1 percent of injected eggs resulting in a transgenic animal (Nottle *et al.*, 2001). This makes it a very expensive technology. Secondly, the conundrum of what trait and which target gene should be manipulated has perplexed the field.

Notwithstanding these issues, pronuclear injection has proven to be robust and lead the way in genetically modified (GM) animals. Two examples will be discussed. The first highlights the issue of what gene for which trait. The second describes the use of transgenic livestock as an alternative route to produce biomedically important proteins. It is worth emphasizing that, to date, the use of transgenic animals has been restricted to experimental studies or a restricted number of biomedicine-targeted applications.

4.   ENHANCING GROWTH

There was much speculation about to what transgenic technology could be applied. It was postulated that transgenesis would allow the generation of farm animals with altered phenotype, with a focus on enhancing growth rate and carcass composition or improving disease resistance. The enhanced growth claim was fuelled by the dramatic growth of mice carrying a transgene encoding for growth hormone (Palmiter *et al.*, 1992). Transgenic pigs were generated that carried the same or similar transgene, but very little growth enhancement was observed. In contrast, these animals were considerably disabled, suffering from widespread deleterious effects, including susceptibility to stress, lameness and reduced fertility (Pursel *et al.,* 1989). Although some progress has been achieved using different transgenes, this approach has not been successful overall, and has not been taken up commercially. Nevertheless, the lasting image of mice the size of rats still motivates thinking on this topic.

5.   ANIMAL BIOREACTOR

An alternative use of transgenesis was to generate animal bioreactors, and it is this application that has seen the most progress (Wilmut and Whitelaw, 1994; Houdebine, 2000). Many of the proteins currently used to treat the vast array of diseases from which man suffers are purified from blood. An alternative production route involves *in vitro* cell culture fermentation systems. The blood route suffers from limited supplies and the attendant risk of contamination with infection agents, while the second is expensive. These limitations encouraged the development of transgenic animals that produced human pharmaceuticals in a body fluid that could be easily and routinely harvested, i.e. milk.

The first demonstration that transgenic livestock could produce a human protein in their milk was more than a decade ago. Although the levels of, in this study, human factor IX (Clark *et al.*, 1989) were low, subsequent reports clearly demonstrated that high levels of a human pharmaceutical protein could be produced in this way. The intervening years have seen this approach being tweaked in several directions, with an ever-increasing list of proteins being produced in the milk of a number of animal species: sheep, cattle, pigs, goats, rabbits and mice (Wilmut and Whitelaw, 1994; Houdebine, 2000). In some cases, very high levels of expression have been achieved. Thus, "Tracy" the transgenic sheep produced more than 30 g/litre of human protease inhibitor α1-antitrypsin in her milk (Wright *et al.,* 1991).

This application is now primarily in the commercial biotechnology arena and success relies as heavily on product development as financial survival. Historically, these studies have fuelled the development of gene transfer technologies in livestock.

6.   VIRAL VECTORS

Viruses are highly evolved gene delivery systems. There is currently much anticipation that gene-delivery vectors based on viral sequences will enable gene therapy to become a reality (Thomas, Erhardt and Kay, 2002; Logan, Lutzo and Kohn, 2002). The approach taken is to remove as much of the viral sequence as possible, thereby reducing the potential for vector-induced disease. Initial attempts to generate transgenic animals utilized oncoretroviruses such as the Mouse Mammary Tumour Virus (MMTV) Chan *et al.*, 1998). Unfortunately, first-generation vectors proved to be poorly expressed, with transgene activity often silenced through generic genomic gene inactivation mechanisms, such as DNA methylation (Jahner and Jaenisch, 1985). Efforts to use adenovirus vectors have worked, but questions regarding vector toxicity and the low rates of success hamper this method (Gordon, 2002). Recently, it has been demonstrated that lentiviral vectors can be used to efficiently introduce foreign DNA into the mouse germline (Lois *et al.*, 2002; Pfeifer *et al.*, 2002).

7.   ALTERNATIVE METHODS

Most transgenic livestock have been produced by pronuclear injection, while only a handful to date have been produced using viral vector technology. Several attempts to develop variations on these methods have been reported. The transgene can be introduced by injection, viral vector or simple co-culture into the germ cells themselves. Limited effort has been applied to manipulating the oocyte. Instead, attention has focused on sperm. In particular, claims that co-culturing sperm and DNA resulted in the generation of transgenic mice (Lavitrano *et al.*, 1989) were initially questioned (Brinster *et al.*, 1989). This work has continued, and there is some anticipation that success will be reported soon. It is worth noting that if it can be demonstrated that sperm-mediated gene transfer (SMGT) is robust, then by combining it with intra-cytoplasmic sperm injection (ICSI) (Chan *et al.*, 2001) this could prove a very attractive route for the generation of transgenic animals.

8.   CELL-BASED METHODS

All the above methods are limited in that they can only add a (trans)gene to the genome. In the mouse, techniques for removing or knocking out genes were established during the 1980s (Smithies *et al.*, 1985; Capecchi, 1989). This development has revolutionized all aspects of modern-day biology by enabling the direct assessment of gene function *in vivo*. This technique is based on the ability of DNA sequences to recombine, albeit very infrequently, with the identical partner in the genome. This technology is now very sophisticated: genes can be knocked out in specific tissues and single base-pair mutations can be introduced into a selected gene. The technology relies upon targeting endogenous genes by homologous recombination in totipotent embryonic stem (ES) cells in culture. These cells are then re-introduced into the early embryo to colonize the tissues of the developing mouse. Unfortunately, to date, no livestock ES cells have been isolated (Stice, 2002).

The lack of methods for gene knock-out in livestock was the driving force behind the development of nuclear transfer technology (Campbell *et al.*, 1996). This technique was made famous through the generation of "Dolly" (Wilmut *et al.*, 1997). Importantly, nuclear transfer used somatic cells grown in culture, and therefore offered a completely new route for precise sequence-directed changes to the germline of livestock. In theory, this would bypass the need for ES cells and therefore, for the first time, offered the potential of gene targeting in livestock.

This has now been shown to be possible (McCreath *et al.*, 2000; Denning *et al.*, 2001a). However, the generation of knock-out transgenic livestock is technically and financially demanding. The techniques utilized efficiently in mice are considerably less effective in farm livestock cells. Secondly, the stringent cell culture phase required for targeting somatic cells reduces their developmental potential (Denning *et al.*, 2001b), dramatically reducing the efficiency of this technology. In addition, nuclear transfer in livestock is beset by losses *in utero* and postnatal (Wilmut, 2002), having both a welfare and an economic cost. Finally, the genetic modification is recessive, unless animals are bred together or a second targeting event is engineered. The latter is possible (Phelps *et al.*, 2003), but requires considerable effort and as of yet lacks precision.

In summary, nuclear transfer enables gene-targeting strategies to be applied to livestock, but, unless there is a conceptual leap in our understanding of the technique, it will not become commonplace.

9.   AGRICULTURAL TRAITS

Both gene addition and gene targeting are inefficient methods for modifying the livestock germline. Perhaps more importantly, the required introgression programme incorporating backcrossing would result in a loss of selection for other traits. The result is that the benefit of the transgene must substantially exceed that which could be achieved by conventional selection. This has been calculated to be at least a factor of 10 (Gamma, Smith and Gibson, 1992), although this value could be reduced through geographical or political issues. Even if this were possible, complex strategies would need to be implemented to avoid inbreeding depression, which would have the effect of reducing the overall productivity of the animals.

As a consequence, gene addition and gene targeting have been restricted to primarily biomedical rather than agricultural applications. For example, although it is possible to generate animals lacking a copy of the scrapie sensitive gene PrP (Denning *et al.*, 2001a), it is difficult to imagine how this could be introgressed and maintained in the homozygous state in large populations.

10. TECHNICAL HURDLES

Transgenic animals can be generated. Indeed, transgenic animals exist, but what are the technical limitations that prevent their widespread use? First, all the current technologies, regardless of the overinflated claims, are inefficient. More efficient methods would reduce cost. Second, the techniques used are highly technical, demanding both expertise and specialized equipment. This is particularly so for gene targeting methods. Simpler methods would result in wider uptake of the technology. Finally, there is the issue of "Which gene?" It is relatively easy to identify appealing traits – growth, fertility, disease – but much more difficult to identify the specific genetic targets to be modified.

Emerging new technologies, however, may soon revolutionize the scope and efficiency of the genetic modification of livestock, which in turn could allow application of transgenic technologies to modify agriculturally significant characteristics of livestock.

Table 2. Major achievements in transgenic livestock.

Year	Event	Reference
1989	First animal bioreactors (for human factor IX)	Clark *et al.*, 1989
1989	Pigs expressing growth hormone	Pursel *et al.*, 1989
1991	High-level protein production in milk	Wright *et al.*, 1991
1992	First attempt at disease resistance (influenza in pigs)	Muller *et al.*, 1992
1994	Pigs expressing human complement inhibitors	Fodor *et al.*, 1994
1996	Transgenic attempt to improve wool production	Damak *et al.*, 1996
1997	'Dolly' the sheep created	Wilmut *et al.*, 1997
2000	First gene targeting in livestock	McCreath *et al.*, 2000
2001	Transgenic pigs produced addressing environmental issue	Golovan *et al.*, 2001

11.   THE FUTURE FOR TRANSGENIC LIVESTOCK

There is considerable effort required to overcome the issues of inefficiency, simplicity and target. At the moment, perhaps the most motivating is the potential that lentiviral vectors (Lois *et al.*, 2002; Pfeifer *et al.*, 2002; Tiscornia *et al.*, 2003) offer in combination with the truly revolutionary molecular tool of RNA interference (RNAi) (McManus and Sharp, 2002; Hannon, 2002).

Lentiviral vectors appear to offer an extremely efficient method to generate transgenic animals. In previous studies, using pronuclear injection, about 70 sheep were required to make just one transgenic founder. In contrast, using lentiviral vectors in combination with *in vitro* matured and fertilized oocytes, we estimate that as few as 5 animals will be required! This technique has been developed in mouse studies, but there is no reason why these efficiencies will not be the same for livestock.

Even more appealing is the simplicity of delivery, abrogating the need for specialized equipment. Lentiviral vectors can be delivered by injection into the perivitelline space of the fertilized egg (Lois *et al.*, 2002) or, after removal of the zona pelucida, simply by incubating the denuded eggs in a viral solution (Pfeifer *et al.*, 2002). Looking to a transgenic future for livestock, this aspect – simplicity – could transfer this technology from the research lab to veterinary practices.

The potential of combining these gene delivery efficiencies with the revolutionary new technology of RNAi is exciting. RNAi enables the sequence-specific destruction of target mRNAs (McManus and Sharp, 2002). This is termed gene knockdown and has long been recognized as a major mechanism of post-transcriptional gene silencing in *C. elegans*, *Drosophila* and plants (Hannon, 2002). The first steps have been accomplished as gene expression knockdown has been demonstrated in transgenic mice (Tiscornia *et al.*, 2003).

This strategy is all the more attractive as it is genetically a dominant trait. In theory, it can be used to knockdown any target mRNA. This could be an endogenous gene controlling growth or fertility. Alternatively, it could be used to combat disease in animals. The easiest targets would be RNA molecules produced by or as part of an infectious agent. In this respect, it is worth noting that the majority of the OIE List A diseases are caused by RNA viruses. The prospect of reducing the incidence of animal diseases is truly exciting. As yet, this is only a desire, but the potential benefit offered to agriculture, and societies, around our world must make it worthy of support.

12.   CONCLUSION

It is now 18 years since the first transgenic livestock were reported. Much has happened in the intervening years. The techniques used to produce transgenic livestock range from conceptually simple pronuclear injection, first developed in mice, to more complex and as yet mechanistically unknown nuclear transfer with, perhaps, a new era in this technology being heralded by the development of viral-vector-based approaches.

Although transgenic livestock have had a high profile, courting much attention from both the pro- and anti- camps, do we have fields filled full of GM animals? No. This is not to say that considerable effort has been employed in developing transgenic technology and numerous proof-of-principle studies performed. Rather, taking the concept from the developmental phase to commercial and breeding reality has not happened. By combining new technology with transgenes that evoke the powerful intracellular machinery involved in RNAi, innovative approaches to increase our understanding of biology through the use of transgenic animals and, perhaps, pioneering application of these technologies to generate animals that are less susceptible to infectious disease, may be possible (Clark and Whitelaw, 2003).

ACKNOWLEDGEMENTS

The author's research is supported by the Biotechnology and Biological Sciences Research Council (BBSRC) (UK). Thanks go to the many colleagues at Roslin Institute who share my enthusiasm for transgenic technology.

REFERENCES

Brinster, R.L., Sandgren, E.P., Behringer, R.R. & Palmiter, R.D. 1989. No simple solution for making transgenic mice. *Cell,* **59**: 239–241.

Campbell. K.H., McWhir, J., Ritchie, W.A. & Wilmut, I. 1996. Sheep cloned by nuclear transfer from a cultured cell line. *Nature,* **380**: 64–68.

Capecchi, M.R. 1989. Altering the genome by homologous recombination. *Science,* **244**: 1288–1292.

Chan, A.W., Chong, K.Y., Martinovich, C., Simerly, C. & Schatten, G. 2001. Transgenic monkeys produced by retroviral gene transfer into mature oocytes. *Science,* **291**: 309–312.

Chan, A.W., Homan, E.J., Ballou, L.U., Burns, J.C. & Bremel, R.D. 1998. Transgenic cattle produced by reverse-transcribed gene transfer in oocytes. *Proceedings of the National Academy of Sciences, USA,* **95**: 14028–14033.

Clark, A.J. & Whitelaw, C.B.A. 2003. A future for transgenic livestock. *Nature Reviews Genetics,* **4**: 825–833.

Clark, A.J., Bessos, H., Bishop, J.O., Brown, P., Harris, S., Lathe, R., McClenaghan, M., Prowse, C., Simons, J.P., Whitelaw, C.B.A. & Wilmut, I. 1989. Expression of human anti-hemophilic factor IX in the milk of transgenic sheep. *Biotechnology,* **7**: 487–492.

Damak, S., Jay, N.P., Barrell, G.K. & Bullock, D.W. 1996. Targeting gene expression to the wool follicle in transgenic sheep. *Biotechnology,* **14**: 18118–18114.

Denning, C., Burl, S., Ainslie, A., Bracken, J., Dinnyes, A., Fletcher, J., King, T., Ritchie, M., Ritchie, W.A., Rollo, M., de Sousa, P., Travers, A., Wilmut, I. & Clark, A.J. 2001a. Deletion of the alpha(1,3)galactosyl transferase (GGTA1) gene and the prion protein (PrP) gene in sheep. *Nature Biotechnology,* **19**: 529–530.

Denning, C., Dickinson, P., Burl, S., Wylie, D., Fletcher, J. & Clark, A.J. 2001b. Gene targeting in sheep and pig primary fetal fibroblasts. *Cloning Stem Cells,* **3**: 205–215.

Fodor, W.L., Williams, B.L., Matis, L.A., Madri, J.A., Rollins, S.A., Knight, J.W., Velander, W. & Squinto, S.P. 1994. Expression of a functional human complement inhibitor in a transgenic pig as a model for the prevention of xenogeneic hyperacute organ rejection. *Proceedings of the National Academy of Sciences, USA,* **91**: 11153–11157.

Gamma, L.T., Smith, C. & Gibson, J.P. 1992. Transgene effects, introgression strategies and testing schemes in pigs. *Animal Production,* **54**: 427–440.

Golovan, S.P., Meidinger, R.G., Ajakaiye, A., Cottrill, M., Wiederkehr, M.Z., Barney, D.J., Plante, C., Pollard, J.W., Fan, M.Z., Hayes, M.A., Laursen, J., Hjorth, J.P., Hacker, R.R., Phillips, J.P. & Forsberg, C.W. 2001. Pigs expressing salivary phytase produce low-phosphorus manure. *Nature Biotechnology,* **19**: 741–745.

Gordon, J.W. 2002. High toxicity, low receptor density, and low integration frequency severely impede the use of adenovirus vectors for production of transgenic mice. *Biology of Reproduction,* **67**: 1172–1179.

Gordon, J.W. & Ruddle, F.H. 1981. Integration and stable germ line transmission of genes injected into mouse pronuclei. *Science,* **214**: 1244–1246.

Hammer, R.E., Pursel, V.G., Rexroad, C.E., Wall, R.J., Bolt, D.J., Ebert, K.M., Palmiter, R.D. & Brinster, R.L. 1985. Production of transgenic rabbits, sheep and pigs by microinjection. *Nature,* **315**: 680–683.

Hannon, G.J. 2002. RNA interference. *Nature,* **418**: 244–251.

Houdebine, L.M. 2000. Transgenic animal bioreactors. *Transgenic Research,* **9**: 305–320.

Jahner, D. & Jaenisch, R. 1985. Terovirus-induced de novo methylation of flanking host sequences correlates with gene inactivity. *Nature,* **315**: 594–597.

Lavitrano, M., Camaioni, A., Fazio, V.M., Dolci, S., Farace, M.G. & Spadafora, C. 1989. Sperm cells as vectors for introducing foreign DNA into eggs: genetic transformation of mice. *Cell,* **57**: 717–723.

Logan, A.C., Lutzko, C. & Kohn, D.B. 2002. Advances in lentiviral vector design for gene-modification of hematopoetic stem cells. *Current Opinions in Biotechnology,* **13**: 429–436.

Lois, C., Hong, E.J., Pease, S., Brown, E.J. & Baltimore, D. 2002. Germline transmission and tissue-specific expression of transgenes delivered by lentiviral vectors. *Science,* **295**: 868–872.

McCreath, K.J., Howcroft, J., Campbell, K.H., Colman, A., Schneike, A.E. & Kind, A.J. 2000. Production of gene-targeted sheep by nuclear transfer from cultured somatic cells. *Nature,* **405**: 1066–1069.

McManus, M.T. & Sharp, P.A. 2002. Gene silencing in mammals by small interfering RNAs. *Nature Reviews Genetics,* **3**: 737–747.

Muller, M., Brenig, B., Winnacker, E.L. & Brem, G. 1992. Transgenic pigs carrying cDNA copies encoding the murine Mx1 protein which confers resistance to influenza virus infection. *Gene,* **121**: 263–270.

Nottle, M.B., Haskard, K.A., Verma, P.J., Du, Z.T., Grupen, C.G., McIlfatrick, S.M., Ashman, R.J., Harrison, S.J., Barlow, H., Wigley, P.L., Lyons, I.G., Cowan, P.J., Crawford, R.J., Tolstoshev, P.L., Pearse, M.J., Robins, A.J. & d'Apice, A.J. 2001. Effect of DNA concentration on transgenesis rates in mice and pigs. *Transgenic Research,* **10**: 523–551.

Palmiter, R.D., Brinster, R.L., Hammer, R.E., Trumbauer, M.E., Rosenfeld, M.G., Birnberg, N.C. & Evans, R.M. 1992. Dramatic growth of mice that develop from eggs microinjected with metallothionein-growth hormone fusion genes. *Nature,* **300**: 611–615.

Pfeifer, A., Ikawa, M., Dyn, Y. & Verma, I.M. 2002. Transgenesis by lentiviral vectors: lack of gene silencing in mammalian embryonic stem cells and pre-implantation embryos. *Proceedings of the National Academy of Sciences, USA,* **99**: 2140–2145.

Phelps, C.J., Koike, C., Vaught, T.D., Boone, J., Wells, K.D., Chen, S.H., Ball, S., Specht, S.M., Poljaeva, I.A., Monahan, J.A., Jobst, P.M., Sharma, S.B., Lamborn, A.E., Garst, A.A., Moore, M., Demetris, A.J., Rudert, W.A., Bottino, R., Bertera, S., Truco, M., Starzl, T.E., Dai, Y. & Ayares, D.L. 2003. Production of alpha 1,3 galactosyl-transferase deficient-pigs. *Science,* **299**: 411–414.

Pursel, V.G., Pinkert, C.A., Miller, K.F., Volt, D.J., Campbell, R.G., Palmiter, R.D., Brinster, R.L. & Hammer, R.E. 1989. Genetic engineering of livestock. *Science,* **244**: 1281–1288.

Smithies, O., Gregg, R.G., Boggs, S.S., Koralewski, M.A. & Kucherlapati, R.S. 1985. Insertion of DNA sequences into the human chromosomal beta-globin locus by homologous recombination. *Nature,* **317**: 230–234.

Stice, S.L. 2002. Opportunities and challenges in domestic animal embryonic stem cell research. pp. 64–71, *in:* A.J. Clark (ed). *Animal Breeding: Technology for the 21st Century.* Switzerland: Harwood Academic Press.

Thomas, C.E., Ehrhardt, A. & Kay, M.A. 2002. Progress and problems with the use of viral vectors for gene therapy. *Nature Reviews Genetics,* **4**: 346–358.

Tiscornia, G., Singer, O., Ikawa, M. & Verma, I.M. 2003. A general method for gene knockdown in mice by using lentiviral vectors expressing small interfering RNA. *Proceedings of the National Academy of Sciences, USA,* **100**: 1844–1848.

Wilmut, I. 2002. Are there any normal cloned mammals? *Nature Medicine,* **8**: 215–216.

Wilmut, I. & Whitelaw, C.B.A. 1994. Strategies for the production of pharmaceuticals in milk. *Reproduction Fertility and Development,* **6**: 625–630.

Wilmut, I., Schnieke, A.E., McWhir, J., Kind, A.J. & Campbell, K.H. 1997. Viable offspring derived from fetal and adult mammalian cells. *Nature,* **385**: 810–813.

Wright, G., Carver, A., Cottom, D., Reeves, D., Scott, A., Simons, P., Wilmut, I., Garner, I. & Colman, A. 1991. High level expression of active human alpha-1-antitrypsin in the milk of transgenic sheep. *Biotechnology,* **9**: 830–834.

DNA POLYMORPHISMS IN THE SAHIWAL BREED OF ZEBU CATTLE REVEALED BY SYNTHETIC OLIGONUCLEOTIDE PROBES

Shashikanth and B.R. Yadav*

*Livestock Genome Analysis Laboratory, National Dairy Research Institute, Karnal – 132 001, India. *Corresponding author.*

Abstract:

Genomic DNA of 15 randomly selected unrelated animals and from two sire families (11 animals) of the Sahiwal breed of Zebu cattle were investigated. Four oligonucleotide probes – $(GTG)_5$, $(TCC)_5$, $(GT)_8$ and $(GT)_{12}$ – were used on genomic DNA digested with restriction enzymes *Alu*I, *Hinf*I, *Mbo*I, *Eco*RI and *Hae*III in different combinations. All four probes produced multiloci fingerprints with differing levels of polymorphisms. Total bands and shared bands in the fingerprints of each individual were in the range of 2.5 to 23.0 KB. Band number ranged from 9 to 17, with 0.48 average band sharing. Probes $(GT)_8$, $(GT)_{12}$ and $(TCC)_5$ produced fingerprinting patterns of medium to low polymorphism, whereas probe $(GTG)_5$ produced highly polymorphic patterns. Probe $(GTG)_5$ in combination with the *Hae*III enzyme was highly polymorphic with a heterozygosity level of 0.85, followed by $(GT)_8$, $(TCC)_5$ and $(GT)_{12}$ with heterozygosity levels of 0.70, 0.65 and 0.30, respectively. Probe GTG_5 or its complementary sequence CAC_5 produced highly polymorphic fingerprints, indicating that the probe can be used for analysing population structure, parentage verification and identifying loci controlling quantitative traits and fertility status.

1. INTRODUCTION

Man has domesticated very few species for his use from the enormous number of animals known to exist. Seven mammalian and two avian species supply most human needs for milk, meat, eggs, animal fibre and draught animal power. In addition, there are about 10 species of domesticated animals with highly specific adaptations for particular habitats, such as

H. P. S. Makkar and G. J. Viljoen (eds.), Applications of Gene-Based Technologies for Improving Animal Production and Health in Developing Countries, 111-120.
© 2005 *IAEA. Printed in the Netherlands.*

camelids, high altitude bovines, canines and elephants. Moreover, within all these species, there are many breeds or types, each having special features and adaptations, which have been developed over thousands of years of domestication in different environments. Since all the breeds within a species can interbreed, there are almost unlimited options for new genetic combinations. The evolution of these livestock is not static but is a dynamic process, with new breeds being continuously generated, although sometimes at the expense of earlier breeds.

India is a vast country extending between 8.4° and 37.6°N and 68.7° and 97.25°E. Consequently, the country spans tropical, sub-tropical and temperate zones. In addition to variation in climatic and topography, there are widely differing ethnic groups with varying social structure and habits, which also adds to the variation in agricultural practices, especially animal breeding. The rich animal genetic resources and the wide variety of livestock of different types, productivity and adaptability have evolved over time in these varying agroclimatic zones. They have endured famines, insect pests, diseases, dry heat, humid heat, and cold. Some breeds are extremely heat tolerant. Zebu cattle have a lower metabolic rate, which suits well in hot climates and also makes them a comparatively better utilizer of low quality roughage. The different breeds are economically well suited to the areas where they exist. India has been recognized as the home for many important breeds of livestock, especially draught cattle, milch buffaloes, carpet-wool sheep and highly prolific goat breeds. There are 26 recognized cattle breeds, 7 buffalo, 40 sheep, 20 goat, 4 camel, 6 horse and 3 pig breeds, together with species like yak and mithun.

The national livestock diversity is currently under threat from several sources. Intensification of agriculture in developed countries and the pursuit of higher production targets have led to a greater reliance on a few breeds and the consequent neglect (and, in some cases, eventual extinction) of the remainder. In developing countries, the major threats to genetic diversity are the extensive use of artificial insemination, intensification of agriculture and indiscriminate cross-breeding of local breeds with less adaptable exotic germplasm. Coupled with natural disasters, frequent wars, new political boundaries and changing technologies, these factors have eroded the indigenous gene pool at a rate that far outstrips its possible regeneration. In this situation there is an urgent need for a centralized data bank for animal genetic resources that would have a key role to play in listing and describing the breeds that currently exist. Documentation and monitoring is essential to identify erosion of resources and to highlight the breeds that are likely to disappear in near future unless suitable conservation programmes are adopted. There is also an urgent need for characterization and identification of unique genes in these breeds and species at the DNA level.

Livestock improvement greatly depends on the exploitation of DNA-level polymorphisms. Specific sequences of DNA are being used as genetic markers to identify loci responsible for expression of complex traits both in man and animals. These markers have short-range uses such as parentage determination, individual identification (Jeffreys *et al.,* 1986), detection of twin zygosity (Plante *et al.,* 1992), etc., and long-range applications such as gene mapping and marker assisted selection (Ashwell *et al.,* 1997; Fries, 1993; Georges *et al.,* 1993; Mitra *et al.,* 1999).

At present, several classes of markers (Caetano-Anolles, 1998) are available, including restriction fragment length polymorphisms (RFLPs), amplified fragment length polymorphisms (AFLPs), variable number tandem repeats (VNTRs), sequence tandem repeats (STRs), single-strand conformational polymorphisms (SSCPs) and single nucleotide polymorphisms (SNPs). DNA sequences harbouring basic repeat motifs of between two and six nucleotides can be synthesized and hybridized to microsatellite sequences from a variety of species to produce multilocus hybridization patterns (Epplen *et al.,* 1991). Several such oligonucleotide sequences have been reported to be useful in producing highly polymorphic DNA fingerprints in a variety of species (Ali, Muller and Epplen, 1986; Buitkamp *et al.,* 1991).

The degree of polymorphism elucidated from a probe or a marker may differ among species. Georges *et al.* (1988) observed a considerable heterogeneity in genetic variation depending on the probe–species combination. It is important to screen DNA markers for their informativeness and polymorphism for various domestic species of animals before considering them for further use. In earlier studies, several synthetic probes having core sequences of (AT), (GT), (GC), (CAC), (GAA), (GGAT), (GACA), (TGG) and (GATA) have been used for DNA fingerprinting of a variety of animal species (Buitkamp *et al.,* 1991; Schwaiger *et al.,* 1991). However, the indigenous zebu cattle, which constitute the major proportion of India's cattle population, have thus far been poorly explored with DNA-based markers. In this study, four different microsatellite probes were screened for their usefulness as markers in zebu cattle.

2. MATERIALS AND METHODS

2.1 Animals

The investigation was carried out on randomly selected unrelated cattle of the Sahiwal breed. The number of animals used for the study of each probe is noted in Table 1. The animals were maintained in a herd at the

National Dairy Research Institute, Karnal. Blood samples were collected aseptically by jugular venepuncture into sterile plastic vials containing 0.5 M EDTA (pH 8.0) at 100 µl/10 ml blood.

2.2 Restriction enzymes

Five oligonucleotide probes – *Alu*I, *Hinf*I, *Mbo*I, *Eco*RI and *Hae*III – in different combinations were used to reveal polymorphism in DNA of Sahiwal cattle.

2.3 Oligonucleotide probes

Four oligonucleotide probes – $(GT)_8$, $(GT)_{12}$, $(TCC)_5$ and $(GTG)_5$ – and sequences were selected on the basis of literature reports, and were obtained in synthesized form from Bangalore Genei. All these probes were used in different combinations with the above-mentioned restriction enzymes.

2.4 Methodology

Genomic DNA was extracted from 10 ml peripheral blood of each animal using the phenol-chloroform extraction method (Clamp *et al.*, 1993), with some modifications. Restriction digestion of the DNA was carried out separately for each restriction enzyme. Five enzymes – *Alu*I, *Hinf*I, *Mbo*I, *Eco*RI and *Hae*III – were used to complete digestion according to the supplier's specifications. Digested DNA samples were subjected to electrophoresis on 0.8 percent agarose gel at 1.5 V/cm overnight or until a 2 kb fragment of lambda *Hind*III marker (Boehringer Mannheim, Germany) travelled more than 15 cm, using 1× tris/acetic acid/EDTA (TAE) as running buffer. The size-separated DNA fragments were transferred from gel to nylon membrane (Hybond^{TM} N$^+$, Amersham, UK) using the alkali transfer protocol of the vacuum transfer method (VacuGene XL, Pharmacia LKB). The DNA on nylon membrane was immobilized by exposing the membrane to ultraviolet radiation (John and Ali, 1997). Oligonucleotide probes were custom synthesized and used after radioisotopic labelling with $(\gamma^{32}P)$ dATP^{32}P using the enzyme polynucleotide kinase (New England Biolab, UK) by the standard procedure (Sambrook, Fritsch and Maniatis, 1989). Hybridization of labelled oligonucleotide probes to genomic DNA on nylon membranes was carried out at 45°C for probes $(GTG)_5$ and $(TCC)_5$, at 43°C for $(GT)_8$, and 65°C for $(GT)_{12}$, in a hybridization cylinder of a hybridization oven (Hybridizer 600, Stratagene). Post-hybridization treatments and autoradiography were carried out as described by earlier workers (Ali,

Muller and Epplen, 1986). The molecular size of each fragment on X-ray film, i.e. the DNA fingerprint, was estimated using GelBase computer software (UVP, UK). The number of total bands and shared bands in the fingerprints of each individual were recorded in the range of 2.3 to 23.0 kb.

2.5   Analysis of band patterns

Number of bands, average band sharing rate , mean allelic frequencies (a) and heterozygosity (h) level were calculated as per Jeffreys *et al.*, 1986.

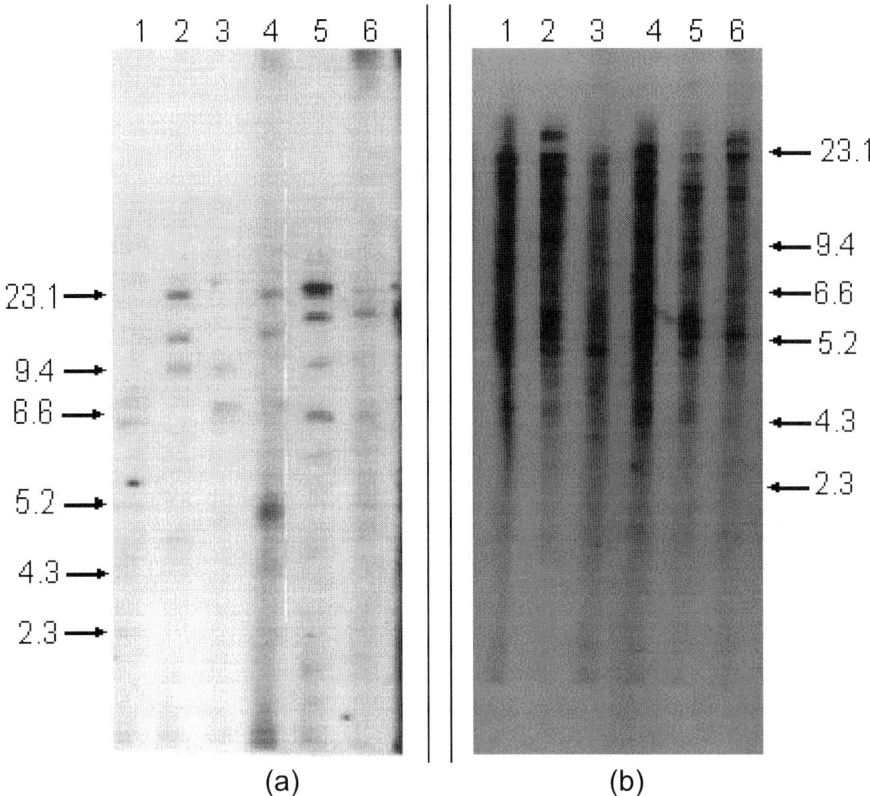

Figure 1. DNA fingerprinting by (GTG)₅ in combination with (a) *Mbo*I and (b) *Eco*RI.

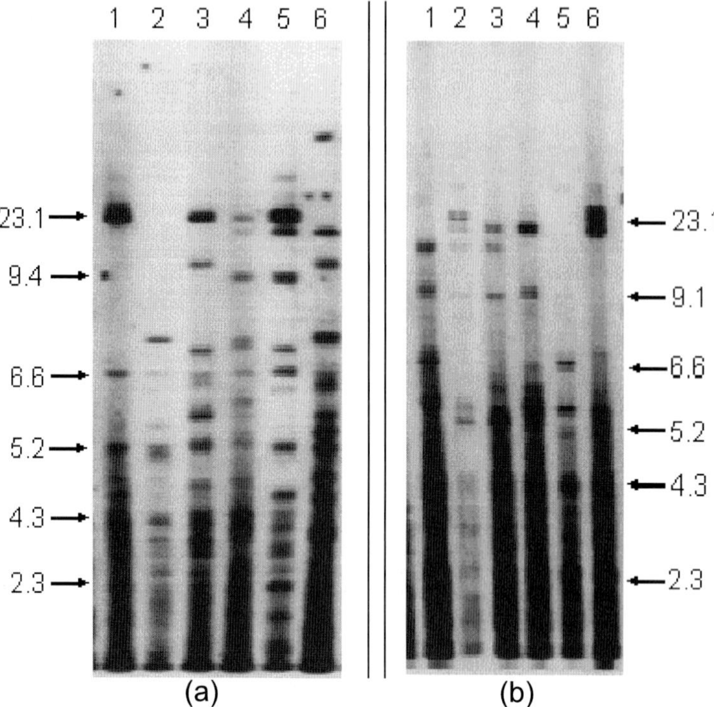

Figure 2. DNA fingerprinting by (GTG)$_5$ in combination with (a) *Alu*I and (b) *Hinf*I.

3.   RESULTS AND DISCUSSION

All the oligonucleotide probes used in the present investigation were of short tandem repeat (STR) nature. The observations identified the presence of these repeats in digested DNA of Sahiwal cattle, in different combinations of enzymes and probes. All the four probes produced multilocus fingerprints. The number and intensity of bands varied in different combinations of enzyme probes with differing levels of polymorphism (See Figures 1 to 3). Restriction enzymes *Mbo*I and *Eco*RI showed fewer bands (Figures 1a and 1b) compared with *Alu*I and *Hinf*I (Figures 2a and 2b) in all the combinations. Restriction enzyme *Hae*III produced a greater number of, and more polymorphic, band patterns in all the combinations (Figure 3).

Table 1. Levels of polymorphism in DNA fingerprints of various probes hybridized to Sahiwal DNA digested with *Hae*III enzyme.

Probe	N	No.	MM	BSR	A	H	Pattern
$(GT)_8$	10	21.28	7	0.71	0.46	0.70	LP
$(GT)_{12}$	5	21.00	19	0.97	0.82	0.30	LP
$(TCC)_5$	10	15.50	1	0.73	0.48	0.65	LP
$(GTG)_5$	10	14.95	0	0.48	0.27	0.85	HP

KEY: N = Number of replicates (randomly selected unrelated cattle); No. = Mean number of bands; MM = Monomorphic bands; BSR = Average band sharing rate; A = Mean allelic frequency; H = Heterozygosity; LP = Low polymorphism band pattern; and HP = High polymorphism band pattern.

Probes $(GT)_8$, $(GT)_{12}$ and $(TCC)_5$ produced fingerprinting patterns of medium to low polymorphism, whereas probe $(GTG)_5$ produced a highly polymorphic pattern. Mean number of bands per individual was highest for GT_8 and lowest for GTG_5. Band sharing averages were highest for GT_{12} and lowest for GTG_5 fingerprints. Probe GTG_5 was highly polymorphic, with heterozygosity level of 0.85, followed by GT_8 and TCC_5, which had heterozygosity levels of 0.70 and 0.65, respectively. Means of number of bands per individual, band sharing rate, allele frequencies and heterozygosity are listed in Table 1.

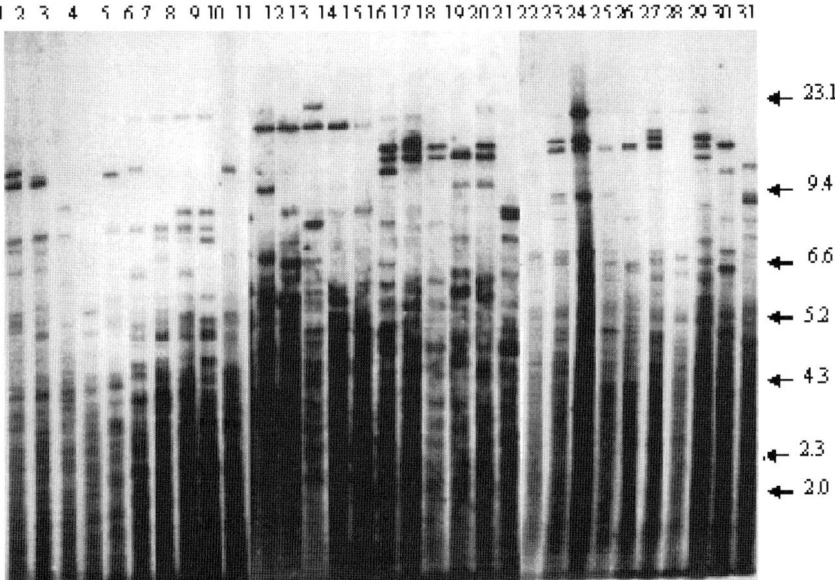

Figure 3. DNA fingerprinting revealed by oligonucleotide probe $(GTG)_5$ in combination with *Hae*III.

The (GT)$_8$ probe produced as many as 32 bands in the resolvable portion of the gel. However, several bands were shared by all the individuals, hence the average band sharing rate was found to be high. The high band sharing values observed is comparable to previous reports (Buitkamp *et al.*, 1991; Schwaiger *et al.*, 1991). The high band sharing rate in this study indicates that the animals examined might be genetically more homogeneous with respect to (GT)$_n$ sequences. Comparison of average number of bands obtained from different probes showed that the (GT)$_8$ probe hybridized to more fragments than the other probes. This result indicates that (GT)$_n$ sequences are more abundant in the zebu cattle genome compared with other sequences studied. It has been reported by earlier workers (Tautz and Renz, 1984) that there are approximately 50,000 (CA)$_n$ arrays in the human genome and that they occur at an average interval of 30,000 bp.

The (GT)$_{12}$ probe produced multilocus fingerprints with a higher band sharing rate compared with (GT)$_8$ fingerprints. The mean number of bands and heterozygosity were low compared with (GT)$_8$ fingerprints. This variation could be due to relatively smaller sample size used for (GT)$_{12}$ fingerprinting or, alternatively, due to variation in the nucleotide constitution of repeat sequences and differences in hybridization and stringency conditions for the same core sequences of differing length. This kind of variation in hybridization pattern of same-core repeats of differing length has been previously reported for TGG core repeat probes in buffaloes (Clamp *et al.*, 1993).

The (TCC)$_5$ probe produced multilocus polymorphic fingerprints. The level of polymorphism was low, as revealed by high mean band sharing value of 0.75. Buitkamp *et al.*, (1991) observed that (TCC)$_5$ detected only two polymorphic bands in *Hinf*I-digested DNA of German Friesian cattle. The reason for this deviation from the present observation could be the variation between the genomes of *Bos taurus* and *Bos indicus*. Alternatively, it is possible that *Hinf*I restriction sites that are adjacent to (TCC)$_n$ sequences are conserved while there may be variation in *Hae*III restriction sites. Few reports are available regarding the conservation of restriction sites adjacent to VNTR sequences. Existence of conserved restriction sites for *Hinf*I and *Mbo*I adjacent to (GACA)$_n$ sequences has been suggested in cattle (Schwaiger *et al.*, 1991). Existence of such restriction sites greatly reduces the level of polymorphism owing to several monomorphic fragments.

The (GTG)$_5$ probe produced highly polymorphic DNA fingerprints. The number of bands ranged between 9 and 17, with average band sharing of 0.48. The (GTG)$_5$ probe or its complement, (CAC)$_5$, produced highly polymorphic fingerprints in a variety of species: man, cattle, pigs, chicken and various other species (Buitkamp *et al.*, 1991; Schwaiger *et al.*, 1991). Thus this probe has been used for studies in a large number of animals and suggested for extensive analysis of DNA polymorphism in the Sahiwal breed and in other cattle breeds of the zebu and taurus groups.

4.   CONCLUSION

Synthetic oligonucleotides were used as probes to test their suitability for analysing genetic variability in the Sahiwal breed of zebu cattle (*Bos indicus*). All the four probes tested – $(GT)_8$, $(GT)_{12}$, $(TCC)_5$ and $(GTG)_5$ – produced multilocus hybridization patterns with varying degrees of polymorphism. The $(GTG)_5$ probe was highly polymorphic, with a heterozygosity level of 0.85, followed by $(GT)_8$ and $(TCC)_5$, which had heterozygosity levels of 0.70 and 0.65, respectively.

The high heterozygosity level obtained in this study and the low level of mutation rate associated with the sequences indicate that the probe can be used for analysing population structure, for parentage verification and as a marker to identify loci controlling quantitative traits, disease resistance, fertility, etc. DNA profiling with highly polymorphic probes such as $(GTG)_5$, which detect loci having better somatic and germline stability, are also useful in getting profiles specific to breeds and to differentiate between breeds, and thus are helpful in conservation of genetic diversity in domestic livestock.

REFERENCES

Ali, S., Muller, C.R. & Epplen, J.T. 1986. DNA fingerprinting by oligonucleotide probes specific for simple repeats. *Human Genetics*, **74**: 239–243.

Ashwell, M.S., Pekroad, C.E. Jr, Miller, R.H., VanRaden, P.M. & Da, Y. 1997. Detection of loci affecting milk production and health traits in an elite US Holstein population using microsatellite markers. *Animal Genetics*, **28**: 216–222.

Buitkamp, J., Zischler, H., Epplen, J.T. & Geldermann, H. 1991. DNA fingerprinting in cattle using oligonucleotide probes. *Animal Genetics*, **22**: 137–146.

Caetano-Anolles, G. 1998. *DNA Markers: Protocols, Applications and Overviews.* New York, NY: Wiley-VCH.

Clamp, P.A., Felters, R., Shalhevet, D., Beever, J.E., Atac, E. & Schook, L.B. 1993. Linkage relationships between ALPL, ENO_1, GP_1, PGD and $TGFB_1$ on porcine chromosome 6. *Genomics*, **17**: 324–329.

Epplen, J.T., Ammer, H., Epplen, C., *et al.* 1991. Oligonucleotide fingerprinting using simple repeat motifs: A convenient, ubiquitously applicable method to detect hypervariability for multiple purposes. pp. 50–69, *in:* T. Burke, G. Dolf, A.J. Jeffreys and R. Welff (eds). *DNA Fingerprinting: Approaches and Applications.* Basel, Switzerland: Birkhauser.

Fries, R. 1993. Mapping the bovine genome: Methodological aspects and strategy. *Animal Genetics*, **24**: 111–116.

Georges, M., Dietz, A.B., Mishra, A., Nielsen, D., Sargeant, L.S., Sorensen, A., Steele, M.R., Zhao, X., Leipold, H., Womack, J.E. & Lathrop, M. 1993. Microsatellite mapping of the gene causing weaver disease in cattle will allow the study of an associated quantitative trait locus. *Proceedings of the National Academy of Sciences, USA*, **90**: 1058–1062.

Georges, M., Lequarre, A.S., Custelli, M., Hamget, R. & Vassart, G. 1988. DNA fingerprinting in domestic animals using four different minisatellite probes. *Cytogenetics and Cellular Genetics*, **47**: 127-131.

Jeffreys, A.J., Wilson, V., Thein, S.L., Weatherall, D.J. & Ponder, B.A.J. 1986. DNA fingerprints and segregation analysis of multiple markers in human pedigrees. *American Journal of Human Genetics*, **39**: 11–24.

John, M.V. & Ali, S. 1997. Synthetic DNA-based genetic markers reveal intra-species DNA sequence variability in *Bubalus bubalis* and related genomes. *DNA and Cell Biology*, **16**: 369–378.

Mitra, A., Yadav, B.R., Ganai, N.A. & Balakrishnan, C.R. 1999. Molecular markers and their applications in livestock improvement. *Current Science*, **77**(8): 1045–1053.

Plante, Y., Schmutz, S.M., Lang, K.D.M. & Moker, J.S. 1992. Detection of leucochimaerism in bovine twins by DNA fingerprinting. *Animal Genetics*, **23**: 295–302.

Sambrook, J., Fritsch, E.F. & Maniatis, T. 1989. *Molecular cloning: A Laboratory Manual*. 2nd Edition. Cold Spring Harbor, NY: Cold Spring Harbor Laboratory Press.

Schwaiger, F.W., Gomolka, M., Geldermann, H., Zischler, H., Buitkamp, J., Epplen, J.T. & Ammer, H. 1991. Oligonucleotide fingerprinting to individualize ungulates. *Applied and Theoretical Electrophoresis*, **2**: 193–200.

Tautz, D. & Renz, M. 1984. Simple sequences are ubiquitous repetitive components of eukaryotic genomes. *Nucleic Acids Research*, **12**: 4127–4138.

GENETIC DIVERSITY AND RELATIONSHIPS OF VIETNAMESE AND EUROPEAN PIG BREEDS

N.T.D. Thuy,[1,2] E. Melchinger,[3] A.W. Kuss,[3] T. Peischl,[3]
H. Bartenschlager,[3] N.V. Cuong[2] and H. Geldermann.[3] *

*1. DAAD fellowship student at the 3. Department of Animal Breeding and Biotechnology, University of Hohenheim, D-70593 Stuttgart, Germany.
2. Institute of Biotechnology (IBT), National Center for Natural Science & Technology, 18 Hoang Quoc Viet Rd, Cau Giay, Hanoi, Viet Nam. * Corresponding author.*

Abstract: Indigenous resources of the Asian pig population are less defined and only rarely compared with European breeds. In this study, five indigenous pig breeds from Viet Nam (Mong Cai, Muong Khuong, Co, Meo, Tap Na), two exotic breeds kept in Viet Nam (Large White, Landrace), three European commercial breeds (Pietrain, Landrace, Large White), and European Wild Boar were chosen for evaluation and comparison of genetic diversity. Samples and data from 317 animals were collected and ten polymorphic microsatellite loci were selected according to the recommendations of the FAO Domestic Animal Diversity Information System (DAD-IS; http://www.fao.org/dad-is/). Effective number of alleles, Polymorphism Information Content (PIC), within-breed diversity, estimated heterozygosities and tests for Hardy-Weinberg equilibrium were determined. Breed differentiation was evaluated using the fixation indices of Wright (1951). Genetic distances between breeds were estimated according to Nei (1972) and used for the construction of UPGMA dendrograms which were evaluated by bootstrapping. Heterozygosity was higher in indigenous Vietnamese breeds than in the other breeds. The Vietnamese indigenous breeds also showed higher genetic diversity than the European breeds and all genetic distances had a strong bootstrap support. The European commercial breeds, in contrast, were closely related and bootstrapping values for genetic distances among them were below 60%. European Wild Boar displayed closer relation with commercial breeds of European origin than with the native breeds from Viet Nam. This study is one of the first to contribute to a genetic characterization of autochthonous Vietnamese pig breeds and it clearly demonstrates that these breeds harbour a rich reservoir of genetic diversity.

H. P. S. Makkar and G. J. Viljoen (eds.), Applications of Gene-Based Technologies for Improving Animal Production and Health in Developing Countries, 121-130.

1. INTRODUCTION

The current genetic composition of a species influences the capacity of its members to adapt to future physical and biotic environments. However, directed selection for economically desirable traits can lead to genetic erosion of individual breeds. A diminished genetic base is likely to limit the genetic gain for production characteristics that may be vital for sustainable agricultural systems also in the future. According to the FAO World Watch List (FAO, 2000), 55% of the world's pigs are located in the Asian and Pacific region, representing 37% of the listed breeds. A number of studies have been conducted on the genetic diversity of European (Laval *et al.*, 2000) and Chinese pig breeds (Li *et al.*, 2000; Kim *et al.*, 2002; Blott *et al.*, 2003). and a new international cooperation project has been started recently to evaluate 50 Chinese breeds in comparison with 59 European breeds (Blott *et al.*, 2003). Little, however, is so far known about the ten Vietnamese indigenous breeds listed in the FAO World Watch List (FAO, 2000).

Polymorphic molecular genetic markers provide a means to assess within- and between-breed genetic diversity. Microsatellite markers, due to their abundant and even distribution throughout the genome, high polymorphism and comparative ease of genotyping, have been used to characterize a wide variety of farm animal breeds in several species, and are recommended for studies of livestock diversity.

This study is concerned with the evaluation of 10 microsatellite markers in five Vietnamese indigenous pig populations with diverse geographical origin, the assessment of their level of genetic diversity and the estimation of genetic distances among them. Samples and data from 317 animals were collected and ten polymorphic microsatellite loci were selected according to the recommendations of the FAO Domestic Animal Diversity Information System (DAD-IS; http://www.fao.org/dad-is/). In comparison with the Vietnamese autochthonous breeds, two Vietnamese exotic breeds of European descent were included in the study, together with three European commercial breeds and the European Wild Boar.

2. MATERIAL AND METHODS

2.1 Population samples and DNA isolation

Blood or sperm samples were collected from five indigenous Vietnamese pig breeds, two exotic breeds from Viet Nam (originating from breeds from Australia, France and the United States of America), three commercial breeds from Germany, and the European Wild Boar (*Sus scrofa scrofa*)

(WB). The Vietnamese breed (VB) samples comprised 31 individuals of Co (CO), 32 of Meo (ME), 32 of Muong Khuong (MK) and 25 of Tap Na (TN) from mountain areas in northern or central Viet Nam, and 32 individuals of Mong Cai (MC). The exotic breeds sampled numbered 22 and 17 for Landrace (LV) and Yorkshire (YV), respectively. MC, LV and YV were kept at breeding stations in Viet Nam. Samples from European commercial breeds included 30 Large White (LW), 32 Pietrain (PI) and 32 German Landrace (LG), and were collected from farms and an artificial insemination station in Baden Württemberg (Germany). Samples of 32 Wild Boar originated from different regions in Germany. DNA from blood and sperm was isolated according to standard protocols.

2.2 Microsatellites, PCR conditions and fragment length analysis

The 10 microsatellites used in this study are listed in Table 1. A 25-µl PCR reaction contained 0.4 mmol/litre of each nucleotide, 100 ng template DNA and 20 pmol of each Fluorescein-labelled primer. Denaturation at 92°C for 135 sec was followed by the appropriate number of cycles, with annealing times between 20 and 40 sec and extension (72°C) times between 35 and 45 sec. The final cycle had an extension interval of 5 min.

Table 1. PCR parameters for microsatellite markers.

Locus	MgCl$_2$ (mmol)	T$_A$[1] (°C)	Forward primer	Reverse primer
SWR345	3.0	60	AACAGCTCCGATTCAACCC	TACTCAGCCTTAAAAGGAAGGG
SW489	3.0	55	CAAGTGTGAAATTTGTGCGG	CGAAGTGCTAACTATAAGCAGCA
IFNG	2.1	58	ATTAGACCCCTAGCCTGGGA	GTTGGTCCTGTTCTCCAATAGG
SW2019	1.9	60	ATGATGCGAACCTGGAACTC	TATGTGTAACTTGGTCCCATGC
TNFB	2.1	60	CTGGTCAGCCACCAAGATTT	GGAAATGAGAAGTGTGGAGACC
SW1083	3.0	50	CCTTGCTGGCCTCCTAAC	CATACTCCAAAATTTCTATGTTGA
SW2410	3.0	56	ATTTGCCCCCAAGGTATTTC	CAGGGTGTGGAGGGTAGAAG
SW957	3.0	54	AGGAAGTGAGCTCAGAAAGTGC	ATGGACAAGCTTGGTTTTCC
S0215	2.5	58	TAGGCTCAGACCCTGCTGCAT	TGGGAGGCTGAAGGATTGGGT
SW2427	3.0	60	GCATGTTATTGAGTTGATGTGTAGG	TCGGAATTCCAGAAAATTGG

NOTE: (1) Annealing temperature.

Fragment length analysis was performed on an Automated Laser Fluorescent Sequencer (A.L.F., Pharmacia, Freiburg) using 5% Hydrolink gels. The 1.5 µl to 3 µl samples of PCR amplificates were mixed with 3 µl of loading buffer (formamide with 10% w/v mixed ion exchanger, 0.6% w/v dextran blue, 19.5 mmol/ml EDTA), with the appropriate length markers, denatured at 92°C for 2 min and cooled on ice prior to loading on the gel. Electrophoreses were carried out for 150 min at 1500 V and 50°C, using 0.6 × TBE (0.06 M Tris/HCl, 49.8 mmol/litre boric acid, 0.06 mmol/litre EDTA, pH 8.2) as running buffer. A 2 mW laser was used to sample every 2 sec.

The actual fragment lengths, based on internal and external length markers, were determined by using the A.L.F. win/Instrument Control and Allele Links software (Pharmacia, Freiburg).

2.3   Cloning and sequencing of microsatellite alleles

Microsatellite amplificates were transferred into the pT7Blue vector using the Perfectly Blunt Cloning Kit (Novagen, Freiburg) according to the instructions of the manufacturer. PCR products from positive clones were purified (GFX PCR DNA and Gel Band Purification Kit, Amersham Pharmacia Biotech, Freiburg) and sequenced by the BIO LUX sequencing Laboratory (Stuttgart) using vector specific primers. Sequence alignments were effected with the GeneDoc software (Nicholas, Nicholas and Deerfield, 1997).

2.4   Statistical analyses

Genetic variability was measured for individual as well as grouped populations. Effective number of alleles (ENA) and polymorphism information content (PIC) were calculated according to Kimura and Crow (1964) and Botstein *et al.* (1980), respectively. The BIOSYS-2 software package (Swofford and Selander, 1989) was used for determination of deviations from Hardy-Weinberg Equilibrium (HWE), F-statistics (Wright, 1951) and standard genetic distances (*Ds*) according to Nei (1972). Standard genetic distances were submitted to 1000 bootstrap re-samplings applying the BOOTDIST routine of BIOSYS-2, and subsequently used for the construction of UPGMA (Unweighted Pair Group Measure using Arithmetic Averages) dendrograms using the NEIGHBOR and CONSENSE routines of the PHYLIP software package (Felsenstein, 1991).

3. RESULTS AND DISCUSSION

3.1 Number of alleles

The number of alleles detected per locus varied widely, with as few as nine at *IFNG* and as many as 18 at *TNFB,* the overall mean number of alleles being 14.4. None of the population samples contained the whole range of alleles for any of the loci, and for locus *S0215* only one allele was found in the exotic breeds. The breed with the greatest mean number of alleles was ME, with 9.5, and the lowest was LG, with 4.0. Among the Vietnamese indigenous breeds, MC, which is kept at an experimental station, had the lowest value for the average allele number (5.7), which was close to the range for the breeds of European descent (4.1–5.1).

As can be seen in Table 2, the group of Vietnamese breeds shows at least a twofold mean number of alleles compared with the breeds of European origin. The ENA of the marker panel was also markedly higher in VB as compared to the groups with European background. Higher ENA values in VB than in the breeds of European background mirror our findings for allele numbers and agree with the results of Li *et al.* (2000), who found increased PIC and ENA values in four indigenous Chinese pig breeds in comparison with one Australian breed. The values we found for breeds with European background are comparable with a previous genetic analysis of Mexican hairless pigs and commercial breeds by Lemus-Flores *et al.* (2001), where the values were 6.8 and 6.7, respectively. Equally low numbers of alleles (3.7 to 5.8) as found by Fan *et al.* (2002) for three Chinese indigenous breeds could be explained by the fact that only populations from livestock farms were investigated. This is supported by our finding for MC (5.7), which is also kept on an experimental station under controlled breeding conditions. Taken together, these results suggest that the Vietnamese indigenous breeds are still highly diverse while the other breeds seem to have a reduced allelic diversity due to breeding influences. This finding is supported by the fact that, while none of the population samples contained the whole range of alleles individually, only VB as a genetic group contains all the observed alleles at four loci (*SW489, SW1083, SW2410* and *SW957,* not shown).

Table 2. Number of alleles per locus and genetic group.

Genetic group	NS	NL	MNA	ENA
VB	120	10	12.8	6.17
XB	39	9	5.4	2.77
EB	64	10	6.4	3.31

KEY: VB = Vietnamese breeds (CO, ME, MK, TN); XB = exotic breeds (LV, YV); EB = European breeds (LG, PI, LW); NS = number of samples; NL = number of polymorphic loci (a locus was considered to be polymorphic if the most frequent allele was <0.99); MNA = mean number of alleles across all loci; ENA = effective number of alleles.

3.2   Polymorphism information content (PIC), heterozygosity and genetic equilibrium

The PIC values, as shown in Table 3, were closely correlated with the effective number of alleles. The heterozygosities observed were lower than expected in all the genetic groups. However, the discrepancies were only small in the groups of European origin (EB, XB) while VB showed a more pronounced difference, having also the lowest number of loci in genetic equilibrium (see also Table 3). It has been suggested that reduced heterozygosities may result from size homoplasy at some loci that cannot be detected by electrophoretic fragment length analyses (Makova, Nekrutenko and Baker, 2000). For example, comparative sequencing at locus *SW498* of individuals from VB, EB and WB breeds of one highly frequent allele revealed a 2 bp deletion in the 5'-flanking region of an allele with 16 repeats in WB (not shown). The deletion caused this particular allele to appear as the 15 bp allele in fragment length analysis.

The number of populations with significant deviation from Hardy-Weinberg Equilibrium was slightly higher than randomly expected. For the Vietnamese local breeds, a certain degree of inbreeding due to the absence of management and to poor infrastructural conditions can be expected. This is reflected by the inbreeding coefficient, F_{IS} (Table 3), for which VB showed the highest value (0.089) while the lowest F_{IS} value (-0.005) was observed in EB. As also shown in Table 3, VB showed an intermediate genetic differentiation between subpopulations (F_{ST}) (0.050), while F_{ST} was highest (0.097) in EB and lowest (0.028) in XB. These F_{ST} values are in a range of 0.005 to 0.15, which is considered to indicate moderate differentiation between breeds (Wright, 1978; Hartl and Clark, 1997). Interestingly, it is the group of European commercial breeds that shows the highest F_{ST} value. This could partly be the result of selection, since when genetic drift and gene flow are weak, the effects of other forces on the gene pool, such as mutation and selection, gain more influence (Neigel, 2002).

Table 3. Heterozygosity, Hardy-Weinberg Equilibrium and F-statistics

Genetic group	PIC	H_E	H_O	NLE	F_{IS}	F_{ST}
VB	0.78	0.80	0.69	7	0.089	0.050
XB	0.50	0.54	0.51	9	0.022	0.028
EB	0.55	0.58	0.53	8	-0.005	0.097

KEY: VB = Vietnamese breeds (CO, ME, MK, TN); XB = exotic breeds (LV, YV); EB = European breeds (LG, PI, LW); PIC = polymorphism information content; H_O = observed heterozygosity (direct-count); H_E = expected heterozygosity (unbiased estimate); NLE = number of loci in Hardy-Weinberg equilibrium (from 10 analysed loci); F_{IS} = inbreeding coefficient of individual in subpopulation; F_{ST} = genetic differentiation between subpopulations.

3.3 Genetic distances and phylogenetic tree

The genetic distances (Nei, 1972) between genetic groups, as well as between individual populations, are indicated in Table 4. The smallest genetic distances were found between the breeds LV and YV (0.06); CO and ME (0.10); and LV and PI (0.07). The largest distances were observed between WB and MC (1.98); and WB and TN or MK (1.89). The genetic distance (0.11) between YV and PI, and LG and LW were identical (Table 4). The distance between TN and MK is larger than the distance between CO and ME by a factor of 2.4; these are already listed as independent breeds in the FAO World Watch List (FAO, 2000), which leads us to the conclusion that TN should also be considered a distinct breed.

Table 4. Matrix of genetic distances between individual populations according to Nei (1972) standard distance (Ds).

Popu-lation	Genetic distance										
	CO	ME	MK	TN	MC	LV	YV	LG	PI	LW	WB
CO	xxx										
ME	0.10	xxx									
MK	0.28	0.30	xxx								
TN	0.27	0.27	0.24	xxx							
MC	0.36	0.28	0.53	0.44	xxx						
LV	0.94	0.62	0.96	0.92	0.88	xxx					
YV	1.06	0.70	1.18	1.21	1.10	0.06	xxx				
LG	1.20	0.84	1.41	1.37	1.08	0.11	0.11	xxx			
PI	0.95	0.63	1.03	0.95	0.80	0.07	0.11	0.14	xxx		
LW	1.02	0.70	1.20	1.01	1.03	0.14	0.11	0.25	0.18	xxx	
WB	1.58	1.25	1.89	1.89	1.97	0.36	0.30	0.30	0.45	0.35	xxx

KEY: CO: Co; ME: Meo; MK: Muong Khuong; TN: Tap Na; MC: Mong Cai; LV: Landrace (Viet Nam); YV: Yorkshire (Viet Nam); LG: Landrace (Germany); PI: Pietrain (Germany); LW: Large White (Germany); WB: European Wild Boar.

Table 5. Matrix of genetic distances between genetic groups according to Nei (1972) standard distance (Ds).

Genetic group	Genetic distance				
	VB	MC	XB	EB	WB
VB	xxx				
MC	0.31	xxx			
XB	0.81	0.96	xxx		
EB	0.85	0.90	0.03	xxx	
WB	1.52	1.96	0.32	0.30	xxx

KEY: VB = Vietnamese indigenous breeds grouped (CO, ME, MK, TN); MC = Mong Cai; XB = exotic breeds grouped (LV, YV); EB = European breeds grouped (LG, PI, LW); WB = European Wild Boar.

Regarding the genetic groups, the smallest genetic distance (0.03) was found between the two groups of European origin (XB and EB), and the largest distances (1.52 and 1.96) between WB and VB or MC (Table 5).

The UPGMA dendrogram for the eleven pig populations, as well as the tree for the genetic groups resulting from these data, are shown in Figure 1. It can be seen that the eleven populations were divided into two branches. Within the branches, the European-based breeds (LV, YV, PI and LG) were very similar and bootstrapping values were below 60%. The Vietnamese indigenous breeds formed a less homogenous cluster and all subclusters had a strong bootstrap support. The dendrogram for the genetic groups (Figure 1) shows high bootstrapping values for all branches and clearly differentiates between EB and XB, as well as between the individual populations (MC, WB) and genetic groups.

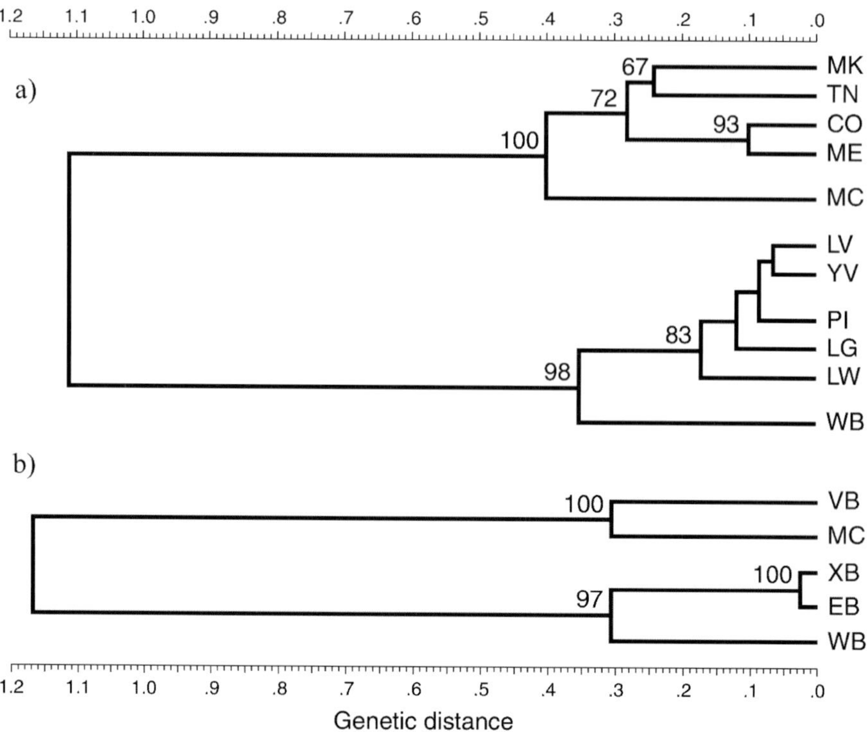

Figure 1. UPGMA dendrograms based on standard genetic distances (Nei, 1972). a) Individual populations. b) Grouped populations compared with MC and WB. Values represent bootstrap results >60%.

The results clearly point to differences between the Vietnamese indigenous breeds, which reflect their geographical distribution. The closely related CO and ME, for example, originate from regions only 150 km apart in Nghe An Province, while MK and TN, which also form a cluster, are both from the northern part of Viet Nam and closer to each other than to any of the other populations. In addition, MC, which has been kept at an experimental station since the 1960s, shows a slightly increased genetic distance relative to all the other local breeds. However, with values below 60%, bootstrapping replicates of the trees did not establish a precise topology among the commercial breeds of European descent. As the BOOTDIST routine of BIOSYS2 (Swofford and Selander, 1989) resamples the locus-specific genetic distance matrices, we consider this to be a consequence of the limited number of loci tested, and so more clarity could be expected from increasing the marker number.

4.   CONCLUSION

This study is one of the first contributions to a genetic characterization of autochthonous Vietnamese pig breeds. It clearly demonstrates that these breeds harbour a rich reservoir of genetic diversity. In comparison with the indigenous breeds listed in the FAO World Watch List (FAO, 2000), the newly tested Tap Na can also be considered to be an independent breed. Further studies on the genetic structure of the Vietnamese pig population are necessary in order to characterize and preserve porcine genetic diversity in this region.

ACKNOWLEDGEMENTS

The study was supported by the Deutsche Forschungsgemeinschaft (DFG) and the German Federal Ministry of Education and Research (BMBF).

REFERENCES

Blott, S., Andersson, L., Groenen, M.A., SanCristobal, M., Chevalet, C., Cardellino, R., Li, Huang, Li, K., Plastow, G. & Haley, C. 2003. Characterization of genetic variation in the pig breeds of China and Europe – The PIGBIODIV2 Project. *Archivos de Zootecnia,* **52**: 207–213.

Botstein, D., White, R.L., Skolnick, M. & Davis, R.W. 1980. Construction of a genetic linkage map in man using restriction fragment length polymorphisms. *American Journal of Human Genetics,* **32**: 314–331.

Fan, B., Wang, Z.G., Li, Y.J., Zhao, X.L., Liu, B., Zhao, S.H., Yu, M., Li, M.H., Chen, S.L., Xiong, T.A. & Li, K. 2002. Genetic variation analysis within and among Chinese indigenous swine populations using microsatellite markers. *Animal Genetics,* **33**: 422–427.

FAO [Food and Agriculture Organization of the United Nations]. 2000. *World Watch List for Domestic Animal Diversity.* 2nd edition. Edited by B.D. Scherf. Rome: FAO.

Felsenstein, J. 1991. PHYLIP, Version 3.57c. Distributed by the author. Department of Genetics, University of Washington, Seattle WA, USA

Hartl, D.A. & Clark, A.G. 1997. *Principles of Population Genetics.* Sunderland, MA: Sinauer Associates Inc.

Kim, K.I., Lee, J.H., Li, K., Zhang, Y.P., Lee, S.S., Gongora, J. & Moran, C. 2002. Phylogenetic relationships of Asian and European pig breeds determined by mitochondrial DNA D-loop sequence polymorphism. *Animal Genetics,* **33**: 19–25.

Kimura, M. & Crow, J.F. 1964. The number of alleles that can be maintained in a finite population. *Genetics,* **49**: 725–738.

Laval, G., Iannuccelli, N., Legault, L., Milan, D., Groenen, M.A., Giuffra, E., Andersson, L., Nissen, P.H., Jorgensen, C.B., Beeckmann, P., Geldermann, H., Foulley, J.-L., Chevalet, C. & Ollivier, L. 2000. Genetic diversity of eleven European pig breeds. *Genetics Selection Evolution,* **32**: 187–203.

Lemus-Flores, C., Ulloa-Arvizu, R., Ramos-Kuri, M., Estrada, F.J. & Alonso, R.A. 2001. Genetic analysis of Mexican hairless pig populations. *Journal of Animal Science,* **79**: 3021–3026.

Li, K., Chen, Y., Moran, C., Fan, B., Zhao, S. & Peng, Z. 2000. Analysis of diversity and genetic relationships between four Chinese indigenous pig breeds and one Australian commercial pig breed. *Animal Genetics,* **31**: 322–325.

Makova, K.D., Nekrutenko, A. & Baker, R.J. 2000. Evolution of microsatellite alleles in four species of mice (genus *Apodemus*). *Journal of Molecular Evolution,* **51**: 166–172.

Nei, M. 1972 Genetic distance between populations. *American Naturalist,* **106**: 283–292.

Neigel, J.E. 2002. Is F_{ST} obsolete? *Conservation Genetics,* **3**: 167–173.

Nicholas, K.B., Nicholas, H.B. Jr, & Deerfield, D.W., II. 1997. GeneDoc: Analysis and Visualization of Genetic Variation. See: http://www.psc.edu/biomed/genedoc

Swofford, D.L. & Selander, R.B. 1989. BIOSYS-2, Version 1.7, Dept. of Genetics and Development, University of Illinois at Urbana Champaign, Urbana, Illinois, USA. 1997 version modified by W.C. Black IV, Dept. of Microbiology, Colorado State University, Fort Collins, Colorado, USA.

Wright, S. 1951. The genetical structure of populations. *Annals of Eugenics,* **15**: 323–354.

Wright, S. 1978. *Evolution and the Genetics of Population, Variability Within and Among Natural Populations.* Chicago IL: University of Chicago Press.

COMBINING GENE-BASED METHODS AND REPRODUCTIVE TECHNOLOGIES TO ENHANCE GENETIC IMPROVEMENT OF LIVESTOCK IN DEVELOPING COUNTRIES

Julius van der Werf and Karen Marshall

School of Rural Science and Agriculture, University of New England, Armidale, Australia.

Abstract: Selection based on DNA markers is most useful for traits that are hard to measure and have low heritability. It allows earlier and more accurate selection, increasing short- and medium-term selection response, and may aid in targeting genotypes for specific production environments or markets. The use of genotypic information in breeding programmes for within-breed selection will generally have limited extra benefit, unless selection based on phenotype is difficult or advanced reproductive technologies are used. Novel reproductive technologies boost reproductive rates of breeding animals and may allow reproduction at juvenile ages. The benefit arises from increased selection intensity, as well as from increased selection accuracy due to larger families and decreased generation interval, as higher reproductive rates result in lower optimal ages for breeding animals. Increased reproductive rates and early selection rely more on between-family selection and potentially decrease effective population size, therefore increasing inbreeding. Selection needs to be optimized with respect to inbreeding and merit. Extra benefit from scenarios with unlimited use of reproductive technologies is restricted by the need to maintain genetic diversity. Benefits from marker assisted selection are higher in breeding programmes that use reproductive technologies, as the value of providing information about genotype is more beneficial for selection of young animals before they have a phenotype. Moreover, genotype information exploits variation within families, which is beneficial in breeding programmes where loss of genetic diversity is to be controlled. In developing countries, use of genotype information is likely to be most useful in marker assisted introgression programmes, where valuable genes are introgressed from one breed into another. A large variety of genetic resources in developing countries exists across breeds and populations, and utilization and management of this variation might greatly benefit from gene technologies.

131

H. P. S. Makkar and G. J. Viljoen (eds.), Applications of Gene-Based Technologies for
Improving Animal Production and Health in Developing Countries, 131-144.
© 2005 *IAEA. Printed in the Netherlands.*

1.   INTRODUCTION

The advent of molecular markers allows determination of actual genotype at gene loci, without error due to random and non-random environmental effects. In the ideal situation, we can directly identify genotypes. This could have important applications in a range of areas, varying from the development of genetically modified organisms, to parentage testing, to marker assisted selection (MAS) in animal breeding programmes. Parentage testing could be a very helpful tool in breeding programmes for low input production systems. However, this paper will mainly explore the benefit of DNA technologies for the purpose of genetic improvement of livestock production efficiency. In that context, the objective is mainly to improve traits related to animal productivity and reproductive efficiency, most of these traits being quantitative by nature. The genetic model underlying such traits is based on a large number of loci affecting genetic variation. Some of these loci contain genes with substantial effects therefore affecting profitability of livestock production. Knowledge about genotype at these loci could have a substantial impact on the way in which we select breeding animals. Selection could be based not only on observed phenotypes but also upon genetic markers that are linked to quantitative trait loci (QTLs).

The actual effect of MAS on the efficiency of breeding programmes depends on a number of characteristics. These are related to the ability to observe phenotype in breeding animals, and the information contained by the genetic marker, being its degree of linkage to the QTL and the actual genotypic effects at the QTL. The conditions under which application of MAS would be favourable are discussed below. It will be shown that the main value of MAS will be realized in cases where phenotypic measurement is either difficult, expensive or late in a breeding animal's life. Although a number of traits would fall in this category, in the context of a complete multi-trait breeding objective the total added value of MAS may be limited, except where production is sex-limited and where specific traits such as disease resistance or adaptability have a major impact on the production system.

Benefits from selection based on genotypic information are potentially higher in breeding programmes where reproductive rates are high. This is for two reasons. First, increased reproductive rates decrease the optimal age of breeding animals, reducing the number of traits that can be measured before the moment of reproduction. MAS is particularly useful in providing information about genotype in juvenile animals. The second advantage is that increased reproductive rates tend to increase inbreeding rates because fewer breeding animals are needed, resulting in smaller effective population size. Information on genotype would provide information about within-family variation, rather than among families, which is advantageous in the long term as genetic improvement can be achieved without eroding genetic variability.

The aim of this paper is to evaluate the benefits of using genotype information in the context of modern animal breeding programmes. In such programmes, high selection intensities and short generation intervals provide high rates of genetic improvement, yet sustainability is maintained through restricting the loss in genetic diversity. Many of the arguments used will be generally valid for livestock improvement programmes. However, a number of specific aspects of the use of new technologies discussed will apply to genetic improvement programmes for low input production systems.

2. USE OF GENETIC MARKERS IN BREEDING PROGRAMMES

The value of genetic markers linked to QTLs accrues from increased selection accuracy. Lande and Thompson (1990) estimated increases of selection accuracy up to threefold for low heritable traits. These estimates were very optimistic and later studies looked at the benefit of MAS evaluated in the context of breeding programmes. In dairy breeding programmes, estimates of improved rates of gain were around 10–20 percent due to pre-selection of young bulls in a progeny testing scheme. Spellman and Garrick (1997) evaluated economic benefits of MAS in a commercial dairy population and found only limited additional gains (up to 2.6 percent) over a 10-year period, mainly attributable to the long pay-back time from initial investment. Meuwissen and Goddard (1996) published a simulation study that looked at the main characteristics determining efficiency of MAS. They varied the size of the QTL effect, the heritability of the trait and whether trait measurement was available at reproductive age. The important results of this study are summarized in Table 1.

Table 1. Additional gain from marker assisted selection (MAS) in a closed nucleus breeding programme, depending on the moment of trait recording, and with varying heritability (h^2) and QTL variance (V_{QTL}) as proportion of additive genetic variance (Meuwissen and Goddard, 1996) after 1 (Gen 1) or 5 (Gen 5) generations of selection.

	Selection *after* recording		Selection *before* recording	
	Gen 1	Gen 5	Gen 1	Gen 5
$h^2 = 0.11$, $V_{QTL}=0.33$	+21%	+6%	+45%	+23%
$h^2 = 0.27$, $V_{QTL}=0.33$	+9%	+2.3%	+38%	+15%
$h^2 = 0.27$, $V_{QTL}=0.11$	+1.3%	+1.3%	+8%	+6%

In addition Meuwissen and Goddard (1996) were able to obtain 64 percent more selection response for carcase traits where information was used on siblings that were pre-selected based on marker information. Their work illustrates the conclusion that MAS is mainly useful for traits where phenotypic measurement is less valuable because of (1) low heritability, (2) sex-limited expression, (3) non-availability before sexual maturity, and (4) unavailable without sacrificing the animal (e.g. slaughter traits). Traits such as feed intake or disease resistance may also be expensive or difficult to measure, and information on genotype might be useful in selection here as well. Selection of animals based on (most probable) QTL genotype will allow earlier and more accurate selection, increasing short- and medium-term selection response.

When selection is based on indirect markers, there is no guarantee of QTL genotype, as marker alleles linked to the preferred QTL allele can be different in different families. In such a case, information about linkage phase needs to be accumulated based on phenotypic and pedigree information (e.g. a progeny test). In their study, Meuwissen and Goddard (1996) assumed selection was on *marker haplotype.* An animal's breeding value at the QTL would be estimated in a genetic evaluation scheme where QTL effects were fitted as random effects. Effects were estimated to be different in different families and information on QTL haplotypes was based on a combination of marker haplotype, pedigree and phenotypic information. Hence, their approach assumes no certainty about individual QTL effects, but knowledge of QTL variance and marker genotyping of all animals in the nucleus. Woolliams, Pong-Wong and Villanueva (2002) pointed out that the result of Meuwissen and Goddard (1996) were too optimistic, and that information from markers is less valuable when markers are linked (as opposed to direct markers). Another consideration is that genotypic selection leads to rapid improvement in the short term, but in the long term this is offset by a decreased polygenic response (Gibson, 1994).

In the context of low input production systems, some questions can be raised concerning the validity and practicality of the simulation studies described. Complete phenotypic and pedigree information is often not available, except in some intensive breeding units. With a lack of such information it would be more difficult to realize the value of the marker information. It would be harder and more expensive to determine linkage phase in the case of using linked markers. Moreover, even if the genetic marker were a direct marker, its effect on phenotype would have to be estimated for the population and the environment in which it is used, as effects of individual genes are likely to be dependent upon the background. Information is currently available for a number of "direct markers" (or actual gene mutations) such as myostatin affecting double muscling in cattle, calipyge doing the same in sheep, and Booroola affecting fecundity. Many of these mutations have a major phenotypic effect. Moreover, it turns out that the genetic model often seems more complex than originally anticipated. It may be that selecting based on actual genotype is less robust than selection based on phenotype. However, we should expect that more and more QTLs will be pinned down to actual gene mutations, such as $DGAT_1$ in dairy cattle, and a number of those will have simple additive inheritance patterns that are consistent across environments.

Another consideration is cost of genotyping, as such services might be available commercially for a fair cost. Marshall, Henshall and van der Werf (2002) looked at strategies to minimize genotyping cost in an extensive sheep breeding programme. Close to maximal gain could be achieved when only high ranking males were genotyped and animals whose marker genotype probability could not be derived with enough certainty based on information on relatives. Marshall (2002) also looked at progeny testing of sires to determine family-specific marker-QTL phase. Again, testing of a limited number of males provided a lot of information about phase in the breeding operation, as progeny-tested sires carry relationships with descendants in the following generations.

Many studies on MAS have taken a single-trait approach and, for some traits, genetic markers would have a large impact on response for individual traits. However, in a multi-trait breeding objective, more response for one trait often appears at the expense of another. For example, genetic markers for carcass traits allow an improved ability to select (i.e. earlier, with higher accuracy) for such traits, but selection emphasis for other traits would be reduced. This is comparable with having to divide selection emphasis between genotypic and polygenic information (Gibson, 1994), or simply multi-trait selection. Therefore, the overall effect of MAS on the breeding programme will generally be smaller than predicted for MAS-favourable cases. An exception may be in the dairy sector, where most traits are sex-limited and therefore favoured by MAS. Meuwissen, Hayes and Goddard (2001) considered a futuristic scenario where massive genotyping becomes cost effective and dense marker maps can be used on selection candidates. They showed that the breeding value of animals without phenotype could be accurately estimated (accuracy ~ 0.7–0.8) based on pedigree and phenotypic information on ancestors, as well as a dense marker map covering the whole genome. As indicated by Woolliams, Pong-Wong and Villanueva (2002), the value of marker information here is that it helps predict variation from Mendelian sampling.

In conclusion, selection based on either direct or indirect DNA markers is currently most useful for traits that are hard to measure, sex limited or have low heritability. Unless the breeding objective is dominated by such traits, the impact of MAS on the rate of genetic gain may be limited in conventional breeding programmes (varying from a few percent to maybe 10 percent extra gain). However, new technologies often lead to other breeding programme designs being more optimal. Genotypic information has extra value in the case of early selection and where within-family variance can be exploited. Reproductive technologies usually lead to early selection and more emphasis on between-family selection. DNA marker technology and reproductive technologies are therefore highly synergistic and complementary. Or, as phrased by Soller and Medjugorac (1999), reproductive technologies might provide the "selection space" which can be exploited when using genotype information.

3.   REPRODUCTIVE TECHNOLOGIES

Novel reproductive technologies boost reproductive rates of breeding males (through artificial insemination (AI)) or of females (through multiple ovulation and embryo transfer (MOET), or through embryo production from harvesting oocytes followed by *in vitro* fertilization (IVEP)). In addition, oocytes can now be harvested in juvenile animals (juvenile *in vitro* fertilization and embryo transfer (JIVET)), which allows a significant reduction in the reproductive age of females. The benefit of new reproductive technologies therefore derives from increased selection intensity (fewer parents are needed for reproduction), as well as from increased selection accuracy (larger families) and decreased generation interval. The shorter generation interval can be a direct consequence of the technology used, as in the case of JIVET, or an indirect effect. For example, MOET causes breeding animals to have more contemporary siblings, allowing more accurate estimation of breeding value. Moreover, higher rates of genetic change also lead to higher probabilities of young animals being genetically superior to the older age classes. Figure 1 illustrates optimization of selection across age class, achieved by simply selecting on BLUP Estimated Breeding Values (EBVs), where the proportion of young animals selected will be larger with higher genetic trends and higher variance of their EBVs. The last-named depends completely on the accuracy of the young sire's EBV, which will be higher if more information (phenotypic or genotypic) can be used for genetic evaluation of young animals.

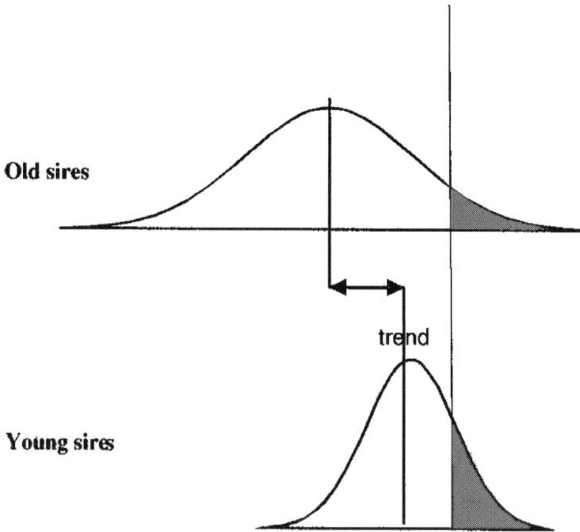

Figure 1. Truncation selection across age classes. Distribution of Estimated Breeding Values (EBVs) for different age classes are shown. The proportion of young animals selected is higher with more genetic trend and with higher accuracy (reflected in standard deviation).

An increased reproductive rate potentially decreases effective population size, and therefore increases inbreeding. Selection for increased merit needs to be balanced against maintenance of sufficient effective population size to ensure long-term sustainability of the breeding programme, and maintenance of genetic diversity. Therefore, selection needs to be optimized such that contributions from selected parents are optimal not only with respect to the next generation, but to future generations as well. Selection methods have been developed, e.g. by Meuwissen (1997), allowing maximizing selection on merit while constraining the rate of inbreeding. Alternative selection schemes need to be compared under optimal selection strategies, keeping the rate of inbreeding constant across schemes. Extra benefit from scenarios with unlimited use of reproductive technologies will then become restricted by the need to maintain genetic diversity (and sufficient effective population size).

To make this point more clear we illustrate this with an application of MOET schemes in beef cattle (Figure 2). When applying conventional MOET, embryos would be flushed from cows at reproductive age (about 15 months) after which many traits can have been measured, e.g. growth, and ultrasound measures of fat and muscling. In this scheme, the only MOET benefit is an increase in female selection intensity. The generation interval would be 24 months, which is similar to the minimum generation interval under normal reproductive schemes.

The overall generation interval might still be reduced as more young cows can be used as nucleus mothers. When juvenile MOET is applied to

 J. van der Werf and K. Marshall

young females at an age of 6 months, using oocyte pickup and *in vitro* fertilization, the generation interval could be reduced to 15 months.

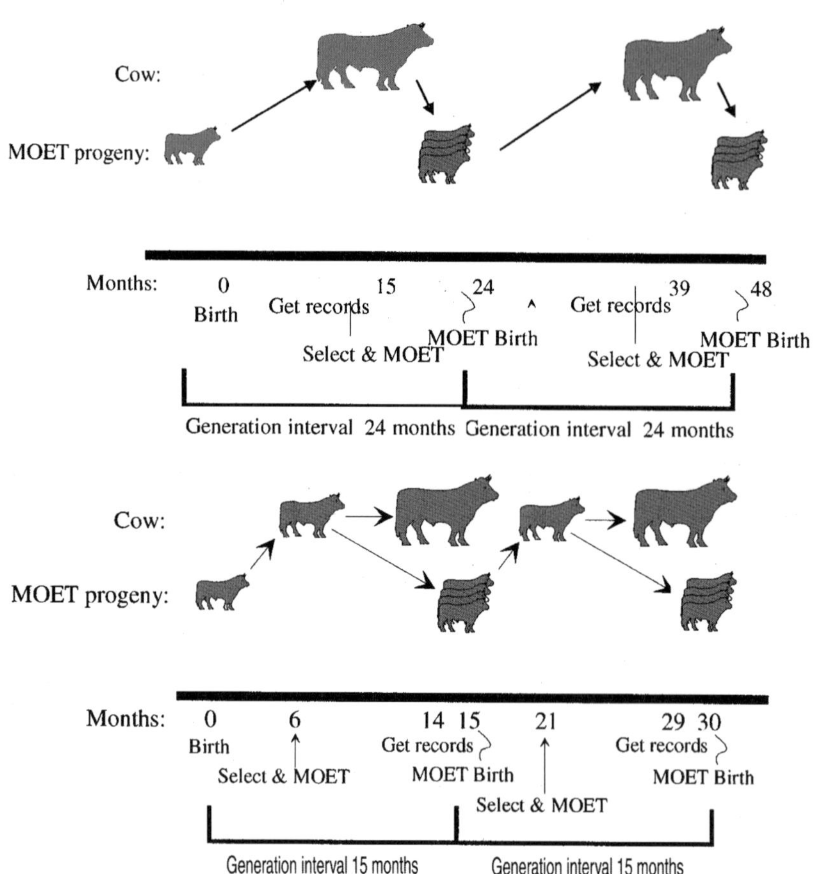

Figure 2. Diagram of MOET schemes in beef cattle where multiple embryos are collected from breeding females either at normal reproductive age (top) or at a juvenile age using oocyte pickup and embryo transfer (bottom) [illustration from Brian Kinghorn].

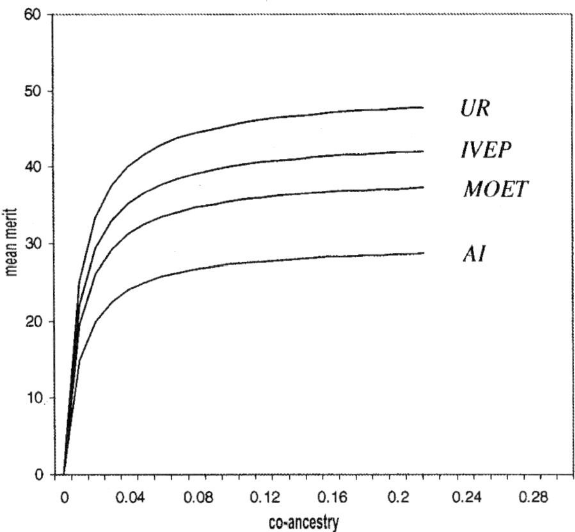

Figure 3. Balance between merit and co-ancestry that can be achieved for various reproductive technologies (AI = Artificial Insemination; MOET = Multiple Ovulation and Embryo Transfer; IVEP = *In vitro* Embryo Production; and UR = Unrestricted reproductive rates

However, the more the inbreeding rate is restricted, the smaller the difference between the various schemes (Figure 3). In practice, schemes that allow selection at earlier age will be restricted most, and relatively more animals need to be selected from a wider number of families, or from older age classes. Animals from older age classes have more within-family information and animals from inferior families have more chance to be selected based on their superior within-family deviation.

As Van Arendonk and Bijma (2003) have pointed out, the generation interval itself is not a design parameter, but results from optimizing merit and diversity in selection. Basically, the benefit of reproductive technologies will result in earlier selection if information about the within-family component is known, i.e. information about the Mendelian sampling component of the breeding value. Such information could be generated by using correlated phenotypes that can be measured early in life, or indeed, DNA marker information to predict QTL genotype(s).

4.   COMBINING GENOTYPE SELECTION AND REPRODUCTIVE TECHNOLOGIES

Marshall (2002) compared Genotype Assisted Selection (GAS) with non-GAS for several reproductive scenarios and use of selection strategies that restrict the rate of inbreeding. In adult schemes, where phenotype measurement can be used at first selection, GAS can increase genetic response initially by up to 10 percent but in the longer term (10 years) the advantage is much smaller, even if major gene and polygenic response are optimally balanced. The initial benefit from applying GAS is lost later because of loss of response from underutilizing the remaining polygenic part of genetic variance.

In the so-called "juvenile schemes" where first selection occurs before the first phenotype has been measured, response based on phenotypic selection is difficult and GAS provides significantly more benefit (Table 2). GAS can nearly double initial response and even after 10 years of optimal selection the superiority of GAS over non-GAS selection can be of the order of 40 percent for juvenile MOET and IVEP schemes.

Table 2. Additional gain (percent) using genotype-assisted selection (GAS) compared with non-GAS from using reproductive technologies in a closed breeding scheme using optimal selection (restricted inbreeding) and a QTL substitution effect of 0.9 of a phenotypic SD.

Selection Round	Adult schemes		Juvenile schemes	
	AI	MOET	Juvenile MOET	IVEP
1	7.8	7.1	63.5	42.7
4	0.0	-3.6	102.0	54.1
10	-3.3	-8.7	41.6	10.4

KEY: MOET = Multiple Ovulation and Embryo Transfer. IVEP = *In vitro* Embryo Production.

5.   BENEFIT FROM SEXING AND CLONING

Sexing of semen or embryos is now possible and semen-sexing services are now available commercially. The technology is being further developed as the amount of semen that can be separated in a given time span is limited. The effect of sexing (semen or embryos) on the rate of genetic improvement is expected to be small: about 5 percent. One could contemplate generating more female offspring in the breeding nucleus to increase female selection intensity. However, this gain is nearly offset by the loss in male selection intensity. In dairy programmes, benefits could be somewhat larger (~10 percent) because of sex-limited progeny testing. The selection intensity in elite females can be increased and more progeny testing capacity can be used for the same cost.

The main benefit of sexing can be expected at the commercial level. The breeding nucleus could generate relatively more males, allowing for some selection intensity in males sold to commercial farmers. This would reduce the genetic lag between the nucleus and the commercial herds. Moreover, commercial progeny can be produced for a specific goal, with female semen used to generate herd replacements, and male semen to generate slaughter progeny. This scenario could be highly profitable at the commercial level in both milk and meat production systems. Gene testing would also be very beneficial in these schemes, as different types of animals are needed for females and males. Similar to using specific matings from specialized male and female lines, as in pig production schemes, MAS would help utilize within-breed variation to target specific genotypes for specific goals, e.g. maternal effects for female replacements, and growth and carcass traits for male progeny.

Animal cloning is currently still very much in an experimental phase, with low efficiency (only a small percentage of clones survive). Applications are mainly for pharmaceutical purposes where high cost and ethical concerns can be easier overcome than in large-scale agricultural applications. As with sexing, the main benefit from cloning is not expected in relation to genetic improvement but in the dissemination of improved breeding animals. After all, clones are copies and cloning multiplies rather than generates improved genotypes. Therefore, the benefit would be that tested clones could be made available to commercial producers. This is not dissimilar to current plant breeding routines. The value of gene technology in the use of cloning could be very high. Gene testing could be useful in the lengthy process of clone selection and testing. Furthermore, specific genotypes could again be targeted for specific environments or markets. This is attractive from a marketing point of view, and will genuinely improve the efficiency of disseminating breeding stock to commercial producers.

6.   INTROGRESSION

Rather than exploiting existing QTLs through within-breed selection, a more likely scenario for developing countries will be that valuable QTLs will be introgressed from one population into another. Especially in developing countries, there is a huge variety across breeds, much of it being useful to exploit in genetic improvement programmes. This includes the variation coming from "exotic" breeds in developing countries. Either indigenous breeds may contain valuable QTLs, but could benefit from upgrading through crossing with superior exotic breeds. Alternatively, valuable QTLs could be introgressed from exogenous breeds. Examples are the Booroola gene in the Garole breed in India (having a moderate and desirable effect on number of lambs weaned), and a number of genes affecting resistance to endemic local diseases. Furthermore, there are many cases of QTLs found in crosses of extreme breeds, and a number of those would be candidates for introgression.

Visscher and Haley (1995) reviewed studies on the efficiency of introgression. A gene marker can be used to introgress a desirable allele into a population, by repeatedly crossing the crossbred animals back to the superior breed that does not contain the desirable allele. The gene marker would help to quickly recover a large proportion of the genetic background without losing the desired QTL allele. DNA markers could also be used to recover the background genes, but the efficiency of this strategy would depend on how much of the variation is within and between populations. Phenotypic selection might be just as efficient when most variation is within populations, or when the between component is very large, but markers would be more efficient for the intermediate cases.

7.   DISCUSSION

The combined application of reproductive and gene technologies in breeding programmes works synergistically. In general, however, use of genotypic information in breeding programmes for within-breed selection will generally have limited extra benefit, unless selection based on phenotype is difficult or advanced reproductive technologies are used. Moreover, MAS is best utilized when embedded in a breeding programme that already relies on extensive recording of phenotypes and pedigree. Phenotypic information will therefore not become redundant in MAS selection programmes. Rather, phenotypic information may be less used in the actual selection, but recording of phenotype will be continuously needed for the purpose of monitoring the QTL effect (retrospectively), and genetic change over time. Therefore, breeding programmes in developing countries are unlikely to benefit much from MAS within populations.

Most breeding programmes in both developed and developing countries struggle to obtain rates of genetic response that are anywhere near to what might be expected based on theoretical considerations. The discrepancy is often due to the lack of control over selection decisions and the lack of clear breeding objectives. Implementation of advanced genetic and reproductive technologies may therefore not be first priority in such programmes.

However, when the gains in response can be significant, they should not be avoided either.

There are a number of useful applications of GAS that would not fit the usual type of evaluation for quantitative traits. Some genes may have significant value in targeting genotypes for specific production environments or markets. Other genes may have very large effects. In many tropical environments, there are specific genes responsible for adaptation or disease resistance (e.g. to external parasites such as ticks or trypanosomes). The value of genetic markers in selection for such genes is large, especially if selection is against recessive genotypes, even though the traits are not easily given a value in a breeding objective context. Such larger benefits from gene technologies may be expected when exploiting variation across populations.

In developing countries, use of genotype information is therefore likely to be more useful in marker-assisted introgression (MAI) compared with selection within breeds. Also in the case of MAI, reproductive technologies will be beneficial because they can help increase the number of animals with the desired genotype. Again, optimal strategies will have to consider genetic diversity as well as risk. Not many case studies have been described for developing countries looking at MAI scenarios in a complete model for livestock production. Such studies would be warranted, as they would probably form an important evaluation framework for the use of genetic technologies in genetic improvement programmes in developing countries.

REFERENCES

Gibson, J.P. 1994. Short-term gain at the expense of long-term response with selection of identified loci. *Proceedings of the 5th World Congress on Genetics Applied to Livestock Production,* 21-201.

Lande, R. & Thompson, R. 1990. Efficiency of marker-assisted selection in the improvement of quantitative traits. *Genetics,* **124**: 743–756.

Marshall, K. 2002. Marker assisted selection in the Australian sheepmeat industry. PhD thesis, University of New England, Armidale, Australia.

Marshall, K., Henshall, J. & van der Werf, J.H.J. 2002. Response from marker-assisted selection when various proportions of animals are marker typed: a multiple trait simulation study relevant to the sheepmeat industry. *Animal Science,* **74**: 223–232.

Meuwissen, T.H.E. 1997. Maximizing the response of selection with a predefined rate of inbreeding. *Journal of Animal Science,* **75**: 934–940.

Meuwissen, T.H.E. & Goddard, M.E. 1996. The use of marker haplotypes in animal breeding schemes. *Genetics Selection Evolution,* **28**: 161–176.

Meuwissen, T.H.E., Hayes, B.J. & Goddard, M.E. 2001. Prediction of total genetic value using genome-wide dense marker maps. *Genetics,* **157**: 1819–1829.

Soller, M. & Medjugorac, I. 1999. A successful marriage: making the transition from QTL mapping to marker-assisted selection. pp. 85–96, *in:* J.C.M. Dekkers, S.J. Lamont and M.F. Rothschild (eds). *From Jay Lush to Genomics: Visions for Animal*

Breeding and Genetics. Proceedings of a conference. Ames, Iowa, 16–18 May 1999. See: http://www.agbiotechnet.com/proceedings/jaylush.asp

Spelman, R. & Garrick, D. 1997. Utilisation of marker assisted selection in a commercial dairy cow population. *Livestock Production Science,* **47**: 139–147.

Van Arendonk, J.A.M. & Bijma, P. 2003. Factors affecting commercial application of embryo technologies in dairy cattle in Europe – a modelling approach. *Theriogenology,* **59**: 635–649.

Visscher, P.M. & Haley, C.S. 1995. Utilizing genetic markers in pig breeding programs. *Animal Breeding Abstracts,* **63**: 1–8.

Woolliams, J.A., Pong-Wong, R., & Villanueva, B. 2002. Strategic optimization of short- and long term gain and inbreeding in MAS and non-MAS schemes. *Proceedings of the 7th World Congress on Genetics Applied to Livestock Production.* Communication No. 23-02. INRA, France.

EVALUATION OF THE UTILITY OF THE *FecB* GENE TO IMPROVE THE PRODUCTIVITY OF DECCANI SHEEP IN MAHARASHTRA, INDIA

Chanda Nimbkar,[1,2] Varsha C. Pardeshi[3] and Pradip M. Ghalsasi[2]

1. Nimbkar Agricultural Research Institute, Phaltan, Maharashtra, India.
2. School of Rural Science and Agriculture, Univ. of New England, Armidale, NSW, Australia.
3. National Chemical Laboratory, Pune, India.

Abstract: Reproductive performance was analysed of ¼ Garole ewes (with ¾ comprising different combinations of Deccani and Bannur genes), 97 of them heterozygous (B+) for the *FecB* (Booroola) gene and 99 non-carrier (++). These ewes belong to the first backcross in the programme of introgression of the *FecB* gene into the Deccani breed, which is one of the main meat-producing breeds in India. Compared with ++ ewes, B+ ewes gave birth to 0.5 more lambs per ewe lambing, weaned 0.3 more lambs and produced 1.1 kg greater lamb weight at 105 days, and this was a significant effect. Of the litters produced by B+ ewes, 49 percent were twins, 2.5 percent triplets and the rest singles. The effect of *FecB* thus appears to be smaller compared with literature reports based on Booroola Merino and its crosses. A moderate effect implies easier manageability. Management efforts are still needed to reduce lamb mortality, especially among multiple-born lambs. Compared with single-bearing ewes, twin-bearing ewes weaned 0.7 more lambs and produced 2.4 kg more weight of lamb at 105 days. Triplet-bearing ewes weaned 0.5 more lambs and 4.7 kg more weight of lamb than twin-bearing ewes. Genes from the small-sized and locally unadapted Garole, with inferior mothering ability, probably had an adverse effect on milk yield, lamb survival and growth. Weaning percentage is likely to improve with further backcrossing. Two unique features of this study are that the gene was introgressed from the Garole and not the Booroola Merino, and all ewes were genotyped using a DNA test rather than classified according to the ovulation rate criterion. The participation of local smallholder sheep owners in the project will improve its chances of successful implementation.

145

H. P. S. Makkar and G. J. Viljoen (eds.), Applications of Gene-Based Technologies for Improving Animal Production and Health in Developing Countries, 145-154.
© 2005 *IAEA. Printed in the Netherlands.*

1. INTRODUCTION

Deccani is a breed of sheep reared in semi-arid areas of the Deccan plateau (16–22°N and 73–77°E) in Maharashtra, Karnataka and Andhra Pradesh States of India. Among Indian sheep breeds, the Deccani and Marwari are considered the most important numerically and the largest contributors to mutton production (Khan *et al.*, 2002). The Deccani is a medium-sized (30 kg mature weight), leggy breed that varies widely in colour and produces a short coarse fleece (600 g/year). There are about 3 million Deccani sheep in Maharashtra, reared in flocks of 30 to 100 breeding ewes by smallholder farmers and landless flock owners. The main (>90 percent) source of income is sale of lambs. All male lambs and most surplus female lambs are sold at about 105 days of age. Deccani ewes usually have only a single lamb; the average litter size is about 1.02. Any improvement in the reproductive rate of the ewe in such a production system would increase the biological and economic efficiency of the system, income and profit earned. A low reproductive rate also limits genetic progress in any breeding scheme. Despite the low heritability of litter size (about 0.1) in sheep, average rates of genetic improvement of 1.3 percent of the mean per year have been realized in Merino, Romney and Galway breeds, mainly because of the high coefficient of variation in these breeds (Land, Atkins and Roberts, 1983). However, the fecundity of the Deccani is extremely low and there is not much variation to exploit.

A diallel cross-breeding programme using the Deccani, Bannur and prolific Garole breeds was conducted at the Nimbkar Agricultural Research Institute (NARI) during 1996–99 (Nimbkar *et al.*, 2003a). Bannur is a non-prolific breed of hair sheep reared for meat production in Karnataka State. One of the reasons for including the Garole breed was to improve the prolificacy of the Deccani by cross-breeding. The project also aimed to determine the genetic basis of the prolificacy of the Garole. Two hypotheses to be tested were, first, that there is a single major gene for prolificacy in the Garole, and, second, that such a major gene is the same as or homologous to the *FecB* (Booroola) gene. Backcross ¼ Garole ewes were therefore produced and their ovulation rates determined (Nimbkar *et al.*, 2003b). The single-gene hypothesis was confirmed because roughly half the daughters of each F_1 sire had an average ovulation rate close to 2, while it was 1 for the other half. The *FecB* mutation was identified in early 2001 (Wilson *et al.*, 2001) and subsequently it was confirmed to exist in the Garole breed (Davis *et al.*, 2002). The *FecB* gene has a large effect on ovulation rate and has been introgressed from the Booroola Merino into several breeds in different countries to improve reproduction rate while maintaining desirable levels of performance for other traits (Meyer *et al.*, 1994; Gootwine *et al.*, 2001). The

Garole breed, being native to the humid, swampy Sunderban region of West Bengal State, is not adapted to a semi-arid environment. The adult weight of Garole ewes is about 15 kg and their milk yield and lamb-rearing ability have been observed to be inferior to comparable breeds. Introgression of the *FecB* gene from the Garole into the Deccani was therefore considered to be the appropriate strategy. It was decided to develop two strains – a fecund Deccani and a composite combining high prolificacy with superior milk yield and mothering ability by introducing the Awassi dairy breed.

This paper reports the effect of the *FecB* gene on the productivity of ¼ Garole ewes in terms of number of lambs weaned, lamb survival and total 105-day weight of lamb produced per ewe lambing.

2.   MATERIALS AND METHODS

2.1   Animals and management

Eleven F_1 rams were produced during 1996–98 by inseminating Deccani and Bannur ewes with fresh diluted semen of 9 Garole rams. Each of the eleven F_1 rams was then single-sire mated to Deccani, Bannur and Deccani × Bannur or Bannur × Deccani ewes to produce ¼ Garole ewe progeny groups of 12 to 27 daughters per sire from 1999 to 2001 (Nimbkar *et al.*, 2003b). In addition, reciprocal backcross ¼ Garole ewes were produced in 2000 by mating 3 Deccani and 3 Bannur rams to F_1 ewes (Garole × Deccani) produced in the diallel crossing programme. Observations on a total of 196 ¼ Garole ewes were available. These ewes had varying proportions (0.0 to 0.75) of Deccani and Bannur genes. All ewes were managed from the age of about 7 months as a single group, in a semi-stall-fed system where they were grazed, herded by shepherds for 8 hours during the day, and were fed mixed dry and green roughage and concentrate at 300 g per head in their pens at night.

The ovulation rate of ewes was determined 4 to 7 days after natural oestrus, twice before and twice after first lambing. They were inseminated with fresh, diluted semen at the third consecutive oestrus. Lambings were therefore spread over the year. In 2000 and 2001, semen of cross-bred rams with proportions of Awassi, Deccani and Garole genes varying from 25 to 50 percent was used. In 2002, the ewes were either inseminated with the semen of Deccani rams or *inter se* mated to 25 percent Garole, *FecB* heterozygote carrier (B+) rams.

Individual lambs were identified with ear tags. Sex, birth type, birth weight and description of lambs and dam identity were also recorded at this time. Thereafter, lambs were weighed once a month. Ewes were given the opportunity to rear all lambs born. Multiple-born lambs of dams without enough milk were also allowed to suckle other ewes losing their lambs or ewes with single lambs and surplus milk. This practice is common in local flocks. Lambs were penned with their dams at night until weaning at approximately 105 days. There was no culling or selection of animals in this study.

2.2   *FecB* genotyping

The 9 Garole grandsires of the ewes whose lambing records are analysed here were assumed to be *FecB* homozygotes on the basis of lambing records of their daughters. Consequently, the F_1 sires of ewes were assumed to be *FecB* heterozygotes (B+). All existing Garole, Deccani and cross-bred rams, ewes and lambs were genotyped using a modified forced restriction fragment length polymorphism (RFLP) Booroola mutation test in November 2001, confirming the above assumptions. In mid-2002, the test was established at the National Chemical Laboratory (NCL) in Pune and fast and accurate *FecB* genotyping of animals became available.

2.3   **Traits studied and statistical analysis**

Lambing records during 2000–02 were available on a total of 196 ¼ G ewes; 97 of them heterozygotes (B+) for the *FecB* gene (159 records) and 99 non-carriers (++) or wild-type homozygotes (162 records). Lambings with at least one live lamb were considered. The distribution of records according to year of lambing and litter size is given in Table 1. The proportions of abortions and stillbirths among ewes of the two genotypes were compared.

The traits of the ewe analysed (computed per ewe per lambing) were:
1. Litter size or number of lambs born alive (NLB).
2. Number of lambs weaned at 105 days (NLW1).
3. Lamb survival to weaning at 105 days (LSURV = NLW1/NLB).
4. Number of weaned lambs surviving to 180 days (NLW2).

Table 1. Distribution of lambing records according to year of lambing, *FecB* genotype of ewe and litter size.

Ewe *FecB* genotype	Year of lambing									Total records		
	2000			2001			2002					
	S	Tw	Tr	S	Tw	Tr	S	Tw	Tr	S	Tw	Tr
Heterozygote (B+)	2	2	–	27	17	1	48	59	3	77	78	4
Non-carrier (++)	6	–	–	51	1	–	103	1	–	160	2	–
Total	8	2	–	78	18	1	151	60	3	237	80	4

KEY to litter size: S = Singles; Tw = Twins; Tr = Triplets.

5. Weight (kg) of lamb weaned at 105 days (LWT1), at which age Deccani lambs are usually slaughtered.

If a lamb died before the age of 105 days, its weight was considered to be zero for the LWT1 trait. Each trait was analysed separately using the ASReml program (Gilmour *et al.,* 2002). The traits NLB, NLW1, NLW2 and LSURV were treated as Poisson variables while LWT1 was treated as a normally distributed variable. All traits were analysed using a mixed model, fitting all significant fixed effects, a random sire effect and the direct permanent environmental effect of the ewe. Two different models were fitted: model A, with the fixed effect of *FecB* genotype of ewe for all five traits; and model B, with the fixed effect of litter size for the traits NLW1, NLW2 and LWT1. Each model also had all other significant fixed effects, as well as random effects. Either *FecB* genotype of ewe or litter size were fitted in the model, but not together because these two effects are highly confounded. Other fixed effects tested were parity of ewe, year of birth of ewe, year-season of lambing, proportion of Deccani and Bannur genes in the ewe, and two-way interactions between all pairs of fixed effects. The proportion of Deccani, Bannur and Awassi genes in the lamb(s) were also tested for all traits other than NLB and found not to be significant.

Model B was investigated to obtain least squares means for single, twin and triplet lambings. Model A allowed the comparison of single-bearing ++ ewes with single- and multiple-bearing B+ ewes, while Model B allowed the comparison of single-bearing ++ and B+ ewes with twin- and triplet-bearing B+ ewes.

3. RESULTS

Parity and year-season of lambing were significant (P<0.001) only for LWT1 in both models. In Model A, *FecB* genotype of ewe had a significant (P<0.01) effect on NLB, NLW1 and LWT1 but not on NLW2 or LSURV. In Model B, litter size had a significant effect (P<0.01) on NLW1, NLW2 and LWT1. Ewe genotype × parity interaction was significant (P<0.05) only for LWT1 in Model A. All other interactions and the proportion of Deccani genes in the ewe were not significant in either of the models.

Least squares means of traits for ewe genotype and litter size for the two models fitted are given in Table 2. It shows clearly that B+ ewes had a significant advantage over ++ ewes for NLB, NLW1 and LWT1. This advantage declined from 53 percent in NLB to 34 percent in NLW1 and 17 percent in NLW2 due to higher mortality of lambs born as twins. It was 13 percent for LWT1. The difference between single- and twin-bearing ewes was much larger, twin-bearing ewes weaning 82 percent more lambs at

105 days (NLW1), 56 percent more at 180 days (NLW2) and producing 28 percent more total weight of lamb at 105 days post-partum (LWT1). Of the lambs born to B+ ewes, 77 percent survived to weaning, compared with 86 percent of lambs born to ++ ewes, but this difference was not significant.

In May-June 2003, there was a severe outbreak of an undiagnosed viral infection among lambs, causing heavy morbidity and mortality and probably leading to reduced growth rates among surviving lambs. Most of the lambs born in the winter of 2002 were around 5 months old at this time.

The proportion of abortions and stillbirths among B+ maiden ewes (0.21) was significantly higher (P<0.05) than that among ++ maiden ewes (0.11). At the second and third parities, the proportion of abortions was similar (0.04 and 0.0, respectively) in both groups.

Table 2. Least squares mean values, standard errors (in brackets) and overall standard error of difference (SED) for ewe productivity traits for the fixed effects of *FecB* genotype of ewe and litter size using models A and B.

	NLB	NLW1	NLW2	LSURV	LWT1 (kg)
Ewe genotype			Model A		
++	1.01 (0.08)	0.88 (0.07)	0.76 (0.07)	0.86 (0.07)	8.79 (0.54)
B+	1.54 (0.10)	1.18 (0.09)	0.89 (0.08)	0.77 (0.07)	9.89 (0.51)
SED	0.10	0.11	0.12	0.12	0.57
Litter size			Model B		
Singles	–	0.84 (0.06)	0.70 (0.06)	–	8.66 (0.46)
Twins	–	1.53 (0.14)	1.09 (0.12)	–	11.05 (0.61)
Triplets[1]	–	2.00 (0.72)	1.57 (0.67)	–	15.75 (2.15)
SED	–	0.30	0.36	–	1.82

NOTE: (1) The means for triplets are based on only 4 sets of triplets.

4.   DISCUSSION

4.1   Benefit of *FecB*

Ovulation rates of B+ ewes in this study have been reported earlier (Nimbkar *et al.*, 2003b) to be about 1 ovum more than ++ ewes. Their ovulation rates were not determined at the oestrus at which they were inseminated. B+ ewes gave birth to 0.5 more lambs per ewe lambing, weaned 0.3 more lambs at 105 days (and 1.1 kg more lamb weight) and had 0.1 more weaned lambs at 180 days than ++ ewes. One copy of *FecB* added 0.9 of a lamb to the litter size of CSIRO Booroola Merino ewes (Piper and Bindon, 1985) while the difference in lambs born per ewe lambing between B+ and ++ ewes among Booroola-Awassi and Booroola-Assaf crosses was 0.66 (Gootwine *et al.*, 2001). In a New Zealand study (Meyer *et al.*, 1994) of

Romney, Perendale and cross-bred ewes, B+ ewes produced 1.1 more lambs at birth relative to ++ ewes and weaned 0.4 more lambs per ewe lambing. Loss due to pre-weaning mortality is comparatively lower (0.2 lambs per ewe lambing) in the present study.

A notable difference between this and other studies of introgression of *FecB* is the low incidence (2.5 percent) of triplet or higher order litters. The majority of lambings of B+ ewes in other studies were three or more lambs (Meyer *et al.,* 1994; Gootwine *et al.,* 2001; Piper and Bindon, 1985). This could be due to the higher average litter size of the breeds into which the *FecB* gene was introgressed, compared with the Deccani.

The effect of the *FecB* gene on prolificacy has been considered to be undesirably large, and higher lamb mortality due to the higher incidence of multiple births in Booroola cross-breds has been widely reported in the literature (Meyer *et al.,* 1994). The damped effect of *FecB* in Garole and ¼ Garole ewes could be due to the lower level and poorer quality of nutrition available to sheep in the tropics. Another possibility is that since the previous studies (with the exception of the Booroola-Assaf, where marker genotypes were used for classification (Gootwine *et al.*, 2001)) were based on determination of ewe genotypes according to phenotypic ovulation rate (Piper and Bindon, 1985), it most probably led to some misclassifications and an overestimation of the effect of the gene. It is also possible that modifier genes in the Garole suppress the expression of the *FecB* gene. The moderate effect of the *FecB* gene found in this study is likely to be more manageable in a shepherd's flock. In Rajasthan State, India, the Central Sheep and Wool Research Institute has introduced Garole sheep into their mutton project, with a view to improving the prolificacy of the local Malpura breed (Sharma, Arora and Khan, 2001).

4.2   Frequency of multiple births among B+ ewes

Twin-bearing ewes in this study weaned 0.7 of a lamb more and 2.4 kg more lamb weight than single-bearing ewes, and had 0.4 of a lamb more at 180 days. On average, 49 percent of the litters produced by B+ ewes in this study were twins and 2.5 percent were triplets while the rest were singles. It seems possible that the more B+ ewes have twin lambs, the higher will be the increase in the productivity of the flock. The average litter size of 52 *FecB* homozygous (BB) Garole ewes at NARI was 2.01 (Nimbkar *et al.,* 2003b). It is therefore possible that cross-bred ewes with two copies of *FecB* might have twins more consistently than ewes with a single copy. The actual proportions of B+ ewes with singles, twins and triplets (0.48, 0.49 and 0.03, respectively) for the average litter size of 1.54 agreed reasonably with the predictions (0.473, 0.520 and 0.007, respectively) made by Z. Gao (pers.

comm.) using the model of Amer *et al.* (1999). For the average litter size of 2.01, this model predicted proportions of ewes with singles, twins and triplets to be 0.21, 0.58 and 0.21, respectively.

4.3   Effect of Garole genes and lamb mortality

This is the first report of the use of a breed other than the Booroola Merino to introduce the *FecB* gene into a non-prolific breed, and the comparison of B+ and ++ ewes based on genotypes obtained using the direct DNA test for the Booroola mutation. All ewes in this study had 25 percent Garole genes. Mortality among lambs of Garole ewes at NARI has been observed to be around 30 percent and there are similar figures reported for the Garole in the literature (Sharma, Arora and Khan, 2001). Overall mortality among single and twin lambs in this study was 28 percent and 44 percent, respectively, while mortality at NARI among Deccani lambs is typically <15 percent (Nimbkar *et al.*, 2003a). Pre-weaning mortality among lambs of B+ ewes was 24 percent compared with 13 percent in lambs of ++ ewes. Garole breeding probably had an adverse effect on the milk yield of ewes and on the growth and adaptability of lambs in this study. There was probably also variation among the ewes in this study due to Mendelian sampling, since one of their parents was cross-bred. It is highly likely that with further reduction in the proportion of Garole genes in subsequent stages of the programme of introgression, survival and growth of lambs would improve significantly. This is because Deccani sheep have been observed to have better ability to rear lambs compared with Garole sheep.

4.4   Introgression of *FecB* into Deccani and development of a composite

A nucleus flock has been established at NARI with purebred Deccani ewes and various cross-bred types generated so far, for the development of a fecund Deccani and a composite with higher prolificacy, lamb-rearing ability and growth. Selection based on an index with growth, reproduction and survival traits will be introduced into this programme and is expected to lead to higher gains. Rigorous evaluation of overall performance of ewes with 0, 1 and 2 copies of *FecB* will be carried out at NARI and in local shepherds' flocks before widespread dissemination is contemplated.

Starting in early 2003, 200 Deccani ewes in 11 local sheep owners' flocks had been mated to or inseminated with semen from B+ rams, and 25 pregnant B+ ewes and 25 ++ ewes, selected on the basis of their reproductive performance, were to be introduced into five more smallholders' flocks. The intention is to assess the sheep owners' ability to manage multiple births, and compare not only the performance of B+ with ++ ewes, but also that of introduced ewes with Deccani ewes in shepherds' flocks. It has been NARI's experience so far that lamb mortality is substantially lower in local shepherds' flocks than the institute's flocks, and lamb growth rates are higher. A geographically dispersed nucleus may be practical to take advantage of shepherds' superior sheep management skills.

5.   CONCLUSION

FecB heterozygous ¼ Garole ewes weaned 0.3 more lambs and yielded 1.1 kg more weight of lamb per ewe lambing than non-carrier ¼ Garole ewes. Twin-bearing ewes, most of which were B+, weaned 0.7 more lambs and 2.4 kg more lamb weight per ewe lambing than single-bearing ewes, despite 40 percent higher pre-weaning mortality among multiple-born lambs. The benefit of *FecB* is moderate because only about half the B+ ewes have multiple births at any one lambing. The small proportion of triplets and absence of litters larger than 3 in B+ ewes implies a more manageable effect under shepherds' flock conditions. The damped expression of *FecB* could be due to genetic or environmental reasons. Cross-bred ewes with two copies of *FecB* are being generated to evaluate their reproductive performance. Reduction in lamb mortality through better management will bring about further improvement in productivity. Weaning percentage is likely to be higher with further backcrossing to Deccani. The *FecB* gene has been introduced into some shepherds' flocks with a view to assessing flock owners' attitude to and management of multiple births, and to develop practical methods to improve lamb survival and growth. Gene technology has been harnessed successfully to utilize the *FecB* gene and appears useful.

ACKNOWLEDGEMENTS

This work was started under project AS1/1994/022 funded by the Australian Centre for International Agricultural Research (ACIAR) and AusAID, and is being continued under ACIAR project AS1/2002/038. The authors acknowledge the contributions of Dr G.D. Gray and Mr B.V. Nimbkar to the planning and design stages of the project, and of NARI staff for animal management and recording. The authors also thank Dr Jill Maddox of the University of Melbourne for her guidance in the establishment of the *FecB* test at NCL.

REFERENCES

Amer, P.R., McEwan, J.C., Dodds, K.G. & Davis, G.H. 1999. Economic values for ewe prolificacy and lamb survival in New Zealand sheep. *Livestock Production Science,* **58**: 75–90.

Davis, G.H., Galloway, S.M., Ross, I.K., Gregan, S.M., Ward, J., Nimbkar, B.V., Ghalsasi, P.M., Nimbkar, C., Gray, G.D., Subandriyo, Inounu, I., Tiesnamurti, B., Martyniuk, E., Eythorsdottir, E., Mulsant, P., Lecerf, F., Hanrahan, J.P., Bradford, G.E. & Wilson, T. 2002. DNA tests in prolific sheep from eight countries provide new evidence on origin of the Booroola (*FecB*) mutation. *Biology of Reproduction,* **66**: 1869–1874.

Gilmour, A.R., Gogel, B.J., Cullis, B.R., Welham, S.J. & Thompson, R. 2002. ASReml User Guide Release 1.0. VSN International Ltd., UK.

Gootwine, E., Zenu, A., Bor, A., Yossafi, S., Rosov, A. & Pollott, G.E. 2001. Genetic and economic analysis of introgression the B allele of the FecB (Booroola) gene into the Awassi and Assaf dairy breeds. *Livestock Production Science,* **71**: 49–58.

Khan, B.U., Sharma, R.C., Kumar, S. & Kushwaha, B.P. 2002. Tailoring genetic improvement to meet the overall livestock development objective. *In: Proceedings of the 7th World Congress of Genetics Applied to Livestock Production.* 19–23 August 2002, Le Corum, Montpellier, France. CD-ROM Communication No. 25-05.

Land, R.B., Atkins, K.B. & Roberts, R.C. 1983. Genetic improvement of reproductive performance. pp. 515–535, *in:* W. Haresign (ed). *Sheep Production.* London: Butterworths.

Meyer, H.H., Baker, R.L., Harvey, T.G. & Hickey, S.M. 1994. Effects of Booroola Merino breeding and the Fec^B gene on performance of crosses with longwool breeds. 2. Effects on reproductive performance and weight of lamb weaned by young ewes. *Livestock Production Science,* **39**: 191–200.

Nimbkar, C., Ghalsasi, P.M., Maddox, J.F., Pardeshi, V.C., Sainani, M.N., Gupta, V. & Walkden-Brown, S.W. 2003b. Expression of the *FecB* gene in Garole and crossbred ewes in Maharashtra, India. pp. 111–114, *in: 50 Years of DNA.* Proceedings of the 15th Conference of the Association for the Advancement of Animal Breeding and Genetics. 6–11 July 2003, Melbourne, Australia.

Nimbkar, C., Ghalsasi, P.M., Swan, A.A., Walkden-Brown, S.W. & Kahn, L.P. 2003a. Evaluation of growth rates and resistance to nematodes of Deccani and Bannur lambs and their crosses with Garole. *Animal Science,* **76**: 503–515.

Piper, L.R. & Bindon, B.M. 1985. The single gene inheritance of the high litter size of the Booroola Merino. pp. 115–125, *in:* R.B. Land and D.W. Robinson (eds). *Genetics of Reproduction in Sheep.* London: Butterworths.

Sharma, R.C., Arora, A.L. & Khan, B.U. 2001. Garole: A prolific sheep of India. *CSWRI [Central Sheep and Wool Research Institute, Rajasthan, India] Research Bulletin.* 17p.

Wilson, T., Wu, X.Y., Juengel, J.L., Ross, I.K., Lumsden, J.M., Lord, E.A., Dodds, K.G., Walling, G.A., McEwan, J.C., O'Connell, A.R., McNatty, K.P. & Montgomery, G.W. 2001. Highly prolific Booroola sheep have a mutation in the intracellular kinase domain of bone morphogenetic protein IB receptor (ALK-6) that is expressed in both oocytes and granulosa cells. *Biology of Reproduction,* **64**: 1225–1235.

EFFECT OF PREGNANCY ON ENDOMETRIAL SEX STEROID RECEPTORS AND ON PROSTAGLANDIN $F_{2\alpha}$ RELEASE AFTER UTERINE BIOPSY IN HEIFERS

Ana Meikle,[1] Daniel Cavestany,[1] Lena Sahlin,[2] William W. Thatcher,[3] Elsa G. Garofalo,[1] Hans Kindahl[4] and Mats Forsberg[4]

1. Faculty of Veterinary Medicine, Uruguay.
2. Karolinska Institutet, Sweden.
3. Animal Sciences, University of Florida, USA.
4. Veterinary Faculty, Swedish University of Agricultural Sciences, Sweden.

Abstract: The effect of pregnancy on oestrogen receptor (ER) and progesterone receptor (PR) endometrial expression in heifers was studied. Holstein heifers were not inseminated (controls, n = 8) or inseminated (n = 21). Endometrial biopsies were taken at Day 17 from the uterine horn ipsilateral to the corpus luteum. Hourly samples were taken on the day of the biopsy in 12 animals (controls = 4 and inseminated = 8) to analyze 15-ketodihydro-$PGF_{2\alpha}$ (PGFM) and progesterone concentrations. Pregnancy determined by ultrasonography diagnosed 6 pregnant cows. The uterine biopsy increased PGFM concentrations, which remained high for 2 to 4 hours, followed by a transient decrease in progesterone concentrations, but the procedure neither provoked luteolysis nor blocked pregnancy. PGFM concentrations were higher in cyclic than in pregnant cows. No differences in PR mRNA expression were observed among groups, but ER mRNA in pregnant heifers tended to be lower than controls, suggesting that this pathway is implicated in maintenance of pregnancy.

1.   INTRODUCTION

Genetic improvement for milk production during the last decades has been associated with decreased reproductive efficiency (Lucy, 2001). Selection for reproductive efficiency in dairy cows has not been attempted worldwide, with the main causes being the time required to evaluate

H. P. S. Makkar and G. J. Viljoen (eds.), Applications of Gene-Based Technologies for Improving Animal Production and Health in Developing Countries, 155-165.
© 2005 *IAEA. Printed in the Netherlands.*

reproductive parameters and the lack of appropriate markers for fertility. One important cause of reproductive failure has been attributed to embryo loss, which has been estimated to be approximately 40 percent (Thatcher *et al.*, 2003), that results from failures in maintaining the life of the corpus luteum. The process whereby the regression of the corpus luteum (luteolysis) is blocked in early gestation in ruminants has been termed maternal recognition of pregnancy (Short, 1969).

In order to inhibit luteolysis, the ruminant embryo delivers a signal (interferon-τ (IFN-τ)) along the uterine horns by elongation. The IFN-τ, acting by modifying uterine gene expression in pathways not completely understood, affects the episodic prostaglandin $F_{2\alpha}$ ($PGF_{2\alpha}$) release that is responsible for the luteal regression. During this period, the embryo is free-living in the uterine lumen and is completely dependent on uterine secretion for all its metabolic needs. Oestrogen and progesterone, acting via their intracellular receptors, ER and PR, are the main regulators of uterine function and have been implicated in the control of luteolysis (Lamming *et al.*, 1995). Maximum concentrations of ER and PR were found around oestrus, while the lowest concentrations of receptors were found at dioestrus (Zelinski *et al.*, 1982; Vesanen *et al.*, 1988). In addition, differences in gene expression were found in heifers with short and normal cycle: cows with short cycle presented higher ER mRNA concentration at day 5 and lower PR mRNA concentration at day 12 (Meikle *et al.*, 2001). These results show that the endometrium of heifers with a short cycle has an altered gene expression during the early luteal phase, which cannot be explained by gene expression at day 0 or by the circulating concentration of steroids between days 0 and 5 (no differences between groups). Heifers that have short-lived corpora lutea and consequently a short cycle are not capable of maintaining pregnancy.

Although progesterone is the principal hormone implicated in the control of embryo development and IFN-τ secretion (Mann and Lamming, 1995), the role of PR in pregnancy remains unclear. Studies consistently report a loss of PR in the uterus around the time of luteolysis in both sheep and cattle (Spencer and Bazer, 1995; Mann *et al.*, 1999; Robinson *et al.*, 2001). These findings are puzzling if we consider the importance of this hormone around the time of maternal recognition of pregnancy. Regarding ER, different responses have been reported in early pregnancy in cattle, such as no changes (Robinson *et al.*, 1999), a decrease in expression within all layers of the endometrium at days 16 and 18 of pregnancy (Robinson *et al.*, 2001), or an increase in uterine glands and stroma but a decrease in luminal epithelium (Kimmins and MacLaren, 2001).

In this study we focused on the effect of the embryo on gene endometrial expression (embryo-to-mother signalling) on sex steroid receptor endometrial expression. As a first step, we investigated if uterine biopsy in heifers provokes $PGF_{2\alpha}$ release, and if it also induces luteolysis or allows pregnancy to be maintained.

2.   MATERIALS AND METHODS

2.1   Experimental design

The experiment was carried out at the INIA experimental farm, Colonia, Uruguay. Twenty-nine Holstein heifers in heat (day 0) were selected after synchronization with two injections of an analogue of $PGF_{2\alpha}$ at an interval of 12 days. Only animals with normal cycle length were analysed. The animals had (mean ±SE) an age of 20.3 ±0.9 months, a body weight of 358 ±8 kg and a body condition score of 2.25 ±0.1 (on a scale of 1 to 5). Eight cows were kept as control and not inseminated, while 21 were inseminated 12 hours after showing standing oestrus. Endometrial biopsies were taken at day 17 from the uterine horn ipsilateral to the corpus luteum, and tissue samples were frozen immediately in liquid nitrogen and stored in a freezer at -80°C until the analysis. Special care was taken in collecting the endometrial samples in order to maintain pregnancy, and the amount of tissue sampled was approximately 0.1 g. Blood samples for progesterone determination were taken daily from day -1 to day +25. An hourly sampling was taken from 5 hours before to 12 hours after the biopsy in 12 animals (four from the control group and eight from the inseminated group) to analyse the $PGF_{2\alpha}$ release by 15-ketodihydro-$PGF_{2\alpha}$. Pregnancy was determined by ultrasonography 35 days after oestrus, and animals were classified in three groups: control, and artificial inseminated (AI) non-pregnant (AI non-pregnant) and pregnant (AI pregnant) heifers.

2.2   Hormone determination

2.2.1   Progesterone

Progesterone (P_4) was determined in plasma using a commercial kit (Coat-a-count, DPC Diagnostic Products Co., Los Angeles CA, USA). In the plasma assay, the intra- and inter-coefficients of variation were 9 percent and 8 percent, respectively. The sensitivity of the assay was 0.04 nmol/litre.

2.2.2   15-Ketodihydro-PGF$_{2\alpha}$

The plasma metabolite of PGF$_{2\alpha}$ was analysed in unextracted plasma by RIA. The detection limit of the assay was between 25 and 30 pmol/litre. The intra-assay coefficient of variation was below 11 percent and the inter-assay coefficient of variation was 14 percent.

2.2.3   Cortisol

Cortisol was determined using a solid phase RIA kit (Coat-a-count, DPC Diagnostic Products Co., Los Angeles CA, USA). The detection limit of the assay was 6 nmol/litre. All samples were determined in the same assay and the intra-assay coefficient of variation for control samples was below 8 percent.

2.3   mRNA determination

A solution hybridization assay of specific mRNAs for ERα and PR was performed in endometrial samples as described previously (Meikle *et al.,* 2001). In short, total nucleic acids (TNA) were prepared and the concentration of DNA in the TNA samples was measured fluorometrically. Probes were synthesized *in vitro* and radiolabelled with ^{35}S-UTP. The probes used for ERα mRNA and PR mRNA determinations were derived from full-length cDNAs containing the whole open reading frame of the human oestrogen and progesterone receptors, respectively. The cross-reactivity between bovine mRNA and human probes for ERα and PR has been demonstrated previously by northern blot (Meikle *et al.,* 2001). Overnight incubation was performed at two different concentrations, and samples were then treated with RNase to digest unhybridized RNA. Labelled hybrids were precipitated with trichloroacetic acid, collected on filters and the radioactivity was determined in a liquid scintillation counter. All the samples from the experiment were determined in the same assay. Receptor mRNA concentrations are expressed as counts per minute (cpm) in relation to DNA content (cpm/µg DNA).

2.4   Statistical analyses

Statistical analysis was carried out using the Statistical Analysis System (SAS Institute Inc., Cary NC, USA). Progesterone, cortisol and 15-ketodihydro-PGF$_{2\alpha}$ concentrations were analysed by a mixed procedure (SAS) and the statistical model included the effects of group (control, AI non-pregnant and AI pregnant heifers), day and the interaction between group and day, and also the random effect of cow within the group. Data of mRNA of ER and PR were analysed by orthogonal contrast using SAS.

3.   RESULTS

3.1   Effect of the uterine biopsy on 15-ketodihydro-PGF2α, progesterone and cortisol levels

Determination of 15-ketodihydro-PGF$_{2\alpha}$, progesterone and cortisol were performed in plasma samples collected from 5 hour before to 12 hour after the performance of the uterine biopsy on day 17 of the oestrous cycle in 12 heifers (controls n = 4; AI n = 8). The inseminated heifers were classified as pregnant (n = 3) or non-pregnant (n = 5) by ultrasound diagnosis at day 35. One AI non-pregnant heifer that had low concentration of P4 on day 17 was already in luteolysis; this animal was excluded from the analyses. The uterine biopsy increased 15-ketodihydro-PGF$_{2\alpha}$ concentration in the first bleeding after performance of the biopsy (p <0.0001), and concentration remained high for the following 2 to 4 hours (Figure 1).

Progesterone concentration increased in the first bleeding after the biopsy (p <0.01) but concentration then decreased 2 to 4 hours after the biopsy, consistent with the 15-ketodihydro-PGF$_{2\alpha}$ peak observed (Figures 1 and 2). No statistical difference in P$_4$ concentration according to physiological status (pregnant vs non-pregnant) could be detected. The significant increase 1 hour after the biopsy was consistent with a cortisol peak at that moment (Figure 2). There was a significant correlation between the two hormones: r = 0.2079 (p = 0.003).

3.2   Outcome of pregnancy

The oestrous cycle in control and inseminated non-pregnant heifers had a duration of 20 ±0.3 and 20.8 ±1.2 days (oestrus to oestrus). Progesterone concentrations from day 0 to day 25 post-oestrus in control (not inseminated), AI non-pregnant and AI pregnant heifers are shown in Figure 3. At day 35, 6 out of 21 inseminated heifers were diagnosed as pregnant. Inseminated non-pregnant cows (n = 15) were classified in two groups according to P$_4$ concentrations at days 21 to 25: luteal (P$_4$ > 18 nmol/litre; AI non-pregnant A; n = 2) or basal (P$_4$ < 3 nmol/litre; AI non-pregnant B; n = 13). Heifers with luteal concentration of P$_4$ at day 25 (AI non-pregnant A) may have suffered early embryonic mortality and were possibly pregnant at day 17; they were therefore considered as pregnant for mRNA analysis (pregnant, n = 8).

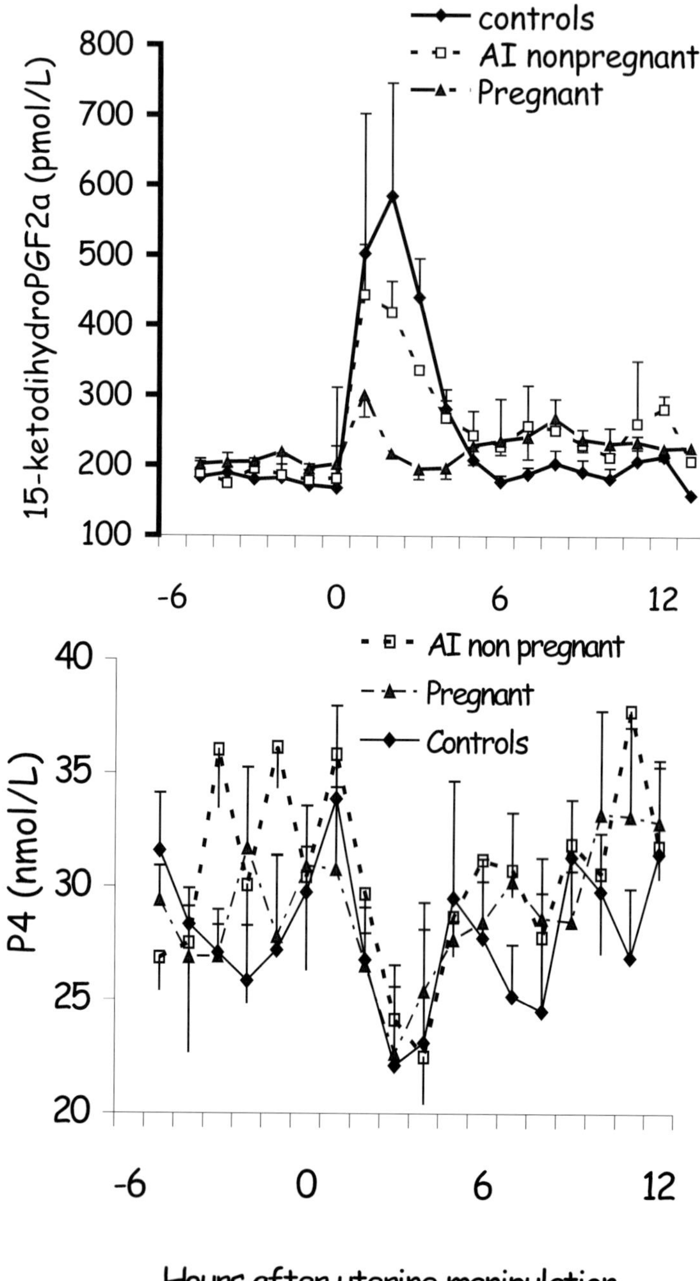

Figure 1. Mean (± standard error of the mean) concentrations of 15-ketodihydro-PGF$_{2\alpha}$ (top panel) and progesterone (bottom panel) in control (n = 4), inseminated non-pregnant (n = 4) and pregnant (n = 3) heifers before and after uterine biopsy.

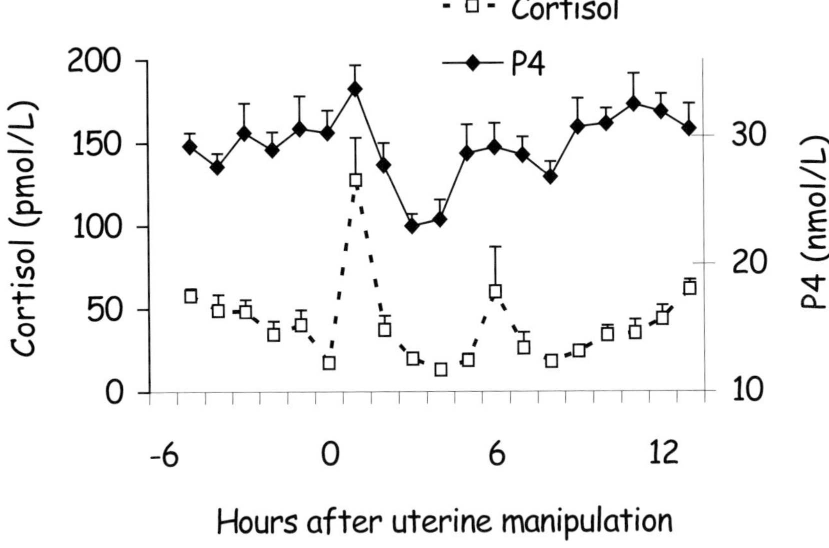

Figure 2. Mean (±SE) concentrations of progesterone and cortisol in heifers before and after uterine biopsy.

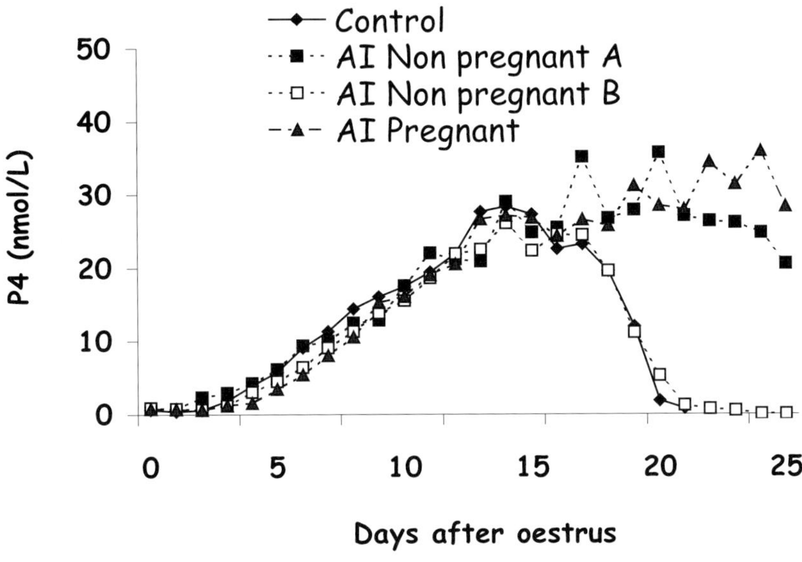

Figure 3. Mean concentrations of progesterone in control, pregnant, non-pregnant heifers with luteal (A) or basal (B) concentration of P4 at day 25 post-oestrus.

3.3   Effect of pregnancy on endometrial mRNA expression of ER and PR

The results of ER mRNA and PR mRNA are shown in Figure 4. No difference could be demonstrated in PR mRNA expression of the different groups. The concentration of ER mRNA in pregnant heifers tend to be lower than control ($p < 0.1$).

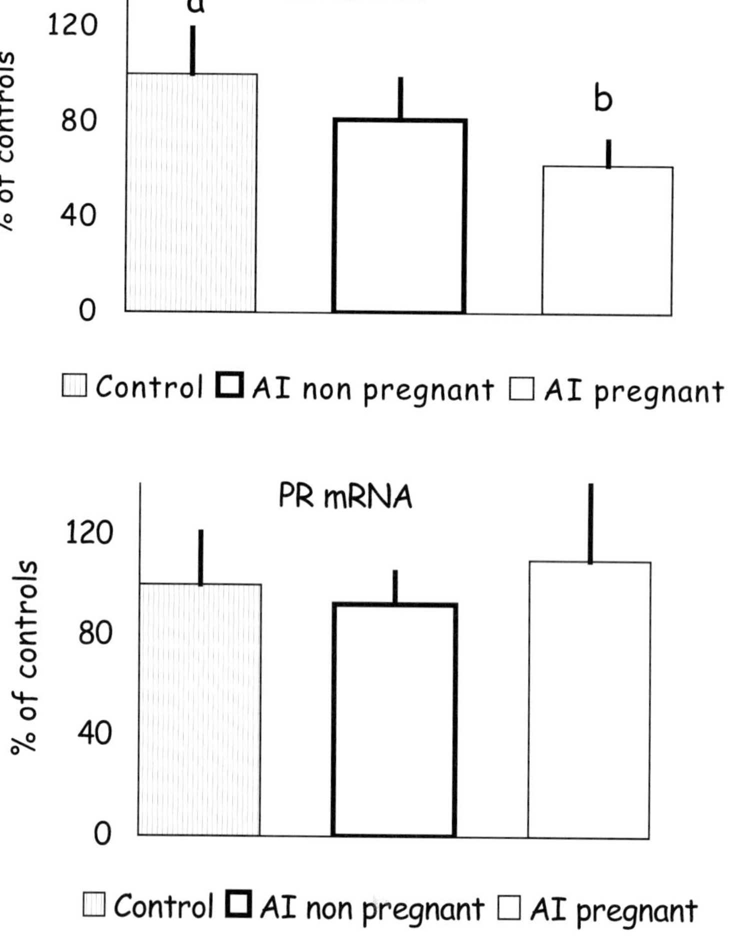

Figure 4. Concentrations of mRNA of oestrogen receptor α (ER mRNA; top panel) and progesterone receptor (PR mRNA; bottom panel) in the endometrial biopsies on day 17 in control, AI non-pregnant and AI pregnant heifers. The results are presented as percentages of control heifers. Bars are least square mean ±SE of a versus b, p <0.10.

4.   DISCUSSION

This is the first study to describe that the effect of an endometrial biopsy on $PGF_{2\alpha}$ release is dependent on the physiological status (pregnant vs cyclic heifers).

Measurement of $PGF_{2\alpha}$ itself in the peripheral circulation is not a suitable parameter since it has an extremely short half-life; 15-ketodihydro-$PGF_{2\alpha}$, a parent compound of $PGF_{2\alpha}$ and having a longer half-life, has been widely used as a good parameter of $PGF_{2\alpha}$ release (Basu *et al.,* 1987). The 3-fold increase of 15-ketodihydro-$PGF_{2\alpha}$ concentration after the uterine biopsy on day 17 is consistent with the known ability of the uterus to secrete high $PGF_{2\alpha}$ concentrations during late dioestrus (Kindahl, Lindell and Edqvist, 1981). There was an effect of the physiological status on 15-ketodihydro-$PGF_{2\alpha}$ concentration; pregnant cows had lower concentration of 15-ketodihydro-$PGF_{2\alpha}$ than control and AI non-pregnant heifers 1 to 3 hour post biopsy. Similar results have been previously reported by Thatcher *et al.* (1995): the magnitude and frequency of $PGF_{2\alpha}$ release from the uterus in non-pregnant cows are higher than that of pregnant cows. This agrees with the final action of the embryo for maternal recognition of pregnancy in ruminants: the embryo signal, interferon-τ, inhibits the episodic $PGF_{2\alpha}$ secretion and in consequence the corpus luteum is maintained.

The uterine biopsy induces a temporary release of $PGF_{2\alpha}$ which is followed by a transient decrease in progesterone concentration, but this procedure does not provoke luteolysis or block pregnancy. Pregnancy was maintained even after entering the horn ipsilateral to the corpus luteum by a transcervical catheter and performing an endometrial biopsy. Thus, transcervical biopsies can be used for studies on the interaction between the conceptus and maternal mRNA endometrial expression. This is useful, avoiding complicated surgery or expensive slaughter. This methodology may allow repeated measurements of gene uterine expression during a biological process.

The progesterone increase observed one hour after the biopsy was surprising, and therefore cortisol was determined. There was an important release of cortisol one hour after the biopsy, which may indicate an important stress for the animals. The correlation observed in progesterone and cortisol concentrations is of no surprise, since progesterone is one of the precursors of cortisol, and may explain the increase found in progesterone concentration one hour after the biopsy (Bolaños, Molina and Forsberg, 1997).

There was a tendency in pregnant heifers to have lower ER mRNA concentration, but no difference was found in PR mRNA concentration. Recently we have further studied the effect of the presence of the embryo on sex steroid receptor expression by different methodologies (Thatcher *et al.*, 2003). Endometrial ER mRNA measured both by northern blot and solution hybridization analysis was higher in cyclic cows at day 17. This finding was substantiated by the increase in ER protein measured by immuno-histochemistry and western blot analyses (Thatcher *et al.*, 2003). Although there was an increased staining intensity of ER protein in luminal epithelium, this cell type was almost devoid of PR staining. At the same time, PR expression in the uterine glands was higher in pregnant cows. The physiological status did not affect PR mRNA concentration, suggesting the embryo may alter PR expression by post-transcriptional pathways.

In summary, as demonstrated for sheep (Spencer and Bazer, 1995), pregnancy affects the expression of ER in cattle and the decrease observed around the time of maternal recognition of pregnancy suggests this pathway is implicated in the inhibition of luteolysis. Although the lack of PR presence in the luminal epithelium is puzzling, the increase of PR staining found in the uterine glands could explain progesterone stimulus to embryo growth.

ACKNOWLEDGEMENTS

The present study received financial support from the International Foundation for Science to A.M. (Grant IFS B/3025-2).

REFERENCES

Basu, S., Kindahl, H., Harvey, D. & Betteridge, K.J. 1987. Metabolites of $PGF_{2\alpha}$ in blood plasma and urine as parameters of $PGF_{2\alpha}$ release in cattle. *Acta Veterinaria Scandinavica,* **28**: 409–420.

Bolaños, J.M., Molina, J.R. & Forsberg, M. 1997. Effect of blood sampling and administration of ACTH on cortisol and progesterone concentrations in ovariectomized Zebu cows (*Bos indicus*). *Acta Veterinaria Scandinavica,* **38**: 1–7.

Kimmins, S. & Maclaren, L.A. 2001. Oestrous cycle and pregnancy effects on the distribution of oestrogen and progesterone receptors in bovine endometrium. *Placenta,* **22**: 742–748.

Kindahl, H., Lindell, J.O. & Edqvist, L.E. 1981. Release of prostaglandin F_{2a} during the oestrous cycle. *Acta Veterinaria Scandinavica,* **77**(Suppl.): 143–148.

Lamming, G.E., Wathes, D.C., Flint, A.F.P., Payne, J.H., Stevenson, K.R. & Vallet, J.L. 1995. Local action of trophoblast interferons in suppression of the development of oxytocin and oestradiol receptors in ovine endometrium. *Journal of Reproduction and Fertility,* **105**: 165.

Lucy, M.C. 2001. Reproductive loss in high-producing dairy cattle: where will it end? *Journal of Dairy Science,* **84**: 1277–1293.

Mann, G.E. & Lamming, G.E. 1995. Progesterone inhibition of the development of the luteolytic signal in cows. *Journal of Reproduction and Fertility*, **104**: 1–5.

Mann, G.E., Lamming, G.E., Robinson, R.S. & Wathes, D.C. 1999. The regulation of interferon-τ and uterine hormone receptors during early pregnancy. *Journal of Reproduction and Fertility*, **54**(Suppl.): 317–328.

Meikle, A., Sahlin, L. Ferraris, A., Masironi, B., Blanc, J.E., Rodriguez-Irazoqui, M., Rodriguez-Piñon, M., Kindahl, H. & Forsberg, M. 2001. Endometrial mRNA expression of oestrogen and progesterone receptors and insulin-like growth factor-I (IGF-I) throughout the bovine estrous cycle. *Animal Reproduction Science,* **68**: 45–56.

Robinson, R.S., Mann, G.E., Lamming, G.E. & Wathes, D.C. 1999. The effect of pregnancy on the expression of uterine oxytocin, oestrogen and progesterone receptors during early pregnancy in the cow. *Journal of Endocrinology*, **160**: 21–33.

Robinson, R.S., Mann, G.E., Lamming, G.E. & Wathes, D.C. 2001. Expression of oxytocin, oestrogen and progesterone receptors in uterine biopsy samples throughout the oestrous cycle and early pregnancy in cows. *Reproduction*, **122**: 965–979.

Short, R.V. 1969. Implantation and the maternal recognition of pregnancy. pp. 2–26, *in:* G. Wolstenholme and M. O'Connor (eds). *Foetal Autonomy.* London: Ciba Foundation, Churchill.

Spencer, T.E. & Bazer F.W. 1995. Temporal and spatial alterations in uterine oestrogen receptor and progesterone receptor gene content during the estrous cycle and early pregnancy in the ewe. *Biology of Reproduction,* **53**: 1527–1543.

Thatcher, W.W., Guzeloglu, A., Meikle, A., Kamimura, S., Bilby, T., Kowalksi, A.A., Badinga, L., Pershing, R., Bartolome, J. & Santos J.E.P. 2003. Regulation of embryo survival in cattle. *Reproduction,* **61**(Suppl.): 253–266.

Thatcher, W.W., Meyer, M.D., Danet-Desnoyers, G. 1995. Maternal recognition of pregnancy. *Journal of Reproduction and Fertility,* **49**: 15–28.

Vesanen, M., Isomaa, V., Alanko, M. & Vihko, R. 1988. Cytosol oestrogen and progesterone receptors in bovine endometrium after uterine involution postpartum and in the estrous cycle. *Animal Reproduction Science,* **17**: 9–20.

Zelinski, M.B., Noel, P., Weber D.W. & Stormshack, F. 1982. Characterization of cytoplasmatic progesterone receptors in the bovine endometrium during proestrus and diestrus. *Journal of Animal Science,* **55**: 376–383.

INVENTORY ANALYSIS OF WEST AFRICAN CATTLE BREEDS

D.M.A. Belemsaga,[1] Y. Lombo,[1] S. Thevenon[1, 2] and S. Sylla[1]
1. Centre international de recherche-développement sur l'élevage en zone subhumide (CIRDES), Bobo-Dioulasso, Burkina Faso.
2. Centre International en Recherche Agronomique pour le Développement (CIRAD), Montpelier, France.

Abstract: The improvement of livestock productivity and the preservation of their genetic diversity to allow breeders to select animals adapted to environmental changes, diseases and social needs, require a detailed inventory and genetic characterization of domesticated animal breeds. Indeed, in developing countries, the notion of breed is not clearly defined, as visual traits are often used and characterization procedures are often subjective. So it is necessary to upgrade the phenotypic approach using genetic information. At CIRDES, a regional centre for subhumid livestock research and development, such studies have been conducted. This paper focuses on cattle breed inventory in seven countries of West Africa as a tool for genetic research on cattle improvement. Data collection was done using a bibliographical study, complemented by *in situ* investigations. According to phenotypic description and concepts used by indigenous livestock keepers, 13 local cattle breeds were recognized: N'dama, Kouri, the Baoule-Somba group, the Lagoon cattle group, zebu Azawak, zebu Maure, zebu Touareg, zebu Goudali, zebu Bororo, zebu White Fulani, zebu Djelli, zebu Peuhl soudanien and zebu Gobra (Toronke). Nine exotic breeds, (American Brahman, Gir, Girolando, Droughtmaster, Santa Gertrudis, Holstein, Montbéliarde, Jersey and Brown Swiss) and five typical cross-breeds (Holstein × Goudali; Montbéliarde × Goudali; Holstein × Azawak; Brown Swiss × Azawak; and Brown Swiss × zebu peuhl soudanien) were also found. From this initial investigation, the areas of heavy concentration of herds and the most important breeds were described. The review has also indicated the necessity for a balance between improving livestock productivity and the conservation of trypanotolerant breeds at risk of extinction in West Africa.

167

H. P. S. Makkar and G. J. Viljoen (eds.), Applications of Gene-Based Technologies for Improving Animal Production and Health in Developing Countries, 167-173.
© 2005 *IAEA. Printed in the Netherlands.*

1.   INTRODUCTION

There is no doubt that livestock play an important role in agriculture and provide opportunities for asset building to support sustainable food production for the increasing population of the developing world, and especially Africa. New and improved technologies are available for increasing agricultural production, such as new cultural techniques, including the use of improved seeds and irradiation technologies for food conservation. Improving animal production by reducing health constraints has become possible with recent developments in biotechnology and subsequent introduction of techniques such as artificial insemination, embryo transfer, cryoconservation of genomes and genes, cross-breeding and marker assisted selection. However, before animals are submitted to any genetic manipulation, breeds have to be characterized in order to determine their genetic composition, a necessary step in the process of conserving animal genetic diversity.

The Centre de Recherche-Développement sur l'Elevage en Zone Subhumide (CIRDES) is a regional research institution with an agenda addressing livestock production and health in West Africa. CIRDES has a long history of work on cattle breed characterization (Moazami-Goudarzi *et al.*, 2001). Genetic research has focused on trypanotolerant livestock as part of a strategy to combat trypanosomiasis, a parasitic disease transmitted by tsetse flies (Berthier *et al.*, 2003). African animal trypanosomiasis is endemic in several parts of the humid and subhumid zones across sub-Saharan Africa. Estimates of economic losses due to the disease are over US$ 500,000 annually in both production losses and the costs of disease control (ILRAD, 1993). In response to the decline in trypanotolerant livestock populations in recent decades, farmers have devised innovative strategies to cope with the changing situation. One such strategy involves crossing local taurine cattle with zebu cattle or exotic breeds in an attempt to increase their productivity. Unabated, this trend could yield undesirable consequences in the long term and weaken incentives for *in situ* conservation of trypanotolerant breeds.

The general objective of the paper is to review available information on West African cattle breeds. A corollary objective of the paper is to make available in one place the results of several important studies on phenotypic traits, geographical localization and breeds under threat of extinction.

2. MATERIALS AND METHODS

2.1 Study area

The study covered all seven countries in West Africa under the mandate of CIRDES, namely Benin, Burkina Faso, Côte-d'Ivoire, Ghana, Mali, Niger and Togo. This area is 3,514,947 km^2, about 11.5% of all Africa (Figure 1). These countries have in common a broad band of territory comprising the subhumid and non-forested areas of the humid zones in which a mixture of trypanotolerant, trypano-susceptible and cross-bred cattle are found.

2.2 Inventory

The inventory of cattle breeds in the CIRDES mandate zone was based on a detailed bibliographical review of available studies. Using available census data, the statistics on cattle population were updated accordingly, using the estimation procedure: $\quad P_n = P_0(1+T)^n$

where: $\quad P_n$ = is the estimated cattle number;

$\qquad$ n = is the projection year;

$\qquad P_0$ = cattle number during the last inventory; and

$\qquad$ T = the growth rate.

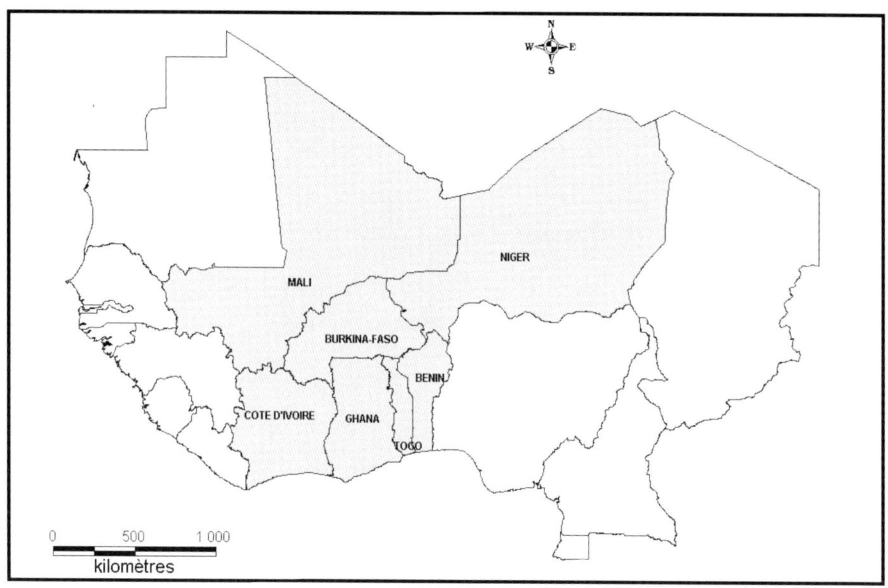

Figure 1. Map of study area in West Africa

     D.M.A. Belemsaga, Y. Lombo, S. Thevenon and S. Sylla

The review was complemented with interviews with livestock assistants, and an exploratory survey of selected farmers to confirm the existence of distinct breeds, based on indigenous knowledge.

3.   RESULTS

3.1   Cattle breed inventory: global results

Table 1 indicates the distribution of local cattle breeds enumerated in the CIRDES mandate zone. As indicated, there are shorthorn and longhorn breeds in both *Bos taurus* and *Bos indicus* groups.

Over the years, natural crossing between zebu and taurine breeds has resulted in the development of stabilized local genotypes such as the Ghana Sanga, Borgou in Benin, or Mere in other countries (Felus, 1995). Some private and commercial breeding schemes and national livestock services have imported exotic breeds to improve the potential of indigenous livestock. These include nine exotic cattle breeds, namely American Brahman, Gir, Girolando, Droughtmaster, Santa Gertrudis, Holstein, Montbéliarde, Jersey and Brown Swiss. The result of crossing with local breeds has produced five cross-breed types: Holstein × Goudali; Montbéliarde × Goudali; Holstein × Azawak; Brown Swiss × Azawak; and Brown Swiss × zebu Peuhl soudanien.

Table 1. Location of available local breeds of cattle in West Africa (CIRDES area).

	Benin	Burkina Faso	Côte d'Ivoire	Ghana	Mali	Niger	Togo
Bos taurus (shorthorn cattle type)							
Baoule-Somba	X	X	X	X			X
Lagoon	X		X	X			X
Bos taurus (longhorn cattle type)							
N'Dama	X	X	X	X	X		X
Kouri						X	
Bos indicus (shorthorn cattle type)							
Azawak		X			X	X	
Maure			X		X		
Touareg					X		
Goudali	X	X		X	X	X	
Bos indicus (longhorn cattle type)							
Bororo	X				X	X	
White Fulani	X	X		X	X		
Djelli						X	
Peuhl soudanien		X	X		X		X
Gobra (Toronké)					X		

In the CIRDES zone countries, the total local cattle population is estimated at 19.2 million head: most (6.6 million) are in Mali (Table 2); while Burkina Faso has the highest density (17 bovines per km^2).

In Figure 2, it can be seen that zebu cattle breeds are predominantly found in Sahelian countries (Niger, Mali, Burkina Faso), while trypano-tolerant livestock outnumber other breeds in countries with a soudanian climate (Côte d'Ivoire, Benin, Ghana, Togo). In all countries, active and passive crossings occur.

4.   DISCUSSION

Adequate knowledge of the animal resources in each country is essential in any genetic improvement programme. A large quantity of information has been collected from the available literature. This review indicates that there is a pressing need for more detailed quantitative data, especially country and regional comparative data on trypanotolerance and other adaptive traits, and special genetic attributes of local breeds (d'Ieteren *et al.*, 1998).

Table 2. Approximate numbers (millions) of local cattle breeds and their density in each country.

	Benin	Burkina Faso	Côte d'Ivoire	Ghana	Mali	Niger	Togo
Number	1.49	4.80	1.33	1.32	6.64	3.31	0.28
Density (no./km^2)	13	17	4	6	5	3	5

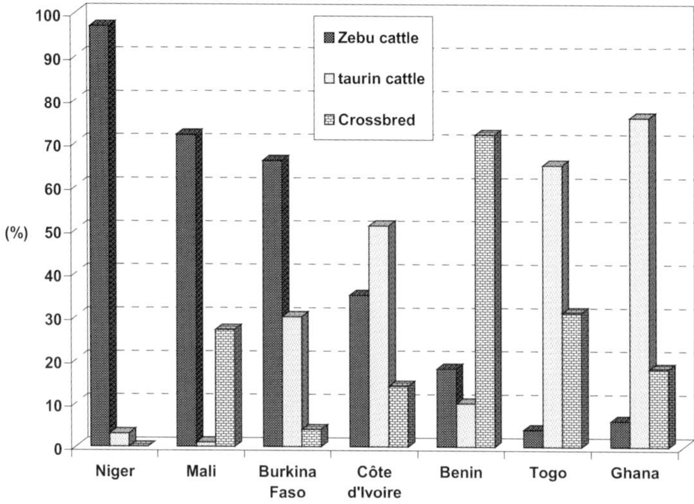

Figure 2. Relative importance of cattle in CIRDES countries

Compared to the cattle population in Africa – 224.3 million (FAO, 2000) – the seven countries of the CIRDES mandate zone (11 percent of Africa's area) account for 8.5 percent of the cattle. This, however, represents more than 70 percent of the total cattle population reared in West Africa. It is interesting to note that the CIRDES zone holds the largest population of trypanotolerant cattle, the most adapted breed in the humid and subhumid zone (Hoste *et al.*, 1988). It has been reported that the Benin Pabli breed has disappeared, that the Lagoon cattle group and the Kouri breed are under threat of extinction, and that the Somba-Baoule group is at risk of gene introgression from Borgou and zebu breeds (Yapi-Gnaore *et al.*, 1996; Youssao *et al.*, 2000). The taurine cattle population is indeed decreasing, compared with the total number of cattle. But the number of Borgou cattle and zebu breeds is on the increase in the zones where trypanosomiasis risk is high. It is clear that indigenous cattle breeds are disappearing not only because of threat from indiscriminate crossing by individual farmers, but also because many schemes for genetic improvement were developed with neither a concern for programmes to preserve local, adapted breeds, nor a built-in plan to conduct routine genetic impact analyses when exotic strains are used (Hall, 1996; Toure, 1993). Even if developing countries need existing and new technologies to improve livestock productions, they first need to put in place firm measures to maintain stocks of breeds at risk of extinction. Indeed, the future of livestock breeding depends on the availability of a large gene pool from which useful traits can be selected.

Measures taken to conserve trypanotolerant breeds of cattle should take into account the diversity in agro-ecological zones and the demand for land use improvement in tsetse-infested but otherwise fertile areas. The question of conserving trypanotolerant breeds should be seen from the perspective of biodiversity and the control of trypanosomiasis. Conservation efforts should emphasize the training of farmers in breeding practices, improved management and primary health care. The efforts of research institutions should be directed at incorporating farmers' preferences for phenotypic and genotype traits. The conservation of trypanotolerant cattle should take a holistic approach that takes into account the strengths and weaknesses of existing livestock production systems.

ACKNOWLEDGEMENTS

Authors gratefully acknowledge breeders in all countries and National Agricultural Research Systems (NARS) for their cooperation in this inventory. Special thanks are due to the European Commission (FED programme N°8 and "Procordel"). They also thank FAO/IAEA for financial support. They would like to record their gratitude to colleagues at CIRDES who contributed significantly in the preparation of this paper.

REFERENCES

Berthier, D., Quere, R., Thevenon, S., Belemsaga, D., Piquemal, D., Marti, J. & Maillard, J.C. 2003. Serial analysis of gene expression (SAGE) in bovine trypano-tolerance: preliminary results. *Genetics Selection Evolution,* **35**(Suppl.1): S35–S47

d'Ieteren, G.D.M., Authie, E., Wissocq, N. & Murray, M. 1998. Trypanotolerance, an option for sustainable livestock production in areas at risk from trypanosomosis. *Revue scientifique et technique de l'Office Internationales des Epizooties,* **17**: 154–175

FAO [Food and Agriculture Organization of the United Nations]. 2000. See: http://www.fao.org/dad-is

Felus, M. 1995. *Cattle breeds – An Encyclopedia.* The Netherlands: Misset Uitgeverij. 799p.

Hall, S.J.G. 1996. Conservation and utilization of livestock breed biodiversity. *Outlook on Agriculture,* **25**: 115–118.

Hoste, C.H., Chalon, E., d'Ieteren, G. & Trail, J.C.M. 1988. Le bétail trypanotolérant en Afrique Occidentale et Centrale. Vol. 3: Bilan d'une décennie. *Etude FAO: Production et Sante Animales,* No. 20/3. 286p.

ILRAD [International Laboratory for Animal Diseases]. 1993. Estimating the costs of animal trypanosomiasis in Africa. *ILRAD Reports,* **11**: 1–6.

Moazami-Goudarzi, K., Belemsaga, D.M.A., Geritti, G., Laloë, D., Fagbohoun, F., Kouagou, N'T., Sidibé, I., Codja, V., Crimelle, M.C., Grosclaude, F. & Touré, S.M. 2001. Caractérisation de la race bovine Somba à l'aide de marqueurs moléculaires. *Revue d'elevage et de medecine veterinaire des pays tropicaux,* **54**: 129–138.

Toure, M.M. 1993. Politique d'amélioration génétique au Mali. pp. 253–255, *in: L'Amélioration génétique des bovins en Afrique de l'Ouest. Etudes FAO : Production et Santé Animales,* No. 110.

Yapi-Gnaore, C.V., Oya, B.A. & Zana, O. 1996. Revue de la situation des races d'animaux domestiques de Côte-d'Ivoire. *Animal Genetic Resources Information (FAO/UNEP),* No. 19: 99–109.

Youssao, A.K.I., Ahissou, A., Touré, Z. & Leroy, P.L. 2000. Productivité de la race Borgou à la ferme de l'Okpara au Bénin. *Revue d'elevage et de medecine veterinaire des pays tropicaux,* **53**: 67–74.

A REVIEW OF GASTROINTESTINAL MICROBIOLOGY WITH SPECIAL EMPHASIS ON MOLECULAR MICROBIAL ECOLOGY APPROACHES

Roderick I. Mackie[1, 2] and Isaac K.O. Cann[1]
1. Department of Animal Sciences. 2. Division of Nutritional Sciences.
Both at University of Illinois at Urbana-Champaign, Urbana IL 61801, USA.

Abstract: All animals, including humans, are adapted to life in a microbial world. Large populations of micro-organisms inhabit the gastrointestinal tract of all animals and form a closely integrated ecological unit with the host. This complex, mixed, microbial culture can be considered the most metabolically adaptable and rapidly renewable organ of the body, which plays a vital role in the normal nutritional, physiological, immunological and protective functions of the host animal.

Bacteria have traditionally been classified mainly on the basis of phenotypic properties. Despite the vast amount of knowledge generated for ruminal and other intestinal ecosystems using traditional techniques, the basic requisites for ecological studies, namely, enumeration and identification of all community members, have limitations. The two major problems faced by microbial ecologists are bias introduced by culture-based enumeration and characterization techniques, and the lack of a phylogenetically-based classification scheme. Modern molecular ecology techniques based on sequence comparisons of nucleic acids (DNA or RNA) can be used to provide molecular characterization while at the same time providing a classification scheme that predicts natural evolutionary relationships. These molecular methods provide results that are independent of growth conditions and media used. Also, using these techniques, bacteria can be classified and identified before they can be grown in pure culture. These nucleic acid-based techniques will enable gut microbiologists to answer the most difficult question in microbial ecology: namely, describing the exact role or function a specific bacterium plays in its natural environment and its quantitative contribution to the whole. However, rather than replacing the classical culture-based system,

H. P. S. Makkar and G. J. Viljoen (eds.), *Applications of Gene-Based Technologies for Improving Animal Production and Health in Developing Countries*, 175-198.
© 2005 *IAEA. Printed in the Netherlands.*

the new molecular-based techniques can be used in combination with the classical approach to improve cultivation, speciation and evaluation of diversity. The study of microbial ecology in gut ecosystems involves investigation of the organisms present (abundance and diversity), their activity (usually determined *in vitro,* but *in vivo* activity or expression of activity is really required), and their relationship with each other and the host animal (synergistic and competitive interactions). This entails the study and measurement of many types of interactions, both beneficial and competitive.

Traditionally, media for isolation of bacteria from natural environments are basically of two types: those that simulate the habitat in broad terms, i.e. habitat-simulating and non-selective media; and those designed to enumerate and isolate bacteria of a particular type or from a specific biochemical niche, i.e. niche-simulating or functional or nutritional group analysis. A third type, less important in the ecological sense, does not simulate the habitat, is often highly selective and is used to isolate specific bacterial groups. Specific nutritional types of bacteria may be isolated by the use of enrichment media. This type of medium, basically a refinement of the habitat-simulating medium, is widely used in environmental microbiology and has been applied with some success to the gut ecosystem. Although estimates of microbial number rely on culture techniques, microscopic examination is a most useful technique for evaluating the efficacy of other enumeration approaches. The combination of microscopy with specific phylogenetic stains or fluorescent antibodies enables bacteria to be specifically detected and enumerated in mixed populations.

The introduction of genetic-based technologies, and in particular those relating to 16S rRNA typing, are rapidly replacing conventional detection and enumeration methods in studies of the mammalian intestinal tract. Although molecular techniques promise a fuller and more accurate description of the true diversity, structure and dynamics of complex microbial communities than the present culturing studies, each technique suffers from its own experimental bias and selectivity. Different methods used for direct molecular detection are reviewed in the paper. Attention is given to molecular characterization of complex communities; in particular, the application of molecular fingerprinting techniques to examine diversity and community structure in complex gut bacterial communities. Denaturing or temperature gradient gel electrophoresis (DGGE/TGGE) methods have been successfully applied to the analysis of human, pig, cattle, dog and rodent intestinal populations. In future, genomics technologies will provide gut microbial ecologists with their best opportunity for a complete or global analysis of the molecular mechanisms involved in metabolism and regulation within the bacterial community and between host and resident microbes. Molecular microbial techniques have been widely and successfully used to study microbial diversity in environmental microbiology but have had limited application in the gastrointestinal ecosystem.

1. INTRODUCTION

All animals, including humans, are adapted to life in a microbial world. The complexity of animal-microbe relationships varies tremendously, ranging from competition to cooperation (Hungate, 1985). The animal alimentary tract has evolved as an adaptation enabling the animal to secure food and limit consumption by other animals. This allows the retention and digestion of ingested food followed by absorption and metabolism of digestion products, whilst feeding and other activities continue. The gastrointestinal tract is a specialized tube divided into various well-defined anatomical regions extending from the lips to the anus. For the purposes of this, and most other, review(s) concerning the ecology, physiology and metabolism of gut microbes, discussion is restricted to the stomach (rumen-reticulum, crop, gizzard), small intestine and large intestine (caecum and colon). The mouth has a characteristic anaerobic bacterial microbiota that is greatly influenced by the practice of coprophagy (the consumption of faecal material). In the context of intestinal function, the mouth determines the physical state and degree of homogenization of the food. This has a considerable effect on the surface area available for the action of enzymes produced by both the host and resident microbes. Large populations of micro-organisms inhabit the gastrointestinal tract of all animals and form a closely integrated ecological unit with the host. This complex, mixed, microbial culture – comprising bacteria, ciliate and flagellate protozoa, anaerobic phycomycete fungi and bacteriophages (at least in the rumen) – can be considered the most metabolically adaptable and rapidly renewable organ of the body, an organ that plays a vital role in the normal nutritional, physiological, immunological and protective functions of the host animal. In monogastric animals, the complex microbiota is simplified by the absence of a large part of the eukaryotic microbiota. It is worthwhile reiterating that the study of microbial ecology in gut ecosystems (Hungate, 1960) involves investigation of the organisms present (abundance and diversity), their activity (usually determined *in vitro,* but *in vivo* activity or expression of activity is really required), and their relationship with each other and the host animal (synergistic and competitive interactions). This entails a study of many types of effects and interactions, competitive and beneficial, dietary, nutritional, physiological and immunological. This emphasizes the urgent need for an integrative, interdisciplinary approach to study digestive physiology, which includes the disciplines of animal physiology, nutrition, biochemistry, microbiology, microbial ecology and molecular biology. However, the first and key part is the description of the microbes present and the measurement of their activities.

Human appreciation for anaerobic life is relatively recent. It was chiefly Pasteur (1858) who alerted the world to actions of microbes in the absence of oxygen and described what we know today as fermentation. Pasteur's demonstration of life without air soon led to the development of many methods for growing such organisms. However, until about 1940 only spore formers and non-spore formers of clinical importance had been isolated and described, probably due to the popularity of the Petri dish and the ease of cultivating aerobic bacteria. Attempts to inoculate and incubate plates under anaerobic conditions were unsuccessful until the anaerobic glove box was perfected (Aranki *et al.*, 1969). Robert E. Hungate is recognized as the father of modern anaerobic microbial ecology. His understanding of the principle of redox potential and achieving low potentials in anaerobic media led to the development of procedures for media preparation enabling enumeration and isolation of anaerobic bacteria. The roll tube technique, with its numerous modifications and improvements since the original description (Hungate, 1950), was considerably superior to other anaerobic methods and contributed much to our understanding of anaerobes. Despite the advent of the anaerobic cabinet, with its many advantages, modifications of the roll tube technique are still widely used and are standard procedures for anaerobe laboratories. Importantly, these techniques have provided the tools that have enabled microbial ecologists, particularly those working in the gut, to advance this field of research considerably.

Bacteria have traditionally been classified mainly on the basis of phenotypic properties. Despite the vast amount of knowledge generated for ruminal and other intestinal ecosystems using traditional techniques, the basic requisites for ecological studies, namely, enumeration and identification of all community members have limitations. The two major problems faced by microbial ecologists include bias introduced by culture-based enumeration and characterization techniques and the lack of a phylogenetically-based classification scheme (Raskin *et al.*, 1997). Modern molecular ecology techniques based on sequence comparisons of nucleic acids (DNA or RNA) can be used to provide molecular characterization while at the same time providing a classification scheme that predicts natural evolutionary relationships. These molecular methods provide results that are independent of growth conditions and media used. Also, using these techniques, bacteria can be classified and identified before they can be grown in pure culture.

In addition, *in situ* hybridization with fluorescently labelled rRNA targeted nucleic acid probes facilitates *in situ* identification and phylogenetic placing of uncultured organisms and provides information on three-dimensional relationships in complex microbial populations (Amann, 1995). Ultimately, genetic capabilities, expression of these capabilities, and taxonomic information are all potentially accessible at the individual cell level using targeted nucleic acid (DNA, mRNA or rRNA) *in situ* hybridization methods. These nucleic acid-based techniques will enable gut microbiologists to answer the most difficult question in microbial ecology: namely describing the exact role or function a specific bacteria plays in its natural environment and its quantitative contribution to the whole (Hungate, 1960). However, rather than replacing the classical culture-based system, the new molecular-based techniques can be used in combination with the classical approach to improve cultivation, speciation and evaluation of diversity. In fact, many of the entries in the SS rRNA database are derived from pure-culture studies. The new research era of "molecular microbial ecology" commands a lot of attention and receives considerable funding support, but it should also be noted that the emphasis should be placed on "ecology" and not "molecular". However, it is also true that microbial ecology cannot be adequately and rigorously pursued without the application of modern molecular techniques.

It could be argued that the technological impetus for major advances in our knowledge of gastrointestinal ecology during the last 40 years has been derived from three major sources: the development of anaerobic culture techniques and their application to the study of the rumen microbial ecosystem by Hungate, Bryant and others; the use of rodent experimental models to define relationships between intestinal bacteria and the host by Dubos, Savage and others; and the development of gnotobiotic technology by which germ-free or defined-microbiota animal models could be derived and maintained. It is already true that the use of molecular ecology and genomics technologies will generate the next major advance in our knowledge and provide, for the first time, not simply a refinement or increased understanding, but a complete description of the gastrointestinal ecosystem.

2.   CLASSICAL CULTIVATION-BASED TECHNIQUES

Physical and chemical conditions within the gut of different animals may differ considerably, but are usually relatively constant in a single species on a given diet. This is the case for homoeothermic animals, in which, allowing for irregularities in food intake, other factors such as temperature, oxygen, acidity and moisture vary little with time. The detailed chemical composition of the gut contents of most animals is extremely complex. The microbial environment in the rumen has, to date, been the most extensively studied and clearly defined and, allowing for variation in the nature and amount of food ingested, serves as a good model for other gut ecosystems, both herbivores and non-herbivores (Mackie, 1997). The hindgut environment is more constant in terms of physical and chemical composition with nutrients for caeco-colic bacteria being provided by undigested dietary polysaccharides and endogenous secretions and tissues such as mucopolysaccharides, mucins, epithelial cells, and enzymes. Many of the properties of rumen contents illustrate the complexities that must be considered in media selection and design in order to cultivate, enumerate and isolate predominant gut bacteria. These properties and conditions form the basis for the formulation of habitat-simulating media used successfully by Hungate, Bryant and others to describe the microbial ecology of the rumen.

Media for isolation of bacteria from natural environments are basically of two types: those that simulate the habitat in broad terms, i.e. habitat-simulating, and non-selective media; and those designed to enumerate and isolate bacteria of a particular type or from a specific biochemical niche, i.e. niche-simulating or functional or nutritional group analysis. A third type, less important in the ecological sense, does not simulate the habitat, is often highly selective and is used to isolate specific bacterial groups. Media in this category include those designed for groups such as clostridia, lactobacilli, bifidobacteria and gram-negative anaerobes. Typically these are commercial general-purpose media and media containing antibiotics or other inhibitory compounds resulting in selective growth conditions. These selective (and erroneously named "specific") media are discouraged during the initial stages of enumeration since they suppress the viable count of bacteria. Specific nutritional types of bacteria may be isolated by the use of enrichment media. This type of medium, basically a refinement of the habitat-simulating medium, is widely used in environmental microbiology and has been applied with some success to the gut ecosystem (Krumholz and Bryant, 1986a, b; Russell, Stroebel and Chen, 1988). A serious criticism of the technique is that it gives no information on the numerical importance of the bacteria isolated and it must be combined with the Most Probable Number (MPN) dilution enumeration procedure to provide a quantitative estimate of numbers and their confidence limits.

Although estimates of microbial number rely on culture techniques, microscopic examination is a most useful technique for evaluating the efficacy of other enumeration approaches. The combination of microscopy with specific phylogenetic stains or fluorescent antibodies enables bacteria to be specifically detected and enumerated in mixed populations (see section describing fluorescence *in situ* hybridization (FISH) in molecular techniques). A few bacteria, such as spirochetes and *Oscillospira*, can be recognized by their unique morphology, but most rods and cocci remain anonymous. Chemical estimation of the mass of bacteria is possible, but also has numerous limitations. Microbial ecologists generally use two methods to quantitate the bacteria present in intestinal samples: the total microscopic count and the total viable count of the numerically dominant groups of bacteria made from serial dilutions of the homogenized sample. The total viable count made on a non-selective solid (agar-based) medium estimates the number of colony forming units (cfu) per gram of faeces or other unit. Since each colony can vary from one cell to a clump or chain of cells it is important to count microscopically in the same way. Comparison of the results using the two methods normally produces the finding that the total viable count is less than the total microscopic count. In the past this was attributed to a large proportion of dead bacteria in the sample. However, this is more likely to be due to an inability to culture the majority of bacteria present. Estimates of the proportion of culturable bacteria in intestinal samples vary from 5–50 percent, but are more likely to be in the range of 5–10 percent. Currently Live/Dead staining kits are commercially available (Molecular Probes, Oregon) and should be a standard component of ecological analyses. With the ability to automate cell counting using specific cell staining procedures, this approach is an essential component of intestinal microbial ecology.

In future, cultivation efforts will intensify with increased attention to:

– the cultivation of previously uncultured bacteria by recreating the chemical environment in their natural setting, resulting in the ability to isolate synergistic cultures, that are unable to grow alone but when in co-culture provide synergistic nutrients or signalling molecules essential for growth of the other partner;

– reduction in carbohydrate/substrate concentrations in medium formulations and longer incubation times to enable slow growing but ecologically and physiologically relevant bacteria that are not fast growers to compete in culture,

– the use of extinction dilution procedures to determine the most abundant bacteria; and

– combining cultivation with molecular approaches, e.g. the use of DGGE/TGGE, to improve isolation and resolution of mixed or synergistic enrichment cultures.

3.   TECHNIQUES USED IN MOLECULAR MICROBIAL ECOLOGY

The introduction of genetic-based technologies, and in particular those relating to 16S rRNA typing, are rapidly replacing conventional detection and enumeration methods in studies of the mammalian intestinal tract

(Tannock, 1999, 2001). The use of ribosomal sequences for classification of micro-organisms in the environment is favoured by several critical attributes. The first is their universal distribution among all cellular forms of life in high copy number. The second is that their essential function in all microbes translates into very slow genetic evolution resulting in high conservation of sequences coding for rRNA. Importantly, the mutation rate of rRNAs corresponds with evolutionary divergence of micro-organisms. Identification and enumeration are basic prerequisites for ecological studies and it is now acknowledged that the most desirable classification scheme should reflect natural evolutionary relationships (Pace, 1997; Woese, 1987). Ribosomal RNA molecules consist of alternating conserved and variable domains making them highly suitable for detection and identification of microbial species and ideal targets for specific DNA probes. By aligning the appropriate 16S rRNA sequences, genus-specific and species-specific sequences can be identified, allowing simultaneous detection and classification. Comparative sequencing of the 16S rRNA molecule has become the most commonly used measure of microbial diversity in the environment (Head, Saunders and Pickup, 1998). The 5S rRNA used in initial studies is rather small (300 nt), giving limited information, while the 16S rRNA, consisting of approximately 1500 nt, provides a large amount of information for phylogenetic inference and is a reasonable size for sequencing. The 23S rRNA (3000 nt) offers substantially more information but requires more sequencing. Thus 16S rRNA has become the established reference and gold standard.

Inferring species identity from genetic data is not without difficulty. Currently identity based on rRNA classification schemes uses <97 percent sequence similarity with any known organism in the database as the rule of thumb for describing a new species. However, in practice, it is more complex. The species concept is receiving considerable attention and the generation of complete genome sequences for prokaryotes of different descent is providing evidence that horizontal gene transfer may be an important factor in bacterial adaptation and speciation (Doolittle, 1999; Woese, 2002). Nevertheless, the state of the art is well founded in the use of rRNA as a phylogenetic marker. Potential uses and limitations of various methods for analysis of complex microbial communities are presented in Tables 1 and 2 (Zoetendal *et al.*, 2004).

Table 1. A summary of molecular techniques used to study complex microbial ecosystems.

Methods	Uses	Limitations
Cultivation	Isolation; traditional approach.	Not representative; slow and laborious.
16S rDNA sequencing	Phylogenetic identification.	Laborious; subject to PCR biases.
DGGE/TGGE/TTGE	Monitoring of community or population shifts; rapid comparative analysis.	Subject to PCR biases; semi-quantitative; identification requires clone library.
T-RFLP	Monitoring of community shifts; rapid comparative analysis; very sensitive; potential for high throughput.	Subject to PCR biases; semi-quantitative; identification requires clone library.
SSCP	Monitoring of community or population shifts; rapid comparative analysis.	Subject to PCR biases; semi-quantitative; identification requires clone library.
FISH	Detection; enumeration; comparative analysis possible with automation.	Requires sequence information; laborious at species level.
Dot-blot hybridization	Detection; estimates relative abundance.	Requires sequence information; laborious at species level.
Quantitative PCR	Detection; estimates relative abundance.	Laborious.
Diversity micro-arrays	Detection; estimates relative abundance.	In early stages of development; expensive.
Non-16S rRNA profiling	Monitoring of community shifts; rapid comparative analysis.	Identification requires additional 16S rRNA-based approaches.

Table 2. Molecular approaches for studying metabolic activities and gene expression in gastrointestinal microbes.

Approach	Target molecule	Is cultivation required?	Is the identity of the target gene required?	Can microbes be identified directly?	Main purpose
BAC vector cloning	Genomic DNA	No	No	No	Identify functional genes.
DNA micro-array	mRNA	No	Yes	No	Obtain tran-scriptional fingerprints.
In situ isotope tracking	Labeled biomarkers	Yes	No	Yes	Identify active microbes.
IVET	Promoter regions	Yes	Yes	Yes	Detect induced promoters.
RT-PCR	mRNA	No	Yes	No	Detect/measure gene expression.
Subtractive hybridization	Genomic DNA	No	No	No	Recovery of unique genes.

4.   DIRECT MOLECULAR DETECTION

4.1   Cell lysis and extraction of nucleic acids

Although molecular techniques promise a fuller and more accurate description of the true diversity, structure and dynamics of complex microbial communities than the present culturing studies, each technique suffers from its own experimental bias and selectivity. The first of these biases is selective nucleic acid extraction. Extraction procedure and protocol is dictated by the application or fingerprinting technique for the isolated DNA. When investigating whole communities a reliable method for extraction and purification of DNA or RNA, or both, from the sample is one of the most critical steps, since all further analyses assume complete and representative presence of accessible nucleic acids. Since not all microbial cells are lysed with equal ease, numerous protocols have been reported for nucleic acid extraction from faecal samples involving enzymatic, chemical and mechanical breaking or disruption of cells. Recently, the more preferred methods involve disruption by bead beating in combination with other treatments (Dore *et al.*, 1998; Simpson *et al.*, 1999; Zoetendal, Akkermans and de Vos, 1998). Optimization of the extraction procedure compatible with the fingerprinting technique applied will obviously provide more confidence in the final results.

4.2   Cloning and sequencing of 16S rRNA and rDNA genes

Ribosomal RNA sequences can be obtained either directly from rRNA or from the encoding genes located at various positions in the genome, i.e. rDNA. In practice, sequences of 16S rDNAs are determined by creating rDNA clone libraries rather than rRNA libraries, for several reasons. Full-length 16S rDNA can be amplified either directly, or after reverse transcription of rRNA, with a set of primers binding to conserved regions of the 16 rDNA. Although this is a routine method for pure cultures, several problems can arise when these techniques are applied to environmental community analysis. These have been reviewed by von Wintzingerode, Gobel and Stackebrandt (1997) and are listed below:
1.   Inhibition of polymerase chain reaction (PCR) amplification by co-extracted substances particularly humic acids or humic substances that strongly inhibit DNA modifying enzymes.

2. Differential PCR amplification of the DNA template from a complex community based on differing abundance of template as well as hybridization and extension efficiency.
3. Formation of artefactual PCR products, such as chimeric molecules, deletion and point mutations.
4. Contaminating DNA, containing the specific target sequence for the PCR reaction involved, leading to amplification in negative controls without added template.
5. 16S rRNA sequence variations due to *rrn* operon heterogeneity that can interfere with the analysis of 16S rDNA clone libraries or gel electrophoresis patterns.

The effect of the number of PCR cycles (10 versus 25 cycles) on the inferred structure of the 16S rDNA library was examined by Bonnet *et al.* (2002). Coverage-based computing, projections and statistical analysis demonstrated that the structures of the two PCR-derived libraries were different and that the 25-cycle rDNA library displayed reduced diversity. Clearly, the number of PCR cycles used for amplification of 16S rDNA genes for phylogenetic diversity studies must be kept as small as possible.

The reverse transcription-polymerase chain reaction (RT-PCR), which converts rRNA into rDNA, is sensitive, requiring high quality template. The 1500 nt size of 16S rRNA, the presence of post-transcriptionally modified ribonucleotides, or processing resulting in fragmentation, can limit the efficiency of the RT reaction, resulting in premature termination of the transcription (Vaughan *et al.*, 2000).

Several programs are available for determining sequence similarity. A comprehensive sequence data set, currently >90,000 small subunit rRNA entries, is available in accessible databases such as GenBank, European Molecular Biology Laboratory (EMBL), Ribosomal Database Project (RDP) (Maidak *et al.*, 1999), and ARB (Strunk and Ludwig, 1995). The last two databases are maintained by specialists and provide services such as alignment of newly submitted sequences, probe check, chimera tests and phylogenetic information (Ludwig *et al.*, 1998).

The number of culture-independent studies performed on human faeces is very limited but demonstrates our limited understanding of bacterial diversity. A pioneering study by Wilson and Blittchington (1996) showed that about 41 percent of all cloned material was not represented by any clonal sequence obtained and only 31 percent of the sequenced colony isolates corresponded with any known species in the databases. The most thorough direct analysis of 16S rDNA clone libraries to date has been generated by the French group (Suau *et al.*, 1999). Their results show that three phylogenetic groups, namely *Bacteroides, Clostridium coccoides* and *C. leptum,* constituted 95 percent of the clones. Comparative sequence analysis again indicated that only 24 percent of the clones corresponded to described species in databases. Thus the vast majority of bacteria in the human intestinal tract have eluded scientific description and our knowledge concerning bacterial diversity is superficial and inadequate. On completion of the review, the same conclusion will be reached concerning the bacterial populations that inhabit the intestinal tract of production animals, except that our knowledge is even more superficial and limited.

5.   MOLECULAR FINGERPRINTING METHODS FOR MICROBIAL COMMUNITIES

It is worthwhile to consider that only some 30 species make up the bulk of the bacterial population in human faeces at any one time, based on the classical cultivation-based approach (Finegold, Attebury and Sutter, 1974; (Moore, and Holdeman, 1974). Thus, it is practical to focus on specific groups of interest within the complex community. These may be the predominant or most active species, specific physiological groups or readily identifiable (genetic) clusters of phylogenetically-related organisms. Several 16S rDNA fingerprinting techniques can be invaluable for selecting and monitoring sequences or phylogenetic groups of interest, and are described below.

Over the past few decades considerable attention has been focused on the identification of pure cultures of microbes on the basis of genetic polymorphisms of DNA encoding rRNA, such as ribotyping, amplified fragment length polymorphism (AFLP), and randomly amplified polymorphic DNA (RAPD) (O'Sullivan, 1999). However, many of these methods require prior cultivation and are less suitable for use in analysis of complex mixed populations, although important in describing cultivated microbial diversity in molecular terms. Much less attention was given to molecular characterization of complex communities. In particular, research into diversity and community structure over time has been revolutionized by the advent of molecular fingerprinting techniques for complex communities (Muyzer, 1999). DGGE/TGGE methods have been successfully applied to the analysis of human (Zoetendal, Akkermans and de Vos, 1998; Zoetendal *et al.*, 2001, 2002), pig (Simpson *et al.*, 1999, 2000), cattle (Kocherginskaya, Aminov and White, 2001), dog (Simpson *et al.*, 2002) and rodent (Deplancke *et al.*, 2000; McCracken *et al.*, 2001) intestinal populations.

5.1   DGGE/TGGE

DGGE is a genetic fingerprinting technique that enables separation of double stranded DNA fragments up to 500 bp in length, utilizing either a denaturing or a temperature gradient gel (Muyzer, 1999; Muyzer *et al.*, 1998). Separation of similar length PCR amplified fragments is achieved by denaturation within discrete melting domains, which results in characteristic banding patterns from PCR product mixtures. Increased resolution of banding patterns is achieved through addition of a GC clamp during PCR amplification (Muyzer *et al.*, 1998). In principle, DGGE can be used for analysis of PCR-amplified ribosomal genes, or functional genes, from mixed microbial communities. For studies on microbial diversity and ecology, or community structure and dynamics, the 16S rDNA is particularly useful due to its mosaic structure comprising highly conserved and hypervariable regions. The application of mixed PCR product to a DGGE gel results in a pattern of bands that corresponds with the predominant species or assemblages (phylotypes) of the microbial community under study. Individual bands, separated by DGGE, can also be identified by direct cloning and sequencing, or by hybridization with group or genus specific DNA probes. This technique is widely used in molecular microbial ecology and has been successfully applied in the authors' laboratory to analyse intestinal and faecal bacterial banding profiles of pigs, rodents and dogs (Deplancke *et al.*, 2000; Simpson *et al.*, 1999, 2000, 2002). Improvements from these studies are optimization of DNA extraction from faecal samples, inclusion of standard DNA fragments from known gut bacteria, which allow more precise gel analysis and between-gel comparisons, as well as image capture and analysis. These improvements have enabled the description of temporal and spatial changes in bacterial populations as a result of diet, dosing of exogenous probiotic bacteria and antibiotic therapy. Importantly, this has demonstrated that each individual animal has a unique but stable and repeatable banding pattern over time. This technique is less labour intensive and biased than traditional cloning and enables rapid estimation of microbial diversity.

Other applications of this technique include identifying 16S rDNA sequence heterogeneity (Nubel *et al.*, 1996); the study of gene diversity, such as tetracycline genes (Aminov, Garrigues-Jeanjean and Mackie, 2001; Aminov *et al.*, 2002); monitoring specific physiological groups; monitoring enrichment; and facilitating isolation (Muyzer, 1999). Similarity indices need to be calculated using numerical methods such as the Shannon-Weaver index (Nubel *et al.*, 1999; Zoetendal *et al.*, 2001) and other indices (Simpson *et al.*, 1999, 2000, 2002). These indices result in a more objective approach to analysing the effect of location, diet or supplementation on gut microbial populations.

5.2   Quantitative Real Time (RT)-PCR of 16S rRNA

Although PCR is the most sensitive technique for detecting sequences that are present in very low concentrations in the environment, many factors can influence the amplification reaction, resulting in misleading results even from quantitative PCR (Vaughan *et al.*, 2000; von Wintzingerode, Gobel and Stackebrandt, 1997). Moreover, the copy number of 16S rRNA genes per genome can vary considerably depending on species and thus competitive PCR procedures with internal standards for 16S rDNA amplification will not accurately reflect bacterial cell numbers or ratios of nucleic acids, but nevertheless give a good estimate.

6.   USE OF DNA PROBES BASED ON rRNA SEQUENCES

6.1   Dot blot hybridization

This technique is useful to measure the amount of a specific 16S rRNA in a mixture relative to the total amount of rRNA. In brief, total RNA is isolated from the sample, bound to a filter using a dot or slot blot manifold device, and hybridized with labelled oligonucleotide probe(s). The amount of label bound to the filter is a measure of the amount of specific rRNA target present, and the relative amount of rRNA may be estimated by dividing the amount of specific probe by the amount of labelled universal probe hybridized under the same conditions. However, the relative amount of rRNA sequence does not reflect the true abundance of the microbe since cells of different species have different ribosome contents and the number of ribosomes within one strain can vary with growth phase. Nevertheless, the relative quantity of rRNA provides a reasonable measure of the relative physiological activity of a specific population. This technique is the most widely used and reliable method for quantitating microbial populations in the intestinal tract of humans and animals (Raskin *et al.*, 1997).

6.2   Fluorescence *in situ* hybridization (FISH)

This method combines the power of molecular techniques with epiflourescent light microscopy (or Confocal Laser Microscopy) for direct visualization of individual cells (i.e. phylogenetic stain) and can be used to determine the relative importance of specific groups or genera detected by cultivation-based and molecular methods. This method can be used to determine if a species, group or cloned sequences indeed comprise a significant part of the original natural community (Amann, Ludwig and Schleifer, 1995). In contrast to most molecular techniques, whole cell fluorescence *in situ* hybridization (FISH) with rRNA-targeted oligonucleotide probes is quantitative on an individual cell basis. Ultimately enumeration of species in the intestinal tract is best addressed using this approach. This technique has become widely used to enumerate various phylogenetic groups that constitute the microbial population of humans (Franks *et al.*, 1998; Harmsen *et al.*, 1999, 2002; Welling *et al.*, 1997). To facilitate enumeration, FISH has been automated and combined with image analysis that is analysed by computer software programs. FISH enables microbial ecologists to address five ecological themes simultaneously:

(i)   to identify subpopulations in natural systems and to locate their niche;

(ii) to obtain information on community structure by using sets of probes;

(iii) to bypass cultivation problems;

(iv) to determine *in situ* cellular rRNA content and "metabolic fitness"; and

(v)   to accurately enumerate defined cell populations (Vaughan *et al.*, 2000).

The lowest level of detection at present is 10^6 cells per gram of faeces.

6.3   Flow cytometry

FISH enumeration methods rely on membrane filter/epifluorescence microscopy and thus flow cytometry may be a more promising enumeration technique. Flow cytometry has several advantages over image cytometry (microscopy), including speed and automation. Several studies have demonstrated the use of flow cytometry in association with FISH, and the general conclusion is that this combination is a very powerful tool for the rapid and automated analysis of mixed microbial communities (Davey and Kell, 1996). The feasibility of this technique has been reported for enumeration of bacterial species in the human gut where bifidobacteria were enumerated with good precision in comparison with microscopic counts (Vaughan *et al.*, 2000).

Direct detection of low numbers of specific bacteria is impeded by high background populations ($10^{10} - 10^{11}$) of other bacteria in faecal samples. Currently, subpopulations constituting at least 0.2–1.0 percent of the total population ($>2\times10^8$ per gram) can be detected. Several species of intestinal microbiota, specifically facultative anaerobes and dosed probiotic bacteria, are well below this level and other specific techniques must be used to enumerate these bacteria. Metabolic parameters such as live/dead can be analysed cytometrically (Davey and Kell, 1996). In order to allow simultaneous detection of several taxa in one sample, multi-parameter processing of samples is required but available and affordable flow cytometers can only excite up to three fluorochromes simultaneously. Thus, the potential for flow cytometry in combination with FISH as a quantitative method for analysing microbial ecology of the intestinal tract still remains to be evaluated.

7.   FUTURE PERSPECTIVES

7.1   Advances in probe and primer technology

Denatured DNA used in dot blots, cell blots or PCR reactions allows easy hybridization with probes and primers, regardless of the hybridization site within the target molecule. However, accessibility of the 16S rRNA in its native form to fluorescently labelled probes varies for different regions of the molecule and different bacteria, due to higher order structure of the native ribosome and the 16S rRNA molecule. Very small (3 bp) shifts in choice of binding site could increase effective fluorescence by a factor of 8.

One of the latest developments in PCR product detection is DNA probes called "molecular beacons", which fluoresce only on hybridization with their target sequence (Tyagi and Kramer, 1996). The molecular beacon has a stem-loop structure that positions the fluor and quench in close proximity and uses Fluorescence Resonance Energy Transfer (FRET) to suppress fluorescence. The technique is solution-based and on hybridization to target the molecular beacon linearizes and a fluorescence signal is generated that is proportional to the amount of target nucleic acid. At present, the application of molecular beacon technology based on phylogenetically-designed beacons is limited by a lack of understanding of the mechanism of different aspects of FISH staining, secondary structure interactions and high background fluorescence based on destabilization of the secondary structure of the beacon by nucleic acid and protein interactions.

Another advance in probe technology is the development of peptide-nucleic acid (PNA) probes (Nielsen, 1999; Stender *et al.*, 2002). These uncharged peptide nucleic acids exhibit much higher target specificity, as well as higher hybridization rates. The chemistry permits a higher level of hydrogen-bonding between probe and target; as a consequence, PNA probes can be shorter (8–15 nt) than conventional nucleotide probes (15–30 nt).

Although primer extension is not possible with PNA, PNA-DNA chimeras have recently been reported that would support PCR applications.

7.2 High-throughput and DNA array technology

The emergence of genomics as a new field of scientific endeavour has re-emphasized the need for rapid and cost-effective molecular biological technologies (Ball, and Trevors, 2002). The concept is to use microfabricated systems using a minimum of reactants that facilitate fully automated massive parallel analyses without a complete supporting laboratory. For example, microlithographic etching of silicon surfaces allows creation of miniaturized fluidic systems in which biomolecules can be mixed, reacted, separated and detected in much the same way as the classical laboratory but on a micro scale. By combining such components on a single chip with a temperature cycling microsystem, PCR amplification and detection of specific DNA can be carried out rapidly. The major advantages of such microfluidic systems will be speed of analysis for tens to thousands of samples (e.g. sequencing or restriction endonuclease (RE) mapping), and the possibility of performing experiments in the field.

Another development is that of DNA micro-arrays, also called biochips, gene chips or DNA chips. DNA micro-arrays are basically glass surfaces spotted with arrays of numerous covalently linked DNA fragments that are available for hybridization. Current applications include monitoring expression of the arrayed genes in mRNA species of growing cells (transcriptional profiling) or detecting DNA sequence polymorphisms or mutations in genomic DNA. Attempts to generate DNA micro-arrays for application to gut ecosystems are in progress (Leser *et al.*, 2002; Wilson *et al.*, 2002). The potential of micro-array technology in microbial ecology studies was demonstrated using microchips containing oligonucleotides complementary to 16S rRNA sequences of nitrifying bacteria that could detect and identify the DNA or RNA isolated from samples containing the target bacteria (Guschin *et al.*, 1997).

7.3 Genomic analysis

High throughput DNA sequencing now offers the potential to obtain a complete blueprint for the lifestyle of a specific microbe, and to assess its genetic potential using comparative and functional genomics approaches. The North American Consortium for Genomics of Fibrolytic Ruminal Bacteria was established in 2000 to promote the sequencing of rumen bacterial genomes, with a central focus on ruminal fibre degradation. The genomes of *Fibrobacter succinogenes* S85 and *Ruminococcus albus* 8 will both be sequenced to closure, and the draft sequence for these bacteria has been searchable via The Institute for Genomics Research (TIGR)'s unfinished genomes Web site (www.tigr.org) since July 2001. A project to sequence the genome of *Prevotella ruminicola* strain 23 to approximately 8× coverage has also been started. At the time of writing, there were only a few physical sequencing gaps remaining in the *F. succinogenes* genome, and annotation was in progress (Nelson *et al.*, 2002), while the *R. albus* project had reached the final stages of random (draft) sequencing.

Despite the unequivocal value associated with whole genome sequencing, it is still cost-prohibitive for many investigators to obtain a similar degree of sequence data for multiple strains of the same microbe, or for related species. Subtractive hybridization (SH) can be used as a means to recover "unique" genomic information from other strains of *Ruminococcus* spp. and *F. succinogenes.* Using these methods, Antonopoulos *et al.* (2004) have generated a set of 384 clones that are unique to *R. flavefaciens* FD-1 with respect to *R. flavefaciens* JM1. These results demonstrate how SH can broaden the scope of functional and comparative genomics in a cost-effective manner, and will facilitate the examination of gene diversity and genome plasticity among closely related ruminal microbes. These methods might also help elucidate which gene(s), as well as other ecological or physiological process(es), are rate-limiting to fibre degradation. By doing so, some of the major "informational" constraints to improving ruminal fibre degradation may ultimately be alleviated, and hypothesis-driven rather than empirical experimental designs will be employed.

7.4   Metagenomic analysis

Cloning large fragments of DNA isolated directly from microbes in natural environments provides a method to access metagenomic DNA. This approach is based on the use of a powerful genomics tool, the Bacterial Artificial Chromosome (BAC). BAC technology has been applied in a limited way to prokaryotic genomics but has recently been used to study *Bacillus cereus* and the soil metagenome (Beja *et al.*, 2000; Rondon, Goodman and Handelsman, 1999; Rondon *et al.*, 2000). BACs can be used to clone complex loci such as biosynthetic pathways, secretion systems and pathogenicity islands, because the average insert size is usually circa 100 kb and because the genes for many bacterial pathways are clustered in the genome. Also, because BAC inserts are large, a relatively small number of clones is required to provide complete coverage of the metagenome, reducing the amount of work required to screen a BAC library. Bacterial BAC libraries can be used to detect gene expression from poorly studied, difficult to manipulate or uncultured isolates. Thus BAC libraries can serve to archive DNA for genomics purposes and can also be used to analyse gene expression as the first step in functional genomics analysis. Importantly, the metagenomic approach enables us to link phylogenetic and functional diversity in the intestinal metagenome of production animals. Several laboratories have initiated studies using this BAC technology.

BAC vectors support the cloning of large DNA inserts (>100 kb) and, combined with their low copy number and the stability of insert DNA, offer new opportunities to conduct a meta-analysis of diversity and metabolic potential in virtually any microbial community. Rondon *et al.* (2000) and Beja *et al.* (2000) successfully constructed BAC libraries of soil and marine environmental DNA, respectively, and they assessed the diversity and metabolic potential recovered in these metagenomic libraries. Such efforts have raised expectations associated with the analysis of microbial communities, to move beyond 16S rRNA-based approaches, and integrate a more functional assessment of the metabolic and degradative potential maintained in microbial communities. Subsequently, a number of initiatives to construct and end-sequence BAC libraries are now underway. Hopefully, these efforts will include rumen and other intestinal microbial consortia.

In conclusion, (gen)omics technologies now provide microbiologists with their best opportunity for a complete or global analysis of the molecular mechanisms involved in bacterial metabolism and regulation from both a comparative and functional perspective.

8.   CONCLUSIONS

This paper can be considered unusual in the present day since it began with an overview of the classical anaerobic cultivation techniques that provide the foundation for our understanding of microbial ecology, physiology and metabolism in gastrointestinal ecosystem. This was followed by a description of the techniques, with some background, and highlighted the tremendous potential, as well as the limitations, of the application of molecular biology to the study of gut microbiology and ecology. It is clear that these modern molecular techniques will not simply result in a refinement of our understanding of the gut microbial ecosystem but will provide the first complete description. The use of nucleic acid probes for specific micro-organisms (rRNA) their genes (rDNA and DNA) and their expression (mRNA) will enable gut microbial ecologists to determine the exact role or function a specific organism has in the gut ecosystem and its quantitative contribution to the whole process – the ultimate goal of the microbial ecologist. This will require a combination of established conventional as well as modern and emerging molecular techniques, which will enable us to elevate this research area into experimental science with the emphasis on hypothesis testing and experimental rigour rather than a descriptive science. Also, with the advent of these techniques based on comparative sequence comparisons it is possible to link ecology and evolution, which was not possible previously, since we have no continuous fossil record or evolutionary history.

The gastrointestinal microbiota can be considered as the most metabolically adaptable and rapidly renewable organ of the body – one that plays a vital role in the normal nutritional, physiological and immunological functions of the host animal. Microbial ecology is defined as the study of microbial life and its interactions with its surrounding biotic and abiotic environment. Although modern molecular techniques have significant advantages over classical cultivation techniques, it is worth remembering that microbial ecology is the study of life, and not of techniques. The emphasis and focus should therefore be more on the study of ecology and less on molecular techniques.

Thus studies on gastrointestinal microbial ecology should focus on:
– diversity (i.e. fingerprinting techniques),
– community structure (i.e. quantitative membrane hybridization and FISH),
– function (i.e. physiology and biochemistry), and
– interactions with the host.

REFERENCES

Amann, R.I. 1995. *In situ* identification of micro-organisms by whole cell hybridization with rRNA-targeted nucleic acid probes. pp. 3.3.6: 1–15, *in:* A.D.L. Akkermans, J.D. van Elsas and F.J. de Bruijn (eds). *Molecular Microbial Ecology Manual*. The Netherlands: Kluwer Academic Publishers.

Amann, R.I., Ludwig, W. & Schleifer, K.H. 1995. Phylogenetic identification and *in situ* detection of individual microbial cells without cultivation. *Microbiological Reviews,* **59**: 143–169.

Aminov, R.I., Garrigues-Jeanjean, N. & Mackie, R.I. 2001. Molecular ecology of tetracycline resistance: development and validation of primers for detection of tetracycline resistance genes encoding ribosomal protection proteins. *Applied and Environmental Microbiology,* **67**: 22–32.

Aminov, R.I., Chee-Sanford, J.C., Garrigues, N., Teferedegne, B., Krapac, I.J., White, B.A. & Mackie, R.I. 2002. Development, validation and application of PCR primers for detection of tetracycline efflux genes of gram-negative bacteria. *Applied and Environmental Microbiology,* **68**: 1786–1793.

Antonopoulos, D.A., Nelson, K.E., Morrison, M. & White, B.A. 2004. Strain-specific genomic regions *Ruminococcus flavefaciens* FD-1 as revealed by combinatorial random-phase genome sequencing and suppressive subtractive hybridization. *Environmental Microbiology,* **6**: 335–346.

Aranki, A., Syed, S.A., Kenney, E.B. & Freter, R. 1969. Isolation of anaerobic bacteria from human gingival and mouse caecum by means of a simplified glove box procedure. *Applied Microbiology*, **17**: 568–576.

Ball, K.D. & Trevors, J.T. 2002. Bacterial genomics: the use of DNA microarrays and bacterial artificial chromosomes. *Journal of Microbiological Methods*, **49**: 275–284.

Beja, O., Suzuki, M.T., Koonin, E.V., Aravind, L., Hadd, A., Nguyen, L.P., Villacorta, R., Amjadi, M., Garrigues, C., Jovanovitch, S.B., Feldman, R.A. & DeLong, E.F. 2000. Construction and analysis of bacterial artificial chromosome libraries from marine microbial assemblage. *Environmental Microbiology,* **2**: 516–529.

Bonnet, R., Suau, A., Dore, J., Gibson, G.R. & Collins, M.D. 2002. Differences in rDNA libraries of faecal bacteria derived from 10- and 25-cycle PCRs. *International Journal of Systematic and Evolutionary Microbiology,* **52**: 757–763.

Davey, H.M. & Kell, D.B. 1996. Flow cytometry and cell sorting of heterogeneous microbial populations: the importance of single-cell analyses. *Microbiological Reviews,* **60**: 641–696.

Deplancke, B., Hristova, K.R., Oakley, H.A., McCracken, V.J., Aminov, R.I., Mackie, R.I. & Gaskins, H.R. 2000. Molecular ecological analysis of the succession and diversity of sulfate-reducing bacteria in the mouse gastrointestinal tract. *Applied and Environmental Microbiology*, **66**: 2166–2174.

Doolittle, W.F. 1999. Phylogenetic classification and the universal tree. *Science*, **284**: 2124–2128.

Dore, J., Sghir, A., Hannequart-Gramet, G., Corthier, G. & Pochart, P. 1998. Design and evaluation of a 16S rRNA-targeted oligonucleotide probe for specific detection and quantitation of human faecal *Bacteroides* populations. *Systematic and Applied Microbiology,* **21**: 65–71.

Finegold, S.M., Attebury, H.R. & Sutter, V.L. 1974. Effect of diet on human fecal flora: comparison of Japanese and American diets. *American Journal of Clinical Nutrition,* **27**: 1456–1469.

Franks, A.H., Harmsen, H.J.M., Raangs, G.C., Jansen, G.J., Schut, F. & Welling, G.W. 1998. Variation of bacterial populations in human feces measured by fluorescent *in situ* hybridization with group-specific 16S rRNA-targeted oligonucleotide probes. *Applied and Environmental Microbiology*, **64**: 3336–3345.

Guschin, D.Y., Mobarry, B.K., Proudnikov, D., Stahl, D.A., Rittman, B.E. & Mirzabekov, A.D. 1997. Oligonucleotide chips as genosensors for determinative and environmental studies in microbiology. *Applied and Environmental Microbiology*, 63: 2397–2402.

Harmsen, H.J.M., Elfferich, P., Schut, F. & Welling, G.W. 1999. A 16S rRNA-targeted probe for the detection of lactobacilli and enterococci in faecal samples by fluorescent *in situ* hybridization. *Microbial Ecology in Health and Disease*, **11**: 3–12.

Harmsen, H.J.M., Raangs, G.C., He, T., Degener J.E. & Welling, G.W. 2002. Extensive set of 16S rRNA-based probes for detection of bacteria in human feces. *Applied and Environmental Microbiology,* **68**: 2982–2990.

Head, I.M., Saunders, J.R. & Pickup, R.W. 1998. Microbial evolution, diversity, and ecology: a decade of ribosomal RNA analysis of uncultivated microorganisms. *Microbiological Ecology,* **35**: 1–21.

Hugenholtz, P., Goebel, B.M. & Pace, N.R. 1998. Impact of culture-independent studies on the emerging phylogenetic view of bacterial diversity. *Journal of Bacteriology*, **180**: 4765–4774.

Hungate, R.E. 1950. The anaerobic mesophilic cellulolytic bacteria. *Bacteriology Review,* **14**: 1–49.

Hungate, R.E. 1960. Symposium: selected topics in microbial ecology. I. Microbial ecology of the rumen. *Bacteriology Review,* **24**: 353–364.

Hungate, R.E. 1985. Anaerobic biotransformations of organic matter. pp. 39–95 (Vol. 1), *in:* E.R. Leadbetter and J.S. Poindexter (eds). *Bacteria in Nature.* New York, NY: Plenum Publ. Corp.

Kocherginskaya, S., Aminov, R.I. & White, B.A. 2001. Analysis of the rumen bacterial diversity under two different diet conditions using denaturing gradient gel electrophoresis, random sequencing, and statistical ecology approaches. *Anaerobe,* **7**: 119–134.

Krumholz, L.R. & Bryant, M.P. 1986a. *Syntrophococcus sucromutans* sp. nov. uses carbohydrates as electron donors and formate, methoxymonobenzenoids or *Methanobrevibacter* as electron acceptor systems. *Archives of Microbiology,* **143**: 313–318.

Krumholz, L.R. & Bryant, M.P. 1986b. *Eubacterium oxidoreducens* sp. nov. requiring H_2 or formate to degrade gallate, pyrogallol, phloroglucinol and quercetin. *Archives of Microbiology,* **144**: 8–14.

Leser, T.D., Boye, M., Amenuvor, J.Z., Jensen, T.K. & Moller, K. 2002. Development of a microarray for structural analysis of pig gastrointestinal bacterial communities. *Reproduction Nutrition Development,* **42**(Suppl. 1): S3.

Ludwig, W., Trunk, O., Klugbauer, S., Klugbauer, N., Weizenegger, M., Neumaier, J., Bachleitner, M. & Schleiffer, K.H. 1998. Bacterial phylogeny based on comparative sequence analysis. *Electrophoresis,* **19**: 554–168.

Mackie, R.I. 1997. Gut environment and evolution of mutualistic fermentative digestion. pp. 13–35 (Vol. 1), *in:* R.I. Mackie and B.A. White (eds). *Gastrointestinal Ecosystems and Fermentations.* New York, NY: Chapman and Hall.

Maidak, B.L., Cole, J.R., Parker, C.T., Garrity, G.M., Larsen, N., Li, T.G., Lilburm, B., McCaughey, M.J., Olsen, G.J., Overbeek, R., Pramanik, S., Schmidt, T.M., Tiedje, J.M. & Woese, C.R. 1999. A new version of the RDP (Ribosomal Database Project). *Nucleic Acids Research,* **27**: 171–173.

McCracken, V.J., Simpson, J.M., Mackie, R.I. & Gaskins, H.R. 2001. Molecular ecological analysis of dietary and antibiotic-induced alterations of the mouse intestinal microbiota. *Journal of Nutrition,* **131**: 1862–1870.

Moore, W.E.C. & Holdeman, L.V. 1974. Human fecal flora: the normal flora of 20 Japanese-Hawaiians. *Applied Microbiology,* **27**: 961–979.

Muyzer, G. 1999. DGGE/TGGE a method for identifying genes from natural ecosystems. *Current Opinions in Microbiology,* **2**: 317–322.

Muyzer, G., Brinkhoff, T., Nübel, U., Santegoeds, C., Schäfer, H. & Wawer, C. 1998. Denaturing Gradient Gel Electrophoresis (DGGE) in Microbial Ecology, pp. 3.4.4: 1–27, *in:* A. Akkermans, J.D. van Elsas and F. de Bruijn (eds). *Molecular Microbial Ecology Manual.* Boston, MA: Kluwer Academic.

Nelson, K., Aminov, R., Forsberg, C., Mackie, R.I., Russell, J.B., White, B.A., Wilson, D.B., Mulligan, S., Tran, K., Carty, H., Khouri, H., Nelson, W., Daugherty, S., Fraser, C. & Morrison, M. 2002. The *Fibrobacter succinogenes* strain S85 genome sequencing project. *Reproduction Nutrition Development,* **42**(Suppl. 1): S44.

Nielsen, P.E. 1999. Application of peptide nucleic acids. *Current Opinions in Biotechnology,* **10**: 71–75.

Nubel, U., Engelen, B., Felske, A., Snaidr, J., Weishuber, A., Amann, R.I., Ludwig, W. & Backhaus, H. 1996. Sequence heterogeneities of genes encoding 16S rRNAs in

Paenibacillus polymyxa detected by temperature gradient electrophoresis. *Journal of Bacteriology,* **187**: 5636–5643.

Nubel, U., Garcia-Pichel, F., Kuhl, M. & Muyzer, G. 1999. Quantifying microbial diversity: morphotypes, 16S rRNA genes, and carotenoids of oxygenic phototrophs in microbial mats. *Applied and Environmental Microbiology*, **65**: 422–430.

O'Sullivan, D.J. 1999. Methods for the analysis of the intestinal microflora, pp. 23–44, *in:* G.W. Tannock (ed). *Probiotics: A critical review.* England: Horizon Scientific Press.

Pace, N.R. 1997. A molecular view of microbial diversity and the biosphere. *Science,* **276**: 734–740.

Raskin, L., Capman, W.C., Sharp, R., Poulsen, L.K. & Stahl, D.A. 1997. Molecular ecology of gastrointestinal ecosystems. pp. 243–298 (Vol. 2), *in:* R.I. Mackie, B.A. White and R.E. Isaacson (eds). *Gastrointestinal Microbes and Host Interactions.* New York, NY: Chapman and Hall.

Rondon, M.R., Goodman, R.M. & Handelsman, J. 1999. The earth's bounty: assessing and accessing soil microbial diversity. *Trends in Biotechnology,* **17**: 403–409.

Rondon, M.R., August, P.R., Bettermann, A.D., Brady, S.F., Grossman, T.H., Liles, M.R., Loiacona, K.A., Lynch, B.A., MacNeil, I.A., Minor, C., Tiong, C.L., Gilman, M., Osburne, M.S., Clardy, J., Handelsman, J. & Goodman, R.M. 2000. Cloning the soil metagenome: a strategy for accessing the genetic and functional diversity of uncultured microorganisms. *Applied and Environmental Microbiology*, **66**: 2541–2547.

Russell, J.B., Stroebel, H.J. & Chen, G. 1988. Enrichment and isolation of a ruminal bacterium with a very high specific activity of ammonia production. *Applied and Environmental Microbiology*, **54**: 872–877.

Suau, A., Bonnet, R., Sutren, M., Goodon, J.-J., Gibson, G.R., Collins, M.D. & Dore, J. 1999. Direct analysis of genes encoding 16S rRNA from complex communities reveals many novel molecular species within the human gut. *Applied and Environmental Microbiology*, **65**: 4799–4807.

Simpson, J.M., Martineau, B., Jones, W.E., Ballam, J.M. & Mackie, R.I. 2002. Characterization of fecal bacterial populations in canines: effects of age, breed and dietary fiber. *Microbiological Ecology,* **44**: 186–197.

Simpson, J.M., McCracken, V.J., White, B.A., Gaskins, H.R. & Mackie, R.I. 1999. Application of denaturant gradient gel electrophoresis for the analysis of the porcine gastrointestinal microbiota. *Journal of Microbiological Methods,* **36**: 167–179.

Simpson, J.M., McCracken, V.J., Gaskins, H.R. & Mackie, R.I. 2000. Denaturing gradient gel electrophoresis analysis of 16S rDNA amplicons to monitor changes in fecal bacterial populations of weaning pigs following introduction of *Lactobacillus reuteri* strain MM53. *Applied and Environmental Microbiology*, **66**: 4705–4714.

Stender, H., Fiandaca, M., Hyldig-Nielsen, J. & Coull, J. 2002. PNA for rapid microbiology. *Journal of Microbiological Methods,* **48**: 1–17.

Strunk, O. & Ludwig, W. 1995. ARB - a software environment for sequence data. Department of Microbiology, Technical University of Munich, Germany.

Tannock, G.W. 1999. Analysis of the intestinal microflora: a renaissance. *Antonie van Leeuwenhoek Journal of Microbiology,* **76**: 265–278.

Tannock, G.W. 2001. Molecular assessment of intestinal microflora. *American Journal of Clinical Nutrition,* **73**(Suppl.): 410S–414S.

Tyagi, S. & Kramer, F.R. 1996. Molecular beacons: probes that fluoresce upon hybridization. *Nature Biotechnology,* **14**: 303–308.

Vaughan, E.E., Schut, F., Heilig, H.G.H.J., Zoetendal, E.G., de Vos, W.M. & Akkermans, A.D.L. 2000. A molecular view of the intestinal ecosystem. *Current Issues in Intestinal Microbiology*, **1**: 1–12.

Welling, G.W., Ellferich, P., Raangs, G.C., Wildeboer-Veloo, A.C., Jansen G.J. & Degener, J.E. 1997. 16S ribosomal RNA-targeted oligonucleotide probes for monitoring of intestinal tract bacteria. *Scandinavian Journal of Gastroenterology*, **222**: 17–19.

Wilson, K.H. & Blitchington, R.B. 1996. Human colonic biota studied by ribosomal DNA sequence analysis. *Applied and Environmental Microbiology*, **62**: 2273–2278.

Wilson, K.H., Wilson, W.J., Radosevitch, J.L., DeSantis, T.Z., Viswanathan, V.S., Kuczmarski, T.A. & Anderson, G.L. 2002. High-density microarray of small-subunit ribosomal DNA probes. *Applied and Environmental Microbiology*, **68**: 2535–2541.

von Wintzingerode, F., Gobel, U.B. & Stackebrandt, E. 1997. Determination of microbial diversity in environmental samples: pitfalls of PCR-based rRNA analysis. *FEMS Microbiological Reviews*, **21**: 213–229.

Woese, C.R. 1987. Bacterial evolution. *Microbiological Reviews*, **51**: 221–271.

Woese, C.R. 2002. On the evolution of cells. *Proceedings of the National Academy of Science of the USA*, **99**: 8742–8747.

Zoetendal, E.G., Akkermans, A.D.L. & de Vos, W.M. 1998. Temperature gradient gel electrophoresis analysis of 16S rRNA from human fecal samples reveals stable and host-specific communities of active bacteria. *Applied and Environmental Microbiology*, **64**: 3854–3859.

Zoetendal, E.G., Akkermans, A.D.L., Akkermans-van Vliet, W.M., Visser, A.G.M. & de Vos, W.M. 2001. The host genotype affects the bacterial community in the human gastrointestinal tract. *Microbial Ecology in Health and Disease*, **13**: 129–134.

Zoetendal, E.G., von Wright, A., Vilponnen-Salmela, T., Ben-Amor, K., Akkermans, A.D.L. & de Vos, W.M. 2002. Mucosa-associated bacteria in the human gastrointestinal tract are uniformly distributed along the colon and differ from the community recovered from feces. *Applied and Environmental Microbiology*, **68**: 3401–3407.

Zoetendal, E.G., Collier, C.T., Koike, S., Mackie, R.I. & Gaskins, H.R. 2004. Molecular ecological analysis of the gastrointestinal microbiota: a review. *Journal of Nutrition*, **134**: 465–472.

GENE-BASED VACCINE DEVELOPMENT FOR IMPROVING ANIMAL PRODUCTION IN DEVELOPING COUNTRIES
Possibilities and constraints

J.R. Egerton
Faculty of Veterinary Science, University of Sydney, Camden, Australia

Abstract: For vaccine production, recombinant antigens must be protective. Identifying protective antigens or candidate antigens is an essential precursor to vaccine development. Even when a protective antigen has been identified, cloning of its gene does not lead directly to vaccine development. The fimbrial protein of *Dichelobacter nodosus,* the agent of foot-rot in ruminants, was known to be protective. Recombinant vaccines against this infection are ineffective if expressed protein subunits are not assembled as mature fimbriae. Antigenic competition between different, but closely related, recombinant antigens limited the use of multivalent vaccines based on this technology.

Recombinant antigens may need adjuvants to enhance response. DNA vaccines, potentiated with genes for different cytokines, may replace the need for aggressive adjuvants, and especially where cellular immunity is essential for protection. The expression of antigens from animal pathogens in plants and the demonstration of some immunity to a disease like rinderpest after ingestion of these, suggests an alternative approach to vaccination by injection.

Research on disease pathogenesis and the identification of candidate antigens is specific to the disease agent. The definition of expression systems and the formulation of a vaccine for each disease must be followed by research to establish safety and efficacy. Where vaccines are based on unique gene sequences, the intellectual property is likely to be protected by patent. Organizations, licensed to produce recombinant vaccines, expect to recover their costs and to make a profit. The consequence is that genetically-derived vaccines are expensive.

The capacity of vaccines to help animal owners of poorer countries depends not only on quality and cost but also on the veterinary infrastructure where

H. P. S. Makkar and G. J. Viljoen (eds.), Applications of Gene-Based Technologies for Improving Animal Production and Health in Developing Countries, 199-210.

they are used. Ensuring the existence of an effective animal health infrastructure in developing countries is as great a challenge for the developed world as providing the next generation of vaccines.

1. INTRODUCTION

The human population has doubled in the last half century, due mainly to the success of public health programmes based on antimicrobial treatment of, and immunization against, infectious diseases (Bloom and Widdus, 1998). However increased human populations exert pressure on the availability of land for animal production. Thus a consequence of population pressure can be land shortage, poorer nutrition of animals and an increased impact of their infectious and parasitic diseases. In developed countries, where there is less likely to be overpopulation, production-limiting animal diseases are managed by a combination of immunization, therapy and change in production systems. The costs of these are recognized as part of the business of farming. Furthermore, profits from commercial farming can be invested in professional advice and the trial of new products.

Where animals are a component of a subsistence agricultural system, however, it is unlikely that owners will have the money to buy vaccines and drugs. It is less likely, also, that these people will have access to advice on animal health and animal husbandry, and the optimal application of those products.

In the case of many endemic, production-limiting diseases, like helminthiasis and acariasis, there continues to be reliance on drugs, for either treatment or prevention: no vaccines exist other than for the lungworm, *Dictyocaulus viviparus*, in sheep. The use of drugs introduces the potential for persistence of harmful residues in human food, market restriction, damage to the environment and the recurrent problem of development of resistance in target organisms. The cost of these drugs restricts their use by owners of animals in the developing world.

There are limits to the efficacy of many of the vaccines currently available. The relative failure of some of these has been due to low potency, short duration of immunity, the need for multiple doses and aggressive adjuvants, and cost. In the case of vaccines based on live, attenuated organisms, there has been continuing concern about inadequate attenuation, reversion to virulence or the persistence of infection in the absence of clinical signs of disease in vaccinated animals. Nevertheless, vaccines currently available have made a major contribution to the control of epidemic diseases like foot-and-mouth disease (FMD) and rinderpest.

Gene technology has the potential to correct the deficiencies of existing animal vaccines and to provide new vaccines where they do not exist. This paper examines this potential and identifies some prerequisites for achieving that goal. It is based on experience with development of recombinant DNA vaccine against foot rot (*Dichelobacter nodosus* infection) and its application in developing countries.

2.   PREREQUISITES FOR APPLICATION OF GENE TECHNOLOGY TO VACCINES

2.1   Identification of protective antigens

The empirical approach to vaccinology entailed the use of either whole culture of organisms, which were delivered as killed preparations, or in an attenuated form. Cell-free toxoids, e.g. some clostridial vaccines, perhaps represented an early move towards exploitation of specific components of micro-organisms. Vaccines based on gene technology, in contrast, require pre-existing knowledge of the essential protective antigens involved in the development of resistance. This knowledge can then be applied to the cloning of the genes responsible and their translation into appropriate expression systems.   Among all the agents of animal disease there are relatively few where this information is complete. Even where this information is available, successful cloning may not lead immediately to an effective product.

Recognition that the fimbrial protein was the principal immunogen of *D. nodosus* (Stewart, 1978), the transmitting agent of foot-rot of ruminants, and its cloning, illustrate this point. Whereas the initial expression host, *Escherichia coli*, elaborated subunits of fimbrial protein, it was found that, although antigenic, they were not protective (Emery, Stewart and Clark, 1984). When the gene responsible was transferred to a fimbrial Type IV bacterium, *Pseudomonas aeruginosa*, the subunits were not only expressed but also assembled as mature fimbriae (Mattick *et al.*, 1987). In this form the recombinant material was a powerful immunogen (Egerton *et al.*, 1987). Because the resistance to fimbrial protein is serogroup specific and because there are at least nine serogroups, recombinant vaccines needed to have representatives of all serogroups included. As further evidence of the difficulties that beset vaccine developers, it has been realized (Raadsma *et al.*, 1994) that antigenic competition between the different components of the multi-serogroup mix limits the extent and duration of antibody response to each component.

                                                    J.R. Egerton

In virus diseases like FMD and blue tongue, where multiple serogroups exist and where immunity is also serogroup specific, it can be anticipated that similar problems will be encountered in the preparation of fully effective multivalent vaccines. It has already been shown that subunit vaccines based on the VP1 protein and other components of FMD do not stimulate full immunity in the face of homologous challenge (Taboga, 1997; Beard, 1999; Brown, 2003).

Much loss of animal production is due to metazoan parasites like helminths and acarids. Earlier reports of the identification of immunogenic proteins from nematode worms (Jasmer and McGuire, 1997; Schallig, Van Leeuwen and Cornellissen, 1997) and the cloning of their genes, has not resulted in a commercial product. Immunity to intestinal nematodes of ruminants includes cell-mediated components, and the stimulation of these, together with adequate levels of antibodies to structural and secretory proteins, appear necessary for effective vaccination (Smith, 1999).

A novel approach was used in the development of a vaccine to control *Boophilus microplus,* a tick that causes chronic losses in tropical and subtropical countries through its blood engorgement and irritation of the host. Furthermore, it transmits serious protozoan parasites. A so-called "hidden" antigen, the protein BP86, isolated from the gut mucosa of ticks, was found to stimulate antibodies lethal for tick larvae when used to immunize cattle (Willadsen, 1992). The genes responsible were used to express the protein in a yeast host. That protein and more recently recognized ones, with an appropriate adjuvant, confer a satisfactory, although incomplete resistance in vaccinated cattle (García-García *et al.,* 1998). The availability of commercial, recombinant vaccines against cattle ticks is a rare realization of the potential of gene technology.

Many successful conventional vaccines were developed before the recognition of the role of cellular immunity in resistance to disease. The detailed understanding of cellular immune responses and how these can be modified by natural and cloned immunomodulators will inevitably improve the formulation of the next generation of vaccines. The challenge now is to establish, for each disease whether both humoral and cellular responses are essential for the resistance of each host species to its specific diseases. Next there is a need to discover antigens convertible to recombinant products that stimulate the optimum immune responses. Accurate methods for measuring these responses *in vitro* will be an essential component of this research. Therefore the identification of candidate antigens, to be cloned and produced as immunogens, implies a detailed understanding of the complete pathogenesis of the diseases associated with individual causal agents. This represents a major departure from the original empirical approach to vaccinology.

All the techniques of molecular biology – genome sequencing; promotion, ablation or inactivation of candidate genes; virulence testing of genetically transformed agents; polymerase chain reaction (PCR) enhancement of putative sequences; micro-arrays – are available and together with proteomics are being applied for many of the agents of animal disease. It is now possible to select candidate vaccine antigens by mining the genome sequence of pathogens (Ariel *et al.*, 2003). Encouraging results continue to emerge. However, because the research is specific to each disease, and funding, when available, is finite, progress is necessarily slow and expensive. Proof of vaccine efficacy may be the rate-limiting step in the future, requiring suitable experimental models of infection and considerable resources and time.

2.2   Selection of expression system

Many options are available experimentally but the system selected for vaccine production will be governed by knowledge of the pathogenesis of the disease against which the final vaccine is directed. Where resistance is based on humoral immunity, expression and harvesting of the antigen from easily grown bacterial or yeast cultures may be the method of choice. The important issue is to ensure that the recombinant product is immunogenic in small doses, preferably without the need for tissue-damaging adjuvants. When resistance depends in part or whole on cellular immunity, recombinant products may need to be incorporated into attenuated organisms for presentation to the host.

Living vaccines, recombinant or otherwise, have several advantages. Avirulent viruses, which replicate sufficiently to generate an immune response are being investigated as vehicles for one or more recombinant antigens (Pastoret *et al.*, 1988; Romero *et al.*, 1993). They often need only one dose and there is the potential for their administration other than by injection. There are disadvantages with living vaccines: a perceived potential to revert to virulence; the possibility of their containing other, masked, agents derived from cell culture lines; and a lack of robustness under field conditions.

DNA vaccines have been shown to stimulate cellular immunity and they overcome real and perceived concerns about using viable, attenuated organisms (Lai and Bennett, 1988; Dufour, 2001). Experimental DNA vaccines directed against components of *Mycobacterium bovis* have been shown to stimulate cellular immunity but do not induce sensitivity to tuberculin (Vordermeier, 2000). Evidence that simultaneous incorporation of cytokine genes with DNA vaccines can enhance immune responses of sheep

(Scheerlinck, 2002) suggests a method for removing the need for aggressive adjuvants.

2.3   Adjuvants

Experience with conventional vaccines is that, other than when delivered in a replicating organism, inoculated immunogens are unlikely to provoke long-term resistance. They require enhancement by adjuvants (Stewart-Tull, 2003). All empirically derived adjuvants result in some degree of irritation and exaggerated host response, both at the site of inoculation and more generally within the immune system. At present, the efficacy of many vaccines is dependent on the use of water-in-oil emulsions as adjuvant. Whereas the tissue damage associated with these may be preferable to the impact of a disease like haemorrhagic septicaemia, there is a need for more refined approaches to manipulation of the immune system. Again, increased understanding of the pathogenesis of diseases, and in particular of the immune response to them, offers the opportunity to enhance the potency of vaccines by inclusion of genetic material to stimulate desirable aspects of the response to individual pathogens (Scheerlinck, 2002; Klavinskis, Hobson and Woods, 2003). The range of immunomodulators available is extensive and the role of each in specific disease systems will need to be defined.

2.4   Methods of administration

Traditionally, killed vaccines have been delivered by subcutaneous or intramuscular injection. There is evidence that intra-dermal and trans-dermal injections may give superior responses in some instances, but for practical purposes they are not used. The possibility of iatrogenic transfer of other diseases is a disadvantage of injection of vaccines in difficult environments. There are circumstances where administration of vaccines other than by injection is desirable. Vaccines for the control of respiratory infections may be more effective if administered intra-nasally or intra-ocularly, such as for Newcastle Disease of poultry. Control of rabies in wildlife is dependent on the success of oral administration of vaccines, recombinant or otherwise (Pastoret *et al.,* 1988).

Application of live vaccines, either natural or genetically engineered, must result in either local or systemic infection to ensure an immune response. It is essential therefore that these products are robust enough to retain their viability in the environments in which they are used. DNA (plasmid) vaccines, which have a number of potential advantages over live-agent products, including greater viability at normal environmental temperatures, also vary in efficacy depending on their route of administration (De Rose, 2002; Babiuk, 2003). The gene gun method of vaccine delivery may enhance immune responsiveness (Lodmell, Ray and Ewalt, 1998) but in its present form it is unlikely to have practical application in the developing world.

Expression in plants of immunogens from pathogenic animal viruses like rinderpest (Carrillo, 1998), and evidence of immunity in animals fed these plants (Khandelwal, Lakshmi and Shaila, 2003), suggests the attractive possibility of vaccines being provided to village animals from crops adapted to and grown locally as a source of both nutrient and immunization.

3.   CONSTRAINTS TO APPLICATION OF GENE TECHNOLOGY

The availability of effective vaccines, whether derived from gene technology or otherwise, will not lead inevitably to the reduction of animal disease and enhanced production in developing countries. Many factors will inhibit this.

3.1   Cost of vaccines

The research necessary to identify candidate antigens, the genes coding for them, their translation to suitable expression systems, their formulation and subsequent evaluation for efficacy and safety is expensive. Scientists in research institutes in the developed world are most likely, at present, to be the inventors of these new vaccines and they or their institutions will protect the intellectual property involved by patenting. The commercialization of these products entails further expense for companies licensed for production under patent protection. All the costs associated with production, marketing and distribution of vaccines need to be recouped. Similar difficulties inhibit the delivery to the third world of vaccines for human diseases (Robbins and Freeman, 1988).

The outcome is that, unless alternative funding is available to subsidize their purchase and application, animal owners in developing countries are unlikely to benefit from gene technology. There is recognition of the need for aid programmes to assist developing countries in the management of epidemic diseases like FMD and rinderpest. Collectively, there is much greater impact on farm animals of the undeveloped world from endemic diseases like the tick- and fly-borne fevers, helminthiasis, and zoonoses like tuberculosis and brucellosis. Mechanisms need to be devised and funding found to ensure that improved vaccines benefit animals and owners in developing countries.

3.2 Animal health services

There are few circumstances where vaccination alone can be expected to resolve a disease problem satisfactorily. A functional national and local veterinary and para-veterinary infrastructure makes it more likely that there will be some understanding of the epidemiology, and economic and social impact, of diseases occurring nationally. Where a veterinary infrastructure exists, but where resources are limited, it will be easier to establish priorities for the use of vaccines improved by gene technology, and the role of those vaccines in the management of individual diseases.

Animal owners at village level need to have confidence in the field staff of their animal health service. Where this confidence exists, it is more likely that new products and new approaches to disease management will be adopted. Unless properly managed, a change as simple as the replacement of alum-precipitated adjuvant with oil emulsion in a vaccine may not be acceptable to animal owners.

The existence of an effective field veterinary service is a by-product of national policy and the priority given to agriculture. Where animal health policy is directed towards intensification and income generation from the sale of the product of animals in urban areas, the needs of subsistence farmers and their animals tend to be overlooked. Consequently, subsistence agriculture systems, which often include dependence on animals to maintain soil fertility, to provide draught power, a form of negotiable asset and some social prestige to their owners, will be decreasingly less able to provide food for the poor who depend on them.

Animal health programmes, which may include the use of effective vaccines, depend also on the level of biosecurity that can be achieved for a nation's farm animals. These programmes will be compromised if national boundaries are not secure from threats emanating from other countries with less stringent standards of animal health. This applies not only to contiguous national borders in Asia, Africa and South America, but also to distant countries from which animals might be imported. Along with control and prevention of disease by vaccination, there is a continuing need for the development and application of policies and protocols for the exclusion of unwanted disease and restriction of its transmission.

Systems of animal production that entail a transhumance component increase the risk of disease transmission within regions, which may include more than one country. In Nepal, absence of appropriate protocols in the importation of sheep, their transmission of foot-rot to indigenous sheep and goats in one village, and the spread of this disease to 40,000 other animals during seasonal migration, illustrates this problem (Ghimire and Egerton, 1996). The successful application of rDNA vaccines to the elimination of this disease has been described (Egerton *et al.*, 2002). The success achieved in Nepal and subsequently in Bhutan (Gurung, 2002) was due as much to the quality of the existing veterinary infrastructure as it was to the vaccine. Although transhumance systems facilitate the spread of disease, they have a number of benefits. In the hill districts of Nepal, the annual spring–summer migration of sheep, goats and cattle from the villages to high altitude alpine pastures allows access to feed that would not otherwise be available. Meanwhile, the people who remain in the villages grow and harvest crops from fields previously fertilized by the migrating animals. The by-products from the grain harvest, which itself provides food for the villagers, are used as an important element of animal feed during the winter months.

The biggest challenge in reducing the poverty of rural communities dependent on animals is to promote and support national agricultural policies that ensure that owners have access to comprehensive advice and support in the application of community-level disease control programmes. These may or may not include conventional or recombinant vaccines.

3.3   Social and cultural constraints

The killing of animals is accepted and used as a method of disease control in many developed countries. Thus whole flock and herd slaughter would follow the incursion of FMD in countries like Australia, the USA or the United Kingdom. Major advances have been made in the reduction of prevalence and even eradication of brucellosis and contagious bovine pleuro-pneumonia (CBPP) of cattle by a combination of vaccination and subsequent test and slaughter of reactors identified by wide-scale serological testing. The test-and-slaughter method, without prior vaccination, has been used successfully in the eradication of bovine tuberculosis from many national herds. Killing of animals is generally unacceptable in some communities, e.g. Bhutan, and, more specifically, the killing of cattle is unacceptable in Hindu societies. Elsewhere, poor farmers, whose only assets may be a few farm animals, are understandably reluctant to dispose of diseased animals. This is especially so when the effects of disease are not obvious and when no compensation is available. There is thus a compelling need for development and application of improved vaccines for the management of animal disease in those societies where the removal of known sources of infection from flocks and herds is unlikely to be adopted. The need for better vaccines is especially strong in the case of the major zoonoses – diseases like tuberculosis, brucellosis and hydatidosis.

3.4   Environmental issues

There is continuing debate about the acceptability of genetically modified crops and food derived from these crops. A similar level of concern about the safety of genetically engineered vaccines for people and animals' (Robbins and Freeman, 1988; Traavik, 1999) is based on the perceived threat of undesirable horizontal transfer of genetic material from vaccines or vaccine vectors to other species. Providing compelling evidence to meet these concerns is a major challenge for the scientific community and those responsible for animal health programmes. Inevitably, providing evidence to support the safety of recombinant vaccines will add to their cost and decrease their availability in developing countries.

4.   CONCLUSIONS

In theory, scientists using gene technology have great scope to develop vaccines for the control and perhaps elimination of diseases of animals. Progress towards these goals must be preceded by extensive research on the pathogenesis of diseases important in different environments. Vaccines developed through understanding of disease pathogenesis and the application of gene technology will be more specific in their action, but are likely to be much more expensive.

DNA vaccines, having a demonstrated capacity to provoke cellular and humoral immunity when the appropriate antigens and immunomodulators are delivered, may replace the need for living vectors or attenuated strains as vaccine vehicles. The possibility that recombinant crops might provide a source of edible vaccines against some animal diseases is encouraging.

The application of better animal vaccines will improve the health and well-being of people in the developing world only where the infrastructure exists to enable their use by animal owners. The development and maintenance of that infrastructure is as big a challenge as making better vaccines.

REFERENCES

Ariel, N., Zvi, K.S., Makarova, E., Chitlaru, E., Elhanany, E., Velan, B., Cohen, S., Friedlander, A.M. & Shafferman, A. 2003. Genome-based bioinformatic selection of chromosomal *Bacillus anthracis* putative vaccine candidates coupled with proteomic identification of surface-associated antigens. *Infection and Immunity,* **71**: 4563–4579.

Babiuk, L.A., Pontarollo, R., Babiuk, S., Loehr, B. & van Drunen Littel-van den Hurk, S. 2003. Induction of immune responses by DNA vaccines in large animals. *Vaccine,* **21**: 649–658.

Beard, C., Ward, G., Rieder, E., Chinsangaram, J., Grubman, M.J. & Mason, P.W. 1999. Development of DNA vaccines for foot-and-mouth disease, evaluation of vaccines encoding replicating and non-replicating nucleic acids in swine. *Journal of Biotechnology,* **73**: 243–249.

Bloom, B.R. & Widdus, R. 1998. Vaccine visions and their global impact. *Nature Medicine,* **4**: 480–484.

Brown, F. 2003. Use of peptides for immunization against foot-and-mouth disease. *Vaccine,* **21**: 3282–3289.

Carrillo, C., Wigdorovitz, A., Oliveros, J.C., Zamorano, P.I., Sadir, A.M., Gómez, N., Salinas, J., Escribano, J.M. & Borcal, M.V. 1998. Protective immune response to foot-and-mouth disease virus with VP1 expressed in transgenic plants. *Journal of Virology,* **72**: 1688–1690.

De Rose, R., Tennent, J.M., McWaters, P., Chaplin, P.J., Wood, P.R., Kimpton, W., Cahill, R. & Scheerlinck, J.P. 2002. Efficacy of DNA vaccination by different routes of immunisation in sheep. *Vaccine,* **20**: 2603–2610.

Dufour, V. 2001. DNA vaccines: new applications for veterinary medicine. *Veterinary Sciences Tomorrow* - Issue 2 - May 2001.

Egerton, J.R., Cox, P.T., Anderson, B.J., Kristo, C., Norman, M. & Mattick, J.S. 1987. Protection of sheep against foot rot with a recombinant DNA-based fimbrial vaccine. *Veterinary Microbiology,* **14**: 393–409.

Egerton, J.R., Ghimire, S.C., Dhungyel, O.P., Shrestha, H.K., Joshi, H.D., Joshi, B.R., Abbott, K.A. & Kristo, C. 2002. Eradication of virulent foot rot from sheep and goats in an endemic area of Nepal and an evaluation of specific vaccination. *Veterinary Record,* **151**: 290–295.

Emery, D.L., Stewart, D.J. & Clark, B.L. 1984. The structural integrity of pili from *Bacteroides nodosus* is required to elicit protective immunity against foot rot in sheep. *Australian Veterinary Journal,* **61**: 237–238.

García-García, J.C., Soto, A., Nigro, F., Mazza, M., Joglar, M., Hechevarría, M., Lamberti, J. & de la Fuente, J. 1998, Adjuvant and immunostimulating properties of the recombinant Bm86 protein expressed in *Pichia pastoris. Vaccine,* **16**: 1053–1055.

Ghimire, S.C. & Egerton, J.R. 1996. Transmission of foot rot in migratory sheep and goats of Nepal. *Small Ruminant Research,* **22**: 231–240.

Gurung, R.B. 2002. The epidemiology of foot rot in Bhutan. MVSc Thesis, University of Sydney.

Jasmer, D.P. & McGuire, T.C. 1997. Protective immunity to a blood-feeding nematode (*Haemonchus contortus*) induced by parasite gut antigens. *Parasite Immunology,* **19**: 447–453.

Khandelwal, A.G., Lakshmi, S. & Shaila, M.S. 2003. Oral immunization of cattle with hemagglutinin protein of rinderpest virus expressed in transgenic peanut induces specific immune responses. *Vaccine,* **21**: 3282–3289.

Kitching, R.P. 1998. A recent history of foot and mouth disease. *Journal of Comparative Pathology,* **118**: 89–108.

Klavinskis, L.S., Hobson, P. & Woods, A. 2003. Incorporation of immunomodulators into plasmid DNA vaccines. *Methods in Molecular Medicine,* **87**: 195–210.

Lai, W.C. & Bennett, M. 1988. DNA Vaccines. *Critical Reviews in Immunology,* **18**: 449–484.

Lodmell, D.,L., Ray, N.B. & Ewalt L.C. 1998. Gene gun particle-mediated vaccination with plasmid DNA confers protective immunity against rabies virus infection. *Vaccine,* **16**: 115–118.

Mattick, J.S., Bills, M.M., Anderson, B.J., Dalrymple, B., Mott, M.R. & Egerton, J.R. 1987. Morphogenetic expression of *Bacteroides nodosus* fimbriae in *Pseudomonas aeruginosa. Journal of Bacteriology,* **169**: 33–41.

Pastoret, P.P., Brochier, B., Languet, B., Thomas, I., Paquot, A., Bauduin, B., Costy, F., Antoine, H., Kieny, M.P., Lecocq, J.P., Debruyn, J. & Desmettre, Ph. 1988. First field trial of fox vaccination against rabies with a vaccinia-rabies recombinant virus. *Veterinary Record,* **123**: 481–483.

Raadsma, H.W., O'Meara, T.J., Egerton, J.R., Lehrbach, P.R. & Schwartzkoff, C.L. 1994. Protective antibody titres and antigenic competition in multivalent *Dichelobacter nodosus* fimbrial vaccines using characterized rDNA antigens. *Veterinary Immunology and Immunopathology,* **40**: 253–274.

Robbins, A. & Freeman, P. 1988. Obstacles to developing vaccines for the Third World. *Scientific American,* **259**: 126–133.

Romero, C.H., Barrett, T.S., Evans, A., Kitching, R.P., Gershon, P.D., Bostock, C. & Black, D.N. 1993. Single capripoxvirus recombinant vaccine for the protection of cattle against rinderpest and lumpy skin disease. *Vaccine,* **11**: 737–742.

Schallig, H.D., Van Leeuwen, M. & Cornellissen, W.C. 1997. Parasite immunity induced by vaccination with two excretory secretory proteins in sheep. *Parasite Immunology,* **19**: 447–453.

Scheerlinck J.P., Casey, G., McWaters, P., Kelly, J., Woollard, D., Lightowlers, M.W., Tennent, J.M. & Chaplin, P.J. 2002. The immune response to a DNA vaccine can be modulated by co-delivery of cytokine genes using a DNA prime-protein boost strategy. *Veterinary Immunology and Immunopathology,* **90**: 55–63.

Smith, W.D. 1999. Prospects for vaccines of helminth parasites of animals. *International Journal of Parasitology,* **29**: 17–24.

Stewart, D.J. 1978. The role of various antigenic fractions in eliciting protection against foot rot in vaccinated sheep. *Research in Veterinary Science,* **24**: 14–19.

Stewart-Tull, D.E. 2003. Adjuvant formulations for experimental vaccines. *Methods in Molecular Medicine,* **87**: 175–194.

Taboga, O., Tami, C., Carrillo, E., Nunez, J.I., Rodriguez, A., Saiz, J.C., Blanco, E., Valero, M.L., Roig, X., Camarero, J.A., Andreu, D., Mateu, M.G., Giralt, E., Domingo, E., Sobrino, F. & Palma, E.L. 1997. Large-scale evaluation of peptide vaccines against foot-and-mouth disease: lack of solid protection in cattle and isolation of escape mutants. *Journal of Virology,* **71**: 2606–2614.

Traavik, T. 1999. An orphan in science: environmental risks of genetically engineered vaccines. Research Report for DN 199-6. Directorate of Nature Management, Trondheim, Norway.

Vordermeier, H.M., Cockle, P.J., Whelan, A.O., Rhodes, S., Chambers, M.A., Clifford, D., Huygenb, K., Tascon, R., Lowrie, D., Colston, M.J. & Hewinson, R.G. 2000. Effective DNA vaccination of cattle with the mycobacterial antigens MPB83 and MPB70 does not compromise the specificity of the comparative intradermal tuberculin skin test. *Vaccine,* **19**: 1246–1255.

Willadsen, P., Kemp, D.H., Cobon, G.S. & Wright, I.G. 1992. Successful vaccination against *Boophilus microplus* and *Babesia bovis* using recombinant antigens. *Memorias do Institute Oswaldo Cruz,* **87** (Suppl.3): 289–294.

CURRENT AND FUTURE DEVELOPMENTS IN NUCLEIC ACID-BASED DIAGNOSTICS

Gerrit J. Viljoen,[1] Marco Romito[2] and Pravesh D. Kara[2]
1. Animal Production and Health Section, Joint FAO/IAEA Division, IAEA Vienna, Austria.
2. Biotechnology Division, Onderstepoort Veterinary Institute, Private Bag X5, Onderstepoort 0110, South Africa.

Abstract: The detection and characterization of specific nucleic acids of medico-veterinary pathogens have proven invaluable for diagnostic purposes. Apart from hybridization and sequencing techniques, polymerase chain reaction (PCR) and numerous other methods have contributed significantly to this process. The integration of amplification and signal detection systems, including on-line real-time devices, have increased speed and sensitivity and greatly facilitated the quantification of target nucleic acids. They have also allowed for sequence characterization using melting or hybridization curves. Rugged portable real-time instruments for field use and robotic devices for processing samples are already available commercially. Various stem-loop DNA probes have been designed to have greater specificity for target recognition during real-time PCR. Various DNA fingerprinting techniques or post amplification sequencing are used to type pathogenic strains. Characterization according to DNA sequence is becoming more readily available as automated sequencers become more widely used. Reverse hybridization and to a greater degree DNA micro-arrays, are being used for genotyping related organisms and can allow for the detection of a large variety of different pathogens simultaneously. Non-radioactive labelling of DNA, especially using fluorophores and the principles of fluorescence resonance energy transfer, is now widely used, especially in real-time detection devices. Other detection methods include the use of surface plasmon resonance and MALDI-TOF mass spectrometry. In addition to these technological advances, contributions by bioinformatics and the description of unique signatures of DNA sequences from pathogens will contribute to the development of further assays for monitoring presence of pathogens. An important goal will be the development of robust devices capable of sensitively and specifically detecting a broad spectrum of pathogens that will be applicable for point-of-care use. Advances in biosensors, the development of integrated systems, such as lab-

H. P. S. Makkar and G. J. Viljoen (eds.), Applications of Gene-Based Technologies for Improving Animal Production and Health in Developing Countries, 211-244.

on-a-chip devices, and enhanced communications systems are likely to play significant future roles in allowing for rapid therapeutic and management strategies to deal with disease outbreaks.

1.   INTRODUCTION

Current developments in nucleic acid-based technologies are continuing to improve the specificity, sensitivity, rapidity, user-friendliness and cost-effectiveness of molecular diagnostics. Nucleic-acid based assays are a direct means of pathogen detection and characterization and are particularly valuable where conventional means of isolation, culture or characterization have proven to be tedious, hazardous, timely, uninformative or even impossible (Van Belkum, 2003). DNA is also more resistant to denaturation than proteins and can survive for long periods under appropriate conditions. Basic principles of cloning, the use of restriction enzymes for plasmid fingerprinting, as well as the use of hybridization probes for use in dot blots as well as Southern and northern blotting all played important contributory roles in the initial detection and/or characterization of nucleic acids of various pathogens. Initially hybridization reactions required the separation of hybrid molecules from unhybridized probes using ultracentrifugation or chromatography and filtration. Immobilization and transfer to membranes was a further development with Southern blotting then dominating the field for two decades. The ability to sequence DNA made further important contributions towards subsequent developments in molecular diagnostics. The ability to synthesize oligonucleotides for use as primers and probes facilitated this process. Initial developments in nucleic acid-based procedures were nevertheless reliant on the cultivation of pathogens so as to obtain sufficient nucleic acids for further analysis. The most important subsequent development was the ability to amplify target DNA, and today polymerase chain reaction (PCR) assays are widely used as diagnostic tools. A host of variations on these principles have been devised in an effort to improve their efficacy and usefulness. Solution-based homogenous hybridization formats using fluorescence resonance energy transfer (FRET)-labelling have also allowed for homogenous diagnostic assays and high throughput multiplex analyses. Apart from important contributions to diagnostic activities, nucleic acid-based approaches are also contributing significantly towards molecular theranostics, where the rapid characterization of pathogens allows for more effective therapeutic interventions e.g. optimal selection of anti-microbials, antitoxins or even antiviral agents (Picard and Bergeron, 2002).

Apart from amplifying nucleic acids, technologies are also being developed for enhancing product separation, signal detection, and instrumentation. Not only must these systems prove to be more sensitive and specific, they should also be relatively simple, rapid and cost-effective if clinical applications are to be realized. In this review, we give a cursory overview of ongoing developments in this field.

2.   BIOINFORMATICS

The design of suitable molecular assays will rely increasingly on available bioinformatical data. Available sequence data is required for suitable probe and primer design. Whole-genome shotgun sequencing together with powerful computational algorithms to facilitate sequence data assembly, gene prediction and functional annotation have played important roles in this process. Primers designed to target conserved sequences of bacteria, e.g. rRNA, have allowed for the development of broad spectrum PCR for detecting bacteria (Picard and Bergeron, 2002). The variable internal rRNA gene regions can be sequenced and the data obtained compared with database sequences to make potential identifications at the species level, as well as contributing to epidemiological studies or for strain typing purposes, so as to determine traits such as antimicrobial resistance and virulence. Sequence data from 140 viruses has been used to design long oligonucleotide DNA micro-arrays with the potential of simultaneously detecting hundreds of viruses (Wang *et al.*, 2002).

Many challenges also exist regarding the storage, analysis, management and integration of generated data. In the case of micro-arrays, pattern recognition tools are becoming increasingly important not only for direct pathogen detection, but also for analysing expression profiles from infected cells (Stenger *et al.*, 2002). Methods of improving internet access to vast amounts of biomedical literature, using improved search engines and data mining programs, will also have important contributions to make to bioinformatics.

Internet and wireless communication technologies are also being harnessed to facilitate diagnostic systems (Cranfield Centre for Analytical Science Institute of BioScience and Technology, Cranfield University). Data that has been remotely collected can now be readily transferred to a central application for processing. Broadband internet, mobile phones and wireless communications technology, among others, will allow for the development of fully integrated, distributed applications across the internet.

3.   SAMPLE COLLECTION AND PROCESSING

Improvements in sample collection, transport and storage are designed to ensure stability of virus and bacteria, especially their nucleic acids, without interfering with the performance of subsequent tests (Picard and Bergeron, 2002). Immunomagnetic separation, using superparamagnetic beads (Dynal Biotech, Norway) is a useful means of concentrating a target organism or virus even at low concentrations in large volumes of complex media (Olsvik *et al.*, 1994; Stark, Reizenstein and Uhlen, 1996). Methods used to extract nucleic acids include lysis of the target using detergents and chaotropes and/or target capture. Mechanical means of disruption have utilized mortar and pestle with samples frozen in liquid nitrogen; homogenization with a Dounce; sonicators; French presses; bead mills; blade homogenizers; and ultrasonic disintegration devices. Enzymatic means make use of proteases. Current improvements are directed towards ensuring more rapid and efficient release of nucleic acids, more effective protection from degradation by nucleases, as well as removal of PCR inhibitors. Currently, commercially available kits and reagents are used by most diagnostic laboratories. A novel sample preparation system, pressure cycling technology (PCT) using the Barocycler^{TM} (Boston Biomedica), has been developed which alternately generates high and low pressure environments, achieving high-level release of DNA, RNA and proteins from a variety of biological samples. Several samples can also be processed simultaneously in separate tubes. The MagNA Lyser (Roche) utilizes a centrifuge and ceramic beads to efficiently process tissue samples.

Methods of nucleic acid capture include the use of paramagnetic particles coated with poly dT or specific probes, silica or a glass matrix which concentrate the target with more effective removal of inhibitors (Manak *et al.*, 2001; Greenfield and White, 1993; O'Meara, 2001). By controlling the pH and salinity of the medium, the cationic character of core-shell magnetic latex particles becomes favourable for nucleic acid adsorption/desorption (Eliassari *et al.*, 2001). This approach is particularly amenable for automation and high-throughput. Robotic devices such as MagnaPure (Roche), are capable of extracting nucleic acids from samples and even preparing PCR reagent mixes in an automated manner. Automated devices have the advantage of being closed systems, which significantly reduce the risks of carry-over contamination and analytical errors associated with manual testing.

Integrated systems, with important contributions by microfluidics, are also under development and combine automated sample processing with amplification and detection in microfabricated devices.

4. NUCLEIC ACID AMPLIFICATION

Target amplification includes PCR, transcription mediated, and strand displacement amplification. Amplification of a probe, however, does not lead to amplification of sequences other than that also found within the probe sequence and includes ligase chain reaction (LCR), rolling circle amplification (RCA), cycling probe technology (CPT) and Qβ replicase assays. Signal amplification does not involve amplification of any sequences found in the target and allows for the use of generic components, e.g. bDNA, SMART and Invader assays.

4.1 Target amplification

4.1.1 PCR

PCR has proven to be particularly effective as a diagnostic tool in allowing for the amplification of minute amounts of target nucleic acids. Since its conception by Mullis in 1983 and the discovery of a thermostable DNA polymerase (Saiki *et al.*, 1985; Mullis and Faloona, 1987; Saiki *et al.*, 1988), the procedure has undergone numerous further developments. Nested, multiple, asymmetric, Hot Start, touchdown PCR, RT-PCR, rapid amplification of cDNA ends (RACE), real-time assays, the use of internal controls, multiplexing, quantitative analyses, improved methods of contamination control and miniaturization of reactions are but some examples of currently established procedures (O'Meara, 2001; Nygren, 2000). In addition, ongoing improvements in instrumentation are increasing the efficiency and through-put of PCR-based testing, including the use of rapid cycling devices such as air-heated microcapillaries, microfabricated silicon-based reaction chambers, and PCR chips integrating amplification and product detection using capillary electrophoresis,

Direct pathogen detection occurs with PCR and it has the potential of detecting less than 10 copies of a specific gene within a complex sample. PCR has also helped to resolve uncertainties where antibody levels remain persistently low, e.g. when ELISA signals remain repeatedly in the "grey zone". It is useful as a means of discriminating between infection and maternal immunity in neonates, detecting viruses that are difficult to culture or are present together with cytotoxic substrates, detecting latent carrier animals, discriminating vaccine from field-strains, characterizing strains (including those with antimicrobial resistance or virulence genes), and performing molecular epidemiological analyses (Whitcombe, Newton and Little, 1998). Apart from PCR, several alternative methods of nucleic acid amplification have been developed and some are described below.

4.1.2   Transcription-based amplification

Several transcription-based amplification assays have been described, including the Qβ replicase system which can exponentially amplify RNA sequences encased within the sequence of MDV-1 RNA, which serves as template for Qβ replicase, an RNA-directed tRNA polymerase of bacteriophage Qβ (Kramer, and Lizardi, 1989). When hybridized to complementary DNA, the recombinant RNA serves as template for RNA synthesis, with a billion-fold increase in product. Nucleic acid sequence-based amplification (NASBA) is an optimized form of self-sustained sequence replication (3SR) that utilizes primers containing sequences for the T7 RNA polymerase promoter that becomes functional once double stranded DNA is formed after reverse transcription. These systems can be used for the detection of RNA targets (Compton, 1991; Guatelli *et al.*, 1990) and also offer a means of detecting viable cells. Amplicon detection is done by gel electrophoresis, but probe hybridization can be used for confirmation. Alternatively, amplicons are detected by enzyme-linked gel electrophoresis (ELGA), where they are hybridized to a horseradish peroxidase-labelled species-specific probes, run on an acrylamide gel, and stained in the gel by immersion in a substrate solution. Molecular beacons can also be used, with real-time monitoring (Cook, 2003; Deiman, Van Aarle and Sillekens, 2002). The commercially marketed TMA assay (Gen-Probe Inc, CA, USA), utilizes a hybridization protection assay and chemiluminescence using acridinium ester-labelled DNA probes to detect transcripts (Hill, 2001). Unbound probe is inactivated chemically. Signal mediated amplification of RNA technology (SMART) is also a transcription-based amplification system using a primer with the T7 promoter sequence, but is a signal and not a target amplification system (see Section 4.3.3).

4.1.3   Strand displacement amplification and loop mediated isothermal amplification assays

Strand displacement amplification (SDA) is another exponential isothermal amplification method based on the nicking of a hemi-modified recognition site by a restriction enzyme followed by the displacement of downstream strands by a polymerase (Walker *et al.*, 1992; Spargo *et al.*, 1996; Spears *et al.*, 1997). The original SDA system proceeds at 37–41°C and takes 2 hours to accomplish 10^8-fold amplification. Recently, a new thermophilic SDA system has been developed which proceeds at 50–60°C and takes only 15 min to accomplish 10^{10}-fold amplification (Spargo *et al.*, 1996). Detection of SDA products has been by using microtitre plate sandwich hybridization or gel electrophoresis. A homogenous system utilizes a fluorescently-labelled detector probe and detection by fluorescence polarization (Spears *et al.*, 1997). The latter system discriminates bound and unbound probe according to differences in tumbling motion. SDA is used in assays for the detection of, for example, *Chlamydia trachomatis* (Spears *et al.*, 1997) and *Mycobacterium tuberculosis* (Walker *et al.*, 1992; Spargo *et al.*, 1996). In loop-mediated isothermal amplification (LAMP), a DNA polymerase with strand displacement activity and a set of four specially designed primers, termed inner and outer primers are used (Notomi *et al.*, 2000; Nagamine *et al.*, 2001; Mori *et al.*, 2001). Initially, a stem-loop (hairpin) DNA structure is formed, in which the sequences of both DNA ends are derived from the inner primer and form the starting material. Subsequently, one inner primer hybridizes to the loop on the product in the LAMP cycle and initiates strand displacement DNA synthesis, yielding the original stem-loop DNA and a new stem-loop DNA with a stem twice as long. The final products are stem-loop DNAs with several inverted repeats of the target DNA and cauliflower-like structures with multiple loops. About 10^9 copies of target can be achieved within an hour. This isothermal technique does not require a denatured DNA template and when combined with reverse transcription, can also effectively amplify RNA. Furthermore, single-stranded DNA can be isolated from LAMP products, which can be used for micro-arrays, where ssDNA is preferable to dsDNA because of its superior hybridization abilities. Another method has been devised that accelerates the LAMP reaction even further by using loop primers (Nagamine, Kuzuhara and Notomi, 2002). These hybridize to the stem-loops, except for the loops that are hybridized by the inner primer, and prime strand displacement DNA synthesis. LAMP is able to amplify more DNA than most other methods. It does, however, require a complicated design of multiple primers. The final product is also a complex mix of stem-loop cauliflower-like DNA structures of various sizes but the advanced method was designed to obtain uniform single-stranded DNA.

4.1.4   Rolling Circle Amplification (RCA)

RCA produces many tandem copies of a circularized single stranded DNA probe (C-probe) under isothermal conditions, using a specific polymerase with high processivity (Ø29 DNA polymerase) and a primer binding to the probe. Alternatively, a linear (padlock) probe is used but the 3' and 5' ends undergo enzymatic ligation only after hybridization and their juxtaposition on the target, producing closed circle DNA (Banér *et al.*, 1998). The polymerase extends the bound forward primer along the C-probe and displaces the downstream forward strand, generating a multimeric single-stranded DNA (ssDNA), as for rolling circle amplification. RCA is a highly sensitive and contamination-resistant assay capable of high multiplexity and can be used in a variety of testing formats without the considerable pre-optimization of many other amplification systems. More than a billion-fold amplification can be achieved in as little as 1–2 hours. Sensitive antigen detection is even possible, employing a RCA primer attached to an antibody with addition of preformed circular DNA template after antibody-antigen interaction (ImmunoRCA) (Demidov, 2002).

In ramification amplification (RAM), multimeric ssDNA formed during RCA serves as template for multiple reverse primers to hybridize and be extended (Zhang *et al.*, 2001). These then displace other strands being generated downstream which in turn become targets for the first primer. A large ramified (branching) DNA complex is generated which continues until all ssDNAs become double-stranded. This results in exponential amplification of the probe, with significant amplification occurring within an hour at 35°C. Advantages include isothermal conditions, generic primers that amplify all probes with equal efficiency, use of both DNA and RNA as target, and the ability to detect single nucleotide polymorphisms (SNPs). In the case of intracellular target detection, it allows for specific localization of the signal while isothermic conditions preserve cell morphology and minimal effect on RAM with fixation of cells.

4.2   Probe amplification

4.2.1   Ligation assays

The ligation chain reaction (LCR) involves exponential amplification of ligated probes using two sets of probes in the presence of a thermostable ligase and thermocycling conditions (Weiguo, 2001). Detection of such probes using FRET-based labelling has now also been described (Schweitzer and Kingsmore, 2001). Amplification using PCR has also been combined with ligation of probes or the ligation detection reaction using a single probe pair. In the case of Gap-LCR, a DNA polymerase is used to seal a gap between the primers and a ligase seals the nick, preventing template-independent ligation that might occur with LCR (Ayyadevara, Thaden and Reis, 2000). Their use has also been extended to DNA micro-arrays where capture of short additional sequences occurs by addressable probes immobilized at unique addressable sites. Padlock probes also undergo ligation and circularization once hybridization of ends to target occurs, and can be used to detect localized sequences, alleles or SNPs (Nilsson *et al.*, 1994). Rolling circle amplification (RCA) can be used with these circular probes and enables very sensitive detection of such single nucleotide differences.

4.3   Signal amplification

4.3.1   Branched DNA

Branched DNA (bDNA; Chiron) is used as a signal amplification strategy to detect and quantitate DNA or RNA in cell lysates and clinical samples (Urdea *et al.*, 1991). Branched DNA or DNA dendrimers provide multiple hybridization sites for detection probes. As few as 10^{-19} M target is detectable (Whitcombe, Newton and Little, 1998). Detection of several DNA and RNA viruses has been reported from clinical samples using bDNA technology (Ross *et al.*, 2002). Direct RNA detection can be performed without the requirement for RNA purification. Specificity of the assay is derived from target-specific oligonucleotides and signal is obtained using oligonucleotide-linked alkaline phosphatase (Wagaman, Leong and Simmen, 2002). In the third generation, several modifications were introduced that enhance signal amplification and reduce assay background, including the use of non-natural synthetic nucleotides in some of the probes.

4.3.2   Invader technology

The Invader[®] assay (Third Wave Technologies, USA) is a homogeneous, isothermal DNA probe-based system for use with either DNA or RNA templates without the requirement for reverse transcription in the latter case (De Arruda *et al.*, 2002). High specificity is achieved through a combination of sequence-specific oligonucleotide hybridization and structure-specific enzymatic cleavage. An upstream invader and downstream probe bind to a ssDNA target such that a one-base overlap occurs. The downstream probe forms a 5′ flap upon annealing, as this region is not complementary to the target. The enzyme Cleavase cleaves immediately downstream of the overlap. The invasive complex does not form if the single nucleotide overlap is not present. The cleaved flaps are then released and form invader probes on a synthetic hairpin oligonucleotide or cassette which is labelled with fluorescent dye and quencher. This FRET cassette has a region that is complementary to the flap and a self-complementary region that mimics both a probe and a target. Cleavage of the cassette leads to fluorescence. The FRET cassette is a generic component and can be employed for detection of any other target. Each cleaved flap generates about 10^3 to 10^4 cleaved cassettes per hour producing linear signal amplification. The invasive complex for the second reaction only occurs after flaps have been generated by the specific recognition of probes with target DNA. They are useful for scanning not only polymorphisms such as SNPs, but also insertions and deletions.

4.3.3   **Signal mediated amplification of RNA technology**

Signal mediated amplification of RNA technology (SMART), is a transcription-based amplification system using the T7 promoter sequence incorporated in a primer (Wharam *et al.*, 2001). Unlike other transcription-mediated systems, signal and not target amplification occurs. Two oligonucleotides are used that form a 3-way junction structure when they anneal adjacently on the target as they share a short complementary sequence. A DNA polymerase extends the shorter probe generating a functional T7 polymerase promoter from where transcription of generic probe sequence occurs in the presence of T7 polymerase. RNA is detected by enzyme-linked oligosorbent assay (ELOSA) using biotinylated capture probes and alkaline phosphatase-labelled detection probes.

5.   REAL-TIME PCR ASSAYS

5.1   Real-time PCR

Real-time systems have the ability to detect the presence of amplification products in real time using fluorescent-labelled probes. This process obviates the need to manipulate product after amplification, thereby significantly reducing time and the risk of carry-over contamination. In addition, real-time PCR possesses high sensitivity, precision and low turn-around times. Further developments include their combination with robotic devices that extract nucleic acids and prepare test reagents, allowing for high throughput. Real-time processing has also greatly facilitated multiplexing PCR assays where different fluorophore-labelled probes can be used to detect various pathogens simultaneously in a single tube. On-line monitoring of the reaction allows for the determination of the exponential phase on the PCR reaction curve. An important application is thus quantitative PCR, as used for establishing viral load during the monitoring of therapy. Instead of using end-point quantification, which is unreliable, real-time assays allow for the determination of the cycle threshold (Ct) where the signal generated is significantly higher than background fluorescence. Ct is directly proportional to amount of input template. Several devices are commercially available, e.g. ABI7700 (Applied Biosystems), MX4000 (Stratagene), Lightcycler (Roche), iCycler (Bio-Rad), SmartCycler (Cepheid) and Robocycler (MJ research). Tube-based or microtitre plate-based platforms are used in these devices, with the exception of the LightCycler. It uses air for heating and cooling, with durable glass capillaries as reaction vessels, allowing for considerably reduced turn-around times because of the high surface-to-volume ratio of the capillaries with ultrarapid thermal cycling. They are also ideal for fluorescence detection, because of the small capillary diameter and concentration of emitted light at the tip of the capillary. The LightCycler also has the ability to perform detailed melting curve analyses of PCR products after amplification. The determination of the melting point of a DNA fragment can depend on its length and guanine cytosine (GC) content, and can therefore be used to characterize PCR products. The signal obtained from one specific PCR product can readily be distinguished from another, allowing for product verification.

5.2 Portable devices

The "Ruggedized Advanced Pathogen Identification Device" (R.A.P.I.D.®, Idaho Technology, Salt Lake City, UT) is a more robust version of the LightCycler, weighing only 22 kg. It is designed for use in military field hospitals for field identification of dangerous pathogens in less than 30 minutes. An even smaller version, termed the "RAZOR", is currently in development; reactions are performed in plastic pouches that are preloaded with freeze-dried reagents. A prototype suitcase-sized analytical thermal cycler instrument fabricated by Lawrence Livermore National Laboratories (Livermore, CA) is capable of, amongst others, single-nucleotide discrimination between orthopoxviruses (Ibrahim *et al.*, 1998). A successive instrument, the advanced nucleic acid analyzer (ANAA) is able to detect target in only 7 minutes (Belgrader *et al.*, 1999). The ANAA relies on silicon and platinum "thermal-cycler" units to perform heating and cooling steps at higher speed than conventional, moulded aluminium or brass blocks. A commercial version, the SmartCycler, is also now available (Cepheid; Sunnyvale, CA). More recently, a notebook-sized, real-time thermocycler was designed using independently programmable thermocycler chambers (Belgrader *et al.*, 2001). The hand-held advanced nucleic acid analyzer (HANAA) (Lawrence Livermore National Laboratory) is a battery-powered portable real-time thermocycler in a handheld format suitable for field use (Higgins *et al.*, 2003). It weighs less than 1 kg and uses silicon and platinum-based thermocycler units to conduct rapid heating and cooling of plastic reaction tubes. It is designed to detect the presence of DNA sequences that are signatures of specific pathogens and reports the results in less than 30 minutes. A micromachined silicon thermal cycler can run continuously for 1.5 hours on a fully charged 12 V battery pack. Fluorescent detection is based on SYBR Green dye or Taqman primers and probes. It is being used to detect biological threat agents, such as *Bacillus anthracis,* drug resistant malaria strains and pathogens in environmental samples.

6.   PROBES

Probes play an important role in confirming target or amplified nucleic acid products. Profluorescent probes utilizing the principles of fluorescence resonance energy transfer (FRET) have been designed to allow for real-time monitoring in homogenous assays. The subsequent development of singly labelled fluorescent probes has made them more practical and less costly to use (Demidov, 2003).

6.1   Hydrolysis probes, beacons, scorpions and smart probes

Hydrolysis or TaqMan probes (Applied Biosystems) consist of a fluorophore which is quenched by a second molecule in close proximity on the opposite end of the probe (Holland *et al.*, 1991; Heid *et al.*, 1996). The probe emits a fluorescent signal after being digested by the 5′ nuclease activity of Taq polymerase, after hybridizing to template during the extension phase of PCR. Molecular beacons are probes containing a target recognition loop flanked by a hairpin with a fluorophore and quencher (Tyagi and Kramer, 1996). Fluorescence occurs once the probe has hybridized with target, inducing separation of quencher and fluorophore. They have the ability to discriminate SNPs effectively and are very suitable for identifying polymorphisms. They have proven to be more specific than linear probes and can also be used in isothermal assays such as SDA, NASBA and RCA. Non-hybridized probes remain dark and probe hybrids need not therefore be isolated. Their self-reporting capacity allows for the detection of hybridization without the labelling of targets, making them more user-friendly and robust for diagnostic applications (Tyagi and Kramer, 1996; Vet, van der Rijt and Blom, 2002). Scorpion probes are variants of molecular beacons and simultaneously act as both primers and probes (Whitcombe, Newton and Little, 1998; Schweitzer and Kingsmore, 2001; Broude, 2002). A hairpin loop is linked to the 5' end via a linker and has a sequence complementary to the flanking target sequence. A quencher and fluorophore are present at the 5' and 3' ends of the loop respectively and are separated during self-annealing, following extension and denaturation. Faster kinetics and better stability of the probe-target complex are achieved in comparison with bimolecular probes. Gold nanoparticles are also being used as quenchers in such probes.

Smart probes are fluorescence-labelled oligonucleotides that report the presence of complementary target sequences by a strong increase in fluorescence intensity. They consist of a fluorescent oxazine dye attached at the terminus of a hairpin oligonucleotide that is efficiently quenched by complementary guanosine residues. Smart probes fluoresce when they undergo a conformational change that forces the fluorescent dye and the guanosine residues apart. Fluorescence bursts from individual smart probes passing through a focused laser beam can be detected using confocal fluorescence microscopy (Knemeyer, Marme and Markus, 2000).

6.2   Binary hybridization probes

Binary hybridization probes consist of two probes labelled at the 3' and 5' ends respectively and are used in LightCycler® (Roche Diagnostics Gmbh, Germany) real-time reactions (Wittwer *et al.*, 1997). The donor activates the acceptor fluorophore by FRET when the probes are adjacent to each other following hybridization. Light is emitted at a longer wavelength that can be distinguished from background fluorescence. The probes are not degraded as with Taqman probes and are thus recycled. Other variations include the use of single labelled probes that binds near a quencher on the target amplicon or alternatively opposite a guanidine, which can act as a quencher.

6.3   Padlock probes

Padlock probes are designed so that when the 5' and 3' ends base pair next to each other on a template strand, they undergo enzymatic ligation, which circularizes the probe and concatenates it to the template (Banér *et al.*, 1998). They are particularly specific and sensitive and remain localized to the site. They are also useful for multiplex reactions. RCA can then be used to amplify the signal. This process can only occur if the probe is bound to a site close to the end of the template strand so as to allow the DNA polymerase to function effectively, or if target is cleaved by nucleases.

6.4   RNA and chimeric probes

Chimeric RNA-DNA probes are used in cycling probe technology (CPT; ID Biomedical Corporation) (Duck *et al.*, 1990). The currently marketed system utilizes fluorescence-labelling at the 5' end, while the 3' end carries a quencher. Fluorescence occurs when the RNA part of the probe is cleaved by RNase H after hybridization with its complimentary PCR product, allowing for real-time monitoring. Assays have been described in conjunction with PCR for detecting *Myobacterium tuberculosis* (Modrusan, Bekkaoui and Duck, 1998) and *Bacillus anthracis* (Cycleave™ PCR, Takara Bio Inc.).

Hybrid Capture (Digene) uses RNA probes to hybridize with target DNA molecules. The RNA–DNA hybrids are captured on a solid phase coated with specific antibodies (Wittwer *et al.*, 1997). Detection is done using antibodies conjugated with alkaline phosphatase and a chemiluminescent substrate. The system has been approved by the FDA for the detection of *Neisseria gonorrhoeae, Chlamydia trachomatis* and cytomegalovirus.

6.5　　Catalytic probes

Catalytic molecular beacons are still in the developmental stage and have potential for detecting nucleic acids without PCR amplification. Only in the presence of target does the beacon module allow the module containing the hammerhead-type deoxyribozymes to hybridize and cleave the quencher/emitter probe. (Broude, 2002; Stojanovic, de Prada and Landry, 2001).

6.6　　Probes using oligonucleotide analogues

Peptide nucleic acid (PNA) probes are synthetic DNA mimics that possess an achiral, neutral polyamide backbone formed by repetitive units of N-(2-aminoethyl) glycine that replaces the negatively charged sugar-phosphate backbone of DNA (Egholm *et al.*, 1993; Nielsen, Egholm and Buchardt, 1994). Individual nucleotide bases are attached to each unit and PNA is thus able to hybridize to complementary nucleic acid sequences. The lack of electrostatic repulsion allows for a more rapid and stronger binding to complementary targets (Egholm *et al.*, 1993). They also hybridize virtually independently of the salt concentration. Additionally, the lack of negative charges affects the physico-chemical behaviour of PNA probes relative to DNA, enabling the development of unique PNA hybridization and PNA/target separation methods. The unnatural PNA backbone also means that PNA is not degraded by ubiquitous enzymes, such as nucleases and proteases (Demidov *et al.*, 1994). High biostability provides better shelf life and PNAs are also of value when used as therapeutic antisense reagents. They are not recognized by polymerases and are hence not incorporated into amplicons. PNA probes can also undergo triplex formation with dsDNA helices as in PD-loop technology, where PNA openers are used to expose a single-stranded region (P-loop complex) within the duplex DNA, that then becomes accessible to DNA or PNA probes (Nielsen, 2001; Demidov, 2001). Molecular beacons can then bind to dsDNA under non-denaturing conditions.

Locked nucleic acids (LNA) are sugar-modified ribonucleotide derivatives with O2' to C4' methylene-links and have exceptional hybridization affinity towards complementary DNA and RNA. They have high capturing efficiencies and allow for unambiguous scoring of single nucleotide polymorphisms. The difference in melting temperature between a perfect match and a single-nucleotide mismatch is larger with LNA than with DNA oligomers (Petersen and Wengel, 2003).

6.6.1   Self-reporting PNA probes

Stemless PNA beacons, such as LightSpeed probes (Boston Probes, Bedford, MA), lack complementary stems, but, due to the polyamine backbone, the hydrophobic interaction between the fluorophore and quencher keeps the structure in a closed form in the absence of target. No target-unrelated sequences are present, improving specificity further, allowing for strong mismatch/match discrimination and fast kinetics. LightUp probes (LightUp Technologies, Göteborg, Sweden) are PNA probes conjugated with thiazole orange, which is inherently non-fluorescent in the unbound state, but becomes fluorescent after hybridization through interactions between the label and the nucleic acid target. While DNA beacons need to be designed with internal complementarity to keep the fluorophore and quencher in close proximity, neither LightSpeed nor LightUp probes have this constraint and are thus inherently self-reporting probes not dependent on interactions other than the hybridization reaction itself.

6.6.2   Q-PNA PCR

In Q-PNA PCR a generic quencher-labelled PNA is used to mask the signal from a fluorescent dye-labelled DNA primer (Fiandaca *et al.*, 2001). A short generic tag sequence and a fluorescent label is included in one of the primers, to which the Q-PNA can bind. This places the quencher near the fluorescent dye label of the primer. During PCR, the Q-PNA is displaced by incorporation of the primer into amplicons and the fluorescence of the dye label is liberated from quenching. Q-PNA PCR is described as being a simple means of converting an existing traditional PCR assay into one for real-time use or end-point analysis.

6.6.3   Blocker probes

Blocker probes are unlabelled PNAs used to hybridize to abundant sequences closely related to the target present in a sample. This prevents unwanted hybridizations by labelled detector probes with these related sequences, thereby increasing signal to noise ratio (Stender *et al.*, 2002).

7.   SIGNAL GENERATION, DETECTION AND ANALYSIS

7.1   Intercalating dyes and fluorescent probes

The production and analysis of signals generated in a reaction can be performed in either a heterogeneous or homogenous manner. The former entails the manipulation of amplified product whilst with the latter, signal generation and detection are performed in a closed tube. Heterogeneous assays include the use of gel electrophoresis (including agarose gel electrophoresis, Sanger sequencing reactions, single-strand conformation polymorphism, denaturing gel electrophoresis and heteroduplex analysis); hybridization studies (Southern blot, reverse line blots) including the use of solid supports such as DNA chips, microtitre plates or magnetic beads; and enzymatic analyses, as in pyrosequencing, restriction fragment length polymorphism, cleavage fragment length polymorphism, minisequencing and chromatographic and mass spectrophotometric analyses (O'Meara, 2001; Kristensen *et al.*, 2001).

Homogenous assays utilize fluorescent dyes and labelled probes for detection within the reaction tube, as used in real-time assays. SYBR Green I (Molecular Probes, Eugene, OR) is a fluorescent intercalating dye that can indicate increasing levels of amplicon, although fluorescence emissions from non-specific amplification products can generate false-positive results. Labelled probes can be used to increase specificity. In the hybridization protection assay, probes labelled with a highly chemiluminescent acridinium ester are used (Nelson *et al.*, 1996). The hybridization of probe with target protects the ester from hydrolysis. Tessera Array Technology (TAT) (Applied Gene Technologies; San Diego) is a non-amplification process involving the labelling of unpurified nucleic acid samples with luminescent compounds, followed by hybridization of such nucleic acids with probes in a multi-array format (Yang *et al.*, 2001).

7.2   Signal detection

Various methods are being employed to increase detection sensitivity without dependence on target amplification. Examples include surface enhanced resonance Raman scattering, which is used to detect chromophore-labelled nucleic acid probes irradiated with laser light at the resonant frequency of the chromophore (Whitcombe, Newton and Little, 1998; Graham *et al.*, 1997). Fluorescence correlation spectroscopy records spatio-temporal correlations among fluctuating light signals, which are associated with the size and shape of the molecule (Whitcombe, Newton and Little, 1998; Eigen and Rigler, 1994). Single molecule electrophoresis measures the time for fluorescent labelled molecules to travel a fixed distance between two laser beams and is also capable of monitoring changes in base composition (Castro and Shera, 1995).

7.3   Non-radioactive labelling systems

The advantages of non-radioactively labelled nucleic acid probes include cost, safety, disposal and stability over radioactive methods. Methods used to produce such include enzymatic processes such as nick translations and PCR labelling for the inclusion of labels such as biotin, fluorescein and digoxigenin-dUTP. Chemical means include psoralen alkylating agents and ULS®(Kreatech) (Van Gijlswijk *et al.*, 2001). Up-converting Phosphor Technology uses submicron-sized phosphor particles attached to DNA that emit visible light after exposure to infrared excitation light. There is a total absence of background autofluorescence from other biological compounds (Corstjens *et al.*, 2003).

7.4   Microparticles, microspheres and nanoparticles

Microspheres with probes attached to their surfaces have also been used for labelling purposes. They contain several fluorescent labels which can then be analysed and serve as barcodes (Mullenix *et al.*, 2001). The PCR-Immunochromatography System uses gold particles coated with anti-DNP antibodies that bind amplicon derived from dinitrophenyl (DNP)-labelled primers (Mitani and Oka, 2001). These amplicons are hybridized to biotin-labelled probes, which allows for their entrapment on a streptavidin line, producing a red colour.

Metal and semiconductor nanoparticles coated with receptor molecules are also proving to be useful alternatives to organic probes. Metal nanoparticles have very high extinction coefficients for labelling in colorimetric and surface plasmon resonance (SPR) assays. They also have high scattering coefficients making them sensitive labels for use in oligonucleotide arrays and dot blots. Metal particles are also being used for thermal probing of specific biomolecular interactions. Gold nanoparticles coated with DNA are being used with micro-array technologies (Taton, 2002). Paramagnetic particles are also finding application in magnetic resonance contrast imaging or with magnetic sensors. Recent advances in nanomaterials have produced a new class of fluorescent labels by conjugating semiconductor quantum dots Qdot™ made of cadmium selenide nanocrystals (Quantum Dot Corporation, USA) with biorecognition molecules (including antibodies and oligonucleotides). They are water-soluble and biocompatible, and provide important advantages over organic dyes and lanthanide probes. By simply changing the particle size and shape, the wavelength emitted by quantum-dot nanocrystals can be continuously tuned, whilst a single light source can be used for simultaneous excitation of all different-sized dots. They are also highly stable against photobleaching and have narrow, symmetric emission spectra. Quantum dots are therefore regarded as being ideal fluorophores for ultrasensitive, multicolour, and multiplexing applications in molecular biotechnology and bioengineering (Taton, 2002; Sakar, Yoon and Sommer, 1992; Chan *et al.*, 2002)

8.   GENOTYPING AND FINGERPRINTING

Molecular probes were initially utilized in the 1980s to characterize various pathogens of importance. Various amplification technologies now play a significant role in typing assays. These now make important contributions to diagnostics, theranostics as well as epidemiology and taxonomy. The characterization of bacterial strains has been done using pulsed field gel electrophoresis (PFGE), chromosomal restriction fragment length polymorphism (RFLP), randomly amplified polymorphic DNA (RAPD), repetitive PCR (rep-PCR), ribotyping or bacterial restriction endonuclease analysis (BRENDA) and post-amplification DNA sequencing of polymorphic genes, which are also utilized for molecular epidemiological analyses. Dideoxy fingerprinting is based on cycle sequencing of PCR-derived amplicon, but with the inclusion of only a single dideoxynucleotide reaction. (Sakar, Yoon and Sommer, 1992) The bands produced are then compared between strains.

Apart from rRNA, the detection of transposable genetic elements (Kamerbeek *et al.*, 1997), polymorphisms in the repeat elements and polymorphisms in spacer regions between direct repeats (spoligotyping) have been described for typing mycobacterial strains, including those with drug resistance (Kamerbeek *et al.*, 1997; Groenen *et al.*, 1993; Aranaz *et al.*, 1998).

Electrophoretic discrimination of genetic polymorphisms can be done using single strand conformational polymorphisms (SSCP) and heteroduplex analyses (Lareu, Swanson and Fox, 1997). The latter has been used to genotype viral strains and is based on its ability to discriminate sequence differences based on motility changes in non-denaturing polyacrylamide gel electrophoresis (PAGE). This is determined by fragment length and secondary conformation as determined by primary structure. It has been especially successful when used with PCR. PFGE is also frequently used to analyse large genomic strands. Other electrophoretic approaches include double strand conformation analysis (DSCA), constant denaturant gel electrophoresis (CDGE), denaturing gradient gel electrophoresis (DGGE) and temperature gradient gel electrophoresis (TGGE) that establish subtle sequence differences between nucleic acid strands.

PCR is also used directly to detect presence or absence of microbial resistance or virulence genes. Further improvements have been made using FRET-probe hybridization studies and real-time detection. Other PCR-based strategies include RAPD, rep-PCR, amplification fragment length polymorphism (AFLP) and PCR-RFLP, sequencing of selected genes, reverse hybridization using strain specific probes and PCR with type-specific primers or probes. RAPD or arbitrarily primed PCR (AP-PCR) allows for qualitative and quantitative differences in the genomes of compared systems using arbitrary primers generating a fingerprint (Williams *et al.*, 1990). No preliminary information is required about the sequence. AFLP is a restriction fragment amplification technique that involves the ligation of adapters to genomic restriction fragments from where primers are used to amplify these fragments. The pattern of bands obtained after PAGE of the amplified products can be highly informative. Polymorphism is due to mutations in the restriction sites of different strains, sequence differences in regions adjacent to the restriction sites that are complementary to the selective primer extensions, and insertions or deletions within the amplified fragments. It possesses a high degree of reproducibility and discriminatory ability and has the advantage over RFLP in that only a small amount of purified genomic DNA is needed. The use of fluorescence-labelled primers has made further contributions and has allowed for its use in automated sequencers (Savelkoul *et al.*, 1999). Reverse hybridization with linear probe arrays is particularly useful for discriminating between several related strains and species.

Methods used for viral genotyping include real-time PCR and FRET probes, DNA sequencing, micro-arrays and reverse hybridization techniques. Micro-arrays have particularly promising possibilities as fingerprinting tools and can be expected to replace many other assays and electrophoretic methods. When containing species-specific gene sequences, micro-arrays can be used to establish phylogenetic relationships between isolates (Sakar, Yoon and Sommer, 1992; Chan *et al.*, 2002; Versalovic and Lupski, 2002; Fox, Lareu and Swanson, 1995). DNA fingerprinting has been reliant on conventional slab gel electrophoresis, capillary electrophoresis and more recently microchip electrophoretic methods (Guttman *et al.*, 2002). Gel microchip electrophoresis combines slab gel and capillary gel electrophoresis with detection achieved in real time. In addition, restriction digestion is performed on small pore-size microfibrous membranes with sample loading and electrophoresis in a multilane format.

The combination of microfluidic-based amplicon fragment separation and fluorescence detection with the compilation of electronic DNA profile libraries (DiversiMap; Bacterial Barcodes, http://www.bacbarcodes.com) has greatly facilitated the identification of bacterial strains. Such strategies will enable sophisticated molecular epidemiological studies and possibly assist in identification of bacterial pathogens.

The RNase A mismatch cleavage method uses RNase A digestion of heteroduplexes to generate band patterns characteristic for each isolate with a given probe (Cristina *et al.*, 1990, 1991, 1998). It has been used as a preliminary screening method to distinguish RNA virus isolates. When a probe is used that recognizes a region that is more variable, the pattern generated consists of small protected bands, compared with the large protected bands generated when using probe binding more conserved regions amongst strains.

9.   SEQUENCING AND ALTERNATIVES

9.1   Conventional approaches

The sequencing of the genomes of pathogens is increasingly playing an important contributory role to molecular diagnostics and epidemiology. Since the first sequencing techniques were described, a number of further improvements have been made, notably adaptation to automation, which has contributed significantly to increasing the throughput of samples sequenced (Sterky and Lundeberg, 2000). Robotic devices for sample preparation, optimized chemistry, engineered sequencing enzymes and dyes with higher sensitivity have also made important contributions. On-line detection of fluorescent-labelled fragments is now routinely performed, with a four-dye system increasing throughput (Applied Biosystems) although one-dye systems yield higher accuracy. Capillary array electrophoresis is now widely used for DNA sequencing, allowing for high throughput and resolution, and detection of smaller amounts of DNA. Commercial devices include those from Applied Biosystems, Beckman Coulter, Molecular Dynamics and SpectruMedix. As a result, the genome maps of many organisms have been completed or are in progress. Further developments include improvements with mechanisms of illumination and detection.

PCR has greatly improved the readiness with which sequencing can be performed. Cycle sequencing utilizes Taq DNA polymerase, a single primer and thermocycling conditions (Murray, 1989). Different means of producing single stranded DNA suitable for sequencing have been devised, including solid phase sequencing.

9.2   Alternative approaches

DNA separation, including the products of restriction enzyme reactions, can also be done in a microchip using polyacrylamide polymerized in a capillary, with single-base resolution for sequencing. Woolley and Mathies (1995) were the first to show that sequencing separations on a chip could reach reasonable read lengths while being much faster than with conventional capillaries. Microchip systems have proved to be capable of producing high-quality separations in seconds in contrast to conventional capillary electrophoresis with separation times of about 10 min.

Micro-addressable arrays or DNA chip arrays can allow for the deduction of the DNA sequence of a sample nucleic acid according to the known sequence of probes and the pattern of hybridization produced. Since nucleotide polymorphisms in routine clinical specimen would be known, there is no advantage in performing extended sequencing. Such arrays have the advantage of speed, ability to analyse complex mixtures of nucleic acids, ability to perform many tests simultaneously and are also more suited to total automated analysis (Anthony, Brown and French, 2001).

Matrix-assisted laser desorption/ionization time-of-flight (MALDI-TOF) mass spectrometry can detect molecules in the subpicomole range (Leusher, 2002). This method has recently also found many applications in nucleic acid analysis. Both dideoxy sequencing and chemical cleavage sequencing have been successfully adapted for MALDI detection, where mass differences between successive fragments indicate the last base present. MALDI-TOF mass spectrometry can be used to measure the molecular weights of different probes simultaneously and in a non-radioactive multiplex manner using spot hybridization on a single spot, using probes with different mass tags (Isola *et al.*, 2001). It has the potential to drastically increase the speed of micro-array hybridization analysis in the future (Leusher, 2002; Isola *et al.*, 2001).

Pyrosequencing™ technology (Pyrosequencing AB, Uppsala, Sweden) is a non-electrophoretic, real-time determination of short sequences according to changes in relative light intensity. Advantages include accuracy, flexibility, parallel processing, and ease of automation. The technique also does not require labelled primers, labelled nucleotides or gel-electrophoresis. Sequencing is done by synthesis and relies on real-time quantification of pyrophosphate (PPi) release during DNA synthesis. Biotinylated PCR template is immobilized onto streptavidin-coated beads, followed by generation of single-stranded template and annealing of a sequencing primer. Nucleotides are added sequentially and, with the generation of PPi, a cascade of enzymatic reactions occur that produces visible light in proportion to the number of nucleotides incorporated, followed by degradation of non-incorporated nucleotides by the action of apyrase. This allows new nucleotides to be dispensed every 65th second. By designing a suitable order for the addition of nucleotides, the user can either determine or confirm the nucleotide sequence of the template. The normal read length is about 30 bases but successful reads of more than 100 have been reported. Uses include identification and characterization of viral and bacterial pathogens, including virulence determination, drug resistance mutations and strain identification (Berg, Sanders and Alderborn, 2002).

Minisequencing is a method identifying only the first proximal base to a hybridized probe by using a labelled ddNTP, thus allowing identification of the extended product. The probe can be immobilized to a microtitre plate (Kristensen *et al.*, 2001; Nikifirov *et al.*, 1994).

Single walled carbon nanotubes with diameters of 0.35–2.5 nm have been use in atomic force microscopy (AFM) (Haffner *et al.*, 2001). They offer high resolution and can detect labelled probes (e.g. streptavidin) bound to target DNA sequences. Multiple AFM tip arrays could allow for high throughput sequencing. Further developments using smaller nanotubes could even allow for direct reading of DNA sequences.

10.   MICRO-ARRAYS AND DNA CHIPS

The positive identification and characterization of nucleic acids, including PCR products, have, apart from the determination of size, included real-time hybridization, Southern blotting and sequencing, and the use of arrays such as dot blots or slot blot reverse hybridizations where probes are immobilized onto a support membrane and the resulting hybridization established visually. Checkerboard hybridization utilizes a series of probes deposited onto a membrane using a slot blotter. The membrane is rotated 90° and hybridized in the same slot blotter to a series of unknown nucleic acids. In contrast to these, micro-arrays and DNA chips are higher density arrays usually produced on glass or silicone (Stenger *et al.*, 2002; Anthony, Brown and French, 2001; Calla, Boruckia and Loged, 2003). They allow for the simultaneous detection of thousands of genetic elements. High density micro-arrays are produced by light-directed combinatorial synthesis of DNA oligomers, with lengths of about 10–30 bases on sites with resolutions of 1 μm^2. Affymetrix has been a leader in this field. Spotted micro-arrays consist of DNA oligonucleotides of 20–100 bases cross-linked to glass slides in 40–200 μm sized spots. Bead arrays consist of polymer beads of 1–5 μm diameter containing up to four fluorophores, and are coated with a DNA target. Flow cytometry or fibre optic technology is used to monitor binding reactions. Micro-electronic chips utilize semiconductor and microfabrication technology and contain micro-electrodes Enhancer molecules attached to hybridized DNA can allow for voltametric monitoring. DNA chips have been used to identify pathogen species and drug-resistant strains. Detection of cell expression profiles using fluorophore-labelled cDNA exposed to chip-immobilized probes can also be used to indicate severity of exposure to a pathogen or toxin. The chip then is scanned and points of fluorescence detected. In most cases, complex array systems are not required for clinical microbiology and the dependency on complex computer-assisted analysis is minimized. In the future, extensive universal diagnostic microbiology arrays will be manufactured so cheaply that they will be used even in assays where much of the chip is redundant (Anthony, Brown and French, 2001).

DNA micro-arrays also offer a means to enhance the detection capabilities of PCR (Stenger *et al.*, 2002; Anthony, Brown and French, 2001). Micro-arrays in these cases can serve as a series of parallel dot-blots either for rapid detection of sequence variation within a given locus, e.g. 16S rDNA, or for the detection of multiple products resulting from multiplex PCR. Since product length is not used for identification, PCR can be used to generate only short products. Because non-specific products are not detected, more primers can be used in multiplex reactions. Direct detection of nucleic acids without PCR, or subtyping (fingerprint) bacterial isolates, or generating new diagnostic markers for PCR can also be done. They have the ability of not only directly detecting bacterial pathogens but also characterizing them for antimicrobial resistance. Micro-arrays have also been used as a broad spectrum screen for viruses (Calla, Boruckia and Loged, 2003). A long oligonucleotide (70-mer) DNA micro-array capable of simultaneously detecting hundreds of viruses has been described (Wang *et al.*, 2002). Related viral serotypes could be distinguished by the unique pattern of hybridization generated by each virus. When micro-array elements from highly conserved regions within viral families were used, individual viruses not explicitly represented on the micro-array could still be detected, raising the possibility that this approach could be used for virus discovery. By using a random PCR amplification strategy in conjunction with the micro-array, multiple viruses could be detected without using sequence-specific or degenerate primers. Developments in microfluidics will allow for integration of purification, amplification and detection steps in a single disposable device.

11.   BIOSENSORS

Biosensors consist of a biological recognition element in intimate contact with a transducer. The latter may include amperometric and potentiometric electrodes, field-effect transistors, magnetoresistive sensors, piezo-electric crystals, optical and optoelectronic devices, as well as miniature cantilevers. Developments in DNA-based biosensors are one of the fastest growing areas in nucleic acid analysis. Nucleic acid probes are used as the biological recognition element although ion channels have also been proposed for detecting base pair composition of DNA, with changes in electron current monitored according to the degree of pore blockage by different bases as ssDNA is pulled through (Kristensen *et al.*, 2001; Meller *et al.*, 2000; Scheller *et al.*, 2001).

Electrochemical methods include the voltametric detection of redox intercalators and the mediated oxidation of guanine within the DNA. One format utilizes signalling probes labelled with the redox agent ferrocene, which bind to target immobilized to capture probe embedded in a self assembling monolayer coating a gold electrode (Umek *et al.,* 2001). Alternatively, amplification of the hybridization event can be done using an enzyme label followed by monitoring of impedance. The conductive properties of a DNA duplex for electrons or holes can allow for monitoring using a redox probe at one end and the electrode on the other. Charge transfer is influenced by distortions with mismatches during hybridization. In the future, ultraminiaturized electrochemical DNA sensors should allow for online monitoring and analysis. Electrochemical methods in combination with microfabrication techniques are likely to play important roles in providing highly sensitive assays.

The Bead ARray Counter (BARC) uses DNA hybridization, magnetic microbeads, and giant magnetoresistive (GMR) sensors to detect and identify biological warfare agents. The current prototype is a table-top instrument consisting of a microfabricated chip (solid substrate) with an array of GMR sensors, a chip carrier board with electronics for lock-in detection, a fluidics cell and cartridge, and an electromagnet. DNA probes are patterned onto the solid substrate chip directly above the GMR sensors, and sample analyte containing complementary DNA hybridizes with the probes on the surface. Labelled, micron-sized magnetic beads are then injected that specifically bind to the sample DNA. A magnetic field is applied, removing any beads that are not specifically bound to the surface. The beads remaining on the surface are detected by the GMR sensors, and the intensity and location of the signal indicate the concentration and identity of pathogens present in the sample. The current BARC chip contains a 64-element sensor array. With recent advances in magnetoresistive technology, however, chips with millions of these GMR sensors will soon be commercially available, allowing simultaneous detection of thousands of analytes (Edelstein *et al.*, 2000).

Piezo-electrical sensors are quartz crystal acoustic sensors that detect changes in mass on the crystal surface and would thus detect DNA-DNA hybridizations. The thickness shear mode type acoustic model is now being used especially for biomedical analysis (Pavey, 2002).

Surface plasmon resonance (SPR) allows for real time monitoring of binding between nucleic acid target and probe on the surface of a gold coated prism. SPR monitors accumulating changes in surface mass following DNA-DNA hybridization and can be used to confirm specificity of a PCR reaction (Bier, Kleinjung and Scheller, 1997; Caruso *et al.*, 1997). ssDNA obtained following asymmetric PCR has been used to hybridize with biotinylated probe attached to the sensor surface (Bianchi *et al.*, 1997).

Cantilevers coated with receptor layers act as force transducers and are being used in microfabricated biosensor devices (Oak Ridge National Laboratories; Graviton Inc; Protiveris Inc; Cantion A/S; IBM). Changes in surface stress, temperature and magnetization can be monitored following receptor-target binding, such as DNA hybridization. Monitoring is done using optical lever, interferometry or beam-bounce techniques. In the case of interdigitated cantilevers, a diffraction pattern is monitored that is determined by cantilever deflection. Capacitor plates or piezo-electrical cantilevers can be used to monitor changes in capacitance or conductivity in response to surface stresses (Bianchi *et al.*, 1997). There is no need to attach fluorescent tags, and parallel operation is possible, as is also real-time monitoring.

Silicon nanowires (Nanosys, CA) and carbon nanotubes (Molecular Nanosystems, CA) have been described that can monitor changes in conductance during binding of biological molecules to the surface (Alivasatos, 2001). DNA labelled with gold nanoparticles can be used as probes on chips, which can then be monitored electrically. Carbon nanotubes, molecular transistors and switches including allosteric and ribozymal nucleic acids, have exciting contributions to make as components of biosensors (Soukup and Breaker, 1999).

12.   MICROFABRICATION, MICROFLUIDICS AND INTEGRATED SYSTEMS

The integration of sample processing, amplification and detection systems is an important goal in achieving suitable point-of-care and on-site devices. Several partly integrated systems have been developed for the detection of biowarfare agents. The Lawrence Livermore and UCL Davies consortium have developed a self-contained system that continuously monitors air samples and automatically reports the presence of specific biological agents. The LLNL Autonomous Pathogen Detection System is an integrated aerosol collector, sample preparation and detection module that detects and identifies pathogens and/or toxins by a combination of an immunoassay and PCR. Nanofabrication and molecular electronics are being used by Nanogen to develop various sample-to-answer devices.

Microfabricated devices are being used to perform PCR for faster cycling using small chambers and integrated heaters. A variation of this approach uses the well of a microchip as the PCR chamber, with the entire chip undergoing thermal cycling. Electrophoretic separation at the level of the wells of the microchips can be achieved by transferring samples in some fashion between a PCR chamber and a well. PCR has also been integrated with detection using fluorogenic DNA probes. Most integrated systems combine the processing and detection phases using a microfluidic platform. Microfluidic systems consist of microchannels and tiny volume reservoirs and utilize electrokinetic or pneumatic mechanisms to transport fluids. The flow rates are in the nanolitre per second range through flow channels with cross-sectional dimensions in the tens of micrometres. Advantages include improved speed of analysis, reproducibility, reduced reagent consumption and the ability to perform multiple operations in an integrated fashion. Further development of this technology is expected to yield higher levels of functionality of sample throughput on a single microfluidic analysis chip.

Integrated systems using microfluidics are usually termed "lab-on-a-chip" technologies. Although still in the developmental phase, they are likely to make dramatic future contributions to molecular diagnostics, especially as point-of-care devices, with important contributions from nanotechnology.

Cepheid has developed a device that consists of disposable cartridge containing reagents and chambers for bacterial cell lysis and test sample preparation for fluorescent-based nucleic acid detection in a device called GeneXpert^{TM}. Infectio Diagnostic developed a rapid sample preparation method for both Gram +ve and -ve bacteria in different sample specimens.

Lin, Burke and Burns (2003) constructed a microfabricated device for separating and extracting double-stranded DNA fragments using an array of micro-electrodes and a cross-linked polyacrylamide gel matrix that is amenable to integration with reaction chambers into a single device for portable genetic-based analysis.

Microfluidic-based laboratory card devices have resulted in credit card-sized card designs suited for processing whole blood for haematological applications, and involves flows of sample reagents and control solutions in microchannels using capillary flow, hydrostatic pressure and fluid adsorption (Micronics Inc. WA) (Bousse *et al.*, 2000; Cronin and Mansfield, 2001). Optical microchips have also been devised whereby a micro-array of optical scanning elements is integrated with microfluidic circuits (Ruano *et al.*,

2003). These allow for fluorescence detection and have the potential of providing for other optical biosensor platforms such as SPR, evanescent field technology and interferometry

13.   DISCUSSION

Nucleic acid based-technologies are making considerable contributions to the field of diagnostics. PCR-based assays are already being utilized routinely by many laboratories and on-going developments are refining as well as expanding their capabilities. The use of real-time PCR and automated sample processing devices have already made significant contributions in reducing contamination whilst improving test consistency, rapidity, sensitivity and throughput. Improving the sensitivity of detection would also obviate the need to perform amplification reactions and its requirement to have suitable primers to amplify the target sequence. Several alternative target, probe and signal amplification systems have been described (LCR, SDA, RCA, bDNA, invasive cleavase). In addition, technologies to enhance separation and detection of nucleic acids have been developed (capillary electrophoresis, mass spectrometry). Labelling and detection methods other than radioactivity are also making important contributions (enzymatic, fluorescence, chemiluminescence and nanoparticle labelling) (Fortina, Surrey and Kricka, 2002). Nevertheless, conventional microbiological assays should be maintained to validate and guide further developments with the newer diagnostic approaches (van Belkum, 2003). Commercial kits for the molecular detection of the most important pathogens are increasingly becoming available. There is also a need to standardize nucleic acid assays through ring tests and the establishment of suitable guidelines and quality control programmes. The availability of lyophilized standards will assist in this process. The need for suitably trained staff to perform and evaluate nucleic acid-based assays, as well as the costs associated with many associated technological platforms is also an important requirement and in some cases an obstacle for their wider application. There is a need for centralized facilities to perform such tests, but developments in integrated systems are likely to allow for future point-of- care testing (Fortina, Surrey and Kricka, 2002). Rapid developments in biosensors are producing more effective biological recognition molecules as well as transducers.   Many of these have the potential of generating signals following the detection of single molecules. Micro-array technologies have the potential for parallel testing for large numbers of pathogens simultaneously, and this could make significant contributions to the diagnostic capabilities of many laboratories. Developments in the integration of sample processing, amplification and analysis and the eventual production of effective commercial testing devices would herald an important achievement in allowing for point-of-care testing. Advances in nanotechnology have potentially important contributions to make in this process, with the likelihood that test results could be obtained within minutes. Suitable wireless communication systems with centralized data banks as well as access to decision-making tools will allow for speedy therapeutic and prophylactic decision-making – a desirable achievement in any effective diagnostics programme.

REFERENCES

Alivasatos, A.P. 2001. Less is more in medicine. *Scientific American,* **285**: 59–65.

Anthony, R.M., Brown, T.J. & French, G.L. 2001. DNA array technology and diagnostic microbiology. *Expert Reviews in Molecular Diagnostics,* **1**: 30–38.

Aranaz, A., Liebana, E., Mateos, A., Dominguez, L. & Cousins, D. 1998. Restriction fragment length polymorphism and spacer oligonucleotide typing: a comparative analysis of fingerprinting strategies for *Mycobacterium bovis. Veterinary Microbiology,* **61**: 311–324.

Ayyadevara, S., Thaden, J.J. & Reis, R.J.S. 2000. Discrimination of primer 3'-nucleotide mismatch by taq DNA polymerase during PCR. *Analytical Biochemistry,* **284**: 11–18.

Banér, J., Nilsson, M., Mendel-Hartvig, M. & Landegrin, U. 1998. Signal amplification of padlock probes by rolling circle amplification. *Nucleic Acids Research,* **26**: 5073–5078.

Belgrader P., Benett, W., Hadley, D., Richards, J., Stratton, P., Mariella, R. Jr & Milanovich, F. 1999. PCR detection of bacteria in seven minutes. *Science,* **284**: 449–450.

Belgrader, P., Young, S., Yuan, B.M, Primeau, M., Christel, L.A., Pourahmadi, F. & Northrup, M.A. 2001. A battery-powered notebook thermal cycler for rapid multiplex real-time PCR analysis. *Analytical Chemistry,* **73**: 286–289.

Berg, L.M., Sanders, R. & Alderborn, A. 2002. Pyrosequencing™ technology and the need for versatile solutions in molecular clinical research. *Expert Reviews in Molecular Diagnostics,* **2**: 361–369.

Bianchi, N., Rutigliano, C., Tomasetti, M., Feriotto, G., Zorzato, F. & Gambari, R. 1997. Biosensor technology and surface plasmon resonance for real-time detection of HIV-1 genomic sequences amplified by polymerase chain reaction. *Clinical and Diagnostic Virology,* **8**: 199–208.

Bier, F.F., Kleinjung, F. & Scheller, F.W. 1997. Real-time measurement of nucleic-acid hybridization using evanescent-wave sensors: Steps towards the genosensor. *Sensors and Actuators B: Chemical,* **38**: 78–82.

Bousse, L., Cohen, C., Nikiforov, T., Chow, A., Kopf-Sill, A.R., Dubrow, R. & Wallace Parce, J. 2000. Electrokinetically controlled microfluidic analysis systems. *Annual Review of Biophysics and Biomolecular Structure,* **29**: 155–181.

Broude, N.E. 2002. Stem-loop oligonucleotides: a robust tool for molecular biology and biotechnology. *Trends in Biotechnology,* **20**: 249–256.

Calla, D.R., Boruckia, M.K. & Loged, F.J. 2003. DNA micro-arrays. *Journal of Microbiological Methods,* **53**: 235–243.

Caruso, F., Rodda, E., Furlong, D.N. & Haring, V. 1997. DNA binding and hybridization on gold and derivatized surfaces. *Sensors and Actuators B: Chemical,* **41**: 189–197.

Castro, A. & Shera, B.E. 1995. Single-molecule electrophoresis. *Analytical Chemistry,* **67**: 3181–3186.

Chan, W.C.W., Maxwell, D.J., Gao, X., Bailey, R.E., Han, M. & Nie, S. 2002. Luminescent quantum dots for multiplexed biological detection and imaging. *Current Opinion in Biotechnology,* **13**: 40–46.

Compton, J. 1991. Nucleic acid sequence-based amplification. *Nature,* **350**: 91–92.

Cook, N. 2003. The use of NASBA for the detection of microbial pathogens in food and environmental samples. *Journal of Microbiological Methods,* **53**: 165–174.

Corstjens, P.L., Zuiderwijk, M., Nilsson, M., Feindt, H., Niedbala, R.S. & Tankea, H.J. 2003. Lateral-flow and up-converting phosphor reporters to detect single stranded nucleic acids in sandwich-hybridization assay. *Analytical Biochemistry,* **312**: 191–200.

Cristina, J., Lopez, J.A., Albo, C., Garcia-Barreno, B., Garcia, J., Melero, J.A. & Portela, A. 1990. Analysis of genetic variability in human respiratory syncytial virus by the RNase A mismatch cleavage method: subtype divergence and heterogeneity. *Virology,* **174**: 126–134.

Cristina, J., Moya, A., Arbiza, J., Russi, J., Hortal, M., Albo, C., Garcia-Barreno, B., Garcia, O., Melero, J.A. & Portela, A. 1991. Evolution of the G and P genes of human respiratory syncytial virus (subgroup A) studied by the RNase A mismatch cleavage method. *Virology,* **184**: 210–218.

Cristina, J., Yunus, A.S., Rockemann, D.D. & Samal, S.K. 1998. Genetic analysis of the G and P genes in ungulate respiratory syncitial viruses by RNaseA mismatch cleavage method. *Veterinary Microbiology,* **62**: 185–192.

Cronin, M. & Mansfield, E.S. 2001. Microfluidic arrays in genetic analysis. *Expert Reviews in Molecular Diagnostics,* **1**: 109–118.

De Arruda, M., Lyamichev, V.I., Eis, P.S., Iszczyszyn, W., Kwiatkowski, R.W., Law, S.M., Olson & Rasmussen, E.B. 2002. Invader technology for DNA and RNA analysis: principles and applications. *Expert Reviews in Molecular Diagnostics,* **2**: 487–496.

Deiman, B., Van Aarle, P. & Sillekens, P. 2002. Characteristics and amplification of nucleic-acid sequence-based amplification (NASBA). *Molecular Biotechnology,* **20**: 163–179.

Demidov, V.V. 2001. PD-loop technology: PNA openers at work. *Expert Reviews in Molecular Diagnostics,* **1**: 343–351.

Demidov, V.V. 2002. Rolling circle amplification in DNA diagnostics: the power of simplicity. *Expert Reviews in Molecular Diagnostics,* **2**: 542–548.

Demidov, V.V. 2003. PNA and LNA throw light on DNA. *Trends in Biotechnology,* **21**: 4–7.

Demidov, V.V., Potaman, V.N., Frank-Kamenetskii, M.D., Egholm, M, Buchard, O., Sonnichsen, S.H. & Nielsen PE. 1994. Stability of peptide nucleic acids in human serum and cellular extracts. *Biochemical Pharmacology,* **48**: 1310–1313.

Duck, P., Alvarado-Urbina, G., Burdick, B. & Collier, B. 1990. Probe amplifier system based on chimeric cycling oligonucleotides. *BioTechniques,* **9**: 142–147.

Edelstein, R.L., Tamanaha, C.R., Sheehan, P.E., Miller, M.M., Baselt, D.R., Whitman, L.J. & Colton, R.J. 2000. The BARC biosensor applied to the detection of biological warfare agents. *Biosensors and Bioelectronics,* **14**: 805–813.

Egholm, M., Buchardt, O., Christensen, L., Behrens, C., Freier, S.M., Driver, D.A., Berg, R.H., Kim, S.K., Norden, B. & Nielsen, P.E. 1993. PNA hybridizes to complementary oligonucleotides obeying the Watson-Crick hydrogen-bonding rules. *Nature,* **365**: 566–568.

Eigen, M. & Rigler, R. 1994. Sorting single molecules: application to diagnostics and evolutionary biotechnology. *Proceedings of the National Academy of Sciences, USA,* **91**: 5740–5747.

Eliassari, A., Rodrigue, M., Meunier, F. & Herve, C. 2001. Hydrophilic magnetic latex for nucleic acid extraction, purification and concentration. *Journal of Magnetism and Magnetic Materials,* **225**: 127–133.

Fiandaca, M.J., Hyldig-Nielsen, J., Gildae, B.D. & Coull, J.M. 2001. Self-reporting PNA/DNA primers for PCR analysis. *Genome Research,* **11**: 609–613.

Fortina, P., Surrey S. & Kricka, L.J. 2002. Molecular diagnostics: hurdles for clinical implementation. *Trends in Molecular Medicine,* **8**: 264–266.

Fox, S.A., Lareu, R.R. & Swanson, N.R. 1995. Rapid genotyping of hepatitis C virus isolates by dideoxy fingerprinting. *Journal of Virological Methods,* **53**: 1–9.

Graham, D., Smith, W.E., Linacre, A.M.T., Munro, C., Watson, N.D. & White, P.C. 1997. Selective detection of deoxyribonucleic acid at ultralow concentrations by SERRS. *Analytical Chemistry,* **69**: 4703–4707.

Greenfield, L. & White, T.J. 1993. Sample preparation methods. pp. 122–137, *in:* D.H. Persing, T.F. Smith, F.C. Tenover and T.J. White (eds). *Diagnostic Molecular Microbiology: Principles and Applications.* Washington, DC: ASM Press.

Groenen, P.M., Bunschoten, A.E., van Soolingen, D. & van Embden, J.D. 1993. Nature of DNA polymorphism in the direct repeat cluster of *Mycobacterium tuberculosis*: application for strain differentiation by a novel typing method. *Molecular Microbiology,* **10**: 1057–1065.

Guatelli, J.C., Whitfield, K.M., Kwoh, D.Y., Barringer, K.J., Richman, D.D. & Gingeras, T.R. 1990. Isothermal, *in vitro* amplification of nucleic acids by a multienzyme reaction modeled after retroviral replication. *Proceedings of the National Academy of Sciences, USA,* **87**: 1874-1878 [See also Correction *ibid.,* **87**: 7797].

Guttman, A., Ronai, Z., Barta, C., Hou, Y.M., Sasvari-Szekely, Wang, X. & Briggs, S.P. 2002. Membrane-mediated ultrafast restriction digestion and subsequent rapid gel microchip electrophoresis of DNA. *Electrophoresis,* **23**: 1524–1530.

Haffner, J.H., Cheung, C.-L., Woolley, A.T. & Lieber, C.M. 2001. Structural and functional imaging with carbon nanotube AFM probes. *Progress in Biophysics & Molecular Biology,* **77**: 73–110.

Heid, C.A., Stevens, J., Livak, K.J. & Williams, P.M. 1996. Real time quantitative PCR. *Genome Research,* **6**: 986–994.

Higgins, J.A, Nasarabadi, S., Karns, J.S., Shelton, D.R., Cooper, M., Gbakima, A. & Koopman, R.P. 2003. A handheld real time thermal cycler for bacterial pathogen detection. *Biosensors and Bioelectronics,* **18**: 1115–1123.

Hill, C.S. 2001. Molecular diagnostic testing for infectious diseases using TMA technology. *Expert Reviews in Molecular Diagnosis,* **1**: 445–455.

Holland, P.M., Abramson, R.D., Watson, R. & Gelfand, D.H. 1991. Detection of specific polymerase chain reaction product by utilizing the 5'→3' exonuclease activity of *Thermus aquaticus* DNA polymerase. *Proceedings of the National Academy of Sciences, USA,* **88**: 7276–7280.

Ibrahim, M.S., Lofts, R.S., Jahrling, P.B., Henchal, E.A., Weedn, V.W., Northrup, M.A. & Belgrader, P. 1998. Realtime microchip PCR for detecting single-base differences in viral and human DNA. *Analytical Chemistry,* **70**: 2013–2017.

Isola, N.R., Allman, S.L., Golovlev, V.V. & Chen, C.H. 2001. MALDI-TOF mass spectrometric method for detection of hybridized DNA oligomers. *Analytical Chemistry,* **73**: 2126–2131.

Kamerbeek, J., Schouls, L., Kolk, A., van Agterveld, M., van Soolingen, D., Kuijper. S., Bunschoten, A., Molhuizen, H., Shaw, R., Goyal, M. & van Embden, J. 1997. Simultaneous detection and strain differentiation of *Mycobacterium tuberculosis* for diagnosis and epidemiology. *Journal of Clinical Microbiology,* **35**: 907–914.

Knemeyer, J-P., Marme, N. & Markus, S. 2000. Probes for detection of specific DNA sequences at the single-molecule level. *Analytical Chemistry,* **72**: 3717–3724.

Kramer, F.R. & Lizardi, P.M. 1989. Replicable RNA reporters. *Nature,* **339**: 401–402.

Kristensen, V.D., Kelefiotis, D., Kristensen, T. & Børresen-Dale, A. 2001. High-throughput methods for detection of genetic variation. *BioTechniques,* **30**: 318–332.

Lareu, R.R., Swanson, N.R. & Fox, S.A. 1997. Rapid and sensitive genotyping of hepatitis virus by single strand conformation polymorphism. *Journal of Virological Methods,* **64**: 11–18.

Leusher, J. 2002. MALDI TOF mass spectrometry: an emerging platform for genomics and diagnostics. *Expert Reviews in Molecular Diagnostics,* **1**: 11–18.

Lin, R., Burke, D.T. & Burns, M. 2003. Selective extraction of size-fractioned DNA samples in microfabricated electrophoresis devices. *Journal of Chromatography A,* **1010**: 255–268.

Manak, M.M., Li, C., Tai, C., Lawrence, N.P., Schumacher, R.T. & Tao, F. 2001. Clinical sample preparation using pressure cycling technology. *Clinical Chemistry,* **47**: 2083.

Meller, A., Nivon, L., Brandin, E., Golovchenko, J. & Brandon, D. 2000. Rapid nanopore discrimination between single polynucleotide molecules. *Proceedings of the National Academy of Sciences, USA,* **97**: 1079–1084.

Mitani, Y. & Oka, T. 2001. PCR-ICS (PCR-Immunochromatography System): A novel detection method for PCR amplicon. *Clinical Chemistry,* **47**: 2079.

Modrusan, Z., Bekkaoui, F. & Duck, P. 1998. Spermine-mediated improvement of cycling probe reaction. *Molecular and Cellular Probes,* **12**: 107–116.

Mori, Y., Nagamine, K., Tomita, N. & Notomi, T. 2001. Detection of loop-mediated isothermal amplification by turbidity derived from magnesium pyrophosphate formation. *Biochemical and Biophysical Research Communications,* **289**: 150–154.

Mullenix, M.C., Feng, X., Sivakamasundari, R., Krishna, R.M. & Piccoli, S.M. 2001. The combination of rolling circle amplification and differentially labelled microspheres enables the development of ultrasensitive multiplexed immunoassays. *Clinical Chemistry,* **47**: 2083.

Mullis, K.B. & Faloona, F.A. 1987. Specific synthesis of DNA *in vitro* via a polymerase-catalyzed chain reaction. *Methods in Enzymology,* **155**: 335–350.

Murray, V. 1989. Improved double stranded DNA sequencing using the linear polymerase chain reaction. *Nucleic Acids Research,* **17**: 88–89.

Nagamine, K., Kuzuhara, Y. & Notomi, T. 2002. Isolation of single-stranded DNA from loop-mediated isothermal amplification products. *Biochemical and Biophysical Research Communications,* **290**: 1195-1198.

Nagamine, K., Watanabe, K., Ohtsuka, K., Hase, T. & Notomi, T. 2001. Loop-mediated isothermal amplification reaction using a non-denatured template. *Clinical Chemistry,* **47**: 1742–1743.

Nelson, N.C., Hammon, P.W., Matsuda, E., Goud, A.A. & Becker, M.M. 1996. Detection of single-base mismatches in solution by chemiluminescence. *Nucleic Acids Research,* **24**: 4980–5003.

Nielsen, P.E. 2001. Peptide nucleic acid: a versatile tool in genetic diagnostics and molecular biology. *Current Opinion in Biotechnology,* **12**: 16–20.

Nielsen, P.E., Egholm, M. & Buchardt, O. 1994. Peptide nucleic acid (PNA). A DNA mimic with a peptide backbone. *Bioconjugate Chemistry,* **5**: 3–7.

Nikifirov, T.T., Rendle, R.B.V., Goelet, P., Rogers, Y.-H., Kotewicz, S., Anderson, S., Trainor, G.L. & Knapp, M.R. 1994. Genetic bit analysis: a solid phase method for typing single nucleotide polymorphisms. *Nucleic Acid Research,* **20**: 4167–4175.

Nilsson, M., Malmgren, H., Samiotaki, M., Kwiatkowski, M., Choudhary, B.P., Lundegren, U. 1994. Padlock probes; circularising oligonucleotides for localized DNA detection. *Science,* **265**: 2085–2088.

Notomi, T., Okayama, H., Masubuchi, H., Yonekawa, T., Watanabe, K., Amino, N. & Hase, T. 2000. Loop-mediated isothermal amplification of DNA. *Nucleic Acids Research,* **28**: e63.

Nygren, M. 2000. Molecular diagnostics of infectious diseases. Doctoral thesis. Department of Biotechnology, Royal Institute for Technology, Stockholm.

O'Meara, D. 2001. Molecular tools for DNA analysis. Doctoral thesis. Department of Biotechnology, Royal Institute for Technology, Stockholm.

Olsvik, Ø., Popovic, T., Skjerve, E., Cudjoe, K.S., Hornes, E., Ugelstad, J. & Uhlén, M. 1994. Magnetic separation techniques in diagnostic microbiology. *Clinical Microbiology Review,* **7**: 43–54.

Pavey, K.D. 2002. Quartz crystal analytical sensors: the future of label-free real-time diagnostics? *Expert Reviews in Molecular Diagnostics,* **2**: 173–186.

Petersen, M. & Wengel, J. 2003. LNA: a versatile tool for therapeutics and genomics. *Trends in Biotechnology,* **21**: 74–81.

Picard, F.J. & Bergeron, M.G. 2002. Rapid molecular theranostics in infectious diseases. *Drug Discovery Today,* **7**: 1092–1101.

Ross, R.S., Viazov, S., Sarr, S., Hoffmann, S., Kramer, S. & Roggendorf, M. 2002. Quantitation of hepatitis C virus RNA by third-generation branched DNA-based signal amplification assay. *Journal of Virological Methods,* **101:** 159–168.

Ruano, J.M., Glidle, A., Cleary, A., Walmsley, A., Aitchison, J.S. & Cooper, J.M. 2003. Design and fabrication of a silica on silicon integrated optical biochip as a fluorescence micro-array platform. *Biosensors and Bioelectronics,* **18**: 175–184.

Saiki, R.K., Gelfand, D.H., Stoffel, S., Scharf, S.J., Higuchi, R., Horn, T., Mullis, K.B. & Erlich, H.A. 1988. Primer-directed enzymatic amplification of DNA with a thermostable DNA polymerase. *Science,* **239**: 487–491.

Saiki, R.K., Scharf, S., Faloona, F., Mullis, K.B., Horn, G.T., Erlich, H.A. & Arnheim, N. 1985. Enzymatic amplification of beta-globin genomic sequences and restriction site analysis for diagnosis of sickle cell anemia. *Science,* **230**: 1350–1354.

Sakar, G., Yoon, H.S. & Sommer, S.S. 1992. Didexoy sequencing as a rapid and efficient search for the presence of mutations. *Genomics,* **13**: 491.

Savelkoul, P.H.M., Aarts, H.J.M., De Haas, J., Dijkshoorn, L., Duim, B., Otsen, M., Rademaker, J.L.W., Schouls, L. & Lenstra, J.A. 1999. Amplified-fragment length polymorphism analysis: the state of an art. *Journal of Clinical Microbiology,* **37**: 3083–3091.

Scheller, F.W., Wollenberger, U., Warsinke, A. & Lisdat, F. 2001. A novel approach is to use ion channels as sensors for detecting base pair composition of DNA. *Current Opinion in Biotechnology,* **12**: 35–40.

Schweitzer, B. & Kingsmore, S. 2001. Combining nucleic acid amplification and detection. *Current Opinion in Biotechnology,* **12**: 21–27.

Soukup, G.A. & Breaker, R.R. 1999. Nucleic acid molecular switches. *TIBTECH,* **17**: 46–476.

Spargo, C.A., Fraiser, M.S., Van Cleve, M., Wright, D.J., Nycz, C.M., Spears, P.A. & Walker G.T. 1996. Detection of *M. tuberculosis* DNA using thermophilic strand displacement amplification. *Molecular and Cellular Probes,* **10**: 247–256.

Spears, P.A., Linn, C.P., Woodard, D.L. & Walker, G.T. 1997. Simultaneous strand displacement amplification and fluorescence polarization detection of *Chlamydia trachomatis* DNA. *Analytical Biochemistry,* **247**: 130–137.

Stark, M., Reizenstein, E. & Uhlen, M. 1996. Immunomagnetic separation and solid phase detection of *Bordetella pertussis. Journal of Clinical Microbiology,* **33**: 778–784.

Stender, H., Fiandaca, M., Hyldig-Nielsen, J.J. & Coull, J. 2002. PNA for rapid microbiology. *Journal of Microbiological Methods,* **48**: 1–17.

Stenger, D.A., Andreadis, J.D., Vora, G.J. & Pancrazio, J.J. 2002. Potential applications of DNA micro-arrays in biodefense-related diagnostics. *Current Opinion in Biotechnology,* **13**: 208–212.

Sterky, F. & Lundeberg, J. 2000. Sequence analysis of genes and genomes. *Journal of Biotechnology,* **76**: 1–31.

Stojanovic, M.N., de Prada, P. & Landry D.W. 2001. Catalytic molecular beacons. *Chembiochemistry,* **2**: 411–415.

Taton, T.A. 2002. Nanostructures as tailored biological probes. *Trends in Biotechnology,* **20**: 276–279.

Tyagi, S. & Kramer, F.R. 1996. Molecular beacons: probes that fluoresce upon hybridization. *Nature Biotechnology,* **14**: 303–308.

Umek, R.M., Lin, S.W., Vielmetter, J., Terbrueggen, R.T., Irvine, B., Yu, C.J., Kayyem, J.F., Yowanto, Blackburn, G.F., Farkas, D.H. & Chen, Y.P. 2001. Electronic detection of nucleic acids. *Journal of Molecular Diagnostics,* **3**: 74–84.

Urdea, M.S., Horn, T., Fultz, T.J., Anderson, M., Running, J.A., Hamren, S., Ahle, D. & Chang, C.A. 1991. Branched DNA amplification multimers for the sensitive, direct detection of human hepatitis viruses. *Nucleic Acids Symposium Series,* **24**: 197–200.

Van Belkum, A. 2003. Molecular diagnostics in medical microbiology: yesterday, today and tomorrow. *Current Opinion in Pharmacology,* **3**: 1–5.

Van Gijlswijk, R.P.M., Talman, E.G., Janssen, P.J.A., Snoeijers, S.S., Killian, J., Tanke, H.J. & Heetebrij, R.J. 2001. Universal linkage system: versatile nucleic acid labeling technique. *Expert Reviews in Molecular Diagnostics,* **1**: 81–91.

Versalovic, J. & Lupski, J.R. 2002. Molecular detection and genotyping of pathogens: more accurate and rapid answers. *Trends in Microbiology,* **10**: S15–S21.

Vet, V.E.T., van der Rijt, B.J.M. & Blom, H.J. 2002. Molecular beacons: colorful analysis of nucleic acids. *Expert Reviews in Molecular Diagnosis,* **2**: 77–86.

Wagaman, P.C., Leong, M.A. & Simmen, K.A. 2002. Development of a novel influenza A antiviral assay. *Journal of Virological Methods,* **105**: 105–114.

Walker, G.T., Fraiser, M.S., Schram, J.J., Little, M.C., Nadeau, J.G. & Malinowski, D.P. 1992. Strand displacement amplification – an isothermal, *in vitro* DNA amplification technique. *Nucleic Acids Research,* **20**: 1691–1696.

Wang, D., Coscoy, L., Zylberberg, M., Avila, P., Boushey, H.A., Ganem, D. & DeRisi, L. 2002. Micro-array-based detection and genotyping of viral pathogens. *Proceedings of the National Academy of Sciences, USA,* **99**: 15687–15692.

Weiguo, C. 2001. DNA ligases and ligase-based technologies. *Clinical and Applied Immunology Reviews,* **2**: 33–43.

Wharam, S.D., Marsh, P., Lloyd, J.S., Ray, T.D., Mock, G.A., Assenburg, R., McPjhee, J.E., Brown, P., Weston, A. & Cardy, D.L.N. 2001. Specific detection of DNA and RNA targets using a novel isothermal nucleic amplification assay based on the formation of a three-way junction structure. *Nucleic Acids Research,* **29**: e54.

Whitcombe, D., Newton, C.R. & Little, S. 1998. Advances in approaches to DNA-based diagnostics. *Current Opinion in Biotechnology,* **9**: 602–608.

Williams, J.G., Kubelik, A.R., Livak, K.J., Rafalski, J.A. & Tingey, S.V. 1990. DNA polymorphisms amplified by arbitrary primers are useful as genetic markers. *Nucleic Acid Research,* **22**: 3257–3258.

Wittwer, C.T., Herrmann, M.G., Moss, A.A. & Rasmussen, R.P. 1997. Continuous fluorescence monitoring of rapid cycle DNA amplification. *Biotechniques,* **22**: 130–138.

Woolley, A.T. & Mathies, R.A. 1995. Ultra-high speed DNA sequencing using capillary electrophoresis chips. *Analytical Chemistry,* **67**: 3676–3680.

Yang, Y., Zaslavskaya, P., Bharadwaj, R., Fu, K., Wu, K., Dattagupt, N., Huang, I., Huang, A. & Wu, Y.-C. 2001. Tessera array technology: a rapid, specific detection of *Mycobacterium tuberculosis* using nucleic acid probe. *Clinical Chemistry,* **47**: 2081.

Zhang, D.Y., Brandwein, M., Hsuih, T. & Li, H.B. 2001. Ramification amplification: A novel isothermal DNA amplification method. *Molecular Diagnosis,* **6**: 141–150.

REVERSE GENETICS WITH ANIMAL VIRUSES
NSV reverse genetics

Teshome Mebatsion
Research and Development, Intervet Inc., 29160 Intervet Lane, Millsboro, DE 19966, USA

Abstract: New strategies to genetically manipulate the genomes of several important animal pathogens have been established in recent years. This article focuses on the reverse genetics techniques, which enables genetic manipulation of the genomes of non-segmented negative-sense RNA viruses. Recovery of a negative-sense RNA virus entirely from cDNA was first achieved for rabies virus in 1994. Since then, reverse genetic systems have been established for several pathogens of medical and veterinary importance. Based on the reverse genetics technique, it is now possible to design safe and more effective live attenuated vaccines against important viral agents. In addition, genetically tagged recombinant viruses can be designed to facilitate serological differentiation of vaccinated animals from infected animals. The approach of delivering protective immunogens of different pathogens using a single vector was made possible with the introduction of the reverse genetics system, and these novel broad-spectrum vaccine vectors have potential applications in improving animal health in developing countries.

1.   INTRODUCTION

The term reverse genetics in virology refers to the use of recombinant DNA technology to convert viral genomes into complementary DNA and generate viruses from the cloned DNA. Since manipulating RNA genomes still remains cumbersome, genetic manipulation of RNA viruses has exclusively relied upon cDNA intermediates to generate biologically active RNA molecules. The first reverse genetics system for an RNA virus was established for polio virus, a positive-strand RNA virus (Racaniello and

H. P. S. Makkar and G. J. Viljoen (eds.), Applications of Gene-Based Technologies for Improving Animal Production and Health in Developing Countries, 245-255.

Baltimore, 1981). For this group of viruses, transfection of a full-length RNA derived from the cDNA into eukaryotic cells results in viral protein expression and replication of the recombinant virus. In contrast, for viruses with a negative-sense RNA genome, the minimum infectious unit is not an RNA molecule, but a core structure called a ribonucleoprotein (RNP). For the formation of functional RNPs, the genome RNAs have to be encapsidated with the nucleoprotein (N) and form a complex with the polymerase (L) and phosphoprotein (P). Due to technical difficulties in reconstituting biologically active RNPs, genetic manipulation of negative-strand RNA viruses (NSVs) has lagged behind that of positive-strand RNA viruses. The first breakthrough occurred when Schnell, Mebatsion and Conzelmann (1994) reported the recovery of rabies virus, a neuropathogenic non-segmented NSV belonging to the family of Rhabdoviridae, entirely from cDNA. Reverse genetics of NSV progressed rapidly in the next decade, as documented by generation not only of non-segmented NSV (Conzelmann, 1998; Roberts and Rose, 1999), but also of segmented NSVs, including Bunyamwera virus (Bridgen and Elliott, 1996) and influenza viruses (Hoffmann *et al.*, 2000; Neumann *et al.*, 1999; Fodor *et al.*, 1999). The focus of this article is on the contribution of reverse genetics of non-segmented NSVs in designing improved vaccines to prevent and control economically important diseases of animals. The reader is directed to excellent recent reviews of different aspects of reverse genetics of NSVs, namely those of Conzelmann (1998), Marriott and Easton (1999), Nagai and Kato (1999), Neumann, Whitt and Kawaoka (2002), Palese *et al.* (1996) and Roberts and Rose (1999).

2.   NON-SEGMENTED NSVs OF VETERINARY IMPORTANCE

Non-segmented NSVs are a large and diverse group of enveloped viruses of both medical and veterinary importance. The single-stranded RNA of a non-segmented NSV possesses individual protein-encoding genes that are separated by regulatory regions comprising a gene-end signal, an intergenic region and a gene-start signal. These signals serve as gene transcription-start and -stop (polyadenylation) signals. The 3' and 5' ends of the RNA are represented by short non-translated sequences that contain the viral promoters for transcription and replication. Transcription of individual mRNAs starts at the 3' end of the genome, giving rise to abundant transcripts, and steadily decreases towards the 5' end, resulting in gradients of transcripts. This modular nature of gene expression facilitates engineering of additional genes into non-segmented NSVs by inserting additional gene-start and gene-end signals.

In veterinary medicine, the use of vaccines is the most effective and inexpensive way of combating infectious disease. Although the great majority of veterinary vaccines are still conventional, the availability of recombinant DNA technology in general, and the recent introduction of reverse genetics in particular, are facilitating a leap forward in designing improved vaccines. Following the first successful recovery of rabies virus, in 1994 (Schnell, Mebatsion and Conzelmann, 1994), vesicular stomatitis virus (VSV) and a number of viruses belonging to the family Paramyxoviridae, including viruses causing important animal diseases, such as rinderpest virus (RPV), canine distemper virus (CDV), bovine respiratory syncytial virus (bRSV), bovine parainfluenza virus (bPIV), Newcastle disease virus (NDV), and infectious haematopoietic necrosis virus (IHNV), have succumbed to genetic manipulation (Table 1).

Table 1. Non-segmented NSVs of veterinary importance generated entirely from cDNA.

Family and Species	Reference
Rhabdoviridae	
Rabies virus (RV)	Schnell, Mebatsion and Conzelmann, 1994.
Vesicular stomatitis virus (VSV)	Lawson *et al.*, 1995; Whelan *et al.*, 1995.
Infectious haematopoietic necrosis virus (IHNV)	Biacchesi *et al.*, 2000.
Paramyxoviridae	
Rinderpest virus (RPV)	Baron and Barrett, 1997.
Bovine respiratory syncytial virus (bRSV)	Buchholz, Finke and Conzelmann, 1999.
Newcastle disease virus (NDV)	Krishnamurthy, Huang and Samal, 2000; Nakaya *et al.*, 2001; Peeters *et al.*, 1999; Romer-Oberdorfer *et al.*, 1999.
Bovine parainfluenza virus type-3 (bPIV3)	Haller *et al.*, 2000.
Canine distemper virus (CDV)	Gassen *et al.*, 2000.

3. GENERATION OF ATTENUATED LIVE VIRUS VACCINE CANDIDATES

The ability to genetically manipulate non-segmented NSV opened a wide range of possibilities for studying virus biology and virus-host interaction. Identification and incorporation of attenuating mutations into recombinant viruses has led to design and generation of improved vaccine candidates. For example, introduction of one of the previously studied mutations into an infectious rabies virus (RV) clone by replacing the arginine at position 333 of RV glycoprotein (G) by an aspartic acid or glutamine resulted in a dramatic attenuation of RV in mice (Mebatsion, 2001; Morimoto *et al.*,

2001). Combination of this mutation with a deletion that eliminates the interaction between RV P-protein and the cytoplasmic dynein light chain (LC8), which is presumably involved in retrograde transport of RV, further attenuates the rabies virus by 30-fold after intramuscular inoculation (Mebatsion, 2001). The resultant recombinant virus may be helpful in developing a highly safe and effective live RV vaccine for oral immunizations of animals.

A new approach to generating live attenuated vaccines against NSVs was demonstrated for VSV by rearranging the gene order. Moving the nucleocapsid protein gene away from the 3' end eliminated the potential of the virus to cause disease (Wertz, Perepelitsa and Ball, 1998). Combining this change with moving the G to a promoter-proximal site yielded a vaccine candidate that protected swine against challenge with wild-type VSV, indicating that gene rearrangement provides a rational strategy for developing attenuated non-segmented NSVs (Flanagan *et al.*, 2001).

Reverse genetics of NSV has also helped in providing important insights into viral pathogenesis. The roles played by many accessory proteins, including V, C and NS proteins of Paramyxoviridae and influenza viruses as interferon antagonists, were studied in detail using infectious clones (Goodbourn, Didcock and Randall, 2000). Since interferon antagonists are important virulence determinants, their identification and modification by knocking them out or reducing their expression should provide opportunities to generate safe attenuated vaccine strains. Like other members of Paramyxovirinae, NDV produces the accessory V protein from the P gene by a process called RNA editing. The association of NDV-V protein with viral pathogenicity and its ability to inhibit alpha IFN was recently demonstrated (Huang *et al.*, 2003; Mebatsion *et al.*, 2001). Introduction of mutation into the editing site was shown to significantly reduce V protein expression and, as a result, the recombinant was highly attenuated in chicken embryos (Mebatsion *et al.*, 2001). Administration of the recombinant NDV with an editing site mutation to 18-day-old chicken embryos did not affect hatchability. Hatched chickens developed high levels of NDV-specific antibodies and were fully protected against lethal challenge, demonstrating the potential use of editing-defective recombinant NDV as a safe embryo vaccine (Mebatsion *et al.*, 2001).

In another example, characterization of recombinant bRSVs lacking either or both of the non-structural (NS) genes, NS1 and NS2, provided similar evidence that the NS genes counteract the antiviral effects of IFN-alpha/beta (Schlender *et al.*, 2000). The NS deletion mutants were shown to replicate to wild-type rBRSV levels in cells lacking a functional IFN-alpha/beta system, but were severely attenuated in IFN-competent cells and in young calves. Immunization of calves with either NS1 or NS2 deletion mutants, however, induced serum antibodies and protected calves against challenge with virulent bRSV (Valarcher *et al.*, 2003). Reverse genetics was also applied to generate recombinant bRSV lacking either or both of the envelope glycoprotein G and SH genes (Karger, Schmidt and Buchholz, 2001). Mucosal immunization of calves with the recombinant bRSV lacking the G led to infection but not to mucosal virus replication. However, it gave protection against challenge with wild-type bRSV, suggesting G may be dispensable in vaccinating calves against bRSV (Schmidt *et al.*, 2002).

4.   GENERATION OF MARKER VACCINES

Recently, intensive vaccination with marker vaccines and stamping-out strategies have been gaining popularity in veterinary medicine where eradication of specific diseases is of national or international interest. A marker vaccine is a vaccine that, in conjunction with a diagnostic test, enables serological differentiation of vaccinated animals from infected animals. A major drawback of all currently used whole-virus-based live and inactivated animal vaccines is that vaccinated animals cannot be distinguished from infected animals with standard serological tests. Approaches to develop marker vaccines include deletion of one or more non-essential but immunogenic proteins. This is mainly applicable for large DNA viruses containing several dispensable genes (e.g. herpesviruses). An alternative approach for the development of a marker vaccine is the use of "subunit vaccines". This approach has been implemented for many antigens involved in inducing protective immunity. The disadvantage of most subunit vaccines is that they are less effective than whole-virus-based live vaccines. For RNA viruses, in which most of the genes are essential, a deleted immunogenic gene may be complemented in *trans* or an additional marker gene may be inserted. Currently, control of RPV, which causes serious economic losses in parts of the developing world, is progressing at an encouraging pace. Effective elimination of this disease may require a genetically marked rinderpest vaccine that allows serological differentiation between vaccinated and infected animals. Recombinant RPVs expressing genetic markers, such as green fluorescent protein (GFP) or the influenza virus HA were generated and were effective in stimulating protective immunity against RPV and antibody responses to the marker protein in vaccinated cattle (Walsh *et al.*, 2000).

Another approach to generate live attenuated marker vaccines is generation of chimeric RNA viruses by replacing a whole immunogenic gene or part of a gene with a corresponding gene from another virus. A recombinant virus that expressed the NDV F protein and a chimeric haemagglutinin protein whose immunogenic globular head was replaced with that of avian paramyxovirus type 4 (APMV4) has been described (Peeters *et al.*, 2001). Neutralizing antibodies are developed against the NDV F protein, while the antibodies developed against the APMV4 HN protein allow a distinction from wild-type NDV isolates. Another approach to generating a marked NDV was to localize a conserved B-cell immunodominant epitope (IDE) on the nucleoprotein (NP) gene and then successfully recover a recombinant NDV lacking the IDE (Mebatsion *et al.*, 2002). In addition, a B-cell epitope of the S2 glycoprotein of murine hepatitis virus (MHV) was inserted in-frame to replace the IDE. Recombinant viruses properly expressing the introduced MHV epitope were successfully generated, demonstrating that the IDE is not only dispensable for virus replication, but can also be replaced by foreign sequences. Chickens immunized with the hybrid recombinants produced specific antibodies against the S2 glycoprotein of MHV and completely lacked antibodies directed against the IDE (Mebatsion *et al.*, 2002). Such marked-NDV recombinants in conjunction with a diagnostic test enable serological differentiation of vaccinated animals from infected animals, and may be useful tools in Newcastle disease eradication programmes.

5.   NSV-BASED VECTOR VACCINES

Non-segmented NSVs are able to accommodate large foreign genes in their envelopes and be able to induce broad-spectrum immune responses. Other important features that make them promising vaccine vectors include their ability to grow to high titre, express the foreign protein at high levels, and absence of DNA intermediates in their replication cycle that makes integration of their genomes into host cell genomes unlikely. The modular nature of the genome of NSVs also facilitates engineering of additional genes from several different pathogens in designing improved vaccines for animals and humans. The approach of delivering protective antigens from different pathogens in NSV vectors demonstrates the manifold possibilities of reverse genetics in designing and developing novel vaccines. Although there are numerous examples of published NSV vectored vaccine candidates of medical importance, few have been investigated for veterinary use. Very recently, the potential of NDV for use as a vaccine vector in expressing the haemagglutinin (HA) gene from avian influenza virus (AIV) was described (Swayne *et al.*, 2003). The recombinant provided only partial protection against lethal challenges of NDV and AIV, indicating the need for further research to improve the immunogenicity of the construct. Since current vaccines to prevent avian influenza rely upon labour-intensive parenteral injection, an NDV-based vaccine that could be administered by mass immunization methods would be an advantage.

Surface glycoproteins of most NSVs are potent inducers of antiviral immune response, which may alone be sufficient to protect the host against challenge with a wild type virus. In an attempt to design multivalent vaccines, recombinant viruses were generated in which the surface glycoproteins were exchanged or added in addition. In this respect, a recombinant RPV expressing the F and H proteins of Peste des petits ruminants virus (PPRV), a morbillivirus causing rinderpest-like disease in sheep and goats, was shown to protect goats against challenge (Das, Baron and Barrett, 2000). Recombinant bRSV, whose G and F proteins were replaced with the HN and fusion (F) proteins of bPIV3, was also recovered (Stope *et al.*, 2001). Reverse genetics has also been described for infectious haematopoietic necrosis virus (IHNV), a non-segmented NSV replicating at low temperatures of 14° to 20°C (Biacchesi *et al.*, 2000). IHNV is the causative agent of a devastating acute, lethal disease in wild and farmed rainbow trout. Based on an IHNV infectious clone, the possibility of exchanging IHNV glycoprotein G with that of viral hemorrhagic septicaemia virus (VHSV), another salmonid rhabdovirus, was demonstrated by recovering a chimeric IHNV that was able to replicate to the level of the wild-type rIHNV in cell culture (Biacchesi *et al.*, 2002).

Reverse genetics has also opened the door to the design of live attenuated human vaccines based on non-segmented NSVs of animal pathogens. A construct of NDV expressing influenza virus haemagglutinin was shown to provide complete protection against a lethal dose of influenza virus challenge in mice (Nakaya *et al.*, 2001). Likewise, recombinant VSVs expressing influenza A virus HA or neuraminidase (NA) proteins protected vaccinated mice from lethal challenge with influenza virus (Roberts *et al.*, 1998).

The potential of VSV to serve as vaccine vector was also demonstrated by generating a recombinant virus expressing human RSV (hRSV) G and F envelope glycoproteins (Kahn *et al.*, 1999). Recombinant bRSV in which the G and F genes are replaced with their hRSV counterparts were also generated and characterized (Buchholz *et al.*, 2000). A similar approach has also explored the use of bPIV3 as live attenuated vaccines by introducing the hPIV3 HN and F genes into a bPIV3 background. The chimeric was shown to protect hamsters from challenge with hPIV3 (Haller *et al.*, 2000). Furthermore, a bivalent vaccine virus containing hPIV3 F and HN genes and the RSV G or F gene in bPIV3 background was demonstrated to induce protection against both hPIV3 and RSV (Schmidt *et al.*, 2001). Non-segmented NSVs have also been harnessed to specifically target and eliminate HIV infected cells, as demonstrated by generation of recombinant

VSV or rabies virus possessing cellular proteins on their surfaces (Mebatsion *et al.*, 1997; Schnell *et al.*, 1997). Recombinant VSV and RV expressing env and/or gag genes of HIV-1 were also generated and characterized in experimental animals to investigate their potential as live-viral vaccines against AIDS (Rose *et al.*, 2001; Schnell *et al.*, 2000).

6.   CONCLUSION

To be successful, genetically engineered vaccines should have an advantage over that of conventional vaccines in at least some aspects such as: improved safety, increased efficacy, prolonged duration of immunity, giving a manufacturing and delivery advantage, or the ability to differentiate vaccinated from infected animals. In "transfer of gene-based technology" to developing countries, the advantages brought by such technology over that of conventional methods should be taken into consideration. In many developing countries, improvement of management practices, effective implementation of conventional vaccines and other disease control measures and assigning of adequate funds and trained manpower to upgrade the agricultural sector are still of paramount importance.

REFERENCES

Baron, M.D. & Barrett, T. 1997. Rescue of rinderpest virus from cloned cDNA. *Journal of Virology,* **71**: 1265–1271.

Bridgen, A. & Elliott, R.M. 1996. Rescue of a segmented negative-strand RNA virus entirely from cloned complementary DNAs. *Proceedings of the National Academy of Sciences, USA,* **93**: 15400–15404.

Biacchesi, S., Thoulouze, M.I., Bearzotti, M., Yu, Y.X. & Bremont, M. 2000. Recovery of NV knockout infectious hematopoietic necrosis virus expressing foreign genes. *Journal of Virology,* **74**: 11247–11253.

Biacchesi, S., Bearzotti, M., Bouguyon, E. & Bremont, M. 2002. Heterologous exchanges of the glycoprotein and the matrix protein in a Novirhabdovirus. *Journal of Virology,* **76**: 2881–2889.

Buchholz, U.J., Finke, S. & Conzelmann, K.K. 1999. Generation of bovine respiratory syncytial virus (BRSV) from cDNA: BRSV NS2 is not essential for virus replication in tissue culture, and the human RSV leader region acts as a functional BRSV genome promoter. *Journal of Virology,* **73**: 251–259.

Buchholz, U.J., Granzow, H., Schuldt, K., Whitehead, S.S., Murphy, B.R. & Collins, P.L. 2000. Chimeric bovine respiratory syncytial virus with glycoprotein gene substitutions from human respiratory syncytial virus (HRSV): effects on host range and evaluation as a live-attenuated HRSV vaccine. *Journal of Virology,* **74**: 1187–1199.

Conzelmann, K.K. 1998. Nonsegmented negative-strand RNA viruses: genetics and manipulation of viral genomes. *Annual Review of Genetics,* **32**: 123–162.

Das, S.C., Baron, M.D. & Barrett, T. 2000. Recovery and characterization of a chimeric rinderpest virus with the glycoproteins of peste-des-petits-ruminants virus: homologous F and H proteins are required for virus viability. *Journal of Virology,* **74**: 9039–9047.

Flanagan, E.B., Zamparo, J.M., Ball, L.A., Rodriguez, L.L. & Wertz, G.W. 2001. Rearrangement of the genes of vesicular stomatitis virus eliminates clinical disease in the natural host: new strategy for vaccine development. *Journal of Virology,* **75**: 6107–6114.

Fodor, E., Devenish, L., Engelhardt, O.G., Palese, P., Brownlee, G.G. & Garcia-Sastre, A. 1999. Rescue of influenza A virus from recombinant DNA. *Journal of Virology,* **73**: 9679–9682.

Gassen, U., Collins, F.M., Duprex, W.P. & Rima, B.K. 2000. Establishment of a rescue system for canine distemper virus. *Journal of Virology,* **74**: 10737–10744.

Goodbourn, S., Didcock, L. & Randall, R.E. 2000. Interferons: cell signalling, immune modulation, antiviral response and virus countermeasures. *Journal of General Virology,* **81**: 2341–2364.

Haller, A.A., Miller, T., Mitiku, M. & Coelingh, K. 2000. Expression of the surface glycoproteins of human parainfluenza virus type 3 by bovine parainfluenza virus type 3, a novel attenuated virus vaccine vector. *Journal of Virology,* **74**: 11626–11635.

Hoffmann, E., Neumann, G., Kawaoka, Y., Hobom, G. & Webster, R.G. 2000. A DNA transfection system for generation of influenza A virus from eight plasmids. *Proceedings of the National Academy of Sciences, USA,* **97**: 6108–6113.

Huang, Z., Krishnamurthy, S., Panda, A. & Samal, S.K. 2003. Newcastle disease virus V protein is associated with viral pathogenesis and functions as an alpha interferon antagonist. *Journal of Virology,* **77**: 8676–8685.

Kahn, J.S., Schnell, M.J., Buonocore, L. & Rose, J.K. 1999. Recombinant vesicular stomatitis virus expressing respiratory syncytial virus (RSV) glycoproteins: RSV fusion protein can mediate infection and cell fusion. *Virology,* **254**: 81–91.

Karger, A., Schmidt, U. & Buchholz, U.J. 2001. Recombinant bovine respiratory syncytial virus with deletions of the G or SH genes: G and F proteins bind heparin. *Journal of General Virology,* **82**: 631–640.

Krishnamurthy, S., Huang, Z. & Samal, S.K. 2000. Recovery of a virulent strain of Newcastle disease virus from cloned cDNA: expression of a foreign gene results in growth retardation and attenuation. *Virology,* **278**: 168–182.

Lawson, N.D., Stillman, E.A., Whitt, M.A. & Rose, J.K. 1995. Recombinant vesicular stomatitis viruses from DNA. *Proceedings of the National Academy of Sciences, USA,* **92**: 4477–4481.

Marriott, A.C. & Easton, A.J. 1999. Reverse genetics of the *Paramyxoviridae. Advances in Virus Research,* **53**: 321–340.

Mebatsion, T. 2001. Extensive attenuation of rabies virus by simultaneously modifying the dynein light chain binding site in the P protein and replacing Arg333 in the G protein. *Journal of Virology,* **75**: 11496–11502.

Mebatsion, T., Finke, S., Weiland, F. & Conzelmann, K.K. 1997. A CXCR4/CD4 pseudotype rhabdovirus that selectively infects HIV-1 envelope protein-expressing cells. *Cell,* **90**: 841–847.

Mebatsion, T., Verstegen, S., De Vaan, L.T., Romer-Oberdorfer, A. & Schrier, C.C. 2001. A recombinant Newcastle disease virus with low-level V protein expression is immunogenic and lacks pathogenicity for chicken embryos. *Journal of Virology,* **75**: 420–428.

Mebatsion, T., Koolen, M.J., De Vaan, L.T., de Haas, N., Braber, M., Romer-Oberdorfer, A., van den Elzen, P. & van der Marel, P. 2002. Newcastle disease virus

(NDV) marker vaccine: an immunodominant epitope on the nucleoprotein gene of NDV can be deleted or replaced by a foreign epitope. *Journal of Virology,* **76**: 10138–10146.

Morimoto, K., McGettigan, J.P., Foley, H.D., Hooper, D.C., Dietzschold, B. & Schnell, M.J. 2001. Genetic engineering of live rabies vaccines. *Vaccine,* **14**: 3543–3551.

Nagai, Y. & Kato, A. 1999. Paramyxovirus reverse genetics is coming of age. *Microbiology and Immunology,* **43**: 613–624.

Nakaya, T., Cross, J., Park, M.S., Nakaya, Y., Zheng, H., Sagrera, A., Villar, E., Garcia-Sastre, A. & Palese, P. 2001. Recombinant Newcastle disease virus as a vaccine vector. *Journal of Virology,* **75**: 11868–11873.

Neumann, G., Whitt, M.A. & Kawaoka, Y. 2002. A decade after the generation of a negative-sense RNA virus from cloned cDNA – what have we learned? *Journal of General Virology,* **83**: 2635–2662.

Neumann, G., Watanabe, T., Ito, H., Watanabe, S., Goto, H., Gao, P., Hughes, M., Perez, D.R., Donis, R., Hoffmann, E., Hobom, G. & Kawaoka, Y. 1999. Generation of influenza A viruses entirely from cloned cDNAs. *Proceedings of the National Academy of Sciences, USA,* **96**: 9345–9350.

Palese, P., Zheng, H., Engelhardt, O.G., Pleschka, S. & Garcia-Sastre, A. 1996. Negative-strand RNA viruses: genetic engineering and application. *Proceedings of the National Academy of Sciences, USA,* **93**: 11354–11358.

Peeters, B.P., de Leeuw, O.S., Koch, G. & Gielkens, A.L. 1999. Rescue of Newcastle disease virus from cloned cDNA: evidence that cleavability of the fusion protein is a major determinant for virulence. *Journal of Virology,* **73**: 5001–5009.

Peeters, B.P., de Leeuw, O.S., Verstegen, I., Koch, G. & Gielkens, A.L. 2001. Generation of a recombinant chimeric Newcastle disease virus vaccine that allows serological differentiation between vaccinated and infected animals. *Vaccine,* **19**: 1616–1627.

Racaniello, V.R. & Baltimore, D. 1981. Cloned poliovirus complementary DNA is infectious in mammalian cells. *Science,* **214**: 916–919.

Roberts, A. & Rose, J.K. 1999. Redesign and genetic dissection of the rhabdoviruses. *Advances in Virus Research,* **53**: 301–319.

Roberts, A., Kretzschmar, E., Perkins, A.S., Forman, J., Price, R., Buonocore, L., Kawaoka, Y. & Rose, J.K. 1998. Vaccination with a recombinant vesicular stomatitis virus expressing an influenza virus hemagglutinin provides complete protection from influenza virus challenge. *Journal of Virology,* **72**: 4704–4711.

Romer-Oberdorfer, A., Mundt, E., Mebatsion, T., Buchholz, U.J. & Mettenleiter, T.C. 1999. Generation of recombinant lentogenic Newcastle disease virus from cDNA. *Journal of General Virology,* **80**: 2987–2995.

Rose, N.F., Marx, P.A., Luckay, A., Nixon, D.F., Moretto, W.J., Donahoe, S.M., Montefiori, D., Roberts, A., Buonacore, L. & Rose, J.K. 2001. An effective AIDS vaccine based on live attenuated vesicular stomatitis virus recombinants. *Cell,* **106**: 539–549.

Schlender, J., Bossert, B., Buchholz, U. & Conzelmann, K.K. 2000. Bovine respiratory syncytial virus nonstructural proteins NS1 and NS2 cooperatively antagonize α/β interferon-induced antiviral response. *Journal of Virology,* **74**: 8234–8242.

Schmidt, A.C., McAuliffe, J.M., Murphy, B.R. & Collins, P.L. 2001. Recombinant bovine/human parainfluenza virus type 3 (B/HPIV3) expressing the respiratory syncytial virus (RSV) G and F proteins can be used to achieve simultaneous mucosal immunization against RSV and HPIV3. *Journal of Virology,* **75**: 4594–4603.

Schmidt, U., Beyer, J., Polster, U., Gershwin, L.J. & Buchholz, U.J. 2002. Mucosal immunization with live recombinant bovine respiratory syncytial virus (BRSV) and recombinant BRSV lacking the envelope glycoprotein G protects against challenge with wild-type BRSV. *Journal of Virology,* **76**: 12355–12359.

Schnell, M.J., Mebatsion, T. & Conzelmann, K.K. 1994. Infectious rabies viruses from cloned cDNA. *EMBO Journal,* **13**: 4195–4203.

Schnell, M.J., Johnson, J.E., Buonocore, L. & Rose, J.K. 1997. Construction of a novel virus that targets HIV-1-infected cells and controls HIV-1 infection. *Cell,* **90**: 849–857.

Schnell, M.J., Foley, H.D., Siler, C.A., McGettigan, J.P., Dietzschold, B. & Pomerantz, R.J. 2000. Recombinant rabies virus as potential live-viral vaccines for HIV-1. *Proceedings of the National Academy of Sciences, USA,* **97**: 3544–3549.

Stope, M.B., Karger, A., Schmidt, U. & Buchholz, U.J. 2001. Chimeric bovine respiratory syncytial virus with attachment and fusion glycoproteins replaced by bovine parainfluenza virus type 3 hemagglutinin–neuraminidase and fusion proteins. *Journal of Virology,* **75**: 9367–9377.

Swayne, D.E., Suarez, D.L., Schultz-Cherry, S., Tumpey, T.M., King, D.J., Nakaya, T., Palese, P. & Garcia-Sastre, A. 2003. Recombinant paramyxovirus type 1-avian influenza-H7 virus as a vaccine for protection of chickens against influenza and Newcastle disease. *Avian Disease,* **47**: 1047–1050.

Valarcher, J.F., Furze, J., Wyld, S., Cook, R., Conzelmann, K.K. & Taylor, G. 2003. Role of alpha/beta interferons in the attenuation and immunogenicity of recombinant bovine respiratory syncytial viruses lacking NS proteins. *Journal of Virology,* **77**: 8426–8439.

Walsh, E.P., Baron, M.D., Anderson, J. & Barrett, T. 2000. Development of a genetically marked recombinant rinderpest vaccine expressing green fluorescent protein. *Journal of General Virology,* **81**: 709–718.

Wertz, G.W., Perepelitsa, V.P. & Ball, L.A. 1998. Gene arrangement attenuates expression and lethality of a nonsegmented strand RNA virus. *Proceedings of the National Academy of Sciences, USA,* **95**: 3501–3506.

Whelan, S.P., Ball, L.A., Barr, J.N. & Wertz, G.T. 1995. Efficient recovery of infectious vesicular stomatitis virus entirely from cDNA clones. *Proceedings of the National Academy of Sciences, USA,* **92**: 8388–8392.

VIRAL SUBVERSION OF THE IMMUNE SYSTEM

Laurent Gillet and Alain Vanderplasschen*
Immunology–Vaccinology (B43b), Department of Infectious and Parasitic Diseases, Faculty of Veterinary Medicine, University of Liège, B-4000 Liège, Belgium.
** Corresponding author.*

Abstract: The continuous interactions between host and viruses during their co-evolution have shaped not only the immune system but also the countermeasures used by viruses.   Studies in the last decade have described the diverse arrays of pathways and molecular targets that are used by viruses to elude immune detection or destruction, or both.   These include targeting of pathways for major histocompatibility complex class I and class II antigen presentation, natural killer cell recognition, apoptosis, cytokine signalling, and complement activation.   This paper provides an overview of the viral immune-evasion mechanisms described to date.   It highlights the contribution of this field to our understanding of the immune system, and the importance of understanding this aspect of the biology of viral infection to develop efficacious and safe vaccines.

1.   INTRODUCTION

The continuous interactions between hosts and viruses during their co-evolution have shaped not only the immune system but also the counter-measures used by viruses. The evasion strategies that viruses have devised are highly diverse, ranging from the passive to the active. Passive evasion strategies comprise hiding inside the infected host cell in a dormant form or creating a broad antigenetic diversity among the progeny virions during each replication cycle (as exploited, for example, by retroviruses), thus evading or staying one step ahead of the immune response. Active mechanisms include interferences with pathways for major histocompatibility complex (MHC)

H. P. S. Makkar and G. J. Viljoen (eds.), Applications of Gene-Based Technologies for Improving Animal Production and Health in Developing Countries, 257-291.

class I and class II antigen presentation, natural killer (NK) cell recognition, cytokine signalling, apoptosis of infected cells, and complement activation. In this review, the authors provide an overview of the different active mechanisms that viruses use to evade host immune responses. Due to space constraints, those mechanisms will be presented concisely in pairs of associated figures and tables. The basic concepts of the components of the immune system targeted by the viruses are described in the figures, while viral strategies are listed in the corresponding tables. To save space, viruses are cited using the abbreviations of the International Committee for Taxonomy of Viruses.

2.   VIRAL INTERFERENCE WITH MHC CLASS I PATHWAY

CD8-positive cells play an important role in immunity against viruses. Just how important these cells are is demonstrated by the evolution of viral strategies for blocking the genesis or the display of viral peptide-MHC class I complexes on the surface of viral infected cells. To enhance the understanding of this field, the manner in which viral proteins are processed for recognition by virus-specific CD8+ T cells is briefly described (Figure 1). In the infected cells, peptides are generated from by-products of proteasomal degradation. Most of the substrates consist of defective ribosomal products (DRiPs). Peptides are then transported into the endoplasmic reticulum (ER) by the TAP protein. Here, MHC class I molecules are folded through the actions of general purpose molecular chaperones working with a dedicated chaperone (Tapasin) that tethers MHC class 1 to TAP. After peptide binding, MHC class I molecules dissociate from TAP, leave the ER and migrate to the plasma membrane through the Golgi complex. As viral peptide-MHC class I complexes accumulate on the cell surface, they have a greater chance of triggering activation by CD8+ T cells with a cognate receptor. Viruses have been shown to interfere with virtually every step of T cell antigen processing and presentation (Figure 1 and Table 1). The viral proteins involved in such mechanisms have been called VIPRs (pronounced "viper") for viral proteins interfering with antigen presentation. They are listed in Table 1 together with their mechanism of action. For an excellent review on this subject, see that of Yewdell and Hill (2002).

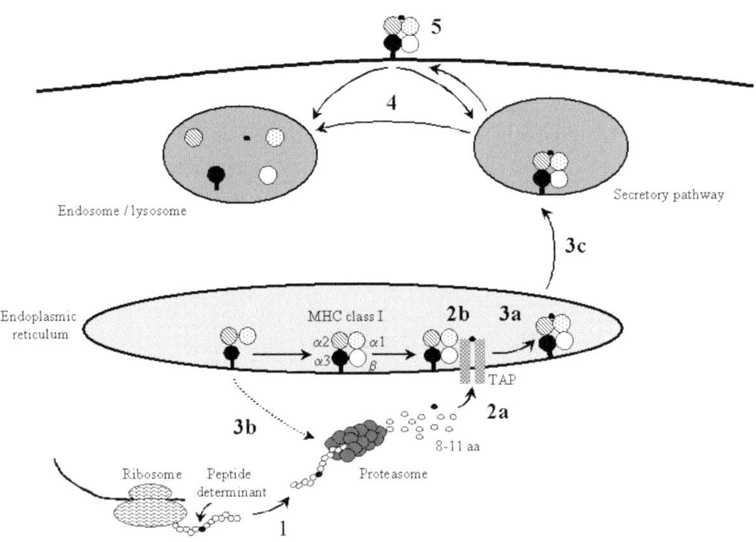

Figure 1. Viral interference with MHC class I pathway.

The classical MHC class I pathway is depicted with reference to viral interfering proteins listed in Table 1. Peptides are derived from DRiPs through the action of proteasomes and transported into the ER by the TAP protein. Nascent MHC class I molecules bind to TAP via tapasin. Binding of peptide to MHC class I molecules releases them from the ER. Peptide-MHC class I molecules then migrate to the cell surface. VIPRs have been shown to interfere with virtually every step of T cell antigen processing and presentation, namely (1) prevention of peptide degradation; (2) inhibition of peptide translocation in the ER, the inhibitory viral protein being either on the cytosolic side (2a) or in the lumen of the ER (2b); (3) retention of MHC class I molecules in the ER (3a) or in the transGolgi network (TGN) (3c), or by targeting of ER MHC class I molecules for degradation by the proteoasomes (3b); (4) reduction of peptide-MHC class I complexes exposed on the cell surface by inhibition of their migration to the cell surface, by increasing their endocytosis from the cell surface and by increasing their degradation into lysosomes; and (5) inhibition of T CD8+ cell recognition of cell surface peptide-MHC class I complexes. The VIPRs acting at those steps are listed in Table 1.

Table 1. Viral interference with the MHC class I pathway.

Site[1]	Virus[2]	Viral gene or protein	Mechanism of action	Source
1	HHV-4	EBNA-1	Contains a sequence that renders it resistant to proteasome degradation and self inhibition of synthesis	[1; 2]
1	HHV-5	UL83	Inhibits generation of antigenic peptides from a 72 kDa transcription factor by phosphorylation of the latter	[3]
2a	HHV-1 HHV-2	ICP47	Prevents peptide translocation by interacting with both TAP 1 and TAP 2 on the cytosolic side of the ER	[4; 5]
2a	BoHV-1	ICP47	Prevents peptide translocation by interacting with both TAP 1 and TAP 2 on the cytosolic side of the ER	[6]
2b	HHV-5	US6	Binds to TAP in the ER lumen and prevents peptide transport	[7; 8; 9]
3a	Adeno-virus	E19	Retains MHC-I in the ER by binding to the α1 and α2 regions (could also inhibit peptide loading of the MHC-I)	[10; 11; 12; 13]
3a	HHV-5	US3	Retains MHC-I in the ER	[14; 15]
3a	MuHV-1	m4	Forms extensive complexes with MHC-I in the ER	[16]
3b	HHV-5	US2, US3	Targets class 1 heavy chains for degradation by the proteasome	[17]
3b	MuHV-4	K3	Targets class 1 heavy chains for degradation by the proteasome and subverts TAP/Tapasin associated class I	[18; 19]
3b	HIV-1	Vpu	Destabilizes newly synthesized class 1 molecules and targets for degradation	[20]
3b	HTLV-1	p12(I)	Targets class 1 heavy chains for degradation by the proteasome	[21]
3c	MuHV-1	m152	Retains MHC-I within the ER-*trans*Golgi intermediate compartment	[22]
4	MuHV-1	m06	Prevents the MHC-I from reaching the cell surface	[23]
4	HIV, SIV	nef	Accelerates endocytosis of class 1 complexes (specific targeting of HLA A and B locus)	[24; 25]
4	EHV-1	?	Enhanced endocytosis of MHC-I from the surface	[26]
4	HHV-8	K3, K5	Targets the MHC-I to lysosomes	[27]
5	MuHV1	m4	Inhibits T CD8+ cell recognition	[28]

NOTES: (1) Site of action. Numbers refer to paths identified in Figure 1. (2) International Committee for Taxonomy of Viruses (ICTV) abbreviations.

SOURCES: [1] Levitskaya *et al.*, 1995. [2] Yin, Maoury and Fahraeus, 2003. [3] Gilbert *et al.*, 1993. [4] Galocha *et al.*, 1997. [5] Ahn *et al.*, 1996. [6] Hinkley, Hill and Srikumaran, 1998. [7] Hengel *et al.*, 1996, 1997. [8] Ahn *et al.*, 1997. [9] Lehner *et al.*, 1997. [10] Cox, Bennink and Yewdell, 1991. [11] Burgert and Kvist, 1987. [12] Jefferies and Burgert, 1990. [13] Bennett *et al.*, 1999. [14] Ahn *et al.*, 1996. [15] Jones, *et al.*, 1996. [16] Kavanagh, Koszinowski and Hill, 2001. [17] Wiertz *et al.*, 1996. [18] Boname and Stevenson, 2001. [19] Lybarger *et al.*, 2003. [20] Kerkau *et al.*, 1997. [21] Johnson *et al.*, 2001. [22] Ziegler *et al.*, 1997. [23] Reusch *et al.*, 1999. [24] Le Gall *et al.*, 1998. [25] Cohen *et al.*, 1999. [26] Rappocciolo, Birch and Ellis, 2003. [27] Hewitt *et al.*, 2002. [28] Kleijnen *et al.*, 1997.

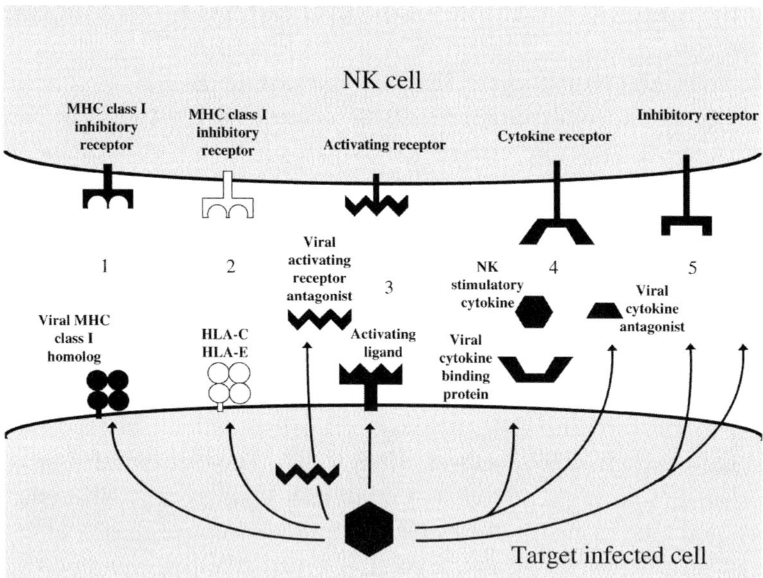

Figure 2. Viral evasion of NK cells.

Viral mechanisms interfering with NK cell functions fall into five categories, namely (1) expression of virally encoded MHC class I homologues that serve as NK cell decoys and ligate inhibitory receptors to block NK cytotoxicity; (2) selective modulation of MHC class I allele expression. Some viruses are able to down-regulate MHC class I molecules that are efficient for presentation of viral peptides to CD8+ cytotoxic T cells (such as HLA-A and HLA-B) without affecting or even increasing the expression of HLA-C and HLA-E, the dominant ligands for NK cell inhibitory receptors; (3) through the various mechanisms listed in Table 2, some viruses are capable of inhibiting the function of NK activatory receptor; (4) other viruses interfere with the cytokine pathways relevant to NK cell activation by producing virally encoded cytokine-binding proteins or cytokine antagonist; and (5) viruses can also directly inhibit NK cells by infecting them or by using viral envelope proteins to ligate NK cell inhibitory receptor.

3.   VIRAL EVASION OF NATURAL KILLER CELLS

NK cells are lymphocytes that, in contrast to B and T cells, do not undergo genetic recombination events to increase their affinity for particular ligands, and are therefore considered as part of the innate immune system. They are capable of mediating cytotoxic activity and producing cytokines after ligation of a variety of germline-encoded receptors. Like CD8+ T cells, NK cells mediate direct lysis of target cells by releasing cytotoxic granules containing perforin and granzymes, or by binding to apoptosis-inducing receptors on the target cells. Several receptors that can activate NK cells have been identified, among which some recognize viral proteins (Orange *et al.*, 2002). Due to the possible consequences of NK cell activation, normal host cells must inhibit NK activity. Various inhibitory receptors are consistently expressed by a subset of NK cells. These receptors bind to host MHC class I molecules and transmit inhibitory signals to the NK cells.

As noted above, many viruses have acquired effective means of avoiding T cell antigen presentation, thus avoiding T cell adaptive immune response. However, by eluding T cells, the viruses might have increased their susceptibility to NK cell-mediated defences. Consequently, in addition to the inhibition of T cell antigen presentation, some viruses have also acquired mechanisms to evade the action of NK cells. These mechanisms fall into five categories, presented in Figure 2; the viruses known to have acquired such mechanisms are listed Table 2. For an excellent review of the viral evasion of NK cells, see Orange *et al.* (2002).

Table 2. Viral evasion of natural killer cells.

Site [1]	Virus [2]	Viral gene	Mechanism of action	Source
1	HHV-5	UL18	Homologue of MHC class I, binds to ILT-2	[1; 2; 3; 4]
1	MuHV-1	m144	Homologue of MHC class I	[5; 6; 7]
1	MuHV-1	m157	Homologue of MHC class I, binds to Ly49-1	[8; 9; 10]
1	MuHV-2	r144	Homologue of MHC class I	[11]
1	MOCV	MC80R	Homologue of MHC class I	[12]
2	HHV-5	US2, US11	Cytosolic degradation of MHC class I, with exception of HLA-C and HLA-E	[13; 14; 15; 16]
2	HHV-5	US2, US3, US6, US11	Degradation or intracellular retention of MHC class I but not IL-18	[17]
2	HHV-5	UL40	Enhances surface expression of HLA-E	[18; 19; 20]
2	MuHV-1	m04	Forms complexes with MHC class I molecules intracellularly and on the cell surface	[21]
2	HIV	Nef	Induces the endocytosis of MHC class I with exception of HLA-C and HLA-E	[22; 23]

Site [1]	Virus[2]	Viral gene	Mechanism of action	Source
2	SIV	Nef	Induces the endocytosis of MHC class I with exception of HLA-C and/or HLA-E	[22; 24]
2	HHV-8	K5	Induces the endocytosis of HLA-A and HLA-B	[19; 25]
3	HHV-5	?	Decreases surface expression of the CD2 ligand LFA-3	[26]
3	HHV-5	UL16	Blocks the interaction of NKG2D-DAP10 and ULBP	[27; 28; 29]
3	MuHV-1	m152	Decreases surface expression of H60 (NKG2D ligand)	[30]
3	HHV-8	K5	Mediates ubiquitination and decreases surface expression of ICAM-1 and B7-2	[25; 31; 32]
3	HIV, HTLV	?	Mediates syalilation of cell surface receptors in infected cells	[33]
3	HIV	Tat	Inhibits LFA-1 mediated Ca^{2+} influx through binding of L-type Ca^{2+} channel	[34; 35]
4	MuHV-1	m131/129	Putative chemokine homologue	[36; 37]
4	HHV-8	vMIP-1, vMIP-2	Chemokine antagonists	[38; 39]
4	HHV-5	UL111a	Viral IL-10 homologue	[40]
4	HHV-4	BCRF1	Viral IL-10 homologue	[41]
4	ECTV	p13	IL-18 binding protein	[42]
4	MOCV	MC54L	IL-18 binding protein	[43]
4	HPV	E6, E7	IL-18 binding protein and antagonistic binding to IL-18 Rα	[44; 45]
4	MuHV-4	hvCKBP	Chemokine binding protein	[46]
4	VACV	vCKBP	Chemokine binding protein	[47]
5	HIV	/	Infects NK cells	[48]
5	HHV-1	/	Infects NK cells	[49]
5	HCV	E2	Binds to CD81	[50; 51]

NOTES: (1) Site of action. Numbers refer to paths identified in Figure 2. (2) International Committee for Taxonomy of Viruses (ICTV) abbreviations.

SOURCES: [1] Beck and Barrell, 1988. [2] Reyburn *et al.*, 1997. [3] Leong *et al.*, 1998. [4] Cosman *et al.*, 1997. [5] Farrell *et al.*, 1997. [6] Kubota *et al.*, 1999. [7] Cretney *et al.*, 1999. [8] Smith, Idris and Yokoyama, 2001. [9] Mandelboim *et al.*, 2001. [10] Arase *et al.*, 2002. [11] Kloover *et al.*, 2002. [12] Senkevich and Moss, 1998. [13] Schust *et al.*, 1998. [14] Gewurz *et al.*, 2001. [15] Machold *et al.*, 1997. [16] Lopez-Botet, Llano and Ortega, 2001. [17] Park *et al.*, 2002. [18] Tomasec *et al.*, 2000. [19] Ishido *et al.*, 2000. [20] Wang *et al.*, 2002. [21] Kavanagh *et al.*, 2001. [22] Le Gall *et al.*, 1998. [23] Cohen *et al.*, 1999. [24] Swigut *et al.*, 2000. [25] Coscoy, Sanchez and Ganem, 2001. [26] Fletcher, Prentice and Grundy, 1998. [27] Sutherland, Chalupny and Cosman, 2001. [28] Kubin *et al.*, 2001. [29] Cosman *et al.*, 2001. [30] Krmpotic *et al.*, 2002. [31] Ishido *et al.*, 2000. [32] Coscoy and Ganem, 2001. [33] Zheng and Zucker-Franklin, 1992. [34] Zocchi *et al.*, 1998. [35] Poggi *et al.*, 2002. [36] Fleming *et al.*, 1999. [37] Saederup *et al.*, 2001. [38] Kledal *et al.*, 1997. [39] Inngjerdingen, Damaj and Maghazachi, 2001. [40] Kotenko *et al.*, 2000. [41] Moore *et al.*, 1990. [42] Born *et al.*, 2000. [43] Xiang and Moss, 1999. [44] Lee *et al.*, 2001. [45] Cho *et al.*, 2001. [46] Parry *et al.*, 2000. [47] Alcami *et al.*, 1998. [48] Chehimi *et al.*, 1991. [49] York and Johnson, 1993. [50] Tseng and Klimpel, 2002. [51] Crotta *et al.*, 2002.

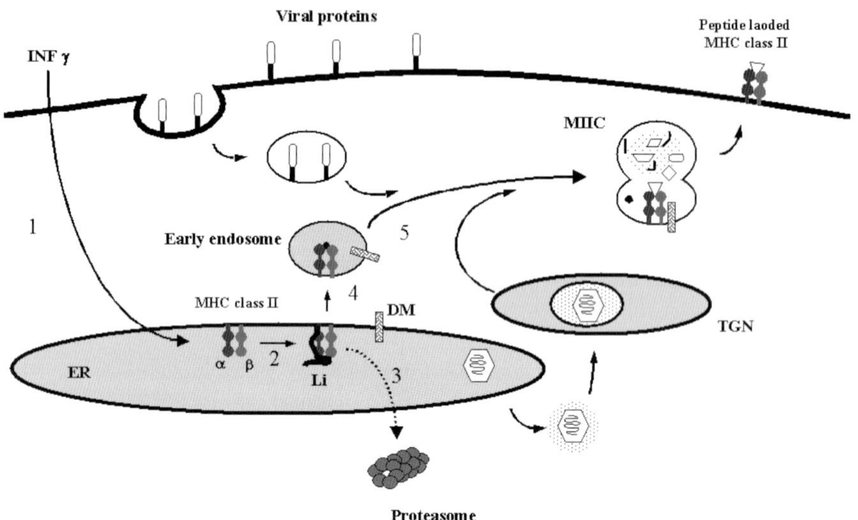

Figure 3. Viral inhibition of MHC class II antigen presentation.

MHC class II α and β chains and the invariant chain (Li) are expressed constitutively or in response to IFN-g stimulation. These molecules assemble in the ER to form the α-β-Li complexes that are then transported from the ER through the Golgi apparatus to the TGN, where the complexes are sorted to endosomes in response to signals present in the cytoplasmic tail of Li. In early endosomes, Li is progressively degraded by low-pH proteases so that fragments of it remain bound to the peptide-binding groove formed by the α-β chains. The MHC class II complexes then traffic into more acidic late endosomes and prelysosomal compartments known as MHC class II loading compartment (MIIC). Viral antigens can reach the endocytic pathway by phagocytosis, endocytosis or recycling of internal vesicules (site of virus assembly). Antigens delivered into the endocytic pathway are degraded by acid-dependent proteases to form peptides that are delivered to MIIC and loaded onto MHC class II α-β dimers. Exchange of peptide antigens for Li fragments occurs in collaboration with class II-like α-β dimers called DM. From the MIIC, peptide-loaded class II moves to the cell surface for presentation to CD4+ T cells. Viral mechanisms interfering with MHC class II antigen presentation fall into 5 categories: (1) inhibition of the IFN-γ transduction cascade leading to the expression of MHC class II; (2) Inhibition of the association of the α and β chains with the Li chains; (3) redirecting the α and b chains and DM for degradation by the proteasome; (4) preventing MHC class II from reaching the endocytic compartment; and (5) interfering with MHC class II processing and acidification of the endosome.

4. VIRAL INHIBITION OF MHC CLASS II ANTIGEN PRESENTATION

CD4-positive cells can recognize viral antigens expressed on virus-infected cells expressing MHC class II molecules to act cytolytically, to produce antiviral cytokines or to coordinate the antiviral immune response. MHC class II molecules are expressed constitutively by thymic epithelial cells, activated T cells and professional antigen-presenting cells, while in other cells, such as fibroblasts, keratinocytes, endothelial, epithelial and glial cells, their expression require IFN-γ stimulation. The latter induces the expression of MHC-II molecules through a complex cascade of factors (reviewed in Hegde, Chevalier and Johnson, 2003).

From the recent literature, it appears that viral inhibition of MHC class II antigen presentation is designed to prevent presentation of endogenous viral antigens in virus-infected cells rather than presentation of exogenous antigens in professional antigen-presenting cells.

Table 3. Viral inhibition of MHC class II antigen presentation.

Site [1]	Virus [2]	Viral gene	Mechanism of action	Source
1	Adeno-virus	E1A	Interferes with MHC class II upregulation (INF γ signal transduction cascade)	[1]
1	HHV-5	?	Interferes with MHC class II upregulation (INF γ signal transduction cascade)	[2; 3]
2	HHV-5	US3	Bounds to α/β subunits of MHC class II complexes in the ER reducing their association with Li	[4]
3	HHV-5	US2	Targets the MHC class II α and DM-α molecules for degradation by the proteasome	[5]
4	HHV-1	?	Redistributes MHC class II molecules away from the endocytic pathway	[6]
4	HIV	Env	Redistributes MHC class II molecules away from the endocytic pathway	[7]
5	HIV	Nef	Interference with MHC class II processing	[8]
5	SIV	Nef	Interference with MHC class II processing	[9]
5	HHV-1	gB	Interference with molecular co-players of MHC class II (DR and DM) processing	[10]
5	HPV/BPV	E5	Interference with MHC class II processing, and acidification of the endosomes	[11; 12]
5	BPV	E6	Interacts with AP-1, the TGN-specific clathrin adaptator complex	[11; 13]

NOTES: (1) Site of action. Numbers refer to paths identified in Figure 3. (2) International Committee for Taxonomy of Viruses (ICTV) abbreviations.

SOURCES: [1] Leonard and Sen, 1996. [2] Miller *et al.*, 1999. [3] Miller *et al.*, 1998. [4] Hegde *et al.*, 2002. [5] Tomazin *et al.*, 1999. [6] Lewandowski, Lo and Bloom, 1993. [7] Rakoff-Nahoum *et al.*, 2001. [8] Stumptner-Cuvelette *et al.*, 2001. [9] Schindler *et al.*, 2003. [10] Neumann, Eis-Hubinger and Koch, 2003. [11] Tortorella *et al.*, 2000. [12] Andresson *et al.*, 1995. [13] Tong *et al.*, 1998.

To enhance the understanding of this field, Figure 3 illustrates how viral peptides are processed for presentation in association with MHC class II molecules on the surface of an infected host cell. Some of the viral mechanisms acquired by viruses to interfere with this process are listed in Table 3. For an excellent review of this topic, see Hegde, Chevalier and Johnson (2003).

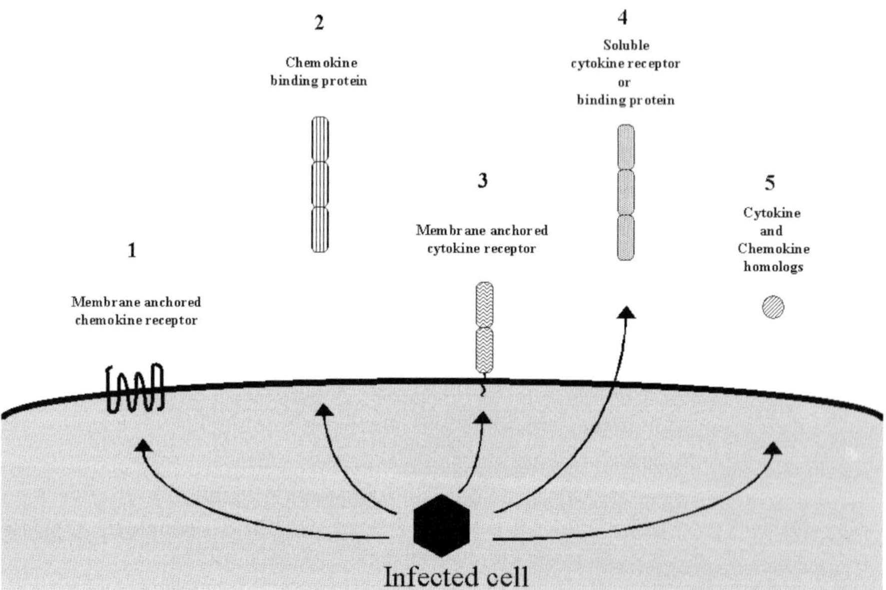

Figure 4. Viral interference with cytokines, chemokines and their receptors.

The strategies acquired by viruses to interfere with or to exploit host cytokines, chemokines and their receptors can be classified into 5 categories: (1) some viruses encode membrane anchored molecules able to bind host chemokine and eventually transduce a signal. Because these viral molecules have sequence similarity with host cellular receptors, they have been called chemokine receptors; (2) other viruses encode soluble proteins capable of binding to chemokines and preventing their action on target cells. Because these viral proteins are not homologues of host cellular proteins, they have been called chemokine binding protein rather than chemokine receptor; similarly, (3) viral encoded membrane anchored cytokine receptors; and (4) soluble cytokine receptors or soluble cytokine binding proteins have been described; (5) viruses are known to encode homologues of cytokines or chemokines.

5. VIRAL INTERFERENCE WITH CYTOKINES, CHEMOKINES AND THEIR RECEPTORS

Viral infection induces the production of cytokines and chemokines playing crucial roles in inducing the migration and activation of immune cells to areas of infection; in immune regulation; in anti-viral defence; as well as in the capacity of target cells to support virus replication. For example, cytokines such as interferons (IFN) and tumour necrosis factor (TNF) induce intracellular pathways that activate an anti-viral state or apoptosis, and thereby contribute to limit viral replication. A very large number of cytokines induce mechanisms that enhance immune recognition, or immune responses that protect against viral infection. Finally, some anti-viral cytokines mediate killing of infected cells by NK cells or cytotoxic T lymphocytes. Therefore, it is not surprising to find that cytokines, chemokines and their receptors are targets of viral immune-evasion strategies. The different strategies developed by viruses to interfere with or to exploit host cytokines, chemokines or their receptors are illustrated in Figure 4. Example of viruses known to have acquired such strategies are listed in the accompanying Table 4. For an excellent review of this topic, see the recent review by Alcami (2003).

Table 4. Viral interference with cytokines, chemokines and their receptors.

Site [1]	Virus [2]	Viral gene	Mechanism of action	Source
1	HHV-8	ORF74	Viral chemokine receptor, induces cell proliferation *in vitro* and tumours in transgenic mice	[1]
1	HHV-5	US28	Viral chemokine receptor	[2]
1	HHV-5	US27	Viral chemokine receptor	[3]
1	SCMV	E3-7	Cluster of five HCMV US28 homologues	[4]
1	MuHV-1	m33	Viral chemokine receptor	[5]
1	HHV-6	U51	Viral chemokine receptor	[6]
1	FWPV	FPV 021, 027, 206	Viral chemokine receptor	[7]
1	SWPV	SPV146	CXCR1 homologue	[3]
1	SPPV	Q2/3L	CC-chemokine receptor	[3]
1	YLDV	7L, 145R	CCR8 homologues	[8]
1	LSDV	LSDV011	CC-chemokine receptor homologue	[9]
2	EHV-1	gG (vCKBP4)	Secreted or membrane anchored C-, CC-, CXC- chemokine binding protein	[10]
2	EHV-3	gG (vCKBP4)	Secreted or membrane anchored C-, CC-, CXC- chemokine binding protein	[10]
2	BoHV-1	gG (vCKBP4)	Secreted or membrane anchored C-, CC-, CXC- chemokine binding protein	[10]
2	BoHV-5	gG (vCKBP4)	Secreted or membrane anchored C-, CC-, CXC- chemokine binding protein	[10]

Site [1]	Virus [2]	Viral gene	Mechanism of action	Source
2	RanHV-1	gG (vCKBP4)	Secreted or membrane anchored C-, CC-, CXC- chemokine binding protein	[10]
2	CapHV-1	gG (vCKBP4)	Secreted or membrane anchored C-, CC-, CXC- chemokine binding protein	[10]
2	CerHV-1	gG (vCKBP4)	Secreted or membrane anchored C-, CC-, CXC- chemokine binding protein	[10]
3	MYXV	vCKBP1	Secreted C-, CC-, CXC- chemokine binding protein	[11]
3	VACV	vCKBP2	Secreted C-chemokine binding protein	[12; 13]
3	CPXV	H5R	Secreted C-chemokine binding protein	[12; 13]
3	MYXV	M-T1	Secreted C-chemokine binding protein	[12; 13]
3	MuHV-4	vCKBP3	Secreted C-, CC-, CXC-, CX_3C- chemokine binding protein	[14]
3	VACV	A41L	vCKBP2 homologue	[15]
4	HHV-5	UL144	Membrane TNFR homologue	[16]
5	CPXV	CrmB	Secreted TNF inhibitor	[17]
5	MYXV	MT-2	Secreted TNF inhibitor	[18]
5	CPXV	CrmC	Secreted TNF inhibitor	[19]
5	CPXV	CrmD	Secreted TNF inhibitor	[20]
5	CPXV	CrmE	Secreted TNF inhibitor, also expressed at the cell surface	[21]
5	VACV	CrmE		[22]
5	LCDV1	ORF167L	Homology to domain of TNFR	[23]
5	SFV	T2	TNF-R homologue	[24; 25]
5	ECTV	E13	Secreted; blocks binding of CD30 to CD30L and induces reverse signalling in cells expressing CD30L	[26]
5	VACV	vCD30	Secreted; blocks binding of CD30 to CD30L and induces reverse signalling in cells expressing CD30L	[27]
5	VACV	B16R	Secreted receptor for interleukin-1 β	[28]
5	MYXV	MT-7	Secreted receptor for IFN-γ	[29]
5	VACV	B8R	Secreted receptor for IFN-γ	[30]
5	VACV	B19R	Secreted and cell surface binding protein for IFN-α/β	[31]
5	HHV-4	BARF1	Secreted binding protein for CSF1	[32]
5	ORFV	GIF	Secreted binding protein for GM-CSF/IL2	[33]
5	MOCV	MC54	Secreted binding protein for IL18	[34]
5	ECTV	E19	Secreted binding protein for IL18	[35]
5	MOCV	MC51, MC53	Secreted binding proteins for IL18	[36]
6	VACV	C11R	Viral epidermal growth factor homologue	[37]
6	ORFV	A2R	Viral vascular endothelial growth	[38]
6	HHV-4	BCRF1	Viral IL-10 homologue	[39]
6	HHV-5	UL111a	Viral IL-10 homologue	[40]
6	ORFV	vIL-10	Viral IL-10 homologue	[41]
6	EHV-2	E7	Viral IL-10 homologue	[42]
6	SaHV-2	ORF13	Viral IL-17 homologue	[43]
6	HHV-8	K2	Viral IL-6 homologue	[44]
6	VACV	A39R	Viral semaphorin, binds semaphorin receptor VESPR	[45]

Site [1]	Virus [2]	Viral gene	Mechanism of action	Source
6	FWPV	FPV080	Viral TGF-β homologue	[7]
6	FWPV	FPV072, FPV076	Viral β-NGF homologue	[7]
6	HHV-8	K6	Viral CR8 agonist	[46]
6	HHV-8	K4	C-, CC-, CXC-, CX3C-chemokine antagonist	[47]
6	HHV-8	K4.1	CCR4 agonist	[48]
6	HHV-6	U83	CC-chemokine agonist	[49]
6	MOCV	MC148	CC-, CXC-chemokine antagonist, CCR8 specific antagonist	[50; 51]
6	MuHV-1	m131/129	CC-chemokine agonist	[52 – 54]
6	HHV-5	UL146	CXCR2 agonist	[55]
6	GaHV-2	MDV003	CXC chemokine	[56]
6	HIV	tat	Partial chemokine similarity	[57]
6	HRSV	gG	Partial chemokine similarity, CX_3CL1 activity	[58]

NOTES: (1) Site of action. Numbers refer to paths identified in Figure 4. (2) International Committee for Taxonomy of Viruses (ICTV) abbreviations.

SOURCES: [1] Arvanitakis *et al.*, 1997. [2] Bodaghi *et al.*, 1998. [3] Murphy, 2001. [4] Alcami, 2003. [5] Davis-Poynter *et al.*, 1997. [6] Milne *et al.*, 2000. [7] Alfonso *et al.*, 1996. [8] Lee, Essani and Smith, 2001. [9] Tulman *et al.*, 2001. [10] Bryant *et al.*, 2003. [11] Mossman *et al.*, 1996. [12] Smith *et al.*, 1997. [13] Graham *et al.*, 1997. [14] Parry *et al.*, 2000. [15] Ng *et al.*, 2001. [16] Benedict *et al.*, 1999. [17] Hu, Smith and Pickup, 1994. [18] Macen *et al.*, 1996. [19] Smith *et al.*, 1996. [20] Loparev *et al.*, 1998. [21] Saraiva and Alcami, 2001. [22] Reading, Khanna and Smith, 2002. [23] Tidona and Darai, 1997. [24] Smith *et al.*, 1990. [25] Smith *et al.*, 1991. [26] Saraiva *et al.*, 2002. [27] Panus *et al.*, 2002. [28] Alcami and Smith, 1992. [29] Upton, Mossman and McFadden, 1992. [30] Alcami and Smith, 1995. [31] Colamonici *et al.*, 1995. [32] Strockbine *et al.*, 1998. [33] Deane *et al.*, 2000. [34] Xiang and Moss, 1999a. [35] Smith, Bryant and Alcami, 2000. [36] Xiang and Moss, 1999b. [37] Twardzik *et al.*, 1985. [38] Meyer *et al.*, 1999. [39] Hsu *et al.*, 1990. [40] Kotenko *et al.*, 2000. [41] Fleming *et al.*, 1997. [42] Rode *et al.*, 1993. [43] Yao *et al.*, 1995. [44] Aoki *et al.*, 1999. [45] Gardner *et al.*, 2001. [46] Boshoff *et al.*, 1997. [47] Kledal *et al.*, 1997. [48] Stine *et al.*, 2000. [49] Zou *et al.*, 1999. [50] Krathwohl *et al.*, 1997. [51] Luttichau *et al.*, 2000. [52] Fleming *et al.*, 1999. [53] Saederup *et al.*, 2001. [54] Saederup *et al.*, 1999. [55] Penfold *et al.*, 1999. [56] Parcells *et al.*, 2001. [57] Albini *et al.*, 1998. [58] Tripp *et al.*, 2001.

6. VIRAL MANIPULATION OF THE CELL DEATH PROGRAMME

Replication of viruses may stimulate suicide of the host cell directly or via recognition by immune effector cells. These cells (cytolytic T cells and NK cells) induce cell death by secretion of cytotoxic cytokines such as TNFs and by processes requiring direct cell-cell contact, such as release of perforin and granzyme. This form of programmed cell death is called apoptosis. Apoptosis is an orchestrated biochemical process that leads ultimately to the demise of the cell, initiated by both internal sensors (intrinsic pathway, mitochondria dependent) and external stimuli (extrinsic pathway, death receptor mediated).

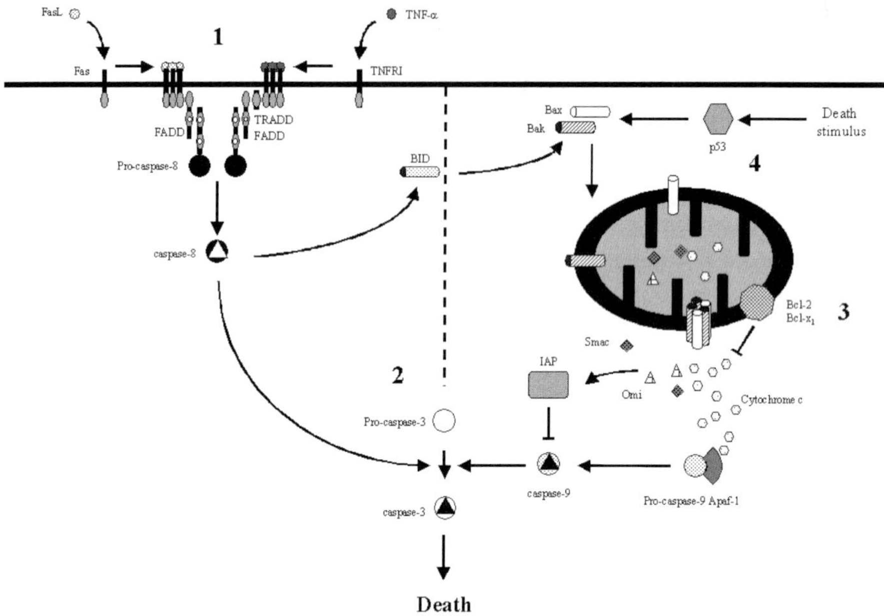

Figure 5. Viral inhibition of apoptosis.

Apoptosis can be initiated by two main pathways. The extrinsic pathway is triggered by death ligands binding to their cognate death receptors. These receptors then multimerize and their death domains (DDs) interact with the DDs of adaptator proteins that bind to pro-caspase 8 and/or pro-caspase 10 to form the DISC. This ends with pro-caspase cleavage in their active form. These caspases can then cleave Bid and activate the effector caspase cascade. On the other end, internal sensors initiate the intrinsic pathway via a process that results in hetero-oligomeric pores formation in the outer membrane of the mitochondria. Factors such as cytochrom c, Smac and Omi are then released in the cytoplasm where cytochrom c promotes formation of the apoptosome, resulting in autocatalytic activation of caspase 9, which initiates the effector caspase cascade. Caspases activation is negatively regulated by IAP, which are counter-balanced by proapoptotic Smac and Omi. Viral mechanisms of apoptosis inhibition fall into 4 main categories: (1) modulating of death receptors signalling; (2) regulation of caspase; (3) mimicking Bcl-2; and (4) blinding the internal sensors.

In the case of replicating viruses, apoptosis can be viewed as an altruistic defence mechanism by which the host infected cell commits suicide in order to prevent virus spread in the infected host. Indeed, premature cell death would enable viruses to maximally replicate or to establish latency. Apoptosis is a complex and highly regulated process. Many viruses have acquired mechanisms to inhibit this important biological process, by targeting different steps. These mechanisms of viral inhibition of apoptosis can be classified into four main classes: modulation of death receptor signalling; caspase regulation; Bcl-2 mimicking; and internal sensors blinding. They are described in Figure 5. Viral proteins inhibiting apoptosis are listed in Table 5, together with their mechanism of action. For an excellent review of this subject, see Benedict, Norris and Ware (2002).

Table 5. Viral inhibition of apoptosis.

Site [1]	Virus[2]	Viral gene	Mechanism of action	Source
1	adeno-virus	E3-6.7	Complexes with 10.4 and 14.5 resulting in downmodulation of TRAIL receptor 1 and 2	[1]
1	adeno-virus	E3-10.4	Inhibits TNF and FasL induced apoptosis	[2; 3]
1	adeno-virus	E3-14.5	Inhibits TNF and FasL induced apoptosis	[2; 3]
1	adeno-virus	E3-14.7	Inhibits TNF induced apoptosis	[4; 5]
1	BoHV-4	ORF71	Inhibits TNF and FasL induced apoptosis (viral homologue of cFLIP)	[6]
1	EHV-2	E8	Inhibits TNF and FasL induced apoptosis (viral homologue of cFLIP)	[7]
1	SaHV-2	ORF71	Inhibits TNF and FasL induced apoptosis (viral homologue of cFLIP)	[8]
1	HHV-8	K13	Inhibits TNF and FasL induced apoptosis (viral homologue of cFLIP)	[9]
1	MOCV	MC159	Inhibits TNF and FasL induced apoptosis (viral homologue of cFLIP)	[7; 10]
1	HHV-5	UL36	Prevents caspase 8 activation	[11]
1	MYXV	MT-2	TNF-R homologue	[12; 13]
1	HHV-4	LMP1	Interacts with TRFAFs and TRADD	[14; 15]
1	SFV	T2	TNF-R homologue	[16; 17]
1	VACV	CrmE	TNF-receptor	[18]
1	CPXV	CrmB	TNF-receptor	[19]
1	CPXV	CrmC	TNF-receptor	[20]
1	CPXV	CrmD	TNF-receptor	[21]
1	CPXV	CrmE	Secreted TNF inhibitor, also expressed at the cell surface	[22]
1	LCDV1	ORF167L	Homology to domain of TNFR	[23]
1	HHV-5	UL144	Membrane TNFR homologue	[24]
2	ASFV	A224L	IAP-related protein	[25; 26]
2	Baculo-virus	P35	Inhibits caspase 1, 3, 6, 8 and 10	[27 – 29]
2	Baculo-virus	IAP	Inhibits caspase 3, 6 and 7	[27; 30]
2	CPXV	CrmA	Inhibits caspase 1, 4, 5 and 11	[31 – 33]
2	VACV	SPI-2	Inhibits caspase 1, 4, 5 and 11	[34]
2	ECTV	SPI-2	Inhibits caspase 1, 4, 5 and 11	[35]

Site [1]	Virus [2]	Viral gene	Mechanism of action	Source
3	Adeno-virus	E1B-19K	Bcl-2-related protein	[36; 37]
3	HHV-4	BHRF1	Bcl-2-related protein	[38; 39]
3	HHV-4	BALF1	Bcl-2-related protein	[40]
3	HHV-8	HHV-8 vBcl-2	Bcl-2-related protein	[41]
3	SaHV-2	ORF16	Bcl-2-related protein	[42; 43]
3	MuHV-4	m11	Bcl-2-related protein	[44]
3	ASFV	A179L	Bcl-2-related protein	[45]
3	HHV-1	US3	Prevents virus induced apoptosis via a post-translational modification of Bad	[46]
3	HHV-1	US5	Cooperates with US3	[46]
3	HHV-5	UL37	Appears to be functionally similar to Bcl-2	[47]
3	HHV-4	LMP1	Up-regulates Bcl-2 and other cell survival proteins	[14; 15]
3	HIV	Nef	Prevents apoptosis via phosphorylation of Bad	[48]
3	HTLV-1	Tax	Activates the Bcl-x_L promoter while repressing transcription of Bax	[49]
4	Adeno-virus	E1B-55K	Binds to p53 and functionally inactivates it	[50]
4	HPV	E6	Targets p53 for degradation	[51 – 53]
4	SV-40	Large T	Binds to p53 and inactivates it	[54; 55]
4	HBV	pX	Complexes p53 and inhibits p53-mediated transcriptional activation	[56]

NOTES: (1) Site of action. Numbers refer to paths identified in Figure 5. (2) International Committee for Taxonomy of Viruses (ICTV) abbreviations.

SOURCES: [1] Benedict *et al.*, 2001. [2] Gooding *et al.*, 1991. [3] Shisler *et al.*, 1997. [4] Gooding *et al.*, 1988. [5] Li, Kang and Horowitz, 1998. [6] Wang *et al.*, 1997. [7] Bertin *et al.*, 1997. [8] Glykofrydes *et al.*, 2000. [9] Sturzl *et al.*, 1999. [10] Shisler and Moss, 2001. [11] Skaletskaya *et al.*, 2001. [12] Macen *et al.*, 1996. [13] Xu, Nash and McFadden, 2000. [14] Kawanishi, 1997. [15] Henderson *et al.*, 1991. [16] Smith *et al.*, 1990. [17] Smith *et al.*, 1991. [18] Reading, Khanna and Smith, 2002. [19] Hu, Smith and Pickup, 1994. [20] Smith *et al.*, 1996. [21] Loparev *et al.*, 1998. [22] Saraiva and Alcami, 2001. [23] Tidona and Darai, 1997. [24] Benedict *et al.*, 1999. [25] Chacon *et al.*, 1995. [26] Nogal *et al.*, 2001. [27] Clem, 2001. [28] Clem, Fechheimer and Miller, 1991. [29] Bertin *et al.*, 1996. [30] Crook, Clem and Miller, 1993. [31] Dbaibo and Hannun, 1998. [32] Tewari and Dixit, 1995. [33] Zhou *et al.*, 1997. [34] Dobbelstein and Shenk, 1996. [35] Turner *et al.*, 2000. [36] Sundararajan and White, 2001. [37] Henry *et al.*, 2002. [38] Henderson *et al.*, 1993. [39] Kawanishi, 1997. [40] Marshall *et al.*, 1999. [41] Sarid *et al.*, 1997. [42] Nava *et al.*, 1997. [43] Derfuss *et al.*, 1998. [44] Wang, Garvey and Cohen, 1999. [45] Afonso *et al.*, 1996. [46] Jerome *et al.*, 1999. [47] Goldmacher *et al.*, 1999. [48] Wolf *et al.*, 2001. [49] Tsukahara *et al.*, 1999. [50] Teodoro and Branton, 1997. [51] Thomas and Banks, 1998. [52] Thomas and Banks, 1999. [53] Pan and Griep, 1995. [54] Lane and Crawford, 1979. [55] Linzer and Levine, 1979. [56] Wang *et al.*, 1995.

7.   VIRUS COMPLEMENT-EVASION STRATEGIES

Complement is part of the innate immune system and is activated in a cascade manner through two main pathways, known as the classical and the alternative, and illustrated in Figure 6.

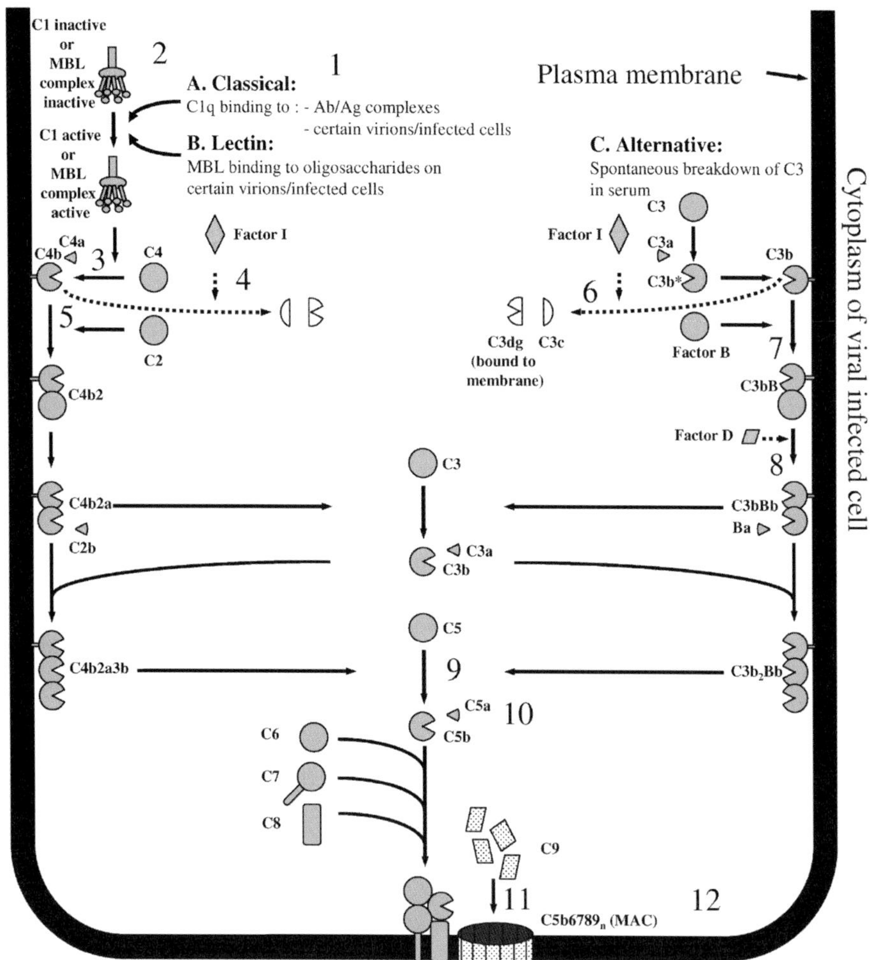

Figure 6. Virus complement evasion strategies.

Complement is part of the innate immune system and is activated in a cascade manner through two main pathways, known as the classical and the alternative pathways. The classical pathway is activated by the recognition proteins C1q or mannose-binding lectin, which bind respectively to charge clusters or neutral sugars on targets. In contrast, activation of the alternative pathway is a default process that proceeds unless down-regulated by specific mechanisms. Complement activation results in cleavage and activation of C3 and deposition of opsonic C3 fragments on surfaces. Subsequent cleavage of C5 leads to assembly of the membrane attack complex (C5b,6,7,8,9), which disrupts lipid bilayers. Viruses have developed different strategies acting at different stages of the complement cascade in order to evade complement-mediated destruction. These are listed in Table 6, and are referred to in this figure. These strategies fall into three main categories: (1) some viruses interfere with the classical pathway by avoiding complement binding to antibody-antigen complexes, either by shedding or internalization of these complexes from the cell surface or by expressing virally-encoded Fc receptor on the cell surface; (2) other viruses encode and express functional homologue of cellular regulators of complement activation (RCA), protecting their lipid envelope and the membrane of the infected cell; and (3) some viruses can incorporate host complement RCA in their envelope and/or up-regulate expression of these proteins in infected cells.

Complement activation on host cells is prevented by several membrane regulators of complement activation (RCA), the activity of which is predominantly restricted to complement of the same species, a phenomenon called homologous restriction. These proteins down-regulate complement activity at two steps in the classical and the alternative pathways: complement receptor 1 (CD35) and decay-accelerating factor (CD55) inhibit the formation and accelerate the decay of the classical pathway and alternative pathway C3-activating enzymes (C3 convertases); complement receptor 1 and membrane cofactor protein (CD46) act as cofactors for Factor I (a serum protease), which catabolizes C4b and C3b, thereby inhibiting formation of the C3 convertases C4b2a and C3bBb; finally, at the end of the complement cascade, CD59 and possibly also homologous restriction factor (C8-binding protein) prevent the formation of the membrane attack complex.

In general, micro-organisms lack mammalian RCA and thus cannot restrict complement deposition and amplification on their surfaces. However, the toxicity of the complement system has selected viruses that have acquired countermeasures. The viral strategies to evade complement-mediated destruction are summarized in Table 6. For a recent review of this topic, see that of Favoreel *et al.* (2003).

Table 6. Virus complement-evasion strategies.

Site [1]	Virus [2]	Viral gene	Mechanism of action	Source
1	SuHV-1	gE-gI	Shedding of viral protein-antibody complexes	[1]
1	SuHV-1	gB-gD	Internalization of viral protein-antibody complexes	[2]
2	HHV-1	gE-gI	Fc receptor activity	[3]
2	HHV-3	gE-gI	Fc receptor activity	[4]
2	SuHV-1	gE-gI	Fc receptor activity	[1]
2	HHV-5	UL118-UL119	Fc receptor activity	[5]
2	HHV-5	TRL11/IRL11	Fc receptor activity	[6]
2	MuHV-1	Fcr1	Fc receptor activity	[7]

Site [1]	Virus [2]	Viral gene	Mechanism of action	Source
2	S	TGEV	Fc receptor activity	[8]
2	S	MHV	Fc receptor activity	[8]
2	S	BCoV	Fc receptor activity	[8]
3	CPXV	IMP	Downregulates chemotactic proteins C3a, C4a, C5a	[9]
4	VACV	VCP	Cofactor for factor I	[10]
4	VARV	SPICE	Cofactor for factor I	[11]
5	VACV	VCP	Binds to C4b	[10]
5	VARV	SPICE	Binds to C4b	[11]
5	SaHV-2	ORF4	Inhibits formation and accelerates decay of classical and alternative C3 convertases	[12]
6	VACV	VCP	Cofactor for factor I	[10]
6	VARV	SPICE	Cofactor for factor I	[11]
6	HHV-4	?	Cofactor for factor I?	[13]
7	HHV-1, HHV-2	gC1, gC2	Binds human C3b	[14]
7	VACV	VCP	Binds to C3b	[10]
7	VARV	SPICE	Binds to C3b	[11]
7	SaHV-2	ORF4	Inhibits formation and accelerates decay of classical and alternative C3 convertases	[12]
7	SuHV-1	gC	Binds species-specific C3b	[15]
7	BoHV-1	gC	Binds species-specific C3b	[15]
7	EHV-1	gC	Binds species-specific C3b	[15]
7	EHV-2	gC	Binds species-specific C3b	[15]
8	HHV-1	gC1	Inhibits Factor D binding	[16]
9	HHV-1	gC1	Inhibits C5 binding	[16]
10	CPXV	IMP	Downregulates chemotactic proteins C3a, C4a, C5a	[9]
11	SaHV-2	ORF15	Homologue of CD59	[17]
12	HHV-5	?	Upregulation of CD55 and CD46	[18]
12	SuHV-2	?	Incorporation of cellular complement regulators	[19]
12	VACV	?	Incorporation of cellular complement regulators	[20]
12	HIV	?	Incorporation of cellular complement regulators	[21]
12	HTLV	?	Incorporation of cellular complement regulators	[22]
12	SINV	?	Incorporation of sialic acids	[23]

NOTES: (1) Site of action. Numbers refer to paths identified in Figure 6. (2) International Committee for Taxonomy of Viruses (ICTV) abbreviations.

SOURCES: [1] Favoreel *et al.*, 1997. [2] Favoreel *et al.*, 1999. [3] Watkins, 1964. [4] Ogata and Shigeta, 1979. [5] Lilley, Ploegh and Tirabassi, 2001. [6] Atalay *et al.*, 2002. [7] Thale *et al.*, 1994. [8] Oleszak *et al.*, 1993. [9] Howard *et al.*, 1998. [10] Kotwal *et al.*, 1990. [11] Rosengard *et al.*, 2002. [12] Fodor *et al.*, 1995. [13] Mold *et al.*, 1988. [14] Friedman *et al.*, 1984. [15] Huemer *et al.*, 1993. [16] Kostavasili *et al.*, 1997. [17] Rother *et al.*, 1994. [18] Spiller *et al.*, 1996. [19] Maeda *et al.*, 2002. [20] Vanderplasschen *et al.*, 1998. [21] Saifuddin *et al.*, 1995. [22] Spear *et al.*, 1995. [23] Hirsch, Griffin and Winkelstein, 1981.

8.   CONCLUSION

During the millions of years they have been co-evolving with their host, viruses have learned how to manipulate host immune control mechanisms. The review of the immune evasion strategies acquired by viruses revealed several fascinating aspects of this field. First, it is remarkable that individual virus families have targeted many common immunological principles. Second, the analysis of viral immunoregulatory proteins revealed that they belong to two classes: those encoded by genes with and those encoded by genes without sequence homology to cellular genes. While the former indicates that viruses have "stolen" genes from the host that were subsequently modified for the benefit of the virus, the latter suggests acquisitions through a mechanism of convergent evolution.

Viruses are obligate parasites that live "on the edge". On the one hand, they need to impair the immune response of their host to be able to replicate and to avoid eradication; but, on the other hand, they need to respect the host immune response in order to ensure their host's (and hence their own) survival. In other words, the perfect adaptation of a virus to its host would represent a virus able to complete its biological cycle without inducing clinical symptoms. Further studies are required to determine the roles of viral immune-evasion mechanisms in this delicate equilibrium. Indeed, most of the studies cited in this review have investigated the ability of viral genes to interfere with the host immune response *in vitro*. However, only *in vivo* experiments will be able to determine the real biological functions of these viral immune-evasion mechanisms. A beautiful example supporting this statement has been provided by the study of vaccinia virus IL-1β receptor. Indeed, while this viral product was thought to contribute to the pathogenicity of the virus, it is interesting to observe that deletion of the corresponding gene enhanced virus virulence and the onset of fever, suggesting that the purpose of a least some of the immune-evasion mechanisms is to reduce immunopathology caused by viral infection (Alcami and Smith, 1996).

In conclusion, this review highlights the complexity and the importance of viral immune-evasion strategies in the host-virus relationship.

ACKNOWLEDGEMENTS

L. Gillet and A. Vanderplasschen are Research Fellow and Senior Research Associate, respectively, of the *Fonds National Belge de la Recherche Scientifique* (FNRS).

REFERENCES

Ahn, K., Gruhler, A., Galocha, B., Jones, T.R., Wiertz, E.J.H.J., Ploegh, H.L., Peterson, P.A., Yang, Y. & Fruh, K. 1997. The EER-luminal domain of the HCMV glycoprotein US6 inhibits peptide translocation by TAP. *Immunity,* **6:** 613–621.

Ahn, K., Meyer, T.H., Uebel, S., Sempe, P., Djaballah, H., Yang, Y., Peterson, P.A., Fruh, K. & Tampe, R. 1996. Molecular mechanism and species specificity of TAP inhibition by herpes simplex virus protein ICP47. *EMBO Journal,* **15:** 3247–3255.

Ahn, K.S., Agulo, A., Ghazal, P., Peterson, P.A., Yang, Y. & Fruh, K. 1996. Human cytomegalovirus inhibits antigen presentation by a sequential multistep process. *Proceedings of the National Academy of Sciences, USA,* **93:** 10990–10995.

Albini, A., Ferrini, S., Benelli, R., Sforzini, S., Giunciuglio, D., Aluigi, M.G., Proudfoot, A.E.I., Alouani, S., Wells, T.N.C., Mariani, G., Rabin, R.L., Farber, J.M. & Noonan, D.M. 1998. HIV-1 Tat protein mimicry of chemokines. *Proceedings of the National Academy of Sciences, USA,* **95:** 13153–13158.

Alcami, A. 2003. Viral mimicry of cytokines, chemokines and their receptors. *Nature Reviews Immunology,* **3:** 36–50.

Alcami, A. & Smith, G.L. 1992. A soluble receptor for interleukin-1-beta encoded by vaccinia virus – a novel mechanism of virus modulation of the host response to infection. *Cell,* **71:** 153–167.

Alcami, A. & Smith, G.L. 1995. Vaccinia, cowpox, and camelpox viruses encode soluble gamma-interferon receptors with novel broad species specificity. *Journal of Virology,* **69:** 4633–4639.

Alcami, A. & Smith, G.L. 1996. A mechanism for the inhibition of fever by a virus. *Proceedings of the National Academy of Sciences, USA,* **93:** 11029–11034.

Alcami, A., Symons, J.A., Collins, P.D., Williams, T.J. & Smith, G.L. 1998. Blockade of chemokine activity by a soluble chemokine binding protein from vaccinia virus. *Journal of Immunology,* **160:** 624–633.

Alfonso, C.L., Neilan, J.G., Kutish, G.H. & Rock, D.L. 1996. An African swine fever virus Bcl-2 homolog, 5-HL, suppresses apoptotic cell death. *Journal of Virology,* **70:** 4858–4863.

Andresson, T., Sparkowski, J., Goldstein, D.J., & Schlegel, R. 1995. Vacuolar H+ATPase mutants transform cells and define a binding site for the papillomavirus E5 on oncoprotein. *Journal of Biological Chemistry,* **270:** 6830–6837.

Aoki, Y., Jaffe, E.S., Chang, Y., Jones, K., Teruya-Feldstein, J., Moore, P.S. & Tosato, G. 1999. Angiogenesis and hematopoiesis induced by Kaposi's sarcoma-associated herpesvirus-encoded interleukin-6. *Blood,* **93:** 4034–4043.

Arase, H., Mocarski, E.S., Campbell, A.E., Hill, A.B. & Lanier, L.L. 2002. Direct recognition of cytomegalovirus by activating and inhibitory NK cell receptors. *Science,* **296:** 1323–1326.

Arvanitakis, L., Geras Raaka, E., Varma, A., Gershengorn, M.C. & Cesarman, F. 1997. Human herpesvirus KSHV, encodes a constitutively active G-protein-coupled receptor linked to cell proliferation. *Nature,* **385:** 347–350.

Atalay, R., Zimmermann, Z., Wagner, M., Borst, E., Benz, C., Messerle, M. & Hengel, H. 2002. Identification and expression of human cytomegalovirus transcription units coding for two distinct Fc gamma receptor homologs. *Journal of Virology,* **76:** 8596–8608.

Beck, S. & Barrell, B.G. 1988. Human cytomegalo-virus encodes a glycoprotein homologous to MHC class-I antigens. *Nature,* **331:** 269–272.

Benedict, C.A., Butrovich, K.D., Lurain, N.S., Corbeil, J., Rooney, I., Schneider, P., Tschopp, J. & Ware, C.F. 1999. Cutting edge: A novel viral TNF receptor superfamily member in virulent strains of human cytomegalovirus. *Journal of Immunology,* **162:** 6967–6970.

Benedict, C.A., Norris, P.S. & Ware, C.F. 2002. To kill or be killed: viral evasion of apoptosis. *Nature Immunology,* **3:** 1013–1018.

Benedict, C.A., Norris, P.S., Prigozy, T.I., Bodmer, J.L., Mahr, J.A., Garnett, C.T., Martinon, F., Tschopp, J., Gooding, L.R. & Ware, C.F. 2001. Three adenovirus E3 proteins cooperate to evade apoptosis by tumor necrosis factor-related apoptosis-inducing ligand receptor-1 and -2. *Journal of Biological Chemistry,* **276:** 3270–3278.

Bennett, E.M., Bennink, J.R., Yewdell, J.W. & Brodsky, F.M. 1999. Cutting edge: Adenovirus E19 has two mechanisms for affecting class-I MHC expression. *Journal of Immunology,* **162:** 5049–5052.

Bertin, J., Armstrong, R.C., Ottilie, S., Martin, D.A., Wang, Y., Banks, S., Wang, G.H., Senkevich, T.G., Alnemri, E.S., Moss, B., Lenardo, M.J., Tomaselli, K.J. & Cohen, J.I. 1997. Death effector domain-containing herpesvirus and poxvirus proteins inhibit both Fas- and TNFR1-induced apoptosis. *Proceedings of the National Academy of Sciences, USA,* **94:** 1172–1176.

Bertin, J., Mendrysa, S.M., LaCount, D.J., Gaur, S., Krebs, J.F., Armstrong, R.C., Tomaselli, K.J. & Friesen, P.D. 1996. Apoptotic suppression by baculovirus P35 involves cleavage by and inhibition of a virus-induced CED-3/ICE-like protease. *Journal of Virology,* **70:** 6251–6259.

Bodaghi, B., Jones, T.R., Zipeto, D., Vita, C., Sun, L., Laurent, L., Arenzan-Seisdedos, F., Virelizier, J.L. & Michelson, S. 1998. Chemokine sequestration by viral chemoreceptors as a novel viral escape strategy: Withdrawal of chemokines from the environment of cytomegalovirus-infected cells. *Journal of Experimental Medicine,* **188:** 855–866.

Boname, J.M. & Stevenson, P.G. 2001. MHC class I ubiquitination by a viral PHD/LAP finger protein. *Immunity,* **15:** 627–636.

Born, T.L., Morrison, L.A., Esteban, D.J., Van den Bos, T., Thebeau, L.G., Chen, N.H., Spriggs, M.K., Sims, J.E. & Buller, R.M.L. 2000. A poxvirus protein that binds to and inactivates IL-18, and inhibits NK cell response. *Journal of Immunology,* **164:** 3246–3254.

Boshoff, C., Endo, Y., Collins, P.D., Takeuchi, Y., Reeves, J.D., Schweickart, V.L., Siani, M.A., Sasaki, T., Williams, T.J., Gray, P.W., Moore, P.S., Chang, Y. & Weiss, R.A. 1997. Angiogenic and HIV-inhibitory functions of KSHV-encoded chemokines. *Science,* **278:** 290–294.

Bryant, N.A., Davis-Poynter, N., Van der Plasschen, A. & Alcami, A. 2003. Glycoprotein G isoforms from some alpha herpesviruses function as broad-spectrum chemokine binding proteins. *EMBO Journal,* **22:** 833–846.

Burgert, H.G. & Kvist, S. 1987. The E3/19K protein of adenovirus type-2 binds to the domains of histocompatibility antigens required for CTL recognition. *EMBO Journal,* **6:** 2019–2026.

Chacon, M.R., Almazan, F., Nogal, M.L., Vinuela, E. & Rodriguez, J.F. 1995. The African swine fever virus IAP homolog is a late structural polypeptide. *Virology,* **214:** 670–674.

Chehimi, J., Bandyopodhyay, S., Prakash, K., Perussia, B., Hassan, N.F., Kawashima, H., Campbell, D., Kornbuth, J. & Starr, S.E. 1991. *In vitro* infection of natural killer cells with different human immunodeficiency virus type 1 isolates. *Journal of Virology,* **65:** 1812–1822.

Cho, Y.S., Kang, J.W., Cho, M., Cho, C.W., Lee, S., Choe, Y.K., Kim, Y., Choi, I., Park, S.N., Kim. S., Dinarello, C.A. & Yoon, D.Y. 2001. Down modulation of IL-18 expression by human papillomavirus type 16 E6 oncogene via binding to IL-18. *FEBS Letters,* **501:** 139–145.

Clem, R.J. 2001. Baculoviruses and apoptosis: the good, the bad, and the ugly. *Cell Death and Differentiation,* **8:** 137–143.

Clem, R.J., Fechheimer, M. & Miller, L.K. 1991. Prevention of apoptosis by a baculovirus gene during infection of insect cells. *Science,* **254:** 1388–1390.

Cohen, G.B., Gandhi, R.T., Davis, D.M., Mandelboim, O., Chen, B.K., Strominger, J.L. & Baltimore, D. 1999. The selective downregulation of class I major histocompatibility complex proteins by HIV-1 protects HIV-infected cells from NK cells. *Immunity,* **10:** 661–671.

Colamonici, O.R., Domanski, P., Sweitzer, S.M., Larner, A. & Buller, R.M.L. 1995. Vaccinia virus B18R gene encodes a type-I interferon-binding protein that blocks interferon-alpha transmembrane signaling. *Journal of Biological Chemistry,* **270:** 15974–15978.

Coscoy, L. & Ganem, D. 2001. A viral protein that selectively downregulates ICAM-1 and B7-2 and modulates T cell costimulation. *Journal of Clinical Investigation,* **107:** 1599–1606.

Coscoy, L., Sanchez, D.J. & Ganem, D. 2001. A novel class of herpesvirus-encoded membrane-bound E3 ubiquitin ligases regulates endocytosis of proteins involved in immune recognition. *Journal of Cell Biology,* **155:** 1265–1273.

Cosman, D., Fanger, N., Borges, L., Kubin, M., Chin, W., Peterson, L., Hsu, M.L. 1997. A novel immunoglobulin superfamily receptor for cellular and viral MHC class I molecules. *Immunity,* **7:** 273–282.

Cosman, D., Mullberg, J., Sutherland, C.L., Chin, W., Armitage, R., Fanslow, W., Kubin, M. & Chalupny, N.J. 2001. ULBPs, novel MHC class I-related molecules bind to CMV glycoprotein UL16 and stimulate KN cytotoxicity through the NKG2D receptor. *Immunity,* **14:** 123–133.

Cox, J.H., Bennink, J.R. & Yewdell, J.W. 1991. Retention of adenovirus-E19 glycoprotein in the endoplasmic reticulum is essential to its ability to block antigen presentation. *Journal of Experimental Medicine,* **174:** 1629–1637.

Cretney, E., Degli-Esposti, M.A., Densley, E.H., Farrell, H.E., Davis-Poynter, N.J. & Smyth, M.J. 1999. m144, a murine cytomegalovirus (MCMV)-encoded major histocompatibility complex class I homologue, confers tumor resistance to natural killer cell-mediated rejection. *Journal of Experimental Medicine,* **190:** 435–444.

Crook, N.E., Clem, R.J. & Miller, L.K. 1993. An apoptosis-inhibiting baculovirus gene with a zinc finger-like motif. *Journal of Virology,* **67:** 2168–2174.

Crotta, S., Stilla, A., Wack, A., D'Andrea, A., Nuti, S., D'Oro, U., Mosca, M., Filliponi, F., Brunnetto, R.M., Bonino, F., Abrignani, S. & Valiante, N.M. 2002. Inhibition of natural killer cells through engagement of CD81 by the major hepatitis C envelope protein. *Journal of Experimental Medicine,* **195:** 35–41.

Davis-Poynter, N.J., Lynch, D.M., Vally, H., Shellam, G.R., Rawlinson, W.D., Barrell, B.G. & Farrell, H.E. 1997. Identification and characterization of a G-protein-coupled receptor homolog encoded by murine cytomegalovirus. *Journal of Virology,* **71:** 1521–1529.

Dbaibo, G.S. & Hannun, Y.A. 1998. Cytokine response modifier A (CrmA): A strategically deployed viral weapon. *Clinical Immunology and Immunopathology,* **86:** 134–140.

Deane, D., McInnes, C.J., Percival, A., Wood, A., Thomson, J., Lear, A., Gilray, J., Fleming, S., Mercer, A. & Haig, D. 2000. Orf virus encodes a novel secreted protein inhibitor of granulocyte-macrophage colony-stimulating factor and interleukin-2. *Journal of Virology,* **74:** 1313–1320.

Derfuss, T., Fickenscher, H., Kraft, M.S., Henning, G., Lengenfelder, D., Fleckenstein, B. & Meinl, E. 1998. Antiapoptotic activity of the herpesvirus saimiri-encoded Bcl-2 homolog: Stabilization of mitochondria and inhibition of caspase-3-like activity. *Journal of Virology,* **72:** 5897–5904.

Dobbelstein, M. & Shenk, T. 1996. Protection against apoptosis by the vaccinia virus SPI-2 (B13R) gene product. *Journal of Virology,* **70:** 6479–6485.

Farrell, H.E., Vally, H., Lynch, D.M., Fleming, P., Shellam, G.R., Scalzo, A.A. & Davis Poynter, N.J. 1997. Inhibition of natural killer cells by a cytomegalovirus MHC class I homologue *in vivo. Nature,* **386:** 510–514.

Favoreel, H.W., Nauwynck, H.J., Halewyck, H.M., Van Oostveldt, P., Mettenleiter, T.C. & Pensaert, M.B. 1999. Antibody-induced endocytosis of viral glycoproteins and major histocompatibility complex class I on pseudorabies virus-infected monocytes. *Journal of General Virology,* **80:** 1283–1291.

Favoreel, H.W., Nauwynck, H.J., Van Oostveldt, P., Mettenleiter, T.C. & Pensaert, M.B. 1997. Antibody-induced and cytoskeleton-mediated redistribution and shedding of viral glycoproteins, expressed on pseudorabies virus-infected cells. *Journal of Virology,* **71:** 8254–8261.

Favoreel, H.W., Van de Walle, G.R., Nauwynck, H.J. & Pensaert, M.B. 2003. Virus complement evasion strategies. *Journal of General Virology,* **84:** 1–15.

Fleming, P., Davis-Poynter, N., Degli-Esposti, M., Densley, E., Papadimitriou, J., Shellam, G. & Farrell, H. 1999. The murine cytomegalovirus chemokine homolog, m131/129, is a determinant of viral pathogenicity. *Journal of Virology,* **73:** 6800–6809.

Fleming, S.B., McCaughan, C.A., Andrews, A.E., Nash, A.D. & Mercer, A.A. 1997. A homolog of interleukin-10 is encoded by the poxvirus orf virus. *Journal of Virology,* **71:** 4857–4861.

Fletcher, J.M., Prentice, H.G. & Grundy, J.E. 1998. Natural killer cell lysis of cytomegalovirus (CMV)-infected cells correlates with virally induced changes in cell surface lymphocyte function-associated antigen-3 (LFA-3) expression and not with the CMV-induced down-regulation of cell surface class I HLA. *Journal of Immunology,* **161:** 2365–2374.

Fodor, W.L., Rollins, S.A., Biancocaron, S., Rother, R.P., Guilmette, E.R., Burton, W.V., Albrecht, J.C., Fleckenstein, B. & Squinto, S.P. 1995. The complement control protein homolog of herpesvirus Saimiri regulates serum complement by inhibiting C3 convertase activity. *Journal of Virology,* **69:** 3889–3892.

Friedman, H.M., Cohen, G.H., Eisenberg, R.J., Seidel, C.A. & Cines, D.B. 1984. Glycoprotein-C of herpes simplex virus-1 acts as a receptor for the C3B complement on infected cells. *Nature,* **309:** 633–635.

Galocha, B., Hill, A., Barnett, B.C., Dolan, A., Raimondi, A., Cook, R.F., Brunner, J., McGeoch, D.J. & Ploegh, H.L. 1997. The active site of ICP47, a herpes simplex virus-encoded inhibitor of the major histocompatability complex (MHC)-encoded peptide transporter associated with antigen processing (TAP), maps to the NH2-terminal 35 residues. *Journal of Experimental Medicine,* **185:** 1565–1572.

Gardner, J.D., Tscharke, D.C., Reading, P.C. & Smith, G.L. 2002. Vaccinia virus semaphorin A39R is a 50–55 kDa secreted glycoprotein that affects the outcome of infection in a murine intradermal model. *Journal of General Virology,* **82:** 2083–2093.

Gewurz, B.E., Wang, E.W., Tortorella, D., Schiust, D.J. & Ploegh, H.L. 2001. Human cytomegalovirus US2 endoplasmic reticulum-lumenal domain dictates association with major histocompatibility complex class I in a locus-specific manner. *Journal of Virology,* **75:** 5197–5204.

Gilbert, M.J., Riddell, S.R., Li, C.R. & Greenberg, P.D. 1993. *Journal of Virology,* **67**: 3461–3469.

Glykofrydes, D., Niphuis, H., Kuhn, E.M., Rosenwirth, B., Heeney, J.L., Bruder, J., Niedobitek, G., Muller-Fleckstein, I., Fleckstein, B. & Ensser, A. 2000. Herpesvirus Saimiri vFLIP provides an antiapoptotic function but is not essential for viral replication, transformation, or pathogenicity. *Journal of Virology,* **74:** 11919–11927.

Goldmacher, V.S., Bartle, L.M., Skaletskaya, A., Dionne, C.A., Kedersha, E.S., Vater, C.A., Han, J.W., Lutz, R.J., Watanabe, S., McFarland, E.D.C., Kieff, E.D., Mocarski, E.S. & Chittenden, T. 1999. A cytomegalovirus-encoded mitochondria-localized inhibitor of apoptosis structurally unrelated to Bcl-2. *Proceedings of the National Academy of Sciences, USA,* **96:** 12536–12541.

Gooding, L.R., Elmore, L.W., Tollefson, A.E., Brady, H.A. & Wold, W.S.M. 1988. A 14,700 mw protein from the E3 region of adenovirus inhibits cytolysis by tumor necrosis factor. *Cell,* **53:** 341–346.

Gooding, L.R., Ranheim, T.S., Tollefson, A.E., Aquino, L., Duerksen-Hughes, P., Horton, T.M. & Wold, W.S.M. 1991. The 10,400-Dalton and 14,500-Dalton proteins encoded by region E3 of adenovirus function together to protect many but not all mouse-cell lines against lysis by tumor-necrosis-factor. *Journal of Virology,* **65:** 4114–4123.

Graham, K.A., Lalani, A.S., Macen, J.L., Ness, T.L., Barry, M., Liu, L.Y., Lucas, A., Clark Lewis, A., Moyer, R.W. & McFadden, G. 1997. The T1/35 kDa family of poxvirus-secreted proteins bind chemokines and modulate leukocyte influx into virus-infected tissues. *Virology,* **229:** 12–24.

Hegde, N.R., Chevalier, M.S., Johnson, D.C. 2003. Viral inhibition of MHC class II antigen presentation. *Trends in Immunology,* **24:** 278–285.

Hegde, N.R., Tomazin, R.A., Wisner, T.W., Dunn, C., Boname, J.M., Lewinsohn, D.M. & Johnson, D.C. 2002. Inhibition of HLA-DR assembly, transport and loading by human cytomegalovirus glycoprotein US3: a novel mechanism for evading major histocompatability complex class II antigen presentation. *Journal of Virology,* **76:** 10929–10941.

Henderson, S., Huen, D., Rowe, M., Dawson, C., Johnson, G. & Rickinson, A. 1993. Epstein-Barr virus-coded BHRF1 protein, a viral homolog of BCL-2, protects human B-cells from programmed cell-death. *Proceedings of the National Academy of Sciences, USA,* **90:** 8479–8483.

Henderson, S., Rowe, M., Gregory, C., Croom-Carter, D., Wang, F., Longnecker, R., Kiefe, E. & Rickinson, A. 1991. Induction of BCL-2 expression by Epstein-Barr-virus latent membrane protein-1 protects infected B-cells from programmed cell-death. *Cell,* **65:** 1107–1115.

Hengel, H., Flohr, T., Hammerling, G.J., Koszinowski, U.H. & Momburg, F. 1996. Human cytomegalovirus inhibits peptide translocation into the endoplasmic reticulum for MHC class I assembly. *Journal of General Virology,* **77:** 2287–2296.

Hengel, H., Koopmann, J.O., Flohr, T., Muranyi, W., Goulmy, E., Hammerling, G.J., Koszinowski, U.H. & Momburg, F. 1997. A viral ER-resident glycoprotein inactivates the MHC-encoded peptide transporter. *Immunity,* **6:** 623–632.

Henry, H., Thomas, A., Shen, Y. & White, E. 2002. Regulation of the mitrochondrial checkpoint in p53-mediated apoptosis confers resistance to cell death. *Oncogene,* **21:** 748–760.

Hewitt, E.W., Duncan, L., Mufti, D., Baker, J., Stevenson, P.G. & Lehner, P.J. 2002. Ubiquitylation of MHC class I by the K3 viral protein signals internalization and TSG101-depenedent degradation. *EMBO Journal,* **21:** 2418–2429.

Hinkley, S., Hill, A.B. & Srikumaran, S. 1998. Bovine herpesvirus-1 infection affects peptide transport activity in bovine cells. *Virus Research,* **53:** 91–96.

Hirsch, R.L., Griffin, D.E. & Winkelstein, J.A. 1981. Host modification of Sindbis virus sialic-acid content influences alternative complement pathway activation and virus clearance. *Journal of Immunology,* **127:** 1740–1743.

Howard, J., Justus, D.E., Totmenin, A.V., Shchelkunov, S. & Kotwal, G.J. 1998. Molecular mimicry of the inflammation modulatory proteins (IMPs) of poxviruses: evasion of the inflammatory response to preserve viral habitat. *Journal of Leukocyte Biology,* **64:** 68–71.

Hsu, D.H., Malefyt, R.D., Fiorentino, D.F., Dang, M.N., Vieira, P., DeVries, J., Spits, H., Mosmann, T.R. & Moore, K.W. 1990. Expression of interleukin-10 activity by Epstein-Barr virus protein BCRF1. *Science,* **250:** 830–832.

Hu, F.Q., Smith, C.A. & Pickup, D.J. 1994. Cowpox virus contains 2 copies of an early gene encoding a soluble secreted form of the Type-II TNF receptor. *Virology,* **204:** 343–356.

Huemer, H.P., Larcher, C., Littel van den Hurk, S.V. & Babiuk, L.A. 1993. Species selection interaction of alphaherpesvirinae with the unspecific immune-system of the host. *Archives of Virology,* **130:** 353–564.

Inngjerdingen, M., Damaj, B. & Maghazachi, A.A. 2001. Expression and regulation of chemokine receptors in human natural killer cells. *Blood,* **97:** 367–375.

Ishido, S., Choi, J.K., Lee, B.S., Wang, C.Y., DeMaria, M., Johnson, R.P., Cohen, G.B. & Jung J.U. 2000. Inhibition of natural killer cell-mediated cytotoxicity by Kaposi's sarcoma-associated herpesvirus K5 protein. *Immunity,* **13:** 365–374.

Ishido, S., Wang, C.Y., Lee, B.S., Cohen, G.B. & Jung, J.U. 2000. Downregulation of major histocompatibility complex class I molecules by Kaposi's sarcoma-associated herpesvirus K3 and K5 proteins. *Journal of Virology,* **74:** 5300–5309.

Jefferies, W.A. & Burgert, H.G. 1990. E3/19K from adenovirus-2 is an immunosubversive protein that binds to a structural motif regulating the intracellular transport of major histocompatibility complex class-I proteins. *Journal of Experimental Medicine,* **172:** 1653–1664.

Jerome, K.R., Fox, R., Chen, Z., Sears, A.E., Lee, H.Y. & Corey, L. 1999. Herpes simplex virus inhibits apoptosis through the action of two genes, Us5 and Us3. *Journal of Virology,* **73:** 8950–8957.

Johnson, J.M., Nicot, C., Fullen, J., Ciminal, V., Casareto, L., Mulloy, J.C., Jacobson, S. & Franchini, G. 2001. Free major histocompatability complex class I heavy chain is preferentially targeted for degradation by human T-cell leukemia/lymphotropic virus type 1 p12(I) protein. *Journal of Virology,* **75:** 6086–6094.

Jones, T.R., Wiertz, E.J.H.J., Sun, L., Nelson, J.A. & Ploegh, H.L. 1996. Human cytomegalovirus US3 impairs transport and maturation of major histocompatibility complex class-I heavy chains. *Proceedings of the National Academy of Sciences, USA,* **93:** 11327–11333.

Kavanagh, D.G., Gold, M.C., Wagner, M., Koszinowski, U.H. & Hill, A.B. 2001. The multiple immune-evasion genes of murine cytomegalovirus are not redundant: m4 and m152 inhibit antigen presentation in a complementary and cooperative fashion. *Journal of Experimental Medicine,* **194**: 967–977.

Kavanagh, D.G., Koszinowski, U.H. & Hill, A.B. 2001. The murine cytomegalovirus immune evasion protein m4/gp34 forms biochemically distinct complexes with class-I MHC at the cell surface and in a pre-golgi compartment. *Journal of Immunology,* **167**: 3894–3902.

Kawanishi, M. 1997. Epstein-Barr virus BHRF1 protein protects intestine 407 epithelial cells from apoptosis induced by tumor necrosis factor alpha and anti-Fas antibody. *Journal of Virology,* **71**: 3319–3322.

Kawanishi, M. 1997. Expression of Epstein-Barr virus latent membrane protein 1 protects Jurkat T cells from apoptosis induced by serum deprivation. *Virology,* **228**: 244–250.

Kerkau, T., Bacik, I., Bennink, J.R., Yewdell, J.W., Hunig, T., Schimpl, A. & Schubert, U. 1997. The human immunodeficiency virus type 1 (HIV-1) vpu protein interferes with an early step in the biosynthesis of major histocompatibility complex (MHC) class I molecules. *Journal of Experimental Medicine,* **185**: 1295–1305.

Kledal, T.N., Rosenkilde, M.M., Coulin, F., Simmons, G., Johnsen, A.H., Alouani, S., Power, C.A., Luttichau, H.R., Gerstoft, J., Clapham, P.R., Clark-Lewis, I., Wells, T.N.C. & Schwartz, T.W. 1997. A broad-spectrum chemokine antagonist encoded by Kaposi's sarcoma-associated herpesvirus. *Science,* **277**: 1656–1659.

Kleijnen, M.F., Huppa, J.B., Lucin, P., Mukherjee, S., Farrell, H., Campbell, A.E., Koszinowski, U.H., Hill, A.B. & Ploegh, H.L. 1997. A mouse cytomegalovirus glycoprotein, gp34, forms a complex with folded class I MHC molecules in the ER which is not retained but is transported to the cell surface. *EMBO Journal,* **16**: 685–694.

Kloover, J.S., Grauls, G.E.L.M., Blok, M.J., Vink, C. & Bruggeman, C.A. 2002. A rat cytomegalovirus strain with a disruption of the r144 MHC class I-like gene is attenuated in the acute phase of infection in neonatal rats. *Archives of Virology,* **147**: 813–824.

Kostavasili, I., Sahu, A., Friedman, H.M., Eisenberg, R.J., Cohen, G.H. & Lambris, J.D. 1997. Mechanism of complement inactivation by glycoprotein C of herpes simplex virus. *Journal of Immunology,* 158: 1763–1771.

Kotenko, S.V., Saccani, S., Izotova, L.S., Mirochinitchenko, O.V. & Pestka, S. 2000. Human cytomegalovirus harbors its own unique IL-10 homolog (cmcIL-10). *Proceedings of the National Academy of Sciences, USA,* **97**: 1695–1700.

Kotwal, G.J., Isaacs, S.N., McKenzie, R., Frank, M.M. & Moss, B. 1990. Inhibition of the complement cascade by the major secretory protein of vaccinia virus. *Science,* **250**: 827–830.

Krathwohl, M.D., Hromas, R., Brown, D.R., Broxmeyer, H.E. & Fife, K.H. 1997. Functional characterization of the C-C chemokine-like molecules encoded by molluscum contagiosum virus types 1 and 2. *Proceedings of the National Academy of Sciences, USA,* **94**: 9875–9880.

Krmpotic, A., Busch, D.H., Bubic, I., Gebhardt, F., Hengel, H., Hasan, M., Scalzo, A.A., Koszinowski, U.H. & Jonjic, S. 2002. MCMV glycoprotein gb40 confers virus resistance to CD8(+) T cells and NK cells *in vivo. Nature Immunology,* **3**: 529–535.

Kubin, M., Cassiano, L., Chalupny, J., Chin, W., Cosman, D, Fanslow, W., Mulberg, J., Rousseau, A.M., Ulrich, D. & Armitage, R. 2001. ULBP1, 2, 3: novel MHC class I-related molecules that bind to human cytomegalovirus glycoprotein UL16, activate NK cells. *European Journal of Immunology,* **31**: 1428–1437.

Kubota, A., Kubota, S., Farrell, H.E., Davis Poynter, N. & Takei, F. 1999. Inhibition of NK cells by murine CMV-encoded class I MHC homologue m144. *Cellular Immunology,* **191**: 145–151.

Lane, D.P. & Crawford, L.V. 1979. T-antigen is bound to a host protein in SV-40-transformed cells. *Nature,* **278:** 261–263.

Le Gall, S., Erdtmann, L., Benichou, S., Berlioz-Torrent, C., Liu, L.X., Benarous, R., Heard, J.M. & Schwartz, O. 1998. Nef interacts with the mu subunit of clathrin adaptor complexes and reveals a cryptic sorting signal in MHC I molecules. *Immunity,* **8:** 483–495.

Lee, H.J., Essani, K. & Smith, G.L. 2001. The genome sequence of Yaba-like disease virus. a yatapoxvirus. *Virology,* **281:** 170–92.

Lee, S.J., Cho, Y.S., Cho, M.C., Shim, J.H., Lee, K.A., Ko, Y.K., Park, S.N., Hoshino, T., Kim, S., Dinarello, C.A. & Yoon, D.Y. 2001. Both E6 and E7 oncoproteins of human papillomavirus 16 inhibit IL-18-induced IFN-gamma production in human peripheral blood mononuclear and NK cells. *Journal of Immunology,* **167:** 497–504.

Lehner, P.J., Karttunen, J.T., Wilkinson, G.W.G. & Cresswell, P. 1997. The human cytomegalovirus US6 glycoprotein inhibits transporter associated with antigen processing-dependent peptide translocation. *Proceedings of the National Academy of Sciences, USA,* **94:** 6904–6909.

Leonard, G.T. & Sen, G.C. 1996. Effects of adenovirus E1A protein on interferon signaling. *Virology,* **224:** 25–33.

Leong, C.C., Chapman, T.L., Bjorkman, P.J., Formankova, D., Mocarski, E.S., Phillips, J.H. & Lanier, L.L. 1998. Modulation of the natural killer cell cytotoxicity in human cytomegalovirus infection: The role of endogenous class I major histocompatibility complex and a viral class I homolog. *Journal of Experimental Medicine,* **187:** 1681–1687. [See also correction: *ibid.,* **188:** 615.]

Levitskaya, J., Coram, M., Levitsky, V., Imreh, S., Steiger-Waldmullen, P.M., Klein, G., Kurill, M.G. & Masucci, M.G. 1995. Inhibition of antigen processing by the internal repeat region of the Epstein-Barr virus nuclear antigen-1. *Nature,* **375:** 685–688.

Lewandowski, G.A., Lo, D. & Bloom, F.E. 1993. Interference with major histo-compatibility complex class II restricted antigen presentation in the brain by herpes-simplex virus type 1 – A possible mechanism of evasion of the immune response. *Proceedings of the National Academy of Sciences, USA,* **90:** 2005–2009.

Li, Y.G., Kang, J. & Horowitz, M.S. 1998. Interaction of an adenovirus E3 14.7-kiloDalton protein with a novel tumor necrosis factor alpha-inducible cellular protein containing leucine zipper domains. *Molecular and Cellular Biology,* **18:** 1601–1610.

Lilley, B.N., Ploegh, H.L. & Tirabassi, R.S. 2001. Human cytomegalovirus open reading frame TRL11/IRL11 encodes an immunoglobulin G Fc-binding protein. *Journal of Virology,* **75:** 11218–11221.

Linzer, D.I.H. & Levine, A.J. 1979. Characterization of a 54 kDalton cellular SV40 tumor-antigen present in SV40-transformed cells and uninfected embryonal carcinoma-cells. *Cell,* **17:** 43–52.

Loparev, V.N., Parsons, J.M., Knight, J.C., Panus, J.F., Ray, C.A., Buller, R.H.L., Pickup, D.J. & Esposito, J.J. 1998. A third distinct tumor necrosis factor receptor of orthopoxviruses. *Proceedings of the National Academy of Sciences, USA,* **95:** 3786–3791.

Lopez-Botet, M., Llano, M. & Ortega, M. 2001. Human cytomegalovirus and natural killer-mediated surveillance of HLA class I expression: a paradigm of host-pathogen adaptation. *Immunological Reviews,* **181:** 193–202.

Luttichau, H.R., Stine, J., Boesen, T.P., Johnsen, A.H., Chantry, D., Gerstoft, J. & Schwartz, T.W. 2000. A highly selective CC chemokine receptor (CCR)8 antagonist encoded by the poxvirus molluscum contagiosum. *Journal of Experimental Medicine,* **191:** 171–180.

Lybarger, L., Wang, X.L., Harris, M.R., Virgin, H.W. & Hansen, T.H. 2003. Virus subversion of the MHC class I peptide-loading complex. *Immunity,* **18:** 121–130.

Macen, J.L., Graham, K.A., Lee, S.F., Schreiber, M., Boshkov, L.K. & McFadden, G. 1996. Expression of the myxoma virus tumor necrosis factor receptor homologue and M11L genes required to prevent virus-induced apoptosis in infected rabbit lymphocytes. *Virology,* **218:** 232–237.

Machold, R.P., Wiertz, E.J.H.J., Jones, T.R. & Ploegh, H.L. 1997. The HCMV gene products US11 and US2 differ in their ability to attack allelic forms of murine major histocompatibility complex (MHC) class I heavy chains. *Journal of Experimental Medicine,* **185:** 363–366.

Maeda, K., Hayashi, S., Tanioka, Y., Matsumoto, Y. & Otsuka, H. 2002. Pseudorabies virus (PRV) is protected from complement attack by cellular factors and glycoprotein C (gC). *Virus Research,* **84:** 79–87.

Mandelboim, O., Lieberman, N., Lev, M., Paul, L., Arnon, T.I., Bushkin, Y., Davis, D.M., Strominger, J.L., Yewdell, J.W. & Porgador, A. 2001. Recognition of haemagglutinins on virus-infected cells by NKp46 activates lysis by human NK cells. *Nature,* **409:** 1055–1060.

Marshall, W.L., Yim, C., Gustafson, E., Graf, T., Sage, D.R., Hanify, K., Williams, L., Fingeroth, J. & Finberg, R.W. 1999. Epstein-Barr virus encodes a novel homolog of the bcl-2 oncogene that inhibits apoptosis and associates with Bax and Bak. *Journal of Virology,* **73:** 5181–5185.

Meyer, M., Clauss, M., Lepple-Wienhues, A., Waltenberger, J., Augustin, H.G., Ziche, M., Lanz, C., Buttner, M., Rziha, H.J. & Dehio, C. 1999. A novel vascular endothelial growth factor encoded by Orf virus, VEGF-E, mediates angiogenesis via signalling through VEGFR-2 (KDR) but not VEGFR-1 (Flt-1) receptor tyrosine kinases. *EMBO Journal,* **18:** 363–374.

Miller, D.M., Rahill, B.M., Lairmore, M.D., Durbin, J.E., Waldman, W.J. & Sedmark, D.D. 1998. Human cytomegalovirus inhibits major histocompatability complex class II expression by disruption of the Jak/Stat pathway. *Journal of Experimental Medicine,* **187:** 675–683.

Miller, D.M., Zhang, Y.X., Rahill, B.M., Waldman, W.J. & Sedmak, D.D. 1999. Human cytomegalovirus inhibits IFN-alpha-stimulated antiviral and immunoregulatory responses by blocking multiple levels of IFN-alpha signal transduction. *Journal of Immunology,* **162:** 6107–6113.

Milne, R.S.B., Mattick, C., Nicholson, L., Devaraj, P., Alcami, A. & Gompels, U.A. 2000. RANTES binding and down-regulation by a novel human herpesvirus-6 beta chemokin receptor. *Journal of Immunology,* **164:** 2396–2404.

Mold, C., Bradt, B.M., Nemerow, G.R. & Cooper, N.R. 1988. Epstein-Barr virus regulates activation and processing of the 3RD component of complement. *Journal of Experimental Medicine,* **168:** 949–969.

Moore, K.W., Viera, P., Fiorentino, D.F., Trounstine, M.L., Khan, T.A. & Mosmann, T.R. 1990. Homology of cytokine synthesis inhibitory factor (IL-10) to the Epstein-Barr virus gene BCRFI. *Science,* **248:** 1230–1234.

Mossman, K., Nation, P., Macen, J., Garbutt, M., Lucas, A. & McFadden, G. 1996. Myxoma virus M-T7. a secreted homolog of the interferon-gamma receptor, is a critical virulence factor for the development of myxomatosis in European rabbits. *Virology,* **215:** 17–30.

Murphy, P.M. 2001. Viral exploitation and subversion of the immune system through chemokine mimicry. *Nature Immunology,* **2:** 116–122.

Nava, V.E., Cheng, E.H.Y., Veliuona, M., Zhou, S.F., Clem, R.J., Mayer, M.L. & Hardwick, J.M. 1997. Herpesvirus Saimiri encodes a functional homolog of the human bcl-2 oncogene. *Journal of Virology,* **71:** 4118–4122.

Neumann, J., Eis-Hubinger, A.M. & Koch, N. 2003. Herpes simplex virus type 1 targets the MHC class II processing pathway for immune evasion. *Journal of Immunology,* **171:** 3075–3083.

Ng, A., Tscharke, D.C., Reading, P.C. & Smith, G.L. 2001. The vaccinia virus A41L protein is a soluble 30 kDa glycoprotein that affects virus virulence. *Journal of General Virology,* **82:** 2095–2105.

Nogal, M.L., de Buitrago, G.G., Rodriguez, C., Cubelos, B., Carrascosa, A.L., Salas, M.L. & Revilla, Y. 2001. African swine fever virus IAP homolog inhibits caspase activation and promotes cell survival in mammalian cells. *Journal of Virology,* **75:** 2535–2543.

Ogata, M. & Shigeta, S. 1979. Appearance of immunoglobulin-GFC receptor in cultured human cells infected with varicella-zoster virus. *Infection and Immunity,* **26:** 770–774.

Oleszak, E.L., Perlman, S., Parr, R., Collisson, E.W. & Leibowitz, J.L. 1993. Molecular mimicry between S peplomer proteins of coronaviruses (MHV, BCV, TGEV and IBV) and Fc receptor. *Advances in Experimental Medicine and Biology,* **342**: 183–188.

Orange, J.S., Fassett, M.S., Koopman, L.A., Boyson, J.E. & Strominger, J.L. 2002. Viral evasion of natural killer cells. *Nature Immunology,* **3:** 1006–1012.

Pan, H. & Griep, A.E. 1995. Temporally distinct patterns of P53-dependent and P53-independent apoptosis during mouse lens development. *Genes and Development,* **9:** 2157–2169.

Panus, J.F., Smith, C.A., Ray, C.A., Smith, T.D., Patel, D.D. & Pickup, D.J. 2002. Cowpox virus encodes a fifth member of the tumor necrosis factor receptor family: A soluble, secreted CD30 homologue. *Proceedings of the National Academy of Sciences, USA,* **99:** 8348–8353.

Parcells, M.S., Lin, S.F., Dienglewicz, R.L., Majerciak, V., Robinson, D.R., Chen, H.C., Wu, Z.N., Dubyak, G.R., Brunovskis, P., Hunt, H.D., Lee, L.F. & Kung, H.J. 2001. Marek's disease virus (MDV) encodes an interleukin-8 homolog (vIL-8): characterization of the vIL-8 protein and a vIL-8 deletion mutant MDV. *Journal of Virology,* **75:** 5159–5173.

Park, B., Oh, H., Lee, S., Song, Y.S., Shin, J., Sung, Y.C., Hwang, S.Y. & Ahn, K. 2002. The MHC class I homolog of human cytomegalovirus is resistant to down-regulation mediated by the unique short region protein (US)2, US3, US6, and US11 gene products. *Journal of Immunology,* **168:** 3464–3469.

Parry, C.M., Simas, J.P., Smith, V.P., Stewart, C.A., Minson, A.C., Efstathoiu, S. & Alcami, A. 2000. A broad spectrum secreted chemokine binding protein encoded by a herpesvirus. *Journal of Experimental Medicine,* **191:** 573–578.

Penfold, M.E.T., Dairaghi, D.J., Duke, G.M., Saederup, N., Mocarski, E.S., Kemble, G.W. & Schall, T.J. 1999. Cytomegalovirus encodes a potent alpha chemokine. *Proceedings of the National Academy of Sciences, USA,* **96:** 9839–9844.

Poggi, A., Carosio, R., Spaggiari, G.M., Fortis, C., Tambussi, G., Dell'Antonio, G., Dal Cin, E., Rubartelli, A. & Zocchi, M.R. 2002. NK cell inactivation by dendritic cells is dependent on LFA-1-mediated induction of calcium-calmodulin kinase II: Inhibition by HIV-1 Tat C-terminal domain. *Journal of Immunology,* **168:** 95–101.

Rakoff-Nahoum, S., Chen, H.C., Kraus, T., George, I., Oei, E., Tyorkin, M., Salik, E., Beuria, P. & Sperber, K. 2001. Regulation of class II expression in monocytic cells after HIV-1 infection. *Journal of Immunology,* **167:** 2331–2342.

Rappocciolo, G., Birch, J. & Ellis, S.A. 2003. Down-regulation of MHC class I expression by equine herpesvirus-1. *Journal of General Virology,* **84:** 293–300.

Reading, P.C., Khanna, A. & Smith, G.L. 2002. Vaccinia virus CrmE encodes a soluble and cell surface tumor necrosis factor receptor that contributes to virus virulence. *Virology,* **292:** 285–298.

Reusch, U., Muranyi, W., Lucin, P., Burgert, H.G., Hengel, H. & Koszinowski, U.H. 199. A cytomegalovirus glycoprotein re-routes MHC class I complexes to lysosomes for degradation. *EMBO Journal,* **18:** 1081–1091.

Reyburn, H.T., Mandelboim, O., Vales Gomez, M., Davis, D.M., Pazmany, L. & Strominger, J.L. 1997. The class I MHC homologue of human cytomegalovirus inhibits attack by natural killer cells. *Nature,* **386:** 514–517.

Rode, H.J., Janssen, W., Rosenwolff, A., Bugert, J.J., Thein, P., Becker, Y. & Darai, G. 1993. The genome of equine herpesvirus type-2 harbors an interleukin-10 (IL10)-like gene. *Virus Genes,* **7:** 111–116.

Rosengard, A.M., Liu, Y., Nie, Z.P. & Jimenez, R. 2002. Variola virus immune evasion design: Expression of a highly efficient inhibitor of human complement. *Proceedings of the National Academy of Sciences, USA,* **99:** 8808–8813.

Rother, R.P., Rollins, S.A., Fodor, W.L., Albrecht, J.C., Setter, E., Fleckenstein, B. & Squinto, S.P. 1994. Inhibition of complement-mediated cytolysis by the terminal complement inhibitor of herpesvirus Saimiri. *Journal of Virology,* **68:** 730–737.

Saederup, N., Aguirre, S.A., Sparer, T.E., Bouley, D.M. & Mocarski, E.S. 2001. Murine cytomegalovirus CC chemokine homolog MCK-2 (m131-129) is a determinant of dissemination that increases inflammation at initial sites of infection. *Journal of Virology,* **75:** 9966–9976.

Saederup, N., Lin, Y.C., Dairaghi, D.J., Schall, T.J. & Mocarski, E.S. 1999. Cytomegalovirus-encoded beta chemokine promotes monocyte-associated viremia in the host. *Proceedings of the National Academy of Sciences, USA,* **96:** 10881–10886.

Saifuddin, M., Parker, C.J., Peeples, M.E., Gorny, M.K., Zollapazner, S., Ghassemi, M., Rooney, I.A., Atkinson, J.P. & Spear, G.T. 1995. Role of virion-associated glycosylphosphatidylinositol-linked proteins CD55 and CD59 in complement resistance of cell line-derived and primary isolates of HIV-L. *Journal of Experimental Medicine,* **182:** 501–509.

Saraiva, M. & Alcami, A. 2001. CrmE, a novel soluble tumor necrosis factor receptor encoded by poxviruses. *Journal of Virology,* **75:** 226–233.

Saraiva, M., Smith, P., Fallon, P.G. & Alcami, A. 2002. Inhibition of type 1 cytokine-mediated inflammation by a soluble CD30 homologue encoded by ectromelia (mousepox) virus. *Journal of Experimental Medicine,* **196:** 829–839.

Sarid, R., Sato, T., Bohenzky, R.A., Russo, J.J. & Chang, Y. 1997. Kaposi's sarcoma-associated herpesvirus encodes a functional Bcl-2 homologue. *Nature Medicine,* **3:** 293–298.

Schindler, M., Wurfl, S., Benaroch, P., Greenough, T.C., Daniels, R., Easterbrook, P., Brenner, M., Munch, J. & Kirchhoff, F. 2003. Down-modulation of mature major histocompatibility complex class II and up-regulation of invariant chain cell surface expression are well-conserved functions of human and simian immunodeficiency virus nef alleles. *Journal of Virology,* **77:** 10548–10556.

Schust, D.J., Tortorella, D., Seebach, J., Phan, C. & Ploegh, H.L. 1998. Trophoblast class I major histocompatibility complex (MHC) products are resistant to rapid degradation of the human cytomegalovirus (HCMV) gene products US2 and US11. *Journal of Experimental Medicine,* **188:** 497–503. [See also correction: *ibid.,* **188:** 497.]

Senkevich, T.G. & Moss, B. 1998. Domain structure, intracellular trafficking, and beta 2-microglobulin binding of a major histocompatibility complex class I homolog encoded by molluscum contagiosum virus. *Virology,* **250:** 397–407.

Shisler, J., Yang, C., Walter, B., Ware, C.F. & Gooding, L.R. 1997. The adenovirus E3-10.4K/14.5K complex mediates loss of cell surface Fas (CD95) and resistance to fas-induced apoptosis. *Journal of Virology,* **71:** 8299–8306.

Shisler, J.L. & Moss, B. 2001. Molluscum contagiosum virus inhibitors of apoptosis: The MC159 v-FLIP protein blocks Fas-induced activation of procaspases and degradation of the related MC160 protein. *Virology,* **282:** 14–25.

Skaletskaya, A., Bartle, L.M., Chittenden, T., McCormick, A.L., Mocarski, E.S. & Goldmacher, V.S. 2001. A cytomegalovirus-encoded inhibitor of apoptosis that suppresses caspase-8 activation. *Proceedings of the National Academy of Sciences, USA,* **98:** 7829–7834.

Smith, C.A., Davis, T., Anderson, D., Solam, L., Beckman, M.P., Jerzy, R., Dower, S.K., Cosman, D. & Goodwin, R.G. 1990. A receptor for tumor necrosis factor defines an unusual family of cellular and viral proteins. *Science,* **248:** 1019–1023.

Smith, C.A., Davis, T., Wignall, J.M., Din, W.S., Farrah, T., Upton, C., McFadden, G. & Goodwin, R.G. 1991. T2 open reading frame from the Shope fibroma virus encodes a soluble form of the TNF receptor. *Biochemical and Biophysical Research Communications,* **176:** 335–342.

Smith, C.A., Hu, F.Q., Smith, T.D., Richards, C.L., Smolak, P., Goodwin, R.G. & Pickup, D.J. 1996. Cowpox virus encodes a second soluble homologue of cellular TNF receptors, distinct from CrmB, that binds TNF but not LT alpha. *Virology,* **223:** 132–147.

Smith, C.A., Smith, T.D., Smolak, P.J., Friend, D., Hagen, H., Gerhart, M., Park, L., Pickup, D.J., Torrance, D., Mohler, K., Scooley, K. & Goodwin, R.G. 1997. Poxvirus genomes encode a secreted, soluble protein that preferentially inhibits beta chemokine activity yet lacks sequence homology to known chemokine receptors. *Virology,* **236:** 316–327.

Smith, H.R., Idris, A.H. & Yokoyama, W.M. 2001. Murine natural killer cell activation/receptors. *Immunological Reviews,* **181:** 115–125.

Smith, V.P., Bryant, N.A. & Alcami, A. 2000. Ectromelia, vaccinia and cowpox viruses encode secreted interleukin-18-binding proteins. *Journal of General Virology,* **81:** 1223–1230.

Spear, G.T., Lurain, N.S., Parker, C.J., Ghassemi, M., Payne, G.H. & Saifuddin, M. 1995. Host cell-derived complement control proteins CD55 and CD59 are incorporated into the virions of 2 unrelated enveloped viruses – human T-cell leukemia/lymphoma virus type-I (HTLV-I) and human cytomegalovirus (HCMV). *Journal of Immunology,* **155:** 4376–4381.

Spiller, O.B., Morgan, B.P., Tufaro, F. & Devine, D.V. 1996. Altered expression of host-encoded complement regulators on human cytomegalovirus-infected cells. *European Journal of Immunology,* **26:** 1532–1538.

Stine, J.T., Wood, C., Hill, M., Epp, A., Raport, C.J., Schweickart, V.L., Endo, Y., Sasaki, T., Simmons, G., Boshoff, C., Clapham, P., Chang, Y., Moore, P., Gray, P.W. & Chantry, D. 2000. KSHV-encoded CC chemokine vMIP-III is a CCR4 agonist, stimulates angiogenesis, and selectively chemoattracts TH2 cells. *Blood,* **95:** 1151–1157.

Strockbine, L.D., Cohen, J.L., Farrah, T., Lyman, S.D., Wagener, F., DuBose, R.F., Armitage, R.J. & Spriggs, M.K. 1998. The Epstein-Barr virus BARF1 gene encodes a novel, soluble colony-stimulating factor-1 receptor. *Journal of Virology,* **72:** 4015–4021.

Stumptner-Cuvelette, P., Morchoisne, S., Dugast, M., Le Gall, S., Raposo, G., Schwartz, O. & Benaroch, P. 2001. HIV-1 Nef impairs MHC class II antigen presentation and surface expression. *Proceedings of the National Academy of Sciences, USA,* **98:** 12144–12149.

Sturzl, M., Hohenadl, C., Zietz, C., Castanos-Velez, E., Wunderlich, A., Ascherl, G., Biberfeld, P., Monini, P., Browning, P.J. & Ensoli, B. 1999. Expression of K13/v-FLIP gene of human herpesvirus 8 and apoptosis in Karposi's sarcoma spindle cells. *Journal of the National Cancer Institute,* **91:** 1725–1733.

Sundararajan, R. & White, E. 2001. E1B 19K blocks Bax oligomerization and tumor necrosis factor alpha-mediated apoptosis. *Journal of Virology,* **75:** 7506–7516.

Sutherland, C.L., Chalupny, N.J. & Cosman, D. 2001. The UL16-binding proteins, a novel family of MHC class I-related ligands for NKG2D, activate natural killer cell functions. *Immunological Reviews,* **181:** 185–192.

Swigut, T., Iafrate, A.J., Muench, J., Kirchhoff, F. & Skowronski, J. 2000. Simian and human immunodeficiency virus Nef proteins use different surfaces to downregulate class I major histocompatibility complex antigen expression. *Journal of Virology,* **74:** 5691–5701.

Teodoro, J.G. & Branton, P.E. 1997. Regulation of p53-dependent apoptosis, transcriptional repression, and cell transformation by phosphorylation of the 55-kiloDalton E1B protein of human adenovirus type 5. *Journal of Virology,* **71:** 3620–3627.

Tewari, M. & Dixit, V.M. 1995. Fas-induced and tumor necrosis factor-induced apoptosis is inhibited by the poxvirus CrmA gene-product. *Journal of Biological Chemistry,* **270:** 3255–3260.

Thale, R., Lucin, P., Schneider, K., Eggers, M. & Koszinowski, U.H. 1994. Identification and expression of a murine cytomegalovirus early gene coding for an FC receptor. *Journal of Virology,* **68:** 7757–7765.

Thomas, M. & Banks, L. 1998. Inhibition of Bak-induced apoptosis by HPV-18 E6. *Oncogene,* **17:** 2943–2954.

Thomas, M. & Banks, L. 1999. Human papillomavirus (HPV) E6 interactions with Bak are conserved amongst E6 proteins from high and low risk HPV types. *Journal of General Virology,* **80:** 1513–1517.

Tidona, C.A. & Darai, G. 1997. The complete DNA sequence of lymphocystis disease virus. *Virology,* **230:** 207–216.

Tomasec, P., Braud, V.M., Rickards, C., Powell, M.B., McSharry, B.P., Gaddola, S., Cerundolo, V., Borysiewicz, L.K., McMicheal, A.J. & Wilkinson, G.W.G. 2000. Surface expression of HLA-E, an inhibitor of natural killer cells, enhanced by human cytomegalovirus gpUL40. *Science,* **287:** 1031–1033.

Tomazin, R., Boname, J., Hegde, N.R., Lewinsohn, D.H., Altschuler, Y., Jones, T.R., Cresswell, P., Nelson, J.A.. Riddell, S.R., & Johnson, D.C. 1999. Cytomegalovirus US2 destroys two components of the MHC class II pathway, preventing recognition by CD4(+) T cells. *Nature Medicine,* **5:** 1039–1043.

Tong, X., Boll, W., Kirchhausen, T. & Howley, P.M. 1998. Interaction of the bovine papillomavirus E6 protein with the clathrin adaptor complex AP-1. *Journal of Virology,* **72:** 476–482.

Tortorella, D., Gewurz, B.E., Furman, M.H., Schust, D.J. & Pleogh, H.L. 2000. Viral subversion of the immune system. *Annual Review of Immunology,* **18:** 861–926.

Tripp, R.A., Jones, L.P., Haynes, L.M., Zheng, H.Q., Murphy, P.M. & Anderson, L.J. 2001. CX3C chemokine mimicry by respiratory syncytial virus G glycoprotein. *Nature Immunology,* **2:** 732–738.

Tseng, C.T. & Klimpel, G.R. 2002. Binding of hepatitis C envelope protein E2 to CD81 inhibits natural killer cell functions. *Journal of Experimental Medicine,* **195:** 43–49.

Tsukahara, T., Kannagi, M., Ohashi, T., Kato, H., Arai, M., Nunez, G., Iwanaga, Y., Yamamoto, N., Ohtani, K., Nakamura, M. & Fujii, M. 1999. Induction of Bcl-x(L) expression by human T-cell leukemia virus type 1 tax through NF-kappa B in apoptosis-resistant T-cell transfectants with tax. *Journal of Virology,* **73:** 7981–7987.

Tulman, E.R., Alfonso, C.L., Lu, Z., Zsak, L., Kutish, G.F. & Rock, D.L. 2001. Genome of lumpy skin disease virus. *Journal of Virology,* **75:** 7122–7130.

Turner, S.J., Silke, J., Kenshole, B. & Ruby, J. 2000. Characterization of the ectromelia virus serpin SPI-2. *Journal of General Virology,* **81:** 2425–2430.

Twardzik, D.R., Brown, J.P., Ranchalis, J.E., Todaro, G.J. & Moss, B. 1985. Vaccinia virus-infected cells release a novel polypeptide functionally related to transforming and epidermal growth factors. *Proceedings of the National Academy of Sciences, USA,* **82:** 5300–5304.

Upton, C., Mossman, K. & McFadden, G. 1992. Encoding of a homolog of the IFN-gamma receptor by myxoma virus. *Science,* **258:** 1369–1372.

Vanderplasschen, A., Mathew, E., Hollinshead, M., Sim, R.B. & Smith, G.L. 1998. Extracellular enveloped vaccinia virus is resistant to complement because of incorporation of host complement control proteins into its envelope. *Proceedings of the National Academy of Sciences, USA,* **95:** 7544–7549.

Wang, E.C.Y., McSharry, B., Retiere, C., Tomasec, P., Williams, S., Borysiewicz, L.K., Braud, V.M. & Wilkinson, G.W.G. 2002. UL40-mediated NK evasion during productive infection with human cytomegalovirus. *Proceedings of the National Academy of Sciences, USA,* **99:** 7570–7575.

Wang, G.H., Bertin, J., Wang, Y., Martin, D.A., Wang, J., Tomaselli, K.J., Armstrong, R.C. & Cohen, J.I. 1997. Bovine herpesvirus 4 BORFE-2 protein inhibits fas- and tumor necrosis factor receptor 1-induced apoptosis and contains death effector domains shared with other gamma-2 herpesviruses. *Journal of Virology,* **71:** 8928–8932.

Wang, G.H., Garvey, T.L. & Cohen, J.I. 1999. The murine gammaherpesvirus-68 M11 protein inhibits Fas- and TNF-induced apoptosis. *Journal of General Virology,* **80:** 2737–2740.

Wang, X.W., Gibson, M.K., Vermeulen, W., Yeh, H., Forrester, K., Sturzbecher, H.W., Hoiejmakers, J.H.J. & Harris, C.C. 1995. Abrogation of P53-induced apoptosis by the hepatitis-B virus-X gene. *Cancer Research,* **55:** 6012–6016.

Watkins, J.F. 1964. Adsorption of sensitized sheep erythrocytes to HeLa cells infected with herpes simplex virus. *Nature,* **202:** 1364–1365.

Wiertz, E.J.H.J., Tortorella, D., Bogyo, M., Yu, J., Mothes, W., Jones, T.R., Rapoport, T.A. & Ploegh, H.L. 1996. Sec61-mediated transfer of a membrane protein from the endoplasmic reticulum to the proteasome for destruction. *Nature,* **384:** 432–438.

Wolf, D., Witte, V., Laffert, B., Blume, K., Stromer, E., Trapp, S., D'Aloja, P., Schurmann, A. & Baur, A.S. 2001. HIV-1 Nef associated PAK and PI3-kinases stimulate Akt-independent Bad-phosphorylation to induce anti-apoptotic signals. *Nature Medicine,* **7:** 1217–1224.

Xiang, Y. & Moss, B. 1999. Identification of human and mouse homologs of the MC51L-53L-54L family of secreted glycoproteins encoded by the molluscum contagiosum poxvirus. *Virology,* **257:** 297–302.

Xiang, Y. & Moss, B. 1999. IL-18 binding and inhibition of interferon gamma induction by human poxvirus-encoded proteins. *Proceedings of the National Academy of Sciences, USA,* **96:** 11537–11542. [See also correction: *ibid.,* **97:** 11673.]

Xu, X.M., Nash, P. & McFadden, G. 2000. Myxoma virus expresses a TNF receptor homolog with two distinct functions. *Virus Genes,* **21:** 97–109.

Yao, Z.B., Fanslow, W.C., Seldin, M.F., Rousseau, A.M., Painter, S.L., Comeau, M.R., Cohen, J.I. & Spriggs, M.K. 1995. Herpesvirus Saimiri encodes a new cytokine, IL-17, which binds to a novel cytokine receptor. *Immunity,* **3:** 811–821.

Yewdell, J.W. & Hill, A.B. 2002. Viral interference with antigen presentation. *Nature Immunology,* **3:** 1019–1025.

Yin, Y., Maoury, B. & Fahraeus, R. 2003. Self-inhibition of synthesis and antigen presentation by Epstein-Barr-virus encoded EBNA1. *Science,* **301:** 1371–1374.

York, I.A. & Johnson, D.C. 1993. Direct contact with herpes simplex virus-infected cells results in inhibition of lymphokine-activated killer cells because of cell-to-cell spread of virus. *Journal of Infectious Diseases,* **168:** 1127–1132.

Zheng, Z.Y. & Zucker-Franklin, D. 1992. Apparent effectiveness of natural killer cells vis-à-vis retrovirus-infected targets. *Journal of Immunology,* **148:** 3679–3685.

Zhou, Q., Snipas, S., Orth, K., Muzio, M., Dixit, V.M. & Salvesen, G.S. 1997. Target protease specificity of the viral serpin CrmA – Analysis of five caspases. *Journal of Biological Chemistry,* **272:** 7797–7800.

Ziegler, H., Thale, R., Lucin, P., Muranyi, W., Flohr, T., Hengel, H., Farrell, H., Rawlinson, W. & Koszinowski, U.H. 1997. A mouse cytomegalovirus glycoprotein retains MHC class I complexes in the ERGIC/cis-Golgi compartment. *Immunity,* **6:** 57–66.

Zocchi, M.R., Rubartelli, A., Morgavi, P. & Poggi, A. 1998. HIV-1 Tat inhibits human natural killer cell function by blocking L-type calcium channels. *Journal of Immunology,* **161:** 2938–2943.

Zou, P., Isegawa, Y., Nakano, K., Haque, M., Horiguchi, Y. & Yamanishi, K. 1999. Human herpesvirus 6 open reading frame U83 encodes a functional chemokine. *Journal of Virology,* **73:** 5926–5933.

THE MOLECULAR BASIS OF LIVESTOCK DISEASE AS ILLUSTRATED BY AFRICAN TRYPANOSOMIASIS

John E. Donelson

Department of Biochemistry, University of Iowa, Iowa City, Iowa 52242, USA

Abstract: African trypanosomes are protozoan parasites, most species of which are transmitted by tsetse flies. They reside in the mammalian bloodstream and evade the immune system by periodically switching the major protein on their surface – a phenomenon called antigenic variation, mediated by gene rearrangements in the trypanosome genome. The trypanosomes eventually enter the central nervous system and cause a fatal disease, commonly called *ngana* in domestic cattle and sleeping sickness in humans. Two sub-species of *Trypanosoma brucei* infect humans (*T. b. rhodesiense* and *T. b. gambiense*) and one sub-species does not survive in humans (*T. b. brucei*) because it is lysed by the human-specific serum protein, apolipoprotein L-I. Wild animals in Africa have other (less well understood) molecular mechanisms of suppressing the number of African trypanosomes in the blood, and some indigenous breeds of African cattle also display a partial "trypanotolerance" whose genetic loci have recently been mapped.

1.   INTRODUCTION

Viral, bacterial, protozoan and helminthic diseases of domestic livestock continue to be serious impediments in the agricultural economies of most developing countries. Many of these livestock pathogens have evolved sophisticated molecular mechanisms for evading or circumventing the mammalian immune system. They also frequently acquire resistance to drugs that are initially effective against them. Among these livestock pathogens are African trypanosomes, which are protozoan parasites that cause the fatal diseases of ngana in cattle, surra in camels and horses, and sleeping sickness in humans. They are the paradigm for a livestock

293

H. P. S. Makkar and G. J. Viljoen (eds.), Applications of Gene-Based Technologies for Improving Animal Production and Health in Developing Countries, 293-311.

pathogen in developing countries about which much is now known, yet little has been achieved in controlling or eliminating the disease.

African trypanosomes were identified as the cause of trypanosomiasis more than 100 years ago (Hide, 1999) and in many ways are ideal pathogens to study in the laboratory. From the perspective of research on the parasites themselves, excellent laboratory rodent models for their infection exist (Hide 1999). They can also be readily grown as free-living organisms in culture (Baltz *et al.*, 1985) and individual colonies can be obtained on agar plates (Carruthers and Cross, 1992). Their main mechanisms of immune evasion are known (Donelson, 2003). The DNA sequence of their genome is nearly determined (El-Sayed *et al.*, 2003). They can be manipulated genetically in the laboratory – genes can be mutated, and deleted from or inserted into their genome (Bellofatto and Cross, 1989). They contain unique organelles and metabolic pathways not found in mammals, which could potentially be exploited for new drug development (Opperdoes and Michels, 1993; Schmidt and Krauth-Siegel, 2002). They are pathogens of humans as well as livestock (Hide, 1999), so they attract the interests and vast resources of the medical research community. From the standpoint of experiments on their animal hosts, advantages also exist. Breeds of domestic cattle that are either "trypanotolerant" or trypanosome-sensitive are well known and animals of each type have been cross-bred (Hanotte, 2003). The molecular basis of trypanosome-tolerance in at least one indigenous wild animal species (the Cape buffalo) has been elucidated (Wang *et al.*, 2002). Finally, the reason some African trypanosome subspecies are killed by human serum, but not by livestock serum, is at least partially understood (Vanhamme *et al.*, 2003).

Despite these extensive molecular characterizations of African trypanosomes and their interactions with their livestock hosts during the past century, African trypanosomiasis remains ranked among the top 10 livestock diseases with a significant negative impact on developing countries in Africa, Asia and South America. In this review, several of the distinctive molecular properties of African trypanosomes will be described. The reasons why this information has not translated into vaccines or better drugs against trypanosomes will be presented, and prospects for more successful applications of gene-based approaches against livestock diseases of developing countries in the future will be examined.

2. ANTIGENIC VARIATION OF AFRICAN TRYPANOSOMES

The phenomenon for which African trypanosomes have become famous amongst molecular biologists the world over is their remarkable ability to undergo antigenic variation during the onslaught of the immune responses directed against them by their mammalian hosts. Antigenic variation, in the case of African trypanosomes, refers to their ability to periodically switch from the expression of about 10 million copies of a single variant surface glycoprotein (VSG) on their surface to 10 million copies of another, immunologically distinct, VSG on their surface. This switch occurs at a rate of 10^{-2} to 10^{-7} switches per doubling time of 5-10 hours (Turner and Barry, 1989; Davies, Carruthers and Cross, 1997; Turner, 1997). The 10 million VSG molecules completely encapsulate the organism's surface and the VSG's only known function is to serve as a protective barrier against the attack of the immune system on the other outer membrane constituents. Thus, antigenic variation permits the trypanosome population as a whole to keep one step ahead of the immune response. The three-dimensional structures of the N-terminal two-thirds of two immunologically distinct VSG molecules have been determined by X-ray crystallography (Blum *et al.*, 1993), and found to be very similar rod-like shapes, despite having quite different amino acid sequences (Figure 1). These rod-like structures permit the 10 million copies of a single VSG to pack very closely together as a dense array on the surface, and suggest that all VSGs share a similar backbone structure from which emerge distinct epitopes derived from the different amino acid side chains. A number of other pathogens have now been shown to display various forms of antigenic variation (Babour and Restrepo, 2000), but the phenomenon is still best understood at the molecular level in African trypanosomes.

The first indications that African trypanosomes could undergo antigenic variation were published in the early 1900s, even though the investigators at the time probably did not realize from their data that they were observing this phenomenon. Figure 2A shows an example of these data, taken from a publication in the medical literature of 1910 (Ross and Thompson, 1910). The authors examined blood smears obtained from a human patient infected with trypanosomes and plotted the number of trypanosomes in the blood versus the probable number of weeks after the patient had been infected via the bite of a tsetse fly. As Figure 2A shows, the infection was characterized by successive waves of parasitaemia.   One week a large number of trypanosomes were in the blood, and a few days later a small number were present, followed by a large number again. This cyclic pattern continued during the ten weeks of sampling.

In their discussion of this observation, the authors suggested that the trypanosomes were periodically "habituating" to the antibodies directed against them (Ross and Thompson, 1910), a deduction remarkably close to what we now understand to be antigenic variation.

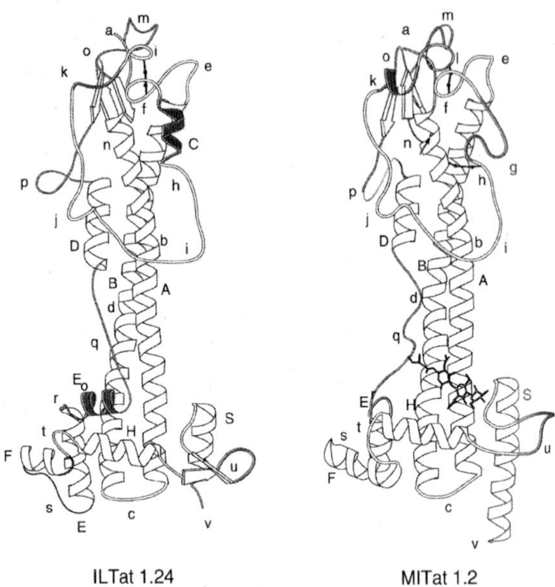

Figure 1. Ribbon diagrams depicting the three-dimensional shapes of the N-terminal two-thirds of two different VSGs (taken from Blum, 1993). The N-termini of both proteins are near the top of the diagram, and the C-termini are near the bottom. The top portions of both proteins are the regions exposed on the outer surface of the trypanosome.

During the next sixty years, not much more was learned about the nature of these waves of parasitaemia, until 1976, when the results were published of a seminal experiment in the understanding of antigenic variation (Bridgen, Cross and Bridgen, 1976). The authors of this work infected a single experimental rabbit with a single trypanosome and then collected parasites from four different peaks of parasitaemia. By the time this work was undertaken, it had been shown that the surface of trypanosomes was covered predominately by multiple copies of a single protein. The authors isolated this surface protein from these four trypanosome populations and subjected the four protein preparations to Edmund degradation to determine their N-terminal amino acid sequences. The results are shown in Figure 2B. The four proteins were found to have completely different N-terminal amino acid sequences that were unrelated to each other by potential frame shift mutations or other possible gene insertions or deletions. In addition to suggesting that these four VSGs were derived from four different genes, or segments of genes, in the trypanosome genome, this finding also provided a hypothesis for why a hallmark of an African trypanosome infection is the successive waves of parasitaemia.

The hypothesis was that, in a given peak of parasitaemia, all or most of the trypanosomes express the same VSG. An immune response is mounted against this VSG, resulting in the elimination of most of the trypanosomes in the population. But, before they are all eliminated, one or more of the trypanosomes switch to the expression of a different VSG and give rise to the next wave of parasitaemia. The immune system must then mount a new response against the new VSG, and again eliminates most of the organisms in this second population. Before they are all gone, however, one or more parasites switches to the expression of a third VSG, giving rise to the third population, and so on. Thus, the occasional switching from one VSG to another VSG by individual trypanosomes and the subsequent multiplication of the "switched" organisms can lead to antigenic variation of the trypanosome population as a whole in the bloodstream. Subsequent experiments during the past 27 years by a number of different laboratories have provided very strong evidence for this hypothesis and have supplied many additional details about the molecular events responsible for antigenic variation. The diagram in Figure 3 summarizes some of these events.

Southern hybridization blots of genomic DNA from African trypanosomes probed with different VSG cDNAs under low hybridization stringency indicate that the genome contains several hundred different VSG genes (*VSGs*) and pseudo-*VSGs* (indicated by the black rectangles labelled A, B and C in upper left corner of Figure 3). The reason *VSGs* cross-hybridize under low hybridization stringency is that they possess some sequence similarities in their C-terminal coding regions, which encode the portion of the rod-shaped protein that lies closest to the membrane. In a given trypanosome, all of these several hundred *VSGs* are transcriptionally silent – they are not expressed – except for one. This expressed *VSG* has usually been duplicated and translocated into a region of the genome located near the physical end of a chromosome, i.e. a telomere, that is called a *VSG* expression site or ES (shown on the left-hand side of Figure 3). This duplicated *VSG* lies downstream of a promoter (the triangular flag in the ES) and several expression site-associated genes (*ESAGs*). When the *VSG* is in this active ES, the *VSG* is transcribed and translated into the VSG protein, called VSG_B in the example shown on the left-hand side of Figure 3.

When a switch from the expression of one *VSG* to another occurs, the most common thing to happen is that another *VSG* at an interior chromosomal site is duplicated and translocated into the ES, displacing the *VSG* already in that ES. The displaced *VSG* is simply degraded and lost from the cell. The new *VSG* in the ES is then transcribed and translated into the new antigenically distinct VSG protein. A third switch involves a third duplicative translocation and the expression of a third *VSG*. Under normal circumstances, one, and only one, *VSG* is expressed at a time in a given trypanosome, although rare exceptions have been noted (Baltz *et al.*, 1986; Esser and Schoenbechler, 1985; Muñoz-Jordán, Davies and Cross, 1996). The unexpressed, "donor" *VSG*s for the duplicative gene transposition are scattered among different chromosomes and are never expressed from their interior chromosomal locations.

Interestingly, sometimes the trypanosome apparently tries to duplicate and express more than one *VSG* simultaneously in the ES. The outcome, as it has been detected experimentally (Thon, Baltz and Eisen, 1989), is the formation of a mosaic, or composite, *VSG* in the ES in which, for example, the first one-third of the gene is derived from one duplicated "donor" *VSG*, while the second one-third comes from another duplicated "donor" *VSG*, and the last one-third from still another. Thus, the trypanosome possesses the potential to create new mosaic *VSG*s in the ES that are not from a single "donor" *VSG* template stored in the genome.

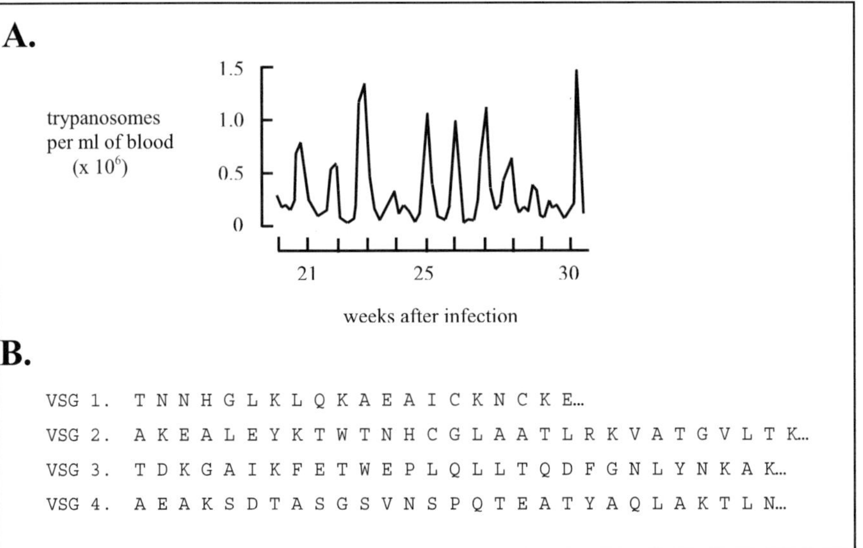

Figure 2. (A) A graph redrawn from the reference by Ross and Thompson (1910) showing the number of trypanosomes per millilitre of blood in a human patient at different times after the probable time of infection. (B) The N-terminal amino acid sequences of VSGs from four different trypanosome populations derived from an experimental rabbit infected with a single trypanosome (taken from Bridgen, Cross and Bridgen, 1976)

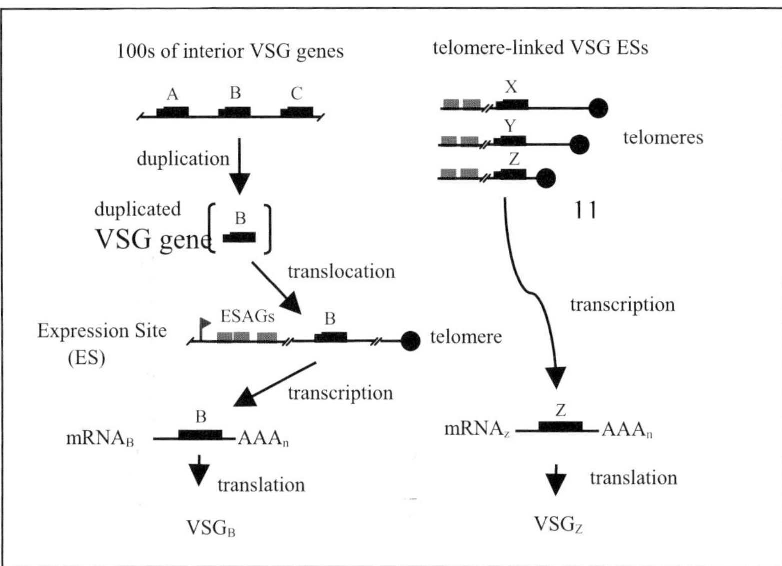

Figure 3. Summary of the events at the DNA and RNA level that are responsible for antigenic variation in African trypanosomes.

Furthermore, examples of "unfaithful" duplications of donor *VSG*s have been detected (Lu *et al.*, 1993). In these cases, the duplicated *VSG* is from a single donor *VSG* template, but single-base replacements are scattered throughout the newly duplicated *VSG* that preserve the open reading frame, but cause amino acid changes. If these amino acid changes occur in the region of the VSG protein exposed to the outer surface of the VSG coat, they can potentially alter the antigenicity of the surface. If the changes alter the structure of the VSG too much, then the VSG may not pack tightly on the membrane and the trypanosome succumbs to the immune system. But that is the price individual organisms pay to accommodate the ability of the trypanosome population as a whole to generate great diversity in its antigenicity. Thus, since the trypanosomes can create both new mosaic *VSG*s and new mutated *VSG*s in the ES, it is likely that the trypanosome population has the potential to generate sequentially a nearly infinite number of VSGs on its surface.

Another layer of complexity in *VSG* expression comes from the fact that multiple telomere-linked ESs exist in the genome. African trypanosomes are diploid and their genome of 35 megabase pairs has eleven pairs of chromosomes, ranging in size from 1 megabase to 6 megabases (Melville *et al.*, 1998). Southern hybridization blots indicated that many of these telomeres have *VSG*s near them and estimates are that as many as 20 telomere-linked expression sites exist (Borst and Ulbert, 2001; Pays and Nolan, 1998; Cross, Wirtz and Navarro, 1998), each of which has a complement of *ESAG*s between the promoter and the *VSG* (indicated in the upper right of Figure 3). The sequences of several telomere-linked ESs of 35–50 kb have now been reported (LaCount *et al.*, 2001; Berriman *et al.*, 2002). Each of these ESs has five to ten *ESAG*s between the promoter and the *VSG*, and a comparison of these *ESAG* sequences reveals that each ES has its own group of *ESAG* isogenes. For example, each ES has its own two isogenes for the two-subunit transferrin receptor. Transferrin is a serum protein that complexes with iron in the blood and is the source of iron for the trypanosome. The transferrin receptors encoded in the different trypanosome ESs are similar, but non-identical, in their amino acid sequences. Binding studies indicate that these different transferrin receptors have differing affinities for transferrins of different mammalian species (Bitter *et al.*, 1998; Gerrits *et al.*, 2002). This observation raises the interesting possibility that multiple ESs provide different assortments of *ESAG* proteins suitable for maximal growth in different environments, i.e. the bloodstreams of different mammals (Bitter *et al.*, 1998; Gerrits *et al.*, 2002). This host adaptation model predicts that different ESs would be favoured by the trypanosomes residing in different mammalian hosts. In addition to the differing transferrin receptor binding affinities, some additional data exist consistent with the model. Bloodstream trypanosomes appear to switch away from expression of ESs encoding a low-affinity receptor for transferrin of the animal serum in which they are cultured *in vitro* (Gerrits *et al.*, 2002). Thus, the 20 ESs potentially provide 20 unique transferrin receptors for the trypanosome to use in its search for the most efficient uptake of iron in a given mammalian bloodstream. The protein products of the other *ESAG*s in a specific ES might also help the trypanosome adapt to a given bloodstream environment (Pays *et al.*, 2001).

3.   WHY DOES *T. BRUCEI RHODESIENSE* INFECT HUMANS AND *T. BRUCEI BRUCEI* DOES NOT?

Three subspecies of *Trypanosoma brucei* have been identified: *T. b. rhodesiense*, *T. b. gambiense* and *T. b. brucei*. The first two subspecies infect humans and cause human sleeping sickness, a fatal infection unless treated with highly toxic drugs. In contrast, the third subspecies, *T. b. brucei*, is lysed by human serum and does not cause sleeping sickness. Indeed, the only known difference between *T. b. rhodesiense* and *T. b. brucei* is their susceptibility to lysis by human serum and there is strong evidence that the two subspecies can intra-convert (Rifkin *et al.*, 1994). All three subspecies can infect a wide range of mammals other than humans, including cattle, rabbits, rats and mice. Much effort has been expended over the years in trying to identify the molecular basis of this difference in susceptibility to lysis by human serum.

Among the first hints about the molecular basis for this difference were the observations in 1978 that a component of the heavy density lipoprotein (HDL) fraction of human serum is responsible for the lysis of *T. b. brucei*, but fails to cause lysis of *T. b. rhodesiense* (Rifkin, 1978). This component was subsequently found to bind to the flagellar pocket of *T. b. brucei*, where nutrients typically enter the cell, after which it is internalized and localized to large lysosome-like vesicles within the trypanosome cell (Hager and Hajduk, 1997). *T. b. brucei* lyses only after the binding, internalization and lysosomal targeting of this component has occurred. In contrast, the resistant *T. b. rhodesiense* also binds the component via receptors in the flagellar pocket, but it is not detected in the lysosomes. The component was initially identified as a haptoglobin-related protein by one group of investigators (Hager and Hajduk, 1997) and later reported by another group to be apolipoprotein L-I (apoL-I) (Vanhamme *et al.*, 2003), a difference yet to be resolved. Both groups agree, however, that the component originally resides in human HDLs and reaches the trypanosome lysosomes prior to cell lysis.

A different experimental strategy for identifying the molecular basis of the human serum resistance has been to search for a gene or genes in the trypanosome genome itself that might correlate with resistance or susceptibility to lysis by human serum. This tactic has also been fruitful. A gene has been found to be expressed in *T. b. rhodesiense*, but not in *T. b. brucei*, that encodes a truncated form of the VSG called the serum resistance associated protein (SRA) (De Greef and Hamers, 1994; Xong *et al.*, 1998). This *SRA* gene is one of the *ESAG*s situated in at least one, but not all, ESs of *T. b. rhodesiense*. If the *T. b. rhodesiense* activates the ES bearing the *SRA*, the cells are resistant to lysis by human serum; if it activates an ES lacking the *SRA*, then the cells are susceptible to lysis. This differential expression of *SRA* is consistent with the earlier observation that *T. b. rhodesiense* (resistant) and *T. b. brucei* (susceptible) are intraconvertible (Rifkin *et al.*, 1994). If during antigenic variation the cells switch from the expression of the *SRA*-bearing ES to expression of an *SRA*-lacking ES, then the cells lose their resistance to human serum. Likewise, it has been shown that when a plasmid bearing the *SRA* gene is transfected into several different *T. b. brucei* isolates, the *T. b. brucei* isolates become human serum resistant (Xong *et al.*, 1998; Wang, Bohme and Cross, 2003), indicating that the SRA protein is the only trypanosome-encoded factor required for the serum resistance. Thus, antigenic variation-linked expression of SRA underlies the resistance of *T. b. rhodesiense* to lysis by human serum.

The SRA of 410 amino acids is smaller by about 80 amino acids than a typical VSG of about 490 amino acids. Nascent VSGs have a C-terminal hydrophobic region that is replaced with a glycosylphosphatidylinositol (GPI) moiety, which anchors the mature VSG to the outer membrane of the trypanosome. Nascent SRA also possesses a canonical C-terminal hydrophobic GPI signal sequence, but the mature SRA does not appear to be GPI-anchored on the cell surface (Wang, Bohme and Cross, 2003). Instead, immunofluorescence assays with anti-SRA antibodies indicate that the SRA accumulates in the lysosomes (Vanhamme *et al.*, 2003). Consistent with this SRA location is the fact that nascent SRA does not have an N-terminal signal peptide-like sequence, as do all nascent VSGs. Thus, the SRA has neither an N-terminal signal sequence to aid its movement across a membrane nor a GPI moiety to anchor it to the outer membrane. In the absence of these signals and post-translational modifications, SRA's default location appears to be the lysosome.

Recent evidence indicates that when both the trypanosome-encoded SRA and the human serum-derived apoL-I are in the lysosomes of *T. b. rhodesiense*, they bind to each other (Vanhamme *et al.*, 2003). The binding affinity between the two is strong enough that when recombinant SRA is covalently attached to Sepharose, it can be used to affinity-purify apoL-I from human serum. Deletion analyses of recombinant versions of both SRA and apoL-I suggest that an amphipathic α-helix near the N-terminus of SRA associates with another amphipathic α-helix near the C-terminus of apoL-I through a coiled-coil protein-protein interaction (Vanhamme *et al.*, 2003). Thus, resistance of *T. b. rhodesiense* to human serum lysis is probably due to an SRA-mediated inhibition of the apoL-I lytic effect within the lysosome. Left unexplained by this model is the question of how apoL-I causes lysis of the cell in the absence of SRA. One possibility is that free apoL-I in the trypanosome lysosome destabilizes the lysosomal membrane, releasing the lysosomal contents into the cytoplasm and causing lysis of the cell. However, if SRA is also in the lysosome, it associates with apoL-I and neutralizes its membrane destabilizing effect. Clearly many details of this SRA-mediated resistance mechanism remain to be elucidated, but if the overall model is correct, it may suggest new ways to approach the prevention and cure of African trypanosome infections in both humans and domestic livestock.

4.   WHY ARE MOST WILD ANIMALS MORE RESISTANT TO AFRICAN TRYPANOSOMES THAN MOST DOMESTIC ANIMALS?

As mentioned above, African trypanosomiasis is typically fatal to domestic animals and humans unless treated with toxic drugs. However, not all mammals succumb to African trypanosomes. Many wild animal species in sub-Saharan Africa live healthy lives in areas inhabited by both trypanosomes and the tsetse flies that transmit them. Indeed, some of the major national wildlife parks in Africa, including the Serengeti in Tanzania and Masai Mara in Kenya, exist today because the early colonial settlers found they could not raise domestic livestock in these regions due to African trypanosomiasis. Analyses of the ingested blood in trapped tsetse flies from these regions reveal that they feed on a number of wildlife species. One of these species is the Cape buffalo and considerable effort has been expended in an effort to determine why these animals display few if any signs of the disease (Wang *et al.*, 2002).

Blood samples collected from wild Cape buffaloes in tsetse fly-endemic regions of Africa often contain antibodies against trypanosomes, demonstrating they have been exposed to the parasites. Furthermore, Cape buffaloes bred and raised in captivity in the absence of tsetse flies are as resistant to trypanosome infection as are wild Cape buffaloes, suggesting that the resistance is not acquired but has a genetic basis (Wang *et al.*, 2002). Experimental infections of Cape buffaloes have shown that trypanosomes can initially grow and multiply in the buffalo, but the initial wave of parasitaemia does not reach as high a level as in a highly inbred domestic cattle breed, such as the Holstein breed. In addition, the subsequent waves of parasitaemia in the Cape buffalo diminish until only a few parasites remain (Figure 4A). These experimentally infected Cape buffaloes do not appear to undergo faster, or higher titre, antibody responses against the VSG than do similarly infected cattle (Reduth *et al.*, 1994), so it is unlikely that the B cell immune response of the Cape buffalo is responsible for its ability to control the infection.

These observations led to an effort to identify the factors in Cape buffalo serum that suppress the initial trypanosome waves of parasitaemia. Fractionation of the serum and subsequent biochemical analyses resulted in the detection of the enzymes xanthine oxidase and catalase as two of these factors (Muranjan *et al.*, 1997). The xanthine oxidase activity in uninfected Cape buffalo serum is about 7-fold higher than in uninfected Holstein serum (Wang *et al.*, 2002). Likewise, the level of catalase activity is much higher in uninfected Cape buffalo serum than in uninfected Holstein serum. However, soon after a Cape buffalo is infected with trypanosomes, its serum catalase activity drops precipitously, whereas the lower level of Holstein catalase activity is unaffected by a trypanosome infection (Figure 4B). The initial level of xanthine oxidase in both uninfected animals remains unchanged by the trypanosome infection.

The reactions catalysed by xanthine oxidase and catalase are summarized in Figure 4C. Xanthine oxidase (XO) catalyses the terminal reactions in the catabolism of excess purines to uric acid. The enzyme's substrates are the nitrogen-containing purine bases hypoxanthine (derived from AMP) and xanthine (derived from either hypoxanthine or GMP). The enzyme utilizes molecular oxygen (O_2) and water to generate uric acid, which is excreted through the kidneys, and hydrogen peroxide (H_2O_2), which is a highly reactive oxygen species. When these xanthine oxidase-catalysed reactions occur in the blood, the serum catalase removes the H_2O_2 by converting it back to O_2 and water. Thus, H_2O_2 as a reactive oxygen species does not have a chance to build up in the blood during the production of uric acid. Furthermore, the high level of catalase in uninfected Cape buffalo serum balances the high amount of xanthine oxidase, and likewise, the low level of catalase in uninfected Holstein serum accompanies the correspondingly low level of xanthine oxidase (Figure 4B).

The decrease in catalase activity in the serum of a trypanosome-infected Cape buffalo dramatically alters the balance in the activities of xanthine oxidase and catalase. This new combination of a high amount of xanthine oxidase and a low level of catalase results in the accumulation of H_2O_2 to a much higher steady state level in the serum than prior to the infection (Figure 4D). It turns out that African trypanosomes are highly sensitive to killing by H_2O_2. They appear to themselves lack catalase activity and can be killed by as little as 1μM H_2O_2 (Wang *et al.*, 2002). This trypanocidal activity of infected Cape buffalo serum can be inhibited by the addition of catalase (Muranjan *et al.*, 1997). Thus, it seems likely that the trypanocidal activity in infected Cape buffalo serum is due at least in part, if not entirely, to the increased level of H_2O_2. Left unresolved by these observations is why African trypanosomes in the blood of a Cape buffalo cause the decrease in its serum catalase activity. A similar decrease in serum catalase activity of Holstein animals does not occur when they are infected with trypanosomes. Likewise, a similar decrease in catalase activity and concomitant increase in H_2O_2 is not seen in the serum of other wild animals (that have been investigated) when they are infected with African trypanosomes. Thus, it appears that this H_2O_2-mechanism of trypanocidal activity has not been exploited by other wildlife species to deal with trypanosome infections – they seem to have arrived at other, still unknown, molecular mechanisms of coping with the infection.

Another major curiosity about this Cape buffalo approach for controlling an African trypanosome infection is that it is relatively short-lived and does not result in complete clearance of the infection (Wang *et al.*, 2002). Within a few weeks of the initial infection, the concentration of the catalase in the blood returns to the pre-infection level and H_2O_2 no longer accumulates to its high trypanocidal level (Figure 4D). The number of trypanosomes in the blood, however, remains suppressed for reasons that are unknown. Perhaps an acquired immune response involving antibodies directed against one or more trypanosome receptors for growth factors may control the organisms at this latent stage of infection (Wang *et al.*, 2002; Black *et al.*, 2001). Clearly, much more remains to be learned about the latent infection stage in the Cape buffalo, and other suppression mechanisms await discovery in other wild animal species that co-exist with trypanosomes and tsetse flies.

5.   WHY ARE A FEW CATTLE BREEDS "TRYPANOTOLERANT" WHILE MOST CATTLE BREEDS ARE NOT?

Trypanotolerance has been defined as the relative capacity of an animal to control the development of African trypanosomes and to limit their pathological effects, the most prominent of which in cattle is anaemia (d'Ieteren *et al.*, 1998). By these criteria, a few indigenous cattle breeds in Africa are more trypanotolerant than others. When these trypanotolerant animals undergo trypanosome challenge, they are better able than susceptible animals to control parasite proliferation, maintain their packed red blood cell volume, mount an immune response and maintain body weight (Hanotte, 2003). For example, N'Dama (*Bos taurus*) cattle of West African origin are a relatively small-sized, "hump-less" breed that is more tolerant of trypanosomiasis than are Boran (*Bos indicus*) cattle of East Africa, which is a breed of larger, humped animals favoured for their growth and productivity by farmers in areas with little threat of trypanosomiasis. Comparison of the immune responses of these two breeds during trypanosome infection shows some small differences in antibody response, complement level and cytokine expression, but these differences do not seem sufficient to account for the dramatic difference in trypanotolerance (Naessens, Teale and Sileghem, 2002).

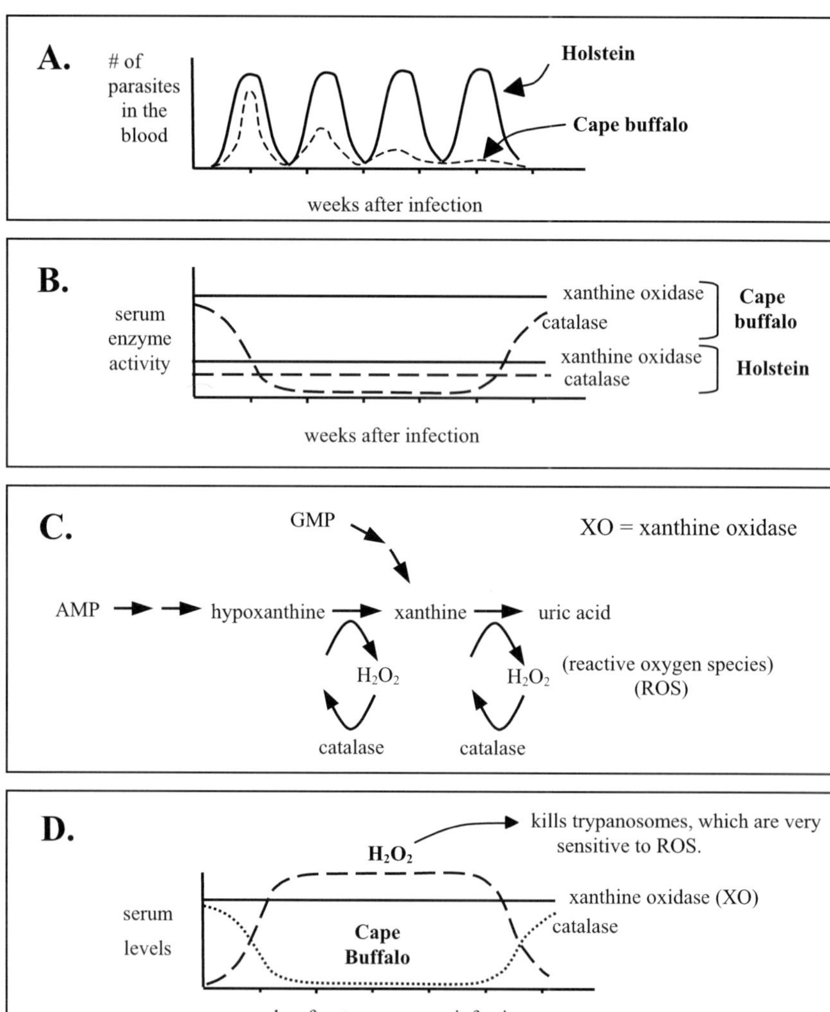

Figure 4. (A) Diagram of the difference in the waves of parasitaemia in a typical Holstein animal and a typical Cape buffalo infected with African trypanosomes. (B) A plot of the relative enzymatic activities of xanthine oxidase (solid line) and catalase (dashed line) in the sera of a Holstein animal and a Cape buffalo infected with African trypanosomes over time (this author's interpretation of data described in Wang *et al.*, 2002). (C) Diagram of the reactions catalysed by xanthine oxidase (XO) and catalase. (D) A plot of the relative serum levels of xanthine oxidase, catalase and H_2O_2 in a Cape buffalo infected with African trypanosomes (this author's interpretation of data described in Wang *et al.*, 2002).

About twenty years ago, in 1983, a research institute in Kenya [now called the International Livestock Research Institute (ILRI)] began an experimental crossbreeding programme between N'Dama and Boran cattle, with the eventual goal of identifying the factors that contribute to the difference in trypanotolerance between the two breeds (Jordt *et al.*, 1986). A total of 177 F_2 offspring and their parents and grandparents have recently been both (i) genotyped at 477 molecular marker loci (mainly microsatellite sequences) distributed among the 29 cattle autosomal chromosomes; and (ii) phenotyped for 16 possible trypanotolerance traits, involving primarily anaemia, body weight and parasitaemia (Hanotte, 2003). The results of this study strongly support a model of multiloci involvement in the inheritance of trypanotolerance. At least one locus correlating with trypanotolerance was found on 17 of the 29 bovine chromosomes and two loci were mapped to still other chromosomes, indicating that at least 19 distinct loci are involved. It will be a major challenge to identify the specific genes corresponding to these 19 loci, but two developments raise hope that this goal can eventually be achieved. First, the sequence of most of the bovine genome will probably be determined in the next few years, which will aid in gene discovery at these loci, and, second, trypanotolerant loci have also been mapped in the mouse genome (Iraqi *et al.*, 2000), whose sequence is nearly completely determined. Thus, some of the mouse trypanotolerant genes, which may be identified before the bovine ones, will probably have a corresponding homologue in cattle that can then be investigated.

Most of the cattle alleles for resistance to trypanosomiasis were found to originate from the trypanotolerant N'Dama parent, as expected, but five of the resistance alleles appeared to come from the trypanosusceptible Boran parent. This surprising result, as the authors of the study point out (Hanotte, 2003), could be due to the fact that the present-day Boran cattle are probably descended from ancestral cattle that inhabited tsetse fly regions and therefore may possess some residual loci associated with trypanotolerance. Interestingly, these residual Boran trypanotolerant loci are different from the trypanotolerant loci found in the N'Dama genome. Independent of the question about why Boran cattle have trypanotolerant loci, this finding raises the possibility that selection within an F_2 cross between the two breeds might produce a new "synthetic" breed that would exceed either parental breed in its level of trypanotolerance. Indeed, some of the 177 F_2 animals investigated possessed higher levels of trypanotolerance than did any of the N'Dama controls (Hanotte, 2003).

6.   CONCLUSIONS

African trypanosomes have evolved a sophisticated and molecularly complex mechanism called antigenic variation for evading the immune response of their mammalian hosts. Built into this duplicative gene transposition mechanism, which occurs at the trypanosome's DNA level, are opportunities for the trypanosomes to create new genes for the variant surface glycoproteins (VSGs) whose sequential expression serves as the foundation of the antigenic variation. Likewise, some mammalian hosts for African trypanosomes have evolved equally distinctive mechanisms for co-existing with, or eliminating, the trypanosome infection. Humans have a protein component in their serum that lyses some, but not all, trypanosome species and subspecies. Cape buffaloes suppress, but do not eliminate, a trypanosome infection by increasing the concentration of H_2O_2 in their blood. Some cattle breeds are more tolerant of trypanosomes than others through a combination of the gene products of as many as 19 different genetic loci in their genome. These studies of African trypanosome infections over the past 100 years indicate that a vaccine against African trypanosomiasis is not feasible. However, better drug therapies based on our improved knowledge of the metabolic pathways and genomic DNA sequence of African trypanosomes is highly likely. The challenge for the future is to translate this improved knowledge about both the parasite and its hosts' defence mechanisms into the design and implementation of these better drug therapies for the livestock owners in tsetse fly-endemic regions of sub-Saharan Africa.

REFERENCES

Babour, A.G. & Restrepo, B.I. 2000. Antigenic variation in vector-borne pathogens. *Emerging Infectious Diseases,* **6**: 449–457.

Baltz, T., Baltz, D., Giroud, C. & Crockett, J. 1985. Cultivation in a semi-defined medium of animal infective forms of *Trypanosoma brucei, T. equiperdum, T. evansi, T. rhodesiense* and *T. gambiense. EMBO Journal,* **4**: 1273–1277.

Baltz, T., Giroud, C., Baltz, D., Roth, C., Raibaud, A. & Eisen, H. 1986. Stable expression of two variable surface glycoproteins by cloned *Trypanosoma equiperdum. Nature,* **319**: 602–604.

Bellofatto, V. & Cross, G.A. 1989. Expression of a bacterial gene in a trypanosomatid protozoan. *Science,* **244**: 1167–1169.

Berriman, M., Hall, N., Sheader, K., Bringaud, F., Tiwari, B., Isobe, T., Bowman, S., Corton, C., Clark, L., Cross, G.A., Hoek, M., Zanders, T., Berberof, M., Borst, P. & Rudenko, G. 2002. The architecture of variant surface glycoprotein gene expression sites in *Trypanosoma brucei. Molecular and Biochemical Parasitology,* **122**: 131–140.

Bitter, W., Gerrits, H., Kieft, R. & Borst, P. 1998. The role of transferrin-receptor variation in the host range of *Trypanosoma brucei. Nature,* **391**: 499–502.

Black, S.J., Sicard, E.L., Murphy, N. & Nolan, D. 2001. Innate and acquired control of trypanosome parasitaemia in Cape buffalo. *International Journal of Parasitology,* **31**: 561–564.

Blum, M.L., Down, J.A., Gurnett, A.M., Carrington, M., Turner, M.J. & Wiley, D.C. 1993. A structural motif in the variant surface glycoproteins of *Trypanosoma brucei. Nature,* **362:** 603–609.

Borst, P. & Ulbert, S. 2001. Control of VSG gene expression sites. *Molecular and Biochemical Parasitology,* **114**: 17–27.

Bridgen, P.J., Cross, G.A. & Bridgen, J. 1976. N-terminal amino acid sequences of variant-specific surface antigens from *Trypanosoma brucei. Nature,* **263**: 613–614.

Carruthers, V.B. & Cross, G.A. 1992. High-efficiency clonal growth of bloodstream- and insect-form *Trypanosoma brucei* on agarose plates. *Proceedings of the National Academy of Sciences, USA,* **89**: 8818–8821.

Cross, G.A.M., Wirtz, L.E. & Navarro, M. 1998. Regulation of *vsg* expression site transcription and switching in *Trypanosoma brucei. Molecular and Biochemical Parasitology,* **91**: 77–91.

d'Ieteren, G.D., Authie, E., Wissocq, N. & Murray, M. 1998. Trypanotolerance, an option for sustainable livestock production in areas at risk from trypanosomosis. *Revue scientifique et technique de L'Office International des Epizooties,* **17**: 154–175.

Davies, K.P., Carruthers, V.B. & Cross, G.A. 1997. Manipulation of the vsg co-transposed region increases expression-site switching in *Trypanosoma brucei. Molecular and Biochemical Parasitology,* **86:** 163–177.

De Greef, C. & Hamers, R. 1994. The serum-resistance associated (SRA) gene of *Trypanosoma brucei rhodesiense* encodes a VSG-like protein. *Molecular and Biochemical Parasitology,* **68**: 277–284.

Donelson, J.E. 2003. Antigenic variation and the African trypanosome genome. *Acta Tropica,* **85**: 391–404.

El-Sayed, N.M., Ghedin, E., Song, J., MacLeod, A., Bringaud, F., Larkin, C., Wanless, D., Peterson, J., Hou, L., Taylor, S., Tweedie, A., Biteau, N., Khalak, H.G., Lin, X., Mason, T., Hannick, L., Caler, E., Blandin, G., Bartholomeu, D., Simpson, A.J., Kaul, S., Zhao, H., Pai, G., Van Aken, S., Utterback, T., Haas, B., Koo, H.L., Umayam, L., Suh, B., Gerrard, C., Leech, V., Qi, R., Zhou, S., Schwartz, D., Feldblyum, T., Salzberg, S., Tait, A., Turner, C.M., Ullu, E., White, O., Melville, S., Adams, M.D., Fraser, C.M. & Donelson, J.E. 2003. The sequence and analysis of *Trypanosoma brucei* chromosome II. *Nucleic Acids Research,* **31**: 4856–4863.

Esser, K.M. & Schoenbechler, M.J. 1985. Expression of two variant surface glycoproteins on individual African trypanosomes during antigen switching. *Science,* **229**: 190–193.

Gerrits, H., Mußmann, R., Bitter, W., Kieft, R. & Borst, P. 2002. The physiological significance of transferrin receptor variations in *Trypanosoma brucei. Molecular and Biochemical Parasitology,* **119**: 237–247.

Hager, K.M. & Hajduk, S.L. 1997. Mechanism of resistance of African trypanosomes to cytotoxic human HDL. *Nature,* **385**: 823–826.

Hanotte, O., Ronin, Y., Agaba, M., Nilsson, P., Gelhaus, A., Horstmann, R., Sugimoto, Y., Kemp, S., Gibson, J., Korol, A., Soller, M. & Teal, A. 2003. Mapping of quantitative trait loci controlling trypanotolerance in a cross of tolerant West African N'Dama and susceptible East African Boran cattle. *Proceedings of the National Academy of Sciences, USA,* **100**: 7443–7448.

Hide, G. 1999. History of sleeping sickness in East Africa. *Clinical Microbiology Reviews,* **2**: 112–125.

Iraqi, F., Clapcott, S.J., Kumari, P., Haley, C.S., Kemp, S.J. & Teale, A. 2000. Fine mapping of trypanosomiasis resistance loci in murine advanced intercross lines. *Mammalian Genome,* **11**: 645–648.

Jordt, T., Mahon, G.D., Touray, B.N., Ngulo, W.K., Morrison, W.I., Rawle, J. & Murray, M. 1986. Successful transfer of frozen N'Dama embryos from the Gambia to Kenya. *Tropical Animal Health and Production,* **18**: 65–75.

LaCount, D.J., El-Sayed, N.M., Kaul, S., Wanless, D., Turner, C.M.R. & Donelson, J.E. 2001. Analysis of a donor gene region for a variant surface glycoprotein and its expression site in African trypanosomes. *Nucleic Acids Research,* **29**: 2012–2019.

Lu, Y., Hall, T., Gay, L.S. & Donelson, J.E. 1993. Point mutations are associated with a gene duplication leading to the bloodstream re-expression of a trypanosome metacyclic VSG. *Cell,* **72:** 397–406.

Melville, S.E., Leech, V., Gerrard, C.S., Tait, A. & Blackwell, J.M. 1998. The molecular karyotype of the megabase chromosomes of *Trypanosoma brucei* and the assignment of chromosome markers. *Molecular and Biochemical Parasitology,* **94**: 155–173.

Muñoz-Jordán, J.L., Davies, K.P. & Cross, G.A.M. 1996. Stable expression of mosaic coats of variant surface glycoproteins in *Trypanosoma brucei. Science,* **272**: 1791–1794.

Muranjan, M., Wang, Q., Li, Y.L., Hamilton, E., Otieno-Omondi, F.P., Wang, J., Van Praagh, A., Grootenhuis, J.G. & Black, S.J. 1997. The trypanocidal Cape buffalo serum protein is xanthine oxidase. *Infection and Immunity,* **65**: 3806–3814.

Naessens, J., Teale, A.J. & Sileghem, M. 2002. Identification of mechanisms of natural resistance to African trypanosomiasis in cattle. *Veterinary Immunology and Immunopathology,* **87**: 187–194.

Opperdoes, F.R. & Michels, P.A. 1993. The glycosomes of the Kinetoplastida. *Biochimie,* **75**: 231–234.

Pays, E. & Nolan, D.P. 1998. Expression and function of surface proteins in *Trypanosoma brucei. Molecular and Biochemical Parasitology,* **91**: 3–36.

Pays, E., Lips, S., Nolan, D., Vanhamme, L. & Perez-Morga, D. 2001. The VSG expression sites of *Trypanosoma brucei*: multipurpose tools for the adaptation of the parasite to mammalian hosts. *Molecular and Biochemical Parasitology,* **114**: 1–16.

Reduth, D., Grootenhuis, J.G., Olubayo, R.O., Muranjan, M., Otieno-Omondi, F.P., Morgan, G.A., Brun, R., Williams D.J.L. & Black, S.J. 1994. African buffalo serum contains novel trypanocidal protein. *Journal of Eukaryotic Microbiology,* **41**: 95–103.

Rifkin, M.R. 1978. Identification of the trypanocidal factor in normal human serum: high density lipoprotein. *Proceedings of the National Academy of Sciences, USA,* **75**: 3450–3454.

Rifkin, M.R., De Greef, C., Jiwa, A., Landsberger, F.R. & Shapiro, S.Z. 1994. Human serum-sensitive *Trypanosoma brucei rhodesiense*: a comparison with serologically identical human serum-resistant clones. *Molecular and Biochemical Parasitology,* **66**: 211–220.

Ross, R. & Thompson, D. 1910. A case of sleeping sickness studied by precise enumerative methods: regular periodic increase of the parasites observed. *Proceedings of the Royal Society of London, Series B,* **82**: 411–415.

Schmidt, A. & Krauth-Siegel, R.L. 2002. Enzymes of the trypanothione metabolism as targets for antitrypanosomal drug development. *Current Topics in Medical Chemistry,* **2**: 1239–1259.

Thon, G., Baltz, T. & Eisen, H. 1989. Antigenic diversity by the recombination of pseudogenes. *Genes Development,* **3**: 1247–1254.

Turner, C.M. 1997. The rate of antigenic variation in fly-transmitted and syringe-passaged infections of *Trypanosoma brucei. FEMS Microbiology Letters,* **153**: 227–231.

Turner, C.M. & Barry, J.D. 1989. High frequency of antigenic variation in *Trypanosoma brucei rhodesiense* infections. *Parasitology,* **99:** 67–75.

Vanhamme, L., Paturiaux-Hanocq, F., Poelvoorde, P., Nolan, D.P., Lins, L., Van Den Abbeele, J., Pays, A., Tebabi, P., Van Xong, H., Jacquet, A., Moguilevsky, N., Dieu,

M., Kane, J.P., De Baetselier, P., Brasseur, R. & Pays, E. 2003. Apolipoprotein L-I is the trypanosome lytic factor of human serum. *Nature,* **422**: 83–87.

Wang, J., Bohme, U. & Cross, G.A. 2003. Structural features affecting variant surface glycoprotein expression in *Trypanosoma brucei. Molecular and Biochemical Parasitology,* **128**: 135–145.

Wang, J., Van Praagh, A., Hamilton, E., Wang, Q., Zou, B., Muranjan, M., Murphy, N.B. & Black, S.J. 2002. Serum xanthine oxidase: origin, regulation, and contribution to control of trypanosome parasitemia. *Antioxidants Redox Signaling,* **4**: 161–178.

Xong, H.V., Vanhamme, L., Chamekh, M., Chimfwembe, C.E., Van Den Abbeele, J., Pays, A., Van Meirvenne, N., Hamers, R., De Baetselier, P. & Pays, E. 1998. A VSG expression site-associated gene confers resistance to human serum in *Trypanosoma rhodesiense. Cell,* **95**: 839–846.

VACCINATION AGAINST TICKS AND THE CONTROL OF TICKS AND TICK-BORNE DISEASE

Peter Willadsen
CSIRO Livestock Industries, Queensland Bioscience Precinct, 306 Carmody Rd., St. Lucia, Queensland 4067, Australia.

Abstract: Economic losses due to ticks and tick-borne disease of livestock fall disproportionately on developing countries. Currently, tick control relies mostly on pesticides and parasite-resistant cattle. Release of a commercial recombinant vaccine against *Boophilus microplus* in Australia in 1994 showed that anti-tick vaccines are a feasible alternative. For vaccines, it is important to understand the efficacy needed for a beneficial outcome. In this, it is relevant that some tick antigens affect multiple tick species; that existing vaccines could be improved by the inclusion of additional tick antigens; and that vaccination against ticks can have an impact on tick-borne disease. Practically, although recombinant vaccine manufacture involves relatively few steps, issues of intellectual property rights (IPR) and requirements for registration of a product may affect economic viability of manufacture. Hence practical vaccines for the developing world will require both successful science and a creative "business solution" for delivery in a cost-effective way.

1. INTRODUCTION

Ticks and tick-borne disease can seriously diminish the well-being of human kind, domestic animals and wild animal populations from the subarctic to the tropics. There is no doubt, however, that the impact, particularly on livestock, is felt most strongly in the tropical and subtropical parts of the world. Hence, the enormous economic losses (de Castro, 1997; Perry *et al.*, 2002) fall disproportionately on the developing countries of the world.

H. P. S. Makkar and G. J. Viljoen (eds.), Applications of Gene-Based Technologies for Improving Animal Production and Health in Developing Countries, 313-321.

The case for anti-tick vaccines and for vaccines against tick-borne disease has been made repeatedly. Globally, the single, most economically important tick is *Boophilus microplus,* the one-host parasite of cattle. This is an endemic parasite of cattle in the major beef producing countries of the world, and problems with pesticide resistance in the tick population, as well as concerns about chemical residues in meat, are two of the factors that drove development of vaccines against this tick species. In many developing countries, the situation is more complex. Domestic animals are likely to be exposed to a number of tick species and these tick species are likely to be multi-host. While the evidence to date is that pesticide resistance is far less problematic in these species than in the one-host *B. microplus,* the simultaneous presence of multiple species is a challenge for a vaccine. Nevertheless, anti-tick vaccines may still be highly desirable in such circumstances because they exhibit a number of other potential advantages, several of which are particularly appropriate to the developing-country situation. Vaccines are applicable to all genotypes. Vaccines are small and transportable; recombinant vaccines have the potential to be highly stable; they have the potential to be cheap technology; and they are non-contaminating for both animal products and the environment. Although the correct application of a vaccine has its own issues, vaccination can in many circumstances be less problematic than the appropriate use of pesticides. There is much evidence that the application of pesticide for tick control is frequently implemented inappropriately or ineffectively.

2.　　FEASIBILITY OF VACCINATION

The feasibility of developing, commercializing and distributing a vaccine against ticks was first demonstrated by the commercial release of the TickGARD recombinant vaccine against *B. microplus* in 1994 in Australia, and the parallel release of the Gavac vaccine against the same tick in Cuba at about the same time. The development of the original vaccine has been described (Willadsen, 1995). A few follow-up studies of the use of the vaccine in Australia have been published (Jonsson *et al.*, 2000; Willadsen, in press), as has some information on its use in Brazil (Hungerford *et al.*, 1995). Considerably more has appeared on the application of the Cuban vaccine, both in that country and in other parts of South America (de la Fuente *et al.*, 1999; Rodriguez *et al.*, 1995a; Rodriguez *et al.*, 1995b). One may ask why, if such vaccines are successful, this remains the only such vaccine nine years after first commercial release. This is an issue this paper seeks to address. Before looking at specific issues in vaccine development, however, it should be remembered that the development of the vaccine against *B. microplus* was almost certainly the only time that a sustained and adequately funded effort was made to develop a vaccine against any tick species, and that the effort was still small, by comparison with the resources poured into pesticide development. Lack of sustained support may well be the single most important factor in the relative lack of progress in this area of research and development.

Issues in the future development of tick vaccines will be discussed under three headings: performance criteria and measures of efficacy; scientific feasibility; and vaccine manufacture and registration. Finally, some comments will be made on the role of tick vaccines in the control of tick-borne disease.

3.   PERFORMANCE CRITERIA AND MEASURES OF EFFICACY

In seeking to develop a vaccine, as for any other form of parasite control, it is both scientifically desirable and commercially important to define the level of efficacy necessary for a beneficial outcome. Consider two examples. *Boophilus microplus* has its impact on cattle production through the transmission of tick-borne disease organisms such as *Babesia bovis, Babesia bigemina* and *Anaplasma marginale,* as well as through a direct effect on animal productivity through blood loss and tick worry. If tick-borne diseases are controlled, then the aim of tick control is to reduce the direct impact on cattle. This impact is typically considered to be directly proportional to the number of ticks engorging (Sutherst *et al.*, 1979). The performance criterion for a vaccine is then relatively straightforward. It must reduce tick numbers to the level where impact on productivity is minimized and the vaccination itself is cost effective. A very different situation arises with a tick whose major impact is on the transmission of a tick-borne disease that causes very high levels of mortality: *Rhipicephalus appendiculatus* and the transmission of *Theileria parva* may be an example. In such circumstances, the performance criterion for an anti-tick vaccine is that it reduces the transmission of tick-borne disease to a tolerable level – a very different and probably vastly more difficult target to achieve.

Unless vaccination against ticks is totally effective, which is unlikely, this question of efficacy needed to have a beneficial effect is both complex and vital. A survey of the literature will show that many of the tick antigens that have been studied induce immunity in a host that results in a 30–70% reduction in tick survival, engorgement or reproduction. The case for taking these observations the next step towards commercial development must depend on demonstrating that such levels of protection are worthwhile for the particular tick–host system of commercial interest.

4.   SCIENTIFIC FEASIBILITY

Scientifically, the identification of increasing numbers of antigens from a variety of tick species has proceeded apace, particularly over the last few years. Feasibility is demonstrated by historical studies of the application of the existing vaccines in the field; by the existence of characterized, recombinant antigens that could be incorporated into alternative vaccine formulations; by the demonstrated ability of some antigens to affect multiple tick species; and by the evidence that vaccination against ticks has an impact on tick-borne disease. Current evidence in each of these areas is summarized below.

Two commercial vaccines against *B. microplus* have been developed, both single antigen recombinant vaccines delivered in an adjuvant. These vaccines are based on essentially the same protein antigen, Bm86 (de la Fuente *et al.*, 1999; Rodriguez *et al.*, 1995b; Willadsen *et al.*, 1989; Rand *et al.*, 1989; Cobon *et al.*, 1996). They have been shown to be efficacious, though with some level of variation, in Argentina, Australia, Cuba, Brazil and Mexico. There are claims that the original Cuban vaccine is ineffective against two strains of *B. microplus* in Argentina (Garcìa-Garcìa *et al.*, 1999). This lack of efficacy is overcome using a minor sequence variant of the original Bm86 (Garcìa-Garcìa *et al.*, 2000), though the evidence for this is still somewhat confusing (Willadsen, in press). The same antigen was reported to be highly efficacious in Egypt (Khalaf, 1999), while a synthetic peptide vaccine based on the Bm86 sequence is efficacious in Colombia (Patarroyo *et al.*, 2002).

In limited trials, the original Bm86 vaccine has shown to be highly efficacious against *Boophilus annulatus* (Fragoso *et al.*, 1998; Pipano *et al.*, 2003), as well as *Hyalomma anatolicum* and *Hyalomma dromedarii* (de Vos *et al.*, 2001). It has a lower level of efficacy against *Boophilus decoloratus,* and little efficacy against *R. appendiculatus* or *Amblyomma variegatum* (de Vos *et al.*, 2001). The 64TRP antigen from *R. appendiculatus* (Trimnell, Hails and Nuttall, 2002) is also efficacious against *Ixodes ricinus* in mice (Labuda *et al.*, 2002). That is, there is now good evidence that cross-species protection by a single antigen may occur, though in an unpredictable fashion.

If one sought to increase the efficacy of the existing vaccines or to protect against other tick species, then availability of alternative tick antigens would be essential. A survey of the literature shows that in addition to Bm86, another five tick proteins have been tested as recombinant antigens and shown to be efficacious, while the results of trials with recombinants are inconclusive for a further four. A further four have been evaluated with positive results as native proteins, and so presumably would be candidates for recombinant expression and testing (Willadsen, 2001, in press).

Anti-tick vaccines would be at their most useful if they also had an impact on the transmission of tick-borne diseases. Tentative evidence to date suggests this may be found in particular cases, as summarized in Section 6, below.

5.   VACCINE MANUFACTURE AND REGISTRATION

Suppose that a prototype anti-tick vaccine of sufficient efficacy has been developed in the laboratory and validated in small-scale trials with a livestock animal of importance. This is the most that can be achieved with the resources typically available to academic research institutions. How can this prototype be converted into a vaccine deliverable to farmers?

It is commonly assumed that commercial development requires the resources and the input of one of the decreasing number of global veterinary product or vaccine manufacturers. Certainly, should such a manufacturer become committed to the development of an anti-tick vaccine, that can be an ideal outcome. It is likely, however, that there will be an immediate incompatibility of interests. Although an anti-tick vaccine may be of great value to the farmers of a developing country, it is very likely that the vaccine will represent a small and unreliable market for the manufacturer – far too small to be commercially attractive. This fact alone may account for much of the lack of development of vaccines over the last decade. The issue, as it relates to the commercial development of human vaccines, has been well described (Batson, 2002). For veterinary vaccines, the situation will be easier in that a broader range of adjuvants may be acceptable and registration requirements may be less demanding, though the markets will be of correspondingly lower value. It is therefore worth considering alternative routes to manufacture. It is important to remember that, in recognition of the importance of market failure in the delivery of vaccines to agriculture, many of the most successful have been manufactured by state or government enterprises. Two current examples include the manufacture of vaccines against babesiosis and anaplasmosis in Australia (Callow, Dalgleish and de Vos, 1997) and against tropical theileriosis caused by *Theileria annulata* in China (Gu *et al.*, 1997; Shirong, 1997). How feasible is this for a recombinant vaccine?

At its simplest, manufacture of a recombinant vaccine involves relatively few steps: production of a recombinant antigen in one of a small number of commercially viable expression systems, most typically in bacteria or yeast, followed by downstream purification and processing of the product to an acceptable level of purity and efficacy, then formulation, typically in an adjuvant, and packaging for distribution. Each step of this relatively simple process relies on usually well-documented methods and each is typically available as a contracted service from local service suppliers. Under the best circumstances, antigen yields of up to 10,000 doses per litre of low cost medium can be achieved, making vaccine manufacture for a local market, even national markets, a small-scale process. Is it then unrealistic to imagine that the process of vaccine manufacture for a developing country market could proceed outside the framework of the major international animal health companies?

The limitations on the feasibility of such an approach are likely to be two-fold: the requirements for registration of a product; and intellectual property rights (IP) restricting freedom to operate. Consider first the requirements for registration. Recombinant veterinary vaccines against parasites are novel products; only one has been fully commercialized to date. They are likely to operate through mechanisms that differ from current, usually chemical, control technologies, and to produce different results. The result is that, lacking guidelines by which the acceptability of such a new product may be judged, regulatory requirements may be excessive, or uncertain. Registration of the TickGARD vaccine against *Boophilus microplus* was preceded by vaccination of approximately 18,000 cattle in a variety of trials (Willadsen, 1995). This is not only a serious deterrent to the development of such products but also a deterrent to their improvement, since improvements are likely to require re-registration.

The issue of IPR now impinges on multiple aspects of biotechnology. Anyone wishing to manufacture a recombinant vaccine for a developing country is likely to find antigen, expression system and adjuvant subject to IPR conditions. This may or may not be a serious deterrent, but the issue must be considered.

6.   TICK-BORNE DISEASE

One of the most common approaches to the control of tick-borne disease is the control of ticks (Kocan, 1995). A tick vaccine is likely to modulate the normal processes of tick-host interaction. The destruction of the gut of *B. microplus* by vaccinating with the Bm86 antigen is a simple example. Since tick-borne pathogens rely on and exploit this tick-host interaction in various and often complex ways in their life cycles, vaccination has the potential to have a direct impact on tick-borne disease. There is tentative experimental evidence for this in three areas. First, field data from Cuba has shown a substantial, though variable, reduction in the incidence of babesiosis and anaplasmosis following vaccination against *B. microplus* (de la Fuente *et al.*, 1998). Second, there is evidence that vaccination of cattle with the Australian TickGARD vaccine prevents transmission of *Babesia bigemina* by *Boophilus annulatus* and reduces the severity of *Babesia bovis* infection (Pipano *et al.*, 2003). Finally, it has recently been reported that vaccination of mice with recombinant forms of the 64 TRP protein from *R. appendiculatus* increased the survival of mice challenged with tick-borne encephalitis virus-infected *I. Ricinus* and partially blocked transmission of virus to uninfected ticks (Labuda *et al.*, 2002).

7.   CONCLUSION

Development and application of successful vaccines against ticks and the control of tick-borne disease will demand collaboration between the research community and those responsible for the delivery of such technologies to farmers of the developing world. The research task is to more accurately define the problem and hence the criteria that must be met for a solution to emerge, and, in parallel, to develop the appropriate vaccines and vaccination strategies. This is still only part of the problem. Without the "business solution" that will manufacture and deliver such vaccines in a cost-effective way, the finest research will fail to achieve a real impact on livestock production. That "business solution" will require a creative approach to product manufacture and registration. Although this represents a challenge, the potential benefits of success are enormous.

REFERENCES

Batson, A. 2002. The costs and economics of modern vaccine development in Orphan Vaccines – Bridging the Gap. *Developmental Biology,* **110:** 15–24.

Callow, L.L., Dalgleish, R.J. & de Vos, A.J. 1997. Development of effective living vaccines against bovine babesiosis – the longest field trial? *International Journal of Parasitology,* **27:** 747–767.

Cobon, G.S., Moore, J.T., Johnston, L.A.Y., Willadsen, P., Kemp, D.H., Sriskantha, A., Riding, G.A. & Rand. K.N. 1996. DNA encoding a cell membrane glycoprotein of a tick gut. US Patent and Trademark Office Application No. 325 071 435/240.2.

de Castro, J.J. 1997. Sustainable tick and tickborne disease control in livestock improvement in developing countries. *Veterinary Parasitology,* **71**: 77–97.

de la Fuente, J., Rodriguez, M., Montero, C., Redondo, M., Garcìa-Garcìa, J.C., Mendez, L., Serrano, E., Valdes, M., Enriquez, A., Canales, M., Ramos, E., Boue, O., Machado, H. & Lleonart, R. 1999. Vaccination against ticks (*Boophilus* spp.): the experience with the Bm86-based vaccine Gavac I. *Genetic Analysis Biomolecular Engineering,* **15**: 143–148.

de la Fuente, J., Rodriguez, M., Redondo, M., Montero, C., Garcìa-Garcìa, J.C., Mendez, L., Serrano, E., Valdes, M., Enriquez, A., Canales, M., Ramos, E., Boue, O., Machado, H., Lleonart, R., de Armas, C.A., Rey, S., Rodriguez, J.L., Artiles, M. & Garcìa, L. 1998. Field studies and cost-effectiveness of vaccination with Gavac against the cattle tick *Boophilus microplus. Vaccine,* **16**: 366–373.

de Vos, S., Zeinstra, L., Taoufik, O., Willadsen, P. & Jongejan, F. 2001. Evidence for the utility of the Bm86 antigen from *B. microplus* in vaccination against other tick species. *Experimental Applied Acarology,* **25**: 245–261.

Fragoso, H., Hoshman-Rad, P., Ortiz, M., Rodgrigues, M., Redondo, M., Herrera, L. & de la Fuente, J. 1998. Protection against *Boophilus annulatus* infestations in cattle vaccinated with the *B. microplus* Bm86-containing vaccine Gavac. *Vaccine,* **16**: 1990-1992.

Garcìa-Garcìa, J.C., Gonzalez, I.L., Gonzalez, D.M., Valdes, M., Mendez, L., Lamberti, J., D'Agostino, B., Citroni, D., Fragoso, H., Ortiz, M., Rodriguez, M. & de la Fuente, J. 1999. Sequence variations in the *B. microplus* Bm86 locus and implications for immunoprotection in cattle vaccinated with this antigen. *Experimental Applied Acarology,* **11**: 883-895.

Garcìa-Garcìa, J.C., Montero, C., Redondo, M., Vargas, M., Canales, M., Boue, O., Rodriguez, M., Joglar, M., Machado, H., Gonzalez, I.L., Valdes, M., Mendez, L. & de la Fuente, J. 2000. Control of ticks resistant to immunization with Bm86 in cattle vaccinated with the recombinant antigen Bm95 isolated from the cattle tick, *B. microplus. Vaccine,* **18**: 2275–2287.

Gu, G., Junkal, S., Mbying, Y., Jukui, Z., Shuun, M., Ali, N., Lixin, T. & Maliyam. 1997. Research on the schizont cell culture vaccine against *Theileria annulata* infection in Xinjiang, China. *Tropical Animal Health and Production,* **29**: 98S–100S.

Hungerford, J., Pulga, M., Zwtsch, E. & Cobon, G. 1995. Efficacy of TickGARD™ in Brazil. *In: Resistencia Y Control en Garrapatas Y Moscas de Importancia Veterinaria.* Seminario Internacional de Parasitologia Animal, Sagar Canifarma, Mexico, 11–13 October 1995. Sagar/Canifarma/FAO/IICA/INIFAP, Acapulco.

Jonsson, N.N., Matschoss, A.L., Pepper, P., Green, P.E., Albrecht, M.S., Hunterford, J. & Ansell, J. 2000. Evaluation of TickGARD(PLUS), a novel vaccine against *B. microplus,* in lactating Holstein-Friesian cows. *Veterinary Parasitology,* **88**: 275–285.

Khalaf, A.S.S. 1999. Control of *B. microplus* ticks in cattle calves by immunization with a recombinant Bm86 glucoprotein antigen preparation. *Deutsche Tierarztliche Wochenschrift,* **106**: 248–251.

Kocan, K.M. 1995. Targeting ticks for control of selected hemoparasitic diseases of cattle. *Veterinary Parasitology,* **57**: 121–151.

Labuda, M., Trimnell, A.R., Lickova, M., Kazimirova, M., Slovak, M. & Nuttall, P.A. 2002. Recombinant tick salivary antigens (64TRP) as a TRANSBLOK vaccine against tick-borne encephalitis virus. p. 51, *in:* Abstracts of the *4th International Conference on Ticks and Tick-Borne Pathogens.* Banff, Canada, 21–26 July 2002.

Patarroyo, J.H., Portela, R.W., de Castro, R.O., Pimentel, J.C., Guzman, F., Patarroyo, M.E., Vargas, M.I., Prates, A.A. & Mendes, M.A.D. 2002. Immunization of cattle with synthetic peptides derived from the *B. microplus* gut protein (Bm86). *Veterinary Immunology and Immunopathology,* **88**: 163–172.

Perry, B.D., Randolph, T.F., McDermott, J.J., Sones, K.R. & Thornton, P.K. 2002. Investing in animal health to alleviate poverty. ILRI [International Livestock Research Institute] Nairobi, Kenya.

Pipano, E., Alekceev, E., Galker, F., Fish, L., Samish, M. & Shkap, V. 2003. Immunity against *Boophilus annulatus* induced by the Bm86 (Tick-GARD) vaccine. *Experimental Applied Acarology,* **29**: 141–149.

Rand, K.N., Moore, T., Sriskantha, A., Spring, K., Tellam, R., Willadsen, P. & Cobon, G.S. 1989. Cloning and expression of a protective antigen from the cattle tick *B. microplus. Proceedings of the National Academy of Sciences, USA,* **86**: 9657–9661.

Rodriguez, M., Massard, C.L., Henrique da Fonseca, A., Ramos, N.F., Machado, H., Labarta, V. & de la Fuente, J. 1995a. Effect of a vaccination with a recombinant Bm86 antigen preparation on natural infestations of *B. microplus* in grazing dairy and beef pure and cross-bred cattle in Brazil. *Vaccine,* **13**: 1804–1808.

Rodriguez, M., Penichet, M.L., Mouris, A.E., Labarta, V., Luaces, L.L., Rubiera, R., Cordoves, C., Sanchez, P.A., Ramos, E., Soto, A., Canales, M., Palenzuela, D., Triiguero, A., Lleonart, R., Herrera, L. & de la Fuente, J. 1995b. Control of *B. microplus* populations in grazing cattle vaccinated with a recombinant Bm86 antigen preparation. *Veterinary Parasitology,* **57**: 339–349.

Shirong, S. 1997. Application and effect of the schizont vaccine against *Theileria annulata* in China. *Tropical Animal Health and Production,* **29**: 101S–103S.

Sutherst, R.W., Norton, G.A., Barlow, N.D., Conway, G.R., Birley, M., Comins, H.N. 1979. An analysis of management strategies for cattle tick (*Boophilus microplus*) control in Australia. *Journal of Applied Ecology,* **16**: 359–382.

Trimnell, A.R., Hails, R.S. & Nuttall, P.A. 2002. Dual action ectoparasite vaccine targeting 'exposed' and 'concealed' antigens. *Vaccine,* **20**: 3360–3568.

Willadsen, P. 2001. The molecular revolution in the development of vaccines against ectoparasites. *Veterinary Parasitology,* **101**: 353–367.

Willadsen, P. In press. Anti-tick Vaccines. *Parasitology* (in press).

Willadsen, P., Bird, P., Cobon, G.S. & Hungerford, J. 1995. Commercialisation of a recombinant vaccine against *B. microplus. Parasitology,* **110**: 43–50.

Willadsen, P., Riding, G.A., McKenna, R.V., Kemp, D.H., Tellam, R.L., Nielsen, J.N., Lahnstein, J., Cobon, G.S. & Gough, J.M. 1989. Immunologic control of a parasitic arthropod. Identification of a protective antigen from *B. microplus. Journal of Immunology,* **143**: 1346–1351.

DEVELOPMENT OF MARKER VACCINES FOR RINDERPEST VIRUS USING REVERSE GENETICS TECHNOLOGY

Satya Parida, Edmund P. Walsh, John Anderson, Michael D. Baron and Thomas Barrett*
*Institute for Animal Health, Pirbright Laboratory, Ash Road, Woking, Surrey GU24 0NF, United Kingdom. *Corresponding author*

Abstract: Rinderpest is an economically devastating disease of cattle (cattle plague), but a live-attenuated vaccine has been very successfully used in a global rinderpest eradication campaign. As a consequence, the endemic focus of the virus has been reduced to an area in eastern Africa known as the Kenya-Somali ecosystem. Although the vaccine is highly effective, it has a drawback in that vaccinated animals are serologically indistinguishable from those that have recovered from natural infection. In the final stages of the eradication campaign, when vaccination to control the spread of disease will only be used in emergencies to contain an outbreak, a marker vaccine would be a very useful tool to monitor possible wild virus spread outside the vaccination area. Marker vaccines for rinderpest, and other viruses with negative-sense RNA genomes, can now be produced using reverse genetics, and the development of such marker vaccines for rinderpest virus is described.

1.   INTRODUCTION

Rinderpest or "cattle plague" is an economically devastating disease that is still endemic in areas of eastern Africa, the aetiological agent being a morbillivirus (RPV) closely related to human measles virus (Barrett and Rossiter, 1999). These are viruses with non-segmented negative-sense RNA genomes that are not infectious as naked RNA. Vaccination is a very effective means of controlling rinderpest since there is only one serotype and

H. P. S. Makkar and G. J. Viljoen (eds.), Applications of Gene-Based Technologies for Improving Animal Production and Health in Developing Countries, 323-333.
© 2005 *IAEA. Printed in the Netherlands.*

a safe attenuated vaccine is available (Plowright and Ferris, 1962), which has been used in a concerted global eradication campaign conducted over the past two decades. The virus now remains only in an ill-defined focus in the Kenya–Somalia ecosystem in eastern Africa. This eradication campaign, known as the Global Rinderpest Eradication Programme (GREP), has as its goal the eradication of the disease by the year 2010. While it is clear that the existing vaccine is highly effective in protecting animals from rinderpest infection, it has a drawback in that vaccinated animals are serologically indistinguishable from those that have recovered from natural infection, and in the final stages of the eradication campaign, when mass vaccination must be discontinued, it would be desirable to use a vaccine that enables this distinction to be made, a so-called "marker vaccine". The ability to unequivocally identify vaccinated animals would allow serological detection of the disease despite vaccination whose history and extent are uncertain.

Marker vaccines for rinderpest, and other viruses with negative-sense RNA genomes, can now be produced using reverse genetics, the process whereby the genomes of RNA viruses can be genetically manipulated through a DNA copy and live virus rescued from the altered DNA. Marker vaccines are increasingly being used for disease eradication programmes, such as for the elimination of Aujeszky's disease and classical swine fever in pigs in Europe (Oirschot *et al.*, 1990; Dewulf *et al.*, 2003). In the final stages of the rinderpest eradication campaign, when vaccination to control the spread of disease will mainly be used in emergency outbreak response, a marker vaccine would be a very useful tool to allow monitoring to check for the spread of wild virus outside the vaccination area.

2.   REVERSE GENETICS

Advances in biotechnology over the past 20 years have revolutionized our approach to new vaccine design, and, in the past decade, reverse genetics has been developed for non-segmented negative-sense RNA viruses, thus allowing the genomes of viruses like those of rinderpest to be genetically modified (Baron and Barrett, 1997; Conzelman *et al.*, 1994; Nagai, 1999). This process allows "site-specific" mutagenesis of virus genomes and makes it possible to add marker genes to the vaccines or to make chimeric vaccines with genes derived from different morbilliviruses, resulting in recombinant viruses that have new immunogenic characteristics.

3.   EXISTING RINDERPEST MARKER VACCINES

The first approach to the production of marker rinderpest vaccines was to incorporate an extra "marker" gene into the Plowright vaccine strain genome. When foreign genes are inserted in the genomes of non-segmented negative strand RNA virus genomes, the positioning of the insertion site determines the level of the gene's expression: the closer the new gene is to the single promoter at the 3' end of the genome, the more mRNA will be produced and, consequently, the more protein (Flanagan *et al.*, 2001). The gradient in mRNA transcription from the first (N) to the last (L) gene results from the possibility that the polymerase will detach from the template at the stop-start signals between each gene. If the polymerase becomes detached, it must return to the 3' promoter and start again, so that the more genes the polymerase has to transcribe the higher the probability of it detaching. This results in abundant transcripts from the 3' proximal genes but low numbers of transcripts from the promoter distal genes.

Since the virus proteins produced from genes closest to the promoter – the nucleocapsid (N) and phospho (P) proteins – are required in large amounts for replication, disrupting these might adversely affect virus growth and so it was decided to insert the marker gene between the P gene and the matrix (M) protein gene, the next in the genome sequence (Walsh *et al.*, 2000a). The authors have now produced versions of the Plowright vaccine that express various forms of the green fluorescent protein (GFP) from modified GFP genes inserted into the cDNA between the P and M genes. Two of these vaccines, rRPV-InsGFP and rRPV-SigGFP, which expressed either intracellular or secreted forms of GFP, were highly effective in protecting animals from virulent rinderpest challenge, although the antibody response to the GFP protein elicited in the vaccinated animals depended on its site of expression. For example, intracellular expression of GFP failed to induce anti-GFP antibody in any of the vaccinated cattle, while secretion of GFP gave rise to a significant anti-GFP antibody response, but only in two of four vaccinated cattle (Walsh *et al.*, 2000a). This indicated that extracellular expression of the marker protein improved the antibody response, but simple secretion of the protein itself was insufficient to enable the generation of a strong and uniform anti-marker humoral antibody response in all vaccinated animals. Since a strong antigenic response to the marker protein is an essential prerequisite for any marker vaccine, a more effective antigen presentation strategy was needed to produce a marker antibody response in all vaccinated cattle. A recombinant RPV vaccine expressing the GFP in a form that was anchored on the outer membrane of infected cells was produced, and this vaccine (rRPV-AncGFP) elicited a humoral antibody response to the marker protein in all vaccinated animals and, at the same time, provided complete immunity to rinderpest virus challenge (Walsh *et al.*, 2000b). Another potential marker protein, the influenza A haemagglutinin (FluHA) protein, was also incorporated as a marker into the RPV vaccine genome and a vaccine, rRPV-InsHA, was produced. A strong antibody response was generated to the FluHA marker protein in all animals vaccinated with the RPV-FluHA recombinant (Walsh *et al.*, 2000b).

An important consideration when developing genetically modified virus vaccines is safety. The possibility that the marker proteins could be incorporated into the virus envelope and thus alter the tropism, and possibly the pathogenicity, of the vaccine produced must be considered, and so, as a safety feature, this vaccine construct expressed a receptor site mutant version of the influenza HA protein in which the receptor binding domain of the HA was mutated to severely reduce its binding efficiency. This precaution was taken in case the HA protein was incorporated into the virus envelope, which could then potentially alter the virus tissue tropism and perhaps the pathogenic potential of the recombinant virus. In general, viruses have mechanisms to exclude the incorporation of foreign proteins into their envelopes, and immunoprecipitation studies using antibodies to the marker proteins failed to precipitate the recombinant vaccine viruses, indicating that the marker is excluded from the virus envelopes. Subsequently, immunoelectron microscopic studies confirmed that budded virus envelopes were free of the marker FluHA protein (Walsh *et al.*, 2000b).

The different vaccines described in this section are all "positively" marked RPV vaccines. They are ideal vaccines for assessing the effectiveness of a vaccination programme and the level of vaccine cover achieved in vaccinated herds. However, they do not allow the identification of vaccinated animals that have subsequently become infected by wild-type virus.

4.   NEW RINDERPEST MARKER VACCINES

An alternative approach is to make "negatively" marked vaccines. These are required if we wish to use serological methods to look for the presence of wild-type virus circulation in the face of a level of vaccination cover that is insufficient to eliminate it. The missing component of the vaccine will only be detected in a vaccinated animal if it subsequently becomes infected. Negative marker DNA vaccines have been used for some time and the marker vaccine for Aujeszky's disease (Oirshot *et al.*, 1990) has a non-essential glycoprotein gene deleted. Vaccinated animals lack a response to the missing protein and so can be distinguished from naturally infected animals. However, the tight constraints imposed by an RNA genome means that no such "luxury" immunogenic gene is available for deletion in these viruses. A way to overcome this problem is to make chimeric vaccines with genes derived from related viruses. For example, to produce a marker vaccine for peste des petit ruminants (PPR), the two surface glycoprotein genes (H and F) of the RPV vaccine were substituted with those from the related PPR virus vaccine and it was shown that the resulting vaccine protects goats from virulent PPR challenge (Das *et al.*, 2000). This is a "negatively" marked PPR vaccine since it lacks many of the normal PPR antigens; in particular, it will have a different response to the highly antigenic N protein of PPR, which in this vaccine is derived from RPV. Two different competitive ELISAs (Anderson and McKay, 1994; Libeau *et al.*, 1995) are available, based on the use of specific monoclonal antibodies directed against either the virus N or H proteins, that can serologically distinguish RPV and Peste des petit ruminants virus (PPRV) infections, and these can be used to identify animals vaccinated with this chimeric vaccine and distinguish them from animals that have been infected with either of the parent viruses.

A similar chimeric RPV vaccine, in which the N protein gene was replaced by that derived from the PPR vaccine, was produced and tested in cattle (rRPV-PPRN) (Parida, Kumar, Baron and Barrett, unpublished). The vaccinated animals can also be distinguished from naturally infected ones using two currently available ELISAs that detect antibodies specific to either the N or haemagglutinin (H) proteins of these two viruses. Vaccinated animals become positive in the PPRV-specific anti-N antibody test and remain negative in the equivalent RPV-specific test. The opposite is true in the case of H protein-specific antibodies. Vaccinated animals that subsequently become infected with wild type virus would become double positive for the N protein antibodies of both viruses (see Table 1).

Table 1. Expected reactivity in diagnostic tests for RP-PPR-N vaccinated animals.

	Specific Monoclonal Antibody cELISA			
Exposure	PPR-N	RPV-N	PPR-H	RPV-H
Vaccinated	+	–	–	+
Infected	–	+	–	+
Vaccinated and infected	+	+	–	+

5.   POSITIVE-NEGATIVE RINDERPEST MARKER VACCINE

In a further development, the authors have introduced a positive marker gene, again the anchored form of GFP, into the rRPV-PPRN genome, a vaccine named (rRPV-PPRN-AncGFP), in an attempt to make a positive-negative marker vaccine. The organization of the different vaccine genomes produced for virus rescue is outlined in Figure 1.

5.1   Construction

Virus was rescued from this cDNA, which, as expected, expressed GFP on the surface of infected Vero cells. GFP expression was demonstrated by autofluorescence of the GFP and using an anti-GFP antibody for immunofluorescence staining of the cells (see Figure 2). Unfortunately, this "positive-negative" marker vaccine, unlike the other marker vaccines produced thus far, did not grow to high titre in Vero cells, reaching a titre of approximately 10^3 TCID$_{50}$/ml instead of the usual titre of over 10^6.

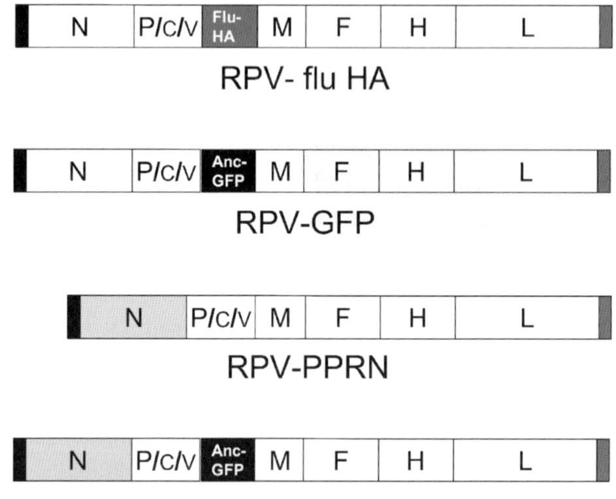

Figure 1. Organization of the different RPV genomes. The gene order of each of the RPV recombinant viruses produced is shown. The normal order of the RPV genome is 3' N, P, M, F, H, L 5' and the Flu HA and Anc GPF genes were inserted between the P and M genes in order to minimize the effect on virus growth and replication. The N gene of PPR is shaded grey in the rRPV-PPRN and rRPV-PPRN-AncGFP chimeras.

5.2   Vaccination

In spite of its poor growth, this virus was used to vaccinate three cattle using a dose of $10^{3.7}$ TCID$_{50}$ of the virus stock, the standard field vaccine dose being 10^2 TCID$_{50}$. The vaccinated animals developed neutralizing antibodies to rinderpest, but at a reduced rate compared to the rRPV-PPR-N chimera. Four weeks after vaccination the cattle were challenged with virulent rinderpest virus (Saudi/81 strain) and were completely protected from clinical disease. They showed an anamnestic response in neutralizing antibody titre, indicating that replication of the challenge virus had occurred (Figure 3). The low titre of neutralizing antibody was reflected in a low response in the cELISA in detection of specific anti-RPV H antibodies following vaccination, compared with that seen with the normal vaccine (Anderson and McKay, 1994). While vaccination with the normal vaccine resulted in the production of high levels of anti-RPV H protein antibody by 4 weeks, the rRPV-PPRN-AncGFP vaccine did not. Again, a strong anamnestic response was seen following challenge, confirming that replication of the challenge virus had occurred in the absence of clinical disease (Figure 4). No antibody response was detected to the GFP protein.

6.   DISCUSSION

New RPV vaccines that allow identification of each vaccinated animal have been developed based on alterations to the genome of the Plowright vaccine (Plowright and Ferris, 1962). This vaccine strain has been used very successfully for vaccination of cattle and other ruminant species for the past 40 years without any adverse effects and has resulted in the near complete eradication of rinderpest from the globe (Rweyemamu and Cheneau, 1995). Positive marker genes (GFP or FluHA) introduced into the genome will allow detection of vaccinated animals in a population, while that with a negative marker (RPV-PPRN chimera) will, in addition, allow wild virus circulation to be detected in the vaccinated population. The next phase will be to test the more promising new RPV marker vaccines – rRPV-Flu HA, rRPV-AncGFP and rRPV-PPRN – in a larger number of cattle. In particular, the antibody responses will need to be studied in African breeds of cattle in order to determine the average level of antibody response to the marker antigens. In addition, the companion diagnostic tests will need to be validated and the minimum vaccine dose that will give solid protection along with an acceptable marker antibody response needs to be established.

However, the rRPV-PPRN-AncGFP vaccine is not suitable for field testing. Its slow growth in tissue culture was reflected in poor growth *in vivo* and both the anti-virus and the anti-GFP antibody responses were too weak to allow accurate detection using ELISA. The availability of a double-marker vaccine (positive and negative markers), although not essential to the success of the eradication campaign, would be desirable since it would allow easier confirmation of an animal's status with respect to vaccination or infection, or both.

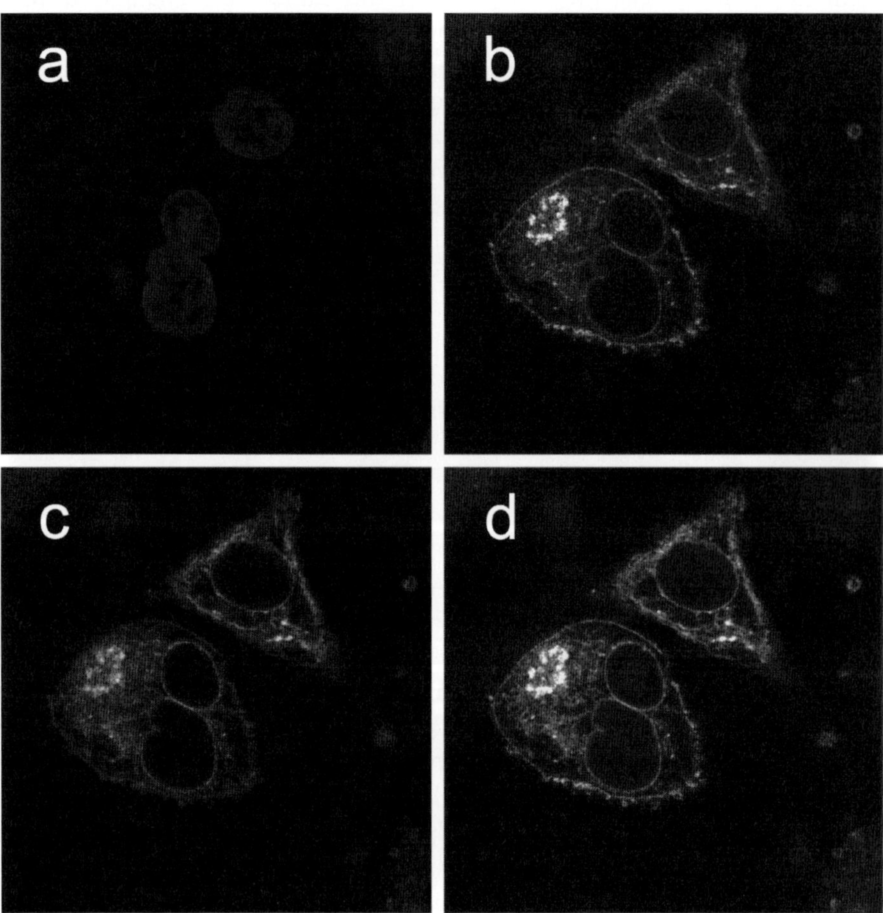

Figure 2. Expression of GFP from rRPV-PPRN-GFP. Panel a: Nuclear staining with DAPI. Panel b: GFP autofluorescence. Panel c: Staining with anti-GFP antibody. Panel d: Merged panels a to c.

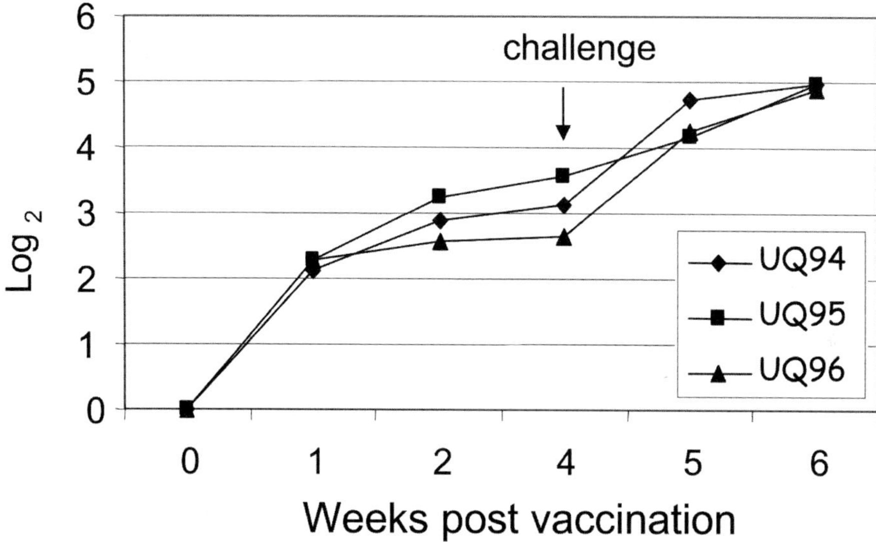

Figure 3. Virus neutralizing antibody responses following vaccination with rRPV-PPRN-GFP. The virus neutralizing antibody responses after vaccination of 3 cattle (animals UQ94, UQ95, UQ96) were determined in a micro-neutralization assay as described in Barrett *et al.*, 1989. The cattle were challenged at 4 weeks post-vaccination with 10^4 TCID$_{50}$ of virulent Saudi/81 strain of rinderpest virus.

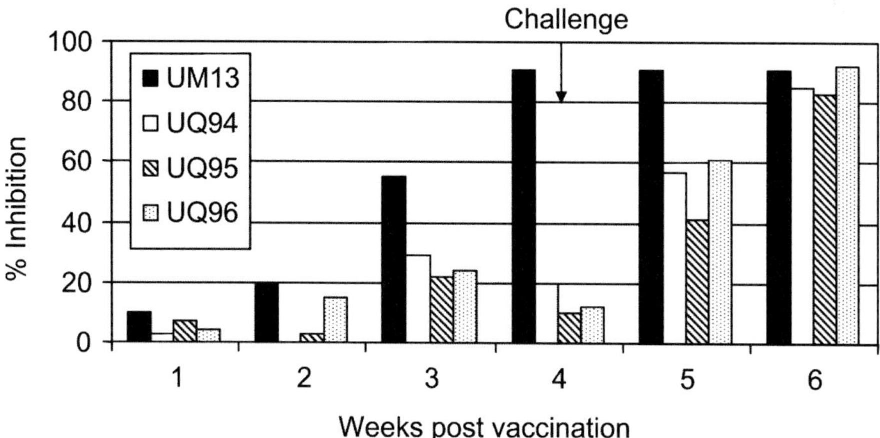

Figure 4. H-specific antibody responses following vaccination with rRPV-PPRN-GFP. The development of specific-anti-H protein antibodies was determined using a cELISA (Anderson and McKay, 1994). The antibody responses after vaccination of 3 cattle (animals UQ94, UQ95, UQ96) were compared with the response following vaccination with the currently used RBOK rinderpest vaccine (animal UM13). The cattle were challenged at 4 weeks post-vaccination with 10^4 TCID$_{50}$ of virulent Saudi/81 strain of rinderpest virus.

The authors are currently investigating other ways to make positive-negative marker vaccines using chimeric H and N protein genes to replace their normal homologues in the vaccine strain. This very powerful new technology for manipulating morbillivirus genomes, in addition to its practical usefulness in allowing the development of marker vaccines, also enables us to investigate, on a rational scientific basis, the molecular determinants of virulence and attenuation in this virus group. This will be an important consideration when assessing the safety of both existing and new vaccines for general use.

ACKNOWLEDGEMENTS

This work was supported by grants from UK Department for International Development (DFID) and the Biotechnology and Biological Sciences Research Council (BBRSC), UK.

REFERENCES

Anderson, J. & McKay, J.A. 1994. The detection of antibodies against Peste des petits ruminants virus in cattle, sheep and goats and the possible implications to rinderpest control programmes. *Epidemiology and Infection,* **112**: 225–231.

Barrett, T., Belsham, G.J., Shaila, M.S. & Evans, S.A. 1989. Immunisation with a vaccinia recombinant expressing the F protein protects rabbits from challenge with a lethal dose of rinderpest virus. *Virology,* **170:** 11–18.

Baron, M.D. & Barrett, T. 1997. Rescue of recombinant rinderpest virus from cloned cDNA. *Journal of Virology,* **71**: 1265–1271.

Barrett, T. & Rossiter, P. 1999. Rinderpest: the disease and its impact on humans and animals. *Advances in Virus Research,* **53**: 89–110.

Conzelmann, K.K. & Schnell, M. 1994. Rescue of synthetic genome analogs of rabies virus by plasmid-encoded DNA. *Journal of Virology,* **68**: 713–719.

Das, S.C., Baron, M.D. & Barrett. T. 2000. Recovery and characterisation of a chimeric rinderpest virus with the glycoproteins of Peste des petits ruminants virus: Homologous F and H proteins are required for virus viability. *Journal of Virology,* **74**: 9039–9047.

Dewulf, J., Laevens, H., Koenen, F., Mintiens, K. & de Kruif, A. 2003. A comparative study for emergency vaccination against classical swine fever with an E2 sub-unit marker vaccine and a C-strain vaccine. pp. 221–231, *in:* Proceedings of the Society for Veterinary Epidemiology and Preventive Medicine. University of Warwick, UK, 1–2 April 2003.

Flanagan, E.B., Zamparo, J.M., Ball, L.A., Rodriguez, L.L. & Wertz, G.W. 2001. Rearrangement of the genes of vesicular stomatitis virus eliminates clinical disease in the natural host: new strategy for vaccine development. *Journal of Virology,* **75**: 6107–6114.

Libeau, G., Perhaud, C., Lancelot, R., Colas, F., Guerre, L., Bishop, D.H.L. & Diallo, A. 1985. Development of a competitive ELISA for detecting antibodies to the Peste des petits ruminants virus using a recombinant nucleoprotein. *Research in Veterinary Science,* **58**: 50–55.

Nagai, Y. 1999. Paramyxovirus replication and pathogenesis: Reverse genetics transforms understanding. *Reviews in Medical Virology,* **9**: 83–99.

Oirschot, J.T., Van Terpstra, C., Moormann, R.J.M., Berns, A.J. & Gielkens, A.L.J. 1990. Safety of an Aujeszky's disease vaccine based on deletion mutant strain 783 which does not express thymidine kinase and glycoprotein 1. *Veterinary Record,* **127**: 443–446.

Plowright, W. & Ferris, R.D. 1962. Studies with rinderpest virus in tissue culture. The use of attenuated culture virus as a vaccine for cattle. *Research in Veterinary Science,* **3**: 172–182.

Rweyemamu, M.M. & Cheneau, Y. 1995. Strategy for the global rinderpest eradication programme. *Veterinary Microbiology,* **44**: 369–376.

Walsh, E.P., Baron, M.D., Rennie, L., Anderson, J. & Barrett, T. 2000. Development of a genetically marked recombinant rinderpest vaccine expressing green fluorescent protein. *Journal of General Virology,* **81**: 709–718.

Walsh, E.P., Baron, M.D., Rennie, L., Monahan, P., Anderson J. & Barrett, T. 2000. Recombinant rinderpest vaccines expressing membrane anchored proteins as genetic markers: evidence for exclusion of marker protein from the virus envelope. *Journal of Virology,* **74**: 10165–10175.

EVALUATION OF DIAGNOSTIC TOOLS FOR EPIDEMIOLOGICAL PURPOSES – APPLICATION TO FMD

Ingrid Bergmann, Viviana Malirat, Erika Neitzert and Eduardo Correa Melo
Pan American Foot-and-Mouth Disease Center, PAHO/WHO.

Abstract
Since the implementation in 1988 of the foot-and-mouth disease (FMD) eradication programme in South America (PHEFA), significant advances have been made in the control of the disease. Particularly relevant has been the progress attained in one of the three subregions, the Southern Cone, where during the late 1990s most areas were recognized as FMD-free by OIE, with or without vaccination. To achieve PHEFA goals, development and implementation of diagnostic tools to trace for potential sources of infection was of utmost importance. Accordingly, priority was given to the identification of genetic links among circulating strains and, in relevant areas, to the systematic sero-epidemiological monitoring of persistent viral activity, regardless of vaccination condition. A genetic database of representative strains from South America, constructed at the Pan American Foot-and-Mouth Disease Center, constituted the basis for the phylogenetic analysis of viruses types O and A recorded recently in endemic regions, and during the emergencies in FMD-free areas of the Southern Cone. The genetic data placed all variants from the Southern Cone in a single lineage, which clearly differed from the evolutionary nodes that included the isolates in the Andean subregion, reflecting two independent production systems and livestock trade circuits. To complement the search for potential sources of infection, serosurveillance became an important adjunct. To this end, new tools to infer viral circulation within and between herds, irrespective of vaccination, were developed and validated. The diagnostic approach, based on an immuno-enzymatic system that detects antibodies against non-capsid proteins, was particularly useful in support of epidemiological investigations after episodes and to confirm absence of viral activity in regions to be recognized as FMD-free, being especially important prior to the suspension of vaccination. Results of samplings in the Southern Cone during the 1990s strongly suggested that emergencies in this subregion did not originate from residual persistent activity in the areas declared FMD-free, but most probably from trade activities involving "hot spots" within the subregion.

H. P. S. Makkar and G. J. Viljoen (eds.), Applications of Gene-Based Technologies for Improving Animal Production and Health in Developing Countries, 335-342.

1.   INTRODUCTION

Foot-and-mouth disease (FMD) is economically the most important disease of domestic livestock. Due to its highly infectious nature, ability to cause persistent infection and long-term effects on the animal species it affects, countries that have the disease have many trade restrictions placed upon them. In 1988, a Continental Program for Eradication of FMD in the Americas (PHEFA) was implemented. Since then, impressive progress has been attained, particularly in the Southern Cone subregion of South America, where, during the 1990s, many areas achieved the FMD-free status, with or without vaccination, as recognized by OIE.

To monitor the progress of FMD control and eradication programmes, laboratory information is critical. In this regard, and since the initial stages of PHEFA, a major challenge for the Pan American Foot-and-Mouth Disease Center (PAFMDC), as a regional OIE reference laboratory, was to update approaches to be able to respond to the new demands for diagnostic precision required in support of the epidemiological transition expected in the region. Regions with advanced eradication campaigns needed to incorporate tools for rapid and precise detection and characterization of the agent during an emergency, and, equally important, for identification of potential sources of infection and dissemination of the virus, circulating either clinically or subclinically.

Reverse-transcription polymerase chain reaction (RT-PCR) and automated genetic sequence determination are valuable tools for constructing the genetic data banks that comprise the basis for phylogenetic analysis to trace for evolutionary linkages among strains. In addition, serosurveys in regions with epidemiological associations with the area where an episode occurred are a convenient approach to support the identification of the potential source of infection and dissemination of the virus, and to further assess the epidemiological links suggested by the phylogenetic analysis. Moreover, serosurveys may open the pathway to find the agent when silent circulation occurs, thus supporting the search for genetic connections.

2.   RESULTS AND DISCUSSION

2.1   Phylogenetic studies

As a result of eradication campaigns and surveillance activities, during the late 1990s, most regions in the Southern Cone of South America attained the status of "free of the disease, with or without vaccination". In 2000, a small number of outbreaks of FMD virus types O and A were registered in the subregion, placing the FMD-free regions of the Southern Cone in a state of emergency. Moreover, in 2001, an important epidemic of virus type A took place, which was relatively rapidly controlled. Since then, isolated cases of type O have been registered in the subregion.

Partial nucleotide sequencing of the gene coding for the capsid polypeptide 1D (VP1) was determined for representative strains of these outbreaks. Pairwise comparisons were carried out with the viral sequences held in the PAFMDC database, constructed from strains of epidemiological relevance in the region and from vaccine virus strains (Figures 1 and 2).

The phylogenetic analysis of the variants demonstrated that, in all cases, they belonged to strains endogenous to the Continent.

All type O variants isolated in the Southern Cone in 2000–2003 were derived from a single lineage, showing homology values of at least 93 percent among them. They present a relatively distant genetic relation with the strain used for vaccine production, O1/Campos/Bra/58, with average divergence values of 15 percent. Another important conclusion was that the isolates were not related to the PanAsia strain.

Molecular analysis of type A isolates from the Southern Cone showed that all variants corresponding to the A-2001 epidemic (22 isolates) were very closely related, with homology values of at least 96 percent among them. They showed a genetic homology of 90 percent with an isolate collected in Argentina in 2000, A/Argentina/2000. When compared with type A strains available in the genetic database, variants responsible for type A outbreaks during 2000–2001 shared a common ancestral source, and were related, although not very closely, to the field strains circulating during the 1990s in the region, with homology values between 91 and 94 percent, and approximately 14 percent genetic divergence from the vaccine strain, A24 Cruzeiro/Bra/55.

Further analysis of the referred O and A strains responsible for the emergencies in the FMD-free areas of the Southern Cone, indicated that there is an evident relationship with strains isolated in the same period in "hot spots" within the subregion, which could still present viral niches and therefore deserve special attention from National Authorities. They were not related to strains circulating in the Andean area between 2000 and 2002, with divergence values higher than 15 percent. This suggested independent circulation of viral variants in both livestock trade circuits, reflecting independent production systems and trade activities for animals and their products. Variants isolated in the Andean region were not restricted to a single genetic lineage.

I. Bergmann, V. Malirat, E. Neitzert and E. Correa Melo

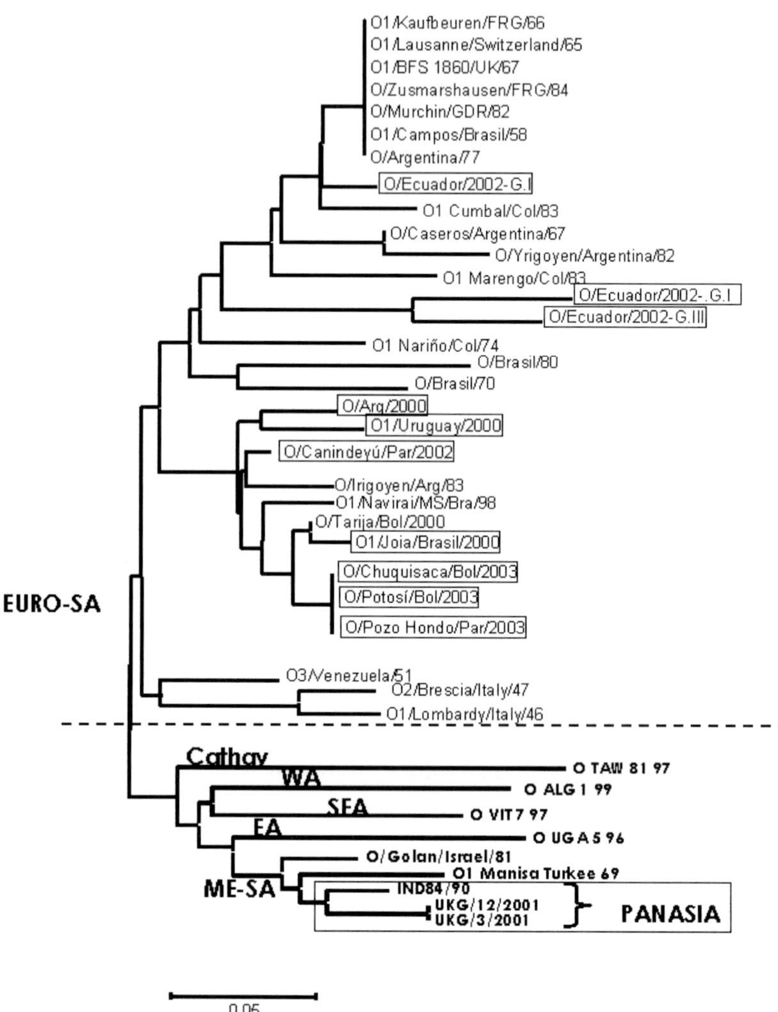

Figure 1. Phylogenetic tree (neighbour joining) showing the genetic relationships of FMD virus type O isolates in South America. Also included were representative strains of six topotypes within type O viruses. P-distances were calculated based on the comparison of the 171 nucleotides of the 3′ end of the VP1 gene. The tree was constructed using the Mega 2.01 program.

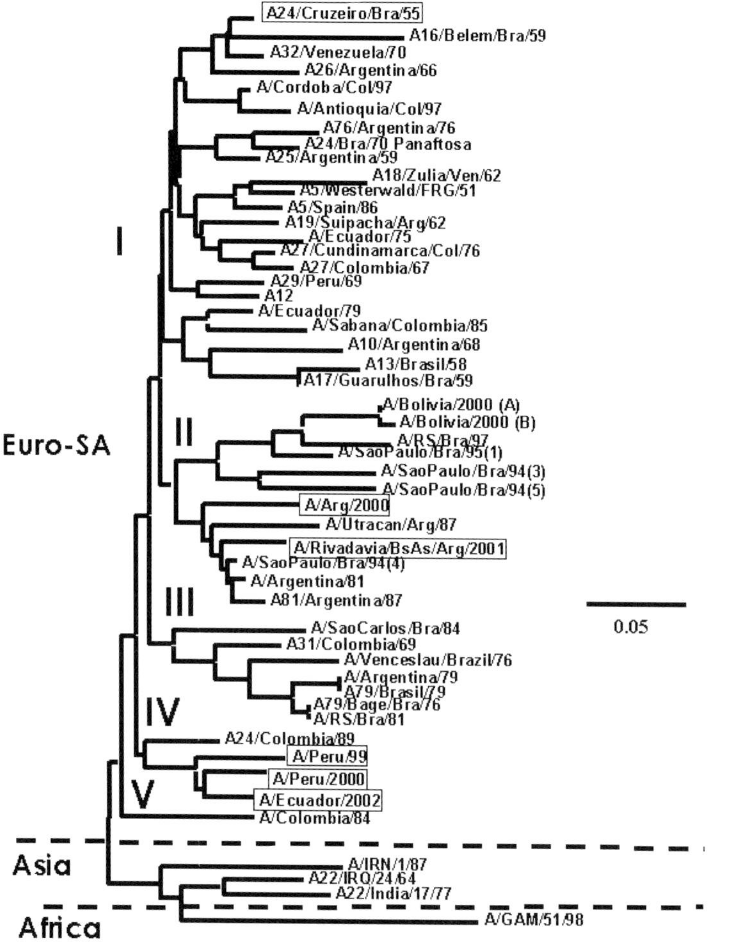

Figure 2. Phylogenetic tree (neighbour joining) showing the genetic relationships of FMD virus type A isolates in South America. Also included were representative strains of the other type A topotypes. P-distances were calculated based on the comparison of the 171 nucleotides of the 3' end of the VP1 gene. The tree was constructed using the Mega 2.01 program.

2.2　Serosurveillance of viral activity

Phylogenetic analysis did not lead to conclusive interpretations on the origin of the infection within the Southern Cone because only one lineage could be established, circulating over several years within the region and with no evident links to other areas. Potential sources of infection that could account for these emergencies might be either "hot spots" within the subregion or by silent infection in areas already declared FMD-free. In this context, serosurveillance of viral circulation constitutes an important adjunct to identify potential sources of infection and spread of the virus.

An immuno-enzymatic serosurveillance system has been developed and validated at PAFMDC, comprising a screening test that detects antibodies against non-capsid polyprotein 3ABC through an indirect ELISA, and an enzyme-linked immuno-electrotransfer blot (EITB) assay that detects antibodies against 5 non-capsid proteins, to confirm I-ELISA 3ABC suspect or positive samples.

The extensive use of this system in South America in regions with various well-documented epidemiological situations has allowed the identification of proper criteria for interpretation of results and for productive exploitation of the data, when applying the system in evaluation of areas with various degrees of risk. Sero-epidemiological investigations in the Southern Cone were used systematically as part of active surveillance activities to follow the progress of the eradication campaigns in the region, and became particularly relevant for confirming absence of viral activity in areas that pursued an FMD-free status.

Results from serosurveys carried out during the 1990s to support recognition of FMD-free status by OIE strongly suggested clearance of viral circulation in those regions recognized as FMD-free with or without vaccination. Moreover, the fact that recurrence of outbreaks was not registered in the free areas for many years after the serosurveys (the last years without vaccination, which increased considerably the number of susceptible cattle) seems to indicate that reappearance of FMD virus in the previously free areas of the subregion was not due to a virus that had lain undetected with the available surveillance tools. Quite the opposite: introduction of FMD into the FMD-free zones may have been caused by movement of clinically or subclinically infected livestock from high-risk zones where sporadic disease occurred, entering the region through convenient trade markets. In this context, serosurveys conducted in regions with trade links to the outbreak area could lead to important information for origin tracing. Border areas should be a priority for joint animal health programmes, which should reinforce serosurveys of "hot spots".

3.   CONCLUSIONS

Reappearance of FMD virus in the Southern Cone of South America clearly indicates the need to further strengthen surveillance activities in the subregion. In fact, success in maintaining an area free of FMD depends greatly on surveillance and prevention. Consequently, achieving FMD-free status magnifies rather than diminishes the importance of having effective laboratories and veterinary services in the field.

The widespread application of phylogenetic techniques has permitted tracing genetic links among circulating viral strains. Such information, together with the results of serosurveys to establish potential viral niches, forms an important input in protecting against re-introductions of the agent in areas considered to be free of the infection. It should be noted that active surveillance activities can become, under certain circumstances, the only reliable tool to search for risk, as genetic-based tracing can only be applied when isolation of the agent has been achieved, and interpretations become conclusive when the variant characterization identifies an evident link with endemic or epidemic regions with reported viral circulation. Moreover, serosurveys became very useful to support epidemiological investigations after episodes, including: serological samplings in regions with epidemiological links to the outbreak area; follow-up of infection as a function of time and space; and ratification of the status of the area.

The application of these new approaches is an important contribution to the understanding of the epidemiology of FMD, which is essential in formulating the most effective control strategies.

REFERENCES

Bergmann, I.E., Malirat, V. & Jeovah, P. 1992. Molecular characterization of foot-and-mouth disease virus strains used for vaccine production in South America. *Bol. Centr. Panam. Fiebre Aftosa,* **58**: 119–127.

Bergmann, I.E., Augé de Mello, P., Neitzert, E. & Gomes, I. 1993. Diagnosis of cattle with persistent aphthovirus infections and their differentiation from vaccinated animals by enzyme-linked immunoelectrotransfer blot using bioengineered nonstructural antigens. *American Journal of Veterinary Research,* **54**: 825–831

Bergmann, I.E., Malirat, V., Neitzert, E., Panizutti, N., Sanchez, C. & Falczuck, A. 2000. Improvement of serodiagnostic strategy for foot-and-mouth disease virus surveillance in cattle under systematic vaccination: A combined system of an indirect ELISA-3ABC with an enzyme-linked immunoelectrotransfer blot assay. *Archives of Virology,* **145**: 473–489.

Bergmann, I.E., Neitzert, N., Malirat, V., Ortiz, S., Colling, A., Sánchez, C. & Correa Melo, E. 2003. Rapid serological profiling by enzyme-linked immunosorbent assay and

                         I. Bergmann, V. Malirat, E. Neitzert and E. Correa Melo

its use as an epidemiological indicator of foot-and-mouth disease viral activity. *Archives of Virology,* **148**: 891–901.

Malirat, V., Bergmann, I..E., Augé de Mello, P., Tiraboschi, B., Beck, E. & Gomes, I. 1994. Genetic variation of foot-and-mouth disease virus during persistent infection in cattle. *Virus Research,* **34**: 31–48.

Malirat, V. & Bergmann, I.E. 2002. Foot-and-mouth disease. Molecular instruments for viral characterization: RT-PCR and cycle sequencing for use in of molecular epidemiology studies on foot-and-mouth disease virus. Pan American Foot-and-Mouth Disease Center, PAHO/WHO, Río de Janeiro, Brazil. *P Series of Teaching Manuals,* No. 17.

Neitzert, E., Beck, E., Augé de Mello, P., Gomes, I. & Bergmann, I.E. 1991. Expression of the aphthovirus RNA polymerase gene in *Escherichia coli* and its use together with other bioengineered nonstructural antigen in detection of late persistent infections. *Virology,* **184**: 799–804.

VIRUS EVOLUTION IN THE FACE OF THE HOST RESPONSE

Esteban Domingo
Centro de Investigación en Sanidad Animal (CISA-INIA), Valdeolmos, Madrid, Spain, and Centro de Biología Molecular "Severo Ochoa" (CSIC-UAM), Universidad Autónoma de Madrid, Cantoblanco, 28049 Madrid, Spain.

Abstract: Microbial infections are highly dynamic. Viruses have evolved two main strategies against the host response: interaction or evasion. Interaction is typical of complex DNA viruses. Their genomes encode a number of proteins that exert modulatory functions that alter the immune response of the host. Evasion strategy is used mainly by RNA viruses, and is based on high mutation rates and quasispecies dynamics. The complexity of viral populations demands research on new antiviral strategies that take into consideration the adaptive potential of viruses, in particular RNA viruses.

1.   INTRODUCTION

Microbial infections are highly dynamic processes in which the invading pathogen must counteract host responses to replicate and to find additional host individuals for long-term survival. Viruses are no exception and display a broad repertoire of strategies to cope with host defences. The presence of viral genomes or virus-like elements in the genomes of differentiated organisms suggests that cells and cellular organisms have undergone a long co-evolutionary process in which functional modules have been frequently exchanged: viruses have probably been active agents of horizontal gene transfer. For reviews of this topic, see Baranowski, Ruiz-Jarabo and Domingo (2001), and Bushman (2002).

In this co-evolutionary process there must have been strong selective pressures to maintain both viable hosts (otherwise this author would not exist to write this review) and infectious viruses (otherwise virus evolution would

H. P. S. Makkar and G. J. Viljoen (eds.), Applications of Gene-Based Technologies for Improving Animal Production and Health in Developing Countries, 343-348.
© 2005 *IAEA. Printed in the Netherlands.*

not be the topic of this review). The process can be imagined in simplified terms. Viruses and other agents of horizontal gene transfer (conjugative plasmids, mobile elements, etc.) contributed to cellular diversification, and possibly also to functional compartmentalization. In turn, viruses acquired specific functions to target subsets of cells. Viruses that killed cells massively would not survive. Cells that were totally resistant to viruses would be deficient in one of the several mechanisms that permit their exploration of new biological properties through genetic change. The result of eons of these interactive processes is a number of mechanisms evolved by viruses to overcome host defences and of hosts to limit the actions of molecular parasites. Dramatically, this evolutionary history is responsible for difficulties we encounter in virus disease prevention and control.

2. TWO VIRAL STRATEGIES: INTERACTION VERSUS EVASION

2.1 The interaction strategy

Complex DNA viruses, such as the poxviruses and the herpesviruses, express a number of genes whose products contribute to evade detection and destruction by the host immune response. These immunomodulatory products include homologues of cytokines, chemokines and their receptors, as well as some that can modulate signal transduction, inhibit apoptosis, the activity of cytotoxic T lymphocytes (CTLs), natural killer (NK) cells, complement, antibody production or down-regulate major histocompatibility complex proteins, among other activities (Alcami, 2003; Seet *et al.,* 2003; Campo, 2002). Interaction strategies extend to some DNA viruses of small genome size, such as the papillomaviruses, and even to some RNA viruses. The important human pathogen hepatitis C virus appears to exploit both interaction and evasion strategies to survive as one of the most successful viral pathogens worldwide (Isaguliants, 2003). By analysing the molecular basis of such interactive mechanisms much can be learned not only of viral infections but also of the host immune system.

2.2 The evasion strategy

In contrast to complex DNA viruses, RNA viruses exploit genetic variation as a means to escape from selective pressures intended to limit their replication. The molecular basis of this adaptive strategy is the high rate of occurrence of mutations during RNA genome replication (Batschelet,

Domingo and Weissmann, 1976; Drake and Holland, 1999). This is due largely to the absence of proofreading-repair activities that are usually associated with cellular DNA polymerases, and also to the absence of post-replicative mismatch correction pathways that can act on double stranded DNA but not on RNA. The result of these critical biochemical differences is that viral RNA replicases and retrotranscriptases display average error rates in the range of 10^{-3} to 10^{-5} misincorporations per nucleotide copied (Drake and Holland, 1999) while for cellular DNA polymerases involved in DNA replication average rates are around 10^{-10} misincorporations per nucleotide copied (Alberts *et al.*, 2002).

A consequence of high error rates is that mutant RNA genomes are generated continuously during replication, even in a single infected cell. Therefore, RNA virus populations consist of complex and dynamic distributions of mutant genomes, termed viral quasispecies (Eigen, 1996; Eigen and Biebricher, 1988). Multiple quasispecies co-exist in infected organisms, even within individual tissues and organs. Quasispecies was developed as a general theory of molecular evolution (Eigen and Schuster, 1979), and it has had a major impact for virology (Domingo, 2000; Domingo *et al.*, 2001). Here, only those features are emphasized related to the evasion strategy of RNA viruses (and, according to recent evidence, also some DNA viruses of small genome size (López-Bueno, Mateu and Almendral, 2003)) that represent a problem for disease prevention and control. In the mutant distributions of viral quasispecies, antibody- or CTL-escape mutants may be present or be generated with high frequency. Their selection may result in progression of infection despite an immune response, and may limit the efficacy of vaccines and other immunotherapeutical treatments. Likewise, inhibitor-escape mutants may be present or generated in mutant spectra, and their selection may result in failure of antiviral treatments. This has been reviewed by Domingo (2003).

High mutation rates may also favour reversion of attenuated virus to virulence, provided the number of genetic lesions involved in reversion is limited (that is, that virulent variants can be present in the mutant spectrum of the attenuated virus), and the virulent form has a selective advantage over the attenuated form in the environment in which replication takes place. More generally, mutants with increased virulence, altered host cell tropism or altered host range may be present or generated in mutant spectra. Their selection may result in more severe disease, or in disease with atypical symptoms, or in viral disease emergence or re-emergence. This has been reviewed by Baranowski *et al.* (2003). The last-named events are often conditioned to alteration in viral traffic prompted by a variety of sociological and ecological factors (Mahy, 1997; Murphy and Nathanson, 1994). Therefore a number of problems relevant to the emergence, prevention and control of viral disease relate directly or indirectly to high mutation rates and quasispecies dynamics of RNA viruses.

3. THE NEED FOR NEW ANTIVIRAL STRATEGIES

The evidence for implication of RNA genome plasticity in current problems related to the control of viral diseases and to disease emergence (Domingo, 2004) suggests that classical approaches to prevention and therapy must be carefully reconsidered. In principle, monotherapy, with either a single antiviral inhibitor or a single monoclonal antibody, is incompatible with the dynamics of rapid mutant generation, competition and selection in viral populations. This is because inhibitor- or antibody-escape mutants are likely to be rapidly selected (Domingo *et al.*, 2001). The frequent (in some case systematic) selection of human immunodeficiency virus (HIV) mutants resistant to antiretroviral agents constitutes a dramatic example. It is remarkable that the mutations that confer resistance to inhibitors have always been compatible with viral function. If resistance mutations entailed lethality, the problem of treatment failure would not occur. The author has speculated that, in their long evolutionary history, viral enzymes had to cope with metabolites that are related to the currently used inhibitors, and that they had to evolve to overcome inhibitory (or other perturbing) effects (Domingo, 2003). According to this proposal, there would be deep evolutionary determinants involved in the current difficulties encountered in suppressing viral replication, in addition to viruses using part of the cellular machinery to complete their replication cycles.

For similar reasons, vaccines that stimulate the host immune system to target a limited number of antigenic determinants of the virus are likely to fail to prevent progress of the infection, and may even promote the selection of antigenic variants of the pathogen (Domingo *et al.,* 2001). These limitations have been amply recognized, and strategies that include use of combination therapy (multiple inhibitors addressed to independent viral targets but administered simultaneously) and multivalent vaccines (with multiple, independent B-cell and T-cell epitopes) have been proposed (Domingo, 2003; Domingo and Holland, 1992). If implemented, these strategies would diminish the frequency of treatment failures. However obvious, these recommendations are not always followed.

Recently, a new antiviral strategy based on virus extinction associated with increased mutagenesis has actively been pursued (Eigen, 2002). It is based on the concept that for any given amount of genetic information to be transmitted during replication there is a maximum error rate compatible with maintenance of the information. Violation of the error threshold should result in loss of virus infectivity, and this has now been documented with several virus-host systems (Domingo, 2000; Eigen, 2002; Graci and Cameron, 2002). This promising new antiviral strategy exploits high mutation rates of viruses for their destruction rather than for their survival.

4. CONCLUDING REMARKS

The picture that has emerged from studies of viral genomes at the population level is one of exceeding complexity, rendering classical strategies for viral disease prevention and control largely obsolete. For many important human and animal diseases, no effective vaccines or antiviral agents are available. Obviously, successful vaccines (to prevent infections by poliovirus, measles virus, influenza viruses, foot-and-mouth disease virus, among others) have had a great positive impact on human and animal health, and they will continue to be used, and will contribute to prevent disease. The key issue here is to make such effective vaccines available to people in need of health care, worldwide. Vaccine efficacy should not impede awareness that even these viruses that can now be controlled harbour enormous potential for genetic change and for causing forms of disease that cannot be anticipated. We know this from experience with the emergence of HIV-1, SARS virus and many other pathogens over the last decades. It is extremely important to view viral pathogens (and in reality any microbial pathogen) as a highly dynamic and flexible entity in its fine genetic identity as well as in its manifestations of an infection (Holland, de La Torre and Steinhauer, 1992). New approaches to treat and prevent viral diseases must be found that take into consideration (rather than ignore) the highly dynamic nature of the pathogens to be controlled.

ACKNOWLEDGEMENTS

Work in the author's laboratory is supported by grants BMC2001-1823-C02-01 (MCyT), QLK2-CT-2002-00825 (EU), 08./0015/2001 (CAM) and Fundación R. Areces. Only a limited number of references have been included for space limitations.

REFERENCES

Alberts, B., Johnson, A., Lewis, J., Raff, M., Roberts, K. & Walter, P. 2002. *Molecular Biology of the Cell*. New York, NY: Garland Science.

Alcami, A.   2003.   Viral mimicry of cytokines, chemokines and their receptors. *Nature Reviews Immunology,* **3**: 36–50.

Baranowski, E., Ruiz-Jarabo, C.M. & Domingo, E.   2001.   Evolution of cell recognition by viruses. *Science,* **292**: 1102–1105.

Baranowski, E., Ruíz-Jarabo, C.M., Pariente, N., Verdaguer, N. & Domingo, E.   2003. Evolution of cell recognition by viruses: a source of biological novelty with medical implications. *Advances in Virus Research,* **62**: 19–111.

Batschelet, E., Domingo, E. & Weissmann, C.   1976.   The proportion of revertant and mutant phage in a growing population, as a function of mutation and growth rate. *Gene,* **1**: 27–32.

Bushman, F.   2002.   *Lateral DNA Transfer. Mechanisms and Consequences.* Cold Spring Harbor, NY: Cold Spring Harbor Laboratory Press.

Campo, M.S.   2002.   Virus escape from immunosurveillance. *Virus Research,* **88**: 1.

Domingo, E.   2000.   Viruses at the edge of adaptation. *Virology,* **270**: 251–253.

Domingo, E.   2003.   Quasispecies and the development of new antiviral strategies. *Progress in Drug Research,* **60**: 133–158.

Domingo, E.   2004.   Microbial evolution and emerging diseases. *In:* C. Power and T. Johnson (eds). *Emerging Neurological Infections.* New York, NY: Marcel Dekker.

Domingo, E. & Holland, J.J.   1992.   Complications of RNA heterogeneity for the engineering of virus vaccines and antiviral agents. *Genetic Engineering (N Y),* **14**: 13–31.

Domingo, E., Biebricher, C., Eigen, M. & Holland, J.J.   2001.   *Quasispecies and RNA Virus Evolution: Principles and Consequences.* Austin TX: Landes Bioscience.

Drake, J.W. & Holland, J.J.   1999.   Mutation rates among RNA viruses. *Proceedings of the National Academy of Sciences, USA,* **96**: 13910–13913.

Eigen, M.   1996.   On the nature of virus quasispecies. *Trends in Microbiology,* **4**: 216–218.

Eigen, M.   2002.   Error catastrophe and antiviral strategy. *Proceedings of the National Academy of Sciences, USA,* **99**: 13374–13376.

Eigen, M. & Biebricher, C.K. (eds).   1988.   Sequence space and quasispecies distribution. Vol. 3, *in: RNA Genetics.* Boca Raton, FL: CRC Press.

Eigen, M. & Schuster, P.   1979.   *The Hypercycle. A Principle of Natural Self-organization.* Berlin: Springer.

Graci, J.D. & Cameron, C.E.   2002.   Quasispecies, error catastrophe, and the antiviral activity of ribavirin. *Virology,* **298**: 175–180.

Holland, J.J., de La Torre, J.C. & Steinhauer, D.A.   1992.   RNA virus populations as quasispecies. *Current Topics in Microbiological Immunology,* **176**: 1–20.

Isaguliants, M.G.   2003.   Hepatitis C virus clearance: the enigma of failure despite an impeccable survival strategy. *Current Pharmaceutical Biotechnology,* **4**: 169–183.

Johnston, J.B. & McFadden, G.   2003.   Poxvirus immunomodulatory strategies: current perspectives. *Journal of Virology,* **77**: 6093–6100.

López-Bueno, A., Mateu, M.G. & Almendral, J.M.   2003.   High mutant frequency in populations of a DNA virus allows evasion from antibody therapy in an immunodeficient host. *Journal of Virology,* **77**: 2701–2708.

Mahy, B.W.   1997.   Human viral infections: an expanding frontier. *Antiviral Research,* **36**: 75–80.

Murphy, F.A. & Nathanson, N.   1994.   The emergence of new virus diseases: an overview. *Seminars in Virology,* **5**: 87–102.

Seet, B.T., Johnston, J.B., Brunetti, C.R., Barrett, J.W., Everett, H., Cameron, C., Sypula, J., Nazarian, S.H., Lucas, A. & McFadden, G.   2003.   Poxviruses and immune evasion. *Annual Review of Immunology,* **21**: 377–423.

RUMEN MICROBIAL GENOMICS

Mark Morrison[1] and Karen E. Nelson[2]
1. Ohio State University, Columbus, Ohio, USA.
2. The Institute for Genomic Research, Rockville, Maryland, USA.

Abstract: Improving microbial degradation of plant cell wall polysaccharides remains one of the highest priority goals for all livestock enterprises, including the cattle herds and draught animals of developing countries. The North American Consortium for Genomics of Fibrolytic Ruminal Bacteria was created to promote the sequencing and comparative analysis of rumen microbial genomes, offering the potential to fully assess the genetic potential in a functional and comparative fashion. It has been found that the *Fibrobacter succinogenes* genome encodes many more endoglucanases and cellodextrinases than previously isolated, and several new processive endoglucanases have been identified by genome and proteomic analysis of *Ruminococcus albus,* in addition to a variety of strategies for its adhesion to fibre. The ramifications of acquiring genome sequence data for rumen micro-organisms are profound, including the potential to elucidate and overcome the biochemical, ecological or physiological processes that are rate limiting for ruminal fibre degradation.

1.   INTRODUCTION

Improving plant cell wall (fibre) degradation remains one of the highest priority goals for all livestock enterprises, from the intensively managed dairy herds in North America and Europe, to the nomadic cattle herds of sub-Saharan Africa. The North American Consortium for Genomics of Fibrolytic Ruminal Bacteria was created in 2000 to promote the sequencing and comparative analysis of rumen microbial genomes. The long-term goal of this consortium is to elucidate the genetics and molecular biology underpinning fibre degradation in ruminants. By doing so, new opportunities to improve feed utilization and animal productivity, as well as to foster a

H. P. S. Makkar and G. J. Viljoen (eds.), Applications of Gene-Based Technologies for
Improving Animal Production and Health in Developing Countries, 349-355.

more positive image for the interface between animal agriculture and the environment in developed countries, should be forthcoming. The authors here outline the scientific merit associated with the application of genomics and related technologies to the study of ruminal microbes, and how such studies might not only advance our knowledge of microbial biology, but also positively affect animal production and health in developing countries.

2.   WHY SEQUENCE THE GENOMES OF RUMINAL BACTERIA?

The technologies employed and the ensuing computational needs for microbial genome sequencing have been outlined in detail elsewhere (Charlebois, 1999; Wren and Dorell, 2002; Fraser, Read and Nelson, 2004) and will not be reviewed here. Suffice it to say that genome sequencing offers the potential to obtain a complete blueprint for the lifestyle of a specific microbe, and to assess its genetic potential in a functional and comparative fashion.

The first rumen bacterium to have its genome sequenced was *Wolinella succinogenes,* because of its phylogenetic placement between *Helicobacter pylori* and *Campylobacter* spp. (Baar *et al.,* 2003). However, the diversity of microbes present in the rumen is great, so how does one select other microbes for genome sequencing? Some prioritization of need is required, and the North American Consortium selected fibre degradation as being an aspect of ruminant microbiology that is of universal relevance to animal health and production. It has long been believed that the major part of fibre degradation is carried out by only a select group of anaerobic bacteria. *Fibrobacter succinogenes, Ruminococcus albus* and *R. flavefaciens* are the predominant cellulolytic species, with *Prevotella ruminicola* and *Butyrivibrio fibrisolvens* the predominant xylanolytic bacteria. Microscopic examinations have shown these bacteria adhere tightly to the plant surface, possess cell bound enzyme complexes, and also show selectivity in adhesion site on the plant surface. However, very little is known about the repertoire of glycoside hydrolases acquired by these bacteria over evolutionary time, factors affecting the expression of these enzymes, and other gene products underpinning the competitiveness of these bacteria to exist within the rumen milieu. Notably, these bacteria were located within the eubacterial tree of life at positions for which little or no genome sequence information was available prior to the initiation of these projects. As such, these projects provide an additional benefit to microbial biologists interested in genome composition and phylogenetics, and expand the opportunities for revealing novel insights into microbial biology and ruminal function. A combination of funds from the United States Department of Agriculture (USDA)'s Initiative for Future Agriculture and Food Systems and the USDA-National Science Foundation (NSF) Microbe Sequencing Program has supported the sequencing of the *F. succinogenes, R. albus* and *P. ruminicola* genomes to closure. More rumen bacterial genomes are now being sequenced: a draft sequence for *R. flavefaciens* is nearing completion (White, pers. comm.) and several other ruminal microbes, including a strain of *B. fibrisolvens*, are currently being sequenced, through the efforts of New Zealand scientists (Attwood, pers. comm.). Even at these early stages, the ramifications of acquiring genome sequence data for rumen micro-organisms are profound. For instance, the North American Consortium projects have revealed that only a small percentage (~25%) of the cellulase- and xylanase-encoding genes had been previously identified, and that novel cellulase systems are employed by at least some ruminal bacteria. With specific reference to *R. albus,* new enzymes rate-limiting to cellulose degradation have been identified (Devillard *et al.*, 2004) and these enzymes possess a novel type of carbohydrate binding module (Xu *et al.*, 2004).

Despite the expediency and cost savings associated with new sequencing technologies, it is still cost prohibitive in many instances to have multiple genomes sequenced for a select group of related bacteria. Therefore, in addition to these sequencing projects, North American Consortium members have used subtractive hybridization methods to characterize the genomic differences among the sequenced genomes and the genomes of additional strains and species of ruminal bacteria. Some of this work with two strains of *R. flavefaciens* has recently been published (Antonopoulos *et al.*, 2004). These studies estimated that almost 10% of the genome content of *R. flavefaciens* strain FD-1 is not present in strain JM-1 (representing almost 400 kilobases of genomic differences between these two strains) and almost half the clones bear little or no resemblance to gene sequences currently deposited in public databases. Accordingly, even closely related strains of bacteria may possess a large amount of genomic information for which little or no knowledge currently exists.

In summation, whole-genome sequencing and subtractive hybridization methods serve as the foundation for a multitude of comparative and functional lines of investigation that seek to understand, at an organismal level, the molecular biology underpinning the lifestyle of a specific microbe. For instance, the potential to elucidate which enzyme(s) or other ecological or physiological process(es) are rate-limiting to fibre degradation by these bacteria, and how this might change relative to dietary composition, are now greatly enhanced. Perhaps more importantly, whole genome sequencing facilitates the opportunity to extend our understanding of gastrointestinal microbiomes beyond the degradative and metabolic characteristics predictably relevant to host animal health and nutrition. The likelihood of productively altering ruminant microbiology is therefore enhanced, to include strategies that alter a bacterium's competitiveness and colonization potential in the rumen milieu, without targeting cellulase and xylanase genes specifically.

3.   WHY STOP AT ONE GENOME – THE ADVENT OF METAGENOMICS

Metagenomics is a term coined with reference to the genetic potential resident within an entire microbial community, and is dependent upon high throughput DNA sequencing, advances in recombinant DNA technologies, and computational biology. It is anticipated that metagenomics will significantly augment the rumen genome studies that are already underway, and allow for the genetic characterization of microbes that cannot currently be cultured in the laboratory. The genetic potential of these uncultured species, which undoubtedly make a significant contribution to the ecology of the rumen environment, have, until now, escaped attention.

Metagenomic libraries can be constructed in several types of vectors, resulting in libraries with insert sizes ranging from only a few kilobases (Knietsch *et al.*, 2003) to ~40 kilobases in fosmid vectors, to more than 150 kilobases in bacterial artificial chromosome (BAC) vectors (Beja *et al.*, 2000; Rondon *et al.*, 2000). The choice of vector can be influenced by the quality of DNA recovered from a microbiome, the types of analyses to be conducted, and the number of clones that can be propagated and archived in the laboratory (see Handelsman *et al.*, 2002, and Morrison *et al.*, in press, for recent reviews). In general terms, the BAC vectors afford the opportunity of recovering large biosynthetic or metabolic operons in their entirety, and also support a more detailed assessment of (meta)genome organization, due to the larger amount of information archived in each clone (Beja *et al.*, 2000). The BAC and fosmid libraries can also be subjected to functional screens that are dependent on heterologous gene expression, most often in *Escherichia coli* (Rondon *et al.*, 2000). Fosmid clones are inherently smaller, but the cloning efficiency of metagenomic DNA tends to be substantially higher with this type of vector, and fosmid vectors also afford the use of more robust DNA extraction methods. Several groups are currently producing metagenomic libraries of rumen microbiomes, and it is anticipated that these libraries will also reveal significant findings relevant to our understanding of microbial biology and fibre degradation.

The BAC and fosmid libraries reported in the literature range from 10,000 to 25,000 clones, representing 0.5 to 1.0 Gigabasepairs of DNA (Beja *et al.*, 2000; Rondon *et al.*, 2000) (1 Gbp = 10^9 bp). Millions of clones are estimated to be required for representative coverage of soil metagenomes (Handelsman *et al.*, 2002) and given that gut microbiomes possess a similar breadth in biodiversity to many soil microbiomes, a comprehensive inventory of the metagenome is currently out of reach for many investigators interested in rumen and gut microbiomes. However, recent successes with less complex microbial communities, such as those present in an acid mine drainage system (Tyson *et al.*, 2004) and in samples from the Sargasso sea (Venter *et al.*, 2004), suggest that the selective enrichment of subpopulations via physical fractionation (e.g. recovering cells tightly adherent to digesta particles), cell sorting or enrichment cultures may expedite the progress that can be made with more complex microbial communities.

4. CONCLUDING REMARKS

The "-omics" technologies offer exciting new opportunities to investigate microbial diversity and physiology in ruminants, other herbivorous animals, and humans. The depth and breadth of information arising from these projects, which might be used to improve herbivore nutrition and livestock productivity, is unprecedented. The availability of genome sequences should also raise the level of scientific exploration, and to increase expectations that the study of ruminal bacteria could make novel contributions to our understanding of microbial biology. Ironically, the increasing opportunities afforded in the genomics era coincide with a relative downturn in the commitment of scientific expertise to the field of ruminal microbiology. There is the potential for this information to remain largely dormant unless systematic and integrated approaches are more widely employed. Such issues were discussed at a recent genomics workshop held in conjunction with the 3rd RRI-INRA Conference on Gut Microbiology in Aberdeen, Scotland, in June 2002. The development of basic reagents and tools, including gene transfer protocols and mutagenesis strategies; sequencing the genomes of more bacteria, including select bacteria from other gastrointestinal microbiomes; and promoting the networking of smaller groups, rather than a mega-consortium approach, were all recommended as viable needs in the future study of rumen microbial genomics. Although some of these activities are already underway, progress still needs to be made in several areas, especially in relation to the development of techniques in bacterial genetics. Hopefully, the current model that has been established by the North American Consortium will be just the beginning, but we are aware that many challenges lie ahead in terms of funding, data acquisition, data mining and data interpretation.

For these reasons, rumen microbiology has reached a crossroads. New opportunities to overcome the existing constraints to ruminal function and herbivore nutrition are most likely to arise from the integration of genomics-enabled studies into research programmes in herbivore nutrition. However, the future vitality of rumen microbiology is also largely dependent on how effectively the study of ruminal microbes contributes new knowledge to our understanding of microbial biology. A balance needs to be retained between research activities that establish new understanding in microbial biology, and the pragmatic requirements of the livestock industries in developed and developing countries. Perhaps for the first time in several decades, microbial genome sequencing affords the opportunity to achieve advances in both fundamental and applied microbial biology. The activities underway within the North American Consortium and in New Zealand will hopefully provide the necessary impetus to ensure this happens to a greater extent than at present. Time will tell.

ACKNOWLEDGEMENTS

The North American Consortium was created through financial support provided by the United States Department of Agriculture's Initiative for Future Agriculture and Food Systems, and the United States Department of Agriculture's and National Science Foundation's Microbial Genome Sequencing Program. The genome sequences described in this article can be obtained through The Institute for Genomic Research (www.tigr.org). The authors also acknowledge the efforts of other principal members of the North American Consortium for Fibrolytic Ruminal Bacteria (W. Nelson, S. Daugherty and E. Mongodin, TIGR; C.W. Forsberg, University of Guelph; D.B. Wilson and J.B. Russell, Cornell University; B.A. White, R.I. Mackie, and I.K.O. Cann, University of Illinois).

REFERENCES

Antonopoulos, D., Nelson, K.E., Morrison, M. & White, B.A. 2004. Strain-specific genomic regions of *Ruminococcus flavefaciens* FD-1 as revealed by combinatorial random-phase genome sequencing and suppressive subtractive hybridization. *Environmental Microbiology,* **6**: 335–346.

Baar, C., Eppinger, M., Raddatz, G., Simon, J., Lanz, C., Klimmek, O., Nandakumar, R., Gross, R., Rosinus, A., Keller, H., Jagtap, P., Linke, B., Meyer, F., Lederer, H. & Schuster, S.C. 2003. Complete genome sequence and analysis of *Wolinella succinogenes. Proceedings of the National Academy of Sciences, USA,* **100**: 11690–11695.

Beja, O., Suzuki, M.T., Koonin, E.V., Aravind, L., Hadd, A., Nguyen, L.P., Amjadi, M., Garrigues, C., Jovanovich, S.B., Feldman, R.A. & DeLong, E.F. 2000. Construction and analysis of bacterial artificial chromosome libraries from a marine microbial assemblage. *Environmental Microbiology,* **2**: 516–529.

Charlebois, R.L. 1999. *Organization of the Prokaryote Genome.* Washington DC: ASM Press.

Devillard, E., Goodheart, D.E., Karnati, S.K.R., Bayer, E.A., Lamed, R., Miron, J., Nelson, K.E. & Morrison, M. 2004. *Ruminococcus albus* 8 mutants defective in cellulose degradation are deficient in two processive endocellulases, Cel48A and Cel9B, both of which possess a novel modular architecture. *Journal of Bacteriology,* **186**: 136–145.

Fraser, C.M., Read, T.D. & Nelson, K.E. 2004. *Microbial Genomes.* Totowa, NJ: Humana Press.

Handelsman, J., Liles, M., Mann, D., Riesenfield, C. & Goodman, R.M. 2002. Cloning the metagenome: culture-independent access to the diversity and functions of the uncultivated microbial world. pp. 241–256, *in:* B. Wren and N. Dorell (eds). *Functional Microbial Genomics.* [Methods in Microbiology, vol. 33.] London: Academic Press.

Knietsch, A., Waschkowitz, T., Bowien, S., Henne A. & Daniel, R. 2003. Construction and screening of metagenomic libraries derived from enrichment cultures: generation of a gene bank for genes conferring alcohol oxidoreductase activity on *Escherichia coli. Applied and Environmental Microbiology,* **69**: 1408–1416.

Morrison, M., Adams, S.E., Nelson, K.E., & Attwood, G.T. 2004. Metagenomic analysis of the microbiomes in ruminants and other herbivores. *In: Methods for Rumen Microbial Ecology Studies.* IAEA Occasional Publication. Vienna, Austria. In press.

Rondon, M.R., August, P.R., Bettermann, A.D., Brady, S.F., Grossman, T.H., Liles, M.R., Loiacono, K.A., Lynch, B.A., MacNeil, I.A., Minor, C., Tiong, C.L., Gilman, M., Osburne, M.S., Clardy, J., Handelsman, J. & Goodman, R.M. 2000. Cloning the soil metagenome: a strategy for accessing the genetic and functional diversity of uncultured micro-organisms. *Applied Environmental Microbiology,* **66**: 2541–2547.

Tyson, G.W., Chapman, J., Hugenholtz, P., Allen, E.E., Ram, R.J., Richardson, P.M., Solovyev, V.V., Rubin, E.M., Rokhsar, D.S. & Banfield, J.F. 2004. Community structure and metabolism through reconstruction of microbial genomes from the environment. *Nature,* **428**: 37–43.

Venter, J.C., Remington, K., Heidelberg, J.F., Halpern, A.L., Rusch, D., Eisen, J.A., Wu, D., Paulsen, I., Nelson, K.E., Nelson, W., Fouts, D.E., Levy, S., Knap, A.H., Lomas, M.W., Nealson, K., White, O., Peterson, J., Hoffman, J., Parsons, R., Baden-Tillson, H., Pfannkoch, C., Rogers, Y.H. & Smith, H.O. 2004. Environmental genome shotgun sequencing of the Sargasso Sea. *Science,* **304**: 66–74.

Wren, B. & Dorell, N. (eds). 2002. *Functional Microbial Genomics.* [Methods in Microbiology, vol. 33]. London: Academic Press.

Xu, Q., Morrison, M., Nelson, K.E., Bayer, E.A., Atamna, N. & Lamed, R. 2004. A novel family of carbohydrate-binding modules identified with *Ruminococcus albus* proteins. *FEBS Letters,* **566**: 11–16.

TRANSGENESIS AND GENOMICS IN MOLECULAR BREEDING OF PASTURE GRASSES AND LEGUMES FOR FORAGE QUALITY AND OTHER TRAITS

German Spangenberg
Plant Biotechnology Centre, Primary Industries Research Victoria, Department of Primary Industries, La Trobe University, Bundoora, Victoria 3083, Australia; and Molecular Plant Breeding CRC.

Abstract: Significant advances in the establishment of the methodologies required for the molecular breeding of temperate forage grasses (*Lolium* and *Festuca* species) and legumes (*Trifolium* and *Medicago* species) are reviewed. Examples of current products and approaches for the application of these methodologies to forage grass and legume improvement are outlined. The plethora of new technologies and tools now available for high-throughput gene discovery and genome-wide expression analysis have opened up opportunities for innovative applications in the identification, functional characterization and use of genes of value in forage production systems and beyond. Selected examples of current work in pasture plant genomics, xenogenomics, symbiogenomics and micro-array-based molecular phenotyping are discussed.

1.   PASTURE PLANT TRANSGENESIS

Gene technology and the production of transgenic plants offers the opportunity to generate unique genetic variation, when the required variation is either absent or has very low heritability. In recent years, the first transgenic pasture plants with simple "engineered" traits have reached the stage of field evaluation. While gaps in our understanding of the underlying genetics, physiology and biochemistry of many complex plant processes are likely to delay progress in many applications of transgenesis in forage plant improvement, gene technology is a powerful tool for the generation of the

H. P. S. Makkar and G. J. Viljoen (eds.), Applications of Gene-Based Technologies for Improving Animal Production and Health in Developing Countries, 357-372.

required molecular genetic knowledge. Consequently, applications of transgenesis to temperate pasture plant improvement are focused on the development of transformation events with unique genetic variation and in studies on the molecular genetic dissection of plant biosynthetic pathways and developmental processes of high relevance for forage production (Spangenberg *et al.*, 2001).

Primary target traits for the application of transgenesis to temperate pasture plant improvement are forage quality, disease and pest resistance, tolerance to abiotic stresses, and the manipulation of growth and development. Some representative approaches and selected examples in temperate forage grasses and legumes are discussed below (Spangenberg *et al.*, 2001).

Molecular breeding based on transgenesis to overcome limitations in forage quality may be targeted to the individual subcharacters involved: dry matter digestibility, water-soluble carbohydrate content, protein content, secondary metabolites, etc. These molecular breeding approaches may include modification of the lignin profile to enhance dry matter digestibility, genetic manipulation of fructan metabolism to increase non-structural carbohydrate content, genetic manipulation of condensed tannin synthesis to develop "bloat-safe" forages, and the expression of "rumen by-pass" proteins to improve the supply of proteins and essential amino acids. Most quality or anti-quality parameters are associated with specific metabolic pathways or the production of specific proteins. This allows target enzymes or suitable foreign proteins to be identified, corresponding genes isolated, and their expression manipulated in transgenic forage plants.

Pathogen and pest infection can considerably lower herbage yield, persistency, nutritive value, and palatability of forage plants. An armoury of genes and strategies for engineering disease and pest resistance in transgenic plants has been developed and tested over the last decade, including chitinases, glucanases, plant defensins, phytoalexins, ribosome-inactivating proteins, viral coat proteins, viral replicase, viral movement proteins, *Bt* toxins, proteinase inhibitors, and α-amylase inhibitors. Some of them have been applied to the development of pasture plants, mainly forage legumes, for enhanced disease and pest resistance (Spangenberg *et al.*, 2001).

Plants can be used to express recombinant heterologous proteins. Transgenic plants may be an attractive alternative to microbial systems for the production of certain biomolecules. The perennial growth habit, the biomass production potential, the capacity for biological nitrogen fixation, and the ability to grow in marginal areas exhibited by forage plants, particularly pasture legumes, make them potential suitable candidates for molecular farming. Advances in genetic manipulation technologies that allow high levels of transgene expression and transgene containment may, in the not too distant future, make it possible to exploit some forage plants as bioreactors for the production, among others, of industrial enzymes, pharmaceuticals, vaccines, antibodies and biodegradable plastics. Multidisciplinary efforts will, however, be needed to identify the most feasible targets, to generate transgenic plants with suitable expression levels, and to develop efficient downstream processing technology that could adapt transgenic forage plants for non-forage uses and make them a cost-effective alternative for molecular farming. Significant progress has been achieved in recent years in the production of value-added proteins in transgenic lucerne (Spangenberg *et al.*, 2001).

Small-scale planned releases of transgenic plants are required to assess the stability of transgene expression and the novel phenotypes under field conditions and to identify transformation events suitable for transgenic germplasm and cultivar development. Only after the transformation events have been thoroughly evaluated for the stability of the novel phenotype outside of the controlled environment of a glasshouse would it be advisable to continue to integrate these in molecular breeding programmes for the development of transgenic cultivars. An illustrative example of design features of such a small-scale field trial can be found in a recent field trial of alfalfa mosaic virus (AMV)-immune transgenic white clover plants (Kalla *et al.*, 2001).

A range of transformation events in forage legumes and grasses with proof of concept for the technology under containment conditions are being developed. The challenge now is how to best deploy these molecular technologies and tools to evaluate their full potential, based on the transgenic transfer of single and multiple valuable genes, to generate novel genetic variability and novel elite transgenic germplasm, and to efficiently incorporate these factors into breeding programmes for the development of improved cultivars.

Efficient strategies for the introgression of transgenes into elite parents for the subsequent production of synthetic cultivars have been developed, ensuring stable and uniform transgene expression in all plants in the population. One such strategy has been applied to the production of AMV-immune transgenic elite white clover plants homozygous for the transgenes. It involves initial top-crosses of transformation events chosen after their field evaluation with elite non-transgenic white clover parental lines; selecting for progeny from the harvested seed carrying the transgene and its linked selectable *npt2* marker gene by antibiotic selection or PCR screening, followed by diallele crosses between the T_1 progeny. The T_2 offspring plants homozygous for transgenes can be directly identified by high-throughput quantitative PCR transgene detection. The elite white clover plants homozygous for the transgenes are then planted in a selection nursery together with elite non-transgenic parental lines for identification of the new parents of transgenic experimental synthetic cultivars and their subsequent multisite evaluation (Spangenberg *et al.*, 2001).

2.   PASTURE PLANT GENOMICS

Forage plant breeding has entered the genome era. The plethora of new technologies and tools now available for high-throughput gene discovery and genome-wide expression analysis have opened up opportunities for innovative applications in the identification, functional characterization and use of genes of value in forage production systems and beyond. Examples of these opportunities include "molecular phenotyping", "symbiogenomics" and "xenogenomics" (Spangenberg *et al.*, 2001).

We have undertaken the discovery of 100,000 Expressed Sequence Tags (ESTs) from the key forage crops of temperate grassland agriculture, perennial ryegrass (*Lolium perenne*) and white clover (*Trifolium repens*), using high-throughput sequencing of randomly selected clones from cDNA libraries representing a range of plant organs, developmental stages and experimental treatments. The DNA sequences were analysed by Basic Local Alignment Search Tool (BLAST) searches, categorized functionally, and subjected to cluster analysis, leading to the identification of unigene sets in perennial ryegrass and white clover corresponding to 14,767 and 14,635 genes, respectively (Spangenberg *et al.*, 2001).

We have further developed high density spotted cDNA micro-arrays with approximately 15,000 unigene sets as a main screening tool for novel ryegrass and clover sequences of unknown function. These EST-based plant micro-arrays will allow the global analysis of gene expression patterns as a main approach for functional genomics and other applications. Novel applications of EST-based forage plant arrays including "molecular phenotyping", i.e. the analysis of global or targeted gene expression patterns using complex hybridization probes from contrasting genotypes or populations and contrasting environments, are now being tested to integrate micro-array data with current conventional phenotypic selection approaches used in temperate pasture plant improvement (Spangenberg *et al.*, 2001).

Comparative sequence and micro-array data analyses from ryegrass and clover with data from complete genome sequencing projects in *Arabidopsis* and rice, as well as from extensive EST discovery programmes in the model forage legume *Medicago truncatula*, have been undertaken to provide insight into conserved and divergent aspects of grass and legume genome organization and function.

3. PASTURE PLANT SYMBIOGENOMICS

Pasture legumes and grasses offer unique and exciting opportunities in genome research to study plant-pathogen interactions, legume-nitrogen-fixing bacteria symbiosis, legume-mycorrhiza associations, and grass-endophyte endosymbiosis, as well as the potential to apply the knowledge gained from these studies to develop resistance to pathogens and to improve beneficial associations in forages.

The author's laboratory has undertaken a gene discovery programme in *Neotyphodium coenophialum* and *N. lolii,* fungal endophytes of tall fescue (*Festuca arundinacea*) and perennial ryegrass, respectively. Approximately 8,500 *Neotyphodium* DNA sequences were generated, analysed by BLAST searches, categorized functionally, and subjected to cluster analyses, leading to the identification of a 3,806 unigene set in *Neotyphodium*. The programme is focused on the discovery of genes involved in host colonization, nutrient supply to the endophytic fungus, and the biosynthesis of active pyrrolopyrazine and pyrrolizidine secondary metabolites (e.g. the insect deterrents peramine and N-formylloline, respectively) and their regulation. It will provide insight into the molecular genetics of the grass endophyte-host interaction, as well as into the physiological mechanisms leading to the increased plant vigour and enhanced stress tolerance. These genomic tools and knowledge will underpin the development of technologies to manipulate grass–endophyte associations for enhanced plant performance, improved grass tolerance to biotic and abiotic stresses, and altered grass endophyte host specificity, to the benefit of the grazing and turf industries (Spangenberg *et al.*, 2001).

4. XENOGENOMICS

Genome research with exotic plant species, i.e. "xenogenomics", includes gene discovery by high-throughput EST sequencing and large-scale simultaneous gene expression analysis with EST-based micro-arrays. Xeno-genomics has opened up opportunities for a "genomic bio-prospecting" of key genes and gene variants from exotic plants. This approach is particularly suited for the discovery of novel genes and the determination of their expression patterns in response to specific abiotic stresses.

The author's laboratory has undertaken a xenogenomic EST discovery focused on selected Australian native and exotic grasses and legumes that show unique adaptation to extreme environmental stresses. Genes that allow certain plant species to tolerate extreme abiotic stresses, including drought, salinity and low fertility soils, are being isolated and characterized. The species targeted in the xenogenomic EST discovery programme include Australian native grasses, such as the halotolerant blown-grasses (*Agrostis adamsonii* and *A. robusta*) and the aluminium-tolerant weeping grass (*Microlaena stipoides*), together with exotic species, such as Antarctic hair-grass (*Deschampsia antarctica*), one of only two vascular plant species native to Antarctica (Spangenberg *et al.*, 2001).

The discovery of novel genes and their functional genomic analysis will facilitate the development of effective molecular breeding approaches to enhance abiotic stress tolerance in forages and other crops.

5. MOLECULAR BREEDING FOR FORAGE QUALITY

5.1 Manipulation of lignin biosynthesis

Dry matter digestibility of forage plants declines markedly (>10 percent) as plants flower and senesce (Buckner *et al.*, 1967; Radojevic *et al.*, 1994; Stone, 1994). The changes in dry matter digestibility contribute significantly to the lowering of nutritive value of forage during summer (Stone, 1994). For example, increasing dry matter digestibility has been ranked as the most important goal in genetic improvement of the nutritive value of forage grasses for dairy pastures (Smith, Reed and Foot, 1997). However, since heritability of dry matter digestibility is low, with a large number of genes involved, the potential for rapid genetic improvement by traditional methods is low (Barnes, 1990). Lignification of plant cell walls has been identified as the major factor responsible for lowering digestibility of forage tissues as they mature (Buxton and Russell, 1988). The inhibitory effects of lignin on forage digestibility depend on lignin monomer composition and functional groups, lignin content, and the extent of cross-linking to cell wall polysaccharides (Jung and Vogel, 1986; Sewalt *et al.*, 1997; Casler and Jung, 1999; Casler and Kaeppler, 2000). Small increases in digestibility are expected to have significant impact on forage quality, and concomitantly on animal productivity. A 1 percent increase in *in vitro* dry matter digestibility has led to an average 3.2 percent increase in mean liveweight gains (Casler and Vogel, 1999). Lignification comprises a highly coordinated and regulated set of metabolic events resulting in the biosynthesis of lignin precursors (monolignols) and lignins (Whetten and Sederoff, 1995; Boudet and Grima-Pettenati, 1996; Boudet, Goffner and Grima-Pettenati, 1996; Campbell and Sederoff, 1996; Dixon *et al.*, 1996).

Molecular breeding for improved digestibility by down-regulating monolignol biosynthetic enzymes through antisense and sense suppression in transgenic forage plants are currently being explored. The main target enzymes being considered are caffeic acid *O*-methyltransferase (COMT), 4-coumarate:CoA ligase (4CL), cinnamoyl CoA reductase (CCR) and cinnamyl alcohol dehydrogenase (CAD). Experiments with the model plants tobacco and poplar have shown that the down-regulation of COMT, CAD and 4CL expression leads to altered lignin composition or reduced lignin content (Halpin *et al.*, 1994; Higuchi *et al.*, 1994; Ni, Paiva and Dixon, 1994; Hibino *et al.*, 1995; Kajita, Katayama and Omori, 1996; Boudet and Grima-Pettenati, 1996; Bernard Vailhé *et al.*, 1996a, 1998; Baucher *et al.*, 1996, 1998; Stewart *et al.*, 1997). A significant improvement in dry matter digestibility of transgenic sense and antisense COMT tobacco plants showing reduced COMT activity has been demonstrated (Bernard Vailhé *et al.*, 1996b; Sewalt *et al.*, 1997). A decrease in syringyl lignin units (Bernard Vailhé *et al.*, 1996b) or a reduction in lignin content (Dwivedi *et al.*, 1994; Sewalt *et al.*, 1997) was observed. Similarly, transgenic antisense CAD tobacco and alfalfa plants with down-regulated CAD activity have been reported to produce chemically more extractable lignin that was altered in composition or structure (Halpin *et al.*, 1994; Higuchi *et al.*, 1994; Hibino *et al.*, 1995; Boudet and Grima-Pettenati, 1996; Bernard Vailhé *et al.*, 1996a, 1998; Baucher *et al.*, 1996, 1998). Furthermore, simultaneous down-regulation of CAD and CCR in transgenic tobacco led to a decrease in lignin content without the alteration in plant development observed in the CCR-down-regulated parental plants. The parental plants had a similar decrease in lignin content (50 percent of the control), which resulted in reduction in size and in collapse of xylem cells (Chabannes *et al.*, 2000). These results indicate that improvement in dry matter digestibility by the introduction of chimeric sense and antisense lignin biosynthetic genes can be achieved, without apparently impairing normal development of the plant.

While the basic effect of transgenically manipulating the expression of some enzymes in the monolignol pathway may be similar to that of native genes coding for decreased enzyme activity in brown mid-rib (*bmr*) mutants (Casler and Kaeppler, 2000), transgenic approaches offer the potential for transgenic plants with highly-unusual, novel lignins; higher frequencies of novel-lignin phenotypes in comparison with natural variation; effective simultaneous down-regulation of multiple enzymes; and highly targeted down-regulation of enzymes through the choice of cell type-specific and developmentally-regulated promoters. Transgenic approaches to genetic manipulation of monolignol biosynthesis to enhance herbage quality are currently being explored in forage legumes (e.g. *Stylosanthes humilis*, *Medicago sativa*) and forage grasses (e.g. *Lolium perenne* and *Festuca arundinacea*) (L. McIntyre, pers. comm.; Guo, Chen and Dixon, 2000; Spangenberg *et al.*, 2000). Genes (cDNAs and genomic clones) encoding the key enzymes COMT, 4CL, CCR and CAD of perennial ryegrass have been isolated, sequenced, characterized and used for the molecular genetic dissection of this biosynthetic pathway in grasses (Heath *et al.*, 1998; Lynch, Lidgett and Spangenberg, 2000; McInnes, Lidgett and Spangenberg, 2000). Once proof of concept and suitable transformation events in forage plants with down-regulation of individual enzymes or simultaneously for multiple enzymes of the monolignol biosynthetic pathway are obtained, a thorough agronomic assessment of these transgenic plants – particularly for vigour and stress tolerances – and hybridization to generate transgenic elite germplasm, with subsequent selection, will be required to produce marketable cultivars.

5.2   Manipulation of fructan metabolism

Fructans are polyfructose molecules produced by many grass species as their main soluble storage carbohydrate form. It has been shown that ryegrass lines that accumulate higher concentrations of soluble carbohydrates do not suffer as great a decline in digestibility during summer (Radojevic *et al.*, 1994). The increased level of soluble carbohydrates appears to offset the decline in digestibility due to lignification. In addition, herbage intake, protein capture in the rumen and liveweight gains may be improved by increasing the concentrations of non-structural carbohydrates in pasture plants (Michell, 1973; Jones and Roberts, 1991; Beever, 1993).

The introduction of the microbial fructosyltransferase *SacB* gene from *Bacillus subtilis* into fructan-devoid and starch-accumulating tobacco and potato plants led to the accumulation of considerable amounts of high molecular weight levan-type fructans (Ebskamp *et al.*, 1994; Van der Meer *et al.*, 1994). These results demonstrate that sucrose, the substrate for fructosyltranferase, can be efficiently routed into a new sink in non-fructan-accumulating species. Furthermore, transgenic tobacco plants that accumulate bacterial levan showed enhanced performance under drought stress (Pilon-Smits *et al.*, 1995). Fructan synthesis in grasses involves the concerted action of at least three enzymes; sucrose:sucrose 1-fructosyltransferase (1-SST); fructan:fructan 1-fructosyltransferase (1-FFT); and sucrose:fructan 6-fructosyltransferase (6-SFT), which synthesizes the more complex mixed-linkage fructans that prevail in grasses and cereals. A number of plant fructan metabolism-related genes, such as barley 6-SFT, onion 6G-FFT and artichoke 1-SST, have been isolated in recent years, and when introduced into native fructan-devoid species they have led to oligofructan accumulation, and to novel fructan production in native fructan-accumulating plants (Sprenger *et al.*, 1995; Hellwege *et al.*, 1997; Vijn *et al.*, 1997). Transgenic approaches for the genetic manipulation of fructan biosynthesis to enhance herbage quality and tolerance to abiotic stresses are being explored in both forage legumes (e.g. *Trifolium repens*, *Medicago sativa*) and forage grasses (e.g. *Lolium perenne* and *Festuca arundinacea*) (Jenkins *et al.*, 2000; Johnson *et al.*, 2000; LePage *et al.*, 2000; Lidgett *et al.*, 2000; Ye *et al.*, 2001). Transgenic Italian ryegrass (*Lolium multiflorum*) plants with altered fructan metabolism brought about by the expression of chimeric bacterial levansucrase genes have been generated (Ye *et al.*, 2001). cDNAs encoding perennial ryegrass fructosyltransferase homologues have been isolated, characterized and are being used for the systematic molecular genetic dissection of fructan biosynthesis in transgenic grasses (Johnson *et al.*, 2000, 2003; Lidgett *et al.*, 2000, 2002; Chalmers *et al.*, 2003). A cDNA encoding 1-SST from tall fescue has been isolated and functionally characterized in transient assays with tobacco protoplasts and in methylotrophic yeast, *Pichia pastoris* (Luescher *et al.*, 2000). Transgenic white clover plants expressing chimeric *Bacillus subtilis SacB* genes for enhanced tolerance to drought have been produced (LePage *et al.*, 2000). Transgenic lucerne (*Medicago sativa*) and white clover plants expressing a fructosyltransferase gene, derived from *Streptococcus salivarius*, have also been generated (Jenkins *et al.*, 2000). The molecular genetic dissection of fructan biosynthesis in key pasture grasses will enhance our knowledge of fructan metabolism and carbohydrate partitioning in grasses and clarify their functional role in tolerance to cold and drought. This knowledge will be instrumental in designing experimental approaches to produce transgenic forage plants with enhanced forage quality and tolerance to abiotic stresses.

5.3   Transgenic expression of "rumen bypass" protein

Sulphur-containing (S-) amino acids, methionine and cysteine, are among the most limiting essential amino acids in ruminant animal nutrition (Ørskov and Chen, 1989). In particular, wool growth in sheep is frequently limited by the supply of S-amino acids under normal grazing conditions (Reis, 1979; Higgins *et al.*, 1989). Rumen fermentation contributes partly to the S-amino acid deficiency, since rumen microflora degrade the feed protein and, in some circumstances, re-synthesize proteins with a lower nutrient value (Rogers, 1990). Post-ruminal supplements of methionine and cysteine have been shown to result in a 16–130 percent increase in the rate of wool growth (Reis and Schinckel, 1963; Langlands, 1970; Pickering and Reis, 1993). There have also been reports of positive effects from feeding protected methionine on milk production in dairy cows and growth rate in beef animals (Buttery and Foulds, 1988). It is therefore predicted that the ingestion of forage containing relatively rumen-stable proteins rich in S-amino acids would enhance the supply of limiting essential amino acids for ruminant nutrition and lead to increased animal productivity, particularly wool growth (Higgins *et al.*, 1989; Rogers, 1990). The production of transgenic forage legumes expressing genes encoding different rumen bypass proteins rich in S-amino acids, such as chicken ovalbumin, pea albumin and sunflower seed albumin, has been reported (Schroeder *et al.*, 1991; Ealing, Hancock and White, 1994; Tabe *et al.*, 1995; Khan *et al.*, 1996). Low expression levels were observed for the ovalbumin gene in transgenic lucerne and the pea albumin gene in transgenic white clover, where accumulation of the proteins was less than 0.01 percent of total cell protein (Schroeder *et al.*, 1991; Ealing, Hancock and White, 1994). Accumulation of sunflower seed albumin up to 0.1 percent of soluble leaf protein was achieved in transgenic lucerne when the gene was driven by an *rbcS* promoter from *Arabidopsis thaliana* (Tabe *et al.*, 1995). In transgenic subterranean clover (*Trifolium subterraneum*), the accumulation of sunflower albumin increased with leaf age, with old leaves of the most highly expressing plants containing 1.3 percent of total extractable protein (Khan *et al.*, 1996). Transgenic tall fescue plants expressing chimeric genes carrying sunflower albumin SFA8 cDNA sequences (Kortt *et al.*, 1991) with the endoplasmic reticulum retention signal KDEL (Lys-Asp-Glu-Leu) (Wandelt *et al.*, 1992), under control of different promoters, were generated through biolistic transformation (Wang *et al.*, 2001). Transgenic tall fescue plants produced the expected transcript and accumulated the methionine-rich SFA8 protein at levels of up to 0.2 percent of total soluble protein. In order to achieve nutritionally useful levels, expression of the sunflower seed albumin may need to reach 2–5 percent of total soluble protein. Strategies for increasing the accumulation levels of foreign proteins in the leaves of forage plants are required if the full potential offered by transgenic approaches to create novel protein-phenotypes of forages is to be captured.

5.4   Manipulation of condensed tannin biosynthesis

Condensed tannins (proanthocyanidins) are polymeric phenyl-propanoid-derived compounds synthesized by the flavonoid pathway. They are agronomically important in a range of forage legumes, where they are regarded as either beneficial or detrimental. At levels above 4–5 percent dry weight, condensed tannins are generally considered to be nutritionally detrimental and act as anti-feedants and anti-nutritional factors for grazing livestock (Barry and Duncan, 1984; Waghorn *et al.*, 1990; Morris and Robbins, 1997), moderate amounts (1–3 percent) improve herbage quality since they reduce bloat in grazing ruminants by disrupting protein foam, decrease the loss of dietary protein by microbial de-amination, and reduce parasitic load (Barry and Duncan, 1984; Howarth *et al.*, 1991; Tanner *et al.*, 1995; McMahon *et al.*, 2000).

Molecular genetic approaches for the manipulation of tannin biosynthesis have been mainly aimed at the introduction of condensed tannins into lucerne and white clover, and at the reduction of tannin content in highly tanniniferous forage legumes. These transgenic approaches, including strategies for increasing tannin content, for modifying tannin structure, molecular weight and tissue distribution, and for novel tannin enzyme and gene discovery, have recently been reviewed (Morris and Robbins, 1997; Gruber, Ray and Blathut-Beatty, 2000).

REFERENCES

Barnes, R.F. 1990. Importance and problems of tall fescue. pp. 2–12, *in:* M.J. Kasperbauer (ed). *Biotechnology in Tall Fescue Improvement.* Boca Raton, FL: CRC Press.

Barry, T.N. & Duncan, S.J. 1984. The role of condensed tannins in the nutritional value of *Lotus pedunculatus* for sheep. 1. Voluntary intake. *British Journal of Nutrition,* **51**: 485–491.

Baucher, M., Chabbert, B., Pilate, G., Van Doorsselaere, J., Tollier, M., Petit-Conil, M., Cornu, D., Monties, B., Van Montagu, M., Inze, D., Jouanin, L. & Boerjan, W. 1996. Red xylem and higher lignin extractability by down-regulating a cinnamyl alcohol dehydrogenase in poplar. *Plant Physiology,* **112**: 1479–1490.

Baucher, M., Bernard Vailhé, M.A., Chabbert, B., Besle, J.M., Opsomer, C., Van Montagu, M. & Botterman, J. 1998. Down-regulation of cinnamyl alcohol dehydrogenase in transgenic alfalfa (*Medicago sativa* L.) and the effect on lignin composition and digestibility. *Plant Molecular Biology,* **39**: 437–447.

Beever, D.E. 1993. Ruminant animal production from forages: present position and future opportunities. pp. 158–164, *in:* M.J. Baker (ed). *Grasslands for Our World.* Wellington, New Zealand: SIR.

Bernard Vailhé, M.A., Cornu, A., Robert, D., Maillot, M.P. & Besle, J.M. 1996a. Cell wall degradability of transgenic tobacco stems in relation to their chemical extraction and lignin quality. *Journal of Agricultural and Food Chemistry,* **44**: 1164–1169.

Bernard Vailhé, M.A., Migné, C., Cornu, A., Maillot, M.P., Grenet, E., Besle, J.M., Atanassova, R., Martz, F. & Legrand, M. 1996b. Effect of modification of *O*-methyltransferase activity on cell wall composition, ultrastructure and degradability of transgenic tobacco. *Journal of the Science of Food and Agriculture,* **72**:385-391.

Bernard Vailhé, M.A., Besle, J.M., Maillot, M.P., Cornu, A., Halpin, C. & Knight, M. 1998. Effect of down-regulation of cinnamyl alcohol dehydrogenase on cell wall composition and on degradability of tobacco stems. *Journal of the Science of Food and Agriculture,* **76**: 505–514.

Boudet, A.M. & Grima-Pettenati, J. 1996. Lignin genetic engineering. *Molecular Breeding,* **2**: 25–39.

Boudet, A.M., Goffner, D.P. & Grima-Pettenati, J. 1996. Lignins and lignification: recent biochemical and biotechnological developments. *Comptes Rendus de l'Academie des Sciences [Paris] – Sciences de la Vie/Life Sciences,* **319**: 317–331.

                                                            G. Spangenberg

Buckner, R.C., Todd, J.R., Burrus, P.B. & Barnes, R.F. 1967. Chemical composition, palatability, and digestibility of ryegrass-tall fescue hybrid, 'Kenwell', and 'Kentucky 31' tall fescue varieties. *Agronomy Journal,* **59**: 345–349.

Buttery, P.J. & Foulds, A.N. 1988. Amino acid requirements of ruminants. pp. 19–33, *in:* W. Haresign and C.J. Cole (eds). *Recent Development in Ruminant Nutrition, 2.* London: Butterworths.

Buxton, D.R. & Russell, J.R. 1988. Lignin constituents and cell-wall digestibility of grass and legume stems. *Crop Science,* **28**: 553–558.

Campbell, M.M. & Sederoff, R.R. 1996. Variation in lignin content and composition. *Plant Physiology,* **110**: 3–13.

Casler, M.D. & Jung, H.G. 1999. Selection and evaluation of smooth bromegrass clones with divergent lignin or etherified ferulic acid concentration. *Crop Science,* **39**: 1866–1873.

Casler, M.D. & Kaeppler, H.F. 2000. Molecular breeding for herbage quality in forage crops. Chapter 10, *in:* G. Spangenberg (ed). *Molecular Breeding of Forage Crops.* Dordrecht, The Netherlands: Kluwer Academic Publishers.

Casler, M.D. & Vogel, K.P. 1999. Accomplishments and impact from breeding for increased forage nutritional value. *Crop Science,* **39**: 19–20.

Chabannes, M., Yoshinaga, A., Ruel, K. & Boudet, A.M. 2000. Simultaneous down-regulation of CAD and CCR genes induces differences in amount, nature and topology of lignins in the cell wall. Abstracts 6th International Congress of Plant Molecular Biology, Quebec. *Molecular Biology Reporter,* **18** (Suppl.): 2, S 04–6.

Chalmers, J., Johnson, X., Lidgett, A. & Spangenberg, G. 2003. Isolation and characterisation of a sucrose : sucrose 1-fructosyltransferase gene from perennial ryegrass (*Lolium perenne* L.). *Journal of Plant Physiology,* **160**: 1385–1391.

Dixon, R.A., Lamb, C.J., Masoud, S., Sewalt, V.J.H. & Paiva, N.L. 1996. Metabolic engineering – prospects for crop improvement through the genetic manipulation of phenylpropanoid biosynthesis and defense responses – a review. *Gene,* **179**: 61–71.

Dwivedi, U.N., Campbell, W.H., Yu, J., Datla, R.S.S., Bugos, R.C., Chiang, V.L. & Podila, G.K. 1994. Modification of lignin biosynthesis in transgenic *Nicotiana* through expression of an antisense *O*-methyltransferase gene from *Populus. Plant Molecular Biology,* **26**: 61–71.

Ealing, P.M., Hancock, K.R. & White, D.W.R. 1994. Expression of the pea albumin 1 gene in transgenic white clover and tobacco. *Transgenic Research,* **3**: 344–354.

Ebskamp, M.J.M., Van Der Meer, I.M., Spronk, B.A., Weisbeek, P.J. & Smeekens, S.C.M. 1994. Accumulation of fructose polymers in transgenic tobacco. *Bio/Technology,* **12**: 272–275.

Gruber, M.Y., Ray, H. & Blathut-Beatty, L. 2000. Genetic manipulation of condensed tannin synthesis in forage crops. Chapter 11, *in:* G. Spangenberg (ed). *Molecular Breeding of Forage Crops.* Dordrecht, The Netherlands: Kluwer Academic.

Guo, D., Chen, F. & Dixon, R.A. 2000. Roles of caffeic acid 3-*O*-methyltransferase and caffeolyl-CoA 3-*O*-methyltransferase in determining lignin content and composition in alfalfa (*Medicago sativa* L.). Abstracts 6th International Congress of Plant Molecular Biology, Quebec, Canada. *Molecular Biology Reporter,* **18** (Suppl.): 2, S 04–12.

Halpin, C., Knight, M.E., Foxon, G.A., Campbell, M.M., Boudet, A.M., Boon, J.J., Chabbert, B., Tollier, M. & Schuch, W. 1994. Manipulation of lignin quality by down-regulation of cinnamyl alcohol dehydrogenase. *Plant Journal,* **6**: 339–350.

Heath, R., Huxley, H., Stone, B. & Spangenberg, G. 1998. cDNA cloning and differential expression of three caffeic acid *O*-methyltransferase homologues from perennial ryegrass (*Lolium perenne*). *Journal of Plant Physiology,* **153**: 649–657.

Hellwege, E.M., Gritscher, D., Willmitzer, L. & Heyer, A.G. 1997. Transgenic potato tubers accumulate high levels of 1-kestose and nystose: functional identification of a sucrose:sucrose 1-fructosyltransferase of artichoke (*Cynara scolymus*) blossom discs. *Plant Journal,* **12**: 1057–1065.

Hibino, T., Takabe, K., Kawazu, T., Shibata, D. & Higuchi, T. 1995. Increase of cinnamaldehyde groups in lignin of transgenic tobacco plants carrying an antisense gene for cinnamyl alcohol dehydrogenase. *Bioscience Biotechnology and Biochemistry,* **59**: 929–931.

Higgins, T.J., O'brien, P.A., Spencer, D., Schroeder, H.E., Dove, H. & Freer, M. 1989. Potential of transgenic plants for improved amino acid supply for wool growth. pp. 441–445, *in:* G.E. Rogers, P.J. Reis, K.A. Ward and R.C. Marshall (eds). *The Biology of Wool and Hair.* London: Chapman and Hall.

Higuchi, T., Ito, T., Umezawa, T., Hibino, T. & Shibata, D. 1994. Red-brown color of lignified tissues of transgenic plants with antisense CAD gene: wine-red lignin from coniferyl aldehyde. *Journal of Biotechnology,* **37**: 151–158.

Howarth, R.E., Chaplin, R.K., Cheng, K.J., Goplen, B.P., Hall, J.W., Hironaka, R., Majak, W. & Radostits, O.M. 1991. Bloat in cattle. *Agriculture Canada Publication,* No. 1858/E. 34p.

Jenkins, C., Simpson, R., Higgins, T.J. & Larkin, P. 2000. Transgenic white clover and lucerne containing bacterial fructan as a novel carbohydrate for grazing animals. p. 32, *in:* Abstracts 2nd International Symposium on Molecular Breeding of Forage Crops. Lorne and Hamilton, Victoria, Australia.

Johnson, X., Lidgett, A., Chalmers, J., Guthridge, K., Jones, E. & Spangenberg, G. 2003. Isolation and characterisation of an invertase gene from perennial ryegrass (*Lolium perenne* L.). *Journal of Plant Physiology,* **160**: 903–911.

Johnson, X., Lidgett, A., Terdich, K. & Spangenberg, G. 2000. Molecular cloning and characterisation of a fructosyl transferase homologue from *Lolium perenne*. p. 36, *in:* Abstracts 2nd International Symposium on Molecular Breeding of Forage Crops. Lorne and Hamilton, Victoria, Australia.

Jones, E.L. & Roberts, E. 1991. A note on the relationship between palatability and water-soluble carbohydrates in perennial ryegrass. *Australian Journal of Agricultural Research,* **30**: 163–167.

Jung, H.G. & Vogel, K.P. 1986. Influence of lignin on digestibility of forage cell wall material. *Journal of Animal Science,* **62**: 1703–1712.

Kajita, S., Katayama, Y. & Omori, S. 1996. Alterations in the biosynthesis of lignin in transgenic plants with chimeric genes for 4-coumarate:coenzyme A ligase. *Plant Cell Physiology,* **37**: 957–965.

Kalla, R., Chu, P., Spangenberg, G. & Kalla, R. 2001. Molecular breeding of forage legumes for virus resistance. pp. 219–238, *in:* G. Spangenberg (ed). *Molecular Breeding of Forage Crops.* [Developments in Plant Breeding, Vol. 10]. Kluwer Academic Publishers.

Khan, M.R.I., Ceriotti, A., Tabe, L., Aryan, A., McNabb, W., Moore, A., Craig, S., Spencer, D. & Higgins, T.J.V. 1996, Accumulation of a sulphur-rich seed albumin from sunflower in the leaves of transgenic subterranean clover (*Trifolium subterraneum* L.). *Transgenic Research,* **5**: 179–185.

Kortt, A.A., Caldwell, J.B., Lilley, G.G. & Higgins, T.J. 1991. Amino acid and cDNA sequences of a methionine-rich 2S protein from sunflower seed (*Helianthus annuus* L.). *European Journal of Biochemistry,* **195**: 329–334.

Langlands, J.P. 1970. The use of sulphur-containing amino acids to stimulate wool growth. *Australian Journal of Experimental Agriculture and Animal Husbandry,* **10**: 665–671.

LePage, C., Mackin, L., Lidgett, A. & Spangenberg, G. 2000. Development of transgenic white clover expressing chimeric bacterial levansucrase genes for enhanced tolerance to drought stress. p. 80, *in:* Abstracts 2nd International Symposium on Molecular Breeding of Forage Crops. Lorne and Hamilton, Victoria, Australia.

Lidgett, A., Johnson, X., Jennings, K., Guthridge, K., Jones, E. & Spangenberg, G. 2002. Isolation and characterisation of a fructosyltransferase gene from perennial ryegrass (*Lolium perenne* L.). *Journal of Plant Physiology,* **159**: 1037–1043.

Lidgett, A., Johnson, X., Terdich, K. & Spangenberg, G. 2000. The manipulation of fructan biosynthesis in pasture grasses. p. 31, *in:* Abstracts 2nd International Symposium on Molecular Breeding of Forage Crops. Lorne and Hamilton, Victoria, Australia.

Luescher, M., Hochstrasser, U., Vogel, G., Aeschbacher, R., Galati, V., Nelson, C.J., Boller, T. & Wiemken, A. 2000. Cloning and functional analysis of sucrose:sucrose 1-fructosyltransferase (1-SST) from tall fescue. *Plant Physiology,* **124**: 1217–1227.

Lynch, D., Lidgett, A. & Spangenberg, G. 2000. Isolation and characterisation of cinnamyl alcohol dehydrogenase homologue cDNAs from ryegrass. *In:* Abstracts 2nd International Symposium on Molecular Breeding of Forage Crops. Lorne and Hamilton, Victoria, Australia.

McInnes, R., Lidgett, A. & Spangenberg, G. 2000. Isolation and characterisation of a cinnamoyl-CoA reductase homologue cDNA from perennial ryegrass. p. 38, *in:* Abstracts 2nd International Symposium on Molecular Breeding of Forage Crops. Lorne and Hamilton, Victoria, Australia.

McMahon, L.R., McAllister, T.A., Berg, B.P., Mayak, W., Acharya, S.N., Popp, J.D., Coulman, B.E., Wang, Y. & Cheng, K.J. 2000. A review of the effects of condensed tannins on ruminal fermentation and bloat in grazing cattle. *Canadian Journal of Plant Science,* **80**: 469–485.

Michell, P.J. 1973. Relations between fibre and water soluble carbohydrate contents of pasture species and their digestibility and voluntary intake by sheep. *Australian Journal of Experimental Agriculture and Animal Husbandry,* **13**: 165–170.

Morris, P. & Robbins, M.P. 1997. Manipulating condensed tannins in forage legumes. pp. 147–173, *in:* B.D. McKersie and D.C.W. Brown (eds). *Biotechnology and the Improvement of Forage Legumes.* Biotechnology in Agriculture Series, No. 17. Wallingford, UK: CAB International.

Ni, W.T., Paiva, N.L. & Dixon, R.A. 1994. Reduced lignin in transgenic plants containing a caffeic acid *O*-methyltransferase antisense gene. *Transgenic Research,* **3**: 120–126.

Orskov, E.R. & Chen, X.B. 1990. Assessment of amino acid requirements in ruminants. pp. 161–167, *in:* S. Hoshino, R. Onodera, H. Minato and H. Itabashi (eds). *The Rumen Ecosystem.* Japan Scientific Societies Press and Springer Verlag.

Pickering, F.S. & Reis, P.J. 1993. Effects of abomasal supplement of methionine on wool growth of grazing sheep. *Australian Journal of Experimental Agriculture,* **33**: 7–12.

Pilon-Smits, E.A.H., Ebskamp, M.J.M., Paul, M.J., Jeuken, M.J.W., Weisbeek, P.J. & Smeekens, S.C.M. 1995. Improved performance of transgenic fructan-accumulating tobacco under drought stress. *Plant Physiology,* **107**: 125–130.

Radojevic, I., Simpson, R.J., St John, J.A. & Humphreys, M.O. 1994. Chemical composition and *in vitro* digestibility of lines of *Lolium perenne* selected for high concentrations of water-soluble carbohydrate. *Australian Journal of Agricultural Research,* **45**: 901–912.

Reis, P.J. 1979. Effects of amino acids on the growth and properties of wool. pp. 223–242, *in:* J.L. Black and P.J. Reis (eds). *Physiological and Environmental Limitations to Wool Growth.* Armidale, Australia: University of New England Publishing Unit.

Reis, P.J. & Schinckel, P.G. 1963. Some effects of sulfur-containing amino acids on the growth and composition of wool. *Australian Journal of Biological Science,* **16**: 218–230.

Rogers, G.E. 1990. Improvement of wool production through genetic engineering. *TIBTECH,* **8**: 6–11.

Schroeder, H.E., Khan, M.R.I., Knibb, W.R., Spencer, D. & Higgins, T.J.V. 1991. Expression of a chicken ovalbumin gene in three lucerne cultivars. *Australian Journal of Plant Physiology,* **18**: 495–505.

Sewalt, V.J.H., Ni, W., Jung, H.G. & Dixon, R.A. 1997. Lignin impact on fiber degradation: increased enzymatic digestibility of genetically engineered tobacco (*Nicotiana tabacum*) stems reduced in lignin content. *Journal of Agricultural and Food Chemistry,* **45**:1977-1983.

Smith, K.F., Reed, K.F.M. & Foot, J.Z. 1997. The relative importance of specific traits for the genetic improvement of nutritive value in dairy pasture. *Grass and Forage Science,* **52**: 167–175.

Spangenberg, G., Baera, K., Bartkowski, A., Huxley, H., Lidgett, A., Lynch, D., McInnes, R. & Nagel, J. 2000. The manipulation of lignin biosynthesis in pasture grasses. p. 41, *in:* Abstracts 2nd International Symposium on Molecular Breeding of Forage Crops. Lorne and Hamilton, Victoria, Australia.

Spangenberg, G., Kalla, R., Lidgett, A., Sawbridge, T., Ong, E.K. & John, U. 2001. Breeding forage plants in the genome era. pp. 1–40, *in:* G. Spangenberg (ed). *Molecular Breeding of Forage Crops.* [Developments in Plant Breeding, Vol. 10] London: Kluwer Academic Publishers.

Sprenger, N., Bortlik, K., Brandt, A., Boller, T. & Wiemken, A. 1995. Purification, cloning, and functional expression of sucrose:fructan 6-fructosyltransferase, a key enzyme of fructan synthesis in barley. *Proceedings of the National Academy of Sciences, USA,* **92**: 11652–11656.

Stewart, D., Yahiaoui, N., Mcdougall, G.J., Myton, K., Marque, C., Boudet, A.M. & Haigh, J. 1997. Fourier-transform infrared and Raman spectroscopic evidence for the incorporation of cinnamaldehydes into the lignin of transgenic tobacco (*Nicotiana tabacum* L.) plants with reduced expression of cinnamyl alcohol dehydrogenase. *Planta,* **201**: 311–318.

Stone, B.A. 1994. Prospects for improving the nutritive value of temperate, perennial pasture grasses. *New Zealand Journal of Agricultural Research,* **37**: 349–363.

Tabe, L.M., Wardley-Richardson, T., Ceriotti, A., Aryan, A., McNabb, W., Moore, A. & Higgins, T.J.V. 1995. A biotechnological approach to improving the nutritive value of alfalfa. *Journal of Animal Science,* **73**: 2752–2759.

Tanner, G.J., Moate, P., Dailey, L., Laby, R. & Larkin, P.J. 1995. Proanthocyanidins (condensed tannins) destabilise plant protein foams in a dose dependent manner. *Australian Journal of Agricultural Research,* **46**: 1101–1109.

Van Der Meer, I.M., Ebskamp, M.J.M., Visser, R.G.F., Weisbeek, P.J. & Smeekens, S.C.M. 1994. Fructan as a new carbohydrate sink in transgenic potato plants. *Plant Cell,* **6**: 561–570.

Vijn, I., Van Dijken, A., Sprenger, N., Van Dun, K., Weisbeek, P., Wiemken, A. & Smeekens, S.C.M. 1997. Fructan of the inulin neoseries is synthesized in transgenic chicory plants (*Cichorium intybus* L.) harbouring onion (*Allium cepa* L.) fructan:fructan 6G-fructosyltransferase. *Plant Journal,* **11**: 387–398.

Waghorn, G.C., Jones, W.T., Shelton, I.D. & McKnabb, W.C. 1990. Condensed tannins and the nutritive value of herbage. *Proceedings of the New Zealand Grassland Association,* **51**: 171–176.

Wandelt, C.I., Khan, M.R.I., Craig, S., Schroeder, H.E., Spencer, D. & Higgins, T.J.V. 1992. Vicillin with carboxy-terminal KDEL is retained in the endoplasmic reticulum and accumulates to high levels in the leaves of transgenic plants. *Plant Journal,* **2**: 181–192.

Wang, Z.Y., Ye, X.D., Nagel, J., Potrykus, I. & Spangenberg, G. 2001. Expression of a sulphur-rich sunflower albumin gene in transgenic tall fescue (*Festuca arundinacea* Schreb.) plants. *Plant Cell Reports,* **20**: 213–219.

Whetten, R. & Sederoff, R. 1995. Lignin biosynthesis. *Plant Cell,* **7**: 1001–1013.

Ye, X.D., Wu, X.L., Zhao, H., Frehner, M., Noesberger, J., Potrykus, I. & Spangenberg, G. 2001. Altered fructan accumulation in transgenic *Lolium multiflorum* plants expressing a *Bacillus subtilis SacB* gene. *Plant Cell Reports,* **20**: 205–212.

INVESTIGATION OF THE RUMEN MICROBIAL COMMUNITY RESPONSIBLE FOR DEGRADATION OF A PUTATIVE TOXIN IN *ACACIA ANGUSTISSIMA*

Erin M.C. Collins,[1] Chris S. Mcsweeney,[2] * Denis O. Krause[2] and Linda L. Blackall[1]

1. School of Molecular and Microbial Sciences, The University of Queensland, St Lucia, Brisbane, Australia.
2. CSIRO Livestock Industries, Queensland Biosciences Precinct, Carmody Rd, St Lucia, Brisbane, Australia.
* Corresponding author.

Abstract: *Acacia angustissima* has been proposed as a protein supplement in countries where availability of high quality fodder for grazing animals is a problem due to extreme, dry climates. While *A. angustissima* thrives in harsh environments and provides valuable nutrients required by ruminants, it has also been found to contain anti-nutritive factors that currently preclude its widespread application. A number of non-protein amino acids have been identified in the leaves of *A. angustissima* and in the past these have been linked to toxicity in ruminants. The non-protein amino acid 4-n-acetyl-2,4-diaminobutyric acid (ADAB) had been determined to be the major non-protein amino acid in the leaves of *A. angustissima*. Thus, in this study, the aim was to identify microorganisms from the rumen environment capable of degrading ADAB. Using an ADAB-containing plant extract, a mixed enrichment culture was obtained that exhibited substantial ADAB-degrading ability. Attempts to isolate an ADAB-degrading micro-organism were carried out, but no isolates were able to degrade ADAB in pure culture. The mixed microbial community of the ADAB-degrading enrichment culture was further examined through the use of pure-culture-independent techniques. Fluorescence *in situ* hybridization (FISH) was employed to investigate the diversity within this sample. In addition two bacterial 16S rDNA clone libraries were constructed in an attempt to further elucidate the members of the microbial population. The clone libraries were constructed from serial dilutions of the enrichment culture, a 10^{-5} dilution where complete degradation of ADAB occurred, and a 10^{-7} dilution where ADAB degradation did not occur. Through the comparison of these two libraries it was hypothesized that clones belonging to the *Firmicutes*

373

H. P. S. Makkar and G. J. Viljoen (eds.), Applications of Gene-Based Technologies for Improving Animal Production and Health in Developing Countries, 373-386.
© 2005 *IAEA. Printed in the Netherlands.*

phylum were involved in ADAB degradation. A FISH probe, ADAB1268, was then designed to target these clones and was applied to the enrichment cultures to investigate their relative abundance within the sample.

1.   INTRODUCTION

Globally there is a requirement for efficient productivity of primary produce and basic commodities, and, as a consequence, livestock nutrition is of great relevance. This is a major issue in developing countries, particularly where dry, harsh environments and rapidly increasing populations prevail. A tropical legume, *Acacia angustissima*, has been proposed as a protein supplement for ruminants in such countries, where limited choice and restricted availability of feedstock is a significant problem in the management of resources. In addition to being an excellent source of dietary protein for ruminants, *A. angustissima* is also advantageous as it grows well in poor soils and exhibits drought tolerance (Gutteridge, 1994). Despite these attributes, the widespread use of *A. angustissima* is currently impeded by the potential toxicity observed during feeding trials with sheep (Odenyo *et al.*, 1997).

Investigations into the potential toxins or anti-nutritive factors of *A. angustissima* have ensued. The anti-nutritive value of tannins in *A. angustissima* has been explored but their potential role in toxicity is unknown (Saarisalo, Odenyo and Osuji, 1999). A number of non-protein amino acids have been found in the leaves and seeds of *A. angustissima* and other similar browse legumes. These include diaminobutyric acid (DABA), 2,3-diaminopropionic acid (DAPA), 4-N-acetyl-2,4-diaminobutyric acid (ADAB), and 2-amino-4N-ureidopropionic acid (albizziine) (Evans *et al.*, 1985, 1993; Shah *et al.*, 1992). ADAB has been found through high performance liquid chromatography (HPLC) and nuclear magnetic resonance (NMR) to be the major non-protein amino acid in the leaves of *A. angustissima* (Reed *et al.*, 2001). Thus, due to the high abundance of ADAB in the leaves of *A. angustissima* and its similarity to well known neurotoxins, such as DABA (O'Neal *et al.*, 1968), ADAB has been implicated in the toxicity of *A. angustissima*.

During a sheep feeding trial it was observed that animals exhibited no toxic effects when their diet was gradually adapted to include a large proportion of *A. angustissima* (Odenyo *et al.*, 1997). It was further illustrated through transfers of rumen contents that the rumen micro-organisms may confer toxin tolerance to naïve sheep, although there were no uninoculated controls in the study (Odenyo *et al.*, 1997).

In light of these findings, the aim of this investigation was to resolve which are the integral members of sheep rumen microbial consortia that facilitate continued subsistence on a diet comprising *A. angustissima*. In order to further define and direct the research, rumen micro-organisms that degrade the non-protein amino acid ADAB were the focus. To this end an approach utilizing a variety of techniques was employed in order to explore the microbial ecology of this complex microbial ecosystem. Initially, the strategy was to selectively enrich for and isolate ADAB-degrading micro-organisms from rumen contents. Serial dilution and anaerobic plating with variations of habitat simulating anaerobic media with an amino acid-containing fraction of *A. angustissima* extract were used in an effort to isolate an ADAB-degrading micro-organism. The resultant enrichment cultures were monitored using assays for ADAB degradation and the diversity of the mixed microbial enrichment cultures was investigated with light microscopy.

Further examination of the diversity of the ADAB-degrading mixed cultures was carried out by fluorescence *in situ* hybridization (FISH). FISH is a method employing phylogenetically based oligonucleotide probes tagged with fluorescent molecules, targeted to the ribosomal RNA within the cells. When observed using fluorescence microscopy, whole cells fluoresce with specific probes and thus the morphology and identity of cells can be determined. By these means the diversity within the mixed microbial ADAB-degrading enrichment cultures was explored.

2. MATERIALS AND METHODS

2.1 Isolation and enrichment

Rumen digesta was taken from a Merino ewe (2 years old; rumen fistulated) that had been fed for 3 weeks on a Rhodes grass (*Chloris gayana*) diet supplemented with 30 percent oven-dried (50°C) *A. angustissima*. Digesta was immediately diluted (1:1) with an anaerobic saline diluent containing 30 percent glycerol (Mackie *et al.*, 1978) and preserved at -70°C. Sub-samples (1 ml) of preserved rumen digesta were inoculated into anaerobic medium for enrichment and isolation of ADAB-degrading bacteria. Enrichment and isolation of micro-organisms capable of degrading ADAB was attempted through serial ten-fold dilution and anaerobic plating experiments using anaerobic medium as described by Menke *et al.* (1979), which contained 20 percent clarified rumen fluid. An amino acid extract recovered from the leaves of *A. angustissima* using the cation ion exchange resin Amberlite IR-120 plus (16–50 mesh) was injected into each tube to give a final concentration of 10 mM ADAB per tube. Carbohydrates, which included soluble starch (0.5 g/litre), cellobiose (0.4 g/litre), xylose (0.4 g/litre), arabinose (0.2 g/litre) and pectin (0.2 g/litre), were added to the medium in some cases.

2.1.1 Spectrophotometric OPA assay for ADAB

A spectrophotometric assay was used to determine ADAB-degrading activity where o-phthaldialdehyde (OPA) was used to derivatize the amino acids and produce a colorimetric adduct that can be measured spectrophotometrically at a wavelength of 472 nm specific for DABA (Rao, 1978). Prior to the assay, the acetyl group was cleaved through hydrolysis, whereby 100 µl of sample was added to 100 µl of 6 M NaOH and heated to 98°C for 50 minutes. Following hydrolysis, the solution was neutralized with a 100 µl aliquot of 6 M HCl. To perform the assay, 40 µl of hydrolysate was added to 400 µl of OPA reagent, and then 2.5 ml of milli Q water was added. The reaction was then allowed to develop for 10 minutes before reading the absorbance at 472 nm. The absorbance of the control sample and current time sample were measured, and a decrease in absorbance correlated to a decrease in DABA present, indicating degradative activity. The OPA reagent consisted of 250 mg OPA dissolved in 5 ml of methanol, 250 µl of 2-mercaptoethanol, and 44.5 ml of 0.4 M potassium borate buffer (pH 9.5) containing 0.1 percent Triton X. The reagent was stored at 4°C for a maximum of 3 days.

2.1.2 High performance liquid chromatography (HPLC) analysis

Amino acids in culture fluid were analysed in the following manner. A 10 µl aliquot of culture fluid was added to 50 µl of phenyl-isothiocyanate (PITC) reagent in a glass vial. The PITC reagent consisted of 20 volumes acetonitrile (ACN), 10 volumes methanol, 4 volumes triethylamine and 1 volume PITC. The solution was allowed to stand for 30 minutes at room temperature, the solution was then flushed with nitrogen at 40°C until dry. Following this, 2 aliquots of 200 µl of ACN were added, and the solution evaporated to produce a dry residue. The resulting residue was then dissolved in 400 µl of mobile phase A, which consisted of 97 parts 50 mM sodium acetate (pH 6.0) and 3 parts ACN. Mobile phase B contained 6 parts ACN and 4 parts water. The derivatized culture solution was separated by HPLC on a TSKGel Super ODS column (4.6 × 100 mm, 2 µm; P/N 18197) using a Waters Novapak C18 guard column, Waters 600 gradient pump system and diode-array 996 detector with UV detection at 254 nm.

2.2   Fluorescence *in situ* hybridization (FISH)

2.2.1   *In situ* hybridization

Fixation of cultures was effected using 4 percent paraformaldehyde/ phosphate buffered saline (PFA/PBS). For storage, the cells were re-suspended in an equal volume of PBS/100 percent ethanol solution for storage at -20°C (Amann, Krumholz and Stahl, 1990). The oligonucleotide probes used for FISH were commercially synthesized with a 5′ fluorescein isothiocyanate (FITC), indotrimethinecyanine 3 (Cy3), or indopentamethinecyanine 5 (Cy5) label (Thermohybaid Interactiva, Ulm, Germany). When performing FISH with the GAM42a probe, unlabelled BET42a was also used to ensure specificity of this probe (Table 1). Fixed samples were applied to wells on teflon coated glass slides and air dried, and were then dehydrated in an ethanol series (50%, 80% and 100%) for a period of 3 minutes in each. Hybridization and washing procedures were carried out as described by Amann and co-workers (1990).

Table 1. Oligonucleotide probe sequences and stringency used in FISH experiments.

Probe name [Reference]	5'-3' sequence	Target organism	rRNA[1] target site	% F[2]
GAM42a [1]	GCCTTCCCACATCGTTT	*γ-proteobacteria*	23S; 1027–1043	35
BET42a [1]	GCCTTCCCACTTCGTTT	*β-proteobacteria*	23S; 1027–1043	35
CF319a [2]	TGGTCCGTGTCTCAGTAC	*Flavobacteria-Cytophaga*	16S; 319–336	35
BAC303 [2]	CCAATGTGGGGGACCTT	*Bacteroides-Prevotella*	16S; 303–319	0
EUBMIX [3]	GCWGCCWCCCGTAGGWGT	Domain *Bacteria*	16S; 338–355	0–60
ARC915 [4]	GTGCTCCCCCGCCAATTCCT	Domain *Archaea*	16S; 915–934	0–60
ADAB1268 [5]	TTCGGGGTTCGCTCCTCCTC	*Firmicutes* – ADAB-degrading clones	16S; 1268–1288	35

NOTES: (1) *E. coli* numbering according to Brosius *et al.* (1981). (2) Optimal probe formamide percentage.
SOURCES: [1] Amann, Krumholz and Stahl, 1990. [2] Manz *et al.*, 1996. [3] Daims *et al.*, 1999. [4] Stahl and Amann, 1991. [5] This study.

2.2.2   Confocal laser scanning microscopy

A Bio-Rad MRC 1024 confocal laser scanning microscope (CLSM) fitted with a 15 mW argon/krypton laser (American Laser Corporation) was used to acquire images and to view slides. The laser excitation peaks were

488 nm (blue), 568 nm (green) and 647 nm (red). The two emission filters used were a 560 nm long-pass emission filter used to distinguish the green signal from the red, and a 640 nm short-pass emission filter that separates the far-red signal (e.g. Cy5) from the near-red signal (e.g. Cy3). Images were captured using a Bio-rad Laser Sharp confocal controller and were compiled using Adobe Photoshop (Adobe Systems, Inc., Calif., USA).

2.3   16S ribosomal DNA clone libraries

2.3.1   DNA extraction and PCR

Enrichment cultures for three replicate dilutions were pooled prior to DNA extraction. DNA extraction for use in cloning was carried out using the FastDNA SPIN Kit for Soil (BIO101 Protocol), and was performed in the manner outlined by the manufacturer. Amplification of the 16S rDNA of bacterial cells or extracted DNA using PCR employed the forward and reverse primers 27f and 1492r (Table 2). Amplification of clone inserts in the pGem-T Easy plasmid vector involved the use of the vector-specific primers SP6 and T7. PCR reactions were performed in 50 µl volumes containing the following: *Tth* plus reaction buffer (Biotech International, Australia), 200 µM dNTPs, 1.5 mM $MgCl_2$, 200 ng of each of a forward and reverse primer, 100–200 ng of DNA template, 1 U of *Tth* plus DNA polymerase (Biotech International, Australia), and volume was made up using sterile milli Q water.

Table 2. Primers used for 16S rDNA PCR, and sequencing.

Primer[1]	Primer Sequence (5′→3′)
27f	GAGTTTGATCCTGGCTCAG
519r	GWATTACCGCGGCKGCTG
530f	GTGCCAGCMGCCGCGG
907r	CCGTCAATTCMTTTRAGTTT
926f	AAACTYAAAKGAATTGACGG
1492r	TACGGYTACCTTGTTACGACTT

Source: (1) Lane, 1991.

Reactions were carried out in a PTC-100 programmable thermal controller (MJ Research, Inc.) where they were subjected to 24 cycles as follows: denaturation at 94°C for 1 minute, annealing at 46°C for 1 minute, and extension at 72°C for 2 minutes, with a final extension of 5 minutes. PCR products were then purified using a QIAquick PCR purification kit protocol (QIAGEN, Australia). PCR products were then examined by agarose gel electrophoresis, using a 1 percent agarose gel containing 0.1 µg/ml ethidium bromide.

2.3.2   Ligation and transformation

Ligation of 80 ng of pure PCR product into the pGEM T Easy plasmid vector was performed overnight at 4°C. The plasmid was then transformed into commercially available competent *E. coli* cells, XL2-Blue ultra-competent cells, as per the manufacturers instruction. The transformed cells were then grown on LB agar with 50 mg/litre ampicillin overnight at 37°C. Blue-white screening was employed, using X-gal and IPTG to determine clones with inserts. Clones were screened for full-length inserts of the 16S rDNA using PCR with the plasmid primers SP6 and T7. Clones containing full-length inserts were grown, and stored in 20 percent glycerol at -80°C.

2.3.3   Restriction fragment length polymorphism (RFLP)

16S rDNA was digested with HinP1 I (New England Biolabs). Reactions of 25 µl consisting of 21.5 µl of PCR product, 2.5 µl of 10× NEBuffer 2, and 1 U of restriction enzyme. Reaction tubes were incubated at 37°C overnight. The digested product was electrophoresed in 2.5 percent agarose gel containing 0.1 µg/ml ethidium bromide for 90 minutes at 40 V. The resultant banding patterns were detected using UV light. Clones exhibiting the same banding pattern were grouped together. Representatives from each RFLP group were sequenced.

2.3.4   Sequencing

Sequencing was carried out in half reactions of Applied Biosystems (ABI) Prism BigDye Terminator (BDT) cycle sequencing ready reaction kit version 3.0. Reaction contained 25 ng of sequencing primer and approximately 400 ng of template DNA, as determined by comparison with Low DNA Mass Ladder (Life Technologies). Sequencing reactions were carried out in the PTC-100 programmable thermal controller (MJ Research Inc.) with an initial denaturation at 96°C for 2 minutes, followed by 29 cycles of 50°C for 15 seconds, 60°C for 4 minutes and 94°C for 30 seconds.

2.3.5 Sequence DNA purification

Sequence DNA was purified using the ethanol precipitation method as recommended by the Australian Genome Research Facility (AGRF). The 10 µl sequencing reaction was added to a microcentrifuge tube with 16 µl water, and 64 µl of 95 percent ultrapure ethanol added, vortexed briefly and allowed to sit at room temperature for 1–2 hour. Tubes were then centrifuged at 13,500 g for 20 minutes. The ethanol was removed immediately without disturbing the DNA pellet, and then the pellet was rinsed with 250 µl of 70 percent ethanol. Tubes were vortexed and centrifuged for a further 10 minutes, the supernatant was carefully aspirated, and the pellet then dried for 10 minutes in a vacuum centrifuge. Sequencing reactions were submitted for analysis to the AGRF.

2.3.6 Sequence analysis

Sequence identity was preliminarily determined by carrying out Basic Local Alignment Search Tool (BLAST) analysis (Altschul *et al.,* 1990). Contiguous sequences were compiled in SeqEd v1.0.3. Sequences were imported to the 16S rDNA database in the ARB software package in Fasta format. Sequences were then aligned automatically using the fast aligner tool. Phylogenetic analysis of sequences was performed in ARB and the robustness of tree topologies was determined through bootstrap analysis performed in PAUP[*] v 4.0b.

3. RESULTS AND DISCUSSION

3.1 Isolation and enrichments

The non-protein amino acid ADAB was degraded in enriched batch cultures from an Australian sheep adapted to *A. angustissima.* Diaminobutyric acid accumulated as an intermediate degradation product in basal medium that did not contain carbohydrate. The DABA that accumulated in culture fluid during the later stages of growth was eventually metabolized (Figure 1). Further investigations into the members of this enriched ADAB-degrading community were undertaken through the use of minimal medium and serial dilution of the enriched mixed culture. An extract from *A. angustissima* containing the amino acid fraction was the source of ADAB, and long incubations were carried out in order to account for slow growing organisms in the community. Attempts were made to isolate in pure culture an ADAB-degrading micro-organism from the rumen environment, using anaerobic spread plating of the enrichment. A number of isolates were obtained and examined for ADAB-degrading ability using an OPA assay or HPLC determination of ADAB concentration. None of these isolates exhibited ADAB degradation in pure culture. Inability to degrade ADAB in pure culture may be due to syntrophic relationships with other members of the community. Therefore, several of these isolates were grown in co-culture to determine whether complex communities are required to metabolize ADAB, but ADAB degradation did not occur.

Dilutions of ADAB-degrading mixed cultures were carried out in 10-fold serial dilution, in an effort to determine the dilutions where ADAB degradation occurred. ADAB concentration in the dilution tubes was assayed and degradation was found to occur at a dilution of 10^{-5} and no degradation was observed at a dilution of 10^{-7}. These dilutions were used in further analysis of the microbial community, for comparison of the microbial communities to facilitate the deduction of those micro-organisms important in the degradation of ADAB.

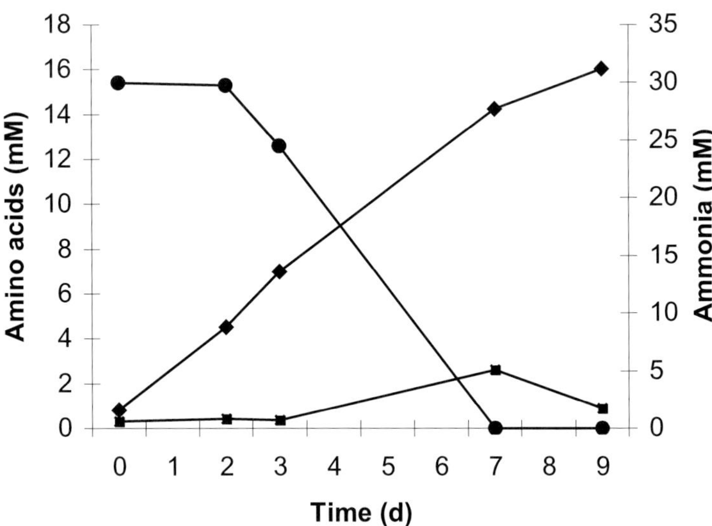

Figure 1. Rates of degradation of ADAB (•) and production of DABA (■) and ammonia (♦) in rumen enrichments on basal medium.

Enrichment of the rumen contents for an ADAB-degrading mixed culture resulted in limited diversity based on monitoring of the cellular morphologies within the sample, using light microscopy. The mixed enrichment cultures consisted mainly of coccobacilli, tiny cocci and long thin rods. The cells with the long thin rod morphology were observed to increase substantially in length over time.

3.2   FISH

FISH is an ideal technique to investigate the degree of diversity and identity of a mixed microbial population (Amann, Fuchs and Behrens, 2001; Moter and Gobel, 2000). Using FISH, cells can be directly visualized *in situ*, thereby demonstrating their relative abundance, and the spatial relationship of cells to each other and plant material can be observed in rumen digesta samples. Other techniques commonly used to quantify members of mixed microbial communities in mixed samples, such as real-time PCR and slot blot hybridization, require DNA extraction or PCR. FISH is not subject to the intrinsic biases of these techniques and more accurately reflects the community composition, and as such is an effective technique to investigate the relative abundance of community members within a mixed sample (Krause, Smith and McSweeney, 2001; von Wintzingerode, Gobel and Stackebrandt, 1997). To further investigate the diversity within the sample, FISH was performed with previously published probes for major groups of bacteria (listed in Table 1). A high proportion of the bacterial cells in the sample hybridized the GAM42a probe, specific for *γ-proteobacteria* (Figure 2). The taxa *Enterobacteriaceae* belongs to the *γ-proteobacteria*, and a number of isolates from this study were determined to be *Escherichia coli* through analysis of sequence data using BLAST analysis.

3.3   16S rDNA clone libraries

The use of FISH demonstrated that the ADAB-degrading mixed cultures consisted of micro-organisms hybridizing the probe specific for the domain *Bacteria*. Consequently, two bacterial 16S rDNA clone libraries were constructed from the ADAB-degrading mixed-culture dilution series, using a degrading (10^{-5} dilution) and a non-ADAB-degrading (10^{-7} dilution) mixed culture. The two bacterial 16S rDNA clone libraries were then analysed and compared. The ADAB-degrading clone library yielded 57 full-length 16S rDNA clones with eight distinct RFLP patterns, while the non-degrading clone library consisted of 62 full-length clones and only four RFLP patterns. Approximately 60 percent of the clones, which were represented by four out of eight RFLP patterns from the ADAB-degrading clone library, belonged to the *Firmicutes,* yet only two clones from the non-ADAB-degrading clone library belonged to this group, and were represented by a single RFLP pattern. Thus, through comparison of the two 16S rDNA clone libraries, it was deduced that the members of the *Firmicutes* were likely to be of importance in ADAB degradation. Phylogenetic analysis was carried out on 16S rDNA clones from the *Firmicutes.* It was found that the ADAB-degrading clones belonging to the *Firmicutes* formed a monophyletic group, which was supported by bootstrap analysis. An oligonucleotide probe, ADAB1268, was designed to specifically target this group for use in FISH.

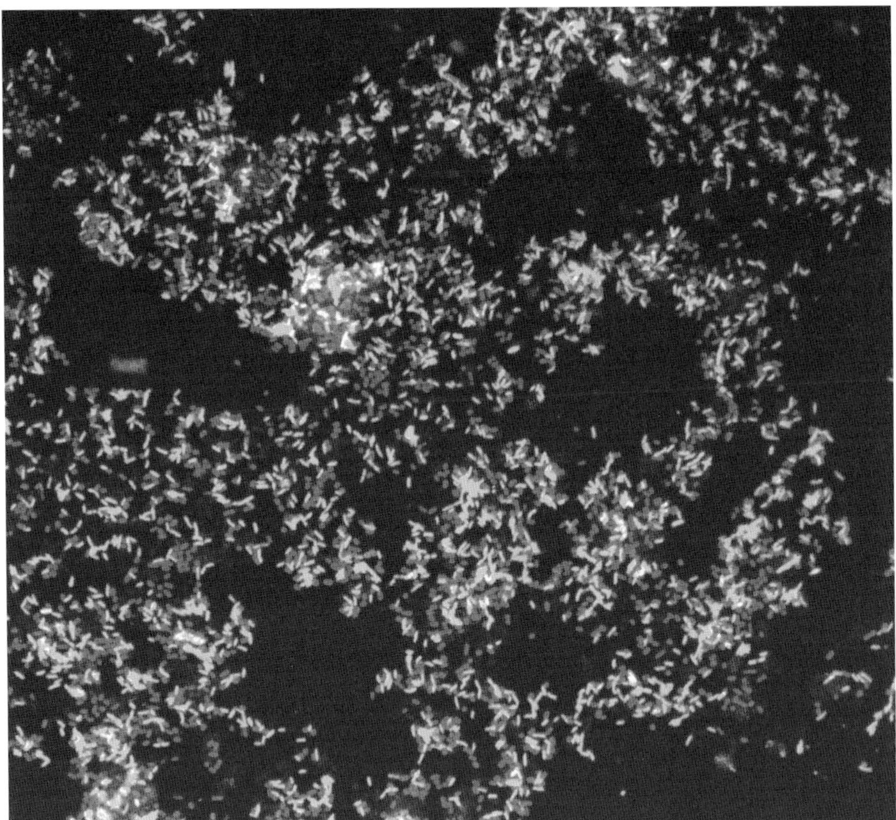

Figure 2. CLSM micrograph of ADAB-degrading mixed culture enrichment where FISH was performed using the probes EUBMIX-Cy5 (blue), GAM42a-Cy3 (red), and ADAB1268-FITC (green) (see Table 1). Cells binding the GAM42a probe for γ-proteobacteria and the EUBMIX probe for bacteria appear magenta, and those binding the ADAB1268 probe in addition to the EUBMIX probe appear cyan in colour.

Figure 3. CLSM micrograph of ADAB non-degrading mixed-culture enrichment where FISH was performed using the probes EUBMIX-Cy5 (blue), GAM42a-Cy3 (red), and ADAB1268-FITC (green) (Table 1).   Cells binding the GAM42a probe for γ-proteobacteria and the EUBMIX probe for bacteria appear magenta, and no cells appear cyan in colour, which would indicate cells binding the ADAB1268 probe in addition to the EUBMIX probe.

The ADAB1268 oligonucleotide probe was subsequently applied to ADAB-degrading enrichment cultures. This probe hybridized a high proportion of long thin rods in the ADAB-degrading mixed culture enrichments as illustrated in Figure 2. It is also of interest to note that in samples of the enrichment cultures at a high dilution, where no ADAB degradation was observed, cells hybridizing the ADAB1268 FISH probe were not present (Figure 3). These results support the hypothesis that these micro-organisms may play an integral role in the degradation of ADAB.

In conclusion, the members of an ADAB-degrading enrichment culture were investigated through a variety of methods. Through the construction of 16S rDNA clone libraries and their analysis, members of the *Firmicutes* were postulated to be of importance in ADAB degradation. FISH probes specific for this group were designed and employed, thereby providing further evidence. However, the significance of these micro-organisms in the mixed microbial community must be determined through linking this information with their function. *In situ* micro-autoradiography (ISMAR), using radiolabelled ADAB incubated with the ADAB-degrading mixed enrichment culture, would allow the direct visualization of the cells responsible for ADAB uptake. Additionally, ISMAR can be used in conjunction with FISH to correlate the identity of these micro-organisms within the mixed culture in relation to their function. This study has shown that when investigating a mixed microbial community it is necessary to combine a number of approaches in order to assess the role and significance of individual members within the community.

ACKNOWLEDGEMENTS

This work was in part supported by the Australian Centre for International Agricultural Research (ACIAR).

REFERENCES

Altschul, S.F., Gish, W., Miller, W., Myers, E.W. & Lipman, D.J. 1990. Basic local alignment search tool. *Journal of Molecular Biology,* **215**: 403–410.

Amann, R., Fuchs, B.M. & Behrens, S. 2001. The identification of micro-organisms by fluorescence *in situ* hybridisation. *Current Opinion in Biotechnology,* **12**: 231–236.

Amann, R.I., Krumholz, L. & Stahl, D.A. 1990. Fluorescent-oligonucleotide probing of whole cells for determinative, phylogenetic, and environmental-studies in microbiology. *Journal of Bacteriology,* **172**: 762–770.

Brosius, J., Dull, T.J., Sleeter, D.D. & Noller, H.F. 1981. Gene organization and primary structure of a ribosomal RNA operon from *Escherichia coli. Journal of Molecular Biology,* **148**: 107–127.

Daims, H., Bruhl, A., Amann, R., Schleifer, K.H. & Wagner, M. 1999. The domain-specific probe EUB338 is insufficient for the detection of all Bacteria: Development and evaluation of a more comprehensive probe set. *Systematic and Applied Microbiology,* **22**: 434–444.

Evans, C.S., Clardy, J., Hughes, P.F. & Bell, E.A. 1985. 2-Amino-4-Acetylaminobutyric Acid, 2,4-Diaminobutyric Acid and 2-Amino-6n-Oxalylureidopropionic Acid (Oxalylalbizziine) in seeds of *Acacia angustissima. Phytochemistry,* **24**: 2273–2275.

Evans, C.S., Shah, A.J., Adlard, M.W. & Arce, M.L.R. 1993. Nonprotein amino-acids in seeds of neotropical species of *Acacia. Phytochemistry,* **32**: 123–126.

Gutteridge, R.C. 1994. Other species of multi-purpose forage tree legume. pp. 97–99, *in:* R.C. Gutteridge and H.M. Shelton (eds). *Forage Tree Legumes in Tropical Agriculture.* Wallingford, UK: CAB International.

Krause, D.O., Smith, W.J. & McSweeney, C.S. 2001. Extraction of microbial DNA from rumen contents containing plant tannins. *Biotechniques,* **31**: 294–295.

Lane, D.J. 1991. 16S/23S rRNA sequencing. pp. 115–175, *in:* E. Stackebrandt and M. Goodfellow (eds). *Nucleic Acid Techniques in Bacterial Systematics.* Chichester, UK: Academic Press.

Mackie, R.I., Gilchrist, F.M.C., Robberts, A.M., Hannah, P.E. & Scwartz, H.M. 1978. Microbiological and chemical changes in the rumen during the stepwise adaptation of sheep to high concentrate diets. *Journal of Agricultural Science,* **90**: 241–254.

Manz, W., Amann, R., Ludwig, W., van Canneyt, M. & Schleifer, K.H. 1996. Application of a suite of 16S rRNA-specific oligonucleotide probes designed to investigate bacteria of the phylum *Cytophaga-Flavobacter-Bacteroides* in the natural environment. *Microbiology,* **142**: 1097–1106.

Menke, K.H., Raab, L., Salewski, A., Steingass, H., Fritz, D. & Schneider, W. 1979. The estimation of the digestibility and metabolizable energy content of ruminant feeding stuffs from the gas production when they are incubated with rumen liquor *in vitro. Journal of Agricultural Science,* **93**: 217–222.

Moter, A. & Gobel, U.B. 2000. Fluorescence *in situ* hybridization (FISH) for direct visualization of micro-organisms. *Journal of Microbiological Methods,* **41**: 85–112.

Odenyo, A.A., Osuji, P.O., Karanfil, O. & Adinew, K. 1997. Microbiological evaluation of *Acacia angustissima* as a protein supplement for sheep. *Animal Feed Science and Technology,* **65:** 99–112.

O'Neal, R.M., Chen, C., Reynolds, C.S., Meghal, S.K. & Koeppe, R.E. 1968. The 'Neurotoxicity' of L-2,4-Diaminobutyric Acid. *Biochemical Journal,* **106**: 699–706.

Rao, S.L.N. 1978. A sensitive and specific colorimetric method for the determination of alpha, beta-diaminopropionic acid and the *Lathyrus sativus* neurotoxin. *Analytical Biochemistry,* **86**: 386–395.

Reed, J.D., Gebre-Mariam, G., Robinson, C.J., Hanson, J., Odenyo, A. & Treichel, P.M. 2001. Acetyldiaminobutanoic acid, a potential lathyrogenic amino acid in leaves of *Acacia angustissima. Journal of the Science of Food and Agriculture,* **81**: 1481–1486.

Saarisalo, E.M., Odenyo, A.A. & Osuji, P.O. 1999. Inoculation with adapted microbes versus addition of polyethylene glycol as methods to alleviate toxicity of *Acacia angustissima* leaves in sheep. *Journal of Agricultural Science,* **133**: 445–454.

Shah, A.J., Younie, D.A., Adlard, M.W. & Evans, C.S. 1992. High-performance liquid-chromatography of nonprotein amino-acids extracted from *Acacia* seeds. *Phytochemical Analysis,* **3**: 20–25.

Stahl, D.A. & Amann, R. 1991. Development and application of nucleic acid probes. pp. 205–248, *in:* E. Stackebrandt and M. Goodfellow (eds). *Nucleic Acid Techniques in Bacterial Systematics.* Chichester, UK: Academic Press.

von Wintzingerode, F., Gobel, U.B. & Stackebrandt, E. 1997. Determination of microbial diversity in environmental samples: pitfalls of PCR-based rRNA analysis. *FEMS Microbiology Reviews,* **21**: 213–229.

APPLICATION OF MOLECULAR MICROBIAL ECOLOGY TOOLS TO FACILITATE THE DEVELOPMENT OF FEEDING SYSTEMS FOR RUMINANT LIVESTOCK THAT REDUCE GREENHOUSE GAS EMISSIONS

G.J. McCrabb,[1] C.S. McSweeney,[2] S. Denman,[2] M. Mitsumori,[3] S. Fernandez-Rivera[1] and H.P.S. Makkar[4]

1. International Livestock Research Institute, PO Box 5689, Addis Ababa, Ethiopia.
2. CSIRO Livestock Industries, Queensland Biosciences Precinct, Carmody Rd, St Lucia, Brisbane, Australia.
3. National Institute of Livestock and Grassland Science, Ikenodai 2, Kukizaki, Ibaraki, 305-0901 Japan.
4. Animal Production and Health Section, Joint FAO/IAEA Division, International Atomic Energy Agency, Vienna, Austria.

Abstract: Ruminant livestock populations in developing countries are increasing in response to increasing demand for meat and milk. These animals are a major global source of methane, a greenhouse gas produced during the degradation of organic matter by micro-organisms in the foregut of ruminant livestock. Chemical inhibition of methanogenic micro-organisms has been reported; however, associated improvements in feed digestion and livestock productivity have not been consistently demonstrated. Gene-based technologies have the potential to contribute new knowledge of the rumen microbial populations involved in these processes, which will assist in identifying feeding practices that lead to methane abatement and improved livestock productivity. For small-scale farmers, feeding interventions that achieve greenhouse gas abatement need also to be associated with improved feed conversion efficiency and enterprise profitability. During the adoption of methane abatement technologies, other regionally important issues such as poverty, food security, sustainable agriculture production systems and environmental management must also be addressed.

H. P. S. Makkar and G. J. Viljoen (eds.), Applications of Gene-Based Technologies for Improving Animal Production and Health in Developing Countries, 387-395.

1.   INTRODUCTION

Ruminant livestock in the developing world are a major global anthropogenic source of methane, an important greenhouse gas. As a result of international concern over climate change induced by rising global greenhouse gas emissions, international policy-makers and national governments are searching for approaches to reduce major sources of greenhouse gas. Methane gas is produced by micro-organisms in the digestive tract of ruminant livestock during feed digestion and released into the environment, and represents a direct loss of digested energy that would more efficiently be used for additional meat or milk production. Gene-based research tools for molecular microbial ecology have only been available recently, permitting changes in the rumen microbial ecosystem to be quantified. Such knowledge can form a sound basis for the management and use of both feed and other natural resources that underpin development of sustainable feeding systems.

2.   CHANGES IN THE GLOBAL RUMINANT LIVESTOCK SECTOR

Dramatic changes have been occurring in the global ruminant livestock sector, including the size of regional livestock populations (Sere, Steinfeld and Gronewold, 1996), the types of feeding and management systems under which ruminant livestock are held (Fernandez-Rivera, Okike and Ehui, in prep.), and the importance of non-livestock issues, including environmental management, human health and the potential for livestock production to improve the livelihoods of small-scale farmers in developing countries (ILRI, 2002).

Global ruminant livestock populations have been increasing over the last two decades and this trend is projected to continue. The increase in numbers of ruminants is being driven by increasing demand for livestock products in the developing world, which has been termed by Delgado *et al.* (1999) as the "Livestock Revolution". The consumption of meat and milk in developing countries increased during 1983–1994, whereas in developed countries consumption was relatively stable. For developing countries, meat and milk consumption are projected to continue increasing annually to 2020 at rates of 2.8 and 3.3 percent, respectively – a trend attributed to continued human population growth, urbanization and income growth (Delgado *et al.*, 1999).

Access to livestock feed resources of sufficient quantity and quality is a major constraint to ruminant production in developing countries. The implications of the "Livestock Revolution" for demand for global livestock feed resources from 2000 to 2020 has been analysed by Fernandez-Rivera,

Okike and Ehui (in prep.). Using a combination of projected annual growth rates for meat and milk consumption (Delgado *et al.*, 1999; FAO, 1998), the number of livestock required to supply both present and projected demands for meat (Fernandez-Rivera, Okike and Ehui, unpublished) and milk (Fernandez-Rivera *et al.*, 2001) were determined. These analyses indicate that in 2000, 53 percent of the global non-dairy cattle and small ruminant populations of 1.8 and 1.3 billion, respectively, were managed in grazing-based and rain-fed mixed crop-livestock systems in developing countries, and 51 percent of the global dairy cattle population of 237 million were managed in rain-fed mixed crop-livestock systems in developing countries. Projections to 2020 indicate that mixed crop-livestock systems will become even more predominant in developing countries, which highlights the need to identify methane abatement strategies for mixed crop-livestock systems in the tropics.

Projections of enteric methane emissions from ruminant livestock in both developing and developed countries have been published by the U.S. Environmental Protection Agency (EPA, 2001, 2002).

Table 1 presents current and projected global emissions from ruminant livestock, and highlights the rapid increase in emissions from developing countries compared with the relatively stable emissions from developed countries. Past experience from developed countries indicates that, in the absence of greenhouse gas mitigation policies, ruminant systems will become more efficient and produce less methane gas production per unit of meat or milk production. Such changes have been occurring in market oriented livestock industries over the last two decades, as reported from Australia (Howden and Reyenga, 1999) and Japan (Terada, 2002).

Table 1. Projections of annual emissions of methane gas (Gg/year) from enteric fermentation of domesticated livestock, from developed[1] and developing[2] countries.

Year	Developing countries	Developed countries
1990	49,300	27,400
2000	58,800	25,100
2010	72,400	26,300
2020	88,800	–

SOURCES: (1) EPA, 2001. (2) EPA, 2002.

3.   GREENHOUSE GAS MITIGATION FROM RUMINANT LIVESTOCK

Livestock systems in developing countries have unique characteristics that render greenhouse gas abatement more difficult than for other major global sources of methane, such as the energy sector or landfills. Ruminant livestock are typically managed by many small-scale farmers, each responsible for a few livestock. Furthermore, grazing land is often common property, and decisions regarding its management are made at the community level. At a global level, impact on enteric methane gas emissions can only be achieved if abatement strategies are adopted by a majority of small-scale farmers and community decision-makers in the developing world.

The options available for greenhouse gas abatement from ruminant livestock have been reviewed by many authors (Reyenga and Howden, 1999; O'Hara, Freney and Ulyatt, 2003; Bates, 2001) and indicate that feed management is a key consideration. In general, mitigation options include improved nutrition through strategic supplementation; improved nutrition through mechanical and chemical feed processing; use of production enhancing agents; improved production through genetic enhancement; improved production efficiency through improved reproduction; and other techniques, such as disease control and the control of product markets and prices. Although full consideration of these options is beyond the scope of the present discussion, it should be noted that in addition to understanding the biology of ruminants, greenhouse gas abatement from ruminants will require consideration of the broader livestock system. All greenhouse gases, including methane, nitrous oxide and carbon dioxide, should be considered when evaluating the impact of mitigation options, because the effects of mitigation of one gas may result in increased production of other greenhouse gases. Furthermore, other non-greenhouse gas issues need to be considered, such as differences in regional and country priorities, the motives for people keeping livestock, management boundaries of livestock systems, and changes in the types of production systems under which livestock are held (McCrabb *et al.*, 2003).

Table 2 summarizes published reports of the effect of chemical inhibition of enteric methane from ruminants on feed digestion and animal performance. This evidence indicates no consistent relationship among methane inhibition, voluntary feed intake, feed digestibility, feed conversion efficiency and liveweight gain. Although feed conversion efficiency is often improved when methane production is inhibited, this is typically due to a reduction in voluntary feed intake rather than increased feed utilization or animal performance *per se*. Clearly, in these situations, knowledge of the associated changes in the rumen microbial populations will assist in identifying feeding practices that will jointly lead to reduced methane gas production and improved production efficiency (McSweeney and McCrabb, 2002).

4.   MONITORING CHANGES IN THE RUMEN MICROBIAL ECOSYSTEM

Until recently, knowledge of rumen microbiology could only be obtained using classical culture-based techniques, such as isolation, enumeration and nutritional characterization – techniques that probably accounted for only 10 to 20 percent of the rumen microbial population.

New gene-based technologies are now being employed to examine microbial diversity through the use of small sub-unit rDNA analysis (e.g. 16S rDNA) to understand the function of the complex rumen microbial ecosystem (Table 3). These technologies have the potential to revolutionize the understanding of rumen function and will overcome the limitations of classical based techniques, including isolation and taxonomic identification of strains critical to efficient rumen function. The future of rumen microbiology research is dependant upon the adoption of these molecular based research technologies.

As gene-based technologies begin to be used, it has become apparent that techniques that optimize the analysis of complex microbial communities rather than the detection of single organisms will additionally need to permit the analysis of many primers and probes in a single sample and the detection of only living microbes.

Several techniques, including real-time quantitative PCR (real-time Q-PCR), have been developed and address some of these issues. The application of real-time Q-PCR methods to monitor rumen microbial ecosystems during dietary interventions will require the development of specific primers to monitor key cellulolytic populations (bacterial and fungal) and broad primers against populations such as methanogenic bacteria, rumen fungi, protozoa and rumen bacteria. In combination, these primers will permit the monitoring of entire populations and their interactions with regard to each other under varying feeding regimes and environmental conditions. Real-time Q-PCR permits the detection and measurement of amplicon product accumulation during each cycle of the PCR. The most common methods employed are 5′ nuclease assays using TaqMan probes (Trei *et al.*, 1971), molecular beacons (Trei, Scott and Parish, 1972) and the DNA intercalating fluorescent dye SYBR green I (Johnson, 1974). Although other methods are available, these three methods are the most commonly adopted, and are now being used routinely to detect and quantify both gene expression and the presence of specific microbial species.

Table 2. A summary of publications reporting the effects of methane inhibition, using bromochloromethane, hemiacetyl of chloral and starch (amichloral) and trichloroacetamide, on feed utilization and animal performance in cattle and sheep.

Animal & Diet	Inhibitor	Treatment duration	Feed intake	Feed digestibility	Daily LWG	FCE	Methane inhibition	Ref.
Cattle Fibrous	BCM	28 days	NS	NS	NS	-	50%	[1]
Cattle Concentrate	HCS	120 days	-9%	Reduced	-4%	+6%	30–40%	[2]
Calves Roughage	HCS	11 weeks	-3%	NS	+10%	+7%	–	[3]
Cattle Roughage	BCM	12 weeks	-10%	–	+5%	+11%	–	[4]
Lambs Not reported	BCM	105 days	–	Increased	-12%	-5%	85%	[5]
Sheep Roughage+ Concentrate	HCS	2 weeks	NS	–	–	–	50–82%	[6]
Lambs Roughage+ Concentrate	TCA	90 days	NS	–	+1%	NS	66%	[7]
Lambs Concentrate	HCS	90 days	-4%	–	+5%	+7%	27–54%	[8]
Lambs Concentrate	HCS	30 days	-13%	Increased	–	–	64%	[9]

KEY: LWG = liveweight gain; FCE = feed conversion efficiency; BCM = bromochloro-methane; HCS = amichloral; TCA = trichloroacetamide; NS = not significantly affected (P>0.05).
SOURCES: [1] Johnson *et al.*, 1972. [2] Cole and McCroskey, 1975. [3] Leibholtz, 1975. [4] McCrabb *et al.*, 1997. [5] Sawyer, Hoover and Sniffen, 1971. [6] Johnson, 1972. [7] Trei *et al.*, 1971. [8] Trei, Scott and Parish, 1972. [9] Johnson, 1974.

Table 3. Currently available gene-based techniques that can be used to describe change in the rumen microbial ecosystem.

Method	Application	Advantages	Disadvantages
Restriction Fragment Length Polymorphisms (RFLPs)	Fingerprinting of isolates and microbial communities.	Rapid screen for identifying and grouping similar organisms.	Broad-based screen that does not provide specific identity.
Oligonucleotide probe hybridization	Enumeration of microbial populations at various levels including domain, phylum, species.	Quantitative and specific. Probes based on phylogenetic sequence database.	Laborious. Quantitation expressed as % of total population and not absolute numbers.
Fluorescence *in situ* hybridization (FISH)	Enumeration of micro-organisms *in situ* within their environment.	Visualize culturable and unculturable micro-organisms spatially in relation to substrate and other community members.	Laborious and quantitation is difficult.
16S rDNA clone libraries	Identify the predominant micro-organisms in a microbial community.	Unculturable and culturable organisms can be identified from sequence analysis of cloned small subunit ribosomal genes.	Laborious and not quantitative. Only predominant organisms are identified.
Denaturing Gradient Gel Electrophoresis (DGGE) and Single-stranded Conformational Polymorphisms	Fingerprint pattern analysis of changes in mixed microbial community composition.	Culturable and unculturable organisms identified. Microbial community composition can be determined by pattern analysis at domain, phylum and species level.	Laborious. Only predominant organisms are identified.
Real time Polymerase Chain Reaction (real-time PCR)	Quantitative estimates of microbial populations at the domain, phylum, species and subspecies level.	Rapid quantitative method for estimating discrete population in a mixed environmental sample.	Technique based on small subunit ribosomal sequence identity of previously sequenced organisms and clone libraries.

5.   CONCLUSION

The application of molecular technologies to monitor rumen microbial populations will allow the effects of various dietary interventions to be described. This knowledge will form a solid foundation for developing sustainable livestock systems based on the efficient use of feed and other natural resources.

Research technologies are available to develop specific primers to monitor key cellulolytic populations (bacterial and fungal) and broad primers against such populations as methanogenic bacteria, rumen fungi and rumen bacteria. In combination with conventional techniques of feed evaluation, these new molecular techniques will allow the monitoring of entire populations and their interactions with each other under varying feeding regimes and environmental conditions. A Coordinated Research Project to address these aspects has been initiated under the auspicious of the FAO/IAEA Joint Division.

For small-scale farmers, feeding interventions that achieve greenhouse gas mitigation also need to be associated with other benefits, such as improved feed conversion efficiency and enterprise profitability, if they are to be adopted broadly in the developing world. Regionally important issues, such as the need for poverty alleviation, food security and sustainable agricultural production, must also be addressed in any attempt to achieve greenhouse gas mitigation in the developing world.

REFERENCES

Bates, J. 2001. *Economic evaluation of emission reductions of nitrous oxides and methane in agriculture in the European Union. Bottom-up analysis.* AEA Technology Environment, Culham, United Kingdom. 99p.

Cole, N.A. & McCroskey, J.E. 1975. Effects of hemiacetyl of chloral and starch on the performance of beef steers. *Journal of Animal Science*, **41**: 1735–1741.

Delgado, C., Rosegrant, M., Steinfeld, H., Ehui, S. & Courbois, C. 1999. *Livestock to 2020: The next food revolution.* Food, Agriculture and Environment Discussion Paper 28. IFPRI [International Food Policy Research Institute], FAO [Food and Agriculture Organization of the United Nations] and ILRI [International Livestock Research Institute].

EPA [United States Environment Protection Agency]. 2001. *Non-CO$_2$ greenhouse gas emissions from developed countries: 1990 to 2010.* US EPA, Washington DC.

EPA. 2002. *Emissions and projections of non-CO$_2$ greenhouse gases from developing countries: 1990 to 2020.* US EPA, Washington, DC.

FAO [Food and Agriculture Organization of the United Nations]. 1998. *FAO Production Yearbook.* Rome: FAO.

Fernandez-Rivera, S., Okike, O. & Ehui, S. In preparation. *The Livestock Revolution: Implications for the demand for feed in developing countries.* Proceedings of an

International Workshop on Forage Adoption. International Livestock Research Institute (ILRI), Addis Ababa, Ethiopia, June 2001.

Fernandez-Rivera, S., van Vuuren, A.M., Okike, I., Ehui, H. & Tegegne, A. 2001. Milk production and feed requirements in dairy systems of developing countries by 2010. pp. 33–35, *in: Tropical Animal Health and Production – Dairy Developments in the Tropics.* Faculty of Veterinary Medicine, Utrecht, University, The Netherlands.

Howden, S.M. & Reyenga, P.J. 1999. Methane emissions from Australian livestock: Implications of the Kyoto Protocol. *Australian Journal of Agricultural Research,* **50**: 1285–1292.

ILRI [International Livestock Research Institute]. 2002. *Livestock – a pathway out of poverty, ILRI's strategy to 2010.* ILRI, Nairobi, Kenya. 24p.

Johnson, D.E. 1972. Effects of a hemiacetyl of chloral and starch on methane production and energy balance of sheep fed a pelleted diet. *Journal of Animal Science,* **35**: 1064–1068.

Johnson, D.E. 1974. Adaptional responses in nitrogen and energy balance of lambs fed a methane inhibitor. *Journal of Animal Science,* **38**: 154–157.

Johnson, D.E., Wood, A.S., Stone, J.B. & Moran, E.T. 1972. Some effects of methane inhibition in ruminants (steers). *Canadian Journal of Animal Science,* **52**: 703–712.

Leibholtz, J. 1975. Ground roughage in the diet of the early-weaned calf. *Animal Production,* **20**: 93–100.

McCrabb, G.J., Berger, K.T., Magner, T., May, C. & Hunter, R.A. 1997. Inhibiting methane production in Brahman cattle by dietary supplementation with a novel compound and the effects on growth. *Australian Journal of Agricultural Research,* **48**: 323–329.

McCrabb, G.J., Fernandez-Rivera, S., Hunter, R.A., Kurihara, M., Terada, F. & Wirth, T. 2003. Managing greenhouse gas emissions from livestock systems. pp. 281–288, *in: Proceedings of the Third International Methane and Nitrous Oxide Mitigation Conference,* Beijing, November 2003. Published by US Environment Protection Agency, Washington DC.

McSweeney, C.S. & McCrabb, G.J. 2002. Inhibition of rumen methanogenesis and its effects on feed intake, digestion and animal production. *In:* J. Takahashi and B.A. Young (eds). *Greenhouse Gases and Animal Agriculture.* Elsevier.

O'Hara, P., Freney, J. & Ulyatt, M. 2003. *Abatement of agricultural non-carbon dioxide greenhouse gas emissions.* Ministry of Agriculture and Forestry, Wellington, New Zealand.

Reyenga, P.J. & Howden, S.M. 1999. *Meeting the Kyoto Target. Implications for the Australian livestock industries.* Bureau of Rural Sciences, Canberra, Australia.

Sawyer, M.A., Hoover, W.J. & Sniffen, C.J. 1971. Effects of a methane inhibitor on growth and energy metabolism in sheep. *Journal of Dairy Science,* **54**: 792–793 (Abstract).

Sere, C., Steinfeld, H. & Gronewold, J. 1996. World livestock production systems: Current status, issues and trends. *FAO Animal Production and Health Paper,* No. 127.

Terada, F. 2002. Global warming and animal agriculture in Japan. *In:* J. Takahashi and B.A. Young (eds). *Greenhouse Gases and Animal Agriculture.* Elsevier.

Trei, J.E., Parish, R.C., Singh, Y.K. & Scott, G.C. 1971. Effect of methane inhibitors on rumen metabolism and feedlot performance of sheep. *Journal of Dairy Science,* **54**: 536–540.

Trei, J.E., Scott, G.C. & Parish, R.C. 1972. Influence of methane inhibition on energetic efficiency of lambs. *Journal of Animal Science,* **34**: 510–515.

EFFECT OF SECONDARY COMPOUNDS IN FORAGES ON RUMEN MICRO-ORGANISMS QUANTIFIED BY 16S AND 18S rRNA

Elizabeth Wina,[1] Stefan Muetzel,[2] Ellen Hoffman,[2] Harinder P.S. Makkar[3] and Klaus Becker[2]

1. Indonesian Research Institute for Animal Production, P.O. Box 221 Bogor, Indonesia.
2. Institute for Animal Production for Tropics and Subtropics (480b), University of Hohenheim, Stuttgart 50793, Germany.
3. Animal Production and Health Section, International Atomic Energy Agency, P.O. Box 100, Wagramerstr. 5, A-1400 Vienna, Austria.

Abstract: A gas syringe method was used to evaluate the effect of secondary compounds from plant materials on *in vitro* fermentation products and microbial biomass. The experiment used *Pennisetum purpureum, Morinda citrifolia* fruit, *Nothopanax scutellarium* leaves, *Sesbania sesban* LS (low saponins type), *Sesbania sesban* HS (high saponins type) and *Sapindus rarak* fruit as substrates. The incubation was conducted with and without polyethylene glycol 6000 (PEG) addition for 24 hours. Gas production and short-chain fatty acids (SCFA) were analysed. Prokaryotic and eukaryotic concentrations were measured by quantifying 16S and 18S rRNA.

The percentage increase in gas production due to PEG was very small (<5%) for all plant materials, which indicated that the biological effect of tannin in these plant materials is limited. TLC analysis revealed that all materials contained saponin, but only *S. rarak*, followed by *S. sesban,* contained a high diversity of saponins. *S. sesban* gave the highest (234 ml/g) while *S. rarak* gave the lowest gas production (115 ml/g). *S. rarak* gave the lowest SCFA production (3.57 mmole/g) and also the lowest ratio of acetate to propionate (1.76), indicating a change in pattern of SCFA production. Total elimination of eukaryotic concentration was evident from the absence of the 18S rRNA band when *S. rarak* and *S. sesban* were used as sole substrates. *S. rarak* also reduced the prokaryotic concentration. To use *S. rarak* as a defaunating agent without affecting prokaryotes, a crude saponin extract was prepared from *S. rarak* for further experiment. Different concentrations of crude saponins in a methanol extract of *S. rarak* fruit dissolved in rumen buffer were added to a

397

H. P. S. Makkar and G. J. Viljoen (eds.), Applications of Gene-Based Technologies for Improving Animal Production and Health in Developing Countries, 397-410.
© 2005 *IAEA. Printed in the Netherlands.*

substrate consisting of elephant grass and wheat bran (7:3 w/w). Microbial biomass yield was quantified by gravimetry and using rRNA as a marker.

Addition of crude saponin extract from *S. rarak* to a high-roughage diet increased microbial biomass (MB) yield to 1.07 and 1.14 times MB yield of the control, estimated by gravimetry and using rRNA as a marker, respectively. A significant, although low, correlation between these methods was found, suggesting that both methods can possibly be used to study the effect of saponin. However, due to the limited correlation between these two methods ($r = 0.5793$), more studies are warranted, using a greater number of samples. Using rRNA as a marker for estimating microbial biomass would be advantageous as the same RNA can be used for further microbial community analysis.

1.   INTRODUCTION

Nutritional quality of forage is determined by the presence and concentrations of both primary compounds, like protein, carbohydrate, fat or fibre, and secondary compounds – tannins, oxalates, glucosinolates, saponins, alkaloids, etc. Of these two broad groups, secondary compounds are more diverse in chemical structure and properties. Many of these secondary compounds are considered as anti-nutritive, or even toxic to animals. Several research workers (Fahey and Jung, 1989; Makkar, 1993; Cheeke, 2000; McSweeney *et al.,* 2001; Wallace *et al.,* 2002) have reviewed the roles of secondary compounds in animal physiology, metabolism and production. Recently, it has been shown that some of these secondary compounds can be utilized to manipulate rumen fermentation towards enhancing ruminant production. For example, saponins from *Yucca schidigera* reduce protozoa (Wang *et al.,* 2000), while tannin-containing feeds under certain conditions may increase microbial protein synthesis (Getachew, Makkar and Becker, 2000).

Various methods are available to study the effect of secondary compounds, but an *in vitro* method is preferred, as it is simple, relatively rapid and comparatively inexpensive. A gas syringe technique (Menke *et al.,* 1979; Makkar, Blümmel and Becker, 1998) is commonly used for conducting *in vitro* experiments, but it has certain limitations. A review published by Getachew *et al.* (1998) considers the advantages and disadvantages of the gas syringe method compared with other available methods. It is important to measure not only gas production, but also to quantify microbial biomass yields (Blummel, Steingass and Becker, 1997). Several methods, including gravimetry (Blummel, Steingass and Becker,

1997), purine (Zinn and Owen, 1986; Obispo and Dehority, 1999), Diamino pimelic acid (DAPA) (Olubobokun, Craig and Nipper, 1988) and ^{15}N incorporation (Getachew, Makkar and Becker, 2000), are available to determine microbial biomass, but recently a new technique using ribosomal RNA (rRNA) has been used, and the method provides simultaneous measurement of prokaryotic and eukaryotic concentration through quantification of 16S rRNA and 18S rRNA (Muetzel, Hoffmann and Becker, 2003). This paper reports the effect of secondary compounds in plant materials on *in vitro* fermentation products and the effect of crude saponin extract on microbial biomass yield quantified by gravimetry and using rRNA as a marker.

2.   MATERIALS AND METHODS

2.1   Substrates

Four plant materials from Indonesia were used, namely elephant grass (Napier grass; *Pennisetum purpureum*), Indian mulberry (*Morinda citrifolia*) fruit, *Polyscias scuttellaria* (syn. *Nothopanax scutellarium*) leaves, and *Sapindus rarak* fruit, together with two types of *Sesbania sesban* – LS (low saponin) and HS (high saponin) – from the International Livestock Research Institute (ILRI) collection, Ethiopia. These were used as substrates for *in vitro* fermentation. *P. purpureum,* a common cultivated grass in Indonesia and which contains no tannin or saponin, was used as a control. *M. citrifolia* and *N. scutellarium* are often used as herbal medicine or as vegetables and *Sapindus rarak* is used as a washing soap. The *in vitro* experiment was done with three replicates.

In a further experiment, a methanol extract of *S. rarak* fruit was prepared and added at different concentrations (0; 0.25; 0.5; 1.0; 2.0; and 4.0 mg/ml in the rumen-buffer solution) to a mixture of elephant grass and wheat bran (7:3 w/w) as a substrate. This experiment was done twice, with duplicate samples.

2.2   Analysis of secondary compounds in plant materials

Biological activity of tannin was determined in terms of percentage increase in gas production due to Polyethylene glycol 6000 (PEG) addition on *in vitro* fermentation, using all plant materials as substrates.

Saponin was identified by thin layer chromatography (TLC). Crude saponin fraction was extracted by methanol (2× 20 ml) from 1 g of plant material. The crude saponin extract was then evaporated to dryness and re-dissolved in 1 ml of methanol, and 4 µl of each extract was spotted on TLC Silica gel 60, and eluted with a chloroform : methanol : water mixture (18:11:2.7) and sprayed with a mixture of ethyl acetate : ethanol : sulphuric acid (5:5:9) (El Gamal *et al.*, 1995). Saponin from *quillaja* bark (Sigma S2149) was used as standard.

2.3   In vitro fermentation

The method used for *in vitro* fermentation was based on the technique described by Menke *et al.* (1979). Feed samples (each of 300 mg) were weighed into 100 ml graduated glass syringes and incubated with 30 ml of buffered rumen liquor at 39°C. Gas production was recorded at regular intervals and samples were drawn after 24 hours of incubation for the first fermentation. The contents were then transferred to pre-weighed centrifuge tubes under continuous stirring and 0.3 ml aliquots were taken for measuring RNA concentration. After centrifugation, supernatant (1 ml) was taken for short-chain fatty acid (SCFA) analysis.

In a further experiment, samples for rRNA analysis were taken after 6, 12, 24 and 48 hours of incubation, and microbial biomass yield was measured by gravimetry, following the method of Blummel, Steingass and Becker (1997).

2.4   Cell lysis and RNA extraction

Cell lysis and RNA extraction were carried out according to the method of Muetzel, Krishnamoorthy and Becker (2001). The mixture of rumen fluid (300 µl), phenol pH 4.9 (600 µl), 10 mM buffer pH 4.9 (270µl), 20% (w/v) SDS solution (30 µl) and 0.1 mm Zirconia-silica beads (1 g) was beaten on a beadmill at maximum speed (~30 Hz) for 2 periods of 2 minutes each. After the cell lysis step, 300 µl chloroform was added to the sample and mixed well, followed by incubation for 2×5 minutes at room temperature. The aqueous phase was separated by centrifugation and transferred into a fresh vial containing ½ volume 7.5 M ammonium acetate and 1 volume of ice-cold isopropanol. The mixture was kept at -20°C overnight and the precipitate was recovered by centrifugation. Supernatant was discarded and the precipitate was washed with 1 ml 75% (v/v) ethanol solution. The RNA pellet was finally dissolved in 200 µl of double-distilled water and stored at -80°C. RNA was separated on 1.4% agarose gel in TBE buffer, and the gel was stained with ethidium bromide (1 µg/ml in TBE buffer). 16S rRNA (for prokaryotes) and 18S rRNA (for eukaryotes) bands were quantified by densitometry, using a series of standards obtained from *Escherichia coli* and *Saccharomyces cerevisiae.*

<u>Calculation of microbial biomass</u>

By gravimetry (Blummel, Steingass and Becker, 1997):

Microbial biomass (mg/g sample) = Apparent undigested residue -True undigested residue (AUDR -TUDR)

By rRNA as a marker (Robinson, Fadel and Ivan, 1996):

Microbial biomass (mg/g sample) = 16S rRNA/0.085 *vol buffer/1000/g sample + 18S rRNA/0.04*vol buffer/1000/g sample

<u>Statistical analysis</u>

The first *in vitro* experiment was done once with triplicate samples and further *in vitro* experiments were carried out twice with duplicate samples for each incubation. Statistical analysis was effected using SAS package V.6.12.

3.   RESULTS

3.1   Secondary compounds in plant materials

Two types of secondary compounds, namely tannins and saponins, were measured qualitatively in the plant materials investigated, except *P. purpureum*, where none occur.

The biological effect of tannin from all substrates was low and this was indicated by the percentage increases in gas production due to PEG addition, which were 4.2, 0, 3.1, 0.3 and 2.0% for *M. citrifolia* fruit, *N. scutellarium* leaves, *S. sesban* leaves (LS), *S. sesban* leaves (HS) and *S. rarak* fruit, respectively. The presence of saponin in plant materials was identified qualitatively by TLC. Figure 1 shows that all tested materials contained saponins, which were indicated by purple-coloured spots. *Quillaja* saponin was used as a standard and it had several light purple bands, which differed from other plant materials tested. *M. citrifolia* showed only one strong purple-coloured spot, while others showed several strong purple spots. *S. rarak* had the highest diversity for saponins. Further evaluation with blood spray (data not shown) revealed haemolytic activity for all purple spots in these plant materials.

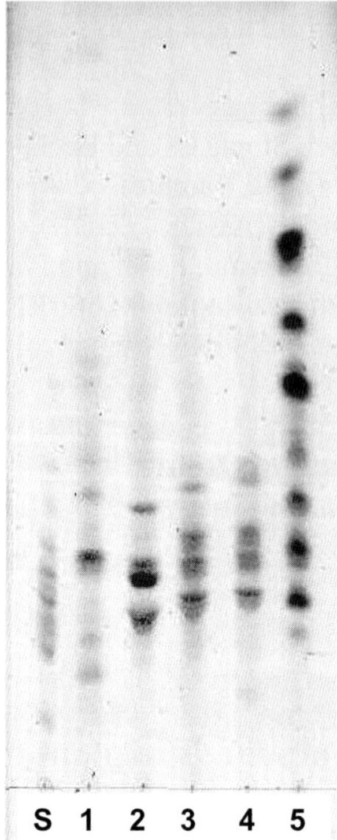

Figure 1. Saponins on TLC Silica gel 60. KEY: S = standard (*Quillaja saponaria*); 1 = *Morinda citrifolia* young fruit; 2 = *Nothopanax scutellarium* leaves; 3 = *Sesbania sesban* leaves LS; 4 = *Sesbania sesban* leaves HS; and 5 = *Sapindus rarak* fruit.

3.2   Effect on fermentation products

The *in vitro* fermentation products of tested materials are presented in Table 1. All materials tested, except for *S. rarak*, gave higher gas production than *P. purpureum*. *S. sesban* LS leaves gave the highest gas production and, hence, the highest SCFA production, but very little SCFA was produced when *S. rarak* was used as a substrate. The molar proportion of individual fatty acids, especially for acetate and propionate, differed very slightly, except for *S. rarak,* which had lower acetate and much higher propionate, and hence a very low ratio of acetate to propionate (1.76).

Table 1. Gas production, short chain fatty acid (SCFA) and ratio of molar proportion of acetate and propionate produced after 24 hours of *in vitro* fermentation for the six substrates.

	Gas production (ml/g)	SCFA (mmol/g)	Acetate : propionate
Pennisetum purpureum	176 [b] (3.4)	4.79 [b] (0.71)	3.27 [d] (0.06)
Morinda citrifolia	200 [d] (3.6)	5.03 [bc] (0.06)	3.16 [c] (0.07)
Nothopanax scutellarium	221 [e] (5.9)	5.41 [cd] (0.16)	3.66 [e] (0.02)
Sesbania sesban (LS)	234 [f] (1.0)	5.71 [d] (0.09)	2.88 [b] (0.05)
Sesbania sesban (HS)	189 [c] (1.7)	4.96 [bc] (0.06)	3.16 [c] (0.03)
Sapindus rarak	115 [a] (2.1)	3.57 [a] (0.04)	1.76 [a] (0.05)

NOTES: Number in bracket is the standard deviation values (n=3); different letters in the same column indicate significant difference (P<0.01).

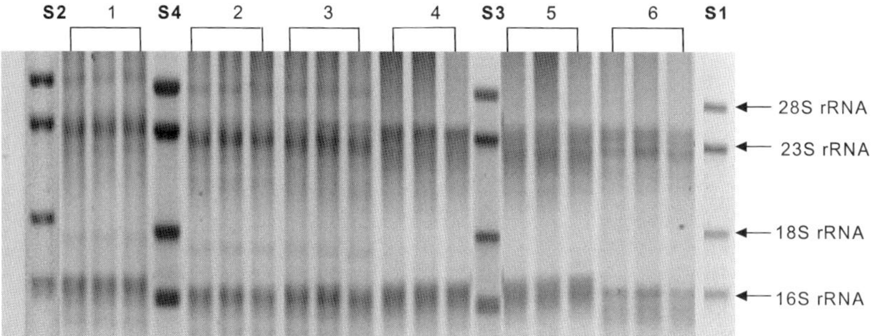

Figure 2. 16SrRNA (prokaryotes) and 18SrRNA (eukaryotes) bands of different substrates on agarose gel. S1–S4 are standards at different concentrations. Substrates: 1 = *Pennisetum purpureum*; 2 = *Morinda citrifolia*; 3 = *Nothopanax scutellarium*; 4 = *Sesbania sesban* (LS); 5 = *Sesbania sesban* (HS); 6 = *Sapindus rarak*.

Figure 2 shows 16S rRNA and 18S rRNA bands of different plant materials on the gel, run together with the standard. With densitometry, prokaryotes (16S rRNA) and eukaryotes (18S rRNA) could be quantified. All plant materials showed lower eukaryotic concentration compared to *P. purpureum* (P<0.05). Lack of an 18S rRNA band on the gel indicated total elimination of eukaryotic concentration when *S. sesban* LS and HS and *S. rarak* were incubated. Prokaryotic concentration was almost 10 times higher than eukaryotic concentration (Figure 3). *N. scutellarium* and *S. sesban* LS have similar prokaryotic concentration (P>0.05), but significantly higher than *M. citrifolia*, *S. sesban* HS and *P. purpureum* (P<0.05), while *S. rarak* gave the lowest concentration.

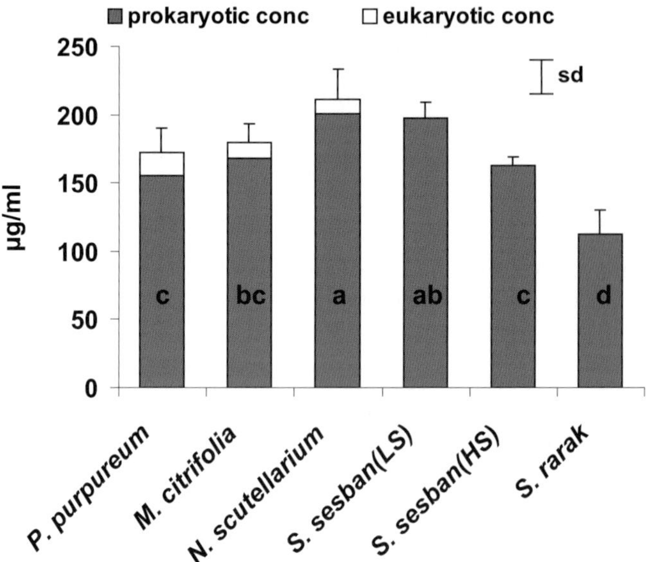

Figure 3. Prokaryotic and eukaryotic concentration in rumen liquor at 24 hr of *in vitro* fermentation quantified by 16S and 18S rRNA concentrations.

3.3   Effect on microbial biomass yield

Further *in vitro* fermentation studies measured microbial biomass (MB) yield by gravimetry and by rRNA as a marker, and the results are presented in Figures 4a and 4b. Inclusion of crude saponin extract from *S. rarak* gave an increase in MB estimated by both methods. MB yield by gravimetry was higher than MB yield estimated using rRNA as a marker (160–180 mg/g vs 80–96 mg/g, respectively). MB yield estimated by gravimetry reached maximum at 2 mg/ml inclusion level of crude *S. rarak* saponin extract (Figure 4a) and was significantly different compared with the control ($P<0.05$) but was not different when compared with 1 or 4 mg/ml extract inclusion. The total MB yield, measured using rRNA (Figure 4b), was not significantly different among treatments, but when analysed separately as prokaryotic and eukaryotic biomass, the differences among treatments were significant ($P<0.05$). The average MB yields due to addition of methanol extract were 1.07 and 1.14 more than MB yield of the control, estimated by gravimetry and rRNA, respectively.

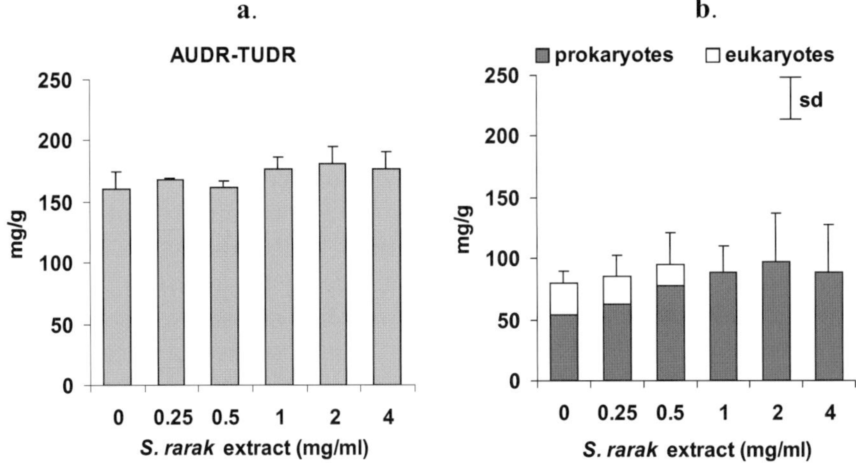

Figure 4. Microbial biomass (MB) yield measured by gravimetry (a) and by rRNA as a marker (b) in the presence of *Sapindus rarak* extract.
NOTE: error bars represent standard deviation (n=3)

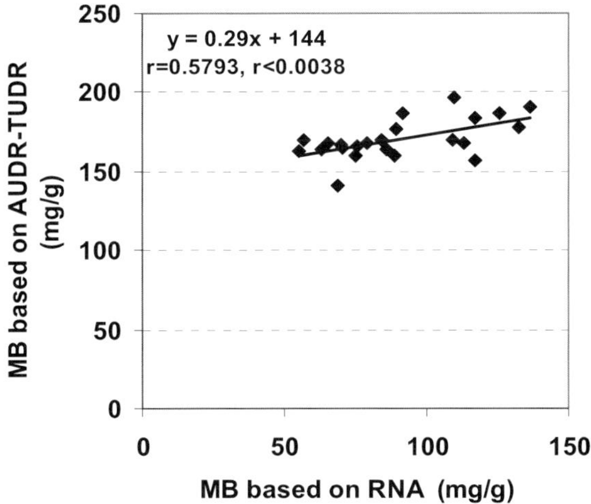

Figure 5. Relationship between Microbial biomass (MB) yield calculated by gravimetry and by RNA as a marker.

4.   DISCUSSION

As PEG addition in *in vitro* fermentation of the plant materials investigated produced very little increase in gas production, the presence of tannin in these plant materials may be too low to have any significant effect on rumen fermentation. Further effects of these plant materials on rumen fermentation, therefore, would not be due to tannin but rather other secondary compounds. TLC analysis shows the presence of saponins, as also shown by their haemolytic activity in all the plant materials used in the present investigation. Haemolytic activity is a characteristic of saponin since saponin can bind the cholesterol of red blood cells. The same reaction occurred when saponin reduced the protozoan concentration. *S. rarak* fruits, compared with other plant materials, have the highest diversity of saponins. They may, however, have different chemical structures. So far, only four acetylated triterpene saponins, which have hederagenin aglycone, have been identified (Hamburger *et al.*, 1992). *S. rarak,* with its diverse saponins, may be a potential source for defaunating agent. Further studies on use of *S. rarak* as animal feed supplement are currently in progress in our laboratory.

With the exception of *S. rarak,* use of the plant materials investigated to supplement *P. purpureum* would be beneficial because the substrates produced more of the SCFA required as an energy source for ruminants and also for microbial growth in the rumen (Table 1). Although these Indonesian plant materials have potential as forages, their utilization as animal feed has never been reported. *S. rarak,* even though it demonstrated a very low SCFA level, changed the pattern of fermentation toward a higher molar proportion of propionate, or a lower ratio of acetate to propionate (A:P). High propionate, a glucogenetic compound, in the rumen is expected in animals fed a high-concentrate diet, and is highly desirable for animal growth. However, one should be careful to look not only at the higher molar proportion or the lower A:P ratio but also at the absolute value of propionate in the rumen. In this experiment, the absolute value of propionate did not change but the absolute value of acetate decreased, and therefore gave a higher molar proportion of propionate. The changes in individual fatty acids are most probably due to changes in microbial community structure in the rumen. This is clearly seen in Figure 3, which shows a variation in prokaryotic and eukaryotic concentrations with different plant materials. As eukaryotic concentration reduces, the prokaryotic concentration is expected to increase, as some protozoa prey on bacteria. Total eukaryote elimination, shown by absence of the 18S rRNA band on the gel, may be due to saponin in LS and HS types of *S. sesban* and *S. rarak.* Saponins in *S. sesban* and *S. rarak* may be at higher concentration, more diverse or more active than *N. scutellarium* and *M. citrifolia,* since the latter two caused only a slight reduction in eukaryotic concentration. Newbold *et al.* (1997) have reported that butanol extract of *S. sesban* that contained saponin caused a large reduction in protozoan concentration. However, this phenomenon was totally in contrast to *in vivo* experiments. Consortia of rumen microbes adapted very quickly to *S. sesban* and protozoa concentration increased rapidly after 7 days (Newbold *et al.,* 1997). Odenyo *et al.* (1997) found that *S. sesban* reduced protozoan concentration almost immediately when added directly to the rumen, but had no effect when fed orally. Therefore, using *Sesbania sesban* as a defaunating agent in the live animal is unlikely to be beneficial.

When *S. rarak* fruit was used as the sole substrate in the fermentation, a reduction occurred not only in eukaryotes but also in prokaryotes. To assess use of *S. rarak* as a defaunating agent without negative effect on rumen bacteria, a further experiment was set up. A crude saponin extract was prepared and added to a substrate that consisted of *P. purpureum* and wheat bran (Figure 4b). Crude saponin extract at 1 mg/ml significantly depressed eukaryotic concentration, and at the highest level (4 mg/ml) only slightly reduced prokaryotic concentration. Therefore, inclusion of crude saponin extract of *S. rarak* in the diet would have a beneficial effect at a level where it affects only eukaryotes, and not prokaryotes. Other studies also showed that the protozoan population was reduced by *S. rarak* saponin extract, measured by direct counting and by the breakdown of ^{14}C-leucine-labelled *Prevotella byrantii* (Ningrat, Garnsworthy and Newbold, 2002). Two fungi, *Neocallimastix frontalis* and *Piromyces rhizinflata,* were negatively affected by saponin in their ability to digest filter paper and in their growth (Wang *et al.*, 2000).

Higher microbial biomass yield was obtained from the inclusion of crude saponin extract of *S. rarak*. The same effect was reported when saponins from quillaja bark (Makkar, and Becker, 1996) and *Yucca schidigera* (Wang *et al.*, 2000) were used. Using a different method to quantify microbial biomass, Makkar and Becker (1996) previously reported an increase in ^{15}N incorporation and efficiency of microbial protein synthesis due to saponins.

In the present study, the absolute amount of MB was higher when calculated by the gravimetric method than when rRNA was used as a marker, but both showed agreement. In both cases, maximum MB yield was obtained at the 2 mg/ml level of extract inclusion. A similar increase in MB yield due to saponin, quantified by both methods, resulted in a significant correlation between these methods (Figure 5), which suggests that both methods can possibly be used to study the effect of saponin. However, the correlation is still low (r = 0.5793) and more studies are warranted, using a larger number of samples.

Estimation of MB yield using the gravimetric method is simple and could be done in any laboratory, without sophisticated equipment. However, it has some limitations. MB yield could be higher at an early stage of fermentation as it might also contain high neutral-detergent-soluble material from feed. At a later stage of fermentation, MB yield could be lower due to microbial lysis (Blummel, Steingass and Becker, 1997). The MB yield was therefore quantified from the residue after 24 hours of fermentation.

Estimation of MB yield using rRNA as marker (Figure 4b) has not been widely used. The concentration of RNA in bacteria and protozoa has to be measured to calculate the total MB yield. RNA concentration in bacteria has been reported to vary from 4 to 24.2 percent (Merry and McAllan, 1983; Hespell and Bryant, 1979), while reports on RNA concentration in protozoa are limited. Robinson, Fadel and Ivan (1996) reported that RNA concentrations in bacteria and in protozoa were 8.5 and 4 percent, respectively, and, using these values, Muetzel, Hoffmann and Becker (2002; 2003) and in the present study calculated MB yields. Using rRNA as marker, the highest MB yield was obtained at the point of maximum microbial growth, which was after 12 hours of fermentation. However, the time for maximum microbial growth might vary with different substrates. Further, high variation of 16S and 18S rRNA concentrations, as found in this experiment at different *in vitro* incubations, highlights a need for further studies using rRNA as a marker in order to quantify MB yield.. This method, however, has the advantage that the same extraction protocol for RNA is used for identification of rumen bacteria or eukaryotes using ribosomal RNA-targeted probes (Muetzel, Hoffmann and Becker, 2003; Muetzel, Krishnamoorthy and Becker, 2001; Odenyo *et al.*, 1994) and for further community analysis (White *et al.*, 1999; Wright and Lynn, 1997).

5.　　CONCLUSION

Sapindus rarak is a potential plant material as a defaunating agent. Addition of crude saponin extract from *S. rarak* increased microbial biomass yield. Both the gravimetric method and rRNA as a marker could be used for quantifying microbial biomass.

ACKNOWLEDGEMENT

The work was done at Hohenheim University and was supported by PAATP, Agriculture Department of Indonesia.

REFERENCES

Blummel, M., Steingass, H. & Becker, K. 1997. The relationship between *in vitro* gas production, *in vitro* microbial biomass yield and N-15 incorporation and its applications for the prediction of voluntary feed intake of roughages. *British Journal of Nutrition*, **77**: 911–921.

Cheeke, P.R. 2000. Actual and potential applications of *Yucca schidigera* and *Quillaja saponaria* saponins in human and animal nutrition. Proceedings of the American Society of Animal Science, Indianapolis. 10p.
See: http://www.asas.org/JAS/symposia/proceedings/0909.pdf.

El Gamal, H.A., Soliman, H.S.M., Karawya, M.S., Mikhova, B. & Duddeck, H. 1995. Isolation of triterpene saponins from *Gypsophila capillaris*. *Phytochemistry,* **38**: 1481–1485.

Fahey, G.C. Jr, & Jung, H-J.G. 1989. Phenolic compounds in forages and fibrous feedstuffs. pp. 123–190, *in:* P.R. Cheeke (ed). *Toxicants of Plant Origin.* Florida, FL: CRC Press .

Getachew, G., Makkar, H.P.S. & Becker, K. 2000. Effect of different amounts and method of application of polyethylene glycol on efficiency of microbial protein synthesis in an *in vitro* system containing tannin-rich browses. Presented at: EAAP Satellite Symposium, *Gas production: fermentation kinetics for feed evaluation and to assess microbial activity.* Wageningen, The Netherlands, 18–19 August 2000.

Getachew, G., Blummel, M., Makkar, H.P.S. & Becker, K. 1998. *In vitro* gas measuring techniques for assessment of nutritional quality of feeds: a review. *Animal Feed Science and Technology,* **72**: 261–281.

Getachew, G., Makkar, H.P.S. & Becker, K. 2000. Effect of polyethylene glycol on *in vitro* degradability of nitrogen and microbial protein synthesis from tannin-rich browse and herbaceous legumes. *British Journal of Nutrition,* **84**: 73–83.

Hamburger, M., Slacanin, I., Hostettmann, K., Dyatmiko, W. & Sutarjadi. 1992. Acetylated saponins in molluscicidal activity from *Sapindus rarak*: Unambiguous structure determination by proton nuclear magnetic resonance and quantitative analysis. *Phytochemical Analysis,* **3**: 231–237.

Hespell, R.B. & Bryant, M.P. 1979. Efficiency of rumen microbial growth: Influence of some theoretical and experimental factors in Y_{ATP}. *Journal of Animal Science,* **49**: 1640–1659.

Makkar, H.P.S., Blümmel, M. & Becker, K. 1998. Application of an *in vitro* gas method to understand the effect of natural plant products on availability and partitioning of nutrients. pp. 147–150, *in:* British Society of Animal Science, *Occasional Publication,* No. 22.

Makkar, H.P.S. 1993. Antinutritional factors in foods for livestock. pp. 69–85, *in:* M. Gill, E. Owen, G.E. Pollott and T.L.J. Lawrence (eds). *Animal Production in Developing Countries.* British Society of Animal Production, *Occasional Publication,* No. 16.

Makkar, H.P.S. & Becker, K. 1996. Effect of quillaja saponins on *in vitro* rumen fermentation. pp. 387–394, *in:* Waller and Yamasaki (eds). *Saponins Used in Food and Agriculture.* New York, NY: Plenum Press.

McSweeney, C.S., Palmer, B., McNeill, D.M. & Krause, D.O. 2001. Microbial interactions with tannins: nutritional consequences for ruminants. *Animal Feed Science and Technology,* **91**: 83–93.

Menke, K.H., Raab, L., Salewski, A., Steingass, H., Fritz, D. & Schneider, W. 1979. The estimation of the digestibility and metabolizable energy content of ruminant feedingstuffs from the gas production when they are incubated with rumen liquor *in vitro. Journal of Agricultural Science,* **93**: 217–222.

Merry, R.J. & McAllan, A.B. 1983. A comparison of the chemical composition of mixed bacteria harvested from the liquid and solid fractions of rumen digesta. *British Journal of Nutrition,* **50**: 701–709.

Muetzel, S., Hoffmann, E.M. & Becker, K. 2002. Effects of sumach (*Rhus typhina*) tannins on rumen fermentation. *Proceedings of the Society of Nutrition Physiology,* **11**: 131. [DLG-Verlags-GmbH, Frankfurt. Germany]

Muetzel, S., Hoffmann, E.M. & Becker, K. 2003. Supplementation of barley straw with *Sesbania pachycarpa* leaves *in vitro*: effects on fermentation variables and rumen microbial concentration structure quantified by ribosomal RNA-targeted probes. *British Journal of Nutrition,* **89**: 445–453.

Muetzel, S., Krishnamoorthy, U. & Becker, K. 2001. Effects of rumen fluid collection site on microbial concentration structure during *in vitro* fermentation of the different substrates quantified by 16S rRNA hybridisation. *Archives of Animal Nutrition,* **55**: 103–120.

Newbold, C.J., El Hassan, S.M., Wang, J.M., Ortega, M.E. & Wallace, R.J. 1997. Influence of foliage from African multipurpose trees on activity of rumen protozoa and bacteria. *British Journal of Nutrition,* **78**: 237–249.

Ningrat, R.W.S., Garnsworthy, P.C. & Newbold, C.J. 2002. Saponin fractions in *Sapindus rarak*: effects on rumen microbes. *Reproduction Nutrition Development,* **42**(Suppl.1): S82.

Obispo, N.E. & Dehority, B.A. 1999. Feasibility of using total purines as a marker for ruminal bacteria. *Journal of Animal Science,* **77**: 3084–3095.

Odenyo, A.A., Mackie, R.I., Stahl, D.A. & White, B.A. 1994. The use of 16S rRNA-targeted oligonucleotide probes to study competition between ruminal fibrolytic bacteria: pure culture studies with cellulose and alkaline peroxide-treated wheat straw. *Applied and Environmental Microbiology,* **60**: 3697–3703.

Odenyo, A.A., Osuji, P.O. & Karanfil, O. 1997. Effect of multipurpose tree (MPT) supplements on ruminal ciliate protozoa. *Animal Feed Science and Technology,* **67**: 169–180.

Olubobokun, J.A., Craig, W.M. & Nipper, W.A. 1988. Characteristics of protozoal and bacterial fractions from microorganisms associated with ruminal fluid or particles. *Journal of Animal Science,* **66**: 2701–2710.

Robinson, P.H., Fadel, J.G. & Ivan, M. 1996. Critical evaluation of diaminopimelic acid and ribonucleic acid as markers to estimate rumen pools and duodenal flows of bacterial and protozoal nitrogen. *Canadian Journal of Animal Science,* **76**: 587–597.

Wallace, R.J., McEwan, N.R., McInntosh, F.M., Teferedegne, B. & Newbold, C.J. 2002. Natural products as manipulators of rumen fermentation. *Asian-Australasian Journal of Animal Sciences,* **15**: 1458–468.

Wang, Y, McAllister, T.A., Yanke, L.J. & Cheeke, P.R. 2000. Effect of steroidal saponins from *Yucca shidigera* extract on ruminal microbes. *Journal of Applied Microbiology,* **88**: 887–896.

Wang, Y., McAllister, T.A., Yanke, L.J., Xu, Z.J., Cheeke, P.R. & Cheng, K.J. 2000. *In vitro* effects of steroidal saponins from *Yucca schidigera* extract on rumen microbial protein synthesis and ruminal fermentation. *Journal of the Science of Food and Agriculture,* **80**: 2114–2122.

White, B.A., Cann, I.K.O., Kocherginskaya, S.A., Aminov, R.I., Thill, L.A., Mackie, R.I. & Onodera, R. 1999. Molecular analysis of Archae, Bacteria and Eucarya Communities in the rumen – Review. *Asian-Australasian Journal of Animal Sciences,* **12**: 129–138.

Wright, A.D.G. & Lynn, D.H. 1997. Phylogenetic analysis of the rumen ciliate family Ophryoscolecidae based on 18S ribosomal RNA sequences with new sequences from Diplodium, Eudiplodium and Ophyroscolex. *Canadian Journal of Zoology,* **75**: 963–970.

Zinn, R.A. & Owen, F.N. 1986. A rapid procedure for purine measurement and its use for estimating net ruminal protein synthesis. *Canadian J. of Animal Science,* **66**: 157–166.

NUTRITION–GENE INTERACTIONS (POST-GENOMICS)

Changes in gene expression through nutritional manipulations

Gregory S. Harper[1], Sigrid A. Lehnert[1] and Paul L. Greenwood[2]

All authors work with Cooperative Research Centre for Cattle and Beef Quality.
1. CSIRO Livestock Industries, Queensland Bioscience Precinct, St. Lucia, Queensland 4067, Australia.
2. NSW Agriculture Beef Industry Centre, University of New England, Armidale, New South Wales 2351, Australia

Abstract: This paper discusses the effects of severe nutritional restriction, both pre- and post-weaning, on development of skeletal muscle in food animals. Given recent predictions about growth in demand for muscle-foods in developing countries, the global community will need to face the food-feed dilemma, and balance efficiency of production against the quality-of-life aspects of local livestock husbandry. It is likely that production animals will be grown in successively more marginal environments and at higher stocking rates on unimproved pastures. Understanding the nutritional limits to animal growth at the level of muscle gene networks will help us find optima for nutrition, growth rate and meat yield. Genomic approaches give us unprecedented capacity to map the networks of control under nutritionally restricted conditions, though the challenges remain of identifying steps that regulate substrate flux. The paper describes some approaches currently being taken to understanding muscle development, and concludes that the genes contributing to two ruminant phenotypes should be mapped and characterized. These are: the capacity to depress metabolic rate in response to nutritional restriction; and the capacity to exhibit compensatory growth after restriction is relieved.

411

H. P. S. Makkar and G. J. Viljoen (eds.), Applications of Gene-Based Technologies for Improving Animal Production and Health in Developing Countries, 411-428.
© 2005 *IAEA. Printed in the Netherlands.*

1.   INTRODUCTION

Several authors have described the developing *Livestock Revolution*; that is, the increased demand for livestock products in the developing world, driven by rapid population growth, real income growth, dietary changes and urbanization. For example, it has been estimated that total global demand for muscle-foods will increase from 185 million tonne in 1993 to 303 million tonne by 2020 (Delgado *et al.*, 1999). This increase will be demand-driven, as opposed to the supply-driven Green Revolution, and muscle-food consumption per capita in the developing world is predicted to increase by more than 40 percent, from 21 kg/year in 1993 to 30 kg/year in 2020. Production growth in China, sub-Saharan Africa and Southeast Asia is predicted to exceed 2.9 percent annually, considerably greater than the expected world average annual growth rate of 1.8 percent. These predictions raise planning and policy issues for the financial, technological, environmental, regulatory and marketing sectors. At the simplest level, demand can be met by increasing growth and production efficiency of existing animals; by increasing the number of animals; increasing the land under production; or by making trade in animal products from the developed world more accessible to people in the developing world. Some of the foreseeable constraints include: the amount of land the global community is willing to commit to animal production; the capacity of that land to produce sufficient nutrients to sustain growth in these animals; the capacity of animal producers to care for their animals; and the efficiency of animals in converting plant protein into animal food products.

This array of issues is intractably broad, so the reader is directed to recent reviews for further consideration (Ruane and Zimmermann, 2001; Hodges, 2000). This paper will focus on muscle-food production and will survey recent publications that reflect on the effects of nutrient restriction on that production. Whilst there are many levels of expression and order at which one could discuss muscle-food production, the focus will be on how muscle phenotypes, genes and gene networks respond to nutrient deprivation. In particular, the paper will emphasize long lasting effects of nutrition on growth of skeletal muscle. The power of genomics will be exemplified by drawing conclusions about developmental, cellular and biochemical mechanisms for production species, from understanding reached in mammalian species.

2. IMPACT OF NUTRITIONAL RESTRICTION IN THE MUSCLE-FOOD PRODUCTION SYSTEM

Muscle-foods are produced in both intensive and extensive production systems. Intensification of livestock production is predicted to continue in the developing world, though ultimately the relative costs of local and imported feed grains will determine the sustainability of this practice. Intensification ensures that the nutritional needs of monogastric species (pigs and chickens) are adequately met and, indeed, approach being optimized. Whilst intensification and grain feeding have real advantages for the production of ruminants, this approach forgoes the potential benefits of browse feeding on native forages of lower nutritive value. Certainly, the production of ruminants in feedlots is expected to increase in areas where extensive production is impractical (Delgado *et al.*, 1999). Intensification also implies industrialization to spread capital costs, and this has significant implications for communities in the developing world, where livestock have historically provided other benefits, including traction and financial management (Kinoti, 2000).

By far, the majority of cattle in the developing world are fed under extensive conditions, and chronic undernutrition of livestock is a recurring feature, as in the more climatically marginal regions of the developed world (Cronje, 1990; Hunter and Buck, 1992). The global impact of this nutritional deficiency for cattle in the developing world relative to the developed world may be 80 kg of beef per head per year (Delgado *et al.*, 1999). The deficiency for small ruminant meat species may be as high as 30 percent compared with developed world growth rates. How can animal genomics help address this deficiency without sacrificing food safety, nutrient density or other aspects of a sustainable system?

3. IMPACTS OF CHRONIC OR SEVERE NUTRITIONAL RESTRICTION ON MUSCLE

Nutritional restriction limits skeletal and muscle growth in ruminants, though muscle in particular is a resilient tissue, with a great capacity to return to a developmental programme once nutrition is repleted (Greenwood and Bell, 2003). The "severe" restriction discussed in this article implies stunting of skeletal development, and hence might include both chronic undernutrition and acute severe undernutrition. The discussion is organized along the lines of animal age, because the animal is more sensitive to the long-term effects of severe nutritional restriction during the pre-weaning phase, including intra-uterine life, than during the post-weaning phase.

3.1   Pre-weaning restriction

In production mammals, foetal tissue mass increases manyfold from mid to late gestation. Growth is associated with greater absolute rates of uptake of oxygen and nutrients by the uterus and umbilical structures, greater foetal whole-body protein synthesis, and greater urea export. These rates do however, decline on a foetal weight-specific basis. Changes in glucose and lipid metabolism, and increased nutrient supply, are key hallmarks of the transition from prenatal to postnatal life. Associated with this increase is a shift in the quality of the supply of nutrients from primarily glucose and amino acids, to less total and more complex carbohydrate and more fat. Plasma glucose concentration in the newborn increases rapidly, and induction of expression of genes for key regulatory enzymes important in gluconeogenesis and lipid metabolism occurs during this transition (Girard *et al.*, 1997).

It is difficult to significantly alter growth of the early- to mid-gestation foetus of livestock species, because foetal nutrient requirements are small and the dam can mobilize body tissues to support its own needs. Nonetheless, when it has been achieved, the results support foetal programming (or epigenetic effects) on metabolic and cellular phenotypes later in life, even though treatments may not change gross foetal phenotype. Corticosteroids have been implicated in foetal programming due to maternal undernutrition, and these findings in ruminants have been reviewed recently (Greenwood and Bell, 2003). A particularly interesting effect is on later life blood pressure homeostasis (Yuen *et al.*, 2003).

The short- and long-term implications of restricted nutrition during early life on growth and development of muscle include (Greenwood and Bell, 2003; Bell, 1992):
- altered structure and cellularity that may influence subsequent capacity for growth;
- reduced strength and capacity to meet energetic needs at birth, resulting in increased mortality and morbidity, as well as lessened competitive capacity; and
- altered nutrient requirements relative to well-grown counterparts, which may influence performance and morbidity later in life.

The magnitude of the effects on the foetus depends on the severity and timing of the restriction relative to ontogeny of organs, peripheral tissues and physiological systems, and is contingent on the relative responses of dam and foetus. Severe intra-uterine growth retardation (IUGR) has been shown to influence body composition in postnatal life, in terms of reduced muscle mass, total protein, total bone or ash content, alone or in combination (Greenwood *et al.,* 1998, 2000). Bone, as indicated by ash content, is less reduced than is total protein content (Greenwood *et al.,* 1998), consistent with bone being slower to respond to restriction or repletion of nutrition than is skeletal muscle. This effect may contribute to the smaller mature size (stunting) of sheep severely undernourished during pregnancy (Schinckel and Short, 1961; Everitt, 1968). Since peripheral tissues develop later than vital organ systems, the long-term effects of severe IUGR on growth and development of these tissues are apparent later (Greenwood *et al.,* 2002). Postnatal environment can also have a multiplicative effect on the direct effects of prenatal nutrition on growth capacity (Greenwood *et al.,* 1998).

Some responses of the neonate to nutritional restriction whilst *in utero* might be considered adaptations to enhance survival during early postnatal life, even though responses like increased fatness and insulin resistance (Hales *et al.,* 1996) may not be advantageous to the animal in the longer term. Greenwood *et al.* (2002) found that small neonatal lambs had elevated levels of growth hormone (GH), depressed levels of insulin-like growth factor-1 (IGF-1), and Rhoads *et al.* (2000a) found down-regulated hepatic expression of the acid-labile subunit (ALS) gene during the early postpartum period. ALS gene expression is GH-dependent and is up-regulated at or soon after birth in normal lambs (Rhoads *et al.,* 2000b). Perhaps nutritionally-restricted foetuses are developmentally delayed, so that neonates express phenotypes reminiscent of the late-gestation foetus, at least in terms of the somatotropic axis. Elevated GH and depressed IGF-1 persisted in small compared to large neonates for the first two weeks of postnatal life (Greenwood *et al.,* 2002), and absolute rates of growth lagged significantly behind normal-birth-weight lambs during this period. However, growth rates subsequently became equal between high- and low-birth-weight groups (Greenwood *et al.,* 1998).

During the postpartum period, small lambs deposit more adipose tissue than their larger counterparts (Greenwood *et al.,* 1998). This results from consumption of more feed on a weight-specific basis, coupled with reduced maintenance energy requirements and limited capacity to deposit muscle (Greenwood *et al.,* 1998, 2000). Circulating concentrations of insulin were also persistently higher in the small neonates during the period to weaning (Greenwood *et al.,* 2002), presumably as a consequence of greater relative feed intake coupled with limited degradation by peripheral tissues during the postpartum period. These observations would be consistent with some degree of insulin resistance.

Table 1. Survey of genes that may be involved in regulating effects of caloric or protein restriction in skeletal muscle pre-weaning, including *in utero.*

Functional pathway and genes involved	Indicative references
Myogenesis and myosatellite cell number and activity	Regulation of myogenesis:
Myogenic Regulatory factors (MyoD, Myf5, Mrf4, Myogenin)	[1–6]
Myocyte Enhancer Factor 2 (Mef2) family of genes	
Growth factors (IGF-I and -II, FGFs, EGFs, TGFs incl. Myostatin, PDGFs, HGF)	Regulation of myosatellite cells:
Hormones (insulin, somatotropin, thyroid hormones, androgens, corticosteroids)	[2, 3, 7–9] Pre-weaning
Cell cycle activators and inhibitors (cyclins, cdks, Rb, p21, p27, p57)	effects on enzyme levels: [10]
Sonic hedgehog (shh)	
Wnt genes (5a & 11)	Developmental
Bone Morphogenic Proteins (2, 4 & 7)	regulators:
Notch cascade	[11–13]
Cytokines (IL6)	
Leukaemia Inhibitory Factor (LIF)	
Neuregulins (NRGs)	
Myofibre hypertrophy or atrophy	*See* Table 2

SOURCES: [1] Buckingham *et al.,* 2003. [2] Kitzmann and Fernandez, 2001. [3] Parker, Seale and Rudnecki, 2003. [4] Rupp, Singhal and Veenstra, 2002. [5] Wigmore and Evans, 2002. [6] Zorzano *et al.,* 2003. [7] Bornemann, Maier and Kuschel, 1999. [8] Goldring, Partridge, and Watt, 2002. [9] Seale and Rudnicki, 2000. [10] Allingham *et al.,* 2001. [11] Amthor, Christ and Patel, 1999. [12] Gustafsson *et al.,* 2002. [13] Anakwe *et al.,* 2003.

As is noted for post-weaning restriction, IUGR directly affects the hypertrophic growth of myofibres, as well as the capacity of muscle for hyperplastic growth through the number and activity of the satellite cells in late prenatal and postnatal life (Greenwood *et al.,* 1999; Jeanplong *et al.,* 2003). Muscle growth is also limited synergistically through cessation of skeleton elongation, and the stretch it induces in muscle (Day *et al.,* 1997).

Table 1 lists some of the genes and gene networks underpinning these physiological responses to early-life nutritional restriction, though clearly the networks lead into other aspects of cell and tissue development.

3.2   Post-weaning restriction

In circumstances of unlimited nutrition, animals grow at a rate and to skeletal dimensions that are determined genetically (Berg and Butterfield, 1976). In intensive production systems, animals are never likely to reach skeletal maturity, or an age at which they begin to lose muscle through sarcopenia (Grounds, 2002), because they are slaughtered young. In extensive beef production systems, restricted nutrition and environmental energy loads limit the animal's capacity to express its full growth capacity.

Prior to skeletal maturity, the animal has the capacity to compensate for the lost growth potential, at least partially. One of us (Harper, 1999) has previously surveyed the literature and found evidence for diverse effects of nutritional restriction on ruminant skeletal muscle, and particularly the meat toughness trait, and concluded that nutritional restriction effects were sometimes confounded by the effects of timing and severity of nutritional restriction. Oddy and Sainz (2002) noted that there is an apparent discontinuity in animal responses to nutritional restriction at around 40 percent skeletal maturity; before that point, restriction led to measurable reductions in the animal's capacity to compensate. After 40 percent maturity, the growth potential of muscle accommodated the restriction and returned to its genetically-determined growth trajectory once nutrition was replete. Given that the restriction of energy *per se* is of greater significance than restriction of any particular amino acid, even in ruminants (Oddy and Sainz, 2002), the literature on caloric restriction in monogastric mammals is pertinent to questions of mechanism and genetic networks of control.

At the structural level, postnatal muscle growth occurs through hypertrophy of existing myofibres rather than development of new fibres (Harper, 1999). Fibre size is influenced by nutritional restriction, decreasing during nutritional restriction, and increasing again once nutrition is replete (Yambayamba and Price, 1991). The proportions of connective tissue in the muscle vary in an opposite sense to fibre size during a weight loss phase (Harper *et al.,* 2002). The proportion of type I fibres increases, due to differential effects of restriction on the various fibre types, and, providing cell death has not occurred, the fibre type profile returns to control values upon repletion (Yambayamba and Price, 1991). In other words, oxidative muscle fibres are preferentially spared during nutritional sarcopenia.

At the biochemical level, nutritional restriction influences enzyme activities characteristic of the myofibrillar component (lactate dehydrogenase and isocitrate dehydrogenase) as well as synthetic products characteristic of the connective tissue (collagen type I) (Harper, Allingham and LeFeuvre, 1999). These changes appear to be transient effects of nutritional restriction in the animal after weaning (Brandstetter, Picard and Geay, 1998).

Functional genomic approaches to nutritional restriction have utilized micro-arrays and the major target species of human and mouse. A myriad of genes are influenced by nutritional restriction, and this is consistent with existing biochemical paradigms for muscle hypertrophy and atrophy (summarized in Table 2). As indicated recently by Glass (2003), muscle atrophy is not necessarily an inverse of muscle hypertrophy, even though many of the same genes are involved. Nutritional restriction initially leads to a dramatic increase in protein degradation, and activation of some unique transcriptional pathways involved in protein turnover. In later phases of restriction, both protein synthesis and degradation are down-regulated intracellularly and extracellularly. The work of Jeanplong *et al.* (2003) suggests that severe nutritional restriction leads to muscle responses at the gene level, which may be differentiated into at least three phases. Down-regulation of intracellular detoxification pathways may explain the intriguing effects of nutritional restriction on life-span in mammals, because autolytic damage of macromolecules is reduced in caloric-protein restriction.

Muscle extracellular matrix genes are known to be responsive to restricted nutrition. The synthesis of collagens (I and III, as well as others) in connective tissues, and specifically by mesenchymal cells, is known to be down-regulated (Laurent, 1987), and, in the longer term, degradatory enzyme systems are also down-regulated. There is some evidence for higher order coordination of groups of genes by the *hcKrox* transcriptional factor (Widom *et al.,* 2001). We have previously found that cycles of weight loss followed by compensatory growth leads to remodelling of connective tissue relative to the myofibrils, and that this has implications for the eating qualities of meat (Harper, Allingham and LeFeuvre, 1999).

Sustained nutritional restriction also leads to adaptive responses by the animal, and these include depression of basal metabolic rate by as much as 20 percent (Guppy and Withers, 1999). As a result, the animal is able to sustain growth rates that are in excess of those predicted from energy intake using the published relationships (ARC, 1980; AFRC 1993; NRC, 1996). Human-twin studies showed that genetic influences on resting metabolic rate were entirely explained by body weight: there was no independent genetic contribution to resting metabolic rate (Hewitt *et al.,* 1991). Metabolic rate under psychological stress, in contrast, showed a significant genetic effect. The allometric exponent (3/4) relating body weight to resting metabolic rate was found to hold, as it does for many animal species, and it is proposed that this represents a genetically determined body weight set point. Certainly aspects of energetic efficiency (feed conversion efficiency) are clearly heritable in production animals (Pitchford, 2004), and hence we predict that basal metabolic rate may also be a heritable trait.

Nutritional restriction leads to a whole-tissue response of muscle. Current research is targeted toward measurements at the gene level, and the complexity of these results probably reflects the responses from various cell types, including myocytes, adipocytes, fibroblasts, pericytes and various precursor cell types, such as satellite cells. Independent measures of cellularity will be required to interpret these integrative results at the metabolic pathway level.

Table 2. Survey of genes implicated in regulating effects of caloric or protein malnutrition in skeletal muscle post-weaning.

Functional pathway and gene	Indicative references

Cellular signalling
  IGF-1 (IGFBP 1-6)
  IRS-1
  Ras
  Raf
  MEK
  ERK
  Phosphoinositides
  PI3K
  Akt1
  mTOR
  p70S6K
  Insulin sensitivity
  800 genes
	Grounds, 2002;
	Glass, 2003;
	Sreekumar *et al.*, 2002;
	Kayo *et al.*, 2001;
	Weindruch *et al.*, 2001;
	Rome *et al.*, 2003.

Energy metabolism
  Cidea
  UCP-1
	Zhou *et al.*, 2003.

Protein turnover
  E1 ubiquitin-activating enzyme
  E2 conjugating enzyme
  E3 ligases
  MuRF1
  MAFbx
	Tawa, Odessey and Goldberg, 1997.

Energy metabolism
  Phosphoinositides PI3K
  GSK3β
  eIF-2β
  LDH & ICDH
	Harper, Allingham and LeFeuvre, 1999;
	Brandstetter, Picard and Geay, 1998.

Extracellular matrix synthesis
  hcKrox
  Collagens I and III
  Osteonectin
  Lysyl oxidase
	Widom *et al.*, 2001;
	Spanheimer *et al.*, 1991;
	Laurent, 1987.

Extracellular matrix degradation MMPs	Streuli, 1999.

Myogenic regulators
  myostatin
  Myf-5
  IGF-1
  myogenin
	Jeanplong *et al.*, 2003.

Unknown and novel genes	Reverter *et al.*, 2003.

4. DEFINING GENE NETWORKS RESPONSIVE TO NUTRITIONAL RESTRICTION

From Tables 1 and 2 it can be seen that some of the genes have been defined that contribute to the maintenance of muscle mass in the face of nutritional restriction, but it is certain that novel genes will be discovered, as will novel functions for known genes. The optimal strategy for gene discovery and functional definition is difficult to define, though one might consider the successful strategies of the past. Muscle growth genes have been found from studies involving:

- genetic extreme production animals, or other species of research interest;
- quantitative trait loci (QTLs) for phenotypic variance in large cross-bred populations;
- comparative genomic approaches based on discoveries in genetically modified animals;
- extreme nutritional treatments; and
- more detailed analysis of known or candidate genes *in vitro*.

The responses of mammalian muscle to nutritional restriction are clearly governed by networks of genes. Understanding even simple linear connections between genotype and phenotype are complex given a myriad of uncontrollable physiological variables, without the complications of regulatory networks. The path for quantification of gene effects underlying monogenic traits is now well trodden. The path for identification of gene effects underlying polygenic traits is less well defined, and has become significantly more complicated since the discovery of unusual inheritance patterns (e.g. polar over-dominance (Georges *et al.,* 2002)) and the effects of imprinting in general. Similarly, working from the phenotype back to the genotypic information via metabolic pathways is difficult and has only recently been approached in single cells (Westerhoff, 2001). What then is a reasonable approach to defining gene networks?

In our case we have made several strategic decisions that we hope will facilitate gene discovery and application. These are:

- Work in production species (sheep and cattle). Whilst this has the disadvantage of a dearth of genomic information at present, it avoids erroneous conclusions reached from species-specific responses to nutritional treatments. The enormous genomic resources available for human and mouse can be utilized to confirm metabolic pathways and regulatory cascades, though not the specifics of substrate fluxes or physiological responses.
- Focus on aspects of nutritional atrophy. We have been performing nutritional studies in cattle focusing on extreme and prolonged nutritional restriction (1) pre-weaning, and (2) post-weaning, where we have

controlled genotype in the experimental herds. Likewise, one of us has published extensively on sheep (Greenwood and Bell, 2003; Greenwood *et al.,* 1998, 1999, 2000, 2002).

– Work within extended collaborative networks. The complexity of genetic networks demands that many measurements be made on tissue samples from a particular animal. This requires large multidisciplinary teams. Through three Cooperative Research Centres (Beef Industry CRC [Meat Quality]; Cattle and Beef Quality CRC; Australian Sheep Industry CRC) and a new Sheep Genomics Programme, we have drawn on the expertise of hundreds of researchers and some 15 research organizations, as well as the financial resources of the Australian Government and private enterprises.

– Build up knowledge within a "systems" framework. We are constructing analytical and logistic frameworks to build from knowledge at the gene level, through to decision-support systems at the industry level. We have found the gap between the gene and the metabolic levels to be particularly difficult to span, and tools that proved valuable include:
  - the Kyoto Encyclopedia of Gene and Genomes (www.kegg.com);
  - Gene Ontology Consortium (http://www.geneontology.org);
  - IUBMB Nicholson minimaps
    (http://www.tcd.ie/Biochemistry/IUBMB-Nicholson); and
  - integrating analysis across experiments (Reverter *et al.,* in press).

– Deliver outcomes using a range of mechanisms. There are many ways that this new knowledge might deliver value in the animal production industries. These include (1) training packages for students and industry practitioners; (2) marker assisted selection for animals with superior growth performance or nutrient efficiency; (3) performance vaccines to modify metabolic rate during periods of restricted nutrition, though this has been extraordinarily difficult to achieve; (4) biologicals or growth modifying substances that can be used to temporarily modulate the animals' metabolism to achieve an energetic efficiency benefit; and (5) transgenic animals that have engineered advantages in growth performance.

These approaches have proven to be appropriate in the context of animal production in the developed world, but this is not to conclude that the same approaches will be productive in the context of the developing world.

5. APPLICATIONS WITH EMPHASIS ON THE DEVELOPING WORLD

As pointed out by Kinoti (2000), livestock make diverse contributions to the quality of life of communities in developing countries, but we will confine our comments to aspects of growth, and the production of meat. Advantageous performance qualities for livestock in a developing world production system would appear to be parasite and disease resistance, and heat tolerance. Muscle growth efficiency and nutritional "resilience" would also seem to be important, where resilience refers to:

- a capacity to reduce metabolic rate in response to nutritional restriction, at a standard physiological age; and
- a capacity to express compensatory growth once nutritional restrictions are lifted, again at standard physiological age.

It is unclear if these phenotypes are mechanistically independent, or related to thrift in humans (Ozanne and Hales, 1999), though that possibility needs to be considered. We recommend measurement of genetic and phenotypic variance in resilience, particularly in animal breeds that express other phenotypes that appear to be well adapted to relatively extreme environments.

Similar phenotypes have been measured before on individual animals, and in experimental groups (Harper, Allingham and LeFeuvre, 1999; Ryan, 1990), but, to our knowledge, they have not been measured on populations of production animals, though a related trait, net feed conversion efficiency, certainly has been measured (Pitchford, 2004). The lack of interest in these traits is understandable, given the current close linkage between agricultural research priorities and developed world industrial priorities.

Knowing the genes and understanding the networks of control may not be sufficient for us to overcome or even manipulate the physiological set points of production species. Evolution of the terrestrial mammals led to animal sizes and body plans that were local optima within environmental, bio-energetic and performance constraints. The extreme genetic manipulations to date include the natural mutants and highly selected breeds, and these do not always cope well with environmental stress. For example, one would need to consider carefully the value of introducing high performance genotypes like double muscled cattle (*mh-/-*) or *callipyge* sheep into developing world production systems with a view to capturing the high yielding characteristics. Dependence on human intervention for reproduction, and parasite susceptibility would be foreseeable problems, along with the probability that these elite animals would not express their yield advantages under nutritionally restricted conditions (Greenwood *et al.,* unpublished data). Similar arguments apply to each of the mammalian production animal species, where choices of species and size have previously been made based on a mixture of indigenous knowledge and historical precedent, and not on full knowledge of the global gene pool of production animals. Sustainable alternative choices can only be made through integration of the available information at the whole-system level.

If nutritional responsiveness and adaptability are linked to feed efficiency, then there is likely to be significant variation within breeds, and perhaps among breeds (Johnston, 2001). A target for future research with relevance to the developing world would be definition of the specific nutritional requirements of high meat yielding breeds, and particularly the optimal timing of supplementary feeding. Increased understanding of early life and epigenetic effects suggests animal producers increase emphasis on feeding the pregnant dam, and often her offspring up to weaning, so that post-weaning performance is predictable and optimal.

An emerging issue will be nutritional management of the transgenic production animal, when they eventually become available. Whilst there are currently no transgenic production animals approved for commercial applications in the developed world, this is an active area of production animal research and development (R&D), and several groups are preparing submissions for the relevant national regulatory bodies (Sang, 2003). Increasing growth efficiency is one of the targets for such R&D (Solomon *et al.*, 1997), along with enhanced phytate degradation (Golovan *et al.*, 2001). Given the significant physiological modifications induced by incorporation of transgenes, these animals might exhibit novel relationships between nutrition and allometric growth and hence require new equations for quantitative nutrition.

6. CONCLUSIONS

Functional genomic studies are mapping the networks of genes responsive to nutritional deprivation in a number of mammalian species. Relatively few studies have targeted the genes that determine structural, cellular and metabolic adaptations to nutritional restriction, and the compensatory growth response subsequent to restriction, though we should not underestimate the challenges in doing this. Given the growing demand for muscle-foods in developing countries, mapping these genes may eventually yield tangible benefit for animal production in the developing world. Animal production in the developed world is progressing inexorably toward higher precision through genetic selection and nutritional management, and gene-based technologies are underpinning this progress. Agricultural research priorities in the developed world are set jointly with industry, and hence tend to ignore the genetics and physiology of undernutrition. International and not-for-profit agricultural research funds will be required to address the post-genomic aspects of nutritional restriction.

ACKNOWLEDGEMENTS

The authors wish to thank Professor Shaun Coffey for catalytic discussions around the social and ethical dimensions of animal production. We also acknowledge the work of a large team of researchers and collaborators within the Cattle and Beef Quality Cooperative Research Centre.

REFERENCES

AFRC [Agricultural and Food Research Council, UK]. 1993. *Energy and Protein Requirements of Ruminants.* An advisory manual prepared by the AFRC Technical Committee on Responses to Nutrients. Wallingford, UK: CAB International.

Allingham, P.G., Harper, G.S., Hennessy, D.W. & Oddy, V.H. 2001. The influence of preweaning nutrition on biochemical and myofibre characteristics of bovine *semitendinosus* muscle. *Australian Journal of Agricultural Research,* **52:** 891–902.

Amthor, H., Christ, B. & Patel, K. 1999. A molecular mechanism enabling continuous embryonic muscle growth – a balance between proliferation and differentiation. *Development,* **126:** 1041–1053.

Anakwe, K., Robson, L., Hadley, J., Buxton, P., Church, V., Allen, S., Hartmann, C., Harfe, B., Nohno, T., Brown, A.M., Evans, D.J. & Francis-West, P. 2003. Wnt signalling regulates myogenic differentiation in the developing avian wing. *Development,* **130:** 503–514.

ARC [Agricultural Research Council, UK]. 1980. *The Nutrient Requirements of Ruminant Livestock.* Slough, UK: Commonwealth Agricultural Bureaux.

Bell, A.W. 1992. Foetal growth and its influence on postnatal growth and development. pp. 111–127, *in:* P.J. Buttery, K.N. Boorman and D.B. Lindsay (ed). *The Control of Fat and Lean Deposition.* Oxford, UK: Heinemann.

Berg, R.T. & Butterfield, R.M. 1976. *New Concepts of Cattle Growth.* Sydney, Australia: Sydney University Press.

Bornemann, A., Maier, F. & Kuschel, R. 1999. Satellite cells as players and targets in normal and diseased muscle. *Neuropediatrics,* **30:** 167–175.

Brandstetter, A.M., Picard, B. & Geay, Y. 1998. Muscle fibre characteristics in four muscles of growing bulls. II. Effect of castration and feeding level. *Livestock Production Science,* **53:** 25–36.

Buckingham, M., Bajard, L., Chang, T., Daubas, P., Hadchouel, J., Meilhac, S., Montarras, D, Rocancourt, D. & Relaix, F. 2003. The formation of skeletal muscle: from somite to limb. *Journal of Anatomy,* **202:** 59–68.

Cronje, P.B. 1990. Supplementary feeding in ruminants – A physiological approach. *Suid Afrikaanse Tydskrif vir Veekunde,* **20:** 110–117.

Day, C.S., Moreland, M.S., Floyd, S.S. & Huard, J. 1997. Limb lengthening promotes muscle growth. *Journal of Orthopedic Research,* **15:** 227–234.

Delgado, C., Rosegrant, M., Steinfeld, H., Ehui, S. & Courbois, C. 1999. *Livestock to 2020: The next food revolution.* International Food Policy Research Institute, Washington DC, USA.

Everitt, G.C. 1968. Prenatal development of uniparous animals, with particular reference to the influence of maternal nutrition in sheep. pp. 131–157, *in:* G.A. Lodge and G.E. Lamming (eds). *Growth and Development of Mammals.* London: Butterworths.

Georges, M., Baraldi, F., Buys, N., Charlier, C., Colette, C., Davis, E., Di Silvestro, F., Moreau, L., Nezer, C., Nguyen, M., Segers, K., Shay, T., Smit, M. & Cockett, N. 2002. On the contribution of imprinted loci to variation in animal body composition. *7th World Congress on Genetics Applied to Livestock Production.* Montpellier, France, 19–23 August 2002.

Girard, J., Chatelain, F., Boillot, J., Prip-Buus, C., Thumelin, S., Pegorier, J.-P., Foufelle, F. & Ferre, P. 1997. Nutrient regulation of gene expression. *Journal of Animal Science,* **75:** 46–57.

Glass, D.J. 2003. Molecular mechanisms modulating muscle mass. *Trends in Molecular Medicine,* **9:** 344–350.

Goldring, K., Partridge, T. & Watt, D. 2002. Muscle stem cells. *Journal of Pathology,* **197:** 457–467.

Golovan, S.P., Meidinger, R.G., Ajakaiye, A., Cottrill, M., Wiederkehr, M.Z., Barney, D.J., Plante, C., Pollard, J.W., Fan, M.Z., Hayes, M.A., Laursen, J., Hjorth, J.P., Hacker, R.R., Phillips, J.P. & Forsberg, C.W. 2001. Pigs expressing salivary phytase produce low-phosphorus manure. *Nature Biotechnology,* **19:** 741–745.

Greenwood, P.L. & Bell, A.W. 2003. Prenatal nutritional influences on growth and development of ruminants. *Recent Advances in Animal Nutrition,* **14:** 57–73.

Greenwood, P.L., Hunt, A.S., Hermanson, J.W. & Bell, A.W. 1998. Effects of birth weight and postnatal nutrition on neonatal sheep: I. Body growth and composition, and some aspects of energetic efficiency. *Journal of Animal Science,* **76:** 2354–2367.

Greenwood, P.L., Slepetis, R.M., Hermanson, J.W. & Bell, A.W. 1999. Intrauterine growth retardation is associated with reduced cell cycle activity, but not myofibre number, in ovine fetal muscle. *Reproduction Fertility and Development,* **11:** 281–291.

Greenwood, P.L., Hunt, A.S., Hermanson, J.W. & Bell, A.W. 2000. Effects of birth weight and postnatal nutrition on neonatal sheep: II. Skeletal muscle growth and development. *Journal of Animal Science,* **78:** 50–61.

Greenwood, P.L., Hunt, A.S., Slepetis, R.M., Finnerty, K.D., Alston, C., Beermann, D.H. & Bell, A.W. 2002. Effects of birth weight and postnatal nutrition on neonatal sheep: III. Regulation of energy metabolism. *Journal of Animal Science,* **80:** 2850–2861.

Grounds, M.D. 2002. Reasons for the degeneration of ageing skeletal muscle: a central role for IGF-1 signalling. *Biogerontology,* **3:** 19–24.

Guppy, M. & Withers, P. 1999. Metabolic depression in animals: physiological perspectives and biochemical generalizations. *Biological Reviews of the Cambridge Philosophical Society,* **74:** 1–40.

Gustafsson, M.K., Pan, H., Pinney, D.F., Liu, Y., Lewandowski, A., Epstein, D.J. & Emerson, C.P. 2002. Myf5 is a direct target of long-range Shh signaling and Gli regulation for muscle specification. *Genes Development,* **16:** 114–126.

Hales, C.N., Desai, M., Ozanne, S.E. & Crowther, N.J. 1996. Fishing in the stream of diabetes: control of fetal organogenesis. *Biochemical Society Transactions,* **24:** 341–350.

Harper, G.S., McKay, M., Johnson, S., Allingham, P.G. & Hunter, R.A. 2002. Perimysium is relatively spared during nutritional atrophy. *Proceedings of the Matrix Biology Society of Australia and New Zealand,* September: 48.

Harper, G.S. 1999. Trends in skeletal muscle biology and the understanding of toughness in beef. *Australian Journal of Agricultural Research,* **50:** 1105–1129.

Harper, G.S., Allingham, P.G. & LeFeuvre, R.P. 1999. Changes in connective tissue of *M. semitendinosus* as a response to different growth paths in steers. *Meat Science,* **53:** 107–114.

Hewitt, J.K., Stunkard, A.J., Carroll, D., Sims, J. & Turner, J.R. 1991. A twin study approach towards understanding genetic contributions to body size and metabolic rate. *Acta Genet. Med. Gemellol. (Rome),* **40:** 133–146.

Hodges, J. 2000. Why livestock, ethics and quality of life? pp. 1–26, *in:* J. Hodges and I.K. Han (eds). *Livestock, Ethics and Quality of Life.* New York, NY: CABI Publishing.

Hunter, R.A. & Buck, N.G. 1992. Nutritive and climatic limits in beef production in semi-arid and tropical zones. pp. 379–392, *in:* R.J. Jarringe and C. Beranger (eds). *World Animal Science, Beef Cattle Production.* Amsterdam, The Netherlands: Elsevier.

Jeanplong, F., Bass, J.J., Smith, H.K., Kirk, S.P., Kambadur, R., Sharma, M. & Oldham, J.M. 2003. Prolonged underfeeding of sheep increases myostatin and myogenic regulatory factor Myf-5 in skeletal muscle while IGF-1 and myogenin are repressed. *Journal of Endocrinology,* **176:** 425–437.

Johnston, D. 2001. Within-breed variation in feed efficiency and associations with other traits. *In:* J.A. Archer, R.M. Herd and P.F. Arthur (eds). *Feed Efficiency in Beef Cattle: Proceedings of the Feed Efficiency Workshop.* University of New England, Armidale, Australia, May 2003. Armidale, Australia: Cooperative Research Centre for Cattle and Beef Quality.

Kayo, T., Allison, D.B., Weindruch, R. & Prolla, T.A. 2001. Influences of aging and caloric restriction on the transcriptional profile of skeletal muscle from rhesus monkeys. *Proceeding of the National Academy of Sciences, USA,* **98:** 5093–5098.

Kinoti, G. 2000. Livestock, ethics, quality of life and development in Africa. pp. 221–242, *in:* J. Hodges and I.K. Han (eds). *Livestock, Ethics and Quality of Life.* New York, NY: CABI Publishing.

Kitzmann, M. & Fernandez, A. 2001. Crosstalk between cell cycle regulators and the myogenic factor MyoD in skeletal myoblasts. *Cellular and Molecular Life Sciences,* **58:** 571–579.

Laurent, G.J. 1987. Dynamic state of collagens: pathways of collagen degradation *in vivo* and their possible role in regulation of collagen mass. *American Journal of Physiology,* **252:** C1–C9.

NRC [National Research Council, USA]. 1996. *Nutrient Requirements of Beef Cattle.* 7th revised edition. Subcommittee on Beef Cattle Nutrition, Committee on Animal Nutrition, Board on Agriculture, National Research Council. Washington, DC: National Academy Press.

Oddy, V.H. & Sainz, R.D. 2002. Nutrition for sheep-meat production. pp. 237–262, *in:* M. Freer and H. Dove (eds). *Sheep Nutrition.* New York, NY: CABI Publishing.

Ozanne, S.E. & Hales, C.N. 1999. The long-term consequences of intra-uterine protein malnutrition for glucose metabolism. *Proceedings if the Nutrition Society,* **58:** 615–619.

Parker, M.H., Seale, P. & Rudnecki, M.A. 2003. Looking back to the embryo: Defining transcriptional networks in adult myogenesis. *Nature Reviews Genetics,* **4:** 497–507.

Pitchford, W.S. 2004. Genetic improvement of feed efficiency of beef cattle: what lessons can be learnt from other species? *Australian Journal of Experimental Agriculture,* **44**: 371–382.

Reverter, A., Byrne, K.A., Bruce, H.L., Wang, Y.H., Dalrymple, B.P. & Lehnert, S.A. 2003. A mixed model-based cluster analysis of DNA microarray gene expression data on Brahman and Brahman-composite steers fed high-, medium- and low-quality diets. *Journal of Animal Science,* **81:** 1900–1910.

Reverter, A., Wang, Y., Byrne, K.A., Tan, S.-H., Harper, G.S., Bruce, H.L. & Lehnert, S.A. In press. Joint analysis of multiple microarray studies via mixed-effects model with null residual covariance structure. *Journal of Animal Science,* in press.

Rhoads, R.P., Greenwood, P.L., Bell, A.W. & Boisclair, Y.R. 2000a. Organization and regulation of the gene encoding the sheep acid-labile subunit of the 150 kDa-binding protein complex. *Endocrinology,* **141:** 1425–1433.

Rhoads, R.P., Greenwood, P.L., Bell, A.W. & Boisclair, Y.R. 2000b. Nutritional regulation of the genes encoding the acid-labile subunit and other components of the circulating insulin-like growth factor system in the sheep, *Journal of Animal Science,* **78:** 2681–2689.

Rome, S., Clement, K., Rabasa Lhoret, R., Loizon, E., Poitou, C., Barsh, G.S., Riou, J.P., Laville, M. & Vidal, H. 2003. Microarray profiling of human skeletal muscle reveals that insulin regulates approximately 800 genes during a hyperinsulinemic clamp. *Journal of Biological Chemistry,* **278:** 18063–18081.

Ruane, J. & Zimmermann, M. 2001. Agricultural Biotechnology for Developing Countries – Results of an Electronic Forum. *FAO Research and Technology Paper,* No. 8.

Rupp, R.A.W., Singhal, N & Veenstra, G.J.C. 2002. When the embryonic genome flexes its muscles – Chromatin and myogenic transcription regulation. *European Journal of Biochemistry,* **269:** 2294–2299.

Ryan, W.J. 1990. Compensatory growth in cattle and sheep. *Nutrition Abstracts and Reviews. Series B, Livestock Feeds and Feeding,* **60:** 653–664.

Sang, H. 2003. Genetically modified livestock and poultry and their potential effects on human health and nutrition. *Trends in Food Science and Technology,* **14:** 253–263.

Schinckel, P.G. & Short, B.F. 1961. The influence of nutritional level during pre-natal and early post-natal life on adult fleece and body characteristics. *Australian Journal of Agricultural Research,* **12:** 176–202.

Seale, P. & Rudnicki, M.A. 2000. A new look at the origin, function, and "stem-cell" status of muscle satellite cells. *Development Biology,* **218:** 115–124.

Solomon, M.B., Pursel, V.G., Campbell, R.G. & Steele, N.C. 1997. Biotechnology for porcine products and its effect on meat products. *Food Chemistry,* **59:** 499–504.

Spanheimer, R., Zlatev, T., Umpierrez, G. & DiGirolamo, M. 1991. Collagen production in fasted and food-restricted rats: Response to duration and severity of food deprivation. *Journal of Nutrition,* **121:** 518–524.

Sreekumar, R., Halvatsiotis, P., Schimke, J.C. & Nair, K.S. 2002. Gene expression profile in skeletal muscle of type 2 diabetes and the effect of insulin treatment. *Diabetes,* **51:** 1913–1920.

Streuli, C. 1999. Extracellular matrix remodelling and cellular differentiation. *Current Opinion in Cell Biology,* **11:** 634–640.

Tawa, N.E., Odessey, R. & Goldberg, A.L. 1997. Inhibitors of the proteasome reduce the accelerated proteolysis in atrophying rat skeletal muscles. *Journal of Clinical Investigation,* **100:** 197-203.

Weindruch, R., Kayo, T., Lee, C.K. & Prolla, T.A. 2001. Microarray profiling of gene expression in aging and its alteration by caloric restriction in mice. *Journal of Nutrition,* **131:** 918S–923S.

Westerhoff, H.V. 2001. The silicon cell, not dead but live! *Metabolic Engineering,* **3:** 207–210.

Widom, R.L., Lee, J.Y., Joseph, C., Gordon Froome, I. & Korn, J.H. 2001. The hcKrox gene family regulates multiple extracellular matrix genes. *Matrix Biology,* **20:** 451–462.

Wigmore, P.M. & Evans, D.J.R. 2002. Molecular and cellular mechanisms involved in the generation of fiber diversity during myogenesis. pp. 175–232, *in:* K.W. Jeon (ed). *International Review of Cytology – A Survey of Cell Biology,* Vol. 216.

Yambayamba, E.S.K. & Price, M.A. 1991. Fibre-type proportions and diameters in the *longissimus* muscle of beef heifers undergoing catch-up (compensatory) growth. *Canadian Journal of Animal Science,* **71:** 1031–1035.

Yuen, B.S.J., Owens, P.C., Muhlhausler, B.S., Roberts, C.T., Symonds, M.E., Keisler, D.H., McFarlane, J.R., Kauter, K.G., Evens, Y. & McMillen, I.C. 2003. Leptin alters the structural and functional characteristics of adipose tissue before birth. *FASEB Journal,* **17:** 1102–1104.

Zhou, Z., Yon Toh, S., Chen, Z., Guo, K., Ng, C.-P., Ponniah, S., Lin, S.-C., Hong, W. & Li, P. 2003. Cidea-deficient mice have lean phenotype and are resistant to obesity. *Nature Genetics,* **35:** 49–56.

Zorzano, A., Kaliman, P., Guma, A. & Palacin, M. 2003. Intracellular signals involved in the effects of insulin-like growth factors and neuregulins on myofibre formation. *Cellular Signalling,* **15:** 141–149.

GENETIC OPPORTUNITIES TO ENHANCE SUSTAINABILITY OF PORK PRODUCTION IN DEVELOPING COUNTRIES: A MODEL FOR FOOD ANIMALS

Cecil W. Forsberg,[1] Serguei P. Golovan,[2] Ayodele Ajakaiye,[2] John P. Phillips,[3] Roy G. Meidinger,[3] Ming Z. Fan,[2] John M. Kelly[4] and Roger R. Hacker.[2]

1. Department of Microbiology.
2. Department of Molecular Biology and Genetics.
3. Department of Animal and Poultry Science.
4. MaRS Landing, Medical and Related Sciences.
All are part of the University of Guelph, Guelph, Ontario, Canada.

Abstract: Currently there is a shortage of food and potable water in many developing countries. Superimposed upon this critical situation, because of the increasing urban wealth in these countries, there is a strong trend of increased consumption of meat, and pork in particular. The consequence of this trend will be increased agricultural pollution, resulting not only from greater use of chemical fertilizer, but also from manure spread on land as fertilizer that may enter freshwater and marine ecosystems causing extensive eutrophication and decreased water quality. The application of transgenic technologies to improve the digestive efficiency and survival of food animals, and simultaneously decreasing their environmental impact is seen as an opportunity to enhance sustainability of animal agriculture without continued capital inputs. Transgenes expressed in pigs that have potential include, for example, genes coding for phytase, lactalbumin and lactoferrin. At the University of Guelph, *Escherichia coli* phytase has been expressed in the salivary glands of the pig. Selected lines of these pigs utilize plant phytate phosphorus efficiently as a source of phosphorus and excrete faecal material with more than a 60 percent reduction in phosphorus content. Because of their capacity to utilize plant phytate phosphorus and to produce less polluting manure they have a valuable trait that will contribute to enhanced sustainability of pork production in developing countries, where there is less access to either high quality phosphate supplement or phytase enzyme to include in the diet. Issues that require continued consideration as a prelude to the introduction of transgenic

429

H. P. S. Makkar and G. J. Viljoen (eds.), Applications of Gene-Based Technologies for Improving Animal Production and Health in Developing Countries, 429-446.
© 2005 *IAEA. Printed in the Netherlands.*

animals into developing countries include food and environmental safety, and consumer acceptance of meat products from genetically modified animals.

1.   INTRODUCTION

Eradication and prevention of evolving exotic diseases, such as HIV/AIDS, malaria and hepatitis, are a major focus of a global investment in developing countries (Kim, 2002). A key factor contributing to the spread of these diseases is widespread debilitating malnutrition (Pimentel and Morse, 2003), which not only predisposes people to infection, but also affects the social and political development of countries. With the world population projected to increase by a factor of 1.5 to over 9 billion by 2050 (Population Reference Bureau, 2003), the problem of malnutrition will only increase.

As a result, developing countries need to significantly increase their food production to maintain the rapidly growing population, yet at the same time they need to do it without degrading their environment, and especially the quality of their water resources. By 2025, almost half of the world population, many of them in developing countries, will live in water-stressed regions. Water will become scarce not only because of the increased demand, but also because of increased pollution (Johnson, Revenga and Echeverria, 2001; Somerville and Briscoe, 2001).

Animal waste is a leading source of phosphorus pollution from agriculture (Jongbloed and Lenis, 1998) and its effect exceeds that of inorganic fertilizers or other anthropogenic fluxes (Smil, 2000). In the United States of America alone, 100 million tonnes of animal manure is produced annually, with the liberation of 1 million tonnes of phosphorus into the environment each year (Walsh, Power and Headon, 1993). Phosphorus pollution is one of the greatest problems in both freshwater and marine environments. Freshwater eutrophication degrades the quality of drinking water, creating an offensive taste and odour (Smil, 2000). Increased nutrient inputs in coastal water are associated with serious environmental impacts, which are becoming a major threat to the coastal environments on which a large number of people in developing countries depend for their survival (Jickells, 1998; Harvell, 1999; Jackson *et al.*, 2001).

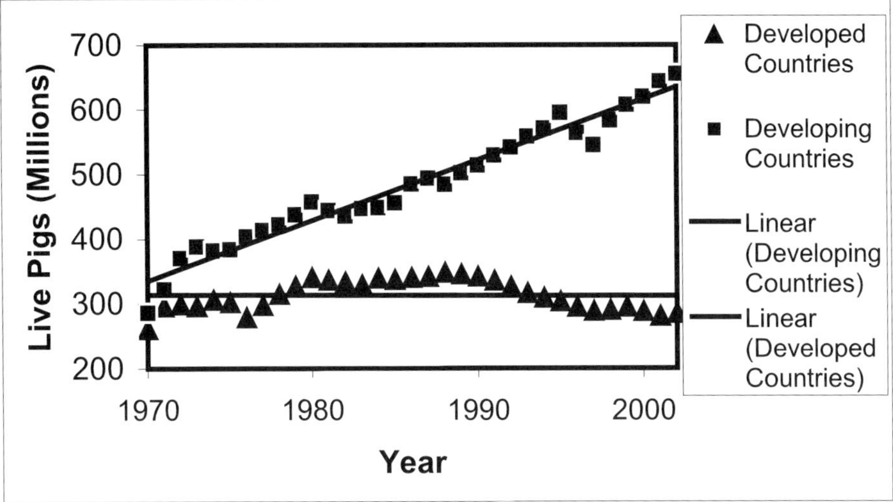

Figure 1. Trends in pig production in developed and developing countries (SOURCE: FAO Statistics, 2002).

The effect of food shortages is compounded by the increasing shortages of potable water due to pollution (Alternative Energy Institute, 2003), and increased shortages in the future will be exacerbated by more intensive agricultural activities (Tilman *et al.*, 2001). A large part of this pollution is expected to be caused by increased production of pigs and poultry, primarily in developing countries (Delgado *et al.*, 1999), but contributions will come from other food animals as well. For example, pig production in developing countries has increased at a linear rate of 10 percent per year since the early 1970s, while pig production in developed countries has remained comparatively constant (Figure 1). Since an increase in the number of monogastric food animals will occur, a major effort should be made to increase their capability to more efficiently utilize cereal grains, and at the same time decrease their deleterious impact on the environment. As with other man-made burdens, the best way to reduce the phosphorus impact from animal agriculture is to minimize the initial inputs.

1.1   Production of food animals

The production of food animals will continue to be a key element in the agricultural economy of developing countries, and, depending upon the particular location, the challenges will include one or all of the following: (i) production of sufficient feed to maintain animals, (ii) prevention of animal diseases, and (iii) development of systems to reduce pollution from animal waste. Meeting these challenges requires a multifaceted approach.

The University of Guelph approach is to develop food animals with novel traits to improve fitness and growth in these challenging environments.

The classical approach to genetic improvement has been through quantitative selection and artificial breeding processes, which have led to a steady genetic improvement of 0.5 to 3 percent per year in North America and Europe (Smith, 1998). A recent innovation in the traditional selection process is genetic marker-assisted selection (Davis and DeNise, 1998). Unfortunately, these approaches often do not allow the separation of desirable from undesirable traits, and, furthermore, they do not enable the transfer of beneficial traits from one species to another. These shortcomings can be circumvented by transgenic techniques.

2.   TRANSGENIC METHODOLOGIES

A variety of methods are now available for the generation of transgenic food animals. These include: (i) classical pronuclear micro-injection (Wall, 1985), (ii) sperm-mediated gene transfer (Chang *et al.*, 2002; Lavitrano *et al.*, 2002), (iii) retroviral vectors (Lois, 2002), and (iv) nuclear transfer methods (Park *et al.*, 2001; Nagashima *et al.*, 2003).

A wide variety of enabling improvements in domestic animals via transgenesis have been proposed, ranging over disease resistance, enhanced digestion, improved milk composition, increased wool growth, enhanced carcass composition and improved reproductive characteristics (Houdebine, 2002). However, until now, only a few transgenic food animals with improved production traits have been developed, and these are listed in Table 1. The slow progress may be attributed to our limited knowledge of suitable promoters, lack of understanding of regulation at the levels of transcription and translation, not to mention chromosome organization. However, many of these issues should be resolved when the genomes of food animals have been sequenced.

Table 1. Transgenes currently under investigation in pigs.

Gene	Effect	Reference
Phytase	Reduced P in manure	Golovan, 2001
IGF-1	Muscle growth	Pursel *et al.*, 1999
α-Lactalbumin	Enhanced piglet growth	Wheeler, Bleck and Donovan, 2002
Lactoferrin	Enhanced piglet health	Lee *et al.*, 2003
Plant oleate desaturase	Improved nutritional value	Irani, Univ. of Kinki [1]

NOTE: (1) Reported in *New Scientist Online News,* 12:30 25 January 2002 (http://www.newscientist.com/hottopics/gm/gm.jsp?id=ns99991841).

Genetically modified animals are probably the most promising tools in creating environment-friendly agriculture in developing countries. While developed countries, with their acquired infrastructure and a large pool of expertise, may choose from a variety of approaches to decrease pollution, many of the same approaches will not be available to developing countries due to the enormous investment required to establish the necessary infrastructure and to gain experience. Genetically modified animals, with the encoded ability to resist diseases and better utilize nutrients from local sources, would not depend on commercial supplements and antibiotics or require an extensive infrastructure (Herrera-Estrella, 1999; Trewavas, 1999; Avery, 2001). Also, genetically modified animals can easily be incorporated into local farm practices; anybody who raises pigs today can raise an Enviropig^{TM} tomorrow with no special training.

3.　　APPLICATIONS OF THE ENVIROPIG^{TM}

Cereal grains, sources of protein supplements of plant origin, and by-products derived from the milling and brewing industries provide the major components of the diets of pigs throughout the world. The energy and protein in these feedstuffs are readily available. However, the third most important nutrient in the diet, phosphorus, has low bioavailability (Table 2) because the bulk of the phosphorus exists in the form of phytic acid (myo-inositol 1,2,3,4,5,6-hexakisdihydrogen phosphate). This compound is not digested by monogastric animals, including the pig, but instead is concentrated 3- to 4-fold in the excreted faecal material. As a consequence, unless diets are supplemented with readily available forms of phosphorus, such as mono or dicalcium phosphate, or animal by-products, such as bone meal, the low level of available phosphorus in the diet restricts the growth of pigs, leading to lameness and stiffness, weakened bone structure and breeding difficulties. Extreme forms of phosphorus deficiency result in rickets in young pigs and osteomalacia in older animals (Ensminger and Parker, 1997). A secondary nutritional effect caused by phytic acid stems from the fact that, in the small intestine at neutral pH, phytic acid complexes minerals, reducing their bioavailability, and instead of being absorbed the minerals are lost along with the phytic acid in the faeces (Kornegay, 2001). Supplementation of cereal-based diets of the pig with the feed additive enzyme phytase is an alternative method by which the dietary requirement for phosphorus can be partially satisfied and mineral absorption improved. For example, inclusion of phytase in the diet at the level of 250 to 1000 units of enzyme per kilogram of feed enhances phosphorus utilization and decreases faecal phosphorus by 25 to 50 percent (Ketaren *et al.,* 1993;

Simons *et al.*, 1990) at a cost of US$ 0.80 to 1.00 in raising a pig to market weight. With higher levels of phytase added, the extent of hydrolysis is improved, but with diminishing returns (Kornegay, 2001). In Europe there is widespread adoption of the feeding of phytase to pigs because of the banning of the feeding of meat and bone meal (Rodehutscord *et al.*, 2002). This legislation was introduced to decrease the potential for spread of serious diseases, such as foot-and-mouth disease (FMD), but there was already a trend to feeding phytase in order to reduce phosphorus excretion in food animals, in response to problems from the extent of phosphorus pollution from manure, which has become quite serious. The outcome of phosphorus pollution of water resources is eutrophication (nutrient enrichment with extensive algal growth, anoxia and toxin production), with the death of fish and other aquatic organisms, and a serious reduction in water quality (Jongbloed and Lenis, 1998; Diaz, 2001).

Table 2. Bioavailability of phosphorus in feed ingredients for pigs.

Feedstuff	Bioavailability of P[1] (%)	Feedstuff	Bioavailability of P[1] (%)
Cereal grains		*Plant protein supplements*	
Maize	14	Canola meal	21
Oats	22	Soybean meal, dehulled	23
Barley	30	Soybean meal, 44% protein	31
Triticale	46	Cotton seed meal[2]	27
Wheat	50	Peanut meal	12
Grain by-products			
Oat groats	13	*Animal protein supplements*	
Maize gluten meal	15	Feather meal	31
Rice bran	25	Meat and bone meal	90
Wheat bran	29	Blood meal	92
Brewers grains	34	Fish meal	94
Wheat middlings	41		
Maize gluten meal	59	*Phosphate supplements*	
Distillers' grains	77	Steamed bone meal	85
Sorghum grain	20	Defluorinated phosphate	90
Millet[2]	37	Dicalcium phosphate	100

NOTES: (1) Bioavailability relative to the availability of P in monosodium/monocalcium phosphate, which equals 100 (NRC, 1998). (2) Based on difference between total and phytate phosphorus (Liao, Sauer and Kies, 2002).

In Canada and the United States of America, there has been an increasing awareness of phosphorus pollution of freshwater systems caused by manure from pigs fed supplemental phosphate (Tilman *et al.*, 2001). This has led to increased replacement of phosphorus supplements by phytase in the diet. Recently, several large pig producing organizations or cooperatives have voluntarily banned the feeding of meat and bone meal to pigs, including Smithfield Foods Inc. in the United States of America, and Maple Leaf Foods Inc. and a provincial pork producers' marketing board, Ontario Pork, in Canada (Greig, 2003). The primary driving force for this change in feeding practice was the demand for safer pork in Japan, a country that accounts for one-third of the world's current imports of pork. Undoubtedly, this practice has led to a further adoption of phytase as a dietary supplement for pig diets.

3.1   Characteristics of the EnviropigTM

The Enviropig^{TM} is the trade name for transgenic pigs that secrete phytase in their saliva. The saliva mixes with feed consumed and the phytase becomes active and digests phytic acid while present in the stomach. Numerous lines of the phytase pigs were developed by classical pronuclear micro-injection of fertilized porcine embryos with a transgene composed of the mouse parotid secretory protein (PSP) promoter linked to the *Escherichia coli* phytase gene (Golovan *et al.*, 2001). We have documented that lines of these pigs efficiently digest phytate present in cereal grains and oil meals, releasing phosphate and thereby eliminating the need to include either supplemental phosphate or added phytase in the diet. This trait is inherited by simple Mendelian genetics, and hemizygous transgenic pigs fed a conventional diet lacking supplemental phosphate utilize the phytate phosphorus and produce faecal material with at least 60 percent less total phosphorus than is present in the faecal material of non-transgenic pigs fed the same diet. As a component of these studies, we have determined the concentration of phosphorus in faecal samples from pigs producing different activities of phytase in the saliva. The data shows that a minimum level of phytase in the saliva for efficient plant phosphorus utilization is approximately 25 units (μmol of phosphate released per minute) per millilitre of saliva (Figure 2). We have also developed homozygous lines of the Enviropig^{TM} that could serve as the terminal sire line for conventional breeding stock. As expected, crossing these homozygous boars with non-transgenic gilts produces offspring that are heterozygous for the phytase trait, with all piglets containing the phytase transgene and all secreting phytase in the saliva. Enviropig^{TM} lines tested exhibit other traits identical to non-transgenic pigs. The enabling phytase trait of the Enviropig^{TM} could be of considerable benefit in both developed and developing countries. It obviously would be best to introduce the trait into indigenous pigs adapted to the unique environment. This could be accomplished by either natural breeding, or more readily by artificial insemination, or by any of the techniques developed for generating transgenic pigs. However introgression of the gene by breeding would be the simplest technique to introduce the gene, and it would avoid the technical expertise, time and cost associated with introducing the transgene into the different breeds of pigs by transgenic methodologies referred to previously. Furthermore, we anticipate having the safety testing completed to satisfy Canadian and United States of America regulatory authorities during the next several years. These standards of safety are comparable to those proposed by the Food and Agriculture Organization of the United Nations (FAO), and, assuming that safety data are approved, there could be swift dispersion of the genetic resource to other countries.

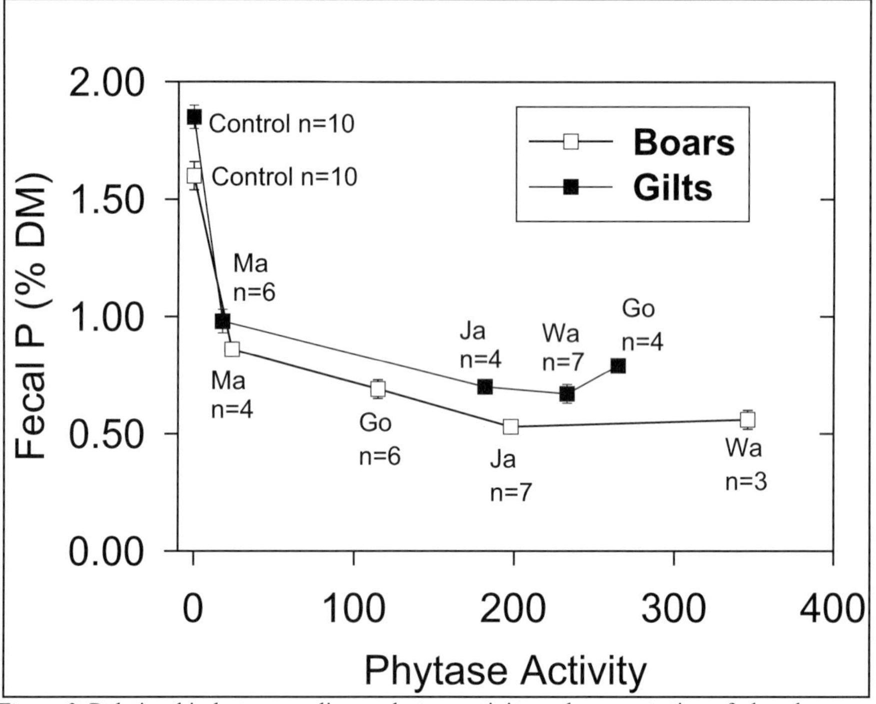

Figure 2. Relationship between salivary phytase activity and concentration of phosphorus on a dry matter (DM) basis in the faecal material collected from different lines of transgenic phytase pigs (n = number of pigs tested; unpublished data).

3.2   Potential benefit of the EnviropigTM trait in pigs raised in selected developing countries

3.2.1   Nigeria and other West African countries

There are approximately 6 million pigs in Western Africa. In Nigeria, about 97 percent of the pigs are free range, with a varied diet, with the remainder raised on large, managed farms (Tewe and Bokanga, 2001). The major sources of energy are maize and cassava (treated to remove most of the cyanogenic glycosides), while the major protein supplements are soybean meal from solvent-extracted soybeans, locally roasted and mechanically extracted soybean cake, and groundnut meal. Wheat and maize offal are also added for additional energy, but they are high in phytate. Dicalcium phosphate, the most common phosphorus supplement in developed countries, must be imported, and when available is three to four times more expensive (on a bioavailable-P basis) than locally produced bone meal. Therefore bone meal, which is primarily from cattle bones, is the major source of phosphorus supplement on managed pig farms. However, there is a systemic problem, with limited availability and quality of bone meal. Often it is not completely incinerated, with chunks of bone marrow remaining, and in other cases it reportedly may be diluted with wood ash, which has the same appearance, and as a consequence contains less phosphate.

Europe has banned the inclusion of meat and bone meal in the diets of animals because it has the potential to containing infectious agents such as the bovine spongiform encephalopathy (BSE) pathogen (Rodehutscord *et al.,* 2002). If there were a problem with BSE or other serious animal disease in West Africa, the reliance on bone meal as the primary source of supplemental phosphate could be catastrophic for the pork industry. If the phytase trait were introduced into West African pigs, the problems associated with the availability and quality of supplemental phosphate in the diet would be circumvented since the plant portion of the diet probably contains sufficient phosphorus to meet the dietary requirement under the managed feeding system. Similarly, free-range phytase pigs would have the capability as they were rooting of utilizing phytate phosphorus from plants, which would substantially improve pig growth, as phosphorus is a key limiting nutrient in the acidic soil in parts of West Africa (Schaffert *et al.,* 2003).

### 3.2.2	Brazil and other South American countries

There are approximately 30 million pigs in Brazil and more than 15 million produced in other South American countries. In Brazil, pigs are generally raised on large farms and are fed 60 percent maize and 40 percent soybean meal as the bulk of the diet, with either supplemental dicalcium phosphate or phytase providing the source of phosphorus. Brazil exports approximately 80 percent of its production to Europe, Russia and Japan, which means that plant protein supplements are included in the diet rather than meat or bone meal because of European import restrictions, and, more recently, Japanese restrictions. The dicalcium phosphate used is imported at a cost of approximately US$ 340 per tonne, and because of this high cost the inclusion of phytase in the diet has become very popular. However, more than 4 million tonnes of non-edible and recyclable by-products are produced during the processing of pork (Bellaver, 2001), and presumably some part of it is used by the pork industry.

In the Brazilian state of Santa Catarina, surveys have shown that swine production and spreading of manure has resulted in 85 percent of water sources being contaminated by faecal organisms manure. This is particularly serious during the rainy season, due to runoff from fields. The Enviropig^{TM} trait would enhance the nutritional efficiency of pigs and reduce the environmental impact of Brazilian pork production, but the option is currently unavailable since Brazilian legislation forbids the use of genetically modified livestock.

### 3.2.3	People's Republic of China

More than half of the world's pigs, some 465 million, are produced in China and they account for 68 percent of meat produced (Skorburg, 2003). Up to four-fifths of Chinese pigs are raised in backyards, with the remainder on farms with over 100 pigs (Skorburg, 2003). Maize and soybean meal are the major sources of energy and protein in the diet. However, depending upon the area, by-products are included as protein supplements, including common groundnut meal, rapeseed meal, cottonseed meal, linseed meal, wheat shorts and rice bran. These are particularly high in phytate. Either dicalcium phosphate or phytase is therefore included in the diet. The dicalcium phosphate is mined within the People's Republic of China, but it is often of low quality, with serious contaminating concentrations of fluoride. Use of meat and bone meal apparently has been banned to avoid any potential for transmission of serious animal diseases such as FMD and BSE, and phytase supplementation is beginning to be widely practised.

The serious nutrient surplus situations and pollution of freshwater systems common in developed countries is now increasingly observed in selected locations in fast growing economies of eastern Asia, including the People's Republic of China (Blackburn, 2003). For example, Lake Taihu in Jiangsu province, the third-largest lake in China, suffers from excessive phosphorus pollution from both raw sewage and non-point source inputs from livestock farms in the lake's catchment (Anon., 2001). Similarly, Taiwan (Province of China) has a serious nutrient excess as a consequence of an intensive pork industry (Chu and Kung, 1995). Based on these observations, the Enviropig^{TM} trait would seem to have the potential to make an important contribution to pork production in Southeast Asia because of both increased efficiency of nutrient utilization and decreased impact on the environment.

4.   TRANSGENIC TRAITS TO ENHANCE HEALTH OF PIGLETS

The first week of a piglet's life is the period of greatest danger, with malnutrition and scours contributing to high morbidity and mortality. The introduction into pigs of transgenes that improve piglet health, with expression of the bovine α-lactalbumin gene (Noble *et al.*, 2002) to increase early lactation and lactose synthesis so as to promote growth of the litter and overall growth of pigs (Mahan and Lepine, 1991), and porcine lactoferrin expressed in the milk (Lee *et al.,* 2003) to protect the piglets against microbial infections are feasible objectives. Both would be of significant benefit in developing countries. Although data is not available on the benefit of recombinant lactoferrin on piglets, feeding of lactoferrin to non-transgenic piglets has a protective effect when piglets are challenged with a lethal dose of bacterial endotoxin (Lee *et al.*, 1998).

5.   SATISFYING GOVERNMENTAL REGULATORY REQUIREMENTS PRIOR TO THE INTRODUCTION OF THE ENVIROPIG^{TM} INTO PORK PRODUCTION SYSTEMS

Genetically modified food animals are considered to be novel organisms and therefore are subject to animal health, environmental and food safety legislation prior to their introduction into the food production system. Regulatory requirements, however, differ for each country. FAO provides a leadership role in the development of safety testing objectives and protocols. In Canada, the regulatory agencies involved include Health Canada, which is concerned with food safety for humans; Environment Canada, with requirements for documentation of "no deleterious effect" on the environment; and the Canadian Food Inspection Agency (CFIA), which has a role in assessing animal welfare issues, and animal feed and fertilizer safety, since by-products of slaughtered animals enter the animal feed and fertilizer paths. In the United States of America, the Food and Drug Agency (FDA), the Environmental Protection Agency (EPA) and the United States Department of Agriculture (USDA) have regulatory roles.

5.1   Safety testing

At present, the only food animal that has been subjected to safety evaluation in North America is transgenic fish by the FDA (Kleter and Kuiper, 2002). The pre-market food safety issues include safety of the insert DNA, potential toxicity of the novel protein product, allergenicity of the protein and the safety from any unintended effects (Chassy, 2002). Other aspects include the equivalence of composition, retention of nutritional value and human dietary exposure. Web sites where information can be found on issues regarding safety testing of transgenic plants and animals include those of Health Canada, FDA and FAO[1].

The chromosomal site of transgene integration must be characterized in the transgenic animal line presented for safety testing. The most desirable case is a single copy inserted at one site on a chromosome in the genome and it must be documented to have no deleterious effect on animal performance in the homozygous state. There is the question of safety of the transgene when humans consume the genetically modified meat. Studies with human volunteers in the UK have shown that no recombinant DNA survived passage through the entire human digestive tract. Although some DNA survived in laboratory created environments that simulated human or animal gastrointestinal tracts, the research concluded that the likelihood of functional DNA being taken up by bacteria in the human or animal gut is extremely low (UK Food Standards Agency, 2002). This conclusion is supported by a thorough review of the previous literature by Jonas *et al.* (2001).

---

1. For Health Canada, see: www.hc-sc.gc.ca/food-aliment/mh-dm/ofb-bba/nfi-ani/e_novel_foods_and_ingredient.html.
    For FDA, see: www.fda.gov/cvm/biotechnology/bio_drugs.html.
    For FAO, see: www.fao.org/es/ESN/food/risk_biotech_papers_en.stm.

To focus on the novel protein, acute oral toxicity testing methods may be conducted using a fixed dose method, the acute toxic class method, or the up-and-down method, as described by the EPA (Acute oral toxicity (Up-down procedure)[2]). Of these methods, the up-and-down protocol described in OECD guideline 425[3] is highly recommended. This method involves testing the novel protein on individual rodents at an upper selected limit – an approach well suited to novel food proteins that should be non-toxic. The advantages of the up-and-down method have been described by Lipnick *et al.* (1995).

Perhaps the greatest concern of regulatory agencies, and indeed, eventually the consumer, is the potential for the expressed novel protein such as the phytase in the Enviropig^{TM} to exhibit allergenic properties when consumed by humans. Currently, the weight-of-evidence approach is taken to assess allergenicity of a novel protein. Aspects considered include, the source of the transgene, sequence homology with known allergens, and physicochemical properties, including heat stability and digestive stability. If a suspected allergenic peptide is identified, additional criteria include immunoreactivity of the novel protein with serum IgE from individuals with know allergies to species that are broadly related to the source of the transferred DNA, and the immunogenicity of the novel protein in appropriate animal models (Taylor, 2002; CAC, 2003).

The other important aspect of the testing is compositional analyses of the meat samples to ensure that characteristics other than the novel protein are equivalent to that of meat from closely related non-transgenic animals.

In our case, given that the transgenic pig is healthy, a useful perspective is the similarity of the porcine species to humans in terms of their nutritional (Miller and Ullrey, 1987), physiological (Higgins and Cordell, 1995; Tumbleson, 1986) and immunological (Helm *et al.,* 2002) characteristics. Therefore a healthy transgenic pig should serve as a persuasive indicator of safe pork.

6. ETHICAL AND CONSUMER ISSUES OF TRANSGENIC FOOD ANIMALS

Ethical concerns about animal biotechnology, according to Etherton *et al.* (2003), can be separated into four groups: (1) impacts on human health and safety, (2) animal welfare, (3) environmental impacts, and (4) fears that

2. www.epa.gov/docs/OPPTS_Harmonized/870_Health_Effects_Test_Guidelines/Series/870-1100.pdf
3. www.unitar.org/cwm/ghs-cd/Documents/cat5/International/OECD_24_JT00111082.pdf

scientists are "playing God". Although not mentioned, a fear of the unknown may also be an important factor. These issues to a degree overlap with consumer issues, including: (1) food safety, (2) taste and healthfulness, (3) reasonable price, (4) whether the meat is from natural domesticated animals, (5) whether the animals were raised with proper care and treatment, and (6) whether production had a deleterious impact on the environment. The issue that does not overlap is whether one is "playing God". However, when a lobby group describes a technology as unnatural, the group almost certainly has concerns about the risk to humans, animals or environment, singly or in combination, and, as suggested by Etherton *et al.* (2003), it is on these concerns that proponents of biotechnology should focus. One cannot help but ask whether hunger influences arguments over ethics. Finucane (2002) has pointed out that risk is a social construct, meaning different things to different individuals. When one thinks of developing countries that were bypassed by the Green Revolution, one has to conclude that innovations that simultaneously help human kind and the environment have distinct benefits.

7.   CONCLUSIONS

The development of transgenic food animals with traits to enhance their digestive repertoire improves the nutritional value of their feed and decreases their impact on the environment, providing a sustainable option for developing countries to achieve a higher standard of living. In the application of this technology, it is recognized that careful consideration must be given to ensuring food and environmental safety and that its adoption must be conscious of, and compatible with, the aspirations of people in these countries.

ACKNOWLEDGEMENTS

Our research programme has been supported by funding from Ontario Pork, Ontario Ministry of Agriculture and Food, Food Systems Biotechnology Center, University of Guelph, Natural Sciences and Engineering Research Council of Canada, Agriculture and Agri-Food Canada and the MaRS-Landing Program supported by the City of Guelph, and the Ontario Government.

REFERENCES

Alternative Energy Institute. 2003. Effects of the current and future population on food supply. See: http://www.altenergy.org/2/population/effects/effects.html.

Anon. 2001. China's third-largest freshwater lake suffers phosphorus pollution. *People's Daily*, 20 April 2001.
http://fpeng.peopledaily.com.cn/200104/20/eng20010420_68189.html

Avery, D.T. 2001. Genetically modified organisms can help save the planet. pp. 205–215. *in*: G.C. Nelson (ed). *Genetically Modified Organisms in Agriculture: Economics and Politics*. San Diego, CA: Academic Press.

Bellaver, C. 2001. Food safety and quality control in the use of ingredients for swine feeding. Second International Virtual Conference on Pork Quality, 5 November– 6 December 2001. See:
http://www.conferencia.uncnet.br/pork/seg/pal/anais01p2_bellaver_en.pdf

Blackburn, H. 2003. Livestock production, the environment and mixed farming systems. See: http://www.fao.org/wairdocs/lead/x6141e/x6141e00.htm.

CAC [Codex Alimentarius Commission]. 2003. Report of the 3rd Session of the Codex Ad hoc Intergovernmental Task Force on Foods Derived from Biotechnology, Yokohama, Japan, 4–8 March 2002. 79p. Paper presented at Jt FAO/WHO Food Standard Programme, Codex Alimentarius Commission, 25th Session, Rome, Italy, 30 June–5 July 2003. Report available at ftp://ftp.fao.org/codex/Alinorm03/al03_34e.pdf.

Chang, K., Qian, J., Jiang, M., Liu, Y.H., Wu, M.C., Chen, C.D., Lai, C.K., Lo, H.L., Hsiao, C.T., Brown, L., Bolen, J. Jr, Huang, H.I., Ho, P.Y., Shih, P.Y., Yao, C.W., Lin, W.J., Chen, C.H., Wu, F.Y., Lin, Y.J., Xu, J. & Wang, K. 2002. Effective generation of transgenic pigs and mice by linker-based sperm-mediated gene transfer. *BMC Biotechnology*, **2**: 5.

Chassy, B.M. 2002. Food safety evaluation of crops produced through biotechnology. *Journal of the American College of Nutrition*, **21**: 166S–173S.

Chu, R. & Kung, C.-M. 1995. Waste management in pig production in Taiwan. *Pig News and Information*, **16**: 31N–33N.

Davis, G.P. & DeNise, S.K. 1998. The impact of genetic markers on selection. *Journal of Animal Science*, **76**: 2331–2339.

Delgado, C.L., Rosegrant, M.W., Steinfeld, H., Ehui, S. & Courbois, C. 1999. The coming livestock revolution. *Choices*, **40**: 40–44.

Diaz, R.J. 2001. Overview of hypoxia around the world. *Journal of Environmental Quality*, **30**: 275–281.

Ensminger, M.E. & Parker, R.O. 1997. *Swine Science*. Danville, Illinois: Interstate Printers & Publishers.

Etherton, T.D., Bauman, D.E., Beattie, C.W., Bremel, R.D., Cromwell, G.L., Kapur, V., Varner, G., Wheeler, M.B. & Wiedmann, M. 2003. Animal agriculture's future through biotechnology. Part 1 – Biotechnology in animal agriculture: an overview. *Council for Agricultural Science and Technology (CAST)*, **23**: 1–12.

Finucane, M.L. 2002. Mad cows, mad corn and mad communities: the role of socio-cultural factors in the perceived risk of genetically-modified food. *Proceedings of the Nutrition Society*, **61**: 31–37.

Golovan, S.P., Hayes, M.A., Phillips, J.P. & Forsberg, C.W. 2001. Transgenic mice expressing bacterial phytase as a model for phosphorus pollution control. *Nature Biotechnology*, **19**: 429–433.

Golovan, S.P., Meidinger, R.G., Ajakaiye, A., Cottrill, M., Wiederkehr, M.Z., Barney, D., Plante, C., Pollard, J., Fan, M.Z., Hayes, M.A., Laursen, J., Hjorth, J.P., Hacker, R.R., Phillips, J.P. & Forsberg, C.W. 2001. Pigs expressing salivary phytase produce low phosphorus manure. *Nature Biotechnology,* **19**: 741–745.

Greig, J. 2003. The meat and bone meal decision. *Ontario Hog Farmer,* **15**: 6–8.

Harvell, C.D., Kim, K., Burkholder, J.M., Colwell, R.R., Epstein, P.R., Grimes, D.J., Hofmann, E.E., Lipp, E.K., Osterhaus, A.D., Overstreet, R.M., Porter, J.W., Smith, G.W. & Vasta, G.R. 1999. Emerging marine diseases – climate links and anthropogenic factors. *Science,* **285**: 1505–1510.

Helm, R.M., Furuta, G.T., Stanley, J.S., Ye, J., Cockrell, G., Connaughton, C., Simpson, P., Bannon, G.A. & Burks, A.W. 2002. A neonatal swine model for peanut allergy. *Journal of Allergy and Clinical Immunology,* **109**: 136–142.

Herrera-Estrella, L. 1999. Transgenic plants for tropical regions: some considerations about their development and their transfer to the small farmer. *Proceedings of the National Academy of Sciences, USA,* **96**: 5978–5981.

Higgins, L.S. & Cordell, B. 1995. Genetically engineered animal models of human neurodegenerative diseases. *Neurodegeneration,* **4**: 117–129.

Houdebine, L.M. 2002. Transgenesis to improve animal production. *Livestock Production Science,* **74**: 255–268.

Jackson, J.B., Kirby, M.X., Berger, W.H., Bjorndal, K.A., Botsford, L.W., Bourque, B.J., Bradbury, R.H., Cooke, R., Erlandson, J., Estes, J.A., Hughes, T.P., Kidwell, S., Lange, C.B., Lenihan, H.S., Pandolfi, J.M., Peterson, C.H., Steneck, R.S., Tegner, M.J. & Davis, R.P. 2001. Historical overfishing and the recent collapse of coastal ecosystems. *Science,* **293**: 629–637.

Jickells, T.D. 1998. Nutrient biogeochemistry of the coastal zone. *Science,* **281**: 217–221.

Johnson, N., Revenga, C. & Echeverria, J. 2001. Ecology. Managing water for people and nature. *Science,* **292**: 1071–1072.

Jonas, D.A., Elmadfa, I., Engel, K.-H., Heller, K.J., Kozianowski, G., König, A., Müller, D., Narbonne, J.F., Wackernagel, W. & Kleiner, J. 2001. Safety considerations of DNA in food. *Annals of Nutrition and Metabolism,* **45**: 235–254.

Jongbloed, A.W. & Lenis, N.P. 1998. Environmental concerns about animal manure. *Journal of Animal Science,* **76**: 2641–2648.

Ketaren, P.P., Batterham, E.S., Dettmann,E.B. & Farrell, D.J. 1993. Phosphorus studies in pigs, 3. Effect of phytase supplementation on the digestibility and availability of phosphorus in soya-bean meal for grower pigs. *British Journal of Nutrition,* **70**: 289–311.

Kim,Y. 2002. World exotic diseases. pp. 331–354, *in:* D. Pimentel (ed). *Biological Invasions: Economic and Environmental Costs of Alien Plant, Animal, and Microbe Species.* Boca Raton, Florida: CRC Press.

Kleter, G.A. & Kuiper, H.A. 2002. Considerations for the assessment of the safety of genetically modified animals used for human food or animal feed. *Livestock Production Science,* **74**: 275–285.

Kornegay, E.T. 2001. Digestion of phosphorus and other nutrients: the role of phytases and factors influencing their activity. pp. 237–271, *in:* M.R. Bedford and G.G. Partridge (eds). *Enzymes in Farm Animal Nutrition.* Marlborough, UK: CABI Publishing.

Lavitrano, M., Bacci, M.L., Forni, M., Lazzereschi, D., Stefano, C.D., Fioretti, D., Giancotti, P., Marfé, G., Pucci, L., Renzi, L., Wang, H., Stoppacciaro, A., Stassi, G., Sargiacomo, M., Sinibaldi, P., Turchi, V., Giovannoni, R., Casa, G.D., Seren, E. & Rossi, G. 2002. Efficient production by sperm-mediated gene transfer of human decay accelerating factor (hDAF) transgenic pigs for xenotransplantation. *Proceedings of the National Academy of Sciences, USA,* **99**: 14230–14235.

Lee, J.-W., Wu, S.-C., Tian, X.C., Barber, M., Hoagland, T., Riesen, J., Lee, K.-H., Tu, C.-F., Cheng, W.T.K. & Yang, X. 2003. Production of cloned pigs by whole-cell intracytoplasmic microinjection. *Biology of Reproduction,* **69**: 995–1001.

Lee, W.J., Farmer, J.L., Hilty, M. & Kim, Y.B. 1998. The protective effects of lactoferrin feeding against endotoxin lethal shock in germ-free piglets. *Infection and Immunity,* **66**: 1421–1426.

Liao, S.F., Sauer, W.C. & Kies, A.K. 2002. Supplementation of microbial phytase to swine diets: Effect on utilization of nutrients. pp. 199–227, *in:* T. Nakano and L. Ozimek (eds). *Food Science and Product Technology.* Kerala, India: Research Signpost.

Lipnick, R.L., Cotruvo, J.A., Hill, R.N., Bruce, R.D., Stitzel, K.A., Walker, A.P., Chu, I., Goddard, M., Segal, L., Springer, J.A., *et al.* 1995. Comparison of the up-and-down, conventional LD_{50}, and fixed-dose acute toxicity procedures. *Food and Chemical Toxicology,* **33**: 223–231.

Lois, C., Hong, E.J., Pease, S., Brown, E.J. & Baltimore, D. 2002. Germline transmission and tissue-specific expression of transgenes delivered by lentiviral vectors. *Science,* **295**: 868–872.

Mahan, D.C. & Lepine, A.J. 1991. Effect of pig weaning weight and associated nursery feeding programs on subsequent performance to 105 kilograms body weight. *Journal of Animal Science,* **69**: 1370–1378.

Miller, E.R. & Ullrey, D.E. 1987. The pig as a model for human nutrition. *Annual Review of Nutrition,* **7**: 361–382.

Nagashima, H., Fujimura, T., Takahagi, Y., Kurome, M., Wako, N., Ochiai, T., Esaki, R., Kano, K., Saito, S., Okabe, M. & Murakami, H. 2003. Development of efficient strategies for the production of genetically modified pigs. *Theriogenology,* **59**: 95–106.

Noble, M.S., Rodriguez-Zas, S., Cook, J.B., Bleck, G.T., Hurley, W.L. & Wheeler, M.B. 2002. Lactational performance of first-parity transgenic gilts expressing bovine alpha-lactalbumin in their milk. *Journal of Animal Science,* **80**: 1090–1096.

NRC [National Research Council]. 1998. *Nutrient Requirements of Swine, 1998.* Washington DC: National Academy Press.)

Park, K.W., Lai, L., Cheong, H.T., Im, G.S., Sun, Q.Y., Wu, G., Day, B.N. & Prather, R.S. 2001. Developmental potential of porcine nuclear transfer embryos derived from transgenic fetal fibroblasts infected with the gene for the green fluorescent protein: comparison of different fusion/activation conditions. *Biology of Reproduction,* **65**: 1681–1685.

Pimentel, D. & Morse, J. 2003. Malnutrition, disease, and the developing world. *Science,* **300**(5617): 251.

Population Reference Bureau. 2003. *World population data sheet.* See: http://www.prb.org/pdf/WorldPopulationDS03_Eng.pdf.

Pursel, V.G., Wall, R.J., Michell, A.D., Elsasser, T.H., Solomon, M.B., Coleman, M.E., DeMayo, F. & Schwartz, R.J. 1999. Expression of insulin-like growth factor-1 in skeletal muscle of transgenic swine. pp. 131–144, *in:* J.D. Murray, G.B. Anderson, A.M. Oberbauer and M.M. McGloughlin (eds). *Transgenic Animals in Agriculture.* New York, NY: CABI Publishing.

Rodehutscord, M., Abel, H.J., Friedt, W., Wenk, C., Flachowsky, G., Ahlgrimm, H.J., Johnke, B., Kuhl, R., & Breves, G. 2002. Consequences of the ban of by-products from terrestrial animals in livestock feeding in Germany and the European Union: alternatives, nutrient and energy cycles, plant production, and economic aspects. *Archives of Animal Nutrition – Archiv fur Tierernahrung,* **56**: 67–91.

Schaffert, R.E., Alves, V.M.C., Parentoni, S.N. & Roghothama, K.G. 2003. Genetic control of phosphorus uptake and utilization efficiency in maize and sorghum under marginal soil conditions. See: http://www.cimmyt.org/ABC/map/research_tools_results/ wsmolecular/workshopmolecular/WSdroughtGeneticcontrol.htm.

Simons, P.C.M., Versteegh, H.A.J., Jongbloed, A.W., Kemme, P.A., Slump, P., Bos, K.D., Wolters, M.G.E., Beudeker, R.F. & Verschoor, G.J. 1990. Improvement of phosphorus availability by microbial phytase in broilers and pigs. *British Journal of Nutrition,* **64**: 525–540.

Skorburg, J. 2003. China briefing book – issue 7. American Farm Bureau Federation. http://www.fb.com/issues/analysis/China_Briefing_Issue7.html.

Smil, V. 2000. Phosphorus in the environment: natural flows and human interferences. *Annual Review of Energy and the Environment,* **25**: 53–88.

Smith, C. 1998. Introduction: Current animal breeding. pp. 1–10, *in:* A.J. Clark (ed). *Animal Breeding: Technology for the 21st century.* Amsterdam, The Netherlands: Harwood Academic.

Somerville, C. & Briscoe, J. 2001. Genetic engineering and water. *Science,* **292**: 2217.

Taylor, S. 2002. Protein allergenicity assessment of foods produced through agricultural biotechnology. *Annual Review of Pharmacology and Toxicology,* **42**: 99–112.

Tewe, O.O. & Bokanga, M. 2001. Post-harvest technologies in Nigeria's livestock industry: Status, challenges and capacities. Presented at a Workshop. Entebbe, Uganda, 17–21 September 2001. See: http://www.foodnet.cgiar.org/Post%20Harvest/Papers/Tewe's%20Final%20paper.htm.

Tilman, D., Fargione, J., Wolff, B., D'Antonio, C., Dobson, A., Howarth, R., Schindler, D., Schlesinger, W.H., Simberloff, D. & Swackhamer, D. 2001. Forecasting agriculturally driven global environmental changes. *Science,* **292**: 281–284.

Trewavas, A. 1999. Much food, many problems. *Nature,* **402**: 231–232.

Tumbleson, M.E. 1986. *Swine in Biomedical Research.* New York NY: Plenum Press.

UK Food Standards Agency. 2002. Extremely low risk of GM transfer. See: http://www.animalbiotechnology.org/archive.asp?news_id=284&mode=showarticle.

Wall, R.J., Pursel, V.G., Hammer, R.E. & Brinster, R.L. 1985. Development of porcine ova that were centrifuged to permit visualization of pronuclei and nuclei. *Biology of Reproduction,* **32**: 645–651.

Walsh, G.A., Power, R.F. & Headon, D.R. 1993. Enzymes in the animal-feed industry. *TIBTECH,* **11**: 424–430.

Wheeler, M.B., Bleck, G.T. & Donovan, S.M. 2002. Transgenic alteration of sow milk to improve piglet growth and health. *Reproduction (Cambridge),* **58**(Suppl.): 313–324.

ETHICAL, SOCIAL, ENVIRONMENTAL AND ECONOMIC ISSUES IN ANIMAL AGRICULTURE

P.C. Kesavan and M.S. Swaminathan

M.S. Swaminathan Research Foundation, Third Cross Street, Taramani, Chennai - 600 113, India

Abstract   Livestock are vital to subsistence farming and sustainable livelihood in most developing countries. Of India's population of one billion people, more than 70 percent live in the rural areas. India also has more than 30 percent of the world's bovine population. This has resulted in not only egalitarian ownership of cattle, but also in an almost inseparable cultural and symbiotic relationship between rural families and their farm animals, particularly large ruminants. It is against this scenario that the ethical, social and environmental issues of gene-based technologies need to be carefully evaluated.

The use of transgenic cows with modified milk composition or for any other purpose has little economic benefit in a system of "production by masses", as typifies India and a few other developing countries, compared with "mass production" systems in developed countries. Rather, the use of rDNA technology for developing drought-resistant fodder and forage crops is likely to bring immediate relief to most regions. Cattle, particularly in India, have poor quality feeds and this results in poor nutrition, with production of large amounts of methane. Immunocastration through biotechnological means would also be advantageous. Developing countries like India need sustainable livelihood security, and, in this regard, gene-based technologies in animal agriculture seem more to raise ethical, social and environmental concerns, rather than being likely to transform "subsistence farming" into vibrant agribusiness. Ethical issues concerning animal welfare, rights and integrity are also discussed, in addition to social, environmental and economic issues.

447

H. P. S. Makkar and G. J. Viljoen (eds.), Applications of Gene-Based Technologies for Improving Animal Production and Health in Developing Countries, 447-462.

1.   INTRODUCTION

It is wonderful that ancient civilizations practised biotechnology, a considerable part of which was based in particular on cow's milk. Examples include the making of curd, butter and cheese. The initial steps in the domestication of farm animals (cows, bullocks, buffaloes and small ruminants) and cultivation of crop plants essentially resulted in biodynamic farming. Animal manure improved the organic content of the soil and its health, while crop-related operations provided feeds for the domestic animals. Since time immemorial, cow dung ash has been used for dusting vegetable crops to deter insect defoliators and borers (Anonymous, 2001). The use of cow dung as an anti-bacterial agent in households and surroundings, and particularly the administration of *Panchgavya* (a mixture of cow's milk, dung, urine, curd/buttermilk and ghee derived from judicious melting of unsalted butter) in the Ayurvedic system of medicine for treating several disorders and assisting the recuperation of women, especially after parturition, represent high levels of traditional knowledge. There was a clear understanding that cow's milk is the nearest to mother's milk for human infants. These all constitute both practical and sentimental reasons for the Hindus and Sikhs, in general, to not accept the slaughter of cows for food.

In the present-day context, the ethical, social, environmental and economic issues in animal agriculture in developing countries are inherently linked with sustainable livelihood and food security for millions of rural poor women and men. In order to break the vicious spiral between the degradation of life support systems (soil, water, forest, biodiversity, etc.) and accentuation of rural poverty, a mixed farming system comprising cows, bullocks, buffaloes, small ruminants and poultry, together with agri-horticultural crops, is fundamental. In countries where large areas of semi-arid wastelands need to be reclaimed using limited external inputs, the role of farm animals in providing draught power as well as manure is obvious. Dung-urine compost enhances soil health and moisture retention.

The vital role of farm livestock in the social, cultural and economic well-being of a large part of humanity in the developing countries contrasts strongly with the situation in developed countries. India, with about one-sixth of the human and about one-third of the cattle (cows and buffaloes) populations of the world, merits particular attention. It is not only because of sheer numbers, but also because of the cultural and social ethos and their mutual reinforcement of livelihood security. India and several countries in South Asia are also rich in microbial, plant and animal biodiversity. Hence, there is need for an objective evaluation of the biosafety and possible environmental risks associated with applications of certain types of gene-based technologies for improvement of livestock and agriculture. Huge populations with fragile livelihood security should not, even unintentionally, be exposed to possible disastrous consequences from inadequately tested – or even as yet untested – gene-based technologies.

2.   ETHICAL ISSUES

The ethical dimensions can be discussed under two heads. One is related to the radical alterations of the genetic endowment of biological organisms far beyond the realms of natural gene exchange through sexual reproduction, and the other concerns animal welfare. The animals used in biological and medical experiments are referred to as "voiceless victims of science" by the animal activists. The extremely painful treatment of animals while being transported and slaughtered in many developing countries, including India, is an antithesis of the professed sentiments and love for animals in general, and cows in particular.

2.1   Gene-based technologies and ethical dimensions

First of all, in the interests of brevity and clarity, and for more focused discussions of applicable ethical aspects, the wide spectrum of gene-based technologies under consideration or already in use for livestock production and management (FAO/IAEA, 2001) can be grouped into three categories.

2.1.1   Disease diagnosis and vaccination

Gene-based technologies here include the expression of a gene product that can be used as a vaccine or to serve as a reagent in a diagnostic assay; molecular epidemiology to pinpoint the source of infection to achieve improved disease control; and production of therapeutic substances through the insertion of specific genes into a variety of living tissues, ranging from single cells to complete organisms.

2.1.2   Genetics and breeding

The least objectionable application, on ethical grounds, would be the use of gene-based technology to identify the genes, including quantitative trait loci (QTLs), that control advantageous productivity traits, and the subsequent use of these QTLs in selection and breeding of superior individuals within breeds, or for selection of appropriate breeds for use in cross-breeding programmes (Cunningham, 1999a). The gene technologies for identifying such markers include selective genome mapping, analysis of microsatellite libraries, restricted fragment length polymorphisms (RFLPs) and single nucleotide polymorphisms (SNPs). So long as these are meant to assist conventional breeding, there can be no valid objections on ethical grounds. However, the use of recombinant DNA technology (rDNA) to insert advantageous genes from a widely unrelated organism into a particular breed or species raises several ethical issues.

An ethical viewpoint that cannot be set aside, even by any strong scientific justification, revolves around humankind and its relationship with nature, especially the commodification of nature by humans, and antagonism between small agrarian communities and large industrial sectors; adverse social and economic impacts, especially economic benefits in "mass production" vis-à-vis "production by masses" (the case of *recombinant bovine somatotropin* (r-BST) is discussed later); biosafety (namely harmful side effects in the recipient animals and the humans who consume the products of these transgenic animals); loss of animal genetic diversity and integrity of animal species; and altering of the natural course of animal evolution. In short, all these considerations need to be subject to critical ethical, social, economic and environmental analyses.

It is agreed that transgenic technologies such as introduction of genes to replace or modify physiological processes and functions (e.g. growth, lactation, reproductive efficiency, immune responses, and resistance to climatic and abiotic stress) through regulatory hormones and other substances (e.g. IGF, prolactin, inhibin, interferons) offer tremendous potential (Cunningham, 1999b; Natarajan and Rasool, 1997). However, it should also be acknowledged that biosafety and environmental risks have not been adequately evaluated in terms of time and space, even for some of the transgenic food crops. The most appropriate precautionary, and perhaps a rightly non-negotiable ethical consideration, ought to be to avoid assumptions of biosafety and environmental safety; it would be prudent to expect and be prepared for any unforeseen ecological events, possibly disasters, specially in the context of biodiversity-rich developing countries.

2.1.3   Nutrition and growth

The consideration here is to improve the nutritional quality of plant feedstuffs and by-products through genetic manipulation (FAO/IAEA, 2001). Examples include changing the leaf : stem ratio; introducing "stay green" traits; increasing the digestibility of nutrients, especially the fibre in tropical forages; decreasing plant fibre and lignin content; increasing soluble carbohydrates in roughages; increasing the sulphur amino acids in leguminous forages; regulating protein and carbohydrate contents and their degradation to achieve maximum microbial protein synthesis in the rumen; and changing the make-up of fatty acids in order to alter the fatty acid composition of milk, meat, etc. A step further is to manipulate the rumen microflora to better degrade fibre and lignin, increase efficiency of nitrogen utilization, and to break down anti-nutritional and toxic factors.

Development and use of molecular techniques for improving ruminant performance through a reduction in methane production has positive implications, both for animal health and for the environment. India's weekly news magazine, *The Week,* on 3 November 2002 carried an article (Sachidananda Murthy, 2002) describing how annually about 8.8 million tonnes of methane is released into the air by the bovine population of India, and that this constitutes 17.6 percent of the gases that damage the ozone layer. Another report (Mitra, *et al.*, 2002) states that methane emission by animals in India is 7.6 Tg/yr (~7.6 million tonnes per year). There is obviously a need to reduce release of methane from the "mobile gas cylinders" that are both large and small ruminants. In the initial stages, molecular probes for quantifying populations of methanogens, fibre-degrading bacteria, fungi and bacteria would be extremely useful. However, non-gene-based approaches, such as inhibitors of methanogens and use of polyunsaturated fatty acids or ingredients containing these acids, might be tested for reducing methane production and enhancing microbial protein and energy supply, as these would be relatively less controversial, and also within the technological and economic capacities of rural communities in the developing countries. Dr Jamie Newbold of the Rowett Research Institute, Aberdeen, states that the amount of methane produced could be cut by almost 50 percent by adding bacteria *Brevibacillus parabrevis* as supplement to feed for farm animals (Nelson *et al.*, 2000). Further, this would not raise objections on ethical grounds. The developing countries, particularly India with its enormous cattle population, would stand to benefit greatly.

2.2   Ethical considerations from the point of view of animal welfare

It is rather unfortunate, but true, that animal welfare (a term including both physical health and behaviour in a natural environment) and animal rights are often violated in most of the developing countries in the matter of their day-to-day existence, as well as just before and during slaughter. Further, neither raising calves in the dark and feeding them with a diet specifically for developing high quality veal, nor battery rearing of broilers, are gene-based, but these violate their rights. When animals are slaughtered for food, modern methods of stunning to prevent them from dying of excruciating pain are more ethical. Cutting the jugular vein to allow the animal bleed to death is certainly not humane, no matter how one tries to justify it. In several cases of animal experiments for science and medicine, a humane code of conduct has been violated. It does not make a difference to the animals whether the unbearable torture and pain to them is caused by our scientific pursuits, or by religious and cultural dogma. However, in recent years, there has been conscious awakening in this regard, and greater adherence to a code of conduct for design of laboratory studies with animals.

There is no denial of the fact that developing countries, particularly India, must learn from developed countries, such as the Netherlands and Sweden, how to handle ethical concerns in practice. Holland has developed legislation in three phases, each recognizing a new dimension of the moral status of animals. These include an anti-cruelty law, an animal protection law and a law recognizing that animals have intrinsic value and are not purely instrumental for humans (Brom and Schroten, 1993).

With particular reference to modern biotechnology in breeding animals, two major ethical considerations relate to animal health and welfare and to the integrity of the animal (Broom, 1998). A case of unethical violation of animal welfare has been the use of biotechnology to produce leaner meat in the "Beltsville pigs" (Pursel and Rexroad, 1993). These pigs contained human growth hormone genes to accelerate growth, but suffered from health problems, such as lameness, ulcers, cardiac diseases and reproductive problems (Rollin, 1997). For broiler chickens, which gain about 2 kg in 40–50 days, the muscles and gut grow faster but the skeleton and cardiovascular system do not keep up, leading to leg problems and heart failure.

Another ethical consideration is animal integrity, which applies when the genetic endowment of the animals is so much altered that it reduces them to being merely instruments for human interests. There is a considered view that genomes should be left intact. Integrity can be considered violated when a change in the composition of the milk in a transgenic cow results, although there might be no risk to the welfare of the animal. It is nature's design that humans consume what animals produce naturally, and humans should not violate nature by making the animals produce what they normally do not. Many individuals hold the view that the superior powers of the humans do not grant them a right on ethical grounds to use the animals in any way they please. There is no reason to believe that animals lack sentiency or the capacity to experience pain and pleasure, and that they are mere automata, incapable of mental or physical experience (Straughan, 2000).

The utilitarianism approach involves weighing different levels of human benefit against different levels of animal suffering in deciding whether what humans do to animals in a particular situation is ethically acceptable or not. For example, the benefits human beings derive from eating leaner pork are unlikely to outweigh the discomfort and diminution of the welfare and integrity of the Beltsville Pigs (Broom, 1998). This situation might, however, have been ethically acceptable if the Beltsville Pigs could save human lives from a life-threatening disease rather than merely serving leaner meat (Mepham, 1993; Christiansen and Sandoe, 2000)!

The legislation in the Netherlands imposes tight constraints on aspects of animal biotechnology, including transgenesis, embryo manipulation, and administration of substances obtained by biotechnology. Using the principles of beneficience, non-maleficence, justice and integrity of the animal, it was concluded that benefits of cloning are relatively small, whereas the ethical drawbacks are considerable (Boer, Brom and Vorstenbosch, 1995).

Related to ethical aspects, the term *telos* – the way of living exhibited by an animal and whose fulfilment results in happiness or whose thwarting results in psychological suffering – has also been brought into the discussion (Straughan, 2000). While genetic engineering may or may not affect the telos of animals, most modern ways of rearing animals for human food certainly do. Hence, Sweden has introduced legislation requiring that farm animals be allowed to live their lives in accordance with their telos, e.g. that cattle have a right to graze, and that chickens and pigs have the right to freedom of movement. Due concerns have also been expressed about the implications for animal welfare of the application to animals of products derived from genetically modified organisms (GMOs) (Boer, Brom and Vorstenbosch, 1995; Juskevich and Guyer, 1990), such as the use of recombinant bovine somatotropin (r-BST) to promote increased milk yield in cows (Juskevich and Guyer, 1990). Use of even non-GM BST has been banned in EU on human health and animal welfare grounds.

The ethical aspects arising from creation of the oncomouse, which is genetically designed to develop cancer, or from the raising of animals for the sole purpose of using their organs for transplantation to humans, are viewed with widely different perceptions and are excluded from consideration in this paper. However, these have been discussed in detail elsewhere (Bach and Ivinson, 2002; NHMRC, 2002; Bramstedt, 1999).

3.   SOCIAL ISSUES

No other branch of biology has received so much support and global importance as modern biotechnology. The time has come for modern biotechnology to fulfil its social contract. While delivering the keynote address on *Science in response to basic human needs* at the World Summit on Science in Budapest in 1999 (UNESCO, 2000), Professor M.S. Swaminathan emphasized that science should serve humanity and must provide basic human needs. The question, therefore, particularly in the context of developing countries, is whether biotechnology in general, and specifically gene-based animal biotechnology, has been or would be able to provide for basic human needs and also to stand up to societal scrutiny. For most developing countries, technological empowerment of rural communities should concurrently address social equities and gender concerns. In the case of India, over 75 million women, as against 15 million men, are involved in feeding cows and buffaloes, milking and selling to cooperatives. The National Agriculture Policy (Department of Animal Husbandry) of the Government of India states that women constitute 71 percent of the labour force in livestock farming. India has vast resources of livestock and poultry, which play a vital role in improving the socio-economic conditions of the rural masses. India has about 16 percent of the cattle, 57 percent of the buffalo, 17 percent of the goat and 5 percent of the sheep population of the world, and in gross production terms ranks first with respect to cattle and buffalo, second for goat and third for sheep. Employment generation in the livestock sector is considerable. At present, about 11 million workers have primary employment, and 8 million have subsidiary status, in the livestock sector.

As mentioned earlier, mixed farming (agricultural crops plus livestock) provides a safety net for livelihood and food security, particularly for the myriad small farm holdings (average ~2 ha), most of which are also monsoon-dependent for cultivation of crops. It thus makes sense that in India and South Asia, the ownership of livestock is more egalitarian than ownership of land. In India, about 70 percent of the livestock is in the hands of small-scale farmers and landless labourers, who control less than 30 percent of the land area.

The production systems are based on traditional low-cost input systems, and many of the livestock owners subsist at or below the poverty line (i.e. an income of US$ 1/day for five family members). Hence, it is emphasized that livestock production and management in the developing countries is but one part of an integrated agricultural system, and therefore modern biotechnological strategies should address the total production system (FAO/IAEA, 2001).

Today, the most serious challenge for health, welfare and sustainable productivity of animals in the developing countries is the inadequate production and supply of fodder and forage crops. Therefore, while considering gene-based technologies, it is necessary to keep in view the social objectives of mixed farming systems. In mixed systems, crops and livestock activities compete for the same scarce resources, such as land, water, labour, capital and technological skills. Consequently, the production levels of livestock in mixed systems (e.g. milk production per day per animal, growth and reproduction rates), in general, are lower than in specialized systems. Mixed farming facilitates use of crop by-products and residues for feed and manure. An example of a traditional association is the one between nomads and crop farmers, whereby livestock of nomads convert crop residues (feed) into manure *in situ* for crop cultivation. Mixed farming assumes enormous significance in the context of reclamation of semi-arid wastelands in the developing countries. Since most of the farm holdings are small and are owned by resource-poor farmers, low external input agriculture (LEIA), where individual farmers have to recycle the resources they have on their own farm, results in on-farm mixing. Under these circumstances, the ideal gene-based technology should first address the problems of abiotic stress (i.e. salinity and drought-induced stress) in fodder and forage crops. The rDNA technology employed by the M.S. Swaminathan Research Foundation, Chennai, to genetically equip crop plants with salinity tolerance genes from mangrove species (*Avicennia marina*) needs to be extended to fodder and forage crops. Similarly, transfer of drought-tolerance genes from drought-resistant species (e.g. *Prosopis juliflora*) through rDNA technology into fodder and forage crops will be prudent. These would be meaningful not only to improve animal health and welfare without raising undue ethical concerns, but also to achieve greater social acceptance of as yet a contentious technology. Production of fodder and forage crops on what have long remained as wastelands would not only improve the well-being of livestock but also reduce the drudgery for women who otherwise have to walk long distances every day to fetch some feed for their cows, buffaloes and bullocks.

4. ENVIRONMENTAL ISSUES

The importance of livestock, particularly farm animals, in enriching the organic content of the soil and improving soil health in on-farm mixing has already been noted. Cow-dung cakes are used for cooking, and this results in saving forest trees from being cut down for fuelwood. However, the excessive production and release of methane that results in depletion of the ozone layer in the atmosphere is a worrisome environmental concern. Hence, appropriate technologies to inhibit methanogens and also to reduce the level of methane production by up to 50 percent in animals fed roughage diets are urgently needed. Also, molecular probes for quantifying populations of methanogens, fibre-degrading bacteria, fungi, etc., will contribute to protection of the environment.

However, the use of rDNA technology to achieve any of the above goals needs to be carefully evaluated from the point of environmental risk. In fact, most of the present regulations and associated risk assessments are focused on plants and not on animals. The consolation seems to be that there are no GM farm animals, other than those used for biopharming, which takes place in controlled and contained premises. Any livestock farming has some environmental impact, but it is not known if GM animals would exert any specific additional impact on the environment. It is more certain that GM fish (Hew, Fletcher and Davies, 1995) and GM insects raise serious environmental concerns. Release – either intentional or inadvertent – of faster-growing GM fish into the marine environment might result in unintended negative consequences for native wild species. And any such negative effects could also have adverse economic consequences. From the point of evolving precautionary principles to ensure environmental protection, cloned farm animals should be subject to the same rigorous standards of control and containment as are now stipulated and enforced for research animals.

It will be prudent to set up an independent Strategic Advisory Body that should take a holistic view of the implications of developments common to different kinds of animals, both in relation to GM and to cloning, particularly where the issues are complex and likely to be of interest to the public and stakeholders. Further, this advisory body should ensure that GM and cloning applications, particularly those in agriculture and environment, are handled in a transparent manner to satisfy public concerns (AEBC, 2002).

It is equally important that public funding for public good should be the guiding principle in ensuring that biosafety and environmental protection standards are absolutely the bottom line. It should be remembered that the crisis caused by bovine spongiform encephalopathy (BSE) and foot-and-mouth disease (FMD) has accelerated public interest in food safety and led to a distrust of regulatory agencies.

5.   ECONOMIC ISSUES

Since the onset of the post-Mendelian era, commercial ventures in the last half-century have generally developed only after the scientific basis for an innovation had been firmly established, its technological premise quite adequately tested, and socio-economic factors well assessed. For instance, radiation preservation technology had its scientific origins in the 1950s, but it took almost three decades of extensive genetic toxicological testing before clearance was given for commercialization in India. There were questions of if nanogram quantities of radiolytic changes in the carbohydrates and proteins per kilogram of grains and meat exposed to around 8–10 kGy could cause damage to the DNA of the consumers. In contrast, *Bt* and herbicide transgenic crops have been given clearance without much fuss, on the basis of substantial equivalence. The fact of the matter is that there was no commercial push in the case of radiation food preservation technology. The argument generally put forward for the aggressive commercial drive in favour of a product generated by modern biotechnology is that it would solve the food and nutritional insufficiencies of the developing countries. It is not exactly clear how this could really happen for most developing countries, especially India, where about 250 million people remain partially or fully hungry, not because there is no food grain in the country, but because they have no purchasing power. Further, experience clearly shows that new technologies have generally resulted in digital, genetic and other divides because of constraints in their diffusion to the rural areas where the bulk of the resource-poor and landless labourers exist, below the poverty line. Further, these new technologies are adopted by only a small sector – with personal financial resources or corporate agencies – and so the rich–poor divide widens further. Monopolistic regimes might result.

In the case of animal biotechnology, this can be illustrated by r-BST. BST, a natural growth hormone secreted by the anterior pituitary of animals, exerts a major effect on the regulation of growth and also milk production. Since the quantities of BST obtained from slaughtered animals are quite small, rDNA technology-based BST (r-BST) is produced and widely used in the USA to increase milk production by 10–20 percent. The bacterium, *Escherichia coli*, which is found in the intestinal tract of humans and animals, can be genetically equipped to produce large amounts of r-BST. The r-BST produced by bacteria is purified and injected into dairy cattle. Exhaustive evaluation tests conducted in the USA have shown that r-BST has no harmful effects in milk (Juskevich and Guyer, 1990), but a high level of milk production makes higher demands on animal physiology, and if adequate food supply is lacking, negative effects are observed on fertility, with other health problems, especially mastitis and ketosis (Jarvis, 1996). As stated earlier, the EU has not approved the use of BST on human health and animal welfare grounds. Currently, neither r-BST nor the complementary adequate nutritious feed is within the reach of millions of resource-poor farming families and nomads owning cattle in India and other developing countries. Jarvis (1996) has also observed that economic factors concentrate its use in sections of high-producing herds and limit its potential use in developing countries. In the context of the Indian scenario, the use of r-BST for boosting milk production in the USA and the lack of resources for wider adoption of this technology by millions of rural women and men in the

Indian dairy sector adds to already existing economic disparity, caused by basically different production systems in these two countries. For instance, in 2001–2002, India produced about 88 million tonnes of milk, involving as many as 150 million cows and buffaloes, owned and managed by about 75 million women and 15 million men. This is a case of production by masses. In contrast, the USA in the corresponding period produced about 69 million tonnes of milk involving only 9.2 million dairy cattle owned and managed by about 200,000 farmers (mostly men). This is a case of mass production (factory farming). It should be obvious that dairy farming is an industry in the USA, whereas in India it is simply an agrarian approach and largely for subsistence. It is against this background that the economic implications of the capacity to afford or not the use of r-BST assume enormous significance.

The scenario can justifiably be further stretched within the purview of the Word Trade Agreement in Agriculture. Crop and animal products resulting from mass production in the developed countries also enjoy considerable domestic support under the green and blue boxes and export subsidies. Since surpluses can easily result from mass production systems, the export subsidies help both in maintaining a high price level for farm produce in the domestic economy and also to promote dumping of the surpluses in outside markets. For example, the actual export subsidy by EEC in 1998 was US$ 1.6/kg (Rs 74) on butter, and the USA subsidized skim milk powder by US$ 1.2/kg (Rs 53). When, as a result of an array of domestic support and export subsidies, crop and dairy products of the developed countries are dumped into developing countries, the livelihood security and consequently the food security of millions and millions of small-scale and resource-poor farming women and men and their children can be devastated. Famine of livelihood is the cause of hunger, and there are already over 250 million people with little economic access (i.e. purchasing power) to food in India.

At best, only a few rich farmers and multinational companies in the developing countries would derive economic benefit from such gene-based biotechnologies if and when these are cleared from the ethical, social and environmental points of view. The various aspects of commercialization of animal biotechnology have been discussed elsewhere (Faber *et al.,* 2003).

6.　CONCLUSIONS

Insofar as the developing countries are concerned, the economic considerations resulting from modern biotechnology require most immediate attention. This matches the statement "poverty is the greatest polluter", made by the late Ms Indira Gandhi, then Prime Minister of India, at the UN Conference on Human Environment in Stockholm in 1972. Nothing else would better explain the grim situation that roughly one-third of the world's cattle population compete for fodder and forage from the small land holdings of millions of resource poor women and men, who also struggle for their own existence based mostly on subsistence agriculture. The on-farm mixing of crops and livestock results in low yield efficiencies for both plant and animal products. And living below the poverty line causes serious hunger-related health problems, for both humans and their livestock. Under these circumstances, the relevant questions are "Which of the biotechnologies would be most relevant?" and "How would these make a difference towards economic well-being?" It should also be borne in mind that the global output of the livestock sector, particularly in developing countries, is expected to double by about 2020.

Research is needed to develop still cheaper and safer methods of embryo transfer in the developing countries. Artificial insemination (AI) is practiced in many rural areas, but molecular technologies to identify bulls with superior genetic endowment for enhancing milk production in the female offspring need to be added. Concurrently, immunocastration (Pell and Aston, 1995) of inferior bulls would prevent the expansion of less productive bovine populations. In general, molecular epidemiology leading to pinpointing sources of infection, and recombinant vaccines and vector-virus-expressed diagnostic reagents would be ethically, socially and environmentally free from major concerns, and economically beneficial.

As recommended by an FAO/IAEA Consultants Meeting (FAO/IAEA, 2001), the first step should be the characterization of the gene pools of livestock, microbes and forages. Capacity building in developing countries to improve ruminant performance through a reduction in methane production will result in economic and environmental benefits.

Improving feed quality through genetic manipulation holds great promise. The biotechnological approach to improving the nutritive value of alfalfa by transferring the sunflower seed storage albumin (identified as being both rich in cysteine and methionine and resistant to degradation by ruminal fluid *in vitro*) is a step in the right direction (Tabe *et al.*, 1995).

The long-term goal of manipulating rumen microflora to better degrade fibre and lignin, increase efficiency of nitrogen utilization, and to break down anti-nutritional and toxic factors needs more careful evaluation from the points of view of both animal health and environmental concerns. The approach to solving problems through rDNA technology should remain as the very last resort. We should bear in mind that if we, the humans, respect our own genetic integrity, we should also respect the integrity, welfare, sentience and telos of all other organisms. After all, no one really knows the very long-term consequences of transgenics for human and ecological health. We should learn from past mistakes. The chemical pesticides that were expected to eradicate a variety of domestic and agricultural pests have largely failed in this regard, but instead are today slowly poisoning us and our children through their toxic residues in soil and groundwater, particularly in the developing countries. Pesticide residues in groundwater are now a major cause of health concern in India and possibly several other agrarian countries in the developing world. The scientists must show courage and honesty in admitting that biosafety and environmental safety are at best assumed, and not yet established through long-term genetic toxicological studies for most of the GM food crops.

Human civilization is under test – whether it should correct the human environment in order to conserve the precious genomes of the human and other species, or else to change the genetic endowment of all organisms, including ourselves, to adapt to the constantly deteriorating environment. The solution is in finding a balance between unacceptable levels of poverty on the one hand and unsustainable lifestyles on the other. The question is whether modern biotechnology can help in bridging the rich-poor divide within developing countries, and between developed and developing countries, without causing ecological harm or compromising biosafety. The answer is not simple and positive. Capacity building in the developing countries should continue, but application should await the emergence of a clearer picture regarding biosafety, environmental effects and economic consequences for developing countries.

REFERENCES

AEBC [Agriculture and Environment Biotechnology Commission]. 2002. Animals and Biotechnology. A Report by the AEBC. 88p.
See: http://www.aebc.gov.uk/aebc/pdf/animals_and_biotechnology_report.pdf

Anonymous. 2001. Report – Cowdung spray has been shown to control bacterial leaf blight disease by the Tamil Nadu Rice Research Institute, Aduthurai. *Pesticide Post*, **9**: 3.

Bach, F.H. & Ivinson, A.J. 2002. A shrewd and ethical approach to xenotransplantation. *Trends in Biotechnology*, **20**(3): 129–131.

Boer, I.J.M., Brom, F.W.A. & Vorstenbosch, J.M.G. 1995. An ethical evaluation of animal biotechnology: the case of using clones in dairy cattle breeding. *Animal Science*, **61**: 453–463.

Bramstedt, K.A. 1999. Ethics and the clinical utility of animal organs. *Biotopic Tibtech,* **17**: 428–429.

Brom, F.W.A. & Schroten, E. 1993. Ethical questions around animal biotechnology: the Dutch approach. *Livestock Production Science,* **36**(1): 99–107.

Broom, D.M. 1998. The effects of biotechnology on animal welfare. pp. 69–82, *in:* A. Holland and A. Johnson (eds). *Animal Biotechnology and Ethics.* London: Chapman and Hall.

Christiansen, S.B. & Sandoe, P. 2000. Bioethics: limits to the interference with life. *Animal Reproduction Science,* **60-61**: 15–29.

Cunningham, E.P. 1999a. Recent developments in biotechnology as they relate to animal genetic resources for food and agriculture. [FAO] Commission on Genetic Resources for Food and Agriculture, Background Study Paper, No. 10.

Cunningham, E.P. 1999b. The application of biotechnologies to enhance animal production in different farming systems. *Livestock Production Science,* **58**: 1–24.

Faber, D.C., Molina, J.A., Ohlrichs, C.L., Van der Zwaag, D.F. & Ferre, L.B. 2003. Commercialization of animal biotechnology. *Theriogenology,* **59**: 125–138.

FAO/IAEA [Food and Agriculture Organization of the United Nations/International Atomic Energy Agency Joint Division]. 2001. Consultants meeting to discuss and make recommendations on significance, suitability and potential applications of gene-based technologies for improving livestock production in developing countries (Report on FAO/IAEA Consultants Meeting, 27 to 30 November 2001.) *(FAO/IAEA) Animal Production and Health Newsletter,* No. 35: 16–18.

Hew, C.L., Fletcher, G.L. & Davies, P.L. 1995. Transgenic salmon: tailoring the genome for food production. *Journal of Fish Biology,* **47**(A): 1–19.

Jarvis, L.S. 1996. *The Potential Effect of Two New Biotechnologies on the World Dairy Industry.* Boulder, CO: Westview Press.

Juskevich, J.C. & Guyer, C.G. 1990. Bovine growth hormone: human food safety evaluation. *Science,* **249**: 875–884.

Mepham, T.B. 1993. Approaches to ethical evaluation of animal biotechnologies. *Animal Production,* **57**: 353–359.

Mitra, A.P., Dileep Kumar, M., Rupa Kumar, K., Abrol, Y.P., Kalra Naveen, Velayuthan, M. & Naqvi, S.W.A. 2002. *Global-Regional Linkages in the Earth System.* (The IGBP Global Change Series, (Series eds: P. Tyson *et al.*)). Berlin: Springer. 198p.

Natarajan, C. & Rasool, T.J. 1997. Gene technology in animals: A boon for livestock production. pp. 10–20, *in:* R. Verma. S.N. Singh and K.R. Shingal (eds). *Biotechnology in Animal Health and Production.* Pune, India: Pune University Press.

Nelson, N., Valdes, C., Hillman, K., McEwan, N.R., Wallace, R.J. & Newbold, C.J. 2000. Effect of methane-oxidising bacterium isolated from the gut of piglets on methane production in Rusitec. *Reproduction Nutrition Development,* **40**: 212 [Abstract].

NHMRC [National Health and Medical Research Council]. 2002. Part 1 – Ethical and Scientific Principles. pp. 9–45, *in:* Draft guidelines and discussion paper on xenotransplantation. Xenotransplantation Working Party. Public consultation 2002. See: http://www.health.gov.au/nhmrc/issues/xeno.pdf

Pell, J.M. & Aston, R. 1995. Principles of immunomodulation. Livestock Production Science, **42**: 122–133.

Pursel, V.G. & Rexroad, C.E. 1993. Status of research with transgenic farm animals. *Journal of Animal Science,* **71**(3): 10–19.

Rollin, B.E. 1997. Send in the clone … don't bother, they are here. *Journal of Agricultural Ethics,* **10**: 25–40.

Sachidananda Murthy. 2002. Quite a lot of hot air. *The Week,* 3 November 2002: 1–3.
Straughan, R. 2000. *Ethics, Morality and Animal Biotechnology.* Report published by The Biotechnology and Biological Science Research Council (BBSRC), UK. See: http://www.bbsrc.ac.uk/tools/download/ethics_animal_biotech/ethics_animal_biotech.pdf
Swaminathan, M.S. 2000. Science is response to basic human needs. Keynote address. pp. 33–40, *in:* A.M. Cetto (ed). *World Conference on Science – Science for the 21st Century: A new Commitment.* Budapest, 1999. Paris: UNESCO.
Tabe, L.M., Wardley-Richardson, T., Ceriotti, A., McNabb, W., Moore, A. & Higgins, T.J.V. 1995. A biotechnological approach to improving the nutritive value of Alfalfa. *Journal of Animal Science,* **73**: 2752–2759.

RISKS OF GENE TRANSFER FROM GMOs TO LIVESTOCK, AND CONSEQUENCES FOR HEALTH AND NUTRITION

Richard H. Phipps,[1] David E. Beever[1] and Ralf Einspanier[2]

1. School of Agriculture, Policy and Development, Department of Agriculture, The University of Reading, Reading RG6 6AR, UK.
2. Institute of Veterinary Biochemistry, Free University of Berlin, Oertzenweg 19b, 14163 Berlin, Germany.

Abstract: There has been rapid uptake of GM crops, with widespread use in livestock production systems. The concept of substantial equivalence as the starting point for the safety assessment of GM crops is discussed, together with the role of compositional and nutritional equivalence. Concerns have been expressed as to the fate of transgenic DNA and the expressed protein, and the safety of milk, meat and eggs derived from animals receiving diets containing GM feeds, and the effects of feed processing, conservation and nutrient digestion on the fate of transgenic DNA and proteins, are presented. Their presence in milk, meat and eggs has never been established in any study to date. There is no evidence to suggest that livestock products from animals receiving GM feed ingredients are anything other than as safe as those produced from conventional feeds. Furthermore, although hypothetically possible, horizontal gene transfer between transgenic DNA and micro-organisms either in the digestive tract of livestock or in the soil has not been established under natural conditions.

1.   INTRODUCTION

In 2002, 6 million farmers in 16 countries grew 59 million hectares of transgenic crops (James, 2003), with further increases predicted for 2003. The increase from 4 million hectares in 1996 to 59 million hectares in 2002 represents the fastest adoption rate of any new technology ever brought to

H. P. S. Makkar and G. J. Viljoen (eds.), Applications of Gene-Based Technologies for Improving Animal Production and Health in Developing Countries, 463-478.

agriculture, with a cumulative total of 235 million hectares grown in this period. The principal genetically modified (GM) crops currently grown are soybean (36.5 million hectares), maize (12.4 million hectares), cotton (6.8 million hectares) and canola oil-seed rape (3.0 million hectares). With few exceptions, these crops have been modified for first-generation agronomic input traits such as herbicide tolerance (*Ht*) or insect protection (*Bt*).

1.1   Use of GM feed resources in livestock production

In many parts of the world, maize grain and soybean meal are the preferred choice of energy or protein supplements, or both, for use in both monogastric and ruminant diets. In 2003, 80 percent of the soybean crop in the United States of America and 97 percent of the crop grown in Argentina was planted to GM cultivars. For the first time, over 50 percent of the world's most popular oilseed crop – soybean – was planted to GM cultivars. Approximately 120 million tonnes of soybean meal are used annually in livestock production; therefore, as 50 percent of the crop is GM, then 60 million tonnes will originate from GM cultivars (Soy Stats, 2001). In spite of the controversy in the EU over the introduction of GM crops, the amount of GM soybeans and soybean meal imported into the EU from the United States of America and Argentina is estimated at 15 million tonnes, and increasing (Soy Stats, 2001). This has in part been due to the banning of meat and bone meal as a nitrogen source for livestock production. In addition, rapeseed and cottonseed meal and maize grain and silage derived from GM cultivars are used extensively in livestock production systems.

The aim of this paper is to review published data relating to the use of GM crops in animal production systems, and to consider the likely implications for the safety of milk, meat and eggs derived from animals fed GM feed ingredients, and the consequences for health and nutrition.

2.   SAFETY ISSUES

A number of excellent reviews have been produced on the issues surrounding the use of the concept of substantial equivalence and the safety assessment of GM cultivars (Aumaitre *et al.,* 2001; Chesson, 2001; Cockburn, 2001; Kuiper, 2001). A very recent, excellent publication by the Organisation for Co-operation and Economic Development (OECD) has also addressed the issue of the safety assessment of animal feedstuffs derived from GM plants (OECD, 2003).

2.1 Concept of substantial equivalence

Substantial equivalence is the starting point for the safety assessment of GM crops, and is based on the concept that existing crops can serve as the basis for comparing the properties of a GM cultivar with an appropriate counterpart considered safe as shown by a long history of safe use (OECD, 2003). Application of the concept is not a safety assessment *per se,* but helps to identify similarities and differences between conventional and GM cultivars. A comparison of phenotypic and agronomic characteristics, together with a compositional analysis of components, including key nutrients and anti-nutritional factors, is the basis for establishing substantial equivalence. While the detection of unintended effects in GM cultivars focuses on chemical analysis of known nutrients and toxicants, it might also involve feeding studies with young, rapidly growing livestock. While the basic concept of substantial equivalence has wide international support, it has, however, been criticised for being subjective, inconsistent and may not reveal unintended effects, as proving a negative is no easy matter (Millstone, Brunner and Mayer, 1999). To this end, the initiative by OECD (OECD, 2001a, b, 2002a, b, c) in developing consensus documents for different crops – where analyses are recommended for use so that the principle of substantial equivalence can be applied more uniformly – is an important development and has been welcomed widely.

2.2 Compositional equivalence

While compositional analysis of feeds has played an important role in the nutritional assessment of conventionally bred cultivars, it should be noted that there are significant differences between the composition of conventionally bred cultivars within crops, and therefore the compositional analysis of GM cultivars must be assessed against not only their near isogenic counterpart but also against the background of the natural variability in conventional cultivars.

Currently, more than 50 biotechnology-derived crop cultivars, mainly with modified agronomic traits, have been assessed by regulatory agencies. The first phase of the evaluation is an assessment of the agronomic and phenotypic characteristics of the new crop variety, which, for maize, includes the following parameters: leaf orientation, plant height, silking date, ear size, height, tip fill, tassel size and colour, dropped ears, early plant vigour, leaf colour, root strength, reaction to pesticides, late season appearance, susceptibility to pathogens, and yield (Cockburn, 2001). These characteristics are sensitive indicators of changes in plant metabolism and genetic pleiotrophy, and as such are robust indicators of equivalence. Thus biotech crops must be phenotypically equivalent to their traditional counterparts.

The next phase of this comparative analysis is based on compositional analyses, which are typically conducted on both GM and conventional cultivars grown under the same field conditions in the same year, as differing environmental conditions may result in significant differences in compositional analyses that are not related to the genetic modification. Again, it is important to have internationally accepted protocols for the collection and analysis of such data.

The analyses conducted to establish compositional equivalence should provide information on macronutrients, micronutrients, anti-nutritive factors and naturally occurring toxins in the feed under evaluation. Guidance on the analyses that should be undertaken in attempting to establish compositional equivalence is provided by OECD (OECD, 2001a, b, 2002a, b, c), but requirements need to be evaluated on a case-by-case basis, as they will vary depending on the specific introduced trait.

If, during these analyses, differences are noted between conventional and GM cultivars, then further assessments are needed to determine the safety and nutritional impact of significant, biologically meaningful differences. Even when statistically significant differences in compositional analyses are noted between the GM cultivar and its near isogenic counterpart and conventional cultivars, these differences should be assessed carefully, because, on their own, they may not indicate the presence of an unintended effect. For example, some of the differences may fall within the natural and often wide variation that exists among currently available commercial cultivars. This emphasizes the importance of comparing the GM cultivar to not just its near isogenic parental line but also to a number of comparable commercially relevant and diverse cultivars.

Compositional data published from the first generation of crops have shown that the food and feed products from these crops were substantially equivalent in terms of composition, except for the introduced traits, to the conventional counterparts. Data have been published for numerous cultivars, including Roundup Ready® soybean (Padgette, 1996); insect-protected maize (Sanders, Lee and Gro, 1998); Roundup Ready® maize (Sidhu *et al.,* 2000); insect-protected cotton (Berberich *et al.*, 1996); Roundup Ready® cotton (Nida, Patzer and Harvey, 1996); insect-protected potatoes (Lavrik, Bartnicki and Feldman, 1995); and insect and virus protected potatoes (Rogan, Bookout and Duncan, 2000). The work conducted by Ridley and colleagues (Ridley *et al.*, 2002) sets a benchmark for others to follow. In their work, over 50 compositional parameters were measured in various plant components from a GM maize cultivar, a control and 15 commercial cultivars, grown at different geographical locations over a two-year period in both replicated and non-replicated trials, and subjected to statistical analysis to confirm compositional equivalence. In all cases, these data have been necessarily combined with specific safety evaluations of the introduced traits, most typically proteins, to assure food and feed safety (Astwood and Fuchs, 2001).

2.3   Nutritional equivalence

Compositional analyses provide only a guideline to the nutritional value of feeds. They cannot provide information on nutrient digestion. In many cases this is an important parameter, and *in vivo* studies are required to determine the bio-availability of nutrients.

In addition, feeding studies with target species have been conducted to establish nutritional equivalence of GM cultivars with their near isogenic counterpart and commercial cultivars, with end-point measurements such as feed intake, nutrient digestion, level of animal performance (milk, meat and egg production), animal health and welfare, and efficacy. Such studies provide valuable information that can be placed in front of the general public to help allay their concerns over the use of GM feed ingredients in the production of staple foods such as milk, meat and eggs.

Numerous recent studies have confirmed nutritional equivalence for a range of GM cultivars when compared with their isogenic counterpart and commercial cultivars. These studies include work with chickens, pigs, sheep, dairy cows and beef cattle, and compared the use of soybean, maize grain, fodder beet, sugar beet, sugar beet pulp and cotton seed modified for herbicide tolerance and insect protection, and are summarized by OECD (OECD, 2003). There is now growing evidence to suggest that once compositional equivalence of GM cultivars has been established, then nutritional equivalence can be assumed and that routine-feeding studies with target species generally add little to a safety and nutritional assessment of these GM constructs.

3.   ENHANCED FEED AND FOOD SAFETY

In many parts of the world, maize is attacked by a variety of insect pests, including the European corn borer (ECB). This pest can significantly reduce crop yield and decrease the quality of food and feed available for both humans and livestock. After the initial insect attack, the maize crop is susceptible to secondary fungal infection as spores enter the wounds left by the insect. Modern biotechnology has produced insect-protected plants that have been genetically enhanced to produce proteins, similar to those produced by the soil bacterium *Bacillus thuringiensis* (*Bt*), which kill the insect pest.

While the majority of reports in the literature show compositional and nutritional equivalence when comparing GM and conventional feed ingredients, reports of increased animal performance in both pigs and poultry when diets contained *Bt* maize grain rather than conventional grain have been noted. The increased performance was attributed to a significant reduction in mycotoxin contamination, an effect of the use of *Bt* maize grain (Piva *et al.*, 2001a, b). Thus, the use of first-generation insect-protected maize cultivars can result in not only more grain but also safer grain for both livestock and humans.

## 4.	FATE OF TRANSGENIC DNA AND GENE PRODUCTS

Although the World Health Organization (WHO) has stated that, given the long history of safe consumption of DNA, the consumption of DNA from all sources – including plants improved through biotechnology – is safe and does not produce a risk to human health, concern has been expressed about the fate of transgenic DNA (tDNA) and protein. Since the introduction of commercial GM crops in the United States of America and other parts of the world, a series of research studies have been undertaken to determine the fate of tDNA and protein in the feed and food chain. In addition to determining the effect of conservation and processing on the fate of tDNA and protein, studies with both ruminants and monogastrics have been conducted to determine the fate of tDNA and protein at key stages in the digestive track, and their possible accumulation in milk, meat and eggs derived from animals receiving GM feed ingredients.

## 4.1	Effect of conservation and feed processing

The ensiling process – chopping of plant tissue and the subsequent lowering of pH by lactic acid fermentation – produces a harsh environment for DNA and will accelerate its degradation. While the origin of silage could be confirmed, it showed that the ensiling process resulted in major fragmentation of tDNA and that the presence of intact, functional genes after an extended time of ensiling was highly unlikely (Hupfer *et al.*, 1999).

While grinding and milling had little effect on DNA fragment size, mechanical extrusion or chemical extraction of oil from seeds can cause extensive fragmentation. Dry heat applied at 90°C appeared to have no effect, while 95°C for 5 minutes caused considerable fragmentation of DNA. Equally, steam at low to moderate pressures caused substantial fragmentation. However, it has been established that processed feed samples may contain DNA fragments large enough to contain functional genes, and thus farm livestock will consume both tDNA and their gene products (Chiter, Forbes and Blair, 2000)

4.2   Digestion of consumed DNA and proteins

It has been estimated (Beever and Phipps, 2001) that the modern dairy cow, consuming on average 24 kg feed DM and producing 40 litres of milk daily, would ingest about 57 g DNA/day, of which 54 µg ($\approx 9.4 \times 10^{-5}$ percent of total intake) was estimated to be of transgenic origin when 60 percent of the diet was supplied as GM maize silage and maize grain. On this basis, it was concluded that the intake of tDNA was extremely small, and posed an almost negligible risk, provided the tDNA did not survive digestion to be absorbed with full functional integrity.

Non-ruminants possess nucleases in salivary and pancreatic secretions, and several studies (Newport and Keal, 1973; D'Mello, 1982; Henderson and Patterson, 1973) have all failed to establish the absorption of intact DNA, or RNA, by the healthy gut. In ruminants, the capacity for nucleic acid digestion is further consolidated by the presence of microbial nucleases within the rumen. The resident population of rumen microbes is responsible for the extensive degradation of dietary fibre, together with varying amounts of dietary protein and starch. There is also evidence (Smith, 1975) to show that dietary nucleic acids were also extensively degraded in the rumen, with 85 percent of the total quantity entering the small intestine being of rumen microbial origin, i.e. synthesized *in situ* within the rumen. This confirmed earlier studies (Razzaque. and Topps, 1972) that demonstrated rapid digestion of plant nucleic acids following feed ingestion, whilst other studies showed little evidence of any direct incorporation of plant nucleotides into rumen microbes (Van Nevel and Demeyer, 1977).

Studies were conducted with growing cattle (McAllan, 1980, 1982) fed hay and concentrates to determine the fate of nucleic acids (primarily of microbial origin) in the post-ruminal section of the alimentary tract. The animals were slaughtered and the data showed that, at the end of the small intestine, over 75 percent of the DNA and RNA entering had been digested, with most of this occurring in the first 30 percent of the small intestine. It was also reported that prior to the terminal ileum there was a true digestibility in excess of 97 percent. In addition, examination of the levels of free adenine, guanine and pyrimidine bases (nucleic acid breakdown products) in intestinal digesta indicated a substantial increase by the first segment of the intestine after the abomasum, followed by their effective removal thereafter, such that the amounts present at the terminal ileum were largely undetectable. While these studies did not actual establish their fate, it is safe to conclude that they would have no functional integrity with respect to genetic coding even if they were absorbed across the intestinal wall.

The processes of protein digestion in both ruminants and monogastrics are well documented, although there has been relatively little attention paid to the digestive fate of individual proteins. In monogastrics, the small intestine is the most active site of protein digestion, followed by extensive absorption of free amino acids and some short-chain peptides. In ruminants, the system is again different, as already alluded to above, with extensive microbial degradation of feed protein occurring in the rumen, together with the associated production of microbial protein. In relation to duodenal protein supply, the quantity may vary with respect to the amount consumed due to the efficiency of capture of degraded dietary protein. In most situations, microbial protein is likely to constitute the major portion of duodenal protein.

Studies have been conducted to examine the fate of proteins produced by the *Ht* and *Bt* constructs. A recent study (Wehrmann *et al.*, 1996) examined the intestinal degradation of PAT (phosphinothricine acetyl transferase) protein over a wide range of pH conditions, using an *in vitro* model system, and established an optimal pH of 6.5. In the absence of pepsin, PAT degradation was 500 g/kg total PAT protein after 10 minutes exposure to the system at optimal pH, taking more than 120 minutes for digestion to exceed 950 g/kg. In contrast, the addition of pepsin to the incubation resulted in almost complete degradation of PAT protein after 5 minutes of exposure. This indicates that, with the currently used GM cultivars, gene products are rapidly degraded in the ruminant and monogastric digestive tract, which is one indicator suggesting that the possibility of allergic reactions from the gene products is remote. It should also be noted that to date there has been no authenticated case of allergic reaction attributable to the consumption of GM foods, which have been consumed by many millions of people for at least seven years.

4.3 Detection of tDNA and protein in food produced from animals fed on GM feeds

Numerous studies have now been conducted with ruminants and monogastrics to determine if either tDNA or protein can be detected in milk, meat or eggs produced by animals receiving GM feeds as part of their diet. These studies, which have been reviewed by OECD (OECD, 2003), included work with dairy cows, beef cattle, poultry (laying hens and broilers) and pigs in which the commonly used GM crops of *Ht* soybean and *Bt* maize formed a significant part of dietary DM. Using highly sensitive Polymerase Chain Reaction (PCR) analyses, neither tDNA nor gene products were found in milk, meat or eggs derived from animals fed GM feed ingredients.

Other recently published studies (Calsamiglia *et al.*, 2003; Jennings *et al.*, 2003; Phipps, Deaville and Maddison, 2003) have confirmed the earlier results. One of these studies used silage with stacked genes for both *Ht* and *Bt*, while another, conducted at this laboratory, examined the fate of tDNA and endogenous plant DNA at key stages of the bovine digestive tract, by analysing samples of rumen fluid, duodenal digesta, faeces, blood and milk of lactating dairy cows. This study failed to detect tDNA in milk produced by cows receiving a total mixed ration containing both GM soybean meal (*cp4epsps* gene) and GM maize grain (*cry1a(b)* gene) or the near isogenic non-GM counterparts. However, tDNA was detected in the solid phase of rumen fluid and duodenal digesta, but not in blood or faeces. In contrast to the single copy transgenes, fragments of the multi-copy rubisco gene were detected in the majority of rumen and duodenal digesta, milk and faeces samples analysed, but rarely in blood. The size of the rubisco gene fragments detected decreased from 1176 bp in rumen and duodenal digesta to 351 bp in faeces. In addition to the 189-bp fragments of the rubisco gene initially detected in milk, fragments of 351 bp, and in a limited number of cases 850-bp fragments, were also detected, which is substantially larger than previously reported. It has been suggested that the presence of rubisco gene fragments in milk is "proof of principle", namely that tDNA may be found in milk. However, the ability to detect fragments of plant multi-copy genes is perhaps a function of their copy number, which is high compared with the single copy of the transgenes present per cell, and may also reflect analytical sensitivity.

5.   TRANSMISSION OF PLANT AND TRANSGENIC DNA

As reviewed recently, the natural flow of nucleic acids is wide and includes DNA from ingested feed or food (Doerfler, 2000). Published reports investigating the stability and transformation activity of tDNA have shown that minute amounts of small DNA fragments are to be expected in the gastrointestinal tract of sheep (Duggan *et al.*, 2000, 2003) and dairy cows (Phipps, Deaville and Maddison. 2003).

Horizontal gene flow has been suggested as a major force in bacterial evolution (Ochman, Lawrence and Groisman, 2000). Although the mammalian genome was initially reported to contain more than 100 genes acquired from bacteria (Lander *et al.,* 2001), this has been refuted (Salzberg *et al.*, 2001). Very recently, a distinct mitochondrial gene transfer between unrelated plants may however imply a frequent horizontal transfer between eukaryotic species (Berezina *et al.*, 2003).

While endogenous plant and animal DNA fragments from multi-copy genes have been detected in livestock products, all published studies to date have shown that tDNA has never been detected in milk, meat or eggs derived from animals receiving diets containing GM feed ingredients (Einspanier *et al.*, 2001; Klotz, Mayer and Einspanier, 2002; Einspanier, 2001). A possible explanation may be that, for example, the chloroplast genome is present in each plant cell in far greater numbers (>200 copies) when compared with the transgenes in GM plant material, which contains only one copy per cell. Consequently the uptake of tDNA from single-copy genes would be a rare event compared with multi-copy chloroplast DNA, and moreover is difficult to detect, even if highly sophisticated analytical techniques are used (Klaften *et al.*, 2004). It should be noted that no general concerns have been expressed over the presence of endogenous plant and or animal DNA fragments in milk, meat or eggs, and numerous organizations, including the World Health Organization, have also stated that they do not consider that the consumption of DNA, including that from GM crops, as inherently unsafe. In addition, the distinct integration of food genes into the genome of farm animals has never been reported (Chambers *et al.*, 2002).

Antibiotic Resistant Marker (ARM) genes have been used in the production of some of the first generation GM crops, as an aid in identifying transformed cells, and are present in a number of commercialized GM cultivars, although having no specific function in that cultivar. The most commonly used ARM gene is *npt-II,* which inactivates the antibiotics neomycin and kanomycin, while another example is the *bla* gene, which confers resistance to ampicillin. It should be noted that these ARM genes were given a full risk assessment and it was concluded that they presented no risk to human and animal health or to the environment and constructs containing them were given regulatory approval for commercialization. Nevertheless concerns were raised that they might be transferred to pathogenic bacteria, conferring resistance to these specific antibiotics. However, it should be noted that the antibiotics in question, such as neomycin and kanomycin, have all but disappeared from medicinal and veterinary use as more efficacious compounds have replaced them, often without the side effects noted in the earlier compounds.

The concerns over the use of ARM genes were expressed in the preamble of the new EU Directive 2001/18 on the deliberate release in the environment of genetically modified organisms (GMOs) and states that "The issue of antibiotic-resistance genes should be taken into particular consideration when conducting the risk assessment of GMOs containing such genes". It goes on to state in Article 4 that:

> "Member States and the Commission shall ensure that GMOs which contain genes expressing resistance to antibiotics in use for medical or veterinary treatment are taken into particular consideration when carrying out an environmental risk assessment, with a view to identifying and phasing out antibiotic resistance markers in GMOs which may have adverse effects on human health and the environment. This phasing out shall take place by the 31 December 2004 in the case of GMOs placed on the market according to part C and by 31 December 2008 in the case of GMOs authorised under part B."

The presence of ARM genes is widespread and is found in many reservoirs in the environment, including soil, water and intestinal bacteria (Scott and Flint, 1995). Microbiologists have considered that the potential for any resistance-gene transfer from plants to bacteria, thus creating a risk to public health, will be nearly zero. Although there is a theoretical chance that it could occur, the probability is extremely low and borders on an evolutionary timescale. Currently, however, the use of ARM genes in the development of new biotech products is under debate, due perhaps more to public perception than to scientific fact. As reported previously, the probability of transformation of intestinal micro-organisms from feed derived from transgenic cultivars appears very low, but still generates concerns about health and safety of such constructs (Nielsen *et al.*, 1998). Thus publications have also dealt with a probable horizontal transmission of such antibiotic resistance gene from GM plants to soil bacteria, as well as to intestinal micro-organisms (Chambers *et al.*, 2002; Gebhard and Smalla, 1999) A recent study monitoring *AmpR* genes within rumen bacteria of cows fed transgenic maize indicates that the ampicillin resistance gene (*bla*) is detectable in animals fed both the isogenic non-GM as well as the transgenic diet (Berezina *et al.*, 2003). These results may lead to the conclusion that bacteria containing the *bla* gene are found ubiquitously in any experimental setting and override any signal potentially derived from recombinant maize material. These findings correspond with earlier reports indicating a naturally occurring ampicillin resistance of rumen bacteria isolated from conventionally fed cattle (Scott and Flint, 1995). In view of the high level of ampicillin, neomycin and kanomycin resistance genes present in such commensal bacteria, the detection of a specific horizontal transmission from forage plants to intestinal bacteria of the mammalian genomes will be extremely difficult to establish.

The conference of Entransfood, the European Network on Safety Assessment of Genetically Modified Food Crops, concluded that "the risk of gene transfer from foods derived from GM crops that are currently commercially available is deemed negligible." It went on to address the role of ARM genes in biotechnology, and stated that "the use of marker genes expressing resistance to antibiotics in use for medicinal or veterinary purposes has been evaluated and that if the antibiotic is widely used or is a tool of last resort, such genes should be avoided." It concluded that "marker genes coding for neomycin (*np-11*) or hygromycin (*hpt*) can be used without the risk of compromising human or animal health."

However, it should be noted that the *npt-II* gene could be removed or replaced by alternative markers, each of which would then need to be assessed for safety.

6.   CONCLUSIONS

The digestive tract of both monogastrics and ruminants are highly efficient at fragmenting both endogenous and transgenic DNA. While both have been found at key stages of the digestive tract, no studies have detected tDNA in milk, meat or eggs produced by animals receiving diets containing GM feed ingredients. There is no evidence to suggest that milk, meat or eggs derived from animals receiving GM feed ingredients are less safe to eat than those produced from conventional feeds. Furthermore, although it is hypothetically possible, horizontal gene transfer between tDNA and either micro-organisms in the digestive tract of livestock or in the soil has not been established under natural conditions. A specific horizontal gene transfer from currently commercialized GM plants to intestinal bacteria or the mammalian nucleus, followed by a functional expression of such foreign genes, must be considered to be very improbable. Future research will increase the transparency of the production process of a product appearing on the food market.

REFERENCES

Astwood, J. & Fuchs, R. 2001. Status and safety of biotech crops. pp. 152–164, *in:* D.R. Baker and N. K. Umetsu (eds). *Agrochemical Discovery: Insect, Weed and Fungal Control.* ACS Symposium Series, No. 774.

Aumaitre, A., Aulrich, K., Chesson, A., Flachowsky, G. & Piva, G. 2001. New feeds from genetically modified plants: substantial equivalence, nutritional equivalence, digestibility and safety for animals and the food chain. *Livestock Production Science,* **73**: 223–238.

Beever, D.E. & Phipps, R.H. 2001. The fate of plant DNA and novel proteins in feeds for farm livestock: a United Kingdom perspective. *Journal of Animal Science,* 79(E suppl).: E290-5.

Berberich, S.A., Ream, J.E., Jackson, T.L., Wood, R., Stiponovic, R., Harvey, P., Patzer, S. & Fuchs, R.L. 1996. Safety assessment of insect-protected cotton: the composition of the cottonseed is equivalent to conventional cottonseed. *Journal of Agricultural and Food Chemistry,* **41**: 365–371.

Berezina, O., Zverlov, V., Rief, S., Einspanier, R. & Schwarz, W. 2003. Bacterial diversity in the rumen of cows fed transgenic maize. VAAM-Symposion. Berlin, March 2003.

Calsamiglia, S., Hernandez, B., Hartnell, G.F. & Phipps, R.H. 2003. Effect of feeding corn silage produced from corn containing MON810 and GA21 genes on feed intake, milk production and composition in lactating dairy cows. *Journal of Dairy Science,* **86**(Suppl. 1): Abstract 247.

Chambers, P.A., Duggan, P.S., Heritage, J. & Forbes, J.M. 2002. The fate of antibiotic resistance marker genes in transgenic plant feed material fed to chickens. *Journal of Antimicrobial Chemotherapy,* **49**: 161–164.

Chesson, A. 2001. Assessing the safety of GM food crops. Reprinted from R.E. Hester and R.M. Harrison (eds). *Food Safety and Food Quality. Issues in Environmental Science and Technology,* No. 15: 1–24.

Chiter, A., Forbes, J.M. & Blair, G.E. 2000. DNA stability in plant tissues: implications for the possible transfer of genes from genetically modified food. *FEBS Letters,* **481**: 164–168.

Cockburn, A. 2002. Assuring the safety of genetically modified (GM) foods: the importance of an holistic, integrative approach. *Journal of Biotechnology,* **98**: 79–106.

D'Mello, J.P.F. 1982. Utilization of dietary purines and pyrimidines by non-ruminant animals. *Proceedings of the Nutrition Society,* **41**: 301–308.

Doerfler, W. 2000. *Foreign DNA in Mammalian Systems.* New Jersey: Wiley-VCH

Duggan, P.S., Chambers, P.A., Heritage, J. & Forbes, J.M. 2000. Survival of free DNA encoding antibiotic resistance from transgenic maize and the transformation activity of DNA in ovine saliva, ovine rumen fluid and silage effluent. *FEMS Microbiology Letters,* **191**: 71–77.

Duggan, P.S., Chambers, P.A., Heritage, J. & Forbes, J.M. 2003. Fate of genetically modified maize DNA in the oral cavity and rumen of sheep. *British Journal of Nutrition,* **89**: 159–166.

Einspanier, R. 2001. Quantifying genetically modified material in food – Searching for a reliable certification. *European Food Research and Technology,* **213**: 415–416.

Einspanier, R., Klotz, A., Kraft, J., Aulrich, K., Poser, R., Schwägele, F., Jahreis, G. & Flachowsky, G. 2001. The fate of forage plant DNA in farm animals: A collaborative case-study investigating cattle and chicken fed recombinant plant material. *European Food Research and Technology,* **212**: 129–134.

Gebhard, F. & Smalla, K. 1999. Monitoring field releases of genetically modified sugar beets for persistence of transgenic plant DNA and horizontal gene transfer, *FEMS Microbiology Ecology,* **28**: 261–272.

Henderson, J.F. & Patterson, A.R.P. 1973. *Nucleotide Metabolism; An Introduction.* London: Academic Press.

Hupfer, C., Hotzel, H., Sachse, K. & Engel, K.H. 1999 The effect of ensiling on PCR based detection of genetically modified *Bt* maize. *European Food Research Technology,* **209**: 301–304.

James, C. 2003. Global status of commercialised transgenic crops 2002. International Service for the Acquisition of Agri-biotech Applications. Ithaca, New York NY, USA.

Jennings, J.C., Albee, L.D., Kolwyck, D.C., Suber, J.B., Taylor, M.L., Hartnell, G.F., Lirette, R.P. & Glenn, K.C. 2003. Attempts to detect transgenic and endogenous plant DNA and transgenic protein in muscle from broilers fed YieldGuard corn borer corn. *Poultry Science,* **82**: 371–380.

Klaften, M., Whetsell, A., Webster, J., Grewal, R., Fedyk, E., Einspanier, R., Jennings, J., Lirette, R. & Glenn, K. 2004. Agricultural biotechnology: challenges and prospects. pp. 83–99, *in:* M.M. Bhalgat, W.P. Ridley, A.S. Felsot and J.N. Seiber (eds). *Development of PCR methods to detect plant DNA in animal tissues.* ACS Symposium Series No. 866. Washington DC: American Chemical Society.

Klotz, A., Mayer, J. & Einspanier, R. 2002. Carry over of feed-DNA into farm animals: first investigations with pigs, chicken embryos and commercial available chicken samples. *European Food Research and Technology,* **214**: 271–275.

Kuiper, H.A., Kleter, G.A., Noteborn, H.P.J.M. & Kok, E.J. 2001. Assessment of the food safety issues related to genetically modified foods. *The Plant Journal,* **27**: 503–528.

Lander, E.S. *et al.* 2001. Initial sequencing and analysis of the human genome. *Nature,* **409**: 860–921.

Lavrik, P.B., Bartnicki, D.E. & Feldman, J. 1995. Safety assessment of potatoes resistant to Colorado potato beetle. pp. 148–158, *in:* K.H. Engel, G.R. Takeoka and R. Teranishi (eds). *Genetically Modified Foods. Safety Issues.* Washington DC: ACS.

McAllan, A.B. 1980. The degradation of nucleic acids in, and the removal of breakdown products from the small intestines of steers. *British Journal of Nutrition,* **44**: 99–112.

McAllan, A.B. 1982. The fate of nucleic acids in ruminants. *Proceedings of the Nutrition Society,* **41**: 309–317.

Millstone, E., Brunner, E. & Mayer, S. 1999. Beyond the substantial equivalence of GM foods. *Nature,* **401**: 525–526.

Newport, M.J. & Keal, H.D. 1973. Artificial rearing of pigs. 10. Effects of replacing dried skim-milk by a single-cell protein (Pruteen) on performance and digestion of protein. *British Journal of Nutrition,* **44**: 161–170.

Nida, D.L., Patzer, S. & Harvey, P. 1996. Glyphosate-tolerant cotton: the composition of the cottonseed is equivalent to conventional cottonseed. *Journal of Agricultural and Food Chemistry,* **44**: 1967–1974.

Nielsen, K.M., Bones, A.M., Smalla, K. & van Elsas, J.D. 1998. Horizontal gene transfer from transgenic plants to terrestrial bacteria – a rare event? *FEMS Microbiology Reviews,* **22**: 79–103.

Ochman, H., Lawrence, J.G. & Groisman, E.A. 2000. Lateral gene transfer and the nature of bacterial innovation. *Nature,* **405**: 299–304.

OECD [Organisation for Economic Co-operation and Development.]. 1993. Safety evaluation of foods derived from modern biotechnology. Paris: OECD.

OECD. 2001a. Consensus document on key nutrients and key toxicants in low erucic acid rapeseed (canola). Series on the Safety of Novel Foods and Feeds, No. 1. Paris: OECD.

OECD. 2001b. Consensus document on compositional considerations for new varieties of soybean: key food and feed nutrients and anti-nutrients. Series on the Safety of Novel Foods and Feeds No.2. Paris: OECD.

OECD. 2002a. Consensus document on compositional considerations for new varieties of sugar beet: key food and feed nutrients and anti-nutrients. Series on the Safety of Novel Foods and Feeds, No. 3. Paris: OECD.

OECD. 2002b. Consensus document on compositional considerations for new varieties of potatoes: key food and feed nutrients, anti-nutrients and toxicants. Series on the Safety of Novel Foods and Feeds, No. 4. Paris: OECD.

OECD. 2002c. Consensus document on compositional considerations for new varieties of maize (*Zea mays*): key food and feed nutrients and anti-nutrients and secondary metabolites. Series on the Safety of Novel Foods and Feeds, No. 6. Paris: OECD.

OECD. 2003. Considerations for the safety assessment of animal feedstuffs derived from genetically modified plants. Series on the Safety of Novel Foods and Feeds, No. 9. Paris: OECD.

Padgette, S.R., Taylor, N., Nider, D., Bailey, M.R., MacDonald, J., Holden, M.R. & Fuchs, R.L. 1996. The composition of glyphosate-tolerant soybean seed is equivalent to that of conventional soybeans. *Journal of Nutrition,* **126**: 702–716.

Phipps, R.H., Deaville, E.R. & Maddison, B. 2003. Detection of transgenic and endogenous plant DNA in rumen fluid, duodenal digesta, milk, blood and feces of lactating dairy cows. *Journal of Dairy Science,* **86**: 4070–4078.

Piva, G., Morlacchini, M., Pietri, A., Piva, A. & Casadei, G. 2001b. Performance of weaned piglets fed insect-protected (MON 810) or near isogenic corn. *Journal of Animal Science,* **79**(Suppl. 1): 106. Abstract 441.

Piva, G., Morlacchini, M., Pietri, A., Rossi, F. & Prandini, A. 2001a. Growth performance of broilers fed insect-protected (MON 810) or near isogenic control corn. *Poultry Science,* **80**: 320.

Razzaque, M.A. & Topps, J.H. 1972. Utilisation of dietary nucleic acids by sheep. *Proceedings of the Nutrition Society,* **31**: 105A–106A.

Ridley, W.P., Sidhu, R.S., Pyla, P.D., Nemeth, M.A., Breeze, M.L. & Astwood, J.D. 2002. Comparison of the nutritional profile of glyphosate-tolerant corn event NK603 with that of conventional corn (*Zea mays* L.). *Journal of Agricultural and Food Chemistry,* **50**: 7235–7243.

Rogan, G.J., Bookout, J.T. & Duncan, D.R. 2000. Compositional analysis of tubers from insect- and virus-resistant potato plants. *Journal of Agricultural and Food Chemistry,* **48**: 5936–5945.

Salzberg, S.L., White, O., Peterson, J. & Eisen, J.A. 2001. Microbial genes in the human genome: lateral transfer or gene loss? Transfer of plasmids between strains of *Escherichia coli* under rumen conditions. *Science,* 292: 1903–1906.

Sanders, P.R., Lee, T.C. & Gro, M.E. 1998. Safety assessment of the insect-protected corn. pp. 241–256, *in:* J.A. Thomas (ed). *Biotechnology and Safety Assessment.* 2nd edition. San Antonio, Texas: University of Texas Health and Science Centre.

Scott, K.P. & Flint, H.J. 1995. Transfer of plasmids between strains of *E. coli* under rumen conditions. *Journal of Applied Bacteriology,* **78**: 189–193.

Sidhu, R.S., Hammond, B.G., Fuchs, R.L., Mutz, J.N., Holden, L.R., George, B. & Olson, T. 2000. Glyphosate-tolerant corn: the composition and feeding value of grain from glyphosate-tolerant corn is equivalent to that of conventional corn (*Zea Mays* L.). *Journal of Agricultural and Food Chemistry,* **48**: 2305–2312.

Smith, R.H. 1975. Nitrogen metabolism in the rumen and the composition and nutritive value of nitrogen compounds entering the duodenum. pp. 399–415, *in:* I.W. McDonald and A.C.I. Warner (eds). *Digestion and Metabolism in the Ruminant.* Armidale, Australia: University of New England publ. Unit.

Soy Stats. 2001. Prepared by American Soybean Association, St Louis, Missouri, USA. See: http://www.soystats.com.

Van Nevel, C.J. & Demeyer, D.I. 1977. Determination of rumen microbial growth *in vitro* from ^{32}P-labelled phosphate incorporation. *British Journal of Nutrition,* **38**: 101–114.

Wehrmann, A., van Vliet, A., Opsomer, C., Botterman, J. & Schulz, A. 1996. The similarities of bar and pat gene products make them equally applicable for plant engineers. *Nature Biotechnology*, **14**: 1274–1278.

REGULATORY AND BIOSAFETY ISSUES IN RELATION TO TRANSGENIC ANIMALS IN FOOD AND AGRICULTURE, FEEDS CONTAINING GENETICALLY MODIFIED ORGANISMS (GMO) AND VETERINARY BIOLOGICS

H.P.S. Kochhar, G.A. Gifford and S. Kahn
Animal Health and Production Division, Animal Products Directorate, Canadian Food Inspection Agency, Ottawa, Ontario. Canada.

Abstract: Development of an effective regulatory system for genetically engineered animals and their products has been the subject of increasing discussion among researchers, industry and policy developers, as well as the public. Since transgenesis and cloning are relatively new scientific techniques, transgenic animals are new organisms for which there is limited information. The issues associated with the regulation and biosafety of transgenic animals pertain to environmental impact, human food safety, animal health and welfare, trade and ethics. To regulate this new and powerful technology predicated on limited background information is a challenge not only for the regulators, but also for the developers of such animals, who strive to prove that the animals are safe and merit bio-equivalency to their conventional counterparts. In principle, an effective regulatory sieve should permit safe products while forming a formidable barrier for those assessed of posing an unacceptable risk.

Adoption of transgenic technology for use in agriculture will depend upon various factors that range from perceived benefits for humans and animals, to safe propagation, animal welfare considerations and integrity of species, as well as effects on bio-diversity. A regulatory framework designed to address the concerns connected with the environmental release of transgenic animals needs to also take into account the ability of genetically modified animals to survive and compete with conventional populations. Regulatory initiatives for biotechnology-derived animals and their products should ensure high standards for human and animal health; a sound scientific basis for evaluation; transparency and public involvement; and maintenance of genetic diversity.

H. P. S. Makkar and G. J. Viljoen (eds.), Applications of Gene-Based Technologies for Improving Animal Production and Health in Developing Countries, 479-498.

Feeds obtained by use of biotechnology have to be evaluated for animal and human safety by using parameters that define their molecular characterization, nutritional qualities and toxicological aspects, while veterinary biologics derived from biotechnology must be shown to be pure, potent, safe and effective when used according to label recommendations.

The Canadian regulatory system relies on the "precautionary principle" in its approach to regulate the "product" instead of the "process". The regulatory framework captures transgenic animals under the *Canadian Environmental Protection Act* (CEPA). Food from transgenic animals is assessed for safety by Health Canada under its *Novel Foods Regulations* of the *Food and Drugs Act.* Feed containing any genetically modified organism is considered Novel *Feed* under the *Feeds Act and Regulations*. The regulation of veterinary biologics, in an effort to prevent and diagnose infectious diseases of animals, relies on effective science-based regulatory controls under the *Health of Animals Act and Regulations*. The Canadian system of regulation for feeds, veterinary biologics and transgenic animals could be useful to developing countries in the process of establishing an effective framework for new regulations.

1.   INTRODUCTION

Modern agricultural biotechnology, as it relates to animal health and production, is a fascinating blend of animal science, veterinary medicine and molecular biology. Biotechnology research has the potential to yield a broad range of improvements in human health and nutrition, animal health and productivity, food safety and environmental protection. While biotechnology has already had a dramatic impact on the biomedical research and human health field, some of the most promising future applications for biotechnology are in the area of animal health and production. In recent years, we have seen impressive strides in the direction of developing new agricultural applications, such as transgenic animals for research and bio-pharmaceutical production, cloning of livestock species, genetically modified (GM) feed ingredients, and recombinant DNA (rDNA)-based vaccines and diagnostic tests.

Before new biotechnology products enter the market, it is essential that they be rigorously evaluated by the appropriate regulatory agencies to ensure their safety and efficacy. Concerns about the potential detrimental effects of release of genetically modified organisms (GMOs) in the environment must be addressed through implementation of science-based regulatory standards to assure that all products are subjected to thorough safety assessments (Kapuscinski *et al.*, 2003).

The term "biosafety" is used to describe measures taken to reduce and eliminate potential risks resulting from biotechnology and its products. Based on the "precautionary principle" approach, it is considered prudent for regulatory agencies to withhold regulatory approvals in circumstances where there is insufficient information to rule out the possibility of significant detrimental effects.   The need to assure the biosafety of genetically engineered micro-organisms and animals is undoubtedly one of the most critical challenges that the agricultural biotechnology industry and regulatory agencies face.   Science-based regulatory agencies are now faced with the challenge of developing and implementing appropriate standards for assessing the safety of an increasingly broad range of novel biotechnology products (Scientists' Working Group on Biosafety, 1998).

The objective of this paper is to summarize the issues and challenges related to the regulation of animal health and production applications of agricultural biotechnology, with a focus on the relevance of the Canadian experience in developing and implementing regulatory standards for transgenic animals, feeds containing genetically modified ingredients, and recombinant DNA-based veterinary vaccines.

2.   TRANSGENIC ANIMALS IN FOOD AND AGRICULTURE

2.1   Background

Transgenic animals first emerged during the mid-1980s with reports of success in mice (Palmiter *et al.*, 1982). The ability to alter the genome of an animal by introducing exogenous DNA is a major technological advance in biotechnology and animal agriculture. Transgenic animals are produced by the introduction of a small fragment of DNA into pre-implantation embryos. Through this technique, scientists have been able to add, delete, silence or partially inactivate genes of interest. Production of transgenic livestock provides a method of rapidly introducing "novel" genes into poultry, cattle, swine, sheep, goats and fish. Transgenic animals are generally produced for four reasons:
– to improve animal health;
– to increase productivity and improve product quality;
– to mitigate environmental impact of food animal production; or
– to produce bio-pharmaceuticals and industrial biochemical products.

2.2 Methodology

The majority of the methods developed for genetic modification of livestock have been based on techniques that have proven successful in mice (Wilmut and Clark, 1991). The first technique used to generate transgenic livestock (Pursel and Rexroad, 1993) was pronuclear injection of oocytes, which was originally developed in mice. In this technique, fertilized eggs are recovered from donor females, a gene construct is injected in one of the pronuclei of each egg, and the manipulated eggs are transferred into recipient females for development to birth. Transgenic animals produced by this method contain copies of injected gene construct incorporated into their chromosomes. However, a significant portion of such transgenic animals do not express detectable quantities of the protein produced by the transgene. The efficiency of the technique may also be limited due to an early onset of mosaicism (Kang *et al.*, 2002). Despite its inherent limitations, micro-injection has allowed commercial application of transgenic animals for biomedical research purposes and for modification of agricultural traits.

A major advance in transgene technology resulted from the isolation of embryonic stem cells that can be manipulated *in vitro* (Schnieke *et al.*, 1997). These cells can be modified by DNA transfection, which can take advantage of gene targeting by homologous recombination. This method can be used to replace, add or delete endogenous genes, and tends to result in a more consistent expression of the transgene than from some other methods.

Nuclear transfer experiments that produce cloned animals have been conducted successfully in many livestock species, and have proven to be a useful tool for scientists (Wilmut *et al.*, 1997). If the cells used for nuclear transfer are genetically modified, the nuclear transfer technique can be used to generate and propagate new transgenic animals

Another approach to the generation of genetically modified animals takes advantage of "artificial chromosomes", which has recently been reported in work involving cattle (Kuroiwa *et al.*, 2002). In this procedure, the chromosome containing a desired transgene is first introduced into bovine cells in culture. The cells are subsequently utilized as nuclear donors to create transgenic cloned foetal cell lines. These foetal cells are then used in a second round of nuclear transfer to produce transgenic calves (Kuroiwa *et al.*, 2002).

The introduction of foreign genes into the sperm cells before fertilization is another possible approach to create transgenic animals. Fertilization is done by incubation of transgenic sperm (Perry *et al.*, 1999; Robl, 1999) with oocytes in culture, or by direct micro-injection of transgenic sperm into oocytes. Although this method has been used with limited success in livestock animals, it may eventually be refined to become more efficient.

2.3   Applications of transgenic animals

In agriculture, transgenic animals are being investigated as a potential way of developing breeding stock to enhance production traits (e.g. growth rate, feed efficiency), mitigate environmental impacts or increase disease resistance. Potential human health applications of this technology include producing biopharmaceuticals, and generating organs, tissues and cells for xenotransplantation.

We have also seen reports of production of transgenic cows with altered casein, and others that have increased resistance to mastitis, with a secondary benefit of reducing dependence on anti-microbials. Growth hormone genes have been inserted into sheep to produce faster growing lambs, with increased feed efficiency for a net increase in productivity. Scientists have developed transgenic pigs by inserting in their genome a phytase gene that is expressed in salivary glands. This allows such swine to digest phosphorus more efficiently, which reduces the excretion of phosphorus into the manure.

Production of biopharmaceuticals (animal "pharming") has achieved by generating transgenic animals expressing novel proteins in their secretions such as milk or seminal plasma. A range of polyclonal and monoclonal human antibodies, industrial bio-chemicals such as spider silk, and other proteins for treatment of human diseases have been generated through animal "pharming". Latest among the achievements has been the "double knock out pig" (Phelps *et al.*, 2003) that may eventually lead to the production of organs and cells, such as pancreatic islet cells to treat diabetes, or kidneys for transplantation into humans.

2.4   Criteria for evaluating transgenic animals

The adoption of transgenic technology for use in agriculture largely depends on the perceived benefits and risks for humans, as well as the potential impact on the overall health and welfare of farm animals. Some of the more pertinent criteria that must be considered in the light of potential safety concerns are discussed below.

2.4.1   Benefits to human health, animal health and environment

A key underlying objective is to enhance the quality of life for humans, without compromising (and preferably enhancing) the health and welfare of domestic livestock. The assessable benefits include reduced antibiotic use, reduced pathogen loads, enhanced metabolic efficiencies and increased disease resistance.

2.4.2   Safe and "ethical" propagation

The criteria for selection of desirable traits to be propagated through transgenic technology are largely based on the demand to augment specific commercially valuable characteristics. When developing and regulating these commercial applications of biotechnology, the animal biotechnology industry and regulatory agencies must concurrently address society's expectations regarding conservation of genetic material, maintenance of diversity and sustainability of agriculture.

2.4.3   Animal welfare considerations

Some aspects of gene transfer also have the potential for unforeseeable impacts, such as infectious disease hazards or impaired reproduction. For example, some transgenic technologies could potentially lead to the activation and recombination of endogenous retroviruses, resulting from transgenic technologies, which could become virulent. Also, using nuclear transfer techniques to propagate genetic modifications may increase risks to the reproductive health and welfare of surrogate dams and transgenic offspring, due to foetal oversize and perinatal mortalities (Young, Sinclair and Wilmut, 1998; Hill *et al.*, 2001).

2.4.4   Transgene sources

Transgenic animals may be generated by the introduction of genes from number of sources, including: animals of the same species; animals of a different species; microbes; human cells; or *in vitro* nucleic acid synthesis. Transgenes originating from within the same food animal species or transgenes of synthetic origin may be preferable, since it can alleviate some potential concerns about incorporation of genetic material from other species into food animals.

2.4.5   Preservation of the integrity of species

Stringent controls will be required to assure the maintenance of biological diversity and the genetic integrity of species. These unaltered germlines may prove to be an invaluable "gene bank" in the event that novel infectious diseases or heritable genetic defects are inadvertently introduced into modified sub-populations as a consequence of such alterations.

2.5   Regulatory considerations

The regulation of products derived from biotechnology can be based on the principles used for conventionally produced animals. Regulations and standards for determining a responsible use of animal biotechnology in food and agriculture are based on principles that take into account criteria such as benefits and risks, scientific basis of biotechnology and effects on the environment, and must also consider animal welfare and social acceptance (Howard, Homan and Bremel, 2001). Transgenic animals may be viewed as most acceptable if the end result of the genetic manipulation applied is to provide better quality of life for humans, or to provide "environmentally friendly" alternatives to "factory farms".

Producers of transgenic food animals may encounter considerable resistance among some sectors of the general public, as well as among producers of conventional or "organic" livestock who are concerned that the presence of transgenic animals among the general livestock population may result in inadvertent mixing with the conventional or "organic" products. Consequently, regulatory initiatives must be founded on principles that encompass careful evaluation and monitoring of products of animal biotechnology to assure appropriate segregation, identification and tracking of transgenic animals and their products.

Possible considerations for developing regulations to ensure responsible introduction of transgenic animals include those discussed below.

2.5.1   Establish high standards for safeguarding human health and animal health and welfare

A key component of animal health and welfare is the evaluation of proposals to generate transgenic animals – the rationale and justification for the transgenesis and the associated welfare issues (CCAC, 1997). Current societal values could preclude employment of any procedures that might significantly increase risk of pregnancy wastage, dystocia or neonatal loss. In developing regulatory frameworks for transgenic livestock, acceptable target animal or human food safety data will require development and validation of assays, based on good laboratory practice. Selectable markers encoding resistance to antibiotics have been identified as a potential concern in developing animals intended for food purposes. Similarly, there is some concern that endogenous retroviruses may be activated in transgenic animals, or may recombine with exogenous viruses, and cause disease in transgenic animals or in other species.

2.5.2   Development of clear technical standards and assessment guidelines

A transgenic animal's health and productivity parameters may serve as indicators of the safety of its products and by-products (Berkowitz, 1993). The evaluation of products from transgenic animals is based on the principle of demonstrating substantial equivalence to products from non-transgenic animals. The necessary data should include molecular characterization, toxicological studies and nutritional similarities for products intended for food use. In animals genetically modified for "pharming" purposes, a principal concern is demonstrating the desired uniform expression of the transgene in the target tissue. For example, in designing expression of a biopharmaceutical in milk, the effort is focused on optimizing the expression of the transgene in mammary epithelium. Unintended expression of transgenic protein in non-target tissues (referred to as "leaking") or mosaicism should be assessed by appropriate evaluation. The possibility of the transgene promoter acting as a vector for diseases must also be considered.

2.5.3   Provision of a sound scientific basis to evaluate associated risk

The risk assessment of any novel organism is a critical component in the evaluation and designing of criteria for regulation of associated products. If the required safety data is not available, the "precautionary principle" may need to be followed in implementing regulatory controls for certain classes of transgenic animals. Risk assessment also overarches the sound evaluation criteria to demonstrate that equivalence in genetically modified livestock may be impossible to demonstrate unequivocally. The demonstration of minimal acceptable risk in a transgenic animal could be a way to prove it safe. The documents and scientific evaluations in support of providing a sound scientific data basis to demonstrate environmental safety (Kappeli and Auberson, 1997) are fundamental to effective regulations.

2.5.4   Consultation and involvement of stakeholders and the general public in the development of regulations

The development of regulations that safeguard public concerns and those that allow the technology to benefit agriculture are most effective if they neither "restrict" nor "facilitate". Public acceptance is one of the key determinants and is best achieved by consultation with all stakeholders, including industry, academia, regulators and developers, as well as with producers of conventional animals. The government of a modern democratic society is obligated to not merely accommodate the moral convictions of its citizens but to treat them with respect and thus devise policies that can command something close to a reflective consensus (Nuffield, 1999).

2.5.5   Building upon existing regulations and technical standards

The existing regulations for conventionally derived animals or recombinant micro-organisms may be adapted to accommodate transgenic animals. It is appropriate to build on existing regulations and use for capacity building the legislative authority of regulatory institutions already involved in similar activities. In situations where overlaps exist in the intended use of the products of transgenic livestock, several acts and regulations can be triggered and the necessary regulatory control can be achieved only by developing regulatory standards in collaboration with the departments and agencies involved.

2.5.6   Maintenance of genetic diversity and conservation of environment

The regulatory framework has to take into account genetic diversity and biosafety and should strive to minimize the environmental impact of introducing transgenic animals. The ability of modified animals to become feral and compete with their wild type counterparts has to be assessed and the regulations should be based on minimizing loss of diversity. Regulations should establish compulsory standards for identification, tracking and segregation of transgenic animals so that they can be differentiated from conventional animals where necessary to satisfy consumer demands and concerns.

2.6   Biosafety considerations

Biosafety is the term used to describe efforts to reduce and eliminate the potential risks resulting from biotechnology and its products. In the light of ongoing advances in biotechnology and of international trade implications, there is a growing need for harmonized regulations and technical standards to facilitate development of safe commercial applications. Appropriate biosafety standards are one of the prerequisites for the realization of the potential benefits of biotechnology research. The benefits of implementing rigorous biosafety standards include reduction of possible human and environmental risks and maintaining public confidence in biotechnology regulatory controls (Maredia, 1998). In developing biosafety standards and guidelines for transgenic animals, regulatory agencies have to balance the potential benefits against the potential risks. In food and agriculture, the major potential negative impact is the risk of contaminating food products. The number of ingredients in prepared foods make it increasingly difficult to trace the source of the food or its ingredients. The potential for adverse reactions associated with food, including food intolerance and allergy, necessitate comprehensive biosafety regulations that take into account these issues. Biosafety regulations and standards must take into account the above factors and facilitate research while minimizing risk to researchers, animals, consumers and the environment.

3. LIVESTOCK FEEDS DERIVED FROM GENETICALLY MODIFIED ORGANISMS

Public concerns about food safety and GMOs have dramatically raised the profile of agricultural biotechnology. With the advancement of biotechnology, more and more animal feed ingredients derived from GMOs are being developed for the market. The crucial link between animal feed and the human food chain necessitates the establishment of rigorous criteria for the evaluation of feeds that contain ingredients derived from GMOs.

In the Canadian *Feed Regulations*, feeds are defined as being

"any substance or mixture of substances manufactured, sold or represented for use or consumption by livestock, for providing the nutritional requirement of livestock or for the purpose of preventing or correcting nutritional disorders of livestock."

Feeds derived from GMOs include ingredients from plant, microbial and animal sources. Currently on the market we can find plant ingredients derived from modified maize, canola, soybean, rice, wheat and cotton. As for ingredients derived from microbial sources, vitamins, amino acids, enzymes are marketed, as well as modified live direct fed microbials. In the future, animal by-products may become available, such as milk replacers, whey, etc., derived from transgenic animals.

When conducting a safety evaluation for feed, different concerns and parameters are taken into consideration, compared with those present in a human food use evaluation. In particular, the assessment of animal feeds must take into account any risk to the health of the animals consuming the feed, as well as any secondary impacts on food safety with regard to the transfer of potential residues ("downstream effects" – meat, milk and eggs). Limited variety in animal diets, the daily feed intake, as well as their production potential are factors in an evaluation. Other important criteria in a feed safety assessment are "bystander" animal exposure and the environmental impact.

The regulatory issues related to feeds containing ingredients derived from GMOs encompass the principles of science-based risk assessment for safety and efficacy. Many livestock feeds make use of the same ingredients used for human food. Consequently, many elements of a safety assessment are common to both. Both assessments require detailed identification and molecular characterization of the introduced genetic element(s), the expression of the novel trait(s) and the impact of these in the newly modified organism. Dietary composition of the modified feed in comparison with an equivalent counterpart, and potential allergenicity and toxicity of the modified feed, are also common elements in assessment for human safety.

In a livestock feed safety assessment, concerns of the differential expression (spatial, temporal) of the introduced trait(s) in the organism may be different from the human food evaluation, since livestock feed makes use of products and by-products or organisms that are not intended for human food use, e.g. cottonseed meal. The introduced DNA or novel protein may be virtually absent, or considerably concentrated in the product or by-product (as in the case of seed meals).

Finally, as with imported food derived from GMOs, imported feed may also be subject to animal and plant health importation requirements. For instance, in cases where animal by-products are used in the manufacture of the feed, the country of origin of the animal by-product will be a factor in the importation of the feed. Countries with BSE prevalence may have a limited market for feed containing products and by-products from animal sources.

4.   VETERINARY BIOLOGICS

Veterinary biologics are products such as vaccines and diagnostic kits that are used for the prevention, treatment or diagnosis of infectious diseases in animals. With the growing use of recombinant DNA techniques in microbiology, many veterinary biological products are now derived through biotechnology. Veterinary biologics can be broadly classified into two classes, depending on the safety aspects and biological characteristics (Sethi, Gifford and Samagh, 1997) (Table 1).

Table 1. Classification of veterinary biologics based on risk.

	Description
Class I	Inactivated vaccines (conventional or rDNA).
(Low risk)	Subunit vaccines (conventional or rDNA).
	Cytokines and monoclonal antibody (hybridoma) products.
	Modified live (attenuated) vaccines or gene-deleted rDNA vaccines.
Class II	Vaccines using a live vector to carry recombinant derived foreign genes.
(High risk)	Live organisms modified by introduction of foreign DNA.

For modified live (attenuated) vaccines, the inherent safety and potential for residual virulence or reversion to virulence is of prime concern for regulatory agencies. In developing modified live (attenuated) vaccines, manufacturers must consider ways to minimize vaccine organism shedding, horizontal transmission and environmental persistence.

The movement of live recombinant DNA (rDNA) veterinary biologics from contained laboratory research to a fully licensed product for distribution and sale, can be accomplished in four stages:
- Preliminary laboratory research.
- Controlled experiments using a prototype vaccine in target and non-target species, under appropriate laboratory containment conditions.
- Limited field trial using target species under confined conditions, with approval from regulatory agencies.
- Submission of complete licensing dossier – including summary test results, supporting documentation and field trial data – to the regulating agency, for approval.

4.1　Regulatory considerations for veterinary biologics

Regulatory issues related to the release and use of veterinary biologics are based on the principles of assessing the purity, potency, safety and efficacy of the product. These veterinary biologics regulations cover a diverse range of products, including vaccines, antibody products and diagnostic kits that are used for the diagnosis, prevention, control or treatment of a wide range of infectious diseases in animals. These products are derived from materials of animal or microbial origin, and may be produced by conventional microbiological methods or by modern biotechnology techniques. The science-based risk assessment that is utilized during the registration of a veterinary biological product is essential to demonstrate the safety of the product that is administered to livestock. Various aspects of this assessment are considered below.

4.1.1　Veterinary biologics manufacturing facility

Veterinary biologics manufacturing establishments must have appropriate facilities, equipment, personnel and quality assurance monitoring systems to assure manufacturing of veterinary biologics. Production and testing records and other related documents may be assessed to monitor compliance with existing regulations and conformance with good manufacturing practices (GMPs) or equivalent standards.

4.1.2   Safety assessments

Human safety (occupational exposure, food safety), target-animal safety and environmental safety are considered when evaluating rDNA vaccines. Target-animal safety is evaluated in both laboratory and field studies. Local and systemic reactions of animals to vaccinations are monitored to assess the safety to livestock. For biotechnology-derived nucleic acid vaccines, potential safety considerations include: potential for integration of injected plasmids into the host genome (somatic cells or gametes); potential for germline transmission if integrated into gametes; and potential for adverse immunological sequelae such as autoimmunity and immune tolerance.

4.1.3   Environmental assessments

Environmental assessments incorporate the following parameters:
- Molecular and biological characterization of plasmids.
- Documentation of human, animal and environmental safety considerations.
- Qualitative and quantitative risk assessment and management options.
- Monitoring strategies for the field use of products.

Regulatory considerations must also take into account any potential hazards associated with novel manufacturing methods or delivery systems. For example, the production of plant-based oral vaccines for livestock diseases, such as a vaccine for porcine transmissible gastroenteritis virus (Tuboly *et al.*, 2000) expressed in maize, will present additional regulatory challenges, due to the fact that the vaccine antigens are produced in maize fields, rather than in a virology laboratory.   Other examples of non-conventional production systems include the production of vaccines in specific-pathogen-free (SPF) eggs that employ embryonated chicken eggs, and immunization of chickens to elicit an immunological response that leads to accumulation of protective antibodies in egg yolks.

5.   REGULATORY FRAMEWORK IN CANADA

In Canada, several federal government statutes define biotechnology as "The application of science and engineering in the direct or indirect use of living organisms or parts or products of living organisms in their natural or modified forms". This broad definition encompasses organisms developed through traditional breeding methods, as well as through newer technologies such as genetic engineering. Any organism that has undergone procedures intended to change its genetic make-up and traits would be considered to be an organism derived from biotechnology. The Canadian regulatory framework is a "product"-based regulatory system where the end products are evaluated and registered on a case-by-case basis, rather than generic approval of a "process". This system is based on the principle that the animate product of biotechnology is evaluated prior to use or release into the environment.

5.1   Transgenic animal regulations in Canada

The term "biotechnology-derived animal" is an extension of the definition of biotechnology. It refers to animals that have been generated through biotechnological methods. This term may include, but not be limited to, the following categories:
- Genetically engineered or modified animals in which genetic material has been added, deleted, silenced or altered to influence expression of genes and traits.
- Clones of animals derived by nuclear transfer from embryonic or somatic cells.
- Chimeric animals that have received transplanted cells from another animal.
- Interspecies hybrids produced by *in vitro* fertilization.

The Canadian Environmental Protection Act, 1999 (known as CEPA 1999), provides the federal government with the authority to regulate substances ranging from chemicals to animate products of biotechnology (i.e. living organisms). CEPA 1999 ensures that new substances, including biotechnology substances not regulated under other legislation, undergo pre-import or manufacture notification and assessment for potential risk to the environment and human health. Currently in Canada, biotechnology-derived animals are regulated through CEPA 1999, principally administered by Environment Canada (EC) and co-administered by Health Canada (HC), in collaboration with the Canadian Food Inspection Agency (CFIA). The *New Substances Notification Regulation* (NSNR), Schedule XIX, section 29.16 of CEPA[4] provides the first trigger, while the *Novel Foods Regulations* under the *Foods and Drugs Act* administered by HC comes into effect if the product is intended for use as a food (EC/HC, 2001).

Environmental assessment notification guidelines for biotechnology-derived animals, which are currently being drafted by CFIA, in consultation with EC and HC, will address issues related to anticipated risks pertaining to mammalian and avian livestock species intended for release outside of research and development facilities. Each genetic modification presents a

---

4. See: http://www2ec.gc.ca/substances/nsb/download/Biogel1201

different set of circumstances and potential risks, which must be assessed on a case-by-case basis.

Health Canada's *Novel Foods Regulations*[5] address the issue of "substantial equivalence" by assessing the three main aspects of evaluation: molecular characterization; nutritional similarities; and toxicological assessment, including allergenicity data. Once the product is evaluated to be "substantially equivalent" to its conventional counterpart, the product is allowed to be released for use. However, if any unintended effects are observed during the post-approval phase, the developers are required to report the findings and the product becomes subject to possible re-evaluation.

Transgenic animals have been developed in Canada for the purpose of biopharmaceutical production (e.g. goats producing "spider silk" in milk, to be used as suture material), and reducing environmental pollution from manure (pigs expressing phytase genes in saliva to reduce phosphorus output in manure). Cattle have been cloned to exploit the genetic potential of valuable breeding animals. In Canada, the developers are required to submit any notification to EC, a federal regulatory agency, for assessment of their animate products of biotechnology. Developers are allowed to continue their development activities under specific exemption criteria of a "research and development" clause in CEPA 1999, provided they have appropriate bio-containment facilities and proper segregation is maintained so that the experimental animals are not allowed to co-mingle with the conventional population.

5.2 Regulations concerning novel feeds in Canada

The *Feeds Act and Regulations* (CFIA, 2003a) requires that, in Canada, all single-ingredient feeds be evaluated prior to their use in livestock feeds. This applies to both imported and domestically manufactured products. The standards and labelling requirements are specified in the legislation. The Feeds Section in CFIA is responsible for administering the *Feeds Act and Regulations* and for approving feeds. In Canada, a feed is considered "novel" if it comprises an organism or organisms, or parts or products thereof, that has a novel trait or has not been previously used in Canada. Novel feeds include microbial products (both viable and non-viable), such as bacteria, yeast, fungi, micro-algae or forage and silage inoculants; plants with novel traits; or fermentation products such as enzymes, biomass proteins, amino acids, vitamins or flavouring ingredients.

---

5. See: http://www.hc-sc.gc/food-ailment/mh-dm/ofbbba/nfiani/ e_novel_foods_and_ingredients.html

Imported and domestically manufactured product applications are reviewed for safety and efficacy. Safety assessments include an evaluation of human (occupational and food safety), animal and environment safety. Labelling must conform to legislated requirements. The data requirements vary with the product type, but may include a complete identification of all ingredients, certificate of analysis, laboratory methodology, quality assurance procedures and a product sample. Toxicity and stability data is also required for safety assessments. Data submissions for novel feeds should include a description of the organism and the genetic modification, the intended use, the environmental fate and a determination of whether the gene products or their metabolic products will reach the human food chain. The importation of a novel feed requires a safety assessment of the feed before entry into Canada is authorized. Applicants must obtain a letter of authorization from the feed section. These products may also be subject to animal and plant health importation requirements. Novel feed products being developed or tested in feeding trials are subject to regulatory requirements.

5.3   Regulations concerning veterinary biologics

The CFIA Veterinary Biologics Section (CFIA, 2003b) is responsible for regulating the manufacturing, importation, testing, distribution and use of veterinary biologics in Canada.   The Veterinary Biologics Section's regulatory controls and standards are implemented under the authority of the *Health of Animals Act and Regulations.*

To meet Canadian licensing requirements, the regulated veterinary biologics must be shown to be pure, potent, safe and effective when used according to the manufacturer's label recommendations. The objective of this regulatory programme is to help protect the health of Canadian animals (including domestic livestock, poultry, companion animals, wildlife and aquatic species) as well as helping to safeguard public health and food safety by controlling indigenous animal diseases and preventing the introduction and dissemination of foreign animal diseases.

In addition to the registration of veterinary biologics for commercial sale in Canada, the Veterinary Biologics Section is also responsible for review and approval of applications for "restricted use" of veterinary biologics in special situations where an appropriate licensed veterinary biologic is not available. This is done through issuance of a *Veterinary Biologics Product Licence, Import Permit,* or *Permit To Release Veterinary Biologics,* in which various restrictions and conditions are applied, depending on the nature of the product, country of origin, target species, intended use (i.e. investigational, educational or emergency purposes) and the associated risks. Under the *Health of Animals Regulations,* manufacturers and importers are required to notify the Veterinary Biologics Section of suspected adverse reactions. Individual animal owners or veterinarians may report suspected adverse reactions by submitting the notification forms through the Canadian licensee, or they may be forwarded directly to the Veterinary Biologics Section.

## 6.	CANADIAN REGULATORY FRAMEWORK AND DEVELOPING COUNTRIES

The Canadian system for regulation of biotechnology (CFIA, 1998) and animate products of biotechnology is based on a regulatory framework developed in 1993. In developing and refining this biotechnology regulatory framework, inputs from stakeholders and the public were taken into consideration, with the objective of ensuring that stakeholders and the general public have a relatively high level of confidence in the regulatory system and the safety of the regulated agricultural commodities.

As the Canadian system undergoes further changes and the criteria for the regulation of transgenic animals under the current legislation is refined, there is wide scope for improvement as technology evolves. The present Canadian regulatory principles may be adaptable for the developing countries, with modification of these regulations. In certain circumstances, countries may find it worthwhile to collaborate through international standard setting bodies, or to pool their resources to assess the risks in introducing GMOs. Many countries currently lack the necessary regulatory framework and specialized enforcement capabilities to ensure that GMOs are not released for unintended purposes. Since the positive or negative effects are not yet known of products derived from genetically modified livestock, it is important that this area of research and development remains in the public domain. Developing countries might benefit from knowledge of the regulatory framework and the experience of other countries.

– The environmental assessment of any genetic modification should form an intrinsic part of the regulatory framework and be further reviewed depending upon the intent of the modification.

– The demonstration of substantial equivalence of a GM food may circumvent the need for a full food safety review for developing countries with limited resources, However, regulations must recognize that the currently available data may not be based on long-term tests and may not have fully evaluated the effects of prolonged consumption.

– Biodiversity and the concept of biosecurity may need to be embedded in the regulatory framework to ensure that the long-term viability and sustainability of conventional species are not compromised.

Genetically modified foods are often viewed as a way to help feed the world's escalating population, especially in the poorest developing countries, where a substantial portion of this growth is expected to occur. However, as a regulator, this context has limited value as safety is the prime issue and the regulatory considerations have to enforce strict compliance to the assessment and evaluation criteria. The development of GM livestock is already an experimental reality and may become a commercial reality. An important question that must be answered is whether or not the current regulatory standards are appropriate for regulating biotechnology-derived animals and their products. As regulatory requirements for transgenic livestock are not yet fully defined, they clearly have potential to positively influence existing regulatory and industry practices in such important areas as animal health, disease diagnosis, international trade certification, and animal identification and traceability. A key policy objective for the animal biotechnology industry and for any regulatory agency should be to build up consumer confidence through development of appropriate regulatory controls (Evans, 1999).

ACKNOWLEDGEMENTS

The authors wish to thank Dr Jan S. Gavora for critical review of this manuscript and comments. The comments provided by Ms Annie Savoie, Dr Donna Hutchings and Dr Bhim Bhatia provided useful additions to the manuscript and are gratefully acknowledged.

REFERENCES

Berkowitz, D.B. 1993. The food safety of transgenic animals: implications from traditional breeding. *Journal of Animal Science,* 7: 43–46.

CCAC [Canadian Council on Animal Care]. 1997. CCAC guidelines on transgenic animals. Canadian Council on Animal Care, Ottawa, Ontario. Available at: http://www.ccac.ca/englishgdlines/transgen/transge1.htm#download.

CFIA [Canadian Food Inspection Agency]. 1998. Session report on "Consultation on regulating livestock animal and fish derived from biotechnology". Available at: http://www.inspection.gc.ca/english/sci/biotech/tech/aniconsulte.shtml

CFIA. 2003a. Canadian Food Inspection Agency – Feeds section. Available at: http://www.inspection.gc.ca/english/anima/feebet/feebete.shtml

CFIA. 2003b. Canadian Food Inspection Agency – Veterinary Biologics section. Available at: http://www.inspection.gc.ca/english/anima/vetbio/vbpbve.shtml

EC/HC [Environment Canada/Health Canada]. 2001. Guidelines for the notification and testing of New Substance organisms, pursuant to "The New Substances Notification

Regulations of the Canadian Environmental Protection Act, 1999". Published by Environment Canada and Health Canada.

Evans, B.R. 1999. The prospects for international regulatory interventions in embryo transfer and reproductive technologies in the next century. *Theriogenology,* **51**: 71–80.

Hill, J.R., Edwards, J.F., Sawyer, N., Blackwell, C. & Cibelli, J.B. 2001. Placental anomalies in a viable cloned calf. *Cloning,* **3**: 83–88.

Howard, T.H., Homan, E.J. & Bremel, R.D. 2001. Transgenic livestock: Regulation and science in a changing environment. *Journal of Animal Science,* **79**(Suppl.): 1–11.

Kang, Y.K., Park, J.S., Koo, D.B., Choi, Y.H., Kim, S.U., Lee, K.K. & Han, Y.M. 2002. Limited demethylation leaves mosaic-type methylation states in cloned bovine pre-implantation embryos. *EMBO Journal,* **21**: 1092–1100.

Kappeli, O. & Auberson, L. 1997. The science and intricacy of environmental safety evaluations. *Trends in Biotechnology,* **15**: 342–349.

Kapuscinski, A.R., Goodman, R.M., Hann, S.D., Jacobs, L.R., Pullins, E.E., Johnson, C.S., Kinsey, J.D., Krall, R.L., La Vina, A.G., Mellon, M.G. & Ruttan, V.W. 2003. Making 'safety first' a reality for biotechnology products. *Nature Biotechnology,* **21**: 599–601.

Kuroiwa, Y., Kasinathan, P., Choi, Y.J., Naeem, R., Tomizuka, K., Sullivan, E.J., Knott, J.G., Duteau, A., Goldsby, R.A., Osborne, B.A., Ishida, I. & Robl, J.M. 2002. Cloned transchromosomic calves producing human immunoglobulin. *Nature Biotechnology,* **20**: 889–894.

Maredia, M.K. 1998. The economics of biosafety: implications for biotechnology in developing countries. *Biosafety,* **3**: 1–10.

Nuffield. 1999. Nuffield Council on Bioethics in Genetically Modified Crops: The ethical and social issues. The Nuffield Foundation, London.

Palmiter R.D., Brinster, R.L., Hammer, R.E., Trumbauer, M.E., Rosenfeld, M.G., Birnberg, N.C. & Evans, R.M. 1982. Dramatic growth of mice that develop with eggs microinjected with matalotheonein-growth hormone fusion genes. *Nature,* **300**: 611–615.

Perry, A.C.F., Wakayama, T., Kishikawa, H., Kasai, T., Okade, M., Toyoda, Y. & Yanagimachi, R. 1999. Mammalian transgenesis by intracytoplasmic sperm injection. *Science,* **284**: 1180–1183.

Phelps, C.J., Koike, C., Vaught, T.D., Boone, J., Wells, K.D., Chen, S.H., Ball, S., Specht, S.M., Polejaeva, I.A., Monahan, J.A., Jobst, P.M., Sharma, S.B., Lamborn, A.E., Garst, A.S., Moore, M., Demetris, A.J., Rudert, W.A., Bottino, R., Bertera, S., Trucco, M., Starzl, T.E., Dai, Y. & Ayares, D.L. 2003. Production of alpha 1,3 galactosyltransferase-deficient pigs. *Science,* **299**: 411–414.

Pursel, V.G. & Rexroad C.E. Jr. 1993. Recent progress in the transgenic modification of swine and sheep. *Molecular Reproduction and Development,* **36**: 251–254.

Robl, J.M. 1999. New life for sperm mediated transgenesis? *Nature Biotechnology,* **17**: 636–637.

Schnieke, A.E., Kind, A.J., Ritchie, W.A., Mycock, K., Scott, A.R., Ritchie, M., Wilmut, I., Colman, A. & Campbell, K.H. 1997. Human factor IX transgenic sheep produced by transfer of nuclei from transfected fetal fibroblasts. *Science,* **278**: 2130–2133.

Scientists' Working Group on Biosafety. 1998. Manual for assessing ecological and human health effects of genetically engineered organisms. Edmonds, WA: Edmonds Institute.

Sethi, M.S., Gifford, G.A. & Samagh, B.S. 1997. Canadian regulatory requirements for recombinant fish vaccines. *Developments in Biological Standardization,* **90**: 347–353.

Tuboly, T., Yu, W., Bailey, A., Degrandis, S., Du, S., Erickson, L. & Nagy, E. 2000. Immunogenicity of porcine transmissible gastroenteritis virus spike protein expressed in plants. *Vaccine,* **18**(19): 2023–2028.

Wilmut, I. & Clark, A.J. 1991. Basic techniques for transgenesis. *Journal of Reproduction and Fertility,* **43**: 265–275.

Wilmut, I., Schnieke, A.E., McWhir, J., Kind, A.J. & Campbell, K.H. 1997. Viable offspring derived from fetal and adult mammalian cells. *Nature,* **385**: 810–813.

Young, L.E., Sinclair, K.D. & Wilmut, I. 1998. Large offspring syndrome in cattle and sheep. *Reviews of Reproduction,* **3**: 155–163.

INTELLECTUAL PROPERTY RIGHTS AND GENE-BASED TECHNOLOGIES FOR ANIMAL PRODUCTION AND HEALTH
Issues for developing countries

Graham Dutfield

Queen Mary Intellectual Property Research Institute, Queen Mary University of London, John Vane Science Building, Charterhouse Square, London EC1M 6BQ.

Abstract: Intellectual property rights (IPR) are legal and institutional devices to protect creations of the mind. With respect to gene-based innovation, the most significant IPR is patents. Appropriate patent regimes have the potential to foster innovation in animal biotechnology and the transfer of gene-based technologies. Inappropriate patent systems may be counter-productive. Indeed, many critics are doubtful that the current international patent standards, based as they are on a combination of the United States of America' and European regimes, can help countries that lack the capacity to do much life science and biotechnology research to become more innovative or contribute to the acquisition, absorption and, where desirable, the adaptation of new gene-based technologies from outside. Present legislation in Europe, North America and internationally is considered, together with the controversies and important policy questions for developing countries, and the choices facing countries seeking to enhance their scientific and technological capacities in these areas.

1. WHAT ARE INTELLECTUAL PROPERTY RIGHTS AND WHY DO THEY MATTER?

Intellectual property rights (IPR) are legal and institutional devices to protect creations of the mind such as inventions, works of art and literature, and designs. They also include marks on products to indicate their difference from similar ones sold by competitors. Over the years, the rather elastic intellectual property (IP) concept has been stretched to include not only

H. P. S. Makkar and G. J. Viljoen (eds.), Applications of Gene-Based Technologies for Improving Animal Production and Health in Developing Countries, 499-520.

patents, copyrights, trademarks and industrial designs, but also trade secrets, plant breeders' rights, geographical indications, and rights to layout-designs of integrated circuits, among others. It is important to note that "intellectual property" does not lend itself to any precise definition that would satisfy everybody. Indeed, a recent document published by the World Intellectual Property Organization expressed some quite reasonable scepticism about its validity:

> "Intellectual property, broadly conceived, may be seen as a misnomer, because it does not necessarily cover 'intellectual works' as such – it covers intangible assets of diverse origins, which need not entail abstract intellectual work; nor need it be defined and protected through property rights alone (the moral rights of authors and the reputation of merchants are not the subject of property, under a civil law concept)."   (WIPO, 2002)

Today's international IP rules require both developed and developing countries to provide unprecedentedly high standards of protection. The 1994 Agreement on Trade-related Aspects of Intellectual Property Rights (TRIPS), one of the main outcomes of the Uruguay Round of the General Agreement on Tariffs and Trade (GATT), which is administered by the Geneva-based World Trade Organization (WTO), is of special importance in that it establishes enforceable global minimum (and high) standards of protection and enforcement for virtually all the most important IPRs in one single agreement.

To proponents, IPRs contribute to the enrichment of society through (a) the widest possible availability of new and useful goods, services and technical information that derive from innovative activity, and (b) the highest possible level of economic activity based on the production, circulation and further development of such goods, services and information. These objectives are supposed to be achieved because owners can seek to exploit their legal rights by turning them into commercial advantages. The possibility of attaining such advantages, it is believed, encourages innovation. But after a certain period of time, these legal rights are extinguished and the now unprotected inventions and works can be freely used by others.

With respect to gene-based innovation, the most significant IPR is patents. Patents provide inventors with legal rights to prevent others from using, selling or importing their inventions for a fixed period, nowadays normally 20 years. Applicants for a patent must satisfy a national or regional patent issuing authority that the invention described in the application is new, susceptible of industrial application (or "useful" in the United States of America), and that its creation involved an inventive step or would be unobvious to a skilled practitioner. For life science firms, which often need to spend heavily to discover new products, develop them and acquire regulatory approval to sell them, patents are essential for achieving high returns on their research and development (R&D) investments.

Patents, as with other kinds of property, provide exclusive rights that allow markets for certain things, in this case valuable information in the form of inventions, to operate where they otherwise could not do so. As Geroski puts it, "patents are designed to create a market for knowledge by assigning propriety rights to innovators which enable them to overcome the problem of non-excludability while, at the same time, encouraging the maximum diffusion of knowledge by making it public". The public goods justification for patents posits that such rights are likely to have various beneficial effects. Several economists and others have countered convincingly that knowledge is not exactly like other public goods and that such justifications need to be qualified (see David, 1993). Persuasive as these counter-arguments may be, let us for the sake of argument accept that knowledge *is* a public good.

Accordingly, two such beneficial effects should be mentioned here. The first is that they encourage investment in invention and the research and development needed to turn inventions into marketable innovations. The second is that partners in collaborative research and development programmes can exploit the market value of an invention in ways they could not do without a patent. How would this work? Ownership of the patent could be shared, or it could be held by one party on behalf of the others by an arrangement that all would benefit from successful appropriation of the patent's market value either by developing a product or selling or licensing the patent to someone else interested in doing so. Since many innovations and patents are dependent upon earlier patents, it is conceivable also that the patent could be used to engage profitably in transactions with follow-on innovators.

It is commonly assumed that patents are for mechanical devices. Mousetraps are frequently mentioned as typical examples. In fact, this has never been accurate. To illustrate this, in 1724, Thomas Greening was granted an English patent for "grafting or budding the English elm upon the stock of the Dutch elm". In 1785, Philip Le Brocq acquired a patent for "rearing, cultivating, training, and bringing to perfection, all kinds of fruit trees, shrubs, and plants; protecting their leaves, blossoms, flowers and fruits".

Nonetheless, patenting activity relating to the structural and functional components of animals and the extension of the patent system to include animals themselves within their ambit are recent phenomena. Before the 1980s non-biological and microbiological processes, including genetic engineering technologies like recombinant DNA, were patentable in Europe and the United States of America, but the situation for life forms as well as their structural and functional components was uncertain to say the least. It was only in 1980 that DNA sequences "first began appearing in patents", and there were only 16 sequences in the whole of that year (Stokes 2001). However, "by 1990 that figure had risen to over 6,000 sequences". Throughout the 1990s the growth in the patenting of sequences expanded exponentially, and one can expect this to continue. "In 2000 over 355,000 sequences were published in patents, a 5000 per cent increase over 1990" (*ibid.*). In 1988, the first patent was granted for an animal, the famous "oncomouse" (see below).

There is now much more clarity about what is and is not patentable in these parts of the world, and the extent to which discoveries arising from gene-based technological processes can be patented, including living things, is now quite substantial. Nonetheless, these and other countries diverge widely with respect to how their patent systems deal with the new biotechnologies, and the current international rules contribute to harmonization in this area only to a moderate degree.

To the extent that developing countries need access to gene-based technologies relating to animal production and health, patents are extremely important. Appropriate patent regimes have the potential to foster innovation in animal biotechnology and the transfer of gene-based technologies. Inappropriate patent systems may do neither, and may even be counter-productive. Indeed, many critics are doubtful that the current international patent standards, based as they are on a combination of the United States of America and European regimes, can help countries that lack the capacity to do much life science and biotechnology research of their own to become more innovative or contribute to the acquisition, absorption and, where desirable, the adaptation of new gene-based technologies from outside (see Dutfield, 2004). The questions facing developing countries are: "What would an appropriate patent system look like?" and "Are the international rules flexible enough to allow sufficient differentiation to fully accommodate each developing country's varied capacities, needs and priorities in the area of animal health and production?"

2.   INTELLECTUAL PROPERTY AND GENE-BASED INVENTION: THE UNITED STATES AND EUROPE

In some respects, the United States of America has played a pioneering role in extending patent law to animal-related inventions. In 1987, the Board of Patent Appeals and Interferences of the United States of America Patent and Trademark Office came up with a groundbreaking ruling in *ex parte Allen*, which concerned a patent application on polyploid oysters. Although the patent was rejected, the ruling established that multicellular organisms were patentable. Soon after, Donald Quigg, then Commissioner of Patents and Trademarks, publicly announced that the United States of America Patent and Trademark Office would examine "claims directed to multicellular living organisms, including animals" as long as they do not include human beings within their scope. He also clarified that "an article of manufacture or composition of matter occurring in nature will not be considered patentable unless given a new form, quality, properties or combination not present in the original article existing in nature in accordance with existing law" (Quigg, 1987). The following year, the first-ever animal patent was granted for "a transgenic nonhuman mammal" containing an activated oncogene sequence. The patent is commonly referred to as the oncomouse patent, since it describes a mouse into which a gene has been introduced which induces increased susceptibility to cancer.

The patent systems of the United States of America and Europe, including their courts, have adopted somewhat different approaches with respect to exceptions and limitations. For example, the United States of America makes no explicit exceptions in biotechnology and the related fields of animal and plant breeding concerning patentable subject matter. Furthermore, the United States of America patent system makes no reference to *ordre public* or morality in determining whether or not an invention should be protected. In addition, the United States of America courts have limited the extent to which non-owners may use a patented invention in research or experimentation without authorization of the owners so far that such activities are almost completely precluded. In contrast, while Europe permits animals and plants to be protected, animal and plant varieties are not patentable. Moreover, patents are not available for inventions the exploitation or publication of which would be contrary to ordre public or morality. Another difference with the United States of America is that most jurisdictions in Europe have a broader research or experimental use exemption.

The importance of the European Patent Convention (EPC), which was signed in 1973, extends well beyond the confines of the European continent. Some of its provisions have been incorporated, albeit with some modifications, into the TRIPS Agreement, and it has served as a model for legislation in other parts of the world. The Convention has the effect not only of facilitating patent coverage throughout the various European national markets, but also of significantly harmonizing patent law in the region as a whole.

The EPC allows a single patent application to be filed with an institution known as the European Patent Office (EPO), where it is examined. Effectively, an EPC application is a bundle of national applications in the chosen countries. Enforcement is a matter for national courts, and it is perfectly possible for a patent to be invalidated in one country but to remain in force elsewhere. The EPO has an appeals system consisting of various boards, the most important being the Enlarged Board of Appeal. While their judgments are not legally binding, national courts tend voluntarily to accept their authority (Paterson, 2002).

According to Article 53,

"European patents shall not be granted in respect of: (a) inventions the publication or exploitation of which would be contrary to ordre public or morality, provided that the exploitation shall not be deemed to be so contrary merely because it is prohibited by law or regulation in some or all of the Contracting States; (b) plant or animal varieties or essentially biological processes for the production of plants or animals; this provision does not apply to microbiological processes or the products thereof."

2.1   *Ordre public* **and morality**

What do these terms mean? In French civil law, *ordre public* has a wider meaning than "public order" and is more akin to "public policy". According to one interpretation of the term,

"although the expression includes 'public order' in so far as this relates to, for example, rioting, the expression primarily covers such matters as good government, the administration of justice, public services, national economic policy and the proper conduct of affairs in the general interest of the state and society".   (UKBOT, 1970)

Legal experts tend to assume that the ordre public and morality exclusions should be construed narrowly on a case-by-case basis rather than applied to broad classes of patents such as life forms in their broadest sense, and this is what has in fact happened (Moufang, 1998).

Opponents of biotechnological patenting, such as Greenpeace, have sought to make use of the morality and ordre public exclusions in Article 53(a) of the EPC and try to expand their application. But they have had mixed success.

In a 1995 case, *Greenpeace v. Plant Genetic Systems*, the EPO Technical Board of Appeal (TBA) deleted 6 of the 44 claims from the patent. Perhaps the most interesting aspect of the case is that the TBA was challenged to apply the morality and ordre public exclusions. The TBA concluded that an invention is "immoral" if the general public would consider it so abhorrent that patenting would be inconceivable. But it provided no clarification on how "abhorrent" should be interpreted, nor how opponents of a patent should demonstrate that the general public regards the invention as immoral. The TBA rejected the evidence of surveys and opinion polls provided by Greenpeace as inadmissible, arguing that "surveys and opinion polls do not necessarily reflect ... moral norms that are deeply rooted in European culture" (Warren-Jones, 2001). With respect to ordre public, the TBA placed the burden of proof on the patent's opponents by requiring convincing evidence that exploitation of the patent would be seriously prejudicial to the environment. The TBA's rather narrow interpretations of the exceptions led them to reject their application to this particular case.

In the well-known oncomouse case, the EPO's Examining Division was initially reluctant to apply any kind of morality criterion but was instructed to do so by the TBA. In 1991, the Examining Division responded by formulating a balancing test for this particular case that would take into account the following: (i) the interest of mankind in providing remedies for dangerous diseases; (ii) protection against uncontrolled dissemination of unwanted genes; and (iii) prevention of cruelty to animals. On this basis, the Examination Division determined that since the potential benefits of the invention outweighed the negative factors, the patent should be granted, and the EPO consequently did so. However, this patent has remained controversial among environmental and animal welfare groups, and opposition proceedings have continued ever since. One of the main outcomes of such opposition has been that the scope of the patent has been narrowed from transgenic nonhuman eukaryotic animals to rodents.

2.2 Definitional issues

One may reasonably wonder where one draws the line between an essentially biological process on the one hand and non-biological and microbiological ones on the other. The European Patent Convention drew upon the Council of Europe's Convention on the Unification of Certain Points of Substantive Law on Patents for Invention, which was adopted in 1963. The negotiating history of the latter Convention is somewhat helpful. It tells us that "essentially biological" replaced the term "purely biological" from an earlier version of the text. The Council's Committee of Experts on Patents, that was responsible for drafting the convention, changed the wording to broaden the exclusionary language to embrace such "essentially biological" processes as varietal selection and hybridization methods even if "technical" devices were utilized to carry out the breeding processes (Bent *et al.*, 1987). It should be noted that, two years earlier, the UPOV Convention had been adopted, which sought to protect plant breeding innovations derived from such biological processes. The singling out of microbiological processes and products was made at the suggestion of an

NGO called the International Association for the Protection of Industrial Property (AIPPI), which pointed out that micro-organisms were commonly used for industrial purposes, such as in brewing and baking (Bent *et al.*, 1987). Since then the European Patent Office (EPO) has sought to further clarify the term. According to the EPO guidelines for examiners,

> "the question whether a process is 'essentially biological' is one of degree depending on the extent to which there is technical intervention by man in the process; if such intervention plays a significant part in determining or controlling the result it is desired to achieve, the process would not be excluded."

In 1995, the EPO TBA in the aforementioned *Greenpeace v. Plant Genetic Systems* case affirmed that "a process ... comprising at least one essential technical step, which cannot be carried out without human intervention and which has a decisive impact on the final result" is not essentially biological and would thus be patentable. At the same time, conventional plant and animal breeding methods, and other techniques such as artificial insemination, would not be patentable (Warren-Jones, 2001).

Distinguishing *plant variety* from *animal variety* is far from easy, especially considering that the EPC has three official versions: English, French and German. Concerning plants, the EPO has come to a modus operandi following a period of uncertainty that persisted until the late 1990s. In *Greenpeace v. Plant Genetic Systems*, the TBA determined that a claim for plant cells contained in a plant is unpatentable since it does not exclude plant varieties from its scope. This implied that transgenic plants per se were unpatentable because of the plant variety exclusion. Consequently, for the next four years, the EPO stopped accepting claims on plants per se. But the situation changed in 1999 when the EPO Enlarged Board of Appeal decided that while genetically modified plant varieties are unpatentable,

> "a claim wherein specific plant varieties are not individually claimed is not excluded from patentability under Article 53(b), EPC even though it may embrace plant varieties".   (EPO, 1998)

As for animals, it is noteworthy that the oncomouse patent was initially rejected on the grounds of being for an animal variety, but was subsequently adjudged not to be subject to this exclusion. However, one of the main definitional complications is that

> "whereas the French version [of the EPC] ('races animals') approaches the English text, the German term 'tierarten' is equivalent to 'species' "
>   (Moufang, 1989).

In 1989, the European Commission, concerned about the legal uncertainties, which, it was felt, could be prejudicial to the future of biotechnology in Europe, and fearing that some European countries might respond to mounting controversy by banning patents on living organisms and genes, drafted a Directive on the Legal Protection of Biotechnological Inventions. The aim was to harmonize patent law relating to biotechnology around high and clearly defined minimum standards, while preventing member states from "backsliding". In 1998, the Directive was finally adopted, although, at the time of writing, most EU Member States had yet to implement it.

Echoing the EPC, explicit exclusions include animal and plant varieties. More specifically, Article 4.2 states that "inventions which concern plants or animals shall be patentable if the technical feasibility of the invention is not confined to a particular plant or animal variety".

Again following the EPC, essentially biological processes for the production of plants and animals are also excluded. Article 2.2 clarifies that "a process for the production of plants or animals is essentially biological if it consists entirely of natural phenomena such as crossing or selection". This definition has been adopted by the EPO.

Also excluded on the grounds that commercial exploitation of them would be contrary to ordre public or morality are: (a) processes for cloning human beings; (b) processes for modifying the germ line genetic identity of human beings; (c) uses of human embryos for industrial or commercial purposes; [and] (d) processes for modifying the genetic identity of animals which are likely to cause them suffering without any substantial medical benefit to man or animal, and also animals resulting from such processes.

A sequence or partial sequence of a gene may be patented as long as an industrial application is disclosed. This is consistent with EPO practice. In a 1995 case (Howard Florey/Relaxin), the Opposition Division of the EPO declared DNA to be "not 'life'", but a chemical substance which carries genetic information", and therefore patentable just as are any other chemicals.

Article 6 states in part that "inventions shall be considered unpatentable where their commercial exploitation would be contrary to *ordre public* or morality", and that on this basis "processes for modifying the genetic identity of animals which are likely to cause them suffering without any substantial medical benefit to man or animal, and also animals resulting from such processes" are unpatentable.

In sum, the European patent system creates legal uncertainties that its United States of America counterpart does not. At the same time, as we will see below, it is far from clear that the latter system is necessarily a better system in terms of encouraging innovation or of addressing public concerns relating to human welfare, animal rights or economic development.

                                           G. Dutfield

3.   ARE PATENTS APPROPRIATE FOR GENE-BASED INVENTIONS?

Biotechnology patenting raises a number of important policy questions (see Dutfield, 2003a). Should the application of patent systems be modified to accommodate new areas of research that are risky and expensive, even if the resulting discoveries appear to lack genuine novelty or inventiveness, as traditionally defined under patent law? Conversely, to what extent should the design and application of rules that allow the patenting of basic research tools take into account the possible damaging effects of such patenting on downstream innovation?

However, concerns about patenting of gene-based inventions are not just about good public policy-making. Those who doubt the appropriateness of patenting biotechnological inventions are likely to argue that the very notion of scientists inventing a life form, seed, cell line, protein or DNA sequence is fundamentally incorrect or even sacrilegious. It is important to examine all these objections closely.

3.1   Fundamental issues

When it comes to fundamental objections, patents on DNA sequences and living organisms challenge some quite fundamental tenets of patent law and jurisprudence. One of the most frequently expressed opinions is that patents on genes and life forms render the invention or discovery distinction meaningless and thereby allow pure discoveries to be patented. In dealing with this issue, we will discuss first DNA and then life forms.

In the case of DNA, it is worth noting that, despite the extent of patenting going on, some biotech and pharmaceutical companies disagree with the patenting of both naturally occurring DNA sequences and cDNA. But others take the view that, if DNA is just a chemical, then complementary DNA (cDNA) sequences at least should be patentable, provided that they fulfil the criteria of novelty, inventive step and industrial applicability. In a 1998 article in *Science*, John Doll, director of biotechnology examination at the United States of America Patent and Trademark Office (USPTO), clarified that DNA sequences such as expressed sequence tags (ESTs) and single nucleotide polymorphisms (SNPs) can be patented if the applicant discloses specific utilities, such as that the sequence is useful for chromosome mapping or identification, gene mapping, tagging genes with known function such as including increasing predisposition to a disease, or forensic identification (Doll, 1998).

However, one should be very sceptical about this in spite of a recent rule change by the USPTO (see below) making the utility requirement more demanding. This is because techniques for isolating and purifying DNA sequences are well known and no longer require a great deal of skill to use. Even if nobody knew about the naturally occurring equivalent, such a claim should still arguably fail for the lack of an inventive step on the basis of the techniques employed being routine. Nonetheless, as we have seen, several countries do allow isolated and purified DNA sequences to be patented as long as a credible use is disclosed.

The problem goes further even than this. It is now understood better than before that Francis Crick's Central Dogma – namely that DNA makes RNA makes protein – is not applicable in every case. All genes have a function, but not all of them are involved in protein-making processes. To make matters even more complicated, different genes may occupy the same strand of DNA to the extent that it may be extremely difficult to determine where one begins and another ends. Further, the whole protein-making process is complex and still to some extent a mystery. What is becoming apparent, though, is that successful protein manufacture requires the involvement of more than one gene. This is because the various processes that need to take place simultaneously are themselves regulated by other proteins, which in turn are coded for by other genes. So granting patents on a gene on the basis that it performs a single function such as coding for a particular protein or that it is associated with a disease is problematic. This is because it simplistically assumes that genes have independent functions. In fact this is conceptually wrong, because genomes should more accurately be seen as consisting largely of multiple intersecting mini-ecosystems rather than as a single collection of separately functioning "LEGO® bricks". Therefore treating genes as patentable inventions on the basis of a single function is more a reflection of ignorance than of insight, and is essentially anti-innovation, since it potentially hinders opportunities for follow-on researchers to carry out further investigations on genes that a company or university had previously patented on the basis that its scientists had discovered one out of possibly numerous functions. Increasingly, company representatives express frustration that so much basic genetic information is being privatized by other companies through the patent system. Typically, the complainers are from the pharmaceutical and life science firms, while those being criticized are small biotech firms.

As for life forms, it is frequently argued that the patenting of genetically modified organisms should be banned on the basis that a living thing is not a human invention but a discovery or a creation of God. Religious objections on principle to the patenting of life forms certainly deserve to be respected, but one should not assume that inventions must be completely human-made to be classed legitimately as inventions. In Europe and North America, which have the most experience in the patenting of apparently natural substances, there has never been any kind of blanket exclusion of certain types of invention on the basis that because they were not 100 percent human made they cannot be patented. This seems perfectly reasonable, though admittedly it is very difficult to consider an animal as a human invention. An organism should probably not be treated by the law as a single chemical, and a more convincing case needs to be made that inserting or breeding a piece of DNA into an organism should entitle one to property rights not only over the whole organism but over all of its progeny too (Funder, 1999). Interestingly, the EPO Examining Division initially rejected claims to the progeny of transgenic non-human mammals included in the oncomouse patent application, on the basis that the applicant was illegitimately seeking to circumvent the essentially biological process exception. But, as Warren-Jones explains (2001),

> "the Technical Board pointed out that, since the claim was fundamentally a product, the issue did not arise. The Board also indicated that, while not an issue for consideration in the case, claims to progeny could not be assumed to be the result of 'biological processes'".

To this author the second argument seems rather far-fetched. This is not because it is less than 100 percent human-made but because the functioning and behaviour of the life form are mostly out of the control of the "inventor".

It is important also to note that patents are not just for invented machines and chemicals but may include new processes to create things that may or may not already exist, and even new uses of existing things. It is the act of invention not just the thing invented that a patent is awarded for. So while one may agree that an oncomouse is not an invention, the bringing into being of such a mouse did result from human endeavour involving activity that probably deserves to be considered as inventive. It seems fair to reward such inventive work in some way. It may after all be far more beneficial to human welfare in its application to cancer research than yet another mousetrap. But how? One argument is that we should allow the procedure for creating it to be patented, but not the mouse itself. Another argument that is just as logical is that the oncomouse could be patented as long as it is produced by the technique described in the patent. In other words, we should allow a product-by-process patent but not a product-by-any process patent. Alternatively, we might come up with another legal protection system in place of patents.

It is undeniable, though, that the invention or discovery distinction has been blurred to the point of becoming almost meaningless. Some patent practitioners may agree with the critics about this, but claim that it is nothing to be concerned about. Discovering something does not mean that no novelty or inventive step is involved, some of them would argue.

3.2   Policy-related issues

To allow biotech companies to appropriate commercially valuable knowledge, industry and "the patent community" – a term coined by Drahos (1999). It consists of "patent attorneys and lawyers, patent administrators, and other specialists who play a part in the exploitation, administration and enforcement of the patent system – have persuaded governments and courts to be permissive in terms of applying (or not applying) certain customary requirements and to expand the availability of IP protection. Sometimes this has been done by relaxing the rules obliging applicants to fully disclose the invention and demonstrate its novelty, industrial applicability or utility, and inventive step, and by making new rules providing greater legal certainty. On balance many if not most of these regulatory changes may have served a useful purpose in fostering biotechnological innovation. But there have been some negative consequences. We deal first with issues of interpretation that have arisen because of the specificities of patenting in the biotechnology fields, and then look at some of these consequences.

One of the ways in which inventors in this field have benefited from liberal policy-making relates to the concept of exhaustion of rights. Once a patent-protected product is sold by the owner or the licensee, his or her rights over that product are usually exhausted unless there is a contract of sale imposing conditions on buyers. When it comes to living things the rights are not exhausted when the "product" is sold but extend to the progeny whether or not the progeny is "manufactured" by the "inventors". In this sense we are making a concession to the biotechnology patent owner in order to make the patent monopoly meaningful. After all, a strict application of the exhaustion doctrine would make the right so weak as to be almost useless.

In addition, inventions described in a patent are normally meant to be repeatable in the sense that technicians "skilled in the art" (to use the common patent terminology) reading the specification should be able to repeat the invention, that is, come to the same result. But biological processes, even human directed ones, are to some extent random, and such techniques as animal cloning and genetic engineering are still not exact sciences despite the claims of many proponents. For every successful outcome there are likely to be hundreds of failures. But again, patent offices and policy-makers tend to give the applicants the benefit of the doubt.

So what we have done is to bend the rules for biotech inventors to meet their demands if not necessarily their genuine needs. This need not be a problem if negative consequences do not arise. But they do. Now, when we consider the issue of perverse consequences, we need to bear in mind first that in longer established industries and technological fields, legislators and patent offices may be experienced and impartial enough to ensure that the extent of the rights available and that are granted are optimal as far as the public interest is concerned. Even then, this is difficult to achieve. The challenge is much greater still with new technologies such as biotechnology, especially if governments are pressured by powerful economic interests, and if patent offices lack the resources to conduct adequate prior art searches and examinations.

In consequence of recent developments in the field of gene-based inventions, two situations have arisen that should be of major concern to policy-makers. First, disproportionately many patents are being granted in relation to the number of commercial products based upon them. This is because of the enormous quantity of patents on genes and gene fragments, whose existence raises the cost of doing research owing to the need to license related parts of the genome that are "owned" by different institutions. Second, the scope of a patent can sometimes be drawn so broadly as to allow monopoly protection to cover a range of possible products including many unforeseen by the applicant. Both situations can create perverse incentives that may reduce the rate of innovation. Let us look at these situations more closely.

In a now well known article in *Science*, Heller and Eisenberg (1998) warned of an emerging IP problem in the United States of America in the field of biomedical research, which they call the "tragedy of the anti-commons". What they refer to is a situation in which the increased patenting of pre-market, or "upstream" research "may be stifling life-saving innovations further downstream in the course of research and product development". One way this can happen is based on the fact that developing future commercial products such as therapeutic proteins or genetic diagnostic tests often requires the use of multiple gene fragments, an increasing number of which are being patented. The cost of research and development will be affected by the existence of so many of these patents because a company intending to develop such products will need to acquire licences from other patent holders, and thus will incur large (and possibly prohibitive) transaction costs. Since the first patent covering ESTs was controversially awarded to Incyte Pharmaceuticals in 1998, this problem could become more serious. One can expect that potential for such anti-commons situations also appears to exist in gene-based animal-related inventions with technology transfers entailing highly complex and expensive bundles of transactions.

There is little that patent offices can do themselves about such "tragedies" except leave it to the patent holders involved either to accept the high transaction costs entailed by their need to acquire or license other firms' patents or to collaborate with their rivals by setting up private collective rights organizations (CROs) to pool their patents. The advantage for members of setting up such a CRO would be to reduce the transaction costs that would otherwise be incurred by the need both to negotiate multiple licensing arrangements among them, and to distribute royalties from non-member technology licensees (see Merges, 1996). Although CROs would presumably reduce transaction costs for non-member licensees as well, there is also a danger that CROs can become too dominant in the market, leading to reduction of competition and the stifling of innovation.

According to Rai (1999: 840–841), the history of patent pools does not give cause for optimism in this regard. First, some of the best known pools were set up only after protracted litigation. Second, past patent pools were sometimes deemed anti-competitive and therefore illegal under antitrust law. Third, in the biotechnology case, the partners would include a diversity of organizations such as universities, government research agencies, small firms and transnationals. Past experience suggests, in her view, that patent pools have most often come about among homogeneous partners that have previous experience of collaborating. This third reason may be overstated, though, since as we have seen, such types of collaboration involving different types of organization do take place in biotechnology. Logically, the best outcome is probably to bring about a change in the rules so that such basic research tools as DNA sequences cannot be patented at all.

The second problem has been with modern biotechnology from the start. Cohen and Boyer's recombinant DNA patent described a method of inserting genes only into *E. coli*, yet it covered applications of the technology for a much wider range of micro-organisms. This did not prove to be a big problem because it was so widely licensed. But because it is a platform technology, refusing to license it would certainly have slowed down innovation. Similarly the Harvard oncomouse patent disclosed a mouse containing a cancer gene (specifically: "a transgenic nonhuman eukaryotic animal (preferably a rodent such as a mouse)"), but its scope embraced all non-human mammals into which the gene may be inserted plus their progeny. The patenting of genes is bound to raise the issue of excessive breadth (in addition of course to the other problems with gene patenting). Since many and perhaps all genes perform several roles, granting a patent on the gene itself to the first person to discover any role is based on bad science and mistaken on public policy grounds because it is likely to discourage others from investigating other roles that may even be more significant. At the same time, with the product protection acquired and a product in the pipeline, the owners may not see any particular need to discover other functions.

It is fair to say that in individual cases solutions can sometimes be found. The law usually allows opponents to file oppositions or re-examination requests with the patent office. Over-broad patents can also be challenged in the courts if opponents have enough money. The problem could also be dealt with through more careful examinations, but this raises another worrying issue. Nowadays, patent offices are required to become more service oriented and financially self-sufficient. They are expected to demonstrate their efficiency by examining patents speedily and avoiding backlogs. The danger is that the proportion both of excessively broad scope patents and of issued patents lacking genuine novelty and inventive step will increase. In fact, this is known to be a serious problem in the United States of America, where patent examiners are not given sufficient time to do their work properly. One problem is that those opposing such trends, and also specific patents, are likely to find that mounting legal challenges to improperly-granted patents requires financial commitments well beyond their means.

However, these problems are to some extent technical, and adjustments may be made in the courts to correct the situation. Moreover, there are signs that the court may be moving in the opposite direction. One such sign from the United States of America is the 1997 decision in *Regents of the University of California v. Eli Lilly and Co.* to invalidate claims on the basis that "the disclosure of a single species of genetic material does not provide an adequate written description necessary to support patent claims to a broad genus of written material" (Sung and Pelto, 1998). This suggests that the situation may improve as courts become more accustomed to dealing with biotechnology patents.

Another example indicating a possible new trend is the November 2000 decision in *Festo v. Shoketsu* to severely constrain the application of the so-called "doctrine of equivalents". The doctrine, which has been adopted in a number of legal jurisdictions (such as the United States of America and Germany), is meant to ensure that the inventor is able to secure a fair remuneration for unforeseen embodiments that would be obvious to somebody skilled in the art (see Grubb, 1999; Lederer, 1999). In essence, it extends the scope of a patent beyond the actual language of the claims to prevent others from reading the patent and inventing around it without doing anything that would not be obvious to a trained technician. Although the case was unrelated to biotechnology, it may well limit patent coverage in this field as in others.

Another sign of a change in direction, this time at the USPTO, is the new rule announced in January 2001 for DNA-related patent examinations. Patent applications disclosing DNA sequences must now provide convincing evidence that their utility is specific, substantial, and credible (see USPTO, 2001). However, many more far-reaching proposals were rejected, and it seems, to this author at least, that DNA sequences should not be patentable at all, at least until scientists have found a way to develop useful proteins that do not exist in nature. We are still some time away from being able to do this.

4. THE INTERNATIONAL PATENT RULES AND GENE-BASED TECHNOLOGIES

According to Article 27.1 of the TRIPS Agreement, which deals with patentable subject matter, "patents shall be available for any inventions, whether products or processes, in all fields of technology, provided that they are new, involve an inventive step and are capable of industrial application". Paragraph 1 also requires that patents be available and patent rights enjoyable "without discrimination as to the place of invention, the field of technology and whether products are imported or locally produced".

To understand the meaning and purpose of this clause, it needs to be divided into the three areas in which it forbids discrimination. The first, relating to place of invention, required the United States of America to change its patent law. The United States of America is now unique in its persistence with a first-to-invent system as opposed to first-to-file, which operates elsewhere in the world. The problem at the time was that when it came to establishing priority of invention in cases of dispute, the United States of America patent system and the courts were barred from accepting foreign evidence of dates of invention and consequently discriminated against foreign inventions (Charnovitz, 1998), which of course tended to be made by non-United States of America nationals. The second area of non-discrimination relates to field of technology. This highlights another difference between the United States of America and Europe. In the United States of America, inventions fulfilling the criteria of novelty, usefulness and non-obviousness can be patented regardless of considerations as to their "technical" nature, including the extent to which they have a physical embodiment. In Europe, inventions must display a "technical effect" in order to be patentable. One consequence is that business methods are not patentable in Europe whereas they are in the United States of America. Critics argue that such a requirement conflicts with TRIPS because it discriminates against certain fields of "technology" such as business methods and also software programs, and causes legal uncertainty since the meaning of the term is fuzzy and arguably anachronistic. The third area means that in countries where governments require inventors to place their patented products, or products manufactured from their patented processes, on the market, this does not mean they have to make them locally in order to enjoy their patent rights. They can simply import them instead (Carvalho, 2002).

TRIPS also permits certain exclusions. Article 27.2 states as follows:

"Members may exclude from patentability inventions, the prevention within their territory of the commercial exploitation of which is necessary toprotect *ordre public* or morality, including to protect human, animal or plant life or health or to avoid serious prejudice to the environment, provided that such exclusion is not made merely because the exploitation is prohibited by their law."

The terms "ordre public" and "morality" are not defined in TRIPS, although references to human, animal or plant life or health and the environment provide some context. The language of Article 27.2 follows very closely that of the 1973 European Patent Convention, yet even in Europe, as we saw, the true meaning and potential extent of the ordre public and morality exclusions remain unclear.

Article 27.3(b) states that members may also exclude from patentability:

"(b) plants and animals other than micro-organisms, and essentially biological processes for the production of plants or animals other than non-biological and microbiological processes. However, Members shall provide for the protection of plant varieties either by patents or by an effective *sui generis* system or by any combination thereof."

With respect to products, plants and animals may be excluded from patentability. As regards processes, essentially biological processes for the production of plants or animals may also be excluded. Patents must be available for micro-organisms as products and for non-biological and microbiological processes for producing plants or animals. Patent protection need not be available for plant varieties but an effective IPR system is still obligatory. This may be an UPOV-type plant breeders' rights system, another *sui generis* alternative, or some combination of systems.

Otherwise, the meaning of Article 27.3(b) is far from straightforward, and is open to different interpretations. Difficulties arise both in the definition of terms and in the extent to which exceptions are allowed.

For example, it is possible to argue that an application relating to a genetically engineered plant is bound to include plant varieties within its scope whether or not the word "variety" even appears in the specification. This is important because in some jurisdictions, plants can be patented but plant varieties cannot. In others neither can be patented, but there may be a separate IP system exclusively for plant varieties. Even terms like "micro-organism" can be interpreted differently from one legal jurisdiction to another (see below).

5. CHOICES FOR DEVELOPING COUNTRIES

Few if any developing countries follow the United States of America, and some of them are extremely restrictive in terms of how far they permit the patenting of biotechnological inventions or even how far they are willing to accept that any kind of natural product or living thing can be classed as an invention at all. The international rules accommodate these differences to some extent, except that micro-organisms and non-biological and microbiological processes must be patentable. TRIPS does not refer directly to DNA sequences, and so the situation is open to more than one interpretation.

While TRIPS does not allow WTO members to exclude biotechnological inventions from their patent systems in any explicit sense, Article 27.3(b) allows them to use their discretion in determining the extent to which inventions in this technological field can be protected. The problem facing developing countries is that if they lack a clear idea of how – and even whether – biotechnology, including gene-based technologies for animal health and production, can benefit their economies and improve the lives of their citizens, then they are in no position to design an IPR system to promote welfare-enhancing biotechnological innovation. Moreover, many of these countries have no biotechnology industries to speak of, and there is every reason to be highly sceptical that such businesses will spring up just because life-forms, DNA sequences and micro- and non-biological processes can be patented.

Another reason why it is difficult for developing countries to come up with a common position on the review of Article 27.3(b) is that they vary so much in their national capacities to generate biotechnological inventions. Policy-makers in the more technologically-advanced developing countries who believe that the new biotechnologies can be beneficial should design their IPR system with the goal of encouraging domestic innovation and technology transfer, and attracting funds for start-up firms. The experience of developed countries suggests that a carefully-designed IPR system could indeed stimulate innovation, although there is a real danger of a carelessly-designed one turning out to be worse than not having one at all, for example, by over-protecting upstream research and thereby inhibiting more applied downstream research, or by allowing large companies to control markets, raise prices, and distort research priorities. But for many, if not most other, developing and least-developed countries, it is difficult to see how strong IPR protection will encourage innovation if the capacity to do the necessary research is barely existent anyway.

Logically, developing countries should take a TRIPS *de minimis* approach for now, excluding plants and animals, construing "micro-organism" narrowly, and opting for a *sui generis* alternative to patents for plant varieties. This is not as straightforward as it may seem. These terms are open to different interpretations. According to the EPO, "micro-organism"

> "includes not only bacteria and yeasts, but also fungi, algae, protozoa and human, *animal and plant cells*, i.e. all generally unicellular organisms with dimensions beneath the limits of vision which can be propagated and manipulated in a laboratory. Plasmids and viruses are also considered to fall under this definition."

Similarly the Japan Patent Office interprets "micro-organism" to include

> "yeasts, moulds, mushrooms, bacteria, actinomycetes, unicellular algae, viruses, protozoa, etc." and also *"undifferentiated animal or plant cells as well as animal or plant tissue cultures"*   (JPO, 2000).

Scientifically speaking, these interpretations seem rather over-expansive. It is not at all obvious that a single cell from a multi-cellular organism is itself an organism even if it has been cultured in a laboratory. There is no reason why developing countries should not define the term in a more restrictive sense if they should consider it advantageous to do so.

Another big problem that is often overlooked is the huge task developing country patent offices face in processing large numbers of lengthy and highly technical patent applications. To give some idea of the potential difficulties here, in 2000, the U.S. Patent and Trademark Office received a biotech patent application that was the equivalent of 400,000 pages long! And courts having the knowledge and experience to adjudicate disputes between different patent holders and to determine the appropriate scope of a biotech patent may simply not exist.

Developing countries are justifiably concerned that TRIPS furthers the interests of the advanced industrialized countries much more than their own (see CIPR, 2002; Dutfield, 2003b; Sell, 2003). Undeniably, it incorporates a combination of European and United States of America IPR rules and standards with little reference to the needs and interests of developing countries. To make matters worse, developing countries also find themselves pressured to raise their national standards even beyond those of TRIPS through bilateral agreements with the United States of America and the EU, and through threats of trade sanctions.

Developed countries must give developing countries time to determine how to respond to the challenges and opportunities of new biotechnologies, even if this means that they delay full implementation of Article 27.3(b) until several years beyond the official deadlines. It is unreasonable to pressure them to speed up implementation before they feel they are ready to introduce legislation that furthers their long-term interests.

Developed countries should also refrain from imposing their own interpretations of Article 27.3(b) based on their own legislation and jurisprudence, and their own economic interests. As long as developing countries see TRIPS as a legal straightjacket rather than a looser-fitting garment with room to move about in, they are bound to feel not only uncomfortable, but also resentful. In the longer term, this is to nobody's benefit.

As for developing countries, both biotechnology and IPRs are highly controversial subjects that have provoked a heated debate and propaganda, and been the focus of highly-committed advocacy campaigns both in favour and against. This is all the more reason for these countries to be sceptical about much of the advice they get from the developed world on both topics, even when its providers claim to be objective and non-partisan.

REFERENCES

Bent, S.A., Schwaab, R.L. Conlin, D.G. & Jeffery, D.D. 1987. *Intellectual Property Rights in Biotechnology Worldwide.* Basingstoke, UK: Macmillan.

Carvalho, N.P. 2002. *The TRIPS Regime of Patent Rights.* London, The Hague and New York: Kluwer Law International.

Charnovitz, S. 1998. Patent harmonization under world trade rules. *Journal of World Intellectual Property,* **1**(1): 127–137.

CIPR [Commission on Intellectual Property Rights]. 2002. *Integrating Intellectual Property Rights and Development Policy. Report of the Commission on Intellectual Property Rights.* London: CIPR.

David, P. 1993. Intellectual property institutions and the panda's thumb: patents, copyrights, and trade secrets in economic theory and history. *In:* M.B. Wallerstein, R.A. Schoen and M.E. Mogee (eds). *Global Dimensions of Intellectual Property Rights in Science and Technology.* Washington, DC: National Academy Press.

Doll, J. 1998. Biotechnology: the patenting of DNA. *Science,* **280**: 689–690.

Drahos, P. 1999. Biotechnology patents, markets and morality. *European Intellectual Property Review,* **21**: 441–449.

Dutfield, G. 2003a. *Intellectual Property Rights and the Life Science Industries: A Twentieth Century History.* Aldershot, UK: Ashgate.

Dutfield, G. 2003b. *Intellectual Property Rights: Implications for Development.* Geneva, Switzerland: ICTSD and UNCTAD.

Dutfield, G. 2004. *Intellectual Property, Biogenetic Resources and Traditional Knowledge: A Guide to the Issues.* London: Earthscan.

EPO [European Patent office]. 1998. EPO Decision G 01/98, See: http://www.european-patent-office.org/dg3/biblio/g980001ex1.htm

Funder, J. 1999. Rethinking patents for plant innovation. *European Intellectual Property Review,* 21: 551–577.

Geroski, P. 1995. Markets for technology: knowledge, innovation and appropriability. *In:* Stoneman (ed). *Handbook of the Economics of Innovation and Technological Change.* Oxford and Malden, UK: Blackwell.

Grubb, P.W. 1999. *Patents for Chemicals, Pharmaceuticals and Biotechnology.* Oxford, UK: Clarendon Press.

Heller, M.A. & Eisenberg, R.S. 1998. Can patents deter innovation? The anti-commons in biomedical research. *Science,* **280**: 698–701.

JPO [Japan Patent Office]. 2000. *Examination Guidelines for Patent and Utility Model.* Tokyo: JPO. See: http://www.jpo.go.jp/infoe/1312-002_e.htm

Lederer, F. 1999. Equivalence of chemical product patents. *International Review of Industrial Property and Copyright Law,* **30**: 275–284.

Merges, R.P. 1996. Contracting into liability rules: intellectual property rights and collective rights organizations. *California Law Review,* **84**: 1293–1386.

Moufang, R. 1989. Patentability of genetic inventions in animals. *International Review of Industrial Property and Copyright Law,* **20**: 823–846.

Moufang, R. 1998. The concept of "ordre public" and morality in patent law. *In:* G. van Overwalle (ed). *Octrooirecht, Ethiek en Biotechnologie / Patent Law, Ethics and Biotechnology / Droit des Brevets, Ethique et Biotechnologie.* Brussels: Bruylant.

Quigg, D.J. 1987. Animals – patentability. *Journal of the Patent and Trademark Office Society,* **69**: 328.

Paterson, G. 2002. The creation of harmonised patent protection in Europe. *In:* H. Lawton-Smith (ed). *The Regulation of Science and Technology.* Basingstoke, UK: Palgrave.

Rai, A.K. 1999. Intellectual property rights in biotechnology: addressing new technology. *Wake Forest Law Review,* **34**: 827–847.

Sell, S.K. 2003. *Private Power, Public Law: The Globalization of Intellectual Property Rights.* Cambridge, UK: Cambridge University Press.

Stokes, G. 2001. Patenting of genetic sequences: on the up and up. *IP Matters,* April See: http://www.derwent.com/ipmatters/2001_01/genetics.html.

Sung, L.M. & Pelto, D.J. 1998. The biotechnology patent landscape in the United States as we enter the new millennium. *Journal of World Intellectual Property,* **1**: 889–901.

UKBOT [United Kingdom Board of Trade]. 1970. *The British Patent System: Report of the Committee to Examine the Patent System and Patent Law ('The Banks Committee').* London: HMSO.

USPTO [United States Patent and Trademark Office]. 2001. Utility examination guidelines. *Federal Register,* **66**: 1092–1099.

Warren-Jones, A. 2001. *Patenting rDNA: Human and Animal Biotechnology in the United Kingdom and Europe.* Witney, UK: Lawtext Publishing.

WIPO [World Intellectual Property Organization]. 2002. Elements of a *sui generis* system for the protection of traditional knowledge. Document No. WIPO/GRTKF/IC/3/8. Prepared by the Secretariat, for the 3rd Session of the Intergovernmental Committee on Intellectual Property, Genetic Resources and Traditional Knowledge. Geneva, 13–21 June 2001. 24p.

ANTIBIOTIC RESISTANCE AND PLASMID CARRIAGE AMONG *ESCHERICHIA COLI* ISOLATES FROM CHICKEN MEAT IN MALAYSIA

Tin Tin Myaing,[1] A.A. Saleha,[2] * A.K. Arifah[2] and A.R. Raha[3]
1. *Pharmacology and Parasitology Dept., University of Veterinary Science, Yezin, Myanmar.*
2. *Faculty of Veterinary Medicine, Universiti Putra Malaysia, 43400 UPM Serdang, Selangor, Malaysia.*
3. *Department of Biotechnology, Faculty of Food Science and Biotechnology, Universiti Putra Malaysia, 43400 UPM Serdang, Selangor, Malaysia.*
* *Corresponding author.*

Abstract: *Escherichia coli* isolates from 131 raw chicken meat samples were tested for susceptibility to 12 antibiotics. Plasmids were isolated from many samples and their DNA molecular weight calculated. An 81.7% plasmid occurrence rate was observed among the isolates, ranging from 0 to 8 in number and with sizes from 1.2 to 118.6 MDa. Plasmids were detected in 93.8% of *E. coli* isolates resistant to all 12 antibiotics, and in 90.5% of *E. coli* isolates resistant to 11. Three (2.8%) isolates harboured 8 plasmids and were resistant to all 12 antibiotics. Antibiotic resistant genes in bacteria are usually carried in extrachromosomal DNA and it is postulated that *E. coli* with a high number of plasmids possesses wider resistance to antibiotics.

1.   INTRODUCTION

Antibiotic resistance is frequently determined by genetic information of plasmid origin. The correlation between antibiotic resistance and plasmid profile may indicate that the genetic information is plasmid borne. Initially, there was a tendency to assume that antibiotic resistance genes appeared only after antibiotics began to be used widely in medicine. However, the genetic diversity within some classes of resistance makes it clear that the genes have been evolving for a much longer time.

H. P. S. Makkar and G. J. Viljoen (eds.), Applications of Gene-Based Technologies for Improving Animal Production and Health in Developing Countries, 521-527.
© 2005 *IAEA. Printed in the Netherlands.*

1.1   Initial comments

The high frequency of plasmid DNA observed among *Escherichia coli* isolates from poultry may result from exposure of the organisms to antibiotics. The accumulation of resistance genes on plasmids and their potential for transfer suggest that plasmids are major vectors in the dissemination of resistance genes throughout the bacterial population (O'Brien *et al.,* 1993). There have been a number of studies on the occurrence of plasmid DNA and antibiotic resistance of *E. coli* isolated from beef, milk, water and village chickens in Malaysia (Son *et al.,* 1997a,b,c, 1998a,b; Samuel, 1999) but limited in broiler and chicken meat.

2.   ANTIBIOTIC SUSCEPTIBILITY TEST

E. coli isolates from 131 raw chicken meat samples were tested for antibiotic susceptibility to twelve antibiotics by the method of Bauer, Kirby and Sheris (1966), using antibiotic discs (OXOID) as follows: ampicillin, 10 µg; cefoperazone, 30 µg; cephradine, 30 µg; ciprofloxacin, 5 µg; chloramphenicol, 30 µg; enrofloxacin, 5 µg; erythromycin, 15 µg; kanamycin, 30 µg; nalidixic, acid 30 µg; tetracycline, 30 µg; trimethoprim, 5 µg; and vancomycin, 30 µg. Bacteria were suspended in saline to give a density of 0.2 MacFarland standard and streaked onto Mueller-Hinton agar. The plates were incubated at 37°C for 18 hours. The diameter of a clear, inhibition zone was measured to the nearest whole millimetre. The diameter of the zone for individual antibiotics was referred to the interpretation table as to whether the *E. coli* isolate was resistant, susceptible or intermediate.

3.   PLASMID PROFILE ANALYSIS

3.1   Stock culture

Stock bacteria (*E. coli*) were cultured in Luria-Bertani (LB) agar medium and incubated overnight at 37°C. A single *E. coli* colony was transferred into 5 ml of LB broth and incubated overnight at 37°C with gentle shaking.

3.2   Isolation of plasmid DNA

The plasmid isolation was carried out according to the method described by Maniatis, Fritsch and Sambrook (1989), with modifications as in the protocol provided by *Taq* Dye Deoxy Terminator Cycle Sequencing Kit (ABI P/N 401150). The newly modified method is a mini alkaline-lysis/PGE precipitation procedure.

3.3   Agarose gel electrophoresis

A horizontal 0.7 percent agarose gel in Tris-Borate+EDTA or TBE buffer was used for staining. The electrophoresis was run at 80 V for about one hour (E-C 105 Apparatus Corporation) and terminated when the tracking dye was about 10 mm from the edge of the gel. The gel was then stained in a solution of ethidium bromide (0.5 µg/ml) for one hour. The plasmid DNA-ethidium bromide that formed bands in the gel was visualized by luminescent detector (BioRad Gel-Doc 2000 model, Video Graphic Printer UP-890MD).

3.4   Determination of plasmid molecular weight

A graphical method of relating the logarithm of the molecular weight of a DNA molecule (log C) to its electrophoretic mobility in gels was used to determine the molecular weight of plasmid. Plasmids of known molecular weight from *E. coli* V517 (Macrina *et al.,* 1978) were used as standards for calibrating the size of plasmid DNA molecules. In this study, DNA fragments are referred to as plasmids.

4.   CORRELATION BETWEEN ANTIBIOTIC RESISTANCE AND PLASMID CARRIAGE AMONG *E. COLI* ISOLATES

The proportion of *E. coli* isolates resistant to the 12 antibiotics is as shown in Table 1. The number of plasmids ranged from 0 to 8, and their sizes from 1.2 MDa to 118.6 MDa. Among the *E. coli* isolates that were resistant to 12 antibiotics, 93.8 percent had plasmids and 6.3 percent lacked plasmids, while of *E. coli* isolates resistant to 11 antibiotics, 90.5 percent had plasmids and 9.5 percent lacked plasmids (Table 2). Plasmid profiling relies on the analysis of the genetic content, and most of typing schemes are based on the phenotypic characteristics of the organisms (Bichler, de Vries and

Rombouts, 1994). Plasmid profiling has been shown to be a valuable epidemiological tool for typing. The typing system in plasmid profiling is based on the size and number of plasmids that form the plasmid profile (Meyers *et al.*, 1976).

The larger the number of plasmids of different sizes, the more useful is this molecular technique in demonstrating the identity of isolates (Tacket, 1989). In the present study, the number of plasmids ranged from 0 to 8 and the sizes of plasmids ranged from 1.2 MDa to 118.6 MDa. *E. coli* isolates resistance to 12, 11 and 10 antibiotics harboured 8, 7 and 6 plasmids, respectively (Table 3). The overall impression from the study is that the plasmids of the 131 isolates were highly diverse, exhibiting different plasmid profiles and different plasmid sizes. High diversity of plasmid content may be due to the variation in sample collection. It is generally agreed that the plasmid content of bacteria sampled from many sources is more diverse (Selanders and Levin, 1980). No apparent correlation was found between the plasmid profiles of the strains and their resistance patterns to the antimicrobial agents. Those *E. coli* isolates resistant to antibiotics were found to harbour more plasmids than those isolates without plasmids. Platt *et al.* (1984) reported that the clinical isolates of *E. coli* in the resistant groups harboured more plasmids than in the sensitive subpopulation, and that the number of plasmids carried by resistant isolates was greater. Blanco González and Blanco (1985) showed that resistance to antibiotics could be related to having more plasmids. The total number of plasmids in any given population can affect the results of the strain analysis. At the same time, the more plasmids an organism contains, the more specific is the interpretation of the results when a large, diverse population is studied. Without the plasmid profiles, one would have thought that the *E. coli* isolates with similar antibiotic resistance patterns were derived from the same ancestral strain.

A further problem in the use of plasmid typing is potential plasmid losses during culturing, which can mislead the interpretation of results. However, in certain cases, strains can be maintained with their original plasmids over several years (Mahony, Ahmed and Jackson, 1992). The high frequency of *E. coli* isolates observed among the poultry carcasses may result from contamination of carcasses during processing. It is similar to the report of Geornaras, Hastings and von Holy (2001), in that the strains isolated from poultry carcasses during processing were genetically diverse.

Table 1. Percentage of *E. coli* isolates resistant to individual antibiotics.

Antibiotics tested	Number and percentage of resistant *E. coli* isolates	
Ampicillin	127	96.2%
Chloramphenicol	100	75.6%
Cephradine	124	93.9%
Cefoperazone	110	83.3%
Ciprofloxacin	103	78.0%
Erythromycin	111	84.0%
Enrofloxacin	95	72.0%
Kanamycin	67	50.8%
Nalidixic acid	125	94.7%
Tetracycline	119	90.2%
Trimethoprim	125	94.7%
Vancomycin	131	100.0%

Table 2. Number of plasmid DNA harboured by *E. coli* isolates in relation to number of antibiotics to which they were resistant.

Number of antibiotics	Number of resistant *E. coli* isolates	*E. coli* isolates			
		With plasmids (n=107)		Without plasmids (n=24)	
12	48	45	93.8%	3	6.3%
11	21	19	90.5%	2	9.5%
10	27	18	66.6%	9	33.3%
9	9	8	88.9%	1	11.1%
8	17	10	58.8%	7	41.2%
7	3	3	100%	0	
6	1	1	100%	0	
5	1	1	100%	0	
4	3	1	33.3%	2	66.7%
2	1	1	100%	0	

NOTE: n = number

Table 3. E. coli isolates resistant to number of antibiotics and harboured plasmids.

No. of antibiotics to which isolate is resistant	Number of plasmids detected					
	3	4	5	6	7	8
0						
1						
2		+				
3						
4	+	+				
5	+					
6			+			
7			+		+	
8		+	+			
9				+		
10			+	+		
11			+	+	+	
12			+		+	+

NOTE: + = resistant to antibiotics and possessed plasmids.

5.　　CONCLUSION

It may be suggested that resistance to antibiotics by *E. coli* isolates is related to plasmid content, and that the number of plasmids harboured by the organism is correlated with the higher number of antibiotics to which the bacteria shows resistance. It was not feasible in this study to serotype or otherwise characterize the *E. coli* isolates because of the large number of strains examined. Further studies with *E. coli* isolates from the present study, or similar investigation on the presence and characteristics of plasmids that harbour resistant genes, may better define risks associated with antimicrobial usage.

ACKNOWLEDGEMENTS

The authors would like to thank Malaysian Technological Cooperation Programme (MTCP), Ministry of Science Technology and Environment (MOSTE) and Universiti Putra Malaysia (UPM) for financial support during this study. The authors would also like to thank the Ministry of Livestock Breeding and Fisheries, Myanmar, for their kind assistance.

REFERENCES

Bauer, A.W., Kirby, W.M.M. & Sheris, J.S. 1966. Antibiotic susceptibility testing by a standard single disc method. *American Journal of Clinical Pathology,* **45**: 493–496.

Bichler, R.R., de Vries, J. & Rombouts, F.M. 1994. Plasmid diversity in *Salmonella* enteritidis of animal, poultry and human origin. *Journal of Food Protection,* **57**: 307–311.

Geornaras, I., Hastings, J.H. & von Holy, A. 2001. Genotypic analysis of *Escherichia coli* strains from poultry carcasses and their susceptibilities to antimicrobial agents. *Applied and Environmental Microbiology,* **67**: 1940–1944.

González, E.A. & Blanco, J. 1985. Relation between antibiotic resistance and number of plasmids in enterotoxigenic and non-enterotoxigenic *Escherichia coli* strains. *Medical Microbiology and Immunology,* **174**: 257–265.

Macrina, F.L., Kopecko, D.J., Jones, K.K., Ayers, D.J. & Mc Gowen, S.M. 1978. A multiple plasmid-containing *Escherichia coli* strain: Convenient source of reference plasmid molecules. *Plasmid,* **1**: 417.

Mahony, D.E., Ahmed, R. & Jackson, S.G. 1992. Multiple typing techniques applied to a *Clostridium perfringens* food poisoning outbreak. *Journal of Applied Bacteriology,* **72**: 309–314.

Maniatis, J., Fritsch, E.F. & Sambrook, J. 1989. *Molecular Cloning. A Laboratory Manual.* 2nd edition. USA: Cold Spring Harbor-Laboratory Press.

Meyers, J.A., Sanchez, D., Elwell, L.P. & Falkow, S. 1976. Simple agarose gel electrophoretic method for the identification and characterization of plasmid deoxyribonucleic acid. *Journal of Bacteriology,* **127**: 1529–1537. [See also correction in *ibid.*, **129**: 1171.]

O'Brien, T.F., Di Giorgio, J., Parsonnet, K.C., Kass, E.H. & Hopkins, J.D. 1993. Plasmid diversity in *Escherichia coli* isolated from processed poultry and poultry processors. *Veterinary Microbiology,* **35**: 243–255.

Platt, D.J., Sommerville, J.S., Kraft, C.A. & Timbury, M.C. 1984. Antimicrobial resistance and the ecology of *Escherichia coli* plasmids. *Journal of Hygiene* (Cambridge), **93**: 181–188.

Samuel, L. 1999. Molecular characterization of *Escherichia coli* isolated from raw milk and village chicken and broiler litter. MSc Thesis. Faculty of Food Science and Biotechnology, Universiti Putra Malaysia, Malaysia.

Selanders, J.R. & Levin, B.R. 1980. Genetic diversity and structure in *Escherichia coli* populations. *Science,* **210**: 545–554.

Son, R., Razali, H., Syamir, A., Zainuri, A., Rusul, G. & Sahila, A.M. 1997a. Multiple antibiotic resistance (MAR) index and plasmids in *Escherichia coli. Jurnal Veterinar Malaysia,* **9**: 23–25.

Son, R., Rozila, A., Rusul, G., Zainuri, E., Raha, A.R. & Sahilah, A.M. 1997c. Antibiotic resistance and plasmid profiles of *Escherichia coli* in raw milk and beef. *Malaysian Applied Biology,* **26**: 77–81.

Son, R., Sahila, A.M., Gulum, R., Rozila, A. & Samuel, L. 1998a. Beef as a source of multi-resistant *Escherichia coli* O157: H7 harbouring transferable plasmid and resistance phenotype. *Journal of Bioscience,* **9**: 20–27.

Son, R., Shahilah, A.M., Gulam, R., Zainori, A., Tadaaki, M., Norio, A., Yung, B.K., Jun, O. & Mitsuaki, N. 1998b. Detection of *Escherichia coli* O157:H7 in the beef marketed in Malaysia. *Applied and Environmental Microbiology,* **64**: 1153–1156.

Son, R., Zainuri, A., Rusul, G., Karim, M.I.A., Ali, A.M. & Harikishna, K. 1997b. Survival of genetically engineered *Escherichia coli* in river water. *Tropical Biomedicine,* **14**: 19–26.

Tacket, C.O. 1989. Molecular epidemiology of *Salmonella. Epidemiologic Reviews,* **11**: 99–108.

COMPARISON OF DNA PROBE, PCR AMPLIFICATION, ELISA AND CULTURE METHODS FOR THE RAPID DETECTION OF *SALMONELLA* IN POULTRY

Jafar A. Qasem, Salwa Al-Mouqati and Gnanaprakasam Rajkumar
Biotechnology Department, Food Resources Division, Kuwait Institiute for Scientific Research , P.O.Box 24885, Safat, Kuwait-13109.

Abstract: The identification of *Salmonella* spp. from poultry meat was studied by comparing bacterial detection using the Gene-Trak colorimetric hybridization method, a PCR amplification kit and an Enzyme Linked Immunosorbent Assay (ELISA), and these methods were compared with the conventional methodology proposed by the United States Food and Drug Administration (US FDA) for detection of Salmonella in food samples. Forty positive and negative samples were studied.

The three methods yielded similar results with levels of Salmonella greater than 10 CFU per sample, even when the samples were highly contaminated with competing bacteria. In contrast, 20 CFU of seed inoculum per sample was the lowest level of Salmonella detectable with all three methods and the standard culture method. The detection limits of the PCR and ELISA assays were 5 CFU/g after enrichment at 37°C for 6 and 9 hours, respectively.

Compared with conventional bacteriology, all three methods here demonstrated high sensitivity and specificity for Salmonella.

1.   INTRODUCTION

Salmonella is a monospecific genus of bacteria belonging to the family Enterobacteriaceae. The species is *Salmonella enterica*, and strains are differentiated by serotyping, and referred to by name. Salmonellae are gram-negative, non-spore-forming, usually motile, bacilli. They are facultative

H. P. S. Makkar and G. J. Viljoen (eds.), Applications of Gene-Based Technologies for Improving Animal Production and Health in Developing Countries, 529-541.

anaerobes and contain three different types of antigens. The somatic (O) antigen is associated with the cell wall and is composed of lipopolysaccharides. The flagellar (H) antigen is associated with the microbe's peritrichous flagella and is proteinaceous. The third type of antigen, a capsular (Vi) antigen, is found only in some *Salmonella* (Bryan, Fanelli and Riemann, 1979; Dolye and Cliver, 1990; Nagaraja, Pomeroy and Williams, 1991a,b; Smith, 1990; Snoeyenbos, 1991a,b; Wilcock and Schwartz, 1992; Ziprin, 1994).

Conventional culture methods for detecting *Salmonella* in food are laborious and time-consuming (requiring 3 to 7 days). Therefore, faster, but reliable, methods of detecting *Salmonella* are needed (Wolcott, 1991). Several rapid methods have been tried, such as rapid culture methods (D'Aoust and Sewell, 1988); various immunological methods (D'Aoust, Sewell and Greco, 1994; D'Aoust and Sewell, 1988; Ripabelli, 1997; Van Poucke, 1990); and DNA methods (Andrews, 1994; Bulte and Jacob, 1995; D'Aoust and Sewell, 1984; Olsen *et al.*, 1995; Rose, Llabres and Bennett, 1991). Although rapid methods are an improvement over traditional culture methods for *Salmonella* detection, their sensitivity is limited, and they still require time consuming pre-enrichment and selective enrichment steps (Curiale *et al.*, 1990).

One of the most promising methods for detecting *Salmonella* is based on polymerase chain reaction (PCR), which combines simplicity with a potential for high specificity and sensitivity in detecting pathogenic bacteria in foods (Aabo, Andersen and Olsen, 1995; Bulte and Jacob, 1995; Fluit *et al.*, 1993; Jones, Law and Bej, 1993; Kwang, Littledike and Keen, 1996; Mahon *et al.*, 1998; McElory, Cohen and Hargis, 1996; Olsen *et al.*, 1995; Pignato *et al.*, 1995; Rahn *et al.*, 1992). Colorimetric assays based on post-PCR hybridization to detect PCR products have been developed and evaluated in food products (Sumet *et al.*, 1997), but they require multiple, time consuming, post-PCR signal development steps. Integration within routine laboratory schedules is a prerequisite to product acceptance.

2.   MATERIAL AND METHODS

2.1   Collection and isolation of specimens

2.1.1   Stock culture and media

The micro-organisms used in this study were reference strains of bacteria obtained from the American Type Culture Collection (ATCC) (Rockville, USA), namely *Salmonella* serotype Poona NCTC 04840, *S.* Enteritidis

ATCC 13076, *S.* Typhimurium ATCC 14028, *S.* Choloraesuis ATCC 29946, *S.* Typhimurium ATCC 07823, *Klebsilla pneumoniae* ATCC 29939, *Staphylococcus aureus* ATCC 25923, *Pseudomonas aeruginosa* ATCC 27853, *Escherichia coli* ATCC 29926 and *Shigella sonnei* ATCC 29930. All test organisms were grown in trypticase broth (BBRL, USA) at 35°C for 24 hr before they were used.

2.1.2   Food samples

A total of 40 food samples were obtained from local retail establishments. These samples comprised chicken from Kuwait [9], chicken from the Kingdom of Saudi Arabia (KSA) [6], chicken from Denmark [2], chicken from Brazil [10], chicken from France [1], beef from Kuwait [6], beef from Turkey [1], lamb from Kuwait [8], lamb from KSA [1], lamb from Australia [1], lamb from China [3], lamb from Egypt [1] and lamb from Lebanon [1].

All of the samples were immediately transported to the Kuwait Institute for Scientific Research (KISR) laboratory in an icebox, and the initial bacteriological analysis was carried out within 2 to 3 hr of purchase. A 25-g analytical aliquot of each sample was homogenized for 1 min with a homogenizer (Elnex Model S#001) along with 225 ml of sterile lactose broth (Difco).

2.2   Preparation of bacterial strains, culturing and purification of cells

Fresh poultry meat, frozen chicken, ground beef and sheep meat were used. Reference strains of bacteria were obtained from ATCC (Rockville, USA) collections of *Salmonella*. The US FDA's *Salmonella* culture protocol was followed for the analysis of a 25-g meat sample in 1 : 9, sample : broth, ratio (see chapter 5 of BAM (FDA, 1995)). The glucose (TSI), lysine decarboxylase (LIA), H_2S (TSI and LIA), urease, indole, simmone citrate, and voges-Proskauer tests and other methods were used as needed. Transia-plate *Salmonella* (TPS) was designed and optimized for use with a two-step enrichment protocol only. The test was performed after an overnight pre-enrichment in buffered peptone water at 37°C, followed by an additional overnight selective enrichment in Rappaport-Vassiliadis soya media at 41.5°C. The samples were boiled for 15-20 minutes and cooled to room temperature prior to the test. The TPS test kit contained ready-to-use reagents and a 96-well microtitreplate. The DNA-Detect Multi-Food 3 PCR kit, which is based on the PCR technique in a multiplex format, was used to detect and identify *Salmonella* food-borne pathogens. The standard amplification protocol was used. The amplification was carried out in a Perkin Elmer 9700 thermal cycler using the following program: 1× denaturation at 94°C for 5 min; 35× denaturation at 94°C for 30 sec; annealing at 59°C for 1 min; elongation at 72°C for 1 min; and 1× final extension at 72°C for 5 min. Samples (15 µl each) of the PCR-amplified DNA were used for electrophoresis in 2% agarose gel.

2.3   Colorimetric hybridization method

The Gene-Trak *Salmonella* Assay DNA hybridization test for detection of *Salmonella* was used. The DNA hybridization test employs *Salmonella*-specific DNA probes and a calorimetric detection system for detection of *Salmonella* strains in food samples following broth culture enrichment. The procedure for *Salmonella* assay was that recommended by the manufacturer and the absorbency was read at 450 nm. The sample was considered non-reactive for the presence of *Salmonella* if the absorbency value obtained was less than or equal to the established cut-off value for the assay.

The laboratory evaluation serology, PCR, hybridization and culture results were tabulated and correlated to sample types, and sensitivity was compared for the most effective method. For quality control of *Salmonella* culture detection and serology, positive and negative control samples from standard references were used with each serological test and hybridization method, and then certified standard reference strains obtained from ATCC were used for serology standardization and comparison. The results were cross-checked with culture and serology results, when available.

2.4   Pre-enrichment

All samples tested (25 g of meat) were homogenized in 225 ml of lactose broth (Difco) in a stomacher laboratory blender 400 (UAC Hos, London, UK) and incubated at 37°C for 24 hr. Each culture was used for further testing with the three methods being compared.

2.5   Standard culture method

After incubation, 1 ml of pre-enrichment broth (lactose broth) was inoculated into 9 ml of selenite-cystine broth (Oxoid), 9 ml of TTB (Difco) and 10 ml of Rappaport-Vassiliadis medium (Oxoid), and incubated at 37°C for enrichment. After incubation, each selective broth was streaked onto Hektoen Eteric agar, bismuth sulphite (BS) agar and xylose lysine deoxycholate (XLD) agar (Difco), and incubated at 37°C for 24 hr.

Suspected *Salmonella* colonies were biochemically and serologically identified using the API 20 E Biomerieux system and commercially available anti-sera.

2.6   Sensitivity of the methods

In order to determine the lowest number of detectable micro-organisms for each of the rapid methods and the culture method, *S.* Enteritidis (ATCC 13076) was cultured in Trypton soy broth (TSB) for 24 hr at 37°C. After incubation, the culture was diluted in buffered peptone water (BPW) and enumerated on Trypton soy agar (TSA) plates. Homogenized chicken breast (tested negative for *Salmonella*) samples (25 g in 225 ml of BPW) were contaminated with serial dilutions of the pathogen, to give an estimated number of *Salmonella* ranging from 2.0×10^4 to 10×10^{-1} per sample. Each sample was then tested using the three methods being compared; one uninoculated aliquot was included in the test as a negative control.

2.7   Competitive tests

The possibility of detecting *Salmonella* in samples contaminated by large numbers of competing bacteria was investigated with the three rapid methods in comparison with the conventional culture method. Homogenized chicken (*Salmonella*-negative) breast samples (25 g in 225 ml of BPW) were inoculated with 1 ml of single and mixed non-*Salmonella* strains in suspension; inoculants ranged from 4.0×10^4 to 5.0×10^6 per sample. Furthermore, each sample was inoculated with similar concentrations of *S.* Enteritidis, as reported in Table 5. In order to ensure that *S.* Enteritidis was present in the highest dilutions, 1 ml of each dilution was inoculated in 10 ml of TSB and processed for isolation. Negative controls were obtained by processing samples with three methods, each containing one competing strain but no *Salmonella*. All samples were processed for *Salmonella* using the three methods being compared.

3.   RESULTS

Various methodologies for *Salmonella* detection were compared in this study. When only *Salmonella* cultures (reference strains) were tested using the type culture, the ELISA, DNA hybridization and PCR methods detected all of the serotypes with no discrepancies. All of the methods employed differentiated non-*Salmonella* organisms (*Klebsilla pneumonia, Staphylococcus aureus, Pseudomonas aeruginosa, E. coli* and *Shigella sonnei*) from *Salmonella* serotypes (*S.* Poona, *S.* Enteritidis, *S.* Typhimurium and *S.* Choloraesuis). Of the 40 food samples and 5 reference strains tested for *Salmonella* using the culture method and detection kits, culturing gave 20 positive samples (50%), PCR gave 20 positive samples (60%), ELISA gave 22 positive samples (73%) and the DNA-hybridization gave 23 positive samples (57.5%). Of 40 different red meat and chicken samples, 4 gave positive results with all of the methods used, and 10 gave positive results with 3 of the four test methods used (Table 1). Seven samples gave high absorbency with ELISA, 2 samples were PCR-positive and 6 samples were hybridization-positive, but all these could not be verified by cultures. These samples could very possibly have contained *Salmonella*, and the organism could have died during selective enrichment but yet have produced enough antigen to be detected, or over-growth of other bacteria on the agar plates, which suppresses the appearance of typical colonies of *Salmonella,* could have caused the discrepancy. Six samples showed positive results with culturing but negative PCR results, and four samples showed positive with culturing but negative hybridization results. One important reason for this could be that the complex composition of food matrices can inhibit PCR amplification (Tables 1 and 2). PCR had a total detection time of less than 24 hr, which is significantly shorter than the conventional technique, which requires 72–96 hr (Table 3). For samples that produced positive results using the rapid methods, culture streaking on Hektoen Eteric, xylose lysine deoxycholate and bismuth sulphite agars should confirm the results, and then analysis can be continued with biochemical and serological identification of presumptive *Salmonella* isolates using standard procedures.

Table 1. The results of four methods testing for the detection of *Salmonella* in food.

No.	Sample	Source	Culture	PCR	ELISA	Hybrid-ization
1	*Salmonella* Poona	NTCC (04840)	+	+	+	+
2	*S.* Enteritidis	ATCC (13076)	+	+	+	+
3	*S.* Typhimurium	ATCC (14028)	+	+	+	+
4	*S.* Choloraesuis	ATCC (29946)	+	+	+	+
5	*S.* Typhimurium	ATCC (07823)	+	+	+	+
6	*Klebsilla pneumoniae*	ATCC (29939)	-	-	-	-

No.	Sample	Source	Culture	PCR	ELISA	Hybrid-ization
7	*Staphylococcus aureus*	ATCC (25923)	-	-	-	-
8	*Pseudomonas aeruginosa*	ATCC (27853)	-	-	-	-
9	*E. coli*	ATCC (29926)	-	-	-	-
10	*Shigella sonnei*	ATCC (29930)	-	-	-	-
11	Naief chicken	Kuwait	+	-	+	+
12	Fakieh chicken	KSA	+	-	+	+
13	Tanmiah chicken	KSA	+	+	+	+
14	Abo Ali chicken	Kuwait	+	+	+	+
15	Al Ghadeer chicken	Kuwait	+	-	+	-
16	Danpo chicken	Denmark	+	-	+	+
17	Silver chicken	Kuwait	+	-	+	+
18	Sultan chicken	Kuwait	-	-	+	+
19	Sadia chicken	Brazil	-	-	+	+
20	Doux chicken	Brazil	-	-	+	+
21	Hilal chicken	Brazil	+	NA	NA	+
22	Rio chicken	France	-	NA	NA	-
23	Alwataniya chicken	KSA	-	NA	NA	-
24	Doux chicken	Brazil	-	NA	NA	-
25	Danpo chicken	Denmark	-	NA	NA	-
26	Halal chicken	Brazil	-	NA	NA	-
27	Fakieh chicken	KSA	+	NA	NA	+
28	Rezande chicken	Brazil	-	NA	NA	-
29	Fresh chicken	Kuwait	+	NA	NA	+
30	Sadia chicken	Brazil	-	NA	NA	+
31	Americana sausage	Kuwait	+	+	+	-
32	Americana sausage	Kuwait	+	+	+	-
33	Khazan beef	Kuwait	-	-	-	-
34	Americana mutton	Kuwait	+	+	+	-
35	Khazan mutton	Kuwait	-	+	+	-
36	Americana beef	Kuwait	+	+	+	+
37	Mubarakiya mutton	Kuwait	-	+	+	-
38	Al Mawashi beef	Kuwait	-	-	+	+
39	Vita mutton	KSA	+	+	+	+
40	Al Barari Mutton	Kuwait	+	+	+	+
41	Mince fresh meat	Kuwait	-	NA	-	-
42	Mince fresh meat	Australia	+	NA	+	+
43	Stewing meat	China	-	NA	-	-
44	Stroganoff meat	China	-	NA	-	-
45	Striplion meat	China	-	NA	+	+
46	Veal with bone	Egypt	+	NA	-	+
47	Beef sausage	Kuwait	-	NA	-	-
48	Lamb sausage	Lebanon	+	NA	+	+
49	Beef sausage	Turkey	-	NA	-	-
50	Lamb neck	Kuwait	-	NA	-	-

KEY: ATCC = American Type Culture Collection. KSA = Kingdom of Saudi Arabia.
NA = Data not available.

Table 2. Time and cost comparison of the four methods used.

Method	+ve results	Time (hours)	Cost	Disadvantages
Culture	20/40	72–96	$$$	Non-viable, overgrow
ELISA	22/30	40–48	$$	Background interference
PCR	10/20	20–24	$	Food matrices interference
Hybridization	23/40	24–30	$$$	Background interference

Table 3. Comparison of the four methods for *Salmonella* detection in food.

Test method	Chicken meat sample		Red meat sample		All tests	
Culture	11/20	(55%)	9/20	(45%)	20/40	(50%)
ELISA	10/10	(100%)	12/20	(60%)	22/30	(73%)
PCR	4/10	(40%)	8/20	(80%)	12/20	(60%)
Hybridization	15/20	(75%)	8/20	(40%)	23/40	(57.5%)

The *Salmonella* detection rate with the rapid detection methods was lower when reference culture organisms were tested than when *Salmonella* was tested from food samples. Of the 40 samples tested, 8 were from open markets, 28 were from supermarkets and 2 were from farms and factory outlet stores. Of the 40 samples tested, 18 were locally produced and 22 were imported. *Salmonella* detection by culture was 41% in imported samples and 55% in local samples. *Salmonella* detection by PCR was 16.6% in imported samples and 57% in local samples, while *Salmonella* detection by ELISA was 69.2% in imported samples and 76.5% in local samples. Hybridization detection of *Salmonella* was 59% for imported samples and 44% for local samples. Except for the hybridization probe, detection of *Salmonella* by all of the other methods showed higher detection of *Salmonella* in local samples than in imported ones.

As can be seen from Table 4, 10–20 CFU of seed inoculum per sample was the lowest level of *Salmonella* detectable for all three methods and the standard culture method (SCM).

The three methods yielded similar results with levels of *Salmonella* greater than 10 CFU per sample, even when the samples were highly contaminated with competing bacteria (Table 5). Samples contaminated with high numbers of *Shigella sonnei* (4×10^6 CFU) and low numbers of *Salmonella* (10 CFU) produced discrepant results. In this situation, only the PCR and ELISA methods were able to detect the pathogen, while the hybridization and SCM gave negative results.

As can be seen from Table 4, only the PCR and ELISA methods were able to detect the pathogen, while the hybridization and SCM gave negative results. As seen from Table 5, a false negative reaction was reported with the SCM in the sample contaminated with a mixed pool of bacteria (2×10^8 CFU) and 10 CFU of *Salmonella*; in this case, positives were obtained with the other methods.

Table 4. Number of *Salmonella* Enteritidis (CFU/sample) detectable with the three methods in artificially contaminated samples.

Salmonella Enteritidis (Seed inocula)	SCM	PCR	ELISA	Hybridization
2.0×10^4	+	+	+	+
2.0×10^3	+	+	+	+
2.0×10^2	+	+	+	+
20	+	+	+	+
10	–	+	+	–
5.0	–	+	+	–
1.0	–	–	–	–

KEY: SCM = standard culture method.

Table 5. Detection of *Salmonella* in samples contaminated with competitive micro-organisms using the four methods.

Competitive micro-organism (CFU/sample)	*S.* Enteritidis (CFU/sample)	Culture	PCR	ELISA	Hybrid-ization
Staphylococcus aureus (4×10^4)	17×10	+	+	+	+
Pseudomonas aeruginosa (5×10^4)	1.3×10	+	+	+	+
E. coli (4×10^6)	1.9×10	+	+	+	+
Shigella sonnei (4×10^6)	1.1×10	–	+	+	–
Klebsiella pneumoniae (5×10^6)	1.0×10	+	+	+	+
Mixed inocula (2×10^8)	1.0×10	–	+	+	+

4. DISCUSSION

Rapid detection methods for use in food samples have been a subject of active research since the early 1980s. In the past 5 years, much interest has been expressed internationally by food microbiologists in adopting new rapid methods (Foster *et al.*, 1992; Fung, 1992), where a rapid test for food microbiology has been defined as a test that takes 4 to 12 hr to complete. (Jarvis, 1984).

The rapid methods have been shown to not only match the conventional techniques in sensitivity and specificity, but also to offer the possibility of considerable cost-saving with their shorter duration for detection. PCR represents a rapid method with both high sensitivity and specificity, but the complex matrices of food samples, which could reduce its efficacy dramatically, could hamper its sensitivity. Positive results need to be confirmed by culture methods. Bacterial DNA extraction from food samples needs to be standardized and made uniform to make better comparison possible.

Although many of the tests are referred to as "rapid methods", nevertheless most of these *Salmonella* detection systems, regardless of the technology or assay form, still rely on culture methods for resuscitation of injured cells and selective amplification of *Salmonella* populations in the broth culture. Therefore, pre-, selective or post-enrichment procedures, or some combination of them, must be used in conjunction with these "rapid" methods to promote better sensitivity and specificity (D'Aoust *et al.*, 1995; D'Aoust, Sewell and Greco, 1994; D'Aoust and Sewell, 1984, 1988).

Pre-enrichment is an initial step in which the food sample is enriched in a non-selective medium to restore injured Salmonellae cells to a stable physiological condition (Bailey, Cox and Blankenship, 1991). Sub-lethal cell damage may have resulted from thermal processing of food, freezing, thawing, osmotic shock or prolonged storage of low-moisture foods at ambient or elevated temperature (Feng, 1992). Satisfactory resuscitation and pre-enrichment generally require a nutritious, non-selective medium (Jarvis, 1984).

The recovery of *Salmonella* from foods presents a problem for food microbiologists, since low numbers of organisms need to be detected in large volumes of product. Most of the assay systems are screening assays and only provide for the presumptive identification of Salmonellae. Negative results, therefore, must be considered for confirmation by conventional methods, and serology must confirm definitive, but presumptive, positive results (Feng, 1992).

The main advantage of the *Salmonella* PCR kit is the speed with which it screens for the presence of *Salmonella*. However, these PCR methods involve electrophoresis, which requires skill and is laborious and time consuming. The number of samples that can be analysed at any one time is also limited. Furthermore, the ethidium bromide used to stain agarose gels is a mutagen and therefore not appropriate for routine use in food-monitoring laboratories. This process requires 8 to 16 hr of pre-enrichment, 1 to 2 hr of DNA extraction, 2 hr of PCR and 20 min to 1 hr of electrophoresis.

4.1.1   Colorimetric DNA hybridization test

A comparison with the conventional Association of Analytical Communities/FDA's Bacteriological Analytical Manual (AOAC/BAM) culture method showed that both methods were equally effective for detecting *Salmonella* in food products (Foster *et al.*, 1992). This modified colorimetric DNA hybridization method (Gene-Trak) was adopted first by the AOAC in 1990 (Andrews, 1994).

According to the data of Rose, Llabres and Bennett (1991), the Gene-Trak *Salmonella* assay appears to be an effective screening procedure for rapid detection of *Salmonella* in meat and poultry products. The major advantage of the colorimetric DNA probe assay is that large numbers of samples can be screened for Salmonellae fairly rapidly (3 to 4 days).

However, all DNA-probe-positive samples should be confirmed culturally. Hybridization requires the presence of at least 10^2 to 10^3 target microbial pathogens in a sample to elicit a positive signal. Without pre-enrichment of the target organism, DNA hybridization does not provide the required sensitivity to detect *Salmonella* at the required levels (Jones, Law and Bej, 1993)

REFERENCES

Aabo, S., Andersen, J.K. & Olsen, J.E. 1995. Detection of *Salmonella* in minced meat by the polymerase chain reaction method. *Letters in Applied Microbiology,* **21**: 180–182.

Andrews, W.H. 1994. Food microbiology (nondairy). *Journal of AOAC International,* **77**: 187–194.

Bailey, J.S., Cox, N.A. & Blankenship, L.C. 1991. A comparison of an enzyme immuno-assay, DNA hybridization, antibody immobilization, and conventional methods for recovery of naturally occurring Salmonellae from processed broilers. *Journal of Food Protection,* **54**: 354–356.

Bryan, F.L., Fanelli, M.J. & Riemann, H. 1979. Salmonella infections. pp. 73–130, *in:* H. Rieman and F.L. Bryan (eds). *Food-borne Infections and Intoxications.* 2nd edition. New York, NY: Academic Press.

Bulte, M. & Jacob, P. 1995. The use of a PCR-generated *InvA* probe for the detection of *Salmonella* spp. in artificially and naturally contaminated foods. *International Journal of Food Microbiology,* **26**: 335–344.

Curiale, M.S., McIver, D., Weathersby, S. & Planer, C. 1990. Detection of *Salmonella* and other Enterobacteriaceae by commercial deoxyribonucleic acid hybridization and enzyme immunoassay. *Journal of Food Protection,* **53**: 1037–1046.

D'Aoust, J.Y., Sewell, A.M., Greco, P., Mozola, M.A. & Colvin, R. 1995. Performance assessment of the GENE-TRAK® colorimetric probe assay for the detection of foodborne *Salmonella* spp. *Journal of Food Protection,* **58**: 1069–1076.

D'Aoust, J.Y., Sewell, A.M. & Greco, P. 1991. Commercial agglutination kits for the detection of food-borne *Salmonella. Journal of Food Protection,* **54**: 725–730.

D'Aoust, J.Y. & Sewell, A.M. 1988. Reliability of the immunodiffusion 1-2 Test system for detection of *Salmonella* in food. *Journal of Food Protection,* **51**: 853–856.

D'Aoust, J.Y. & Sewell, A.M. 1984. Efficacy of an international method for detection of *Salmonella* in chocolate and cocoa products. *Journal of Food Protection,* **47**: 111–113.

Dolye, M.P. & Cliver, D.O. 1990. *Salmonella.* pp. 185–204, *in:* D.O. Cliver (ed). *Food-borne Diseases.* San Diego, CA: Academic Press, Inc.

FDA [US Food and Drug Administration]. 1995. *Bacteriological Analytical Manual* (BAM). 8th edition.

Feng, P. 1992. Commercial assay systems for detecting food-borne *Salmonella*: A review. *Journal of Food Protection,* **55**: 927–934.

Fluit, A.C., Widjojoatmodjo, M.N., Box, A.T.A., Torensma, R. & Verhoef, J. 1993. Rapid detection of Salmonellae in poultry with magnetic immunopolymerase chain reaction assay. *Applied and Environmental Microbiology,* **59**: 1342–1346.

Foster, K., Garramone, S., Ferraro, K.& Groody, E.P. 1992. Modified colorimetric DNA hybridization method and conventional culture method for detection of *Salmonella* in food – Comparison of methods. *Journal of AOAC International,* **75**: 685–692.

Fung, D.Y.C. 1992. Historical development of rapid methods and automation in microbiology. *Journal of Rapid Methods and Automation in Microbiology,* **1**: 1.

Jarvis, B. 1984. A philosophical approach to rapid methods for industrial food control. *In:* K.O. Habermehl (ed). *Rapid Methods and Automation in Microbiology and Immunology.* New York, NY: Springer-Verlag.

Jones, D. D., Law, R. & Bej, A.K. 1993. Detection of *Salmonella* spp. in oysters using polymerase chain reaction (PCR) and gene probes. *Journal of Food Science,* **58**: 1191–1197.

Kwang, J., Littledike, E.T. & Keen, J.E. 1996. Use of the polymerase chain reaction for *Salmonella* detection. *Letters in Applied Microbiology,* **22**: 46–51.

Mahon, J., Murphy, C.K., Jones, P.W. & Barrow, P.A. 1998. Comparison of multiplex PCR and standard bacteriological methods of detecting *Salmonella* on chicken skin. *Letters in Applied Microbiology,* **19**: 169–172.

McElory, A.P., Cohen, N.D. & Hargis, B.M. 1996. Evaluation of the polymerase chain reaction for the detection of *Salmonella* enteritidis in experimentally inoculated eggs and eggs from experimentally challenged hens. *Journal of Food Protection,* **59**: 1273–1278.

Nagaraja, K.V., Pomeroy, B.S. & Williams, J.E. 1991b. Arizonosis. pp. 130–137, *in: Diseases of Poultry.* 9th edition. Ames, Iowa: Iowa State University Press.

Nagaraja, K.V., Pomeroy, B.S. & Williams, J.E. 1991a. Paratyphoid infections. pp. 99–130, *in: Diseases of Poultry.* 9th edition. Ames, Iowa: Iowa State University Press..

Olsen, J.E., Aabo, S., Hall, W., Notermans, S., Wernars, K., Granum, P., Popovic, T., Rasmussen, H.N. & Olsvik, O. 1995. Polymerase chain reaction for detection of food-borne bacteria pathogens. *International Journal of Food Microbiology,* **28**: 1–78.

Pignato, S., Marino, A.M., Emanuele, M.C., Iannotta, V., Caracappa, S. & Giammanco, G. 1995. Evaluation of new culture for rapid detection and isolation of *Salmonella* in food. *Applied and Environmental Microbiology,* **61**: 1996–1999.

Rahn, K., De Grandis, S.A., Clarke, R.C., McEwen, S.A., Galan, J.E., Ginocchio, C., Curtiss, R. & Gyles, C.L. 1992. Amplification of an *InvA* gene sequence of *Salmonella* Typhimurium by polymerase chain reaction as a specific method of detection of *Salmonella. Molecular and Cellular Probes,* **6**: 271–279.

Ripabelli, G., Sammarco, M.I., Ruberto, A., Lannitto, G. & Grasso, G. 1997. Immunomagnetic separation and conventional culture procedure for detection of naturally occurring *Salmonella* in pork sausage and chicken meat. *Letters in Applied Microbiology,* **24**: 493–499.

Rose, B.E, Llabres, C.M. & Bennett, B. 1991. Evaluation of a colorimetric DNA hybridization test for detection of Salmonellae in meat and poultry products. *Journal of Food Protection,* **54**: 127–130.

Smith, B.P. 1990. Salmonellosis. pp. 818–822, *in:* B.P. Smith (ed). *Large Animal Internal Medicine.* St Louis, MO: The C.V. Mosby Company.

Snoeyenbos, G.H. 1991a. Salmonellosis: Introduction. p. 72, *in: Diseases of Poultry.* 9th edition. Ames, Iowa: Iowa State University Press.

Snoeyenbos, G.H. 1991b. Pullorum diseases. pp. 73–86, *in: Diseases of Poultry.* 9th edition. Ames, Iowa: Iowa State University Press.

Sumet, C., Ermel, G., Salvat, G. & Colin, P. 1997. Detection of *Salmonella* spp. in food products by polymerase chain reaction and hybridization assay in microplate format. *Letters in Applied Microbiology,* **24**: 113–116.

Van Poucke, L.S.G. 1990. Salmonella-Tek: A rapid screening method for *Salmonella* species in food. *Applied and Environmental Microbiology,* **56**: 924–927.

Wilcock, B.P. & Schwartz, K.L. 1992. Salmonellosis. pp. 570–583, *in:* D.L. Allen, B.E. Straw, W.I. Mengeling *et al.* (eds). *Diseases of Swine.* 7th edition. Ames, Iowa: Iowa State University Press.

Wolcott, M.J. 1991. DNA-based rapid methods for the detection of foodborne pathogens. *Journal of Food Protection,* **54**: 387–401.

Ziprin, R.L. 1994. Salmonella. pp. 253–318, *in:* Y.H. Hui, J.R. Gorham, K.D. Murrell *et al.* (eds). *Foodborne Disease Handbook: Diseases Caused by Bacteria.* New York, NY: Marcel Dekker, Inc.

CONTROL OF BOVINE SPONGIFORM ENCEPHALOPATHY BY GENETIC ENGINEERING: POSSIBLE APPROACHES AND REGULATORY CONSIDERATIONS

J.S. Gavora,[1] H.P.S. Kochhar[2] and G.A. Gifford[2]

1. Consultant, 51 Forest Hill Avenue, Ottawa, ON K2C 1P7 Canada.
2. Animal Biotechnology Unit, Veterinary Biologics Section, Canadian Food Inspection Agency, 59 Camelot Drive, Ottawa, ON K1A 0Y9 Canada

Abstract: Transmissible spongiform encephalopathies (TSE) include bovine spongiform encephalopathy (BSE), scrapie in sheep and Creutzfeldt-Jakob disease (CJD) in humans. A new CJD variant (nvCJD) is believed to be related to consumption of meat from BSE cattle. In TSE individuals, prion proteins (PrP) with approximately 250 amino acids convert to the pathogenic prion PrP^{Sc}, leading to a dysfunction of the central neural system. Research elsewhere with mice has indicated a possible genetic engineering approach to the introduction of BSE resistance: individuals with amino acid substitutions at positions 167 or 218, inoculated with a pathogenic prion protein, did not support PrP^{Sc} replication. This raises the possibility of producing prion-resistant cattle with a single PrP amino acid substitution. Since prion-resistant animals might still harbour acquired prion infectivity, regulatory assessment of the engineered animals would need to ascertain that such possible 'carriers' do not result in a threat to animal and human health.

1.   INTRODUCTION

Since bovine spongiform encephalopathy (BSE), also known as "mad cow disease", was first observed in 1986 (Willesmith *et al.*, 1998), over 180,000 cattle have been diagnosed in the United Kingdom (Wrathall *et al.*, 2002). Subsequently, BSE has been detected in several European countries and Japan. In 2001, the incidence of BSE, expressed as the number of cases diagnosed per million bovine animals aged over 24 months, ranged from 258

543

H. P. S. Makkar and G. J. Viljoen (eds.), Applications of Gene-Based Technologies for Improving Animal Production and Health in Developing Countries, 543-550.
© 2005 *IAEA. Printed in the Netherlands.*

in the UK, down to 1 in Austria and Japan (OIE, 2002)[6]. The number of human deaths from definite or probable new variant Creutzfeldt-Jakob disease (nvCJD) in humans had reached 117 as of November 2002 (CJD Surveillance Unit, 2002).

The emergence of the new transmissible spongiform encephalopathies (TSE) – BSE, and particularly nvCJD in humans – is not only a serious threat to human and animal health but also a cause of significant economic losses to the livestock industries. While preventative measures, such as a ban on the use of meat and bone meal, seem to be effective in controlling the spread of BSE, there is currently no effective treatment for other TSEs.

Genetic modification of meat animal species, particularly sheep and cattle, might provide a long-term solution and prevention of TSE (Parry, 1979; Perrier *et al.*, 2002). In sheep, as will be discussed below, a naturally occurring genotype at the relevant wild-type prion (wtPrP) gene makes the homozygote carriers of such a gene resistant to a sheep TSE, scrapie, because the "resistant" gene does not support the propagation of prion protein, the causative agent of scrapie. The resistant genotype is being used in conventional breeding to produce scrapie-resistant sheep.

An analogous resistance to BSE has not been detected in cattle. Recent research work on genetic engineering of the prion gene in laboratory mice (Perrier *et al.*, 2002) demonstrated that a substitution of a single amino acid in the relevant gene may prevent propagation of the prion protein and thus render the engineered animals resistant to TSE. The purpose of this paper is to theoretically examine the possibility that such genetically engineered modification might be accomplished in cattle and to discuss some of the potential regulatory challenges associated with the use of marker assisted selection and genetic engineering for control of TSE in livestock animals.

2.   BACKGROUND

The TSEs, including BSE, scrapie and chronic wasting disease, are a unique group of diseases in which affected animals exhibit abnormal neurological behaviour associated with accumulation of prion protein in nervous system tissues. It is currently believed that the diseases are caused by alterations in a protein (PrP^{C}) produced by the individual's (animal or human) own gene, that changes its folding pattern to form a misshapen protein (PrP^{Sc}), named "prion" by Stanley Prusiner, as an abbreviation of

---

6. Since original compilation of this communication, one case of BSE has been diagnosed in Canada and one in USA.

"**pro**teinaceous **in**fectious particles". The PrP superscripts C and Sc stand for cellular and for scrapie cellular, respectively. Stanley Prusiner was awarded the 1997 Nobel Prize for medicine for first proposing the hypothesis that prion diseases are caused by misfolded proteins, and for elucidating the mechanism by which prions bring about the amyloid plaques in the brain (Prusiner, 1991). Table 1 lists the neurological disorders in humans and animals attributable to prion diseases (Prusiner, 1997).

Scrapie, a TSE disease in sheep, was first described in the eighteenth century. Other prion diseases (Table 1), described more recently, generally occur sporadically and at low frequencies. The importance and intensity of the investigation of prion diseases was strongly magnified after the emergence of BSE in the 1980s in England, which appeared to be the consequence of feeding meat and bone meals from sheep carcases to cattle. Reports of an atypical nvCJD appeared in the scientific press in 1995, and it was suggested that its causative agent might be bovine prions (for review, see Prusiner, 1997).

Table 1. Prion diseases.

Disease	Pathogenesis
Human diseases	
Kuru (Fore people of South Pacific)	Infection through ritualistic cannibalism.
Creutzfeldt-Jakob disease	Infection from prion-contaminated human growth hormone, dura mater grafts, etc.
Familial Creutzfeldt-Jakob disease	Germline mutations in PrP gene.
Variant Creutzfeldt-Jakob disease	Infection from bovine prions.
Gertstmann-Sträussler-Scheinker disease	Germline mutations in PrP gene.
Fatal familial insomnia	Germline mutations in PrP gene.
Sporadic Creutzfeldt-Jakob disease	Somatic mutation or spontaneous conversion of PrP^C into PrP^{Sc}.
Animal diseases	
Scrapie (sheep)	Infection in genetically susceptible sheep.
Bovine spongiform encephalopathy	Infection from prion-contaminated meat and bone meals.
Transmissible mink encephalopathy (mink)	Infection from prions from sheep or cattle.
Chronic wasting disease (mule deer, elk)	Unknown.
Feline spongiform encephalopathy (cats)	Infection from prion-contaminated meat and bone meals.
Exotic ungulate encephalopathy (greater kudu, nyala, oryx)	Infection from prion-contaminated meat and bone meals.

SOURCE: Modified from Prusiner, 1997.

Recently, new insights have been gained into the evolution and normal physiological function of the wild-type prion proteins. It was suggested that, in some yeasts, prions provide a selective advantage under adverse conditions, possibly by producing phenotypic variants, and that prions may have been retained because they aid evolution (Csaba, 2001). It now appears that wild-type, normal prion proteins play a role in the transport and regulation of copper in body tissues: prion proteins bind copper in the domain that contains a series of eight amino acids, octarepeats, repeated four or more times (Burns *et al.*, 2000). Such metal binding is sensitive to changes in acidity and increased acidity leads to copper release. Thus PrP^C seems to play a role in copper metabolism, particularly in neural tissues.

Prion proteins consist of approximately 250 amino acids. DNA sequences of the PrP protein are available for humans and a number of animal species in GenBank: for example (Accession Number/Number of bases) for humans U29185/35,522, mouse U29186/38,418, and sheep U67922/31,412. The conversion of a normal, wild-type prion protein PrP^C into the pathogenic prion protein PrP^{Sc} involves a change in protein folding through which the alpha-helical content in PrP^C is reduced and the amount of beta-sheet folding pattern is increased (Pan *et al.*, 1993). The presence of the aberrantly folded isoform PrP^{Sc} of the normal, cellular prion PrP^C stimulates the conversion of PrP^C to PrP^{Sc} and the accumulation of PrP^{Sc} leads to dysfunction of the central neural system and neuronal dysfunction (Perrier *et al.*, 2002). It appears that the conversion of PrP^C to PrP^{Sc} may be mediated by a chaperone protein, "protein X" (Telling *et al.*, 1995). Recent investigations point toward a possible involvement of plasminogen in the re-folding process (Fischer *et al.*, 2000).

The conversion of PrP^C to PrP^{Sc} is accompanied by significant changes in the properties of the protein, rendering the beta-sheet prion virtually indestructible (Prusiner, 1997). PrP^C is soluble in non-denaturing reagents, whereas PrP^{Sc} is not; PrP^C is readily digested by proteases, whereas PrP^{Sc} is resistant (Prusiner, 1997). Prions are not affected by normal sterilization techniques or high doses of gamma irradiation, and survive incineration in ash form. It is hypothesized that the virtual indestructibility of prion may be crucial in the induction of the damage that it causes to the central nervous system and that the spongiform change is brought about by cytoskeletal disruption of neuronal processes caused by liberation of hydrolytic enzymes from lysosomes overloaded by PrP^{Sc} (Laszlo *et al.*, 1992). The same authors hypothesize that the processing of normal PrP into the abnormal isoform takes place in lysosomes. More recently, it was suggested that neurotoxicity in prion disorders is mediated by a complex pathway involving transmembrane prion protein and not by deposits of aggregated and proteinase resistant PrP^{Sc} alone (Gu *et al.*, 2000).

3.   GENETIC RESISTANCE TO PRION DISEASES, NATURAL AND ENGINEERED

Based on clinical manifestations of naturally occurring scrapie, Parry recognized that the disease has a genetic basis and proposed elimination of scrapie by genetic selection programmes (Parry, 1979). It has been shown that the naturally occurring scrapie-resistant sheep that Parry proposed using in his breeding programme have the "ARR" genotype, i.e. amino acids alanine, arginine and arginine in positions 136, 154 and 171. Sheep with this configuration of the PrP gene are resistant to scrapie, and also show resistance to intracranial injection with BSE prions (Baron *et al.,* 2000). Breeding programmes, taking advantage of the resistance phenomenon in sheep based on DNA analyses, are now in progress in several countries, including Canada. In humans, substitution of lysine for glutamine at position 219 of the PrP amino acid chain results in resistance to sporadic CJD (Shibuya *et al.,* 1998).

An attempt to demonstrate the possibility of preventing TSE by genetically engineering the PrP in mice, recently undertaken at the University of California in San Francisco (Perrier *et al.,* 2002), indicated a possible avenue for a biotechnology approach to the introduction of BSE resistance, since, as was mentioned above, natural resistance to BSE has not been demonstrated.

Perrier *et al.* (2002) based their investigation on the association of the above-mentioned amino acid substitutions in sheep and humans with scrapie and sporadic CJD resistance, respectively. At mouse PrP amino acid position 167 (which corresponds to sheep position 171), arginine was substituted for glutamine, and at mouse PrP amino acid position 218 (that corresponds to human position 219), lysine was substituted for glutamine. For each mouse amino acid positions 167 and 218, the genes containing the substitutions were expressed in a homozygous state in one group of mice, so that the mice produced only the mutated prion protein, and in a heterozygous state in another group of mice, so that these mice expressed the mutated protein in combination with a normal wtPrP.

The results of Perrier *et al.* (2002), summarized in Table 2, demonstrated that the mice with the "dominant-negative PrP" prion gene, i.e. mice homozygous for the PrP gene that had either one of the two resistance-associated single amino acid substitutions but did not produce wtPrP, did not support prion propagation in this line of mice. The presence of the modified PrP gene was not associated with any detectable deleterious effects, indicating that the modification did not affect the normal physiological function of wtPrP.

Table 2. Resistance of transgenic mice to prions. One substitution of a single amino acid at a time in the PrP gene was tested in mice that carried such substitution in either heterozygous or homozygous state.

PrP Gene		Gene products present		Status of the transgenic mice after inoculation with prions
Amino acid No.	Substitution*	wtPrP	Prion	
167	Glutamine/**Arginine**	Yes	Yes	At 300 d, vacuoles and glyosis, some PrP^{Sc} in brains
	Arginine/Arginine	No	Yes	>550 d, no neuronal dysfunction, no PrP^{Sc}
218	Glutamine/**Lysine**	Yes	Yes	>300 d, no neuronal dysfunction, no PrP^{Sc}
	Lysine/Lysine	No	Yes	>300 d, no neuronal dysfunction, no PrP^{Sc}

NOTE: * The amino acid substituted in transgenic mice for that in wild-type PrP (wtPrP) is shown in boldface.
SOURCE: Based on Perrier *et al.,* 2002.

Perrier *et al.* (2002) suggested that the inability of their mice carrying the modified PrP genes to support prion replication raises the possibility of producing prion-resistant livestock that would express PrP with a single amino-acid substitution. The suggestion is supported by the observation that sheep with the desirable substitutions at codons 136, 154 and 171 were resistant to BSE and did not have any neurological signs of PrP^{Sc} in their brains (Baron *et al.,* 2000). The San Francisco group further recommend that the mice shown to be resistant to prion formation in their study need to be further tested by inoculation with other strains of prions (Baron *et al.,* 2000). Overall, they consider the introduction of point mutations in livestock that corresponds to naturally occurring resistance to prion diseases in sheep to be probably more acceptable to consumers of livestock products than complete disruption of the prion gene (Prusiner *et al.,* 1993), which might be associated with undesirable consequences.

A problem that will need to be considered in attempts to introduce resistance to BSE by genetic engineering of the prion gene is the possibility that genetically engineered prion-resistant animals might still harbour scrapie infectivity. Such infectivity may result from the presence of small quantities of abnormal prion material, even though it may not accumulate to detectable levels in the usual target tissues used for sampling for diagnostic or regulatory screening purposes. It is also possible that the abnormal prion proteins may exhibit different tissue tropism, susceptibility to proteinase digestion and pathogenic effects, and that the prion material's relative infectivity for other species may be changed. This would have significant implications for detection of animals harbouring abnormal prion proteins, since infected animals might be asymptomatic or have altered disease manifestations. Prion proteins might accumulate at unanticipated sites, such as muscle, which could hamper detection.

This possibility was demonstrated when hamster scrapie agent was injected into the brains of mice that either did or did not express the mouse gene that encodes the normal version of the PrP protein (Race and Cheesboro, 1998). Mice are known to be highly resistant to the hamster scrapie strain and none of the inoculated mice developed any clinical symptoms of scrapie. However, brain and spleen tissue of the mice that expressed the normal mouse prion gene, obtained between 204 and 782 days after inoculation, contained the scrapie agent, which was capable of infecting hamsters but not mice. By contrast, tissues of mice that did not have a functional PrP gene were only rarely able to transmit the disease to hamsters. It appears that expression of normal mouse PrP gene made it possible for the hamster scrapie agent to persist, even though there was no evidence of any replication of the hamster agent in the mice. Based on these observations, the authors suggest that the prion infectivity might persist in various "resistant" species exposed to BSE (prion)-contaminated feed (Race and Cheesboro, 1998).

Those attempting to produce BSE-resistant cattle by introducing point mutation(s) in wtPrP by genetic engineering have to take the possibility of "resistant carriers" of prions into consideration, both in the design and in the testing of the biotechnology approaches to the resolution of the TSE problem. Regulatory approaches to the approval of such engineered animals would need to ascertain that the possible existence of such "carriers", with differential sensitivity to different strains of TSE, does not result in a threat to human and animal health.

The impact of biotechnology-derived BSE-resistant cattle on the world cattle population could be rather rapid and widespread if the wtPrP gene in somatic cells from elite bulls were to be modified as suggested above and used to produce clones. It is conceivable that such clones, instead of the elite bulls themselves, might then be used in breeding by artificial insemination to rapidly introduce BSE resistance into cattle populations.

REFERENCES

Baron, T.G., Madec, J.Y., Calavas, D., Richard, Y. & Barillet F. 2000. Comparison of French natural scrapie isolates with bovine spongiform encephalopathy and experimental scrapie-infected sheep. *Neuroscience Letters,* **284**: 175–178.

Burns, C.S., Aronoff-Spencer, E., Dunham, C.M., Lario, P., Avdevich, N.I., Antholine, W.E., Olmstead, M.M., Vrierlink, A., Gerfen, G.J., Peisach, J., Scott, W.G. & Millhauser, G.L. 2000. Molecular features of the copper binding sites in the octarepeat domain of the prion protein. *Biochemistry*, **41**: 3991–4001.

Csaba, P. 2001. Yeast prions and evolvability. *Trends in Genetics*, **17**: 167–169.

Fischer, M.B., Roeckl, C., Parizek, P., Schwartz H.P. & Aguzzi, A. 2000. Binding of disease-associated protein to plasminogen. *Nature*, **408**: 479–483.

Gu, J., Fujioka, H., Mishra, R.S., Li, R. & Singh, N. 2000. Prion peptide 106-126 modulates the aggregation of cellular prion protein and induces the synthesis of potentially neurotoxic transmembrane PrP. *Journal of Biological Chemistry*, **277**: 2275–2286.

Laszlo, L., Lowe, J., Self, T., Kenwards, N., Landon, M., McBride, T., Farquhar, C., McConnell, I., Brown, J., Hope, J. & Mayer, R.J. 1992. Lysosomes as key organelles in the pathogenesis of prion encephalopathies. *Journal of Pathology,* **166**: 333–341.

OIE [Office International D'Epizooties]. 2002. Available at: http://www.oie.int/eng/info/en_esbinicidence.htm

Pan, K.-M., Baldwin, M., Nguyen, J., Gasset, M., Serban, A., Groth, D., Mehlhorn, I. & Hunag, Z. 1993. Conversion of alpha-helices into beta-sheets features in the formation of scrapie prion protein. *Proceedings of the National Academy of Sciences, USA,* **90**: 10692–10966.

Parry, H.B. 1979. Elimination of natural scrapie in sheep by sire genotype selection. *Nature*, **277**: 127–129.

Perrier, V., Kaneko, K., Safar, J., Vergara, J., Tremblay, P., Dearmond, S.J., Cohen, F.E., Prusiner, S.B. & Wallace, A.C. 2002. Dominant-negative inhibition of prion replication in transgenic mice. *Proceedings of the National Academy of Sciences, USA,* **99**: 13079–13084.

Prusiner, S.B. 1997. Prion diseases and the BSE crisis. *Science*, **278**: 245–251.

Prusiner, S.B. 1991. Molecular biology of prion diseases. *Science*, **252**: 1515–1522.

Prusiner, S.B., Groth, D., Serban, A., Koehler, R., Foster, D., Torchia, M., Burton, D., Yang, S.-L., & Dearmond, S.J. 1993. Ablation of the prion (PrP) gene in mice prevents scrapie and facilitates production of anti-PrP antibodies. *Proceedings of the National Academy of Sciences, USA,* **90**: 10608–10612.

Race, R. & Cheesboro, B. 1998. Scrapie infectivity found in resistant species. *Nature*, **392**: 770.

Shibuya, S., Higuchi, J., Shin, R.-W., Tateishi, J. & Kitamoto, T. 1998. Codon 219 Lys allele of PRNP is not found in sporadic Creuztfeldt-Jakob disease. *Annals of Neurology,* **43**: 826–828.

Telling, G.C., Scott, M., Mastrianni, J., Gabizon, R., Torchia, M., Cohen, F.F., Dearmond, S.J. & Prusiner S.B. 1995. Prion propagation in mice expressing human and chimeric PrP transgenes implicates interaction of cellular with PrP with another protein. *Cell,* **83**: 79–90.

CJD Surveillance Unit. 2002. CJD Surveillance Unit, University of Edinburgh, Scotland, UK. Data as of November 2002. See: www.cjd.ed.ac.uk/figures/htm

Willesmith, J.W., Wells, G.A.H., Cranwell, M.P. & Ryan, J.M.B. 1998. Bovine spongiform encephalopathy: epidemiological studies. *Veterinary Record,* **123**: 638–644.

Wrathall, A.E., Brown, K.F.D, Sayers, A.R., Wells, G.A.H, Simmons, M.M., Farrelly, S.S.J., Bellerby, P., Squirrell, L.J., Spencer, Y.I., Wells, M., Stack, M.J., Bastiman, B., Pullar, D., Scatcherd, J., Heasman, L., Parker, J., Hannam, D.A.R., Helliwell, D.W., Chree, A. & Fraser, H. 2002. Studies of embryo transfer from cattle clinically infected by bovine spongiform encephalopathy (BSE). *Veterinary Record,* **150**: 365–378.

GENETICALLY MODIFIED ORGANISMS IN NEW ZEALAND AND CULTURAL ISSUES

Robin McFarlane[1] and Mere Roberts[2]
1. Animal and Food Science Division, Lincoln University, New Zealand.
2. School of Geography and Environmental Science, University of Auckland, New Zealand.

Abstract   One of the ironies of the current debate in New Zealand about genetic modification is that it highlights the age-old conflict between science and religion, and in so doing demonstrates that modern society is still caught in the dilemma posed by these two views of the world. Two case studies are presented that demonstrate the distance between proponents and opponents of genetic modification (GM), and the difficulty of resolution within the secular-based framework of quantitative risk assessment applied by the Environmental Risk Management Authority (ERMA) and decision-making committees. Alternative frameworks suggested by Maori are beginning to emerge, and along with the results of several government-funded research projects in this area, should make a valuable contribution to a new framework that more equitably incorporates the fundamental principles of both knowledge systems. If this aim is achieved, it will be of considerable interest to other indigenous peoples in the world who are also faced with real and perceived threats to their cultural beliefs and values originating from new biotechnologies

1.   INTRODUCTION

Ko Aorangi toku maunga
Ko Waitaki toku awa
Ko Takitimu toku waka
Ko Ngaitahu me Te Ati Awa oku iwi
Ko Tuahuriri me Te Rakiamoa oku hapu
Ko Pokiri toku tipuna
Ko Robin McFarlane taku ingoa
No riera tena koutou tena koutou tena koutou katoa

551

H. P. S. Makkar and G. J. Viljoen (eds.), Applications of Gene-Based Technologies for Improving Animal Production and Health in Developing Countries, 551-561.

I wish to introduce myself by identifying with my mountain, river, canoe, tribe, sub-tribe and family affiliation. We would also like to acknowledge the contribution of colleagues and state that this paper is not *the* Maori view of genetic modification but rather one of many accounts and analyses within this contemporary area.

Development of regulatory frameworks to oversee research on recombinant DNA technologies have undergone gradual evolution from largely voluntary guidelines adopted in the 1970s, to formal legislation covering the importation and development of new organisms, including genetically modified organisms (GMOs) (Hope, 2002). All activities regulating GMOs in New Zealand are now covered under the Hazardous Substances and New Organisms (HSNO) Act, 1996, (hereinafter referred to as the HSNO Act), and compliance is compulsory. Many of the regulatory elements were derived from other nations, especially the United Kingdom and Australia. Authorization for experiments with GMOs is carried out by a central authority, the Environmental Risk Management Authority (ERMA), but responsibility for enacting the approval and monitoring of low risk experiments – which make up the vast majority of applications – is delegated to institutional biosafety committees (IBSCs). The remaining applications are considered by the Authority – a group of eight people appointed by the Minister for the Environment – which provides policy advice to ERMA, and also make decisions concerning those applications that go direct to ERMA. These include applications for a field trial; for conditional or full release into the environment; and any others that ERMA considers should require public notification, such as the two case studies considered below. Biosafety aspects of a putative project are based on scientific risk assessment, analysis and management. Before the introduction of the HSNO Act, risk assessment was based largely on the Brenner classification (GMAG, 1978), with consideration given to the factors of GMO access, expression and damage. Latterly, this classification has been widened to include environmental considerations. In addition, a more extensive risk analysis is demanded where deemed necessary, such as where field releases may be contemplated.

Opposition to the development or use, or both, of GMOs is widespread within New Zealand society, as indicated from surveys among the public, various interest groups (Macer, Bezar and Gough, 1991; Macer, 1998) and farmers (Fairweather *et al.*, 2003). The first-named surveys indicated that New Zealanders were generally more positive about the use of modern technology, including the use of GMOs, than their European Union, Canadian or Japanese counterparts. The proportion of people worried about genetic engineering increased during the 1990s, and, by 1997, ethical considerations were specifically mentioned as a point of concern. Within the last-named survey, a wide range of farmer groups were in agreement with using GMOs for medical benefits, but were split over the use of the technology in agriculture and food production – a topic of major importance for a country dependent on farm production for its economic survival.

During the assessment process for low-risk applications, limited consideration is given to moral, ethical and spiritual issues compared with attention given to biological safety issues, despite the fact that moral, ethical and spiritual issues resound deeply with many groups of New Zealanders (Roberts and Wills, 1998), especially Maori. However, Section 5(b) of the HSNO Act 1996 includes the need to

> "recognize and provide for … the maintenance and enhancement of the capacity of people and communities to provide for their own economic, social and cultural well-being and for the reasonably foreseeable needs of future generations."

Two further sections of the Act deal specifically with the potential effects of a hazardous substance or new organism (e.g. a GMO) on Maori. One, Section 6(d), requires applicants and decision-makers to

> "take into account … the relationship of Maori and their culture and traditions with their ancestral lands, water, sites, waahi tapu (sacred sites) valued flora and fauna and other taonga (valued possessions)."

Another, Section 8, notes the necessity to take into account the principles of the Treaty of Waitangi, a fundamental document signed between Maori Chiefs and the British Crown in 1840. Article 2 of this Treaty guarantees to Maori their *tino rangatiratanga* or right to retain authority over their natural resources and all other taonga. Since 1975 and the creation of a Waitangi Tribunal to hear Maori grievances against the Crown, the courts have defined a number of principles emanating from the Treaty, which are recognized as morally binding on the Crown (i.e. the elected Government) and its delegated agents, e.g. ERMA. One of the principles that is relevant here is that of partnership between Crown and Maori, which carries with it the responsibility to act in good faith towards each other and to engage in meaningful consultation. Another principle is for the Crown to actively protect Maori interests.

In practice, however, these sections of the Act have proved to be problematic when attempting to equate perceived risks as identified by scientific risk assessment processes, with Maori knowledge systems, and how these two assessments might be weighed in the decision-making process.

In this paper, we provide two case studies to illustrate these problems. We then discuss the findings of an independent body established by the New Zealand Government to report on GM in New Zealand; comment on the major issues that still await resolution in the ERMA process, identified by Maori; and suggest some new ways forward.

2.   CASE STUDIES

A summary of applications (July 1998–September 2003) for the development or importation of New Organisms, including GMOs, is given in Table 1.

The two case studies that follow are part of approved experiments to field test GMOs in containment. Both examples involve the production of transgenic livestock incorporating human genes.

Table 1. Application to develop or import new organisms, New Zealand, 1998–2003.

Activity	Applications	Approved	Initially withdrawn	Declined
Import or develop GMO	148	133	12	0
Field test GMO in containment	16 (4 animals)	13	2	0
Field release GMO	0	0	0	0
Import new organism (non-GMO)	33	28	4	0
Release new organism (non-GMO)	7	5	1	1

2.1   Case study 1

In September 1998, PPL Therapeutics applied to ERMA for approval to develop transgenic sheep in laboratory containment by the insertion of human alpha-1-antitrypsin (hAAT) into the sheep DNA. The donor nucleic acid originated from an ovine β-lactoglobulin gene isolated from sheep material, and the human gene coding for AAT isolated from a Caucasian cell-line. The sheep were then to be multiplied (by natural mating) within a securely fenced enclosure, for the purpose of producing human alpha-1-antitrypsin (hAAT) in their milk. Following purification from harvested milk, this substance would be used in human clinical trials, as a potential treatment for cystic fibrosis or congenital deficiency of hAAT. ERMA considered the views of two Maori *iwi* (tribes) with *mana whenua* (rights and responsibilities to exercise *kaitiakitanga* or guardianship) over the land where the experimental farm was located. Representatives of PPL held several discussions with local Maori communities, including the Pouakani II Trust and Marae Committee, and the Mangakino Township Incorporation. Despite some reservations, each of these groups decided that the potential benefits to human kind outweighed their concerns. However, they stipulated that transgenic sheep must not be allowed to enter the food chain; and that they must be disposed of by incineration followed by deep pit burial to prevent contamination of nearby waterways – especially the Waikato river, a *taonga* to their iwi. *Ngati Raukawa* considered that such cross-species transfer (transgenics) constituted an unacceptable breach of a sacred belief which could not be mitigated, but accepted a compromise in consideration of the proposed benefits to human health, and therefore did not support or oppose the application. However, they also requested that these sheep not be allowed to enter the food chain and that if no benefits from this trial emerged and the programme were terminated, then a *whakanoa* (ritual cleansing ceremony) of the site be undertaken.

Further advice was offered by *Ngai Kaihautu Tikanga Taiao*, a Maori advisory committee to ERMA. In their opinion, the transfer of genes (particularly human genes) from one species to another is in conflict with the *tikanga* (traditional beliefs, values and customs) of Maori – particularly those of *whakapapa* (history of descent and relationships of things) and *mauri* (the essential essence of a thing).

The Authority approved the application but recognized that it involved an activity that conflicted with Maori spiritual and cultural beliefs, and therefore had the potential to adversely affect the relationship of Maori and their *taonga*.

2.2   Case study 2

The second case involved an application to ERMA in December 1998 by AgResearch (a Crown-owned research institute) to field-test cattle that were transgenic for human myelin basic protein (MBP). This human gene construct had already been inserted into the genome of a cow embryo some years before the application to ERMA (prior to the HSNO Act 1996 coming into force). In this application, cloning and surrogacy techniques were to be followed by conventional breeding in contained grazing conditions so that animals could be tested under controlled field conditions.

The proposed research was very similar to that for the PPL sheep application and involved consultation with several different groups belonging to the Tainui tribal confederation who claim to speak for Maori living in the Waikato region, where AgResearch is located. The Waikato Raupatu Lands Trust advised that the use of human genes in animals was inappropriate according to Maori cultural and spiritual beliefs, as is their consumption. However, they neither supported nor opposed the application but reserved their position until they felt able to make a better-informed decision (letter 23 April 1999). *Ngâti Wairere*, the *hapu* (sub-tribe) that claims *mana whenua* status over the actual land on which AgResearch is situated was adamantly opposed to the application. In their submission to ERMA in 1999 they stated that

> "the mixing of whakapapa by humans between different species is ... inherently against their tikanga…. The Authority has a statutory duty under the Treaty of Waitangi to afford protection of all Maori taonga … and should decline the application".

They were also concerned about the potential for this research to cause "metaphysical imbalance" for people in the area, that could continue over many generations. Furthermore, in order to reduce the potential for contamination of their land and waterways by the transgenic material, they requested that, if the application were to be approved, that Ngati Wairere be invited to determine a culturally appropriate protocol for disposal of the research animals.

In reaching a decision, the committee was divided; the minority (a Maori expert in tikanga) supported Ngati Wairere's views and opposed granting the application. The majority noted that no previous cases before the Courts and the Waitangi Tribunal had addressed the need to actively protect Maori spiritual beliefs, such as whakapapa and mauri, in contrast to tangible, i.e. physical, taonga that have spiritual significance. Accordingly, the majority concluded that spiritual taonga, which may be defined variously and from time to time, are not amenable to active protection in the same way as more tangible taonga.

> " It is one thing to take every effort to respect Maori spiritual beliefs, it is another to ask the whole community to accept them as arbiters of whether genetic research should proceed under the HSNO Act."

Accordingly, the application was approved (ERMA decision GMF98009, 23 May 2001).

3.   ROYAL COMMISSION ON GENETIC MODIFICATION

In a speech from the Throne at the opening of Parliament on 21 December 1999, the new Labour Government announced its intention to set up a Royal Commission on Genetic Modification to enquire into and report on the issues surrounding genetic modification in New Zealand. It also announced a moratorium on all applications to field test or release GMOs for the period June 2000 to August 2001 (this was later extended to 31 October 2003). The Commission held more than 50 meetings, received more than 10,000 written submissions and held 12 weeks of formal hearings, before reporting to the Government in July 2001. Of the 107 submissions received from organizations given "interested persons" status, 7 were from Maori organizations. An additional consultation process for Maori included 28 workshops and 10 regional *hui* (meetings on *Marae* or traditional gathering places) culminating in a national hui. Major issues to Maori regarding GM technology that were highlighted as being of concern are listed below.

– Whakapapa. Submitters emphasized the necessity of protecting the integrity of *whakapapa* by avoiding the mixing of genes between species, especially the mixing of human genes with other species. This constituted both spiritual and physical "contamination" of the organisms as well as the environment.

– Taonga. Recognition that Maori should have exclusive control over their *taonga* as specified under the Treaty of Waitangi, that these include both tangible and intangible taonga, and that Maori have the right to decide what constitutes taonga.

– Mauri. Respect for the spiritual essence that pervades all things, and recognition of the role of Maori as *kaitiaki* to maintain the purity and good health of the *mauri* of organisms and their environment.

– Relationships. Acknowledgement of the need to maintain the appropriate balance in relationships (*whanaungatanga*) between all things, including humans and their environmental kinsfolk.

– Precaution. This "precautionary principle" is expressed by Maori as *kia tupato*, or "be careful", particularly with regard to the potential effects on the environment and on future generations.

– Intellectual property rights. A number of submissions opposed the patenting of naturally occurring life forms, particularly of native flora and fauna. Many others wanted better recognition and protection of traditional *matauranga* or intellectual knowledge.

– Process issues. A majority of submitters had concerns about the relationship between the Crown and Maori, arguing that the current ERMA process did not reflect the Treaty principle of a partnership. Maori wanted more involvement in and control over the process, and for the process to more adequately reflect *tikanga* values. There were also concerns about inadequate information on which to make informed decisions, and of the time constraints in which to become informed.

The Commission's recommendation was that New Zealand should reserve its opportunities with GMOs, GM products and innovations, and proceed with caution. In addition it suggested another 44 recommendations on a range of issues, including intellectual property rights, liability, and various areas of GM research. While several of these specifically addressed some of the concerns raised by Maori, in general the Commission avoided giving specific advice to aid in the resolution of cultural, ethical and spiritual issues. Instead it suggested the setting up of *Toi te Taiao* (a Bio-ethics Council) to promote ongoing dialogue on these matters.

4. COMMON THREADS RESISTING RESOLUTION

Maintenance of the relationship between Maori, their culture and traditions, and their ancestral lands and taonga is critical to the survival of their culture. But just how this relationship is perceived and interpreted varies according to *hapu* and *iwi,* as the two case studies above indicate. Maori knowledge is dynamic and changes with respect to time and space. However, common core values are apparent from the submissions made to the Royal Commission. Paramount among these is the concept of whakapapa. As noted by Roberts and Wills (1998),"to know something is to know its whakapapa." Whakapapa acts as a cognitive template upon which all things known in this world are located and described in space and in time, and in terms of their relationships to other things. For humans it acts as more than a pedigree containing the origins and descent lines of each individual Maori within the whanau (family), hapu (sub tribe) and iwi (tribe); it enables people to relate themselves to the mountains, rivers and other physical features within the tribal area wherein they reside. Concerning plants and animals, whakapapa act as folk taxonomies in addition to providing a mental map of the ecosystem of the object under consideration (Roberts *et al.,* 2004). And as all whakapapa derive ultimately from *atua* (gods), it provides an unbroken link and chain of descent between the spiritual world and the material; the inanimate and the animate. Because it is derived from and under the protection of atua, to interfere with whakapapa is to interfere with things sacred or *tapu.*

To compromise the integrity of whakapapa by altering the gene structure of species (particularly human) through unnatural modification is therefore perceived by many Maori as posing a threat to the traditional relationships embodied in whakapapa, and hence to the cultural values and beliefs such as mauri, that derive from it.

In an effort to find common ground between science and Maori knowledge concerning genetic modification, the Authority has attempted to dissect some of these values and beliefs into their component parts and ask, what are the physical aspects of the genetic code that constitute (for example) mauri? One such attempt was outlined above (Case 2). The Authority uses a clearly reductionist assumption that the mauri of an organism is reducible to individual component parts such that "the only mauri present (in a gene) is the mauri of its constituent bases" (ERMA decision GMD02028 30 September 2002). A contrary view to this is that mauri is an emergent property of matter, as is life itself; i.e. it is the sum total of more than its constituent parts. Two eminent writers on Maori culture both comment that the mauri is not located in any organ of the body, and thus, by inference, cells or genes (Firth, 1973). Clearly the decision-making committees and processes of the Authority have great difficulty assessing the weight to give spiritual or "intangible" taonga rather than to something physical to which spiritual values are attached, e.g. a sacred site. Elsewhere, Durie (2003) also points out that the Authority

> "overlooked the fundamental starting point upon which Maori world views are built – the relationships that confer coherence within the natural world. While the scientific method often dissects the whole into smaller parts in order to find the truth, Maori philosophical methods work in the opposite direction; truth is a function of wider relationships and higher order synergies."

For such reasons, a reductionist approach is unlikely to provide satisfactory answers to questions that arise out of a distinctly different mental paradigm.

In fact, ever since the emergence of modern science from the constraints of religion in the sixteenth century, the gap between the two – science and religion – has widened to the point reached today, as highlighted by the GM debate, in which both sides have increasing difficulty in identifying any common ground. Indeed public discourse in the West has emphasized secular, culture-free, "rationalistic" themes due a lurking fear of religion and a desire to bypass it. Thus, although Maori have been conferred certain rights as partners under the Treaty of Waitangi, which appear to "legitimize" their knowledge, in practice the scientifically-based process of assessing the risks is incommensurate with Maori values and beliefs.

Is there a solution to this conundrum? Several ways forward have been suggested by Maori scholars. Based on matauranga Maori, Mead (2001) proposed several criteria or "tests" that might be applied by Maori in order to assess the acceptability or not of a research application to genetically modify an organism. Tapu involves assessing whether an organism's sanctity is to be breached; *take-utu-ea* – understanding the cause (*take*), purpose, purported benefits or recompense, especially if things go wrong; precedent in the culture; and the consideration of a set of Maori principles. A second proposal by Durie (2003), outlined in Table 2, presents an alternative "Maori methodology" based on three fundamental concepts: the natural environment; the human condition; and procedural certainty.

This author suggests that the application of this framework might include a fourth column of "indicators" that could be used to measure the potential of an application to achieve the desired research outcomes identified in column three (Table 2).

> Whatever the outcome of these suggestions, as Mead (2001) concludes,
> "the tests identified here could be useful to families confronted by the dilemma of having to decide to participate in new technologies, new cures for medical problems, and new ways of doing things. There are steps to

take and questions to be asked and answered. The resulting decision would at least be based on a process that would generate much soul searching, many discussions, and result in a greater understanding of tikanga Maori."
Clearly these approaches need to be used alongside currently used risk assessment procedures to be acceptable to the greater community.

Table 2. A Research Potential Framework.

Domain	Maori Value/Concept	Desired Research Outcome
The Natural Environment	*Mauri* (Integrity)	Research that contributes to the integrity of ecological systems.
	Whanaungatanga (Relationships)	Research that contributes to strengthening relationships between people, between people and the natural environment, and between all organisms.
	Kaitiakitanga (Guardianship)	Research that contributes to resource sustainability.
The Human Condition	*Wairua* (Spirituality)	Research that contributes to human dignity within physical and metaphysical contexts.
	Tapu (Safety)	Research that contributes to human survival and safety.
	Hau (Vitality)	Research that contributes to the maintenance of human vitality.
	Whakapapa (Intergenerational transfers)	Research that contributes to the standing of future generations.
Procedural Confidence	*Tikanga* (Protocols)	Research that contributes to the development of protocols to address new environments.

REFERENCES

Durie, M. 2003. *Mana Tangata: Culture, Custom and Transgenic Research.* Report to the New Zealand Bioethics Council (Toi te Taiao).

Fairweather, J.R., Maslin, C., Gossman, P. & Campbell, H.R. 2003. *Farmer views on the use of genetic engineering in agriculture.* Agribusiness and Economics Research Unit - Report 258. Lincoln University, Christchurch.

Firth, R. 1973 (reprint). Chapter VII-appendix, *in: Economics of the New Zealand Maori.* Wellington, NZ: A.R. Shearer, Government Printer.

GMAG [Genetic Manipulation Advisory Group]. 1978. Report. Genetic manipulation: new guidelines for UK. *Nature,* **276:** 104–108.

Hope, J. 2002. A history of biotechnology in New Zealand. *New Zealand Journal of Environmental Law,* **6:** 1–42.

Macer, D. 1998. *Public perception of biotechnology in New Zealand and the international community: EuroBarometer 46.1.* Christchurch, New Zealand: Eubios Ethics Institute.

Macer, D., Bezar, H. & Gough, J. 1991. *Genetic engineering in New Zealand: science, ethics and public policy.* Centre for Resource Management, Lincoln University, Christchurch.

Mead, H.M. 2001. *"Nga ahi e ngiha mai nei: the fires that flare up."* Recommendations of the hui wannga on tikanga Maori and GM.

Roberts, R.M. & Wills, P.R. 1998. Understanding Maori epistemology: a scientific perspective. pp. 43–77, *in:* H. Wautischer (ed). *Tribal epistemologies: Essays in the philosophy of Anthropology.* Aldershot, UK: Ashgate.

Roberts, R.M., Haami, B., Benton, R., Satterfield, T., Finucane, M.L. & Henare, M. 2004. *Whakapapa as a mental construct of Māori, and its implications for genetic engineering.* The Contemporary Pacific University of Hawaii.

OBJECTIVES, CAPABILITIES AND DANGERS IN THE ROLE OF INTERNATIONAL ORGANIZATIONS AND FUNDING AGENCIES IN PROMOTING GENE-BASED TECHNOLOGIES FOR LIVESTOCK IN DEVELOPING COUNTRIES

John Hodges
Lofererfeld 16, A-5730 Mittersill, Austria.

Abstract: Gene-based technologies offer the world unprecedented opportunities for improving quality of life, or for reducing it in irreversible ways. The basic question addressed in this paper is the position and response of international bodies and donors on whether or not to provide gene-based technologies to developing countries. It will not be easy to attain a responsible and coherent answer to this challenging question. Gaining an objective understanding of the essential issues is hard when controversy rages across the supposedly neutral scientific facts.   Nevertheless, the outcome of the discussion is of prime importance at a global level. This paper seeks to bring light into this arena.

After the Introduction, three principle concerns are examined which should be at the top of the agenda of these international institutions. Following this, short reviews of the critical issues are presented covering: the scientific characteristics and uncertainties associated with gene-based technologies; the nature of target areas in which they may be applied; and the considerable disquiet in society generally. These short outlines highlight the possible benefits and dangers associated with the critical issues. It is concluded that the objectives, capabilities, opportunities and dangers cannot be evaluated at the scientific level alone; they must be evaluated as matters of high policy by all stakeholders before gene-based technologies are implemented on the ground.

In view of these perspectives, at the end of the paper it is proposed that scientists should place a moratorium on the development of gene-based technologies for the development of transgenic animals. It is also proposed

H. P. S. Makkar and G. J. Viljoen (eds.), Applications of Gene-Based Technologies for Improving Animal Production and Health in Developing Countries, 563-584.
© 2005 *IAEA. Printed in the Netherlands.*

that, during the moratorium, the United Nations should carry out a global referendum on the desirability of gene-based technologies being applied to the food chain. Meanwhile it is recommended that international organizations and funding bodies should not promote these techniques.

1.   INTRODUCTION

1.1   The role of international organizations and funding agencies

Gene-based technologies appear to offer substantial opportunities for lifting the quality of life. They also arouse conflicting views because of fears of unknown results. Discussion of the objectives, capabilities and dangers reveals a wide divergence of perspectives as the positive and negative consequences are debated in Europe and North America. The concept of transferring this advanced technology to developing countries for use with livestock presents a profound set of challenges. In the developed West these strong positive and negative feelings do not hinge around the detailed techniques, but rather over the visions and policy of how these new gene-based methods are to be used and their impact upon the quality and meaning of life. The international bodies, funding agencies and national governments must take these alternative perspectives into their deliberations.

The brief adopted for this paper is to view dispassionately the objectives, capabilities and dangers that now face the international bodies and donors. Evaluation is not easy since the potentials both for good and harm are speculative and vast. The potential targets of use and the possible beneficial outflows from gene-based technologies are recorded in detail in other papers in this publication and it is not considered appropriate to repeat them in this paper. Here, therefore, the author seeks to provide a balance in the publication as a whole by presenting scientifically-argued information on the uncertain aspects at this time of using gene-based technologies on livestock in the tropics and semi-tropics. This approach does not imply that the author is unaware or sceptical of the possible benefits. Rather it is recognition that the development and application of gene-based technologies for such widespread use is also accompanied by risks of major consequences for humanity. A comprehensive balance of perspectives is particularly important for international organizations and funding agencies as they examine not only the technologies themselves but also decide on the policies they will adopt on this crucially important step in the future of mankind, livestock, food and the environment.

International bodies and donors are central to this debate since, for many decades, they have played a key role in the transfer of technologies in food and agriculture to developing countries. The routine approach of these bodies is to choose priorities, decide upon the level and supply of funds, and then follow through to monitor progress cooperatively with the recipient governments. However, the issues arising with gene-based technologies extend beyond traditional mandates and routines.

The role of the private sector in gene-based technologies for the tropics and sub-tropics is also immensely important. To date their self-interest has been limited to crop agriculture and human health and they have little involvement or apparent interest in gene-based technologies for animal agriculture in developing countries. This posture, based upon commercial judgements, is itself significant as it indicates the extent of the problems to be overcome and the hazards likely to be encountered.

1.2   The powerful nature of gene-based technologies

The frontiers of biology now being breached by gene-based technologies have profound significance for mankind and for the whole planet whose resources humanity manages. Dualism confronts us. There is potential for so much good and the risk of so much harm. This endeavour has the atmosphere of a Faustian bargain at a new global level of benefit and cost. Can we know the consequences with certainty and control them with confidence? The use of this powerful gene-based technology has clear and defined purposes, namely to modify existing life forms of animals, animal feed, microbes, insect parasites and animal diseases. The overall objective is to improve the quality and quantity of food and of services provided to humans by animals. This strategy involves probing a highly complex network of biological genomes with the intention of changing some specific components. Inevitably this type of intervention must affect the interactions of different domestic plant and animal species with each other, with the natural environment, with wild species of plants and animals and with more lowly organisms. Because food is the main target, humanity is also part of this whole genetic network. Further, since the genomes of mankind and mammalian livestock are so similar there are possibilities of unforeseen and undesirable consequences in the human genome.

It is imperative to know before starting such a daring and magnificent venture whether the promotion of gene-based technologies in the tropics and sub-tropics offers guaranteed prospects for long-term success for mankind and for the natural resources of the earth without the risk of permanent damage. The scale of any irreversible damage is so great that new parameters of caution are essential. Therefore it is now, rather than later, that the values and assumptions of those who favour and those who question the strategy need to be examined.

1.3   Assumptions concerning risk and benefit with gene-based technologies

A common philosophy is that mankind has to take risks to make progress. True. But steps toward successive stages of progress need to be evaluated sensibly. That principle means the decision-makers should take account of the interests of all stakeholders, anticipate the consequences if things go wrong, consider the scale of use and define the point beyond which there is no return. Certainly mankind has to take risks to move ahead; but ethical behaviour requires that any negative consequences from taking risks are limited to the decision-makers alone, at any rate in the speculative stages. Linkage of consequences to the risk-taker is a basic ethical assumption in capitalist and civilized societies. Statements by some leading scientists seeking to reassure the public that genetic modification is harmless because it is not new but has been practised for a 1,000 years may be soothing but, by stretching the facts, are misleading. Genetic progress using conventional methods within domestic plant and animal species might be defined as "modification" but the new technology producing "Genetically Modified" products enters a fundamentally new and different dimension of human interventions in biology. Gene-based technologies involve transfer of genetic material between species, which is a novel technique, and scientists have but a few years of experimentation behind them.

Gene-based technologies are so powerful and collateral damage may be so extensive and permanent that the concept of gaining a major benefit of this type is already questioned by informed people and may not be acceptable to the stakeholders. Benefits could fade into insignificance in a world faced with abnormal organisms. Society may eventually ask, "Why did we ever breach these biological boundaries?"

In the physical sciences the application of nuclear energy presents a similar Faustian bargain: enormous energy and benefits with dreadful consequences for humanity if either the technology goes wrong or human management fails. Consequently, society decided early in the development of nuclear technology that research and use must be restricted in location and ownership must remain under strict security. Nuclear technology does not pass out of the hands of scientists and qualified engineers and is not available for unmonitored use by the public. Nuclear material remains always and everywhere under governmental scrutiny and is closely monitored by national governments and by international agencies and agreements. With nuclear energy we all know that one mistake is one too many.

Beyond the similarities of nuclear engineering and gene-based technologies applied to food and the natural environment, there is also a marked contrast. Governments and people can decide whether they wish to use nuclear power – that is their option – and some have chosen nuclear-free zones. By contrast everyone has to eat. Whereas a country can decide to be non-nuclear, as Austria has done, it is impossible to opt out of eating. Further, it is possible, at great expense, to close down nuclear power-plants, to seal off the Chernobyl sarcophagus and to bury nuclear waste deep in the earth. Even Hiroshima, Nagasaki and Chernobyl – dreadful in their local and sub-regional effects – did not affect the whole earth permanently. If things go wrong with gene-based technologies and deleterious organisms appear in the food chain or in the environment, control would be a nightmare. We do not want biological Chernobyls. The challenge of eliminating modified biological organisms that bring tragedy to the human food chain or in natural resources can scarcely be contemplated in physical or economic terms.

1.4   The confidence of scientists over gene-based technologies

The normal scientific position on the question of risk is that science has a whole array of proven techniques available to test new transgenic biological organisms, to assess probabilities of things going wrong and to ensure that the probability of any known harm is anticipated and avoided. The problem with this new scientific intervention into the complex biological system of the food chain is that the door remains open to unknown risks, which science cannot assess in advance. This inability to predict serious negative consequences in livestock subject to gene-based technologies was well argued in 2002 by the National Research Council of the National Academies in the USA (NRC, 2002). Statistical probabilities of success or failure are inadequate means of providing confidence to the world's stakeholders when only one negative biological event could be multiplied without control.

The case of Bovine Spongiform Encephalopathy (BSE), also known as Mad-Cow Disease, in the UK is a prime example of how difficult it is to diagnose an unpredicted novel condition before it is dispersed throughout a main component of the food chain – in this case the UK national cattle population. Instituting measures to remove BSE from the bovine population and to prevent the spread to other countries proved to be extremely difficult even in a country such as the UK with highly developed infrastructure. The problem is entirely different from containing a previously known pathogen which is quickly identified and then isolated and eradicated using proven procedures with experienced staff. This stark contrast was seen in the way that the UK Foot and Mouth Disease (FMD) outbreak was handled in 2000. Although that FMD epidemic was terrible and spread rapidly, identification of the strain of the FMD virus was easy and elimination routines were successful within the year. In contrast, the BSE epidemic, which was first recognized in 1986, continues to perplex scientists, and 18 years later still erupts into new problems with both cattle and humans. Science has no institutional memory and experience to handle BSE. Such a situation occurs when novel and unexpected biological shocks arise from gene-based technologies.

We need to learn from BSE how difficult it is to control and to eliminate novel organisms. BSE and the human form of BSE, known as variant Creutzfeldt-Jakob Disease (vCJD), have not yet been contained in either the cattle or the human populations. BSE was neither expected nor apparent until some years after it had happened – and then a mammoth and expensive diagnosis and cleanup ensued. In 2004 it remains impossible to assess, anticipate or control the continuing new complications in the UK human population. For example, in late 2003, a fresh scenario arose. Blood donors who, years before, were unaware they were incubating vCJD and who later died from vCJD are thought to have been the vectors passing the condition to blood recipients who later also died of vCJD. The tracing process so far has revealed 15 individuals who had received such blood; consequently these people live under the cloud of probably having vCJD. The ramifications continue because such individuals must also now be excluded from donating blood. The BSE/vCJD problem is not simple since effective diagnostic methods are not available before fatal symptoms appear. The BSE saga shows the difficulty of anticipating consequences when molecules specific to one species cross into other species.

The introduction of gene-based technologies to livestock in developing countries presents challenges that could involve a high proportion of the world's population. The negative impacts, if and when they come, will be beyond feasible recall. The recognized ethical principle of Prior Informed Consent is being violated. Although nobody knows the possible consequences, human populations would be unknowingly committed to participation. Adequate supervision and testing of the technologies and their products within the infrastructure of the developing world is immeasurably more difficult than within the laboratory system of the West.

This paper next considers the major positive and negative issues relating to the objectives, capabilities and dangers facing international bodies and donors. In the following part of the paper some critical technical and socio-economic aspects are briefly reviewed. The paper closes with conclusions, and proposals for action.

2. MAJOR POLICY ISSUES – POSITIVE AND NEGATIVE

2.1 What are the good prospects – a renewed renaissance?

Gene-based technologies appear to offer exciting new prospects for dealing with some ancient problems blocking the road to improved quality of human life. With livestock, some hitherto intractable problems may be overcome with these emerging technologies (Hodges, 1986). That vision is truly exciting. For scientists, who by nature are always thrilled by the prospect of original research, it is a remarkable time of challenge and opportunity. Gene-based technologies break open a whole new world of knowledge that lies near the heart of biological life and human society. This is comparable, if one may coin a new term, to a Renewed Renaissance – Western civilization being born a third time. This frontier of science and the application of newly discovered knowledge could become a breakthrough into a new epoch capable of replacing the worn and compromised patterns of society by launching humanity into a new era of progress.

The excitement among scientists is palpable. The thrill is enhanced by the field of application, namely to provide more and better food and services in the tropics and semi-tropics. This objective, seen through the idealism of early hope, can easily be viewed as a new and fundamental way to break the bands of hunger and of poverty and to diminish the gap between rich and poor. These aims have been the long-term agenda of international developmental and funding bodies together with inter-governmental agencies and national governments for decades – but the scourge of poverty remains. Poverty is an ever-present blight upon Renaissance hopes, now four centuries old, that ignorance would flee before rationality and science and thus enable humanity to shake off poverty and its associated ills. Today poverty still reigns undiminished in some sectors of all cultures, from the most advanced to the most lowly and primitive. Two billion people, one third of the world's population, live on a dollar a day.

The best visions for using gene-based technologies offer the high moral ground upon which scientists, governments and international bodies often take their stand and find their justification in that these innovative technologies will solve the world's food problems. In contrast, there are informed critics whose studies question the validity of this position (Food Ethics Council, 2003).

2.2   What are the negative prospects?

They derive from three sources:
– the biological uncertainties and risks;
– the human socio-economic and political mechanism for delivering gene-based technologies to developing countries; and
– the fear that gene-based technologies are fruitful ground for terrorist interests in biological weapons of mass destruction.
Each of these three anxieties is now briefly considered.

2.3   The biological uncertainties and risks

The genetic structure of life, which currently is defined by species boundaries, may be stirred by gene-based technologies into unpredictable dimensions with unknown activities. The molecular architecture of life is remarkably stable, having been honed over huge periods of time into genomes with great constancy, resilience and strength. Gene-based technologies are used to intervene in this complexity with the aim, initially at any rate, of fine tuning genomes to human advantage. As confidence and knowledge grow, these early minor modifications are likely to metamorphose within scientific research and business ambitions into larger and potentially endless agendas to redesign biological life.

Fears exist that interventions that over-ride long-established biological boundaries that have been inviolate for aeons before and during the whole of human civilization may unleash new forces hostile to progress and may even impair current qualities of life. This fundamental step of gene transfer, which can be seen so positively, has also a darker side in the thoughtful human imagination. The details of that darker side are difficult to measure in scientific terms. But it is a strong intuition shared by many people whose professional lives embrace a large vision of life and who are not driven by the inevitably blinkered excitement of scientific research and business development.

The shared fears do not relate only to the scientific method which, properly applied under controlled laboratory conditions in the West, follows rigorous, even pedantic routes of checking and counter-checking. Some of these scientific and research protocols are discussed by Pardey and Koo (2002). Nevertheless, trepidation arises from knowing that decisions to introduce gene-based technologies to the tropics and sub-tropics will not, in fact, be made by objective scientists. Scientists design technologies and products. They prepare work plans with probabilities of biological success carefully documented. However, the decisions about funding and implementing are always made elsewhere than in the scientific realms. The implications of the decision-making hierarchy and process are huge and are linked with anxieties about the competence of international bodies to promote gene-based technologies.

2.4 The human socio-economic and political mechanism for delivering gene-based technologies to developing countries

Decisions about the use of gene-based technologies in the tropics and semi-tropics will be made largely by the public bodies and governments in the global arena. These activities will cover research, transfer of technology and products, education and training, project identification for application, structuring, and monitoring.

The track-record of international public bodies and governments of developing countries is not characterized by clear, rational and scientifically based thought. Decisions are often compromises, based upon political expediency and financial incentives, with short-term interests high on the agenda.

Anxiety concerning the administrative competence of the international bodies to handle such sensitive new technologies would not arise if the biological risk were non-existent and every technique and application were guaranteed to be thoroughly tested and benign even in failure. But the biological uncertainties are great, and the consequences of scientific imprecision or management bungling could be catastrophic.

A parallel fear is already evident in the developed world, where the engine driving gene-based technologies into the public domain is the private sector. Business is a legitimate part of the modern socio-economic system, and its mandate to make a profit and to enhance shareholder value is recognized within market economy capitalism. Business activities are generally regulated by national governments in developed Western society, in the interests of health, safety, environment and fair trading. In contrast, the transfer of gene-based technologies to many developing countries raises fears about the absence of equivalent enforceable regulatory mechanisms at the global level.

Returning now to the role of international agencies, national governments and donors for gene-based technologies for livestock, one must look at the track record of these organizations with transfer technology for the genetic improvement of livestock. It is not encouraging. My own experience of being responsible for animal breeding and genetic resources in the Food and Agriculture Organization of the United Nations (FAO) showed that, as a profession, animal geneticists from the West often fail to understand the difficulties of transplanting proven, temperate, genetic technologies to the tropics. Analyses of the results of cattle breeding projects over many decades studied by Payne and Hodges (1997) indicate that the rate of failure to achieve the proposed objectives has been high. Principal reasons for failure that had not been anticipated were the socio-economic and cultural systems in developing countries. The resulting economic loss that occurred, especially among small farmers with ruminant animals, was serious at the family level but localized and had limited negative impact, which livestock owners were eventually able to overcome. However, failures to achieve objectives accompanied by negative fall-out effects from gene-based technologies are likely to remain and to multiply in the natural, social and economic environments.

2.5 Bioterrorism

The possibility is real that gene-based technologies transferred to developing country locations will provide fruitful ground for terrorist interests in biological weapons of mass destruction. Placing gene-based technology resources into locations away from the supervision of countries that, at the moment, are the identified targets of terrorism is regarded by many people, not least some politicians, as utterly foolhardy.

3. CRITICAL ISSUES IN ASSESSING OPPORTUNITIES, CAPABILITIES AND CHALLENGES

In this next part of the paper, brief highlights are given of some of the critical issues facing international bodies and funding agencies as they decide policy and practice concerning gene-based technologies for livestock. The issues include technical and socio-economic factors.

Although many critical points are examined in the following section, they all bear upon three key issues, which appear to be central to the discussion. First, there is the fear that interventions into the foundational building blocks of species will open the way to unforeseen restructuring of life, thus damaging natural biological resources. Second, there is the possibility that the close genetic relationship between livestock and humans could result in damage to humanity through the food chain. The third issue is the idea that these biotechnologies will reduce the gap between rich and poor, alleviate poverty and increase world food production to match the expected

50 percent expansion in world human population, to ten billion people by 2050. The critical points are now reviewed in a series of short outlines.

3.1 Difficulties of transferring genetic projects for livestock

In the past, funding agencies for programmes and projects to improve food and agriculture using livestock have dealt in "technology transfer". The techniques to be transferred had generally been proven to work in the developed West, with measurable results, defined and controlled risk, and economic success. Transferring them to the tropics or semi-tropics has always required adaptation to different environmental conditions, animal genotypes, animal disease and feed supplies.

We have learned the hard way that ignorance of socio-economic and cultural values in tropical and semi-tropical countries has been a major cause of failure of animal genetic projects. We have also learned that it is easy to bring equipment and products such as live animals, semen, embryos and breeding programmes to the tropics – but it is frequently impossible to find or to create the needed infrastructure to administer, record, control and decide on successive steps to harvest genetic progress. Past experiences have been based upon use of stable genetic resources from temperate zones. Gene-based technologies introduce a new dimension and involve an entirely different set of scientific techniques and resources compared with conventional breeding methods. Thus, the process of bringing these exceptional technologies to developing countries will confront the international bodies and donors with many new hurdles.

3.2 The existing technologies and products are not yet stable

At present – early 2004 – gene-based technologies for livestock are nearly all from research in the developed world. In that advanced scientific arena, these technologies are far from proven. They cannot be lifted off the shelf and used anywhere with predictable results. The vision of possibilities has already far outrun the ability to produce the products. In gene-based technologies, the discard rate of animals *in vitro*, in pre-term and in full-term pregnancies, in peri-natal deaths and in subsequent premature deaths or malfunction announces powerfully to those who will listen that undesirable genetic consequences of unknown origin occur. Dolly was the only successful cloned sheep from more than three hundred attempts. Such attempts are expensive and uncertain.

A further factor is the need to study successive generations of transgenic animals or those treated with transgenic products with the aim of monitoring normality. Few tests for biological stability over long periods of time have been completed. Where sparse data does exist over successive generations of transgenic animals, interpretation is often difficult, because the search for unknown abnormalities and for cause of death is fraught with possibilities of error. Animal generation intervals are long – a disadvantage compared with use of gene-based technologies for plant science. Based upon existing experience it must be concluded that the transfer of gene-based technologies in livestock to the tropics and semi-tropics will encounter much failure in both biological and economic terms.

3.3 The scientific resources devoted to gene-based technologies are largely committed to temperate region problems

In the developed world, the majority of gene-based technology work is directed towards temperate breeds and species of animals with the object of promoting more intensive animal production. The animal genetic resources of the tropics and semi-tropics are markedly different from those of temperate regions. Production systems are generally low-input/low-output. Animals have genetic adaptation providing them with physiological and metabolic processes suited to harsh and challenging environments, coarse and intermittent feed supplies and with resistance against endemic diseases and insects. Thus much fundamental research is needed before genes can be successfully transferred.

3.4 Animals in the West are usually selected for only one food product whereas in developing countries multiple uses are usually required

This point is obvious to all concerned with livestock and does not need elaborating here. But it does pose a new set of questions that gene-based technologies have yet to address. The success with transgenes in one trait may have a negative effect upon another trait. This problem is well known in conventional animal selection programmes. This old problem presents both new opportunities for solution and new challenges to be overcome.

3.5 Lack of scientific backup in developing countries

Developing countries do not yet have – and it may be very difficult to develop – the intensive scientific backup and consultation network available in the West to monitor the extensive and expensive trials that gene-based technologies require. Gene-based technologies with livestock are subject to highly specific standards of control in the West. High levels of scientific and management supervision plus government permission are required in many Western countries for research, development and application in the field. The costs and feasibility of providing such levels for each developing country are unrealistic. Therefore the possibilities of centres of excellence should be examined. In view of the early stages of this type of work with livestock, the idea of one or two global or regional centres for gene-based technologies applied to livestock presents real benefits.

3.6 Gene-based technologies have not yet gained precision in application

Intellectual visions and agendas are far ahead of bench technologies. It is therefore questionable whether the limited financial public resources available to serve food and agriculture together with poverty alleviation should be devoted to gene-based technologies at this stage. Investments in conventional technology rather than gene-based technology approaches may be far more successful and rewarding.

3.7 The disposal of discard animals

Because gene-based technologies with livestock are not efficient, the repeated attempts to gain a specific goal produces many discard animals. These carcases need controlled disposal so that they do not enter the food chain – a practice currently banned in Europe and North America. Such controlled disposals are expensive. Further there is always the temptation to use the abnormal animal or the carcase illegally in the food supply to uninformed consumers. This may have unknown, uncontrolled and serious consequences. The temptations also include passing of live animals produced by gene-base technologies, known or unknown to be abnormal, into rural herds where ultimately the meat will enter the food chain. The control of and recovery from such tragedies is much harder in poorer societies.

3.8   Livestock in some areas of developing countries are subject to predators; the effect of transgene-animals entering the food chain of wild animals is unpredictable

Interspecies transfer of genes and their products lies at the heart of gene-based technologies. When perfected in the developed West, with all possibilities of dangers examined, tested and overcome, it may be feasible to release such transgenic animals into the commercial livestock sector with strict controls. However, doing this in the natural environment in which many ruminants are kept in developing countries will expose them to predation by wild animals. The potentials for unknown biological, genetic and pathogenic dangers are very great. It is impossible to test all such cross-species transfer in advance.

The example of Bovine spongiform encephalopathy (BSE) in the UK, the origin of which remains unknown, is a reminder of the dangers. Some theories on the origin of BSE involve inter-species transfer by ingestion by bovines of tissues from sheep or exotic species discarded from wild-life parks. Certainly variant Creutzfeldt-Jakob Disease (vCJD) is passed from the bovine species to the human species by ingestion.

3.9   Poverty alleviation

The reduction and removal of poverty is a major objective of international agencies and donors. Small-scale subsistence farmers with limited sales of animal products are particularly susceptible to poverty. These farmers are widespread in many parts of the developing world. Estimates show that half the world's population live in this way. Gene-based technologies are likely to be applied in more intensive production units. Thus it is immediately questionable whether gene-based technologies can be quickly and easily used for poverty alleviation. Considerable study will be needed upon the individual products of gene-based technologies to determine their socio-economic impact in poverty reduction.

3.10   Decision-making

Who will be the decision-makers in transferring gene-based-technologies? Will it be the donor or the recipient? Will it be governments, research institutions, NGOs, individual scientists, rural development specialists or investors? How will stakeholders be involved?

3.11   Beneficiaries

The wealth of benefits anticipated from gene-based technologies with livestock are but concepts and ideologies at present. In this stage of visions it is easy to be overwhelmed by enthusiasm and to conclude that gene-based technologies offer a new phase in the battle against world hunger. In the West, that vision is clearly misplaced, since the West already has too much food. Therefore enthusiasts shape their vision in terms of feeding the developing world and providing food for the 800,000 million people who are starving, hungry or malnourished. This concept must be brought down to earth. The vision needs thorough in-depth analyses of how it will be done and who will be the real beneficiaries among the players.

3.12   Global or local technologies

A major question is the extent to which technologies might bring benefit wherever they are applied, compared with those technologies which need to be designed in the research stage for specific problems, local environments and cultures. Problem identification and design of research strategies is an extremely complex and lengthy process. It is naïve to assume that technologies designed to serve temperate zone problems can be transferred successfully. New strategies based upon local practices, assumptions and values must be designed with local leaders.

3.13   Intellectual property rights

The issue of intellectual property rights (IPR) is already embedded in the scenario of gene-based technologies in the West. Moving these technologies to developing countries will not automatically change the situation. Use of information, biological resources and methods frequently means paying for use of IPR. There is likely to be some resistance from donors over making payments to developed country investors (OECD, 2002).

3.14   Are gene-based technologies a public good?

The above issue of IPR raises the question of whether gene-based technologies are truly a public good, which ought to be promoted with public funds. IPR tends to give an exclusive designation to such technologies and their products which removes them from being a public good and has a negative impact on attempts to use them for poverty alleviation.

A further issue for international bodies, donors and governments is the question of whether and by whom the fruits of gene-based technology in developing countries can or should be patented when donors have funded the work. Protocols will be needed well in advance to avoid bitter legal and expensive disputes in cases of success.

3.15   Liability and insurance

A parallel issue to IPR is the question of liability and of insurance when things go wrong in work funded by international bodies and governments. The appalling experience following the explosive release of poisonous chemicals from a manufacturing plant for agricultural chemicals in Bhopal, India, was not limited to the thousands of deaths and hundreds of thousands suffering permanent sickness – it extends also into the awful, hostile and unresolved processes of litigation.

3.16   Key infrastructure

Gene-based technologies require specific infrastructure, which is often not available in developing countries. Establishing the economic, communication, transport, supervisory, educational and other infrastructure is itself a major aspect of development. International bodies and donors may have to ask whether the introduction of gene-based technologies should be postponed until physical, economic and social infrastructure is in place.

3.17   Public health and safety using gene-based technologies for food and agriculture

There are major issues of health and safety of animal feed and human food that recipient governments need to address, working together with donor governments. In the West, it has proved necessary to create new governmental or para-governmental agencies to monitor the rapid and novel changes now appearing in food as a result of gene-based technologies, although so far mainly in foods of plant origin.

3.18   Uses of gene-based technologies for human health and for food are different

It is relatively easy to make a case for the use of gene-based technologies for human health in developing countries, of which HIV is an example. Further cases arise in the development of vaccines and medical drugs for treatment of tropical diseases, which is a positive use of gene-based technologies. In these cases, the laboratories and products remain largely under medical supervision in development, production and use, whether in the West or in the recipient country. The contrast between gene-based technologies in humans and in livestock is fundamental, and must be taken on board by international bodies and donors. The central fact of the contrast is choice by the human patient.

The long-term vision of gene-based technologies for livestock typically sees large-scale application to animal populations in the field to bring about higher productivity, or in the prevention and treatment of disease by genetic means using the model adopted in developed countries for artificial breeding and embryo transfer with cattle. Eventually, this model results in animals that are themselves transgenic or have been treated with transgene products. By contrast, use of gene-based technologies in the human health field does not visualize the creation of genetically modified humans on a grand scale. Gene therapy in humans is currently an uncertain field for application.

The targets with livestock include not only transgenic animals and products but also genetically modified microbes in the rumen and genetically modified plants for animal feed. Without doubt, the aim will be to release these transgenes into the natural environment, where the majority of ruminant livestock live in the tropics and semi-tropics.

Another contrast between humans and animals is the fact that individual choice is offered to patients over use of gene-based vaccines and treatments for human health. Any negative consequences are explained before the patient chooses. The contrast is vivid. Gene-based technologies applied to food in developing countries will not allow choice. Even the European Union requirements introduced in 2004 for labelling Genetically Modified Food would not work among the urban poor and subsistence farmers of the developing world.

3.19 Damage to the genome and its effects in the human population

Gene-based technologies in animals to be used as human food open the prospect of damage to the human population. Why? Because the aims and basic methods of gene-based technology involve moving genetic material across species boundaries. The mammalian farm animals identified as targets have genomes that are closely related to the human genome. Further, farm animals ingest and excrete massive amounts of biological material drawn from the natural environment, which is also the source of much human food of plant origin.

The example of BSE is a formidable warning of the type of damaging activities at the gene and protein levels that can occur while the danger is unknown and unanticipated by science, yet multiplied by scientifically approved systems of producing animal feed and human food. (Phillips, Bridgeman and Ferguson-Smith, 2000; Horn *et al.*, 2001). By the time it was identified in 1986, BSE had already caused great damage to the UK cattle population. The repercussions continue, with the first cases of BSE in Canada in 2003 and in the USA in 2004. The impact of BSE also continues through variant Creutzfeldt-Jakob Disease (vCJD) to humans – approximately 150 people had died by early 2004. Estimates of who will yet die vary considerably because of the lack of knowledge of this condition: dosage, incubation period, genetic susceptibilities, etc. Further, the costs to the livestock industry were and remain enormous, as shown in the report of a Working Party of the European Association for Animal Production (Cunningham, 2003), which estimated the discounted present loss to the beef industry at Euros 92 billion. The legacy of BSE remains, leading to an uncertain future, since much infective meat and bone meal was exported from Europe to developing countries before and after the dangers were identified.

3.20   Levels of wealth

The growth of gene-based technologies in the developed West has been possible because of the wealth of the West in finance, intellectual resources, education and the existence of a highly structured economic system offering a route for application. This has been true of research in both the public and the private sectors. Most research to date has been within the context of wealthy states, which are likely to be made richer economically or in quality of life by prudent use of the new technologies. In contrast, many people in developing countries are poor. It is not clear that gene-based technologies will automatically improve the wealth or the quality of life in developing countries. Further, these technologies, combined with current World Trade Organization (WTO) trade policies for agricultural products, have great potential to increase the gap between rich and poor, both between and within countries.

3.21   Locations of food production in the future

Clearly the potential impact of the WTO is immense on the future locations of food production for the expected population booms in developing countries. The WTO takes precedence over the Cartagena Protocol on Biosafety. The use of gene-based technologies in the West will continue to exceed their application in developing countries. The West will compete for the urban markets in developing countries with imported Western foods. This scenario will result in negative effects upon domestic producers in the hinterlands of the tropics and semi-tropics who will lose their domestic markets and be unable to compete internationally, thus sliding into further poverty.

3.22 Cultural and religious preferences

As learned earlier in the failures of animal genetic projects with cattle, cultural and religious values are often the major factors affecting the possibilities of change in any community. On this issue the West must not fall into the trap of thinking that only people in developing countries have cultural and religious values. Western society is not free of values and assumptions that are embedded in the culture. Even though science is value-free, scientists are not. Western agricultural scientists have values that are largely tied to economics and efficiency (Hodges, 2003), while consumers and citizens show growing interest in other values, especially health and safety, as additional features of choice beyond the lowest compatible cost. The issue of cultural imperialism is a very sensitive topic in many developing countries.

3.23 Assessment of gene-based technologies for quality of life

The most critical question is "How can gene-based technologies be used by international bodies and funding agencies to empower local farmers and not the opposite – to remove their markets and existing livelihood?" Underlying these issues is a basic question "Are gene-based technologies for animals cost-effective when the whole picture is examined and do they have a positive effect upon lifestyles?"

4. CONCLUSION

Every thinking scientist and knowledgeable person in society grants the premise that mankind now has found in gene-based technology another powerful tool that, on the grand scale, is potentially capable of healing and enhancing or abusing and destroying human life and its environment as it now exists. The question is whether gene-based technology for animal production and health purposes should be widely applied, together with its products, throughout large areas of the tropics and subtropics.

Gene-based technologies, without question, present the world with the possibility of dangerous effects in developing countries due to the biological uncertainties and the weakness of human delivery systems. The consequences of things going wrong extend from the biological to the socio-economic. The possibilities are very slim that mankind has sufficient knowledge to anticipate science-based concerns or sufficient wisdom to avoid rupturing human social, cultural, religious, ethical and economic values.

Public organizations and funding agencies must equip themselves in creative ways to be effective in facing these new situations. There are three significant issues concerning the role of international organizations and public funding:

1. the value and safety of the gene-based techniques themselves;
2. responsible handling of risks and the impacts upon the social, economic, cultural and ethical aspects of life in the tropics and semi-tropics; and
3. the level of priority that ought and might be given to gene-based technologies in this era of reduced international and public financial support for research and development in agriculture.

The dilemma is clear. The answers are not. It is easy to be fascinated by gene-based technologies simply because the research strategies, programmes and projects are designed with positive results as targets. It is equally easy to neglect the potential negative consequences and subsidiary effects that might multiply out of control in the target species, in other species, in the infrastructure of human society or in the natural environment. The Cartagena Protocol on Biosafety requires the collaboration of three broad groups: client countries, providers and donors (ISNAR, 2003). While the Cartagena Protocol provides a framework for these principal parties when they are implementing programmes and gene-based-technology research, the larger policy issue for international bodies and donors is whether they should venture at all into funding gene-based technologies for livestock, given other priorities, public uncertainties and the inability to identify some significant classes of risk in advance.

There are helpful parallels between nuclear power and gene-based technologies. Both are recognized as capable of enormous good and also of bad effects upon society on a massive scale. The story of nuclear energy offers some lessons.

The nuclear power near-tragedy in 1979 at Three Mile Island in the USA was averted by swift action. In contrast, in 1986 at Chernobyl in the USSR, the tragedy resulted from human error and caused widespread damage to the environment, to human lives and generations unborn at the time, to food safety and to economic development.

In the 1970s, the Austrian government built a nuclear power station, situated not far from Vienna. In 1978, after it was completed but before it was activated, the government recognized that the public had strong doubts and fears. They therefore held a referendum in which the Austrian people voted to make Austria a non-nuclear zone. The complete and new nuclear power facility has never been switched on.

Public opinion in Europe is accurately, objectively and regularly measured by the European Union EuroBarometer, which shows a consistent pattern of about 75% of citizens who want Europe to be free of genetic modification, especially in food. The State of Paraná in Brazil has already declared itself free of Genetically Modified Food.

The most important task of the international bodies and donors is to find out what people of the world want before they start to implement gene-based technologies on a global scale. This step would be truly democratic and would avoid accusations that a scientific elite is making life-changing decisions that affect the whole world. In the meanwhile, caution is the only responsible policy.

5.　　PROPOSALS

It is proposed:
1. That scientists should place a voluntary moratorium on the development of gene-based technologies for the development of transgenic animals, in view of the close genetic relationships of man and animals and because livestock are integrated in the food chain with the natural environment. This step would be comparable to the voluntary moratorium placed by scientists upon recombinant DNA when the techniques were first developed in the 1970s.
2. That during the moratorium, the United Nations should carry out a global referendum on the desirability of gene-based technologies being used for food.
3. That until the results of the global referendum are clear, international organizations and funding agencies should not promote these techniques.

Support for such a novel approach when life changing, global scenarios emerge is provided by Václav Havel in his speech in New York in September 2002, on the occasion of his last visit as President of the Czech Republic:

"If humanity is to survive and avoid new catastrophes, then the global political order has to be accompanied by a sincere and mutual respect among the various spheres of civilization, culture, nations or continents … . If we examine all the problems facing the world today, be they economic, social, ecological or general problems of civilization, we will always come up against the problem of whether a course of action is proper or not, or whether, from the long-term planetary point of view, it is responsible."

REFERENCES

Cunningham, E.P. 2003. After BSE – A future for the European livestock sector. Report by a Working Party of the European Association for Animal Production (EAAP). EAAP Publication. No. 108. Wageningen, The Netherlands: Wageningen Academic Press.

Food Ethics Council. 2003. *Engineering Nutrition: GM crops for global justice.* Food Ethics Council, 39–41 Surrey Street, Brighton, England, UK.

Hodges, J. 1986. Prospects for using biotechnology for livestock improvement. *World Animal Review,* No.56: 2–10.

Hodges, J. 2003. Livestock, ethics and quality of life. *Journal of Animal Science,* **81**(11): 2887–2894.

Horn G., Bobrow, M., Bruce, M., Goedert, M., Mclean, A. & Webster, J. 2001. Review of the origin of BSE. Report to UK Government Department of Environment, Food and Rural Affairs. Norwich, UK: UK Government Stationery Office.

ISNAR [International Service for National Agricultural Research]. 2003. A Framework for Biosafety Implementation. Report of a meeting by ISNAR in Washington DC, July 2001. The Hague: ISNAR.

NRC [National Research Council]. 2002. *Animal Biotechnology: Science-Based Concerns.* Washington DC: National Research Council of the National Academies.

OECD [Organisation for Economic Co-operation and Development]. 2002. Report on genetic inventions, intellectual property rights and licensing practices: Evidence and policies. Report of an OECD Working Party, Berlin, 2002.

Pardey, P. & Koo, B. 2002. Biotechnology and genetic resource policies. International Food Policy Research Institute (IFPRI), Washington DC, USA.

Payne, W.J.A. & Hodges, J. 1997. *Tropical Cattle. Origins, Breeds, and Breeding Policies.* Oxford, UK: Blackwell Science Ltd.

Phillips, Lord, Bridgeman, J. & Ferguson-Smith, M. 2000. Report, evidence and supporting papers of the Inquiry into the emergence and identification of Bovine Spongiform Encephalopathy (BSE) and variant Creutzfeldt-Jakob Disease (vCJD) and the action taken in response to it up to 20 March 1996. Norwich, UK: UK Government Stationery Office.

SUITABILITY OF BLOOD PROTEIN POLYMORPHISMS IN ASSESSING GENETIC DIVERSITY IN INDIGENOUS SHEEP IN KENYA

Joram. M. Mwacharo,[1] Charles. J. Otieno[2] and Mwai A. Okeyo[1]

1. Department of Animal Production, University of Nairobi, P.O. Box 29053, Nairobi, Kenya.
2. Iowa State University, Dept of Animal Science, 2255 Kildee Hall, Ames, Iowa 50011, USA.

Abstract Knowledge of genetic diversity is important as it forms the basis for designing breeding programmes and making rational decisions on sustainable utilization of animal genetic resources. This study was designed to assess the efficiency of blood protein polymorphism as a rapid tool for assessing genetic diversity, using seven blood proteins (transferrin, albumin, haemoglobin, esterase A, esterase C, carbonic anhydrase and X-protein) and 457 indigenous fat-tailed (351) and fat-rumped (106) hair sheep in Kenya from 7 populations, with 40 Merino as controls. Transferrin was analysed using polyacrylamide gel electrophoresis and starch gel electrophoresis was used to analyse the other six loci. Of the seven loci analysed, two – carbonic anhydrase and X-protein – could not be interpreted. The five interpretable markers, however, showed low levels of polymorphism in allele numbers and heterozygosity. Multilocus mean F_{ST} values of 0.083 indicated a moderate genetic differentiation between the populations analysed. The Dm and Da genetic distance estimates showed the indigenous sheep populations in Kenya to be closely related genetically, with the dendrogram failing to resolve indigenous sheep into fat-tailed sheep and fat-rumped hair sheep. Due to its costs and modest equipment demands, blood protein polymorphism can be used as a rapid tool to assess genetic diversity and prioritize breeds to be analysed by microsatellite DNA markers.

1. INTRODUCTION

Assessment of genetic diversity within and between indigenous domesticated animal populations has of late become the subject of intense research. Information on the inherent genetic diversity is important in the

585

H. P. S. Makkar and G. J. Viljoen (eds.), Applications of Gene-Based Technologies for
Improving Animal Production and Health in Developing Countries, 585-591.
© 2005 IAEA. Printed in the Netherlands.

design of breeding improvement programmes, making rational decisions on sustainable utilization and conservation of animal genetic resources. Lost genetic diversity is irreplaceable and thus there is need to conserve it. This is especially true for indigenous breeds in developing countries, which are being replaced at a fast rate by exotic high producing breeds, in spite of the local breeds' excellent adaptation to prevailing stressful environmental conditions. The danger looms of losing valuable genes for adaptation to extreme environments and diseases, which might be of value in the future.

Several techniques for assessing genetic diversity exist (Avise, 1994). With the discovery of DNA-based markers, traditional techniques such as protein polymorphism have been ignored, arguing that they are less informative. This study was designed to assess the efficiency of blood protein polymorphism as a rapid tool for assessing genetic diversity, using as a test case indigenous sheep populations in Kenya that had not been characterized.

2.   MATERIALS AND METHODS

The study was carried out using two types of indigenous sheep populations in Kenya: fat-tailed (from Kwale, Makueni, Siaya, Kakamega and Kajiado Districts) and fat-rumped hair sheep (from Isiolo District). Merino sheep were included in the study as a reference breed. A total of 497 genetically unrelated animals (fat-tailed = 351 (Kwale = 61; Makueni = 65; Siaya = 60; Kakamega = 77; Kajiado = 88), fat-rumped = 106 and Merino = 40) were sampled from village flocks. Peripheral blood was collected from each animal into 10 ml EDTA vacutainer tubes and processed within 24 hours of collection by centrifuging at 3000 rpm for 20 minutes. The plasma and red blood cells were each pipetted into separate, clearly labelled 2-ml vials and stored at -20°C until electrophoresis.

Seven blood protein coding loci (transferrin (Tf), albumin (Al), esterase A (EsA), esterase C (EsC), haemoglobin (Hb), carbonic anhydrase (Ca) and X-protein (X)) were analysed, using polyacrylamide gel for Tf and starch gel electrophoresis for the others. Known standards were included on each gel to ensure consistency of genotype scoring. A detailed description of the analytical technique is published elsewhere (Mwacharo, 2000). Exact significance probabilities (Elston and Forthofer, 1977) were used to determine allele frequency deviations from the expected Hardy-Weinberg Equilibrium. Allele diversity and expected heterozygosity were computed as measures of genetic variability, while the Wright (1978) F-Statistic was used to analyse genetic differentiation. The minimum (Dm) and arc (Da) genetic distances of Nei (1972) and Cavalli-Sforza and Edwards (1967) were computed as measures of genetic relationships and were used for phylogenetic analysis. All computations were done using the BIOSYS-1 computer program (Swofford and Selander, 1989).

3.   RESULTS AND DISCUSSION

Five loci (Tf, Al, Hb, EsA & EsC) out of seven gave interpretable results. The resolution of Ca and X could not be resolved. This may be attributed to the long period between sample collection and analysis (2 years), which may have affected the quality of the samples. Braend and Khanna (1968) observed weaker Tf zones for samples kept at room temperature for several days compared with corresponding zones in fresh or freshly frozen samples. Additionally, most proteins – even the most stable, such as lactate dehydrogenase – begin to degrade in activity and resolvability after prolonged storage, under even the lowest temperatures (May, 1992). These observations demonstrate one clear disadvantage of protein markers, namely the stringent requirement for fresh tissue samples and for appropriate storage conditions. This may be a major drawback when sampling populations at remote sites.

The five interpretable markers showed low levels of polymorphism or variability in the number of alleles per locus and heterozygosity values of less than 50% (Table 1) compared with those observed using microsatellites (Arranz, Bayon and San Primitivo, 1996, 1998). This may be attributed to the low polymorphisms and variability of protein-based markers, which tend to underestimate the levels of genetic diversity within and between populations. However, blood protein polymorphisms have been used extensively to evaluate genetic diversity in sheep (Clarke, Turker and Osterhoff, 1989; Missohou *et al.*, 1999), goats (Tunon, Gonzalez and Vallejo, 1989) and cattle (McHugh *et al.*, 1998). Protein electrophoresis does not distinguish all the genetic variability at a given structural gene locus (May, 1992) due to the low levels of variation in the coding sequences relative to the other regions of the genome (Arranz, Bayon and San Primitivo, 1996). Routine electrophoresis detects only the amino acid substitutions that result in differences in the net charge of proteins (Ayala and Kiger, 1980). This represents only a third of all the amino acid substitutions, as only approximately 30 percent of all amino acid substitutions on the exterior of the molecule results in a charge shift (May, 1992). Additionally, only the variability in the coding portions of the genome (which constitutes about 10 percent of the total eukaryotic genome) can be sampled using protein electrophoresis.

The significant multilocus F_{ST} value indicated a moderate genetic differentiation, implying a relatively low degree of gene flow between the populations analysed (Table 1). The F_{ST} value observed compares favourably with those observed in sheep (Arranz, Bayon and San Primitivo, 1998) and cattle (McHugh *et al.*, 1998) studied using microsatellite markers.

The genetic distance estimates showed the indigenous sheep populations to be closely genetically related (Table 2). However, the topology of the dendrogram showed poor consistency with the morphological classification based on the localization of fat deposits (Figure 1). Blood proteins have been used successfully to differentiate between the indigenous African *Bos taurus* (Muturu and N'dama) breeds and zebu (*Bos indicus*) breeds (Braend and Khanna, 1968) and between other sheep populations (Zanotti Casati, Gandini and Leone, 1990; Nguyen *et al.*, 1992).

A comparison of the average genetic distances obtained between the indigenous sheep populations in Kenya with those obtained from previous studies of recognized local breeds of sheep (Missohou *et al.*, 1999; Nguyen *et al.*, 1992; Ordas and San Primitivo, 1986) reveal that the genetic differentiation of indigenous sheep populations in Kenya is of the same order of magnitude of that among well recognized and established breeds of sheep.

Table 1. Genetic variability measures and coefficient of genetic differentiation in indigenous sheep populations in Kenya.

Locus	Expected heterozygosity	Observed heterozygosity	Average no. of alleles per locus	F_{ST}	F_{IS}
Albumin	0.069	0.068	1.29	0.046	-0.059
Haemoglobin	0.164	0.162	1.43	0.268	-0.025
Transferrin	0.562	0.614	4.14	0.034	-0.170
Esterase A	0.340	0.394	2.00	0.122	1.00
Esterase C	0.157	0.209	2.00	0.082	1.00
Mean value	–		–	0.083	0.318

NOTES: F_{ST} is correlation between random gametes within subdivisions/subpopulations relative to the total population. F_{IS} is average over all subdivisions/subpopulations of the correlation between uniting gametes relative to their own subdivision/subpopulation.

Table 2. Matrix of D_m and D_a genetic distance measures between the indigenous sheep populations under study.

Populations	1	2	3	4	5	6	7
1. Kwale		0.002	0.004	0.015	0.030	0.006	0.077
2. Makueni	0.045		0.004	0.007	0.018	0.005	0.056
3. Siaya	0.060	0.065		0.020	0.037	0.011	0.072
4. Kakamega	0.112	0.076	0.117		0.007	0.013	0.047
5. Kajiado	0.146	0.114	0.170	0.104		0.018	0.030
6. Isiolo	0.091	0.092	0.122	0.132	0.114		0.052
7. Merino	0.280	0.259	0.286	0.250	0.192	0.243	

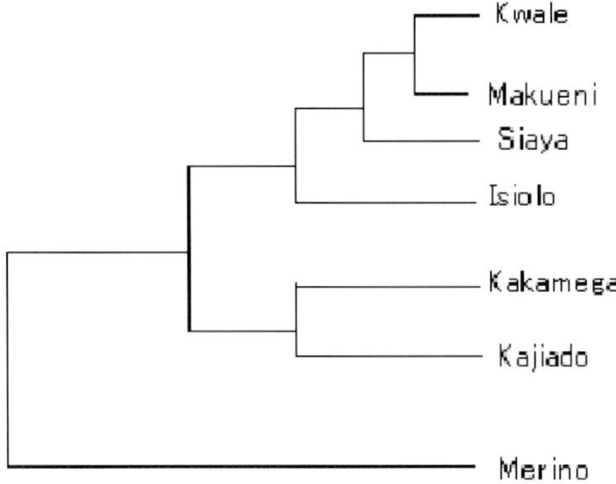

Figure 1. UPGMA tree constructed using the Da genetic distance for the seven populations analysed in the study.
NOTES: Co-phenetic correlation coefficient = 0.984.

Higher estimates of genetic distances have been obtained in other breeds of sheep using microsatellite markers (Arranz, Bayon and San Primitivo, 1998). To improve the accuracy of using protein polymorphism in population genetic studies, a minimum of 30 individuals in each population should be surveyed for approximately 30 loci (Nei, 1976). This is because the number of loci used in this type of study can represent less than one thousandth of the structural genes of an individual's genome (Dobzhansky *et al.*, 1980).

4.   CONCLUSION

The value of protein polymorphism lies in its emphasis on DNA sequences that are expressed phenotypically and when the protein coding loci are sufficiently variable to distinguish the population genetic structure. It offers a rapid, cheap, easy, expedient, reliable and legally compelling means of investigating simultaneously the genetic variation of several functional genes in the same individual. Its equipment demands are modest and personnel can be trained quickly, although the interpretation of gel patterns requires considerable experience. The loci under study are unambiguously named and the functions of their products are known. Their use in population genetics is strengthened by the fact that they are inherited as Mendelian dominant characters and attempts have been made to associate them with quantitative traits (Dally *et al.*, 1980; Vicovan and Rascu, 1989) and as aids to selection. DNA markers give larger genetic distances and higher heterozygosity values and thus are of value when analysing closely related breeds or species. However, for populations whose genetic status is unknown, protein polymorphism may be used first to verify the degree of genetic relationship and to prioritize breeds for subsequent analysis using microsatellites.

REFERENCES

Arranz, J.J., Bayon, Y. & San Primitivo, F. 1996. Comparison of protein markers and microsatellites in differentiation of cattle populations. *Animal Genetics,* **27**: 415–419.

Arranz, J.J., Bayon, Y. & San Primitivo, F. 1998. Genetic relationships among Spanish sheep using microsatellites. *Animal Genetics,* **29**: 435–440.

Avise, J.C. 1994. *Molecular Genetic Markers, Natural History and Evolution.* 1st ed. New York, NY: Chapman and Hall.

Ayala, F.J. & Kiger, J.A. 1980. Gel electrophoresis. *In:* F.J. Ayala and J.A. Kiger (eds). *Modern Genetics.* Menlo Park, CA: Benjamin/Cummings Publishing Co.

Braend, M. & Khanna, N.D. 1968. Haemoglobin and transferrin types of some West African cattle. *Animal Production,* **10**: 129–134.

Cavalli-Sforza, L.L. & Edwards, A.W.F. 1967. Phylogenetic analysis: models and estimation procedures. *Evolution,* **21**: 550–570.

Clarke, S.W., Turker, E.M. & Osterhoff, D.R. 1989. Blood groups and biochemical polymorphisms in the Namaqua sheep breed. *Animal Blood Groups and Biochemical Genetics,* **20**: 279–286.

Dally, M.R., Hohenboken, W., Thomas, D.L. & Craig, A.M. 1980. Relationship between haemoglobin type and reproduction, lamb, wool and milk production and health related traits in crossbred ewes. *Journal of Animal Science,* **50**: 418–427.

Dobzhansky, Th., Ayala, F.J., Stebbins, G.L. & Valentine, J.W. 1980. *Evolucion.* Barcelona: Omega.

Elston, R.C. & Forthofer, R. 1977. Testing for Hardy-Weinberg Equilibrium in small samples. *Biometrics,* **33**: 536–542.

May, B. 1992. Starch gel electrophoresis of allozymes. *In:* A.R. Hoelzel (ed). *Molecular Genetic Analysis of Populations: A practical approach.* London: IRL Press at Oxford University Press, Oxford.

McHugh, D.E., Loftus, R.T., Cunningham, P. & Bradley, D.G. 1998. Genetic structure of seven European cattle breeds assessed using 20 microsatellite markers. *Animal Genetics,* **28**: 333–340.

Missohou, A., Nguyen, T.C., Sow, R. & Gueye, A. 1999. Blood polymorphisms in West African breeds of sheep. *Tropical Animal Health and Production,* **31**: 175–179.

Mwacharo, J.M. 2000. *Characterisation of the genetic diversity of Kenyan indigenous sheep breeds using blood biochemical polymorphism.* MSc Thesis. University of Nairobi, Nairobi, Kenya. 119p.

Nei, M. 1972. Genetic distance between populations. *American Naturalist,* **106**: 283–292.

Nei, M. 1976. Mathematical models of speciation and genetic distance. pp. 723–766, *in:* S. Karlin and E. Nevo (eds). *Population Genetics and Ecology.* New York, NY: Academic Press.

Nguyen, T.C., Morera, L., Llanes, D. & Leger, P. 1992. Sheep blood polymorphisms and genetic divergence between French Rambouillet and Spanish Merino: Role of genetic drift. *Animal Genetics,* **23**: 325–332.

Ordas, J.G. & San Primitivo, F. 1986. Genetic variations in blood proteins within and between Spanish dairy sheep breeds. *Animal Genetics,* **17**: 255–266.

Swofford, D.L. & Selander, R.B. 1989. BIOSYS-1. A computer program for the analysis of allelic variation in population genetics and biochemical systematics. Release 1.7. Illinois Natural History Survey, Campaign IL. 43p.

Tunon, M.J., Gonzalez, P. & Vallejo, M. 1989. Genetic relationships between fourteen native Spanish breeds of goats. *Animal Genetics,* **20**: 205–212.

Vicovan, G. & Rascu, D. 1989. Types of haemoglobin in sheep related to environmental adaptation. *Archiva Zootechnica,* **1**: 33–44.

Wright, S. 1978. *Evolution and the Genetics of Populations. IV. Variability within and among Natural Populations.* Chicago IL: University of Chicago Press.

Zanotti Casati, M., Gandini, G.C. & Leone, P. 1990. Genetic variation and distances of five Italian native sheep breeds. *Animal Genetics,* **21**: 87–92.

PRELIMINARY INVESTIGATION OF GENETIC CHARACTERIZATION OF NATIVE AND ENDEMIC FOWL TYPES OF SRI LANKA

L.P. Silva[1] and W.R.A.K.J.S. Rajapaksha[2]
1. Dept. of Animal Science, Faculty of Agriculture, University of Peradeniya, Sri Lanka.
2. Veterinary Research Institute, Gannoruwa, Sri Lanka.

Abstract: The Red Jungle Fowl (*Gallus gallus*) is generally considered to be main ancestor of the domestic fowl (*Gallus domesticus*). However, it is also believed that other wild *Gallus* species might have contributed to the modern genetic make-up of the domestic fowl, one wild species being the Ceylon Jungle Fowl (*Gallus lafayetti*), endemic to Sri Lanka, which could have contributed to the domestic stock of Sri Lankan native poultry. The present study was conducted in order to investigate the origin of native fowl in Sri Lanka and to establish genetic relationships among them and the Ceylon Jungle Fowl.

Morphological characters of endemic, indigenous and exotic fowl types were recorded. These included Ceylon Jungle fowl; eleven types of native chicken from Sri Lanka; and two exotic chicken breeds (Cornish and Rhode Island Red). Blood samples were collected for DNA extraction. Randomly Amplified Polymorphic DNA (RAPD) analysis was carried out using sixteen non-specific primers.

The results of morphological characterization revealed many variations in plumage and colour pattern. Single and pea comb types were found in both native and exotic types of chicken. A prominent yellow colour marking on a red comb was a unique feature in Ceylon Jungle fowl. The presence of white spots in red earlobes was a distinguishing feature of all native chicken types.

Sixteen non-specific primers were used in the study, and produced 22 polymorphic bands ranging from 500 to 1960 bp. Genetic similarity indices ranged from 0.5 to 1.1 in average genetic distance scale, indicating a broad genetic base in the samples studied. Cluster analysis revealed a clear separation of Ceylon Jungle Fowl from all other types studied, indicating that

593

H. P. S. Makkar and G. J. Viljoen (eds.), Applications of Gene-Based Technologies for Improving Animal Production and Health in Developing Countries, 593-604.
© 2005 *IAEA. Printed in the Netherlands.*

there was early separation and divergent evolution of Ceylon Jungle Fowl from all the domestic races studied.

1.   INTRODUCTION

The evolution of the domestic fowl has a history that extends back to about 6000 BC (West and Zhou, 1989). The main question concerning the origin of the domestic fowl is whether it was domesticated from a single wild species (monophyletic origin) or whether more than one species contributed to the current domestic stock (polyphyletic origin). Stevens (1991) indicated that there are four putative progenitor species from the genus *Gallus*. These four species are native to south and east Asia, namely *G. gallus* (Red Jungle Fowl), *G. sonnerati* (Grey Jungle Fowl), *G. varius* (Java or Green Jungle Fowl) and *G. lafayetti* (Ceylon Jungle Fowl). The most probable theories are that the present-day domestic fowl is entirely derived from a single species, or alternatively from two or more of the four species. An additional possibility is that certain present-day breeds and varieties are derived from one species and other breeds and varieties from another species (Stevens, 1991). These theories are yet to be confirmed.

Of four putative ancestral species, two inhabit the Indian subcontinent, namely *G. gallus* and *G. sonnerati*. *G. lafayetti* inhabits Sri Lanka as an isolated and closed population. Most of the evolutionary investigations documented so far have been based on morphological and behavioural patterns of present-day breeds and their putative progenitors. Few comparisons have been reported at the molecular level. Blood and egg-white protein polymorphism in wild and domestic chicken have been compared by several research groups (Baker and Manwell, 1972; Baker *et al.*, 1971; Nishida *et al.*, 1986). Results were inconclusive regarding a monophyletic or polyphyletic origin for the domestic fowl.

Sri Lanka is a small tropical island, with numerous and various wild and domesticated animals. By virtue of its location, at the crossroads of sea trading between South and Southeast Asia, the country has been a recipient of a variety of livestock and domesticated plant species, including chicken. These domesticated fowl types could have interbred with native ones and evolved as a distinct native population, showing morphological characters typical of the red jungle fowl (*G. gallus*) (Silva *et al.*, 2002). A few observations and studies have reported differences in behaviour (Hutt, 1949; Munechika and Nishiwaki, 1982), morphology (Henry, 1971) and protein and chromosomal polymorphism (Nishida *et al.*, 1986; Okamoto *et al.*, 1986; Watanabe, 1982) between *G. lafayetti* and other wild fowl, but there has been only one investigation (Hashiguchi *et al.*, 1986) reported so far using protein polymorphism to examine the relationship of *G. lafayetti* with native fowl types in Sri Lanka. The present study was conducted in order to investigate, at the DNA level, the origin of native fowl in Sri Lanka, and also to determine whether there is a genetic relationship between native fowl types and the Ceylon Jungle Fowl, *G. lafayetti.*

2.   MATERIALS AND METHODS

Samples from three categories of chicken, namely Ceylon Jungle Fowl (CJF), Sri Lankan Native Fowl (SNF) and Commercial Exotic Fowl (EF), were included in the study (Table 1).

Morphological characterization was based on six characters: plumage colour patterns; type of comb; colour of the ear lobe; colour of the skin; colour of the shank; and type of shank (feathered vs non-feathered).

Blood samples were collected from a wing vein of each bird into a sterilized Eppendorf tube containing 200 μl of 0.129 mM sodium citrate solution. Samples were stored at 4°C until DNA extraction. The DNA extraction was carried out using a commercially available genomic DNA isolation reagent, DNA_{ZOL} (Life Technologies, USA), as follows: 50 μl of blood was washed with 500 μl of 0.9% NaCl and pelletted by centrifuging at 4000 *g* for 10 minutes at 4°C. The supernatant was discarded and 1 ml of DNA_{ZOL} was added to each tube and mixed gently. The lysate was precipitated by adding 0.5 ml of 100% ethanol, mixed, and stored at room temperature for 1 to 4 minutes.

Table 1. Chicken types and strains used in the study.

Number	Category	Chicken type
1	SNF	Village chicken type 1 (Barred)
2	SNF/EF	Naked neck heavy cross
3	SNF	Village chicken type 2 (Red country chicken)
4	SNF/EF	Cornish cross
5	SNF/EF	Naked neck light cross
6	SNF	Village chicken type 3 (White tail)
7	SNF	Village chicken type 4 (Feathered shank)
8	SNF	Village chicken type 5 (Black tail)
9	EF	Cornish
10	EF	Rhode Island Red
11	CJF	Ceylon Jungle Fowl
12	CJF	Ceylon Jungle Fowl

The DNA precipitated was recovered by centrifugation. The DNA pellet was then washed twice with 0.8–1.0 ml of 75% ethanol and centrifuged. The resultant pellet was air dried by storing in an open tube for 5–15 minutes, and the DNA was then dissolved in 300 µl of Tris-EDTA buffer and stored at -20°C.

Sixteen non-specific primers (University of British Colombia, Canada) were used to generate Randomly Amplified Polymorphic DNA (RAPD) profiles (Table 2). The amplification was done (Amplitron II, Thermoline, USA) using 19 µl of reaction cocktail (2 µl 10× Buffer; 2.0 µl 25 mM $MgCl_2$; 0.5 µl 10 mM dNTP; 0.25 µl *Taq* polymerase (1.25 u); 1.0 µl of 20 mM primer and 13.25 µl of sterilized water) and 1 µl of DNA. An initial denaturation step of 2 min at 97°C, was followed by 32 thermal cycles of 95°C for 30 sec, 32–34°C for 1 min and 72°C for 2 min, with a final elongation step at 72°C for 10 min. The PCR products were separated using a 1% agarose gel electrophoresis slab and visualized under UV light after staining with ethidium bromide. Table 2 lists the sixteen primers used for the analysis.

Table 2. Details of the primers used in the RAPD analysis.

Number	Primer sequence
10	^{5}GGG GGG ATT A^3
11	^{5}CCC CCC TTT A^3
26	^{5}TTT GGG CCC A^3
27	^{5}TTT GGG GGG A^3
30	^{5}CCG GCC TTA G^3
31	^{5}CCG GCC TTA C^3
36	^{5}CCC CCC TTA G^3
41	^{5}TTA ACC GGG G^3
42	^{5}TTA ACC CGG C^3
45	^{5}TTA ACC CCG G^3
54	^{5}GTC CCA GAG C^3
58	^{5}TTC CCG GAG C^3
70	^{5}GGG CAC GCG A^3
78	^{5}GAG CAC TAG C^3
90	^{5}GGG GGT TAG G^3
100	^{5}ATC GGG TCC G^3

Cluster analysis was carried out on the morphological information and RAPD profiles. Morphological information on the plumage, comb and shank were used in cluster formation, and all scorable bands were included in the RAPD cluster analysis (Song *et al.*, 2000). The RAPD similarity index was calculated (Nei, 1987) using the MINTAB computer package (Minitab Inc., Release 13.31).

3. RESULTS AND DISCUSSION

3.1 Morphological characterization

The results of morphological characterization revealed many variations in plumage colour pattern. Single and pea comb types were found in both native and exotic types of chicken. The prominent yellow colour marking on the middle of the red serrated single comb was unique to CJF. This observation has been recorded elsewhere (Nishida *et al.,* 1986; Silva *et al.,* 2002; Henry, 1971). Presence of white spots in the red earlobes was a common character observed in all native fowl types. Tables 3 and 4 summarize the observations made on morphological characterization and Figure 1 illustrates the results of the cluster analysis

Cluster analysis showed two distinct clusters, separating two types of SNF and two crosses of SNF and EF from the rest of the tested population. It is interesting to note that according to the morphological characters considered in the present study, the CJF has no clear separation from the other chicken types tested until about the 6.66 distance level (Figure 1). However, morphological analysis showed that the chicken types tested were very divergent in their morphology.

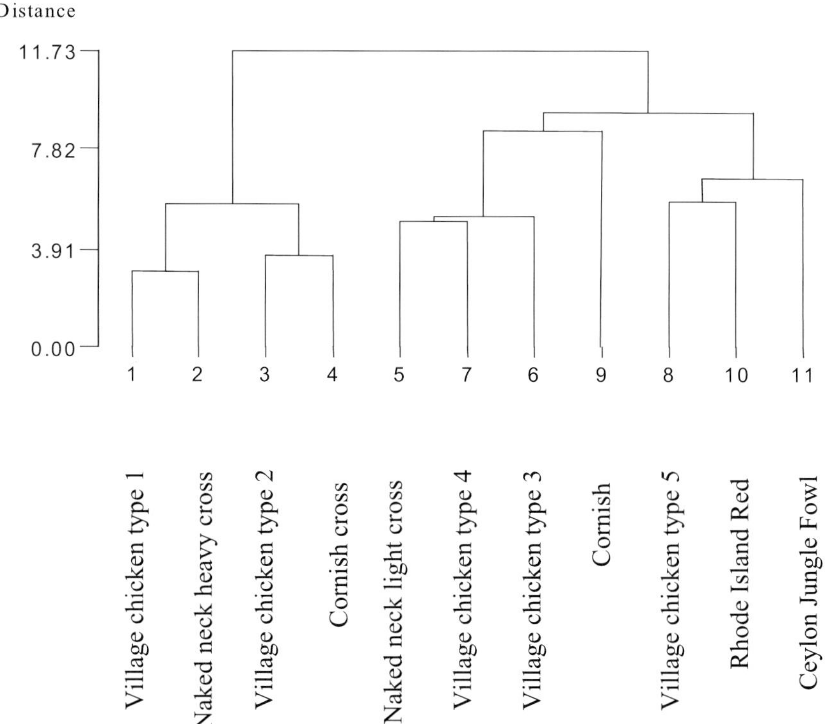

Figure 1. Dendrogram showing the results of cluster analysis on morphological differences of different chicken types. NOTE: Samples 11 and 12 were considered together, as both were Ceylon Jungle Fowl with very similar morphology.

Table 3. Morphological differences in comb, skin, shank and ear lobes of different fowl types.

Chicken type	Comb type	Skin colour	Shank colour	Ear lobe colour
Village chicken type 1	Pea	White	Yellow	Red
Naked neck heavy cross	Single	White	Reddish white	Red (not drooping)
Village chicken type 2	Single	White	Reddish yellow	Red with white spots
Cornish cross	Pea	White	Yellow	Red
Naked neck light cross	Single	White	Reddish yellow	Red
Village chicken type 3	Single	White	Yellow	Red with white spots
Village chicken type 4	Pea	White	Yellowish pink	Red with white spots
Village chicken type 5	Single	White	Yellowish pink	Red with white spots
Cornish	Pea	Yellow	Yellow	Red
Rhode Island Red	Single	Yellow	Yellow	Red
Ceylon Jungle Fowl	Single	White	Reddish yellow	Red

3.2 Genetic characterization

Only eight primers produced polymorphic bands, out of the sixteen non-specific primers used in the study. Of those eight, three produced profiles with very low resolution, and these were considered non-scorable bands. The remaining five primers (numbers 26, 30, 54, 70 and 78) produced 22 polymorphic bands with high resolution. According to the profiles, these bands were distributed within the range of 500 bp to 1960 bp. There were two monomorphic bands common to all chicken types tested (600 bp and 1500 bp; Figure 2). Figures 2, 3 and 4 show the RAPD profiles produced using primers 78, 54 and 70, respectively. Figures 3 and 4 show clearly distinguished banding patterns for CJF.

Genetic similarity indices ranged from 0.5 to 1.1, indicating a wide genetic spread among the chicken types included in the present study. The results of cluster analysis done on the RAPD profiles showed a clear separation of CJF from the other chicken types (Figure 5). This supports an early separation and divergent evolution of CJF from all the domestic chicken types investigated in the present study. This observation is similar to the observation reported by Hashiguchi *et al.* (1986), who compared the blood protein of SNF with four progenitor fowl species. It appears that the contribution of CJF in Sri Lanka Native Fowl (SNF and SNF/EF crossed) is inexistent or very marginal.

Despite limited sample size, a high genetic variation was observed in the SNF types in the present study. This observation supports a diverse origin for the domesticated native fowl in Sri Lanka. Therefore, RAPD can be used to effectively and accurately establish genetic relationships of different fowl types of Sri Lanka.

Table 4. Morphological differences in plumage pattern of different fowl types.

| | | | Feathering | | | |
Neck hackle	Saddle hackle	Breast feathers	Sickle feathers	Tail feathers	1° & 2° wing feathers	Body feathers
Village chicken type 1						
White with black stripes	Yellowish white	White	Black	Black	White & black	Light black & white
Naked neck heavy cross						
–	White + orange bars	White + black bars	White + light black tips	White + black bars	White, some with black bars and some with orange bars	White + black bars
Village chicken type 2						
Brownish orange	Bright brownish orange	Brownish orange + black laced	Bluish black + white tips	Bluish black	Brownish orange & bright brown	Brownish orange + black laced
Cornish cross						
Dark brownish orange	Dark brownish orange	Brown + black lace	Bluish black	Black	Brown & black coloured	Light brown
Naked neck light cross						
–	White	White	White	White	White	White
Village chicken type 3						
Whitish orange	Orange + white laced	Brown & white	White	White	Dark brown & white	White
Village chicken type 4						
Brownish orange	Brownish orange	Black, some with brown laced	Bluish black	Bluish black & some grey	Bright brown & black	Black
Village chicken type 5						
Yellowish white with black stripes	Black & some with yellowish white stripes	Black	Bluish black	Bluish black	Black	Black
Cornish						
White	White	White	White	White	White	White
Rhode Island Red						
Brownish red	Dark brownish red	Brownish red	Bluish black	Bluish black	Brownish red	Light brown
Ceylon Jungle fowl						
Yellow and brownish orange with black stripes	Blackish purple	Brownish orange + black stripes	Blackish purple	Bluish black	Bluish black	Bluish black

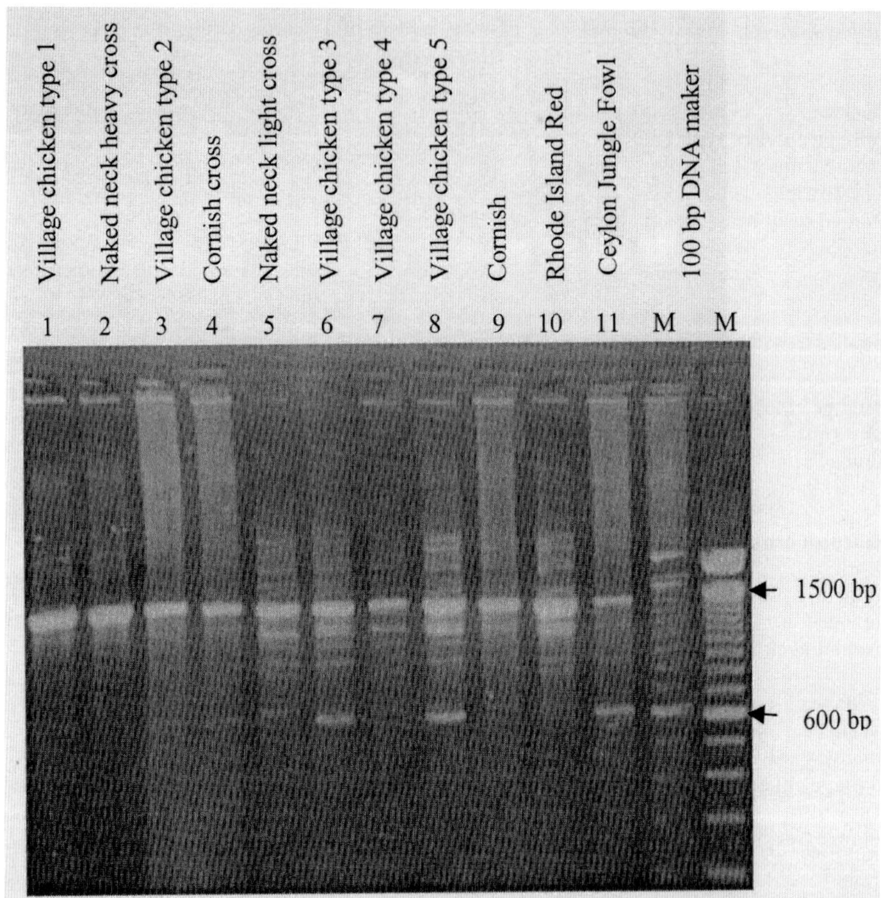

Figure 2. RAPD-PCR profile generated using primer number 78 and electrophoresis with 1% agarose gel.

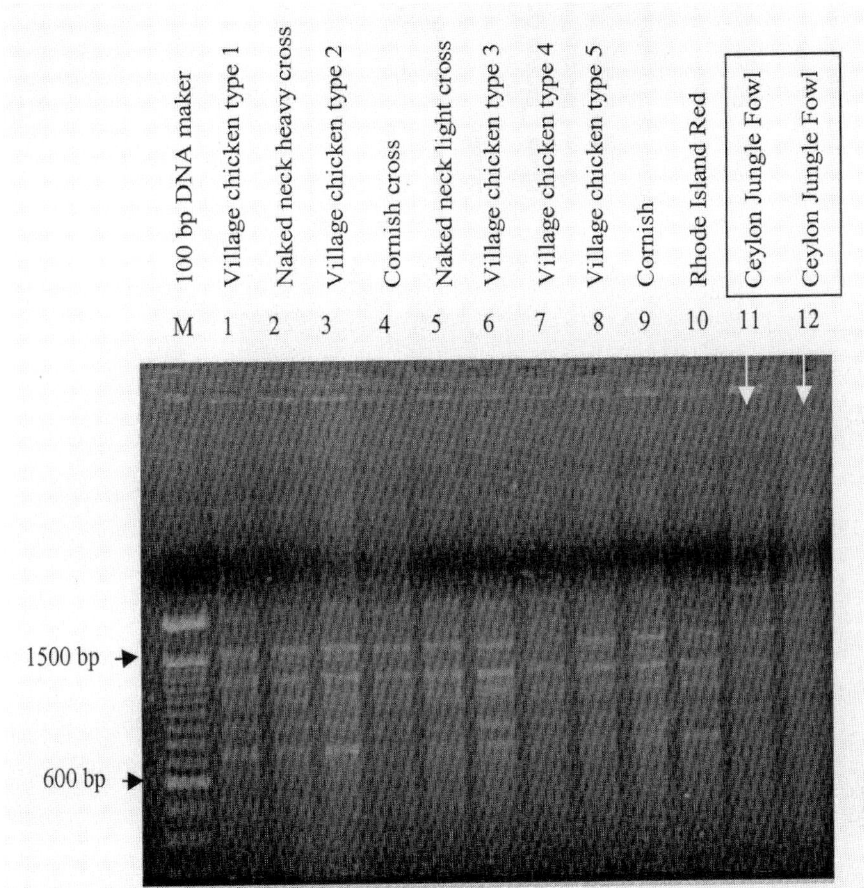

Figure 3. RAPD-PCR profile generated using primer number 54 and electrophoresis with 1% agarose gel.

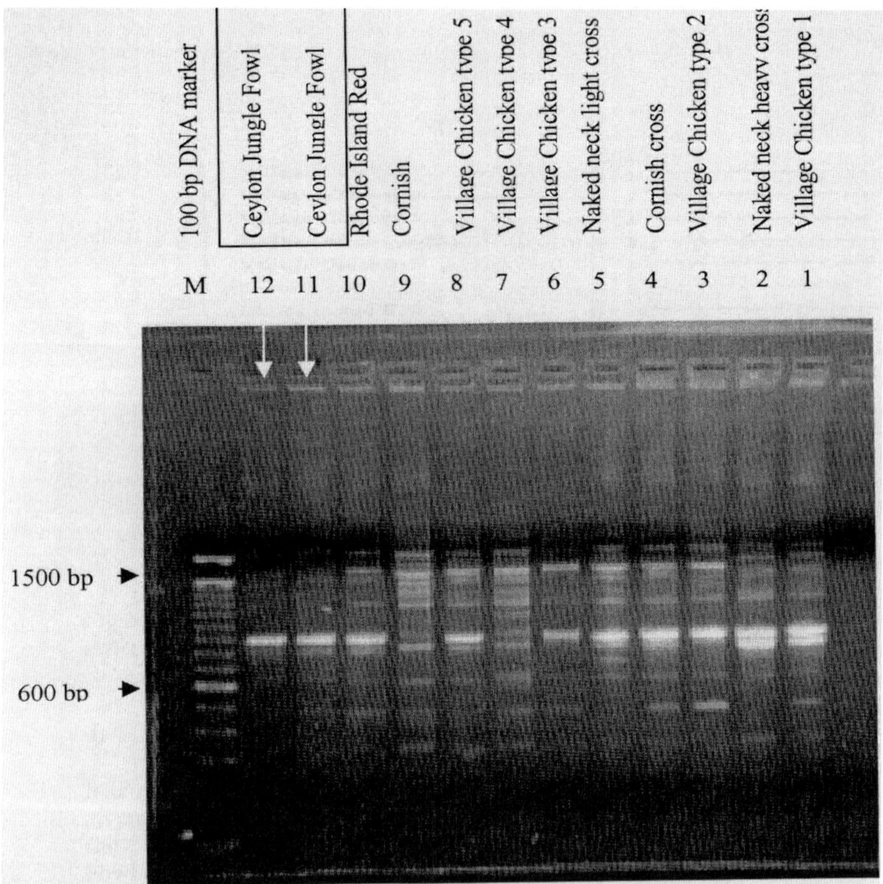

Figure 4. RAPD-PCR profile generated using primer number 70 and electrophoresis with 1% agarose gel.

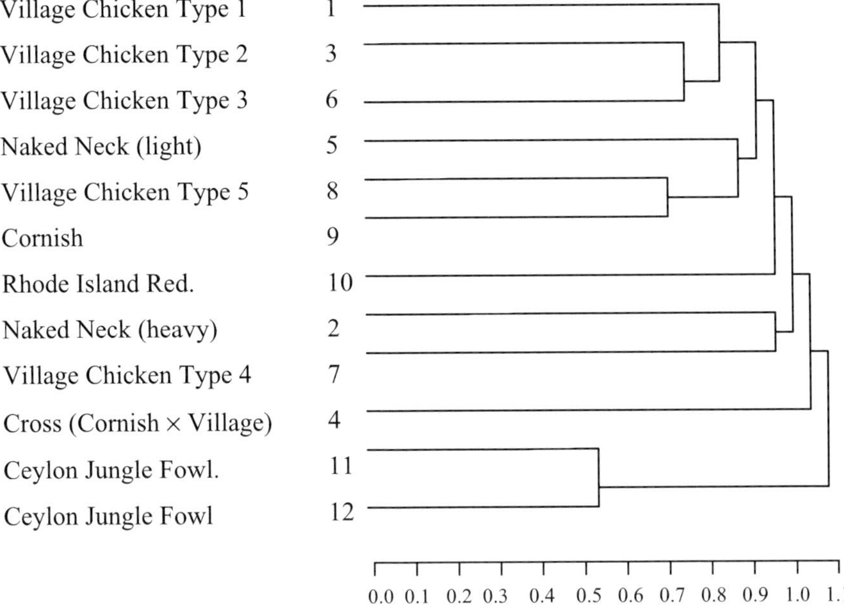

Figure 5. Dendrogram showing the results of cluster analysis of the RAPD profiles of different chicken types.

4.   CONCLUSIONS

Our morphological and RAPD analyses indicate that it is probable that the Sri Lankan Native Fowl population has evolved separately from the endemic putative wild progenitor species of fowl found in Sri Lanka, the Ceylon Jungle Fowl (*Gallus lafayetti*). There is also no evidence of crossbreeding between *G. lafayetti* and the native and exotic fowl types analysed in this study.

ACKNOWLEDGEMENTS

The authors wish to acknowledge the University of Peradeniya, Sri Lanka, for providing funds under the University Research Grant scheme (RG/2000/09/Ag), and thank Mr Suneth Sooriyapathirana, Department of Biotechnology, Faculty of Science, University of Peradeniya, for his contribution in data analysis, and the Director and staff, National Zoological Gardens, Sri Lanka, for their help in sampling Ceylon Jungle Fowl.

REFERENCES

Baker, C.M.A. & Manwell, C. 1972. Molecular genetics of avian proteins. XI. Egg proteins of *Gallus gallus, G. sonnerati* and hybrids. *Animal Blood Groups and Biochemical Genetics,* 3: 101–107

Baker, C.M.A., Manwell, C., Jayaprakash, N. & Francis, N. 1971. Molecular genetics of avian proteins. X. Eggwhite protein polymorphism of indigenous Indian chickens. *Comparative Biochemistry and Physiology,* **40B**: 147–153.

Hashiguchi, T., Okamoto, S., Nishida, T., Hayashi, Y., Goto, H. & Cyril, H.W. 1986. Blood protein variations of the Ceylon Jungle Fowl and native fowl in Sri Lanka. *Report of the Society for Researches on Native Livestock,* **11**: 193–208.

Henry, G.M. 1971. *A Guide to the Birds of Ceylon.* (2nd Ed.) Oxford, UK: Oxford University Press.

Hutt, F.B. 1949. *Genetics of Fowl.* New York NY: McGraw Hill

Munechika, I. & Nishiwaki, M. 1982. Ecological studies on Grey Jungle Fowl (*Gallus sonnerati*) and Ceylon Jungle Fowl (*Gallus lafayetti*). Physiological and Ecological Studies on Jungle Fowls (Tokyo University of Agriculture.) (1982). 35 – 48.

Nei, M. 1987. *Molecular Evolutionary Genetics.* New York NY: Colombia University Press

Nishida, T., Hayashi, Y., Nozawa, K., Okamoto, S., Goto, H. & Cyril, H.W. 1986. Morphological and ecological studies on the Ceylon Jungle Fowl (*Gallus lafayetti*). *Report of the Society for Researches on Native Livestock,* **11**: 179–192.

Okamoto, S., Nishida, T., Goto, H., Hayashi, Y., Hashiguchi, T. & Cyril, H.W. 1986. Analysis of chromosomes of Ceylon Jungle Fowl in Sri Lanka. *Report of the Society for Researches on Native Livestock,* **11**: 207–213.

Silva, L.P., Rajapaksha, W.R.A.K.J.S., Priyangika, P.K. & Himali, S.M.C. 2002. Morphological and genetic characterization of native fowl in Sri Lanka. *Proceedings of the Annual Research Sessions, University of Peradeniya, Sri Lanka.* 7: 17.

Song, B.K., Clyde, M.M., Wikeneswari, R & Noemah, M.N. 2000. Genetic relatedness among *Lansium domesticum* accessions using RAPD markers. *Annals of Botany,* **86**: 297–307.

Stevens, G. 1991. *Genetics and Evolution of the Domestic Fowl.* Cambridge, UK: Cambridge University Press, UK.

Watanabe. S. 1982. Studies on polymorphism of protein and isozymes in the three species of jungle fowls. Physiological and Ecological Studies on Jungle Fowls (Tokyo University of Agriculture.) (1982). 9 – 20

West, B. & Zhou, B.X. 1989. Did chicken go north? New evidence for domestication. *World's Poultry Science Journal,* **45**: 205–218.

TRANSGENIC RABBITS AS A MODEL ORGANISM FOR PRODUCTION OF HUMAN CLOTTING FACTOR VIII

Dusan Vasicek,[1] Peter Chrenek,[1] Alexander Makarevich,[1] Miroslav Bauer,[1] Rastislav Jurcik,[1] Karin Suvegova,[1] Jan Rafay,[1] Jozef Bulla,[1] Ladislav Hetenyi,[1] Jeff Erickson[2] and Rekha K. Paleyanda[2]

1. Research Institute of Animal Production, Nitra, Slovak Republic.
2. American Integrated Biologics, Inc., E. Woodstock, CT, USA.

Abstract: Human clotting factor VIII (hFVIII) is a very complex and large protein whose expression is difficult, as hFVIII requires extensive post-translational modification to be biologically active. This paper reports the generation of transgenic rabbits as a model species for testing the expression of hFVIII in the mammary gland. For micro-injection, a fusion gene construct was used, consisting of 2.5 kb murine whey acidic protein (mWAP) promoter, 7.2 kb cDNA of hFVIII, and 4.6 kb of 3' flanking sequences of the mWAP gene. from 130 micro-injected zygotes transferred into recipients, 30 offspring were delivered. The pups were screened for the transgene by PCR, using DNA isolated from the ear, and results were confirmed by Southern blot analysis. The transgene was identified in one female founder animal, and it was transmitted to the offspring in a Mendelian fashion, thus demonstrating stable integration of the gene construct into the germline of the transgenic rabbits.

1. INTRODUCTION

Transgenic technology focused on production of recombinant proteins of therapeutic value in the milk of mammals has been increasingly successful in recent years. The technology for using the mammary gland as a bioreactor has been developed to the point that pharmaceuticals derived from the milk of transgenic farm animals are currently in the advanced stages of clinical trials. The time required to generate a transgenic animal with high expression levels and to deliver a product to the market are the major

605

H. P. S. Makkar and G. J. Viljoen (eds.), Applications of Gene-Based Technologies for Improving Animal Production and Health in Developing Countries, 605-611.
© 2005 *IAEA. Printed in the Netherlands.*

drawbacks of large animal transgenic technology. Transgenic rabbits offer an attractive alternative to large dairy animals because of their large litter size and short generation interval (Dove, 2000). Rabbits are easily milked and the milk naturally contains 2.5 times as much protein as sheep milk and 4.8 times that of goat milk, as reported by Jennes (1974).

Human clotting factor VIII (hFVIII) is a very complex and large protein, whose expression is difficult, as the hFVIII requires extensive post-translational modification to be biologically active. First transgenic pigs, where the expression of the hFVIII cDNA was targeted to the mammary gland, produced 0.62 U/ml of recombinant hFVIII (rhFVIII) in their milk (Paleyanda *et al.*, 1997). Expression of rhFVIII in transgenic sheep was also achieved, albeit at low levels (Niemann *et al.*, 1999). Recently, Hiripi *et al.* (2003) reported expression of active hFVIII in the mammary gland of transgenic rabbits. Here we report the generation of transgenic rabbits as a model species for testing the expression of hFVIII in the mammary gland.

## 2.	MATERIALS AND METHODS

## 2.1	Gene construct

The construct used for micro-injection consisted of a 2.5 kb murine whey acidic protein (mWAP) promoter, 7.2 kb cDNA of hFVIII and 4.6 kb of 3' flanking sequences of the mWAP gene.

## 2.2	Egg collection and micro-injection

Three days before mating, New Zealand White rabbits were treated with PMSG (Werfaser), followed 72 hr later by hCG (Werfachor). At 19 to 20 hr after mating, the pronuclear stage eggs were flushed with PBS from the oviducts of the animals. After evaluation of the flushed ova, eggs with pronuclei were subjected to micro-injection in CIM+10% FCS (Gibco) medium. The eggs were fixed by suction with a holding pipette, and 5 µg/ml of the gene construct was micro-injected by air pressure into one (male) pronucleus. Swelling of the pronuclei indicated successful micro-injection. The eggs were cultured under 5% CO_2 at 39°C up to blastocyst or 4–6-cell stage, then 10–14 embryos were transferred into the synchronized recipients, as described by Chrenek *et al.* (1998).

2.3   Analysis of transgene integration

Total DNA was isolated from tissue of newborn rabbit. Conditions of PCR analysis (Figure 1) of the amplification hFVIII transgene were reported by Paleyanda *et al.* (1997), using primers hFVIII-F: 5'-GTA GAC AGC TGT CCA GAG GAA-3' and hFVIII-R: 5'-GAT CTG ATT TAG TTG GCC CAT C-3', which define a 587-bp region of human FVIII cDNA. Positive newborn rabbits were confirmed by Southern blot analysis (Figure 2), as described by Paleyanda *et al.* (1997).

2.4   Analysis of transgene expression by RT-PCR

Mammary gland biopsies were taken from lactating founder and her female offspring rabbits on day 21 of lactation. Total RNA was isolated from tissue biopsies of transgenic and non-transgenic lactating females using TRI-Reagent, and 1 µg of total RNA was used for reverse transcription (RT) using oligo(dT)$_{15}$ and ImProm-II Reverse Transcriptase (Promega). Following reverse transcription, 3 µl of the synthesized cDNA was diluted in 27 µl of water, and 1 µl of dilution was used as a template for PCR. PCR was performed using hFVIII-F and hFVIII-R primers. As a control, PCR was also performed using two primers – R1_G3PDHF: 5'-ACG ACC ACT TCG GCA TTG TG-3' and R1_G3PDHR: 5'-TCC ACC ACC CTG TTG CTG TA-3' – of rabbit glyceraldehyde-3-phosphate dehydrogenase (GenBank accession number L23961) that define a 519-bp region of rabbit GAPDH cDNA (Figure 3).

2.5   Analysis of transgene expression by western blotting

Milk samples were taken from lactating founder and her female offspring rabbits on day 20 of lactation. De-fatted milk samples were diluted 1:40 in electrophoresis buffer containing BME, and heated at 95°C for 5 min. Proteins were resolved on 10% SDS-PAGE gel and transferred onto nitrocellulose membrane (ECL Hybond). Blots were blocked 30 min in gelatine (1%), 30 min in rabbit serum (5%), and 15 min in hydrogen peroxide (3%). Blots were probed with goat anti-human Factor VIII at a dilution of 1 : 800. Secondary a/b donkey anti-goat IgG(dilution 1 : 10,000) conjugated with HRP in combination with the detection reagent ECL-plus (Amersham) was used for visualization of positive bands. As a molecular weight marker, a pre-stained protein ladder ~10–180 kDa from Fermentas MBI (SM0671, lot 201) was used (Figure 4).

3.   RESULTS

A total of 130 micro-injected zygotes were transferred into recipients and 30 (26%) offspring were delivered. The pups were screened for the transgene by PCR (Figure 1) of DNA isolated from the ear, and results were confirmed by Southern blot analysis (Figure 2). The transgene was identified in one (3.3%) female founder animal, and it was transmitted to the offspring in a Mendelian fashion. These data demonstrate stable integration of the gene construct into the germline of transgenic rabbits.

In biopsies of mammary gland tissue from a female transgenic rabbit, expression of transgene was demonstrated by RT-PCR as a 587-bp product. Control PCR was performed using rabbit GAPDH cDNA specific primers (Figure 3).

In milk samples from lactating females, a 50 kDa rhFVIII-specific band was detected by western blotting analysis, but was never found in negative controls. The strongest signal was observed in the second lactation (Figure 4). Only a light chain of rhFVIII was detected. Results obtained suggest that another type of rhFVIII-specific antibody should be used for immunoblot analysis.

4.   CONCLUSIONS

Rabbits are a good model for the rapid production of small amounts of certain proteins. The integration frequency of transgenes in this species is generally lower than in mice.

Another factor that may influence integration efficiency, such as the type of construct, is currently being studied with the use of a GFP reporter gene.

The concentration and activity of recombinant hFVIII in milk in further lactations of transgenic females is currently under investigation.

REFERENCES

Chrenek, P., Makarevich, A., Vasicek, D., Laurincik, J., Bulla, J., Gajarska, T. & Rafay, J. 1998. Effects of superovulation, culture and micro-injection on development of rabbit embryos *in vitro. Theriogenology,* **50**: 659–666.

Dove, A. 2000. Milking the genome for profit. *Nature Biotechnology,* **18**: 1045–1048.

Hiripi, L., Makovics, F., Halter, R., Baranyi, M., Paul, D., Carnwath, J.W., Bösze, Z.S. & Niemann, H. 2003. Expression of active human blood clotting factor VIII in the mammary gland of transgenic rabbits. *DNA and Cell Biology,* **22**: 41–45.

Jennes, R. 1974. The composition of milk. pp. 3–105, *in: Lactation, A Comprehensive Treatise.* New York, NY: Academic Press.

Niemann, H., Halter, R., Carnwath, J.W., Herrmann, D., Lemme, E. & Paul, D. 1999. Expression of human blood clotting factor VIII in the mammary gland of transgenic sheep. *Transgenic Research*, **8**: 137–149.

Paleyanda, R.K., Velander, W.H., Lee, T.K., Scandella, D.H., Gwazdauskas, F.G., Knight, J.W., Hoiyer, L.W., Drohan, W.N. & Lubon, H. 1997. Transgenic pigs produce functional human factor VIII in milk. *Nature Biotechnology*, **15**: 971–975.

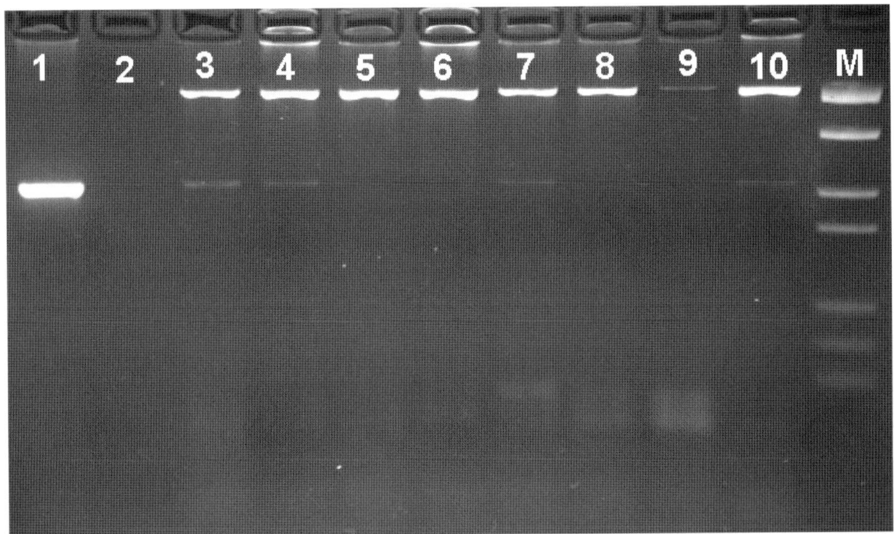

Figure 1. Representative results of hFVIII integration in rabbit genome by PCR analysis. Lane 1: gene construct (+control); 2: negative control; 3, 4, 7, 10: transgenic rabbits; 5, 6, 8, 9: non-transgenic rabbits after micro-injection; M: PhiX174/*Hae*III.

 D. Vasicek et al.

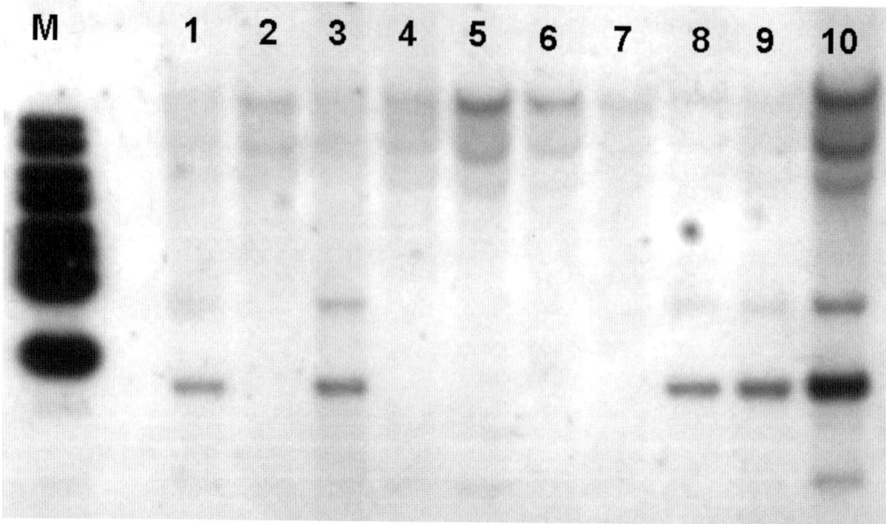

Figure 2. Southern blot detection of hFVIII transgenic rabbits. Lane M: 1 kb DNA Ladder; 1, 3, 8, 9, 10: transgenic rabbit, after micro-injection; 2, 4, 5, 6, 7: non-transgenic rabbit. Analysis showed two FVIII-specific fragments of 2.7 and 1.7 kb, confirming the presence of FVIII transgene in transgenic rabbits.

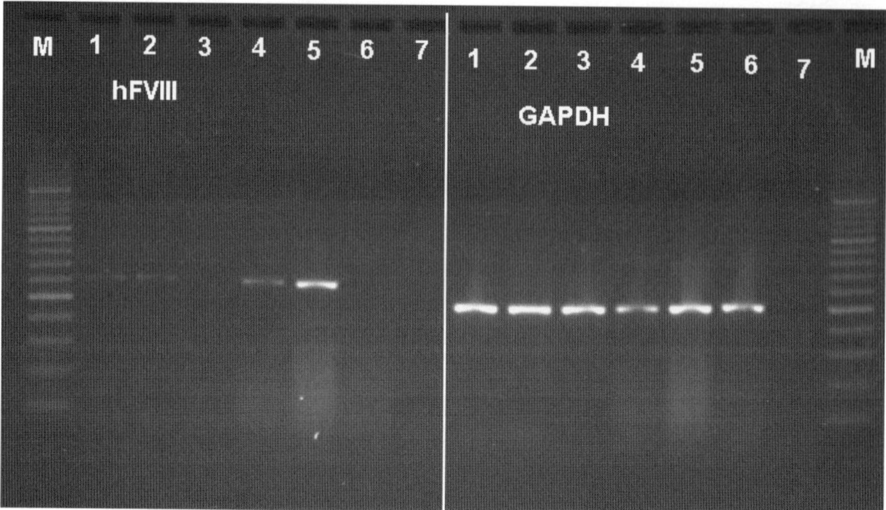

Figure 3. Representative results of RT-PCR of total RNA from mammary gland tissue biopsies of transgenic and non-transgenic rabbits. Lanes 1, 2, 4, 5: transgenic rabbits, first lactation; 3, 6: non-transgenic rabbits; 7: negative control; M: O'RangeRuler™ 100 bp DNA ladder (Fermentas). The specific PCR product for hFVIII cDNA is 587 bp, and for rabbit GAPDH it is 519 bp.

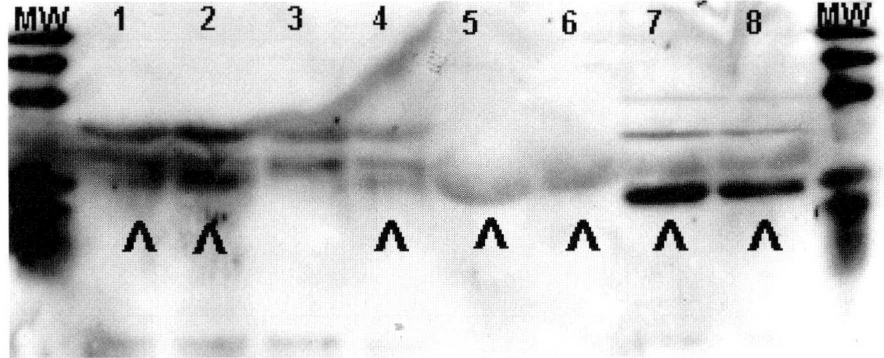

Figure 4. Detection of hFVIII in the milk of transgenic rabbits by western blot. Lanes 1, 4: transgenic rabbits, first lactation; 2: transgenic rabbits, second lactation; 3: non-transgenic rabbit (negative control); 5: rhFVIII 50 µg/lane; 6: rhFVIII 25 µg/lane; 7: human plasma, 1:40 dilution; 8: human plasma, 1:60 dilution; MW: pre-stained protein ladder ~10–180 kDa from Fermentas MBI (SM0671, lot 201: 180, 130, 100, 73, 54, 48, 35, 24, 16 and 10 kDa).

PARENTAGE DETERMINATION IN THREE BREEDS OF INDIAN GOAT USING HETEROLOGOUS MICROSATELLITE MARKERS

N.A. Ganai and B.R. Yadav*
*Livestock Genome Analysis Laboratory, National Dairy Research Institute, Karnal–132 001, India. *Corresponding author.*

Abstract: Parentage verification in Indian goat breeds addresses dubious parentage of three types: 1, exclusion of a putative parent when the genotype of one parent and offspring are known; 2, exclusion of a putative parent when the genotype of the other parent is not available; and 3, exclusion of both the parents of an offspring if falsely recorded. The investigation used 116 unrelated goats and six pedigreed families of three breeds of goat (Jamnapari, Barbari and Sirohi). A set of 12 bovine microsatellite markers was analysed for parentage determination in goats for different types of misidentifications. For Type 1 dubious parentage, the exclusion probability for each marker varied widely, from as low as 13.4% (locus BM-5004 in Jamnapari) to as high as 67% (locus BMS-1237 in Sirohi). For type 2, the values of probability of exclusion ranged from 5% (locus BMS-1237 in Barbari) to 50.1% (locus BMS-1237 in Sirohi). For Type 3, exclusion values ranged from 21.6% to 84%. The exclusion probabilities of falsely recorded parents were estimated for different combinations of 5 markers sets with 12, 8, 6, 5 and 4 markers, respectively.

1.   INTRODUCTION

Animal breeding programmes rely on the accurate estimation of the breeding worth of the animals, using biometrical approaches like REML, BLUP, BLUE and, more recently, the Animal Model. The identification of proven sires, using these methodologies, has been of central importance in animal improvement programmes. Failure to record correct parentage can introduce bias in sire evaluation, by inducing errors in estimates of

H. P. S. Makkar and G. J. Viljoen (eds.), Applications of Gene-Based Technologies for Improving Animal Production and Health in Developing Countries, 613-620.
© 2005 *IAEA. Printed in the Netherlands.*

heritability and breeding values. Using parental determination procedures, 4 to 23 percent parental misidentifications have been reported in several investigations on cattle, even in highly organized improvement programmes (Geldermann, Pieper and Weber, 1986). Error in recording the sire of a progeny can occur at many stages, including handling of semen during freezing, transportation and artificial insemination. Errors can also occur during open-range grazing of animals, particularly sheep and goats. Misidentification rates are potentially much higher in Indian conditions, where recording systems are not properly maintained, especially in field conditions. Correct parentage within breeding stocks is therefore a prerequisite for an efficient breeding programme. Different types of markers have been used for parentage determination, such as blood groups, serum proteins, red cell enzymes and lymphocyte antigen systems; however, frequently these tests do not allow a definitive conclusion.

DNA assays developed for paternity testing include RFLP (Botstein *et al.*, 1980); DNA fingerprinting using multilocus minisatellite and oligonucleotide synthetic probes (Jeffreys, Brookfield and Semeonoff, 1985; Ali, Muller and Epplen, 1986); and PCR-based amplification of minisatellites (Jeffreys *et al.*, 1991) and microsatellites (Litt and Luty, 1989). Each of the different markers used for paternity testing has advantages and disadvantages associated with it. Microsatellite markers not only overcome many of the difficulties, but also have added features that make them markers of choice for paternity verification and individual identification, owing to their high heterozygosity, Mendelian codominant inheritance, ubiquity throughout the genome and ease of scoring by PCR.

The heterologous microsatellite markers are often used in genetic studies because there is significant genome conservation across different related species, which extends to microsatellite loci as well (Vaiman *et al.*, 1996; Ganai and Yadav, 2001). This circumvents the need for obtaining a suitably large panel of polymorphic markers for each species, especially those that are less common and less studied. Cross-species utilization of microsatellite loci not only saves time and effort in the laboratory, but also enables the construction of comparative maps between related species for use in genetic distance studies (Ganai and Yadav, 2001) and parentage determination. The present study was carried out to study the suitability of bovine microsatellite markers for parentage verification in Indian goat breeds in cases of dubious parentage, and for three situations in particular:

1. Exclusion of a putative parent, when the genotype of one confirmed parent and offspring are known.
2. Exclusion of a putative parent, when the genotype of the other parent is not available.
3. Exclusion of both the parents of an offspring if falsely recorded.

2.   MATERIAL AND METHODS

2.1   Animals

The investigation was carried out on 116 unrelated goats in six families of three breeds of goat (Jamnapari, Barbari and Sirohi). The random group of unrelated animals was typed for microsatellite markers to generate the allele frequency data, polymorphic information content (PIC) and heterozygosity for different markers in each breed (Ganai and Yadav, 2001), and estimation of the exclusion probabilities of false parents. Animals from the families were analysed to verify the usefulness of the microsatellite markers for parentage determination.

2.2   Microsatellite markers

A panel of 12 bovine microsatellite markers was used for parentage determination in goats to resolve different types of misidentifications. Microsatellite primers, grouped in sets, are shown in Table 1. The primers were used in five different sets so as to establish the association between accuracy and number of primers used.

Table 1. Microsatellite markers in different sets used to reveal error in identification.

Set No.	No. of markers used	Marker designation
1	12	BMC-5221, BMS-357, BM-7160, BMS-1237, BMS-585, BM-5004, MB-068, BMS-332, BMS-820, BR-6027, BM-7228, MB-045
2	8	BMS-357, BM-7160, BMS-1237, BMS-585, BMS-332, BMS-820, BR-6027, BM-7228
3	6	BM-7160, BMS-1237, BMS-585, BMS-820, BR-6027, BM-7228
4	5	BM-7160, BMS-1237, BMS-585, BR-6027, BM-7228
5	4	BM-7160, BMS-1237, BR-6027, BM-7228

2.3   Methodology

The protocols and conditions for DNA isolation, PCR amplification of microsatellite loci, polyacrylamide gel electrophoresis and detection by autoradiography were the same as reported earlier (Ganai and Yadav, 2001).

2.4   Statistical analysis

Genotypes were recorded directly from the autoradiographs. Allele frequencies were estimated by direct count. Probabilities of exclusion of wrong parentage were calculated from allele frequency data of microsatellite

DNA analysis. The exclusion probabilities were calculated for three types of dubious parentage, using the following formulae. In all the equations:

- P_{El} = probability of exclusion of wrong parent at l^{th} locus;
- p_i = frequency of the ith allele at l^{th} locus;
- p_j = frequency of jth allele at l^{th} locus;
- $i = 1...n$ (number of alleles at i^{th} locus); and
- $j = 1...n-1$.

2.4.1 Exclusion of a putative parent, when the genotypes of a confirmed parent and offspring are known

The formula is derived from Jamieson (1994).

Probability of exclusion of wrong parent at one locus with n alleles is:

$$P_{El} = \sum_i p_i (1-p_i)^2 - \sum_{i>j=1} (p_i\, p_j)^2 [4 - 3(p_i + p_j)]$$

This formula has also been expressed in terms of the powers of allelic frequencies (Jamieson, and Taylor, 1997) to simplify the calculations, becoming:

$$P_{El} = 1 - 2\sum p_i^2 + \sum p_i^3 + 2\sum p_i^4 - 3\sum p_i^5 - 2(\sum p_i^2)^2 + 3\sum p_i^2 + 3\sum p_i^2 \sum p_i^3$$

2.4.2 Exclusion of a putative parent, when the genotype of the other parent is not known

The formula is derived from Gerber, and Morris (1983).

$$P_{El} = \sum p_i^2 (1 - p_i^2) + \sum_{i>j=1} 2p_i\, p_j (1 - p_i - p_j)^2$$

In terms of the powers of allelic frequencies, the formula became:

$$P_{El} = 1 - 4\sum p_i^2 + 2(\sum p_i^2)^2 + 4\sum p_i^3 - 3\sum p_i 4$$

2.4.3 When both the parents of the offspring are putative (doubtful)

The formula is derived from Grundel and Reetz (1981).

$$P_{El} = 1 + \sum [p_i^2 (2 - p_i)]^2 - 2[\sum p_i^2 (2 - p_i)]^2 + 4(\sum p_i^3)^2 - 4\sum p_i^6$$

In terms of the sums of powers of allelic frequencies, the formula became:

$$P_{El} = 1 + 4\sum p_i^4 - 4\sum p_i^5 - 3\sum p_i^6 - 8(\sum p_i^2)^2 + 8(\sum p_i^2)(\sum p_i^3) + 2(\sum p_i^3)^2$$

2.4.4 Combined probability of exclusion over several loci

The combined probability of exclusion over several loci was calculated based on the approach of Weir (1996), namely:

$$P = 1 - \Pi(1 - P_{El})$$

where P_{El} is the probability of exclusion at l^{th} locus, and Π is the product over l loci.

3. RESULTS AND DISCUSSION

The findings indicate that the probability exclusion values varied between different markers in all the three breeds for all three types of dubious paternity. For Type 1 dubious parentage, the exclusion probability for each marker varied widely, from as low as 13.4% (locus BM-5004 in Jamnapari) to as high as 67% (locus BMS-1237 in Sirohi). The markers with exclusion probabilities of >30% in 3 breeds were BMS-357, BM-7160, BMS-1237, BMS-585, BMS-332, BMS-820, BR-6027 and BM-7228. For Type 2 dubious parentage, the values of probability of exclusion ranged from 5% (locus BMS-1237 in Barbari) to 50.1% (locus BMS-1237 in Sirohi). For Type 3, these values ranged from 21.6% to 84%.

The parentage determination of the offspring by markers BM-7160 in two families of Jamnapari breed is shown in Figure 1. The Figure depicts the exclusion of one putative sire and confirmation of the real sire by genotyping a half-sib family for this marker.

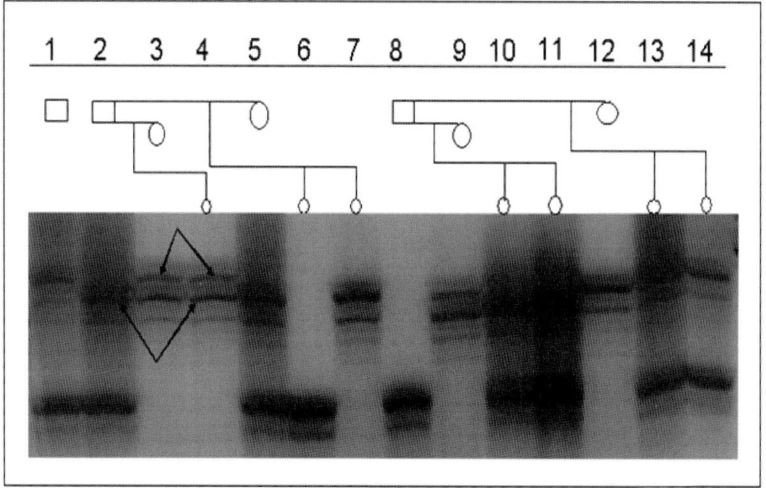

Figure 1. Exclusion of the putative sire in Lane-1 by parentage determination in two families of Jamnapari goats using microsatellite marker BM-7160.

The exclusion probabilities of falsely recorded parents were also estimated for different combinations of markers (Set 1 to Set 5). The numbers of markers in Set 1 to Set 5 were 12, 8, 6, 5 and 4, respectively (Table 1). Among the three breeds, these values were lower in Barbari breed for all sets of markers in all the three types of misidentification. In Type 1 misidentification, the exclusion probability of a false parent was 99.9%, when all 12 markers were considered together (Set 1). Using only 4 markers (Set 5), the exclusion probability was 94.5%. The probability of exclusion of the false parent decreased for all sets of markers when the information on one of the parents was absent (Type 2 misidentification; 98.9% for Set 1, to 80% for Set 5). In Barbari goats, use of Set 5 could exclude only 65% of false misidentifications. For Type 3 misidentification, the probability estimates of excluding both the falsely recorded parents were highest of all the three types of misidentifications for any marker set.

Correct parentage identification among breeding stock is a prerequisite for any efficient breeding programme; its lack can introduce bias in estimation of genetic parameters in sire evaluation and may have serious consequences in long-term improvement programmes. The conventional methods of parentage testing (blood groups, serum proteins, red cell enzymes, lymphocyte antigen systems) do not allow unambiguous exclusion of falsely recorded parents, and therefore should be extended by applying additional genetic marker systems.

Table 2. Exclusion probability of falsely recorded parent(s) for different combinations of markers in three types of parentage misidentifications in three goat breeds.

Marker Set	Sirohi	Jamnapari	Barbari	Overall
Type 1				
Set 1	0.999	0.997	0.995	0.999
Set 2	0.995	0.993	0.985	0.998
Set 3	0.976	0.973	0.947	0.990
Set 4	0.957	0.953	0.931	0.982
Set 5	0.922	0.918	0.862	0.945
Type 2				
Set 1	0.979	0.963	0.945	0.989
Set 2	0.953	0.936	0.899	0.978
Set 3	0.876	0.862	0.796	0.928
Set 4	0.831	0.815	0.762	0.900
Set 5	0.763	0.750	0.649	0.801
Type 3				
Set 1	0.999	0.999	0.999	0.999
Set 2	0.999	0.999	0.998	0.999
Set 3	0.998	0.997	0.986	0.999
Set 4	0.995	0.994	0.978	0.999
Set 5	0.986	0.985	0.935	0.994

From all DNA assays useful for paternity testing (RFLPs, DNA fingerprinting using multilocus minisatellite and oligonucleotide synthetic probes and PCR-based amplification of minisatellites and microsatellites), microsatellites are the markers of choice for paternity verification and individual identification (Caetano-Anolles, 1998).

In the present study, suitability of 12 bovine microsatellite markers was analysed for parentage determination in goats for different types of misidentifications. For excluding more than 99% of the falsely recorded parents, a set of six markers would be needed when the genotype of one parent is known (Type 1). Otherwise, if the genotype of one of the parents were missing (Type 2), then a set of 12 markers would be needed to exclude the other, falsely recorded, parent with the same accuracy (99%). If both the parents are doubtful (Type 3), than a set of 4 markers would exclude both of them with a probability of 99%. The exclusion probabilities of false parents are lower for the Barbari breed than for other two breeds included in the study. This could be due to lower genetic variability in Barbari goats than in other breeds (Ganai and Yadav, 2001).

4. CONCLUSION

More markers are needed in order to exclude false parents in small populations, particularly where there is less variability due to inbreeding depression or bottleneck effect.

REFERENCES

Ali, S., Muller, C.R. & Epplen, J.T. 1986. DNA fingerprinting by oligonucleotide probes specific for simple repeats. *Human Genetics,* **74**: 239–243.

Botstein, D., White, R.L., Skolnick, M. & Davis, R.W. 1980. Construction of genetic linkage maps in man using restriction fragment length polymorphisms. *American Journal of Human Genetics,* **32**: 314–331.

Caetano-Anolles, G. 1998. *DNA Markers: Protocols, Applications and Overviews.* New York NY: Wiley-VCH.

Ganai, N.A. & Yadav, B.R. 2001. Genetic variation within and among three breeds of goats using heterologous microsatellite markers. *Animal Biotechnology,* **12**: 121–136.

Geldermann, H., Pieper, U. & Weber, W.E. 1986. Effect of misidentification on the estimation of breeding value and heritability in cattle. *Journal of Animal Science,* **63**: 1759–1768.

Gerber, R.A. & Morris, J.W. 1983. General equation for the average power of exclusion for genetic systems of n codominant alleles in one parent and in no-parent cases of disputed parentage. pp. 277–280, *in: Inclusion of Probabilities in Parentage Testing.* Virginia, USA: American Association of Blood Banks.

Grundel, H. & Reetz, I. 1981. Exclusion probabilities obtained by biochemical polymorphisms in dogs. *Animal Blood Groups & Biochemical Genetics,* **12**: 123–132.

Jamieson, A. 1994. The effectiveness of using co-dominant polymorphic allelic series for (1) Checking pedigrees and (2) Distinguishing full-sib pair members. *Animal Genetics,* **25**(Suppl.1): 37–44.

Jamieson, A. & Taylor, S.C.S. 1997. Comparisons of three probability formulae for parentage exclusion. *Animal Genetics,* **28**: 397–400.

Jeffreys, A.J., Brookfield, J.F.Y. & Semeonoff, R. 1985. Positive identification of an immigration test-case using human DNA fingerprints. *Nature,* **317**: 818–819.

Jeffreys, A.J., MacLeod, A., Tamaki, K., Neil, D.L. & Monckton, D.C. 1991. Minisatellite repeat coding as a digital approach to DNA typing. *Nature* (London), **354**: 204–209.

Litt, M. & Luty, J.A. 1989. A hypervariable microsatellite revealed by *in vitro* amplification of a dinucleotide repeat within the cardiac muscle actin gene. *American Journal of Human Genetics,* **44**: 397–401.

Vaiman, D., Schibler, L., Bourgeris, F., Oustry, A., Amigues, Y. & Cribiu, E.P. 1996. A genetic linkage map of the male goat genome. *Genetics,* **144**: 279–305.

Weir, B.S. 1996. *Genetic data analysis II.* Sunderland MA: Sinauer.

DNA POLYMORPHISM OF ARABIAN, THOROUGHBRED AND ANGLO-ARAB HORSES IN MOROCCO

Application to identification and parentage verification of individual horses

Lahoussine Ouragh

Laboratoire d'Analyses Génétiques Vétérinaires (LAGEV), Institut Agronomique et Vétérinaire Hassan II, BP 6202 – Instituts, 10101 Rabat, Morocco.

Abstract: New techniques of molecular biology used in analysing DNA polymorphism give access to the whole genetic variability of a given individual, while traditional blood typing (red cell typing and biochemical polymorphisms) gives access only to the transcribed fraction, which is then translated to protein. In addition, this fraction represents only a tiny part (5 to 10%) of the genome's coding fraction. One of the newer testing methods in identifying horses is a DNA-based test using microsatellite marker analysis.

The objective of this work was to evaluate the efficacy of this new technology in the identification and parentage verification of Arabian, Thoroughbred and Anglo-Arab horses in Morocco. The Anglo-Arab horse is a crossbreed between Arabian and Thoroughbred.

Three samples from the three breeds were analysed for 12 microsatellites (HMS2, HMS3, HMS6, HMS7, HTG4, HTG6, HTG7, AHT4, AHT5, VHL20, HTG10 and ASB2). Blood samples were gathered from a total of 1541 horses: 804 Arabians, 559 Thoroughbreds and 178 Anglo-Arabs. Allelic frequencies of the 12 loci studied were calculated in the three groups. The results allowed the determination of intra-population genetic parameters: heterozygosity ratio (h), probability of identification (P_i) and probability of exclusion (P_e).

Based on mean heterozygosity values, variability was relatively lower in Thoroughbred horse (0.7036), while it was almost the same in Arabian and Anglo-Arab horses (respectively 0.7217 and 0.7232). Probabilities of exclusion obtained with the 12 systems were greater than 99.9% for the three populations studied, and probabilities of identification of individual horses

621

H. P. S. Makkar and G. J. Viljoen (eds.), Applications of Gene-Based Technologies for Improving Animal Production and Health in Developing Countries, 621-629.
© 2005 *IAEA. Printed in the Netherlands.*

were 15.4×10^{-12}, 3.5×10^{-12} and 3.2×10^{-12} in the Thoroughbred, Arabian and Anglo-Arab breeds, respectively.

These results indicate that the test using microsatellite marker analysis constitutes a highly efficient and reliable alternative for the identification and parentage verification of individual horses and so it is a useful tool for horse breeding and horse registries.

1.   INTRODUCTION

The Moroccan horse population is estimated AT 160,000 horses, of which about 90 percent are Arab-Barb horses resulting from cross-breeding of the first North African horse, the Barb, with stallions of the Arabian breed, and the cross-breeding of the progeny amongst themselves. The remaining 10 percent comprise Arabian, Barb, Thoroughbred, Anglo-Arab, Anglo-Arab-Barb and saddle horses. The Anglo-Arab horse is a cross-breed between Arabian and Thoroughbred. The populations of Arabians, Thoroughbreds and Anglo-Arabs are estimated at 1,600, 1,500 and 500 horses, respectively.

Despite the modernization of agriculture, the horse, in Morocco holds a significant place in the rural socio-economic matrix. The Arab-Barb and Barb horses are used in agriculture, light traction, riding and leisure activities, notably the fantasia. To this can be added their growing use in equestrian sports and regional racing events. The Arabian, Thoroughbred, Anglo-Arab-Barb and Anglo-Arab horses are mainly used in racing.

The development of systematic breeding requires rationalization and therefore the use of modern management techniques. In order to meet these requirements, the National Horse-Breeding Administration has developed a computerized system called MINISIRE, similar to the SIRE system adopted in France. This system is a rigorous management and horse selection tool, and implementation of MINISIRE requires modern identification means and parentage verification. As a result, the Veterinary Genetics Laboratory (LAGEV) was created in 1990 at the Hassan II Institute of Agronomy and Veterinary Medicine.

Initially, the laboratory used classical gene markers (red cell factors and electrophoretic variants) for blood typing. Since 1998, new gene markers have been introduced. One of the newer testing methods in identifying horses is a DNA-based test using the analysis of short tandem repeat (STR) loci or microsatellites. This technique has the advantage of giving access to the whole genetic variability of a given individual, and not only the transcribed fraction, which represents only a tiny part (5 to 10%) of the genome's coding fraction.

This research studied DNA polymorphism in Thoroughbred, Arabian and Anglo-Arab horses in Morocco, to evaluate its efficacy in the identification and parentage verification of individual horses.

2.   MATERIAL AND METHODS

The analyses were carried out on blood samples from 1541 horses of both sexes, from different regions of Morocco, and comprising 804 Arabian, 559 Thoroughbred and 178 Anglo-Arab.

PCR was used to amplify microsatellites using StockMarks for Horses Equine Paternity PCR Typing Kit (PE Applied Biosystems, Foster City, CA, USA), which includes 12 previously reported loci: ASB2 (Breen *et al.*, 1997); AHT4 and AHT5 (Binns *et al.*, 1995); HMS2, HMS3, HMS6 and HMS7 (Guérin *et al.*, 1994); HTG4 and HTG6 (Ellegren *et al.*, 1992); HTG7 and HTG10 (Marklund *et al.*, 1994); and VHL20 (van Haeringen *et al.*, 1994). Amplification reactions were performed in two multiplex PCR (eight-plex and four-plex). These reactions were performed with reagents supplied in the kit and according to manufacturer instructions.

The pooled PCR products were diluted and mixed with 20 µl formamide and 0.5 µl GeneScan-350 ROX internal size standard, and analysed on an ABI PRISM 310 DNA Sequencer (PE Applied Biosystems) equipped with GeneScan and Genotyper software.

Computation of allele frequencies and mean heterozygosity values (h) were obtained using the GENETIX programme (Belkhir *et al.* 2000). Theoretical probability of identity (P_i) and probability of exclusion (P_e) were estimated according to the formulas of Hanset (1976) and Rendel and Gahne (1961).

3.   RESULTS

Allele frequencies of the 12 microsatellites studied are presented in Table 1 (at end of paper). Each allele is designated by its size expressed in base pairs (bp). The largest number of alleles was found in the Arabian horse (n = 115) and the lowest in the Anglo-Arab horse (n = 85).

Mean heterozygosity values (h), theoretical P_i and P_e are presented in Table 2 (at end of paper). Generally, more than 7 individuals of 10 are heterozygous for all loci. Based on mean heterozygosity value, variability was lower in Thoroughbred (0.7036), but very similar in Arabian and Anglo-Arab (0.7216 and 0.7232, respectively).

Concerning P_i and P_e, Table 2 shows that combined loci present high efficacy for identification and parentage verification of individual horses ($P_i < 10^{-12}$ and $P_e > 0.999$).

4. DISCUSSION

Parameters of intra-population genetics obtained with microsatellite loci confirm those obtained with classical gene markers in Arabian (Ouragh, 1990) and Thoroughbred horses (Benjelloun, 1992). In fact, in those studies, the lowest mean heterozygosity values were observed in the Thoroughbred breed.

The current investigation shows also the efficacy of the microsatellites in identification and parentage verification of individual horses. For the 3 populations, P_e was higher than 0.999, while with classical gene markers they were 0.96 in Thoroughbred and 0.97 in Arabian (Bowling and Clark, 1985).

In using classical gene markers, the probability of drawing two horses from the same breeding population that have identical blood type for all systems under test is 4×10^{-5} in Thoroughbred and 5×10^{-6} in Arabian (Stormont, 1988). In contrast, for microsatellites, this probability is 15.4×10^{-12} and 3.5×10^{-12}, respectively, in Thoroughbred and Arabian, thus proving higher efficiency for identification of individual horses.

5. CONCLUSION

These results indicate that the test using microsatellite marker analysis constitutes a highly efficient and reliable alternative for the identification and parentage verification of individual horses, and hence a useful tool for horse breeding and horse registries. The competent authorities of the equine industry have shown an interest in extending this test to all Moroccan horse breeds in the country, notably Arab-Barb horses.

REFERENCES

Belkhir, K., Borsa, P., Chikhi, L., Raufaste, N. & Bonhomme, F. 2000. GENETIX, logiciel sous Windows TM pour la génétique des populations. Laboratoire Génome, Populations, Interactions, CNRS UMR 5000, Université de Montpellier II, Montpellier, France.

Benjelloun, A. 1992. Analyse génétique des chevaux Pur Sang Anglais et Anglo-Arabes-Barbes du Maroc. Thèse de doctorat vétérinaire. Institut Agronomique et Vétérinaire Hassan II, Rabat, Morocco.

Binns, M.M., Holmes, N.G., Holliman, A. & Scott A.M. 1995. The identification of polymorphic microsatellite loci in the horse and their use in thoroughbred parentage testing. *British Veterinary Journal,* **151:** 9–15.

Bowling, A.T. & Clark, R.S. 1985. Blood Groups and protein polymorphism gene frequencies for seven breeds of horses in the United States. *Animal Blood Groups and Biochemical Genetics,* **16:** 93–108.

Breen, M., Lindgren, G. & Binns, M.M. 1997. Genetical and physical assignments of equine microsatellites – first integration of anchored markers in horse genome mapping. *Mammalian Genome,* **8**: 267–273.

Ellegren, H., Johansson, M., Sandberg, K. & Andersson, L. 1992. Cloning of highly polymorphic microsatellites in the horse. *Animal Genetics,* **22**: 133–142.

Guérin, G., Bertaud, M. & Amigues, Y. 1994. Characterization of seven new horse microsatellites: HMS1, HMS2, HMS3, HMS5, HMS6, HMS7 and HMS8. *Animal Genetics,* **25:** 62.

Hanset, R. 1976. Probabilité d'exclusion de paternité et de monozygotie. Probabilité de similitude. Généralisation à n allèles codominants. *Annales de Médecine Véterinaire,* **119:** 71–80.

Marklund, S., Ellegren, H., Eriksson, S., Sandberg, L. & Andersson, L. 1994. Parentage testing and linkage analysis in the horse using a set of highly polymorphic microsatellites. *Animal Genetics,* **25:** 19–23.

Ouragh, L. 1990. Etude des marqueurs génétiques sanguins des chevaux Arabes, Arabes-Barbes et Barbes au Maroc. Thèse de Doctorat ès Sciences Agronomiques. Institut Agronomique et Vétérinaire Hassan II, Rabat, Morocco.

Rendel, J. & Gahne, B. 1961. Parentage test in cattle using erythrocyte antigens and serum transferrins. *Animal Production,* **3:** 307–314.

Stormont, C. 1988. Identification and parentage verification of individual horses by blood typing tests. *Equine Veterinary Science,* **8**: 176–179.

Van Haeringen, H., Bowling, A.T., Stott, M.L., Lenstra, J.A. & Zwaagstra, K.A. 1994. A highly polymorphic horse microsatellite locus: VHL20. *Animal Genetics,* **25:** 207.

Table 1. Allele frequencies of the 12 microsatellites studied.

Loci	Alleles (bp)	Thoroughbred	Arabian	Anglo-Arab
HTG7	121	0.1747	0.3146	0.2570
	123	–	0.0006	–
	125	0.0107	–	0.0112
	127	0.3262	0.0420	0.1564
	129	0.4822	0.6415	0.5754
	131	0.0062	0.0012	–
AHT4	142	0.0027	0.0114	0.0029
	144	0.2359	0.1128	0.2356
	146	0.0036	0.0684	0.0029
	148	0.2823	0.3536	0.3420
	150	0.1621	0.1458	0.1236
	152	–	0.0006	–
	154	0.0009	0.0659	0.0345
	156	0.0128	0.0114	–
	158	0.2832	0.2256	0.2471
	160	0.0164	0.0044	0.0115
AHT5	127	0.0045	0.0019	0.0028
	129	0.2043	0.2224	0.1833
	131	0.3590	0.1046	0.2389
	133	–	0.0032	–
	135	0.2495	0.1033	0.1528
	137	0.1447	0.4563	0.3306
	139	0.0371	0.1014	0.0917
	143	0.0009	0.0070	–
VHL20	84	0.0009	0.0050	–
	86	0.2757	0.0791	0.2068
	88	0.0009	0.0006	–
	90	0.0009	0.0088	–
	92	0.2081	0.1884	0.2130
	94	0.2838	0.1696	0.1574
	96	0.2261	0.1633	0.2623
	98	0.0018	0.0013	0.0062
	100	0.0018	0.0119	–
	102	–	0.0333	0.0062
	104	–	0.3379	0.1481
	106	–	0.0006	–

Loci	Alleles (bp)	Thoroughbred	Arabian	Anglo-Arab
VHL20	84	0.0009	0.0050	–
	86	0.2757	0.0791	0.2068
	88	0.0009	0.0006	–
	90	0.0009	0.0088	–
	92	0.2081	0.1884	0.2130
	94	0.2838	0.1696	0.1574
	96	0.2261	0.1633	0.2623
	98	0.0018	0.0013	0.0062
	100	0.0018	0.0119	–
	102	–	0.0333	0.0062
	104	–	0.3379	0.1481
	106	–	0.0006	–
ASB2	218	0.0046	0.0033	–
	220	0.0185	0.0229	–
	222	0.0023	0.0336	–
	224	0.0127	0.0008	0.0041
	232	–	0.0016	–
	234	–	0.0426	0.0124
	236	0.0255	0.0090	0.0083
	238	0.1782	0.0663	0.1570
	240	0.0428	0.0074	0.0289
	242	0.1123	0.0786	0.0702
	244	0.0880	0.0213	0.0909
	246	0.0833	0.1326	0.0992
	248	0.0451	0.0254	0.0207
	250	0.2558	0.5376	0.3678
	252	0.1308	0.0172	0.1405
HMS2	219	0.0454	0.0539	0.0234
	221	0.0019	0.0704	0.0146
	223	0.0548	–	0.0351
	225	0.2505	0.0235	0.0994
	227	0.5709	0.3847	0.5585
	229	0.0605	0.2610	0.1754
	231	0.0028	0.0083	–
	233	0.0047	0.0511	0.0029
	237	0.0028	0.0076	0.0058
	239	0.0057	0.1395	0.0848
HMS3	150	–	0.0006	–
	152	0.4858	0.2052	0.4410
	158	0.0009	–	–
	160	0.0027	0.0083	–
	162	0.1263	0.1760	0.1180
	164	0.0463	0.2662	0.1096
	166	0.1121	0.0591	0.0590
	168	0.2260	0.2681	0.2500
	170	–	0.0165	0.0225

Loci	Alleles (bp)	Thoroughbred	Arabian	Anglo-Arab
HMS6	149	–	0.0006	–
	151	0.0009	0.0050	–
	153	0.0009	0.0006	–
	155	0.0018	0.0031	–
	157	0.1099	0.1055	0.1186
	159	0.0441	0.3417	0.1667
	161	0.3757	0.1256	0.2825
	163	0.0009	0.0157	0.0113
	165	0.0099	0.0653	0.0254
	167	0.4559	0.3354	0.3955
	169	–	0.0013	–
HMS7	169	0.0010	0.0080	–
	171	0.1524	0.1725	0.1469
	173	0.0029	0.2172	0.0875
	175	0.1524	0.3543	0.2781
	177	0.2667	0.0414	0.1531
	179	0.2381	0.0862	0.1469
	181	0.1848	0.1190	0.1875
	183	0.0010	0.0013	–
	185	0.0010	–	–
HTG4	125	–	0.0136	0.0056
	127	0.5000	0.2823	0.3792
	129	0.0081	0.1700	0.0534
	131	0.4445	0.4231	0.5084
	133	0.0367	0.0763	0.0309
	135	0.0036	0.0019	–
	137	0.0072	0.0310	0.0225
	139	–	0.0019	–
HTG6	79	0.3757	0.2736	0.3596
	81	0.0009	0.0025	–
	83	–	0.0006	–
	85	0.4365	0.2469	0.3680
	91	0.0063	0.0081	–
	93	–	0.0025	0.0028
	95	0.1637	0.4347	0.2500
	97	0.0036	0.0056	–
	101	0.0134	0.0255	0.0197

Table 2. Parameters of intra-population genetics (h, P_i and P_e).

Microsatellite	Parameter	Thoroughbred N = 559	Arabian N = 804	Anglo-Arab N = 178
HMS2	h	0.6026	0.7533	0.6382
	P_i	0.2058	0.0948	0.1634
	P_e	0.3720	0.5494	0.3161
HMS3	h	0.6823	0.7803	0.1731
	P_i	0.1432	0.0836	0.1225
	P_e	0.4449	0.5662	0.4825
HMS6	h	0.6369	0.7393	0.7212
	Pi	0.2004	0.1092	0.1237
	P_e	0.3647	0.5075	0.4781
HMS7	h	0.7916	0.7742	0.8132
	P_i	0.0764	0.0828	0.0610
	P_e	0.5861	0.5745	0.6284
HTG4	h	0.5509	0.7055	0.6724
	P_i	0.3017	0.1343	0.1757
	P_e	0.2611	0.4665	0.3951
HTG6	h	0.6413	0.6744	0.6724
	P_i	0.2004	0.1669	0.1757
	P_e	0.3711	0.4030	0.3951
HTG7	h	0.6304	0.4877	0.5783
	P_i	0.2069	0.3457	0.2411
	P_e	0.3494	0.2178	0.3075
AHT4	H	0.7577	0.7808	0.7499
	P_i	0.1008	0.0772	0.1044
	P_e	0.5220	0.5845	0.5211
AHT5	h	0.7448	0.7104	0.7683
	P_i	0.1076	0.1216	0.0904
	P_e	0.5188	0.4869	0.5505
VHL20	h	0.7490	0.7872	0.7963
	P_i	0.1092	0.0746	0.0733
	P_e	0.5054	0.5864	0.5936
HTG10	h	0.8026	0.7893	0.8390
	P_i	0.0620	0.0736	0.0467
	P_e	0.6272	0.5955	0.6762
ASB2	h	0.8534	0.6778	0.8390
	P_i	0.0371	0.1237	0.0467
	P_e	0.7091	0.4908	0.6762
	h	0.7036	0.7217	0.7233
	P_i	15.404×10^{-12}	3.545×10^{-12}	3.215×10^{-12}
	P_e	0.9997	0.9998	0.9997

IN VITRO CULTURE OF SKIN FIBROBLAST CELLS FOR POTENTIAL CLONING BY NUCLEAR TRANSFER

S.C. Gupta, N. Gupta, S.P.S. Ahlawat, A. Kumar, R. Taneja, R. Sharma, S. Sunder and M.S.Tantia
National Bureau of Animal Genetic Resources, PO Box 129, Karnal-132001, India.

Abstract: Donor cell lines were developed from skin tissue for the conservation of the endangered Jaiselmeri camel breed of India. Average cell proliferation rates varied from 0.82 to 0.69 in different passages, and population doubling time from 29.3 h to 34.8 h. Around 15 population doublings were accomplished during this culturing. Cell viability was 97 to 99% in different passages. Growth curves of cells from the JC-5 cell line reached a plateau on day 7, while the slower-growing cultures of JC-3 showed elevation even on day 10, possibly due to donor age differences. Cell proliferation rates by both cell count and MTT absorbance showed similar patterns, with a correlation coefficient of 0.79. MTT assay, a colorimetric method, can handle large samples in somatic cell cultures. Diploid chromosomal counts in passages 1, 3 and 5 were normal (2N=74, XY) in 97% of the cells. Occasional metaphase plates showed polyploidy. The present baseline data on standard growth curve, linear relationship in colorimetric assay for estimation of cell proliferation rate, and normal ploidy and karyological levels in camel skin fibroblast cells in multiplication could be useful in developing competent donor somatic cell lines for conservation now and revival of this camel breed by cloning in the future.

1.   INTRODUCTION

After the first successful cloning of a domestic animal using adult somatic cells (Wilmut, 1997), opportunities to select and multiply animals of special merits have increased manyfold (Cunningham, 1999). Cloning holds the promise of bypassing conventional breeding procedures to allow creation

H. P. S. Makkar and G. J. Viljoen (eds.), Applications of Gene-Based Technologies for Improving Animal Production and Health in Developing Countries, 631-640.

of copies of genetically superior or engineered animals in a single generation (Mehra, 2001). Due to increasing commercial pressure and mechanization of agriculture and transport, traditional species, such as camel, are under threat of extinction. Because of remote locations and difficult environmental conditions of desert and arid ecosystems, sampling and storage of adequate number of samples of semen and embryos is not practical. Clonal samples of genetic material, especially the skin cells from unique animals, may be very important for conservation of the available genetic diversity of threatened animal genetic resources (Cunningham, 1999; Kubota *et al.*, 2000).

Somatic cells, which donate the entire genetic material in nuclear transfer (NT) reconstructed embryos and clones, hold the key (Campbell, 1999). Various types of cells originating from different body tissues have successfully been used for production of viable clones (Wilmut, 1997; Wells *et al.*, 1998; Kubota *et al.*, 2000; Cibelli *et al.*, 1998; Kato *et al.*, 1998; Wakayama *et al.*, 1998; Onishi *et al.*, 2000). According to Kubota *et al.* (2000), skin fibroblasts, being easy to collect by non-invasive methods, hardy in culture and amenable to freezing, are the material of choice for conservation of animal biodiversity using cloning techniques.

The aim of this study was to develop donor cell lines from skin tissue for the conservation of the endangered Jaiselmeri camel, establishing culture parameters and determining proliferation rate, viability, population doubling time and karyological profile.

2.   MATERIALS AND METHODS

2.1   Development of cell lines

The skin tissue samples (0.25 cm^2) were collected by ear biopsy from male camels of different age groups from the National Research Centre on Camel, Bikaner, Rajasthan, and transported to the Laboratory at 4°C. The skin tissues were properly cleansed to remove epidermal tissue. Tissue pieces were further cut into small pieces (1 mm^2) and seeded in tissue culture (TC) flask (Nunc) in complete culture medium (DMEM+Ham's F12 culture media containing 10% bovine foetal serum (BFS) and antibiotics). The cell cultures were incubated at 37°C under 5% CO_2 and 95% relative humidity. Primary fibroblast cells were harvested using the trypsin+EDTA protocol described previously (Polejaeva, 2000). The cells were cultured up to 5th passage (15 population doublings) and cryopreserved at -80°C in an ultra-deep-temperature freezer (Forma, USA) in freezing medium (DMEM+Ham's F12, 10% FBS and 10% DMSO (Sigma), penicillin (50 IU) and streptomycin (50 µg/dl).

2.2   Cell proliferation rate

The primary skin fibroblast cells were harvested from tissue explants at 80% confluency. Cell viability, seeding density and cell population at harvest in each passage was estimated using a Neubauer haemocytometer. The population doubling and cell proliferation rates were calculated using the following formulae:

Cell proliferation rate (r) = 3.32 × (Log N_H - Log N_I) / (T_2 - T_1)

Where: N_H = Cells harvested

N_I = Cells initially seeded

T_1 = Initial time

T_2 = Time at harvesting

Population Doubling time (PD) = 24/r

2.3   Growth curve studies

The growth pattern of fibroblast cells at the 4th passage was studied to draw a standard growth curve. For the growth curve assessment, an initial 0.4×10^5 cells were seeded in 25 cm^2 tissue culture flasks. For each cell line (JC-3 and JC-5), a series of 10 cultures were set up. One flask was harvested every 24 hours for a period of 10 days, and cells were counted.

2.4   MTT assay

Cells from the JC-5 cell line were cultured for 6 days on 96-well microtitre plates in 100 µl DMEM+Ham's F-12 medium with 10% BFS. Cells were seeded at 4,000 cells per well (A to J column wells), while K and L column wells contained only media without cells, as negative controls. Each day, 20 µl of MTT (3-(4.5-dimethylthiozol-2yl) 2.5 diphenyl-tetrazolium bromide – a filtered solution of 5 mg/ml MTT in phosphate buffered saline (PBS)) solution were added in each well of 1–4 rows per day and incubated at 37°C for 5 hours. Medium was removed and 200 µl of DMSO was added. The formazan crystals formed were dissolved and the resulting coloured solution was analysed using a multi-well scanning spectrophotometer. Cells of rows 5–8, left without MTT treatment, were harvested each day by the usual trypsin-EDTA protocol described previously and a cell count was made for each well. The data of MTT optical density (OD) values and cell count were entered into spreadsheet software for graphic display.

2.5   Chromosomal preparations

Karyotyping was carried out on the JC-5 skin fibroblast cultures at 1, 3 and 5 passages. The cell suspension of cycling cells (0.40×10^4 cells/cm^2) was seeded in 25-cm^2 TC flasks, and 2.5 µg/ml colchicine (Sigma) was added after observing 50–60% confluency. Effects of colchicine treatment were observed using an inverted microscope, with rounding of cells after incubation for 2–3 hours at 37°C under a 5% CO_2 atmosphere. The cells were harvested by the trypsin-EDTA protocol. Chromosomal slides were prepared using the air-drying protocol described previously (Gupta *et al.*, 2002).

3.   RESULTS

3.1   Cell proliferation

The emergence of fibroblasts from ear tissue explants seeded in the TC flask was observed after 12–15 days of culturing. The cells attained nearly 80% confluency after 3–5 days of removal of tissue pieces from tissue culture flask. Occasional colonies of epithelial cells were observed in culture, but by the 4th passage the fibroblast cultures were found to be free from epithelial cells. Data on cell proliferation and viability of skin fibroblasts is presented in Table 1.

The average cell proliferation rates varied from 0.82 to 0.69 in different passages. The population doubling time varied from 29.26 hr to 34.78 hr. Around 15 population doublings were accomplished during this culturing. The cell viability was nevertheless good, ranging from 97 to 99%.

3.2   Growth curve

Figure 1 shows the growth curves established for cell lines JC-3 and JC-5. The cells from the JC-5 cell line reached a plateau on day 7, while in the slower-growing cultures of JC-3, the curve showed elevation even on day 10. The JC-5 cell line showed a growth pattern similar to standard growth patterns recorded by our group in other species. The age of the JC-5 donor was 4 years, while that of JC-3 was 8 years.

3.3   Cell proliferation index

The data on cell proliferation by MTT assay vis-à-vis the cell count after every 24 hr is presented in Table 2.

Figure 2 shows cell population growth in a 96-well plate. After an initial decrease on day 2, cell growth showed a typical sigmoid curve up to day 6. The initial population of 4,000 cells had increased to around 22,000 at the end of six days. The growth of cells was similar to that of the standard growth curve.

Figure 3 shows the MTT absorbance values from day 1 to day 6. There was a steady increase in OD value, except on day 2. From both experiments, it was evident that out of 4,000 cells seeded in each well, around 10% could not attach to the surface of the culture vessel on day 1 and resume further cell divisions. The cell proliferation curves from both cell count and MTT absorbance showed similar patterns, but the increase in MTT value was not equal to the cell count (Table 2). In cycling cell populations, some cells might have been in an inactive phase, and hence would not have absorbed the formazan (MTT) crystals. The correlation coefficient (r) between the MTT absorbance value and cell was 0.79.

Table 1. Average cell proliferation rate and viability of cultured camel skin fibroblasts.

Passage No.	Population doublings	Cell viability (%)	Proliferation rate (r)[1]	Population doubling per day[1]
1	3	98.50	0.73 ±0.06 (8)	32.80 ±0.47 (8)
2	2	99.00	0.82 ±0.09 (8)	29.26 ±0.34 (8)
3	3	98.00	0.74 ±0.11 (8)	32.43 ±0.61 (8)
4	4	98.50	0.78 ±0.17 (5)	30.76 ±0.84 (5)
5	3	97.00	0.69 ±0.15 (5)	34.78 ±1.14 (5)
Average	3.0 per passage	98.67	0.76 ±0.06 (7)	32.46 ±0.54 (7)

NOTE: (1) Figures in parenthesis indicate the number of animals.

Table 2. Cell proliferation rates for cell count and MTT absorbance in camel skin fibroblasts.

Days of culture	Cell count by haemocytometer		OD value of MTT absorbance at 540 nm	
	Cell concentration	Growth over day zero (x)	OD value	Increase in OD over day zero values (x)
1	4,000	0.000	0.0160	0.000
2	3,600	-0.100	0.0109	-0.110
3	9,200	2.430	0.0275	1.720
4	12,300	3.075	0.0309	1.930
5	14,600	3.650	0.0410	2.560

NOTE: Correlation coefficient r = 0.79.

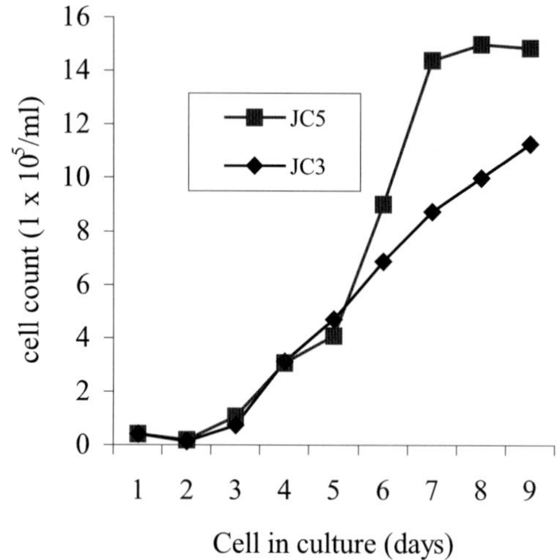

Figure 1. Growth curves of two cell lines from Jaiselmeri camel.

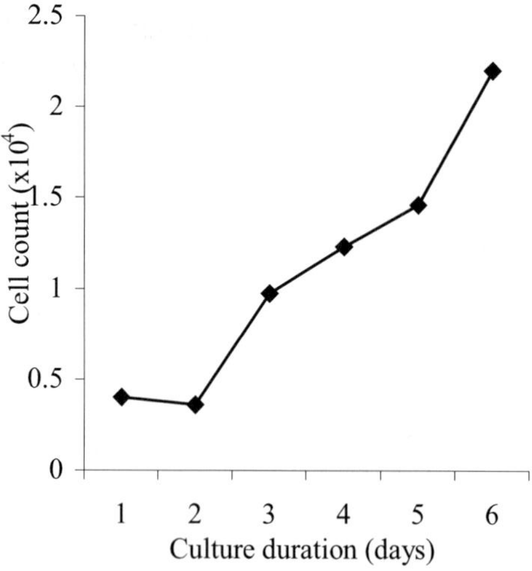

Figure 2. Cell proliferation index by count.

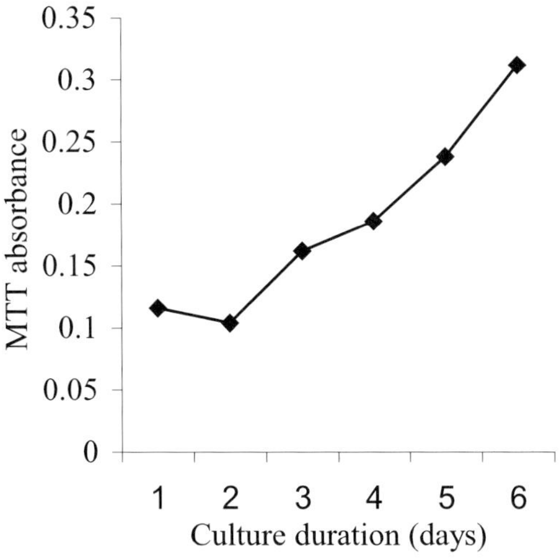

Figure 3. Cell proliferation index by MTT assay.

3.4 Karyological Profile

The 2N chromosomes count in passages 1, 3 and 5 was found to be normal (2N=74,XY) in 97% of the cells. Occasional metaphase plates with polyploidy were observed.

4. DISCUSSION

In this study, skin fibroblast cells from male camels were isolated and cultured, and multiplied in the laboratory for a number of passages before being cryopreserved. Potentially, these cells can later be used for the revival of this breed using nuclear transfer and animal cloning. The importance of skin or other somatic cells isolation and long-term cryopreservation for such remotely located species and livestock breeds has been described (Cunningham, 1999). Cloning using skin fibroblast cells offers the advantage of having easy accessibility and non-invasiveness without animal sex and age limitations (Kubota *et al.*, 2000).

To date, the overall cloning efficiency using various somatic cells has been rather low, with reported efficiency ranging from 0 to 10% (Wilmut,

1997; Wells *et al.*, 1998; Kubota *et al.*, 2000). This is primarily due to higher loss at pre- and post-implantation stages of pregnancy (Wells *et al.*, 1998). The exact mechanism of these embryonic losses are still not clear, although incomplete re-programming of the donor cell genome in the currently used cloning techniques may in part be responsible (Kubota *et al.*, 2000). Monitoring of cell cycle stage and proliferation capacity of donor cells may be very important in understanding the re-programming of cells for the production of NT reconstructed embryos (Campbell, 1999).

In the present study, cell growth measured by the proliferation indices in different passages were recorded, so normally proliferating cell lines with a stable genome could be developed for their ultimate use in animal cloning. The first outgrowth from the tissue was recorded 12–15 days after tissue seeding in culture. These results are comparable with previous studies in cattle and rabbit skin fibroblast cultures (Kubota *et al.*, 2000; Dinnyes *et al.*, 2001). In this study, the standard growth curve of the JC-5 cell line showed a true sigmoid pattern with proper log phase for 3–7 days and then attaining a plateau after 8–10 days of continuous culturing. In comparison with the JC-5 cell line, the JC-3 line deviated from the standard growth curve, especially during log phase. This could be due to differences in donor age. Differences in growth potential and cell proliferation rates of skin fibroblasts from donors of different ages have been reported elsewhere (Dinnyes *et al.*, 2001).

The multiple passaging of fibroblasts in culture helps in improving their totipotency (Kubota *et al.*, 2000), but all cells in culture have a definite lifespan, after which they stop dividing and enter senescence. Recording the proliferation rates at regular intervals enables easy monitoring of the induction of senescence in cycling cells, using routine haemocytometer counting at seeding and at harvesting. However, more reliable colorimetric methods are now available (Maghni, Nicolescu and Martin, 1999). The MTT assay has been designed for the spectrophotometric quantification of cell growth and viability without the use of radioactive isotopes (Maghni, Nicolescu and Martin, 1999; Wemme *et al.*, 1992).

Cell multiplication rates may vary considerably under different culture conditions, and slight change in environment or composition of medium may affect the proliferation rate significantly. For camel skin fibroblast cells, the standard multiplication rate and the population doubling time was not known earlier. We have analysed the proliferation indices of the growing cells using a 96-well plate-based MTT assay, and compared with actual cell counts from replicate wells. In this assay, the dividing and viable cells take up and retain MTT and colour is developed. The intensity of colour was measured by a multiwell scanning spectrophotometer at 540–570 nm. In our study, the cycling cells in culture exhibited a linear relationship between MTT absorbance OD values and the actual cell count ($r = 0.79$).

During long-term culture, cells are likely to develop some type of chromosomal abnormalities. It must be ensured that the cells in different passages be checked for normal karyotype so that viable clones could be developed from them. The important reason for failure of successful cloning of Argali sheep in inter-species nuclear transfer was due to higher frequency of aneuploidy and other structural chromosomal abnormalities in cycling donor cell lines between 7–10 passages (White *et al.*, 1999). Chromosomal aberration (10–15%) was also observed in donor cattle skin fibroblasts (Kubota *et al.*, 2000). Hare and Singh (1979) reported that 90% of spontaneous abortions were due to some type of chromosomal abnormalities carried by the developing embryos. In the present study, the karyological status of cycling skin fibroblast cells in culture was monitored at different passages and it was found that a normal karyotype of 2N=74 was recorded in 97% of cells up to 15 population doublings.

5.　CONCLUSION

The present study is a preliminary attempt to develop skin fibroblast cell lines from an endangered breed of camel known for its excellent riding characteristics. The baseline data on standard growth curve, linear relationship in colorimetric assay for estimation of cell proliferation rate, and normal ploidy and karyological level in camel skin fibroblast cells in multiplication could be useful in developing competent donor somatic cell lines for conservation now and future revival of this camel breed by cloning.

ACKNOWLEDGEMENTS

We thank the Director of the National Bureau of Animal Genetic Resources, Karnal, for providing facilities for this study. This work was supported by a financial grant from the World Bank under the NATP Animal Genetic Resource Biodiversity (Mission Mode) project. We also thank the Director of the National Research Centre on Camel, Bikaner, for providing the samples.

REFERENCES

Campbell, K.H.S. 1999. Nuclear equivalence, nuclear transfer, and the cell cycle. *Cloning,* **1**: 3–15.

Cibelli, J.B., Stice, S.L, Golucke, P.J., Kane, J.J., Jeny, J., Blackwell, C., Ponce de Leon, F.A. & Robl, J.M. 1998. Cloned transgenic calves produced from non-quiescent fetal fibroblasts. *Science,* **280**: 1256–1258.

Cunningham, E.P. 1999. Recent developments in biotechnology as they relate to animal genetic resources for food and agriculture. FAO Commission on Genetic Resources for Food and Agriculture. *Background Study Paper,* No. 10.
See: ftp://ext.ftp.fao.org/ag/cgrfa/BSP/bsp10E.pdf

Dinnyes, A., Dal, Y.P., Barber, M., Liu, L., Xu, J., Zhou, P.L. & Yang, X.Z. 2001. Development of cloned embryos from adult rabbit fibroblasts: effect of activation treatment and donor cell preparation. *Biology of Reproduction,* **64**: 257–263.

Gupta, N., Sharma, R., Taneja, R., Sisodia, B.S., Singla, S.K., Singh, G. & Gupta, S.C. 2002. Growth curve and ploidy levels in multiple passaging of buffalo skin fibroblast cells in culture. *In: Proceedings 4th Asian Congress on Buffaloes.* New Delhi, 22–23 February 2003.

Hare, W.C.D. & Singh, E.L. 1979. *Cytogenetics in Animal Reproduction.* Wallingford, UK: Commonwealth Agricultural Bureaux.

Kato, Y., Tani, T., Solomaru, Y., Kurokawa, A., Kato, J., Doguchi, H., Yasue, H. & Tsunoda, Y. 1998. Eight calves cloned from somatic cells of a single adult. *Science,* **282**: 2095–2098.

Kubota, C., Yamakuchi, H., Todoroki, J., Mizoshita, K., Tabara, N., Barber, M. & Yang, X. 2000. Six cloned calves produced from adult fibroblast cells after long-term culture. *Proceedings of the National Academy of Sciences, USA,* **97**(3): 990–995

Maghni, K., Nicolescu, O.M. & Martin, J.G. 1999. Suitability of cell metabolic colorimetric assays for assessment of CD4+ T cell proliferation: comparison to 5-bromo-2-deoxyuridine (BrdU) ELISA. *Journal of Immunological Methods,* **223**(2): 185–194.

Mehra, K.L. 2001. Animal Biotechnologies: benefits and concerns. *In: Proceeding of National Workshop on Conservation and Management of Genetic Resources of Livestock.* National Academy of Agricultural Sciences, New Delhi.

Onishi, A., Iwamoto, M., Akita, T., Mikawa, S., Takeda, K., Awata, T., Hanada, H. & Perry, A.C. 2000. Pig cloning by microinjection of fetal fibroblast nuclei. *Science,* **289**(5482): 1188–1190.

Polejaeva, I.A., Chen, S.H., Vaught, T.D., Page, R.L., Mullins, J., Ball, S., Dai, Y., Boone, J., Walker, S., Ayares, D.L., Colman, A. & Campbell, K.H.S. 2000. Cloned pigs produced by nuclear transfer from adult somatic cells. *Nature,* **407**: 86–90.

Wakayama, T., Perry, A.C.F., Zuccott, M., Johnson, K.R. & Yangimachi, R. 1998. Full term development of mice from enucleated oocytes injected with cumulous cell nuclei. *Nature,* **394**: 369–374.

Wells, D.N., Misica, P.M., Tervit, H.R. & Vivanco, W.H. 1998. Adult somatic cell nuclear transfer is used to preserve the last surviving cow of the Enderby Island cattle breed. *Reproduction Fertility and Development,* **10**: 369–378.

Wemme, H., Pfeifer, S., Heck, R. & Muller-Quernheim, J. 1992. Measurement of lymphocyte proliferation: critical analysis of radioactive and photometric methods. *Immunobiology,* **185**(1): 78–79.

White, K.L., Bunch, T.O., Mitalipov, S. & Reed, W.A. 1999. Establishment of pregnancy after the transfer of nuclear embryos produced from the fusion of argali (*Ovis ammon*) nuclei in to domestic sheep (*Ovis aries*) enucleated oocytes. *Cloning* **1**(1): 47–54.

Wilmut, I., Schnieke, A.E., Mcwhir, J., Kind, A.J. & Campbell, K.H. 1997. Viable offspring derived from fetal and adult mammalian cells. *Nature,* **385**: 810-813.

GENETIC DIVERSITY IN ALGERIAN SHEEP BREEDS, USING MICROSATELLITE MARKERS

Souheil Gaouar,[1] Nacera Tabet-Aoul,[1] Amal Derrar,[1]
Kathayoun Goudarzy-Moazami[2] and Nadhira Saïdi-Mehtar[1]
1. Department of Biotechnology, University of Oran Es-Senia, CRSTRA, Oran, Algeria.
2. Department of Animal Genetics, INRA, Jouy-en-Josas, France.

Abstract: Two breeds – Ouled-Djellal and Hamra (85 animals) – were genotyped for 12 microsatellites using PCR and sequencing. Allele number and frequency were calculated, and 141 different alleles were found for these microsatellites, reflecting high genetic variability within these breeds. This study is being extended to other Algerian breeds to estimate variability and genetic distances between them. In parallel, blood samples from the various breeds are being collected to build up a DNA bank. The results should support establishment of a strategy to promote the use and development of locally adapted sheep resources.

1. INTRODUCTION

Blood protein polymorphisms have been previously used to study biodiversity and genetic relationships among sheep breeds (Emerson and Tate, 1993). More recently, microsatellite markers have been used to determine allele frequencies in different populations.

In order to define genetic variability of two Algerian sheep breeds, Hamra and Ouled-Djellal, we analysed their DNA using Polymerase Chain Reaction (PCR) with 12 microsatellites localized on 10 different ovine chromosomes, followed by automatic genotyping using sequencing.

H. P. S. Makkar and G. J. Viljoen (eds.), Applications of Gene-Based Technologies for
Improving Animal Production and Health in Developing Countries, 641-644.

2.   MATERIAL AND METHODS

2.1   Animals

Two breeds were analysed in this study:
- Hamra breed. This breed is from in the northwest of Algeria, and 35 unrelated animals were used.
- Ouled-Djellal breed. This breed is found in northeast and central Algeria, and 50 unrelated animals were used.

2.2   Markers

12 microsatellite markers were studied (ovine chromosomal localization is indicated in parenthesis): INRA49 (01); OarFCB20 (02); OarFCB11 (02); MCM527 (05); ILSTS05 (07); CSSM66 (09); MCM42 (09); TGLA53 (12); MAF65 (15); OarCP49 (17); OarHH56 (20); and MAF36 (22).

2.3   Methods

Genomic DNA was purified by protease K digestion and a salting-out procedure (Miller, Dykes and Polesky, 1988). Microsatellites were amplified using fluorescence labelling primers followed by electrophoresis on acrylamide sequencing gel (7%). Allele identification was performed by the GENEPOP 3.1 programme (Raymond and Rousset, 1995).

3.   RESULTS

Figure 1 shows the numbers of alleles and their frequencies for two breeds for microsatellite MCM527.

The results showed a total of 141 different alleles for the 12 microsatellites, reflecting high genetic variability among these breeds. There were 125 alleles among the Ouled-Djellal animals sampled and 104 alleles among the Hamra animals sampled.

A comparison of allele frequencies was carried out to define the use of these microsatellites in population and individual identification in these breeds. The microsatellites studied were divided into two groups.

The first group included ten markers: INRA49, OarFCB20, OarFCB11, MCM527, ILSTS05, CSSM66, MCM42, MAF65, OarCP49 and MAF36. These markers could be used in population identification because they presented different frequencies for the same allele. Some of them (INRA49,

OarFCB20, MCM527, ILSTS05 and MAF36) showed specific alleles for each breed and could be useful for individual identification.

The second group comprised two microsatellites, TGLA53 and OARHH56, which had, for the majority of observed alleles, nearly the same allelic frequencies. So they could not be used for population identification.

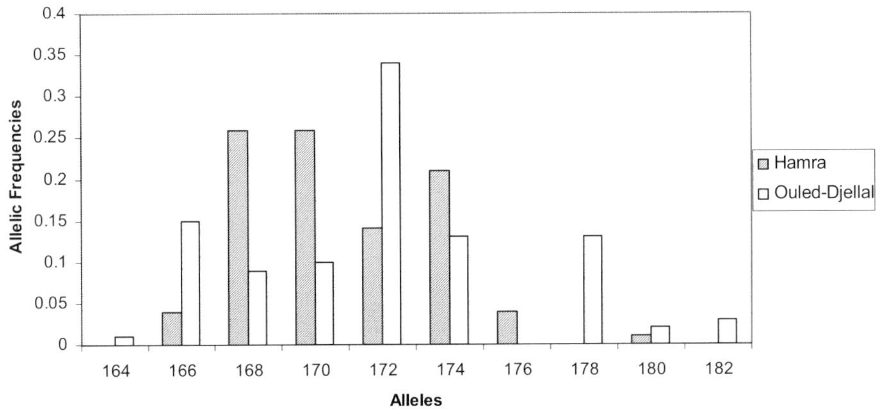

Figure 1. Histogram of allelic frequencies per breed for the microsatellite MCM527.

4.   CONCLUSION

Our results, considered as preliminary, provided the first study of genetic variability of two Algerian ovine breeds, based on DNA polymorphism. Microsatellite genotyping in sheep appears to provide a useful tool for examining genetic variability.

This study will be extended to six other Algerian breeds – Rambi, Taadmit, Barbarine, D'men, Sidaoun and Berber – to estimate their variability and the genetic distances among them.

The results will help in the establishment of a strategy to preserve, promote the use of and develop our locally-adapted sheep breeds. Thus, rational exploitation of domestic animal biodiversity could be realized through a global strategy for the management of Algerian sheep genetic resources.

Moreover, it is hoped to extend this study to other ovine breeds from neighbouring Maghreb countries (Tunisia and Morocco) and other Mediterranean countries (France, Spain, Italy, etc.) to definitively determine evolutionary relationships among these breeds and to indicate their origin and history.

REFERENCES

Emerson, B.C. & Tate, M.L. 1993. Genetic analysis of evolutionary relationships among deer (Subfamily Cervinae). *Heredity,* **84**: 266–273.

Miller, S.A., Dykes, D.D. & Polesky, H.F. 1988. A simple salting out procedure for extracting DNA from human nucleated cells. *Nucleic Acids Research,* **12**: 319–325.

Raymond, M. & Rousset, F. 1995. GENEPOP (version 1.2): Population genetics software for exact tests and ecumenicism. *Journal of Heredity*, **86**: 248–249.

SOMATIC CELL BANKING – AN ALTERNATIVE TECHNOLOGY FOR THE CONSERVATION OF ENDANGERED SHEEP BREEDS

N. Gupta,[1] S.C. Gupta,[1] S.P.S. Ahlawat,[1] R. Sharma,[1] K. Gupta[2] and R. Taneja[1]

1. National Bureau of Animal Genetic Resources, Karnal 132001, India.
2. CSK Himachal Pradesh Krishi Visvavidalya, Palampur, India.

Abstract: Skin samples from ear pinna of 10 male and 10 female sheep were collected and cultured in DMEM+Ham's F12 nutrient medium. Cell viability was 95 to 100% in different cultures. Mean cell proliferation rates were 0.94–0.67 and 1.15–0.56 for males and females in different passages, respectively. Cell proliferation rates were highest in first passage and then showed an age-related decline. Average cell doubling time was 30 h in males and 29.6 h in females. Skin fibroblast cell growth curves were in lag phase for the first 2 days, entered log phase (3rd to 7th days) and plateaued on day 8. Diploid chromosomal counts in proliferating cells up to the 5th passage were normal (2N=54), with no gross chromosomal aberrations recorded. Cells frozen from cycling cells at 80–90% confluency showed superior post-thaw growth compared with cells from overconfluent cultures. DMSO at 10% (v/v) in freezing media was optimal. Controlled-rate freezing at -1°C/min showed better post-thaw cell viability and growth potential. Direct plating of thawed cells without removing DMSO and other contents of the freezing medium gave better post-thaw survival and proliferation rates.

1. INTRODUCTION

Somatic cell nuclear transfer (NT) offers new opportunities for genetic engineering, genome preservation and tissue regeneration. Sheep were the first species in which somatic cell nuclear transfer was successful, with the birth of Dolly in 1997 (Wilmut *et al.,* 1997). Since then, somatic cell NT has

H. P. S. Makkar and G. J. Viljoen (eds.), Applications of Gene-Based Technologies for
Improving Animal Production and Health in Developing Countries, 645-657.
© 2005 *IAEA. Printed in the Netherlands.*

succeeded in various other species, including cattle (Wells *et al.*, 1997, 1998; Cibelli *et al.*, 1998; Kubota *et al.*, 2000), goats (Baguisi *et al.*, 1999), mice (Wakayama *et al.*, 1998), pigs (Onishi *et al.*, 2000; Polejaeva *et al.*, 2000) and gaur (Lanza *et al.*, 2000). Cloned animals have been produced using as the karyoplast donor somatic cells from organs as diverse as mammary gland epithelium, cumulus/granulose cells, oviduct, ear, skin, tail, muscle, liver and sertoli cells. However, it is still not clear which cell type or cell origin is most successful for mammalian cloning (Dinnyes *et al.*, 2001).

Live calves have been produced successfully from multiplied, passaged, frozen and thawed skin fibroblasts (Kubota *et al.*, 2000; Lanza *et al.*, 2000). Yang's group (Jerry Yang, 1999, pers. comm.) has produced 10 cloned calves from a 13-year-old dairy cow using skin fibroblast cells. According to Kubota *et al.* (2000), the skin fibroblasts are now widely used for the conservation and re-generation of unique animals or livestock breeds through animal cloning, because they can be easily collected by tissue biopsy using non-invasive methods, and are easy to culture for long duration before they actually enter into a senescence phase. This study adopted skin fibroblasts from ear tissue for culture, multiple passaging and cryobanking, as an alternative technology for long-term preservation of Gaddi and other endangered indigenous sheep breeds.

The principle is that each cell of an animal's body contains the full genetic code for the whole animal, and that nuclear transfer provides a way of converting cells to whole animal. Cells from endangered breeds – collected by biopsy or from scrapings of soft skin or ear tissue or from hair follicle – can be grown and multiplied in the laboratory and can then be indefinitely stored frozen at -196°C in liquid nitrogen.

2.   MATERIALS AND METHODS

2.1   Sample size, collection and transport

Skin samples from the ear pinna of 10 male and 10 female sheep were collected from random breeding populations in the main centre of the breeds. An area of ear for biopsy was selected, avoiding major veins, and shaved clean on both sides. The surface was cleaned with 70% ethanol and a small piece of tissue (1 cm^2) was cut using a biopsy punch. The tissue was washed several times in Dulbecco's phosphate buffered saline (DPBS) containing 10% foetal bovine serum (FBS) and antibiotic-antimycotic solution. The samples were transferred into complete medium (DMEM+HamsF12 with 10% FBS and antibiotic-antimycotic) within 2 hr and transported in a Thermos flask at 4°C to the laboratory within 72 hr of collection.

2.2   Development of primary cultures

Skin tissue was cleansed thoroughly with a scalpel to remove epidermal layers from both sides, and cut into small pieces (1 mm^2). Skin pieces were seeded about 1 mm apart from each other in a 25-cm^2 flask containing a wetting layer of culture medium (DMEM+Ham's F12, 15% FBS, penicillin, streptomycin and L-glutamine). An additional 2.5 ml of equilibrated media under 5% CO_2 at 37°C was added after 24 hr when the skin pieces were attached to the solid surface of the culture vessel. Tissue pieces were removed when fibroblast colonies appeared around them and the cultures were allowed to grow till nearly 80% confluence.

2.3   Purification and subculturing of skin fibroblast cells

Any epithelial cells visible in primary cultures were removed enzymatically to leave only pure fibroblasts. In trypsin-EDTA treatment, the fibroblasts were collected first, and were pipetted off carefully without disturbing the epithelial colonies. For this, 100 µl of cold trypsin-EDTA (TE) solution was added to make a wetting layer and incubated for 5–10 min at 37°C. The flasks were gently tapped to detach the cells from the surface and then an additional 2.5 ml of DPBS containing 10% FBS was added. The cell suspension was transferred into a 15-ml centrifuge tube by gentle pipetting and centrifuged at 1000 rpm for 10 min. For cell counting and viability testing, a cell pellet was suspended in 1 ml of culture media, from which 10 µl of cell suspension was mixed with an equal volume of 0.4% trypan blue. The blue-stained cells (dead) and unstained (live) were counted using Neubar's haemocytometer. For subculturing and further passaging, the cells were re-seeded at 3.2×10^3 cells/cm^2 in a 25 cm^2 flask containing 2.5 ml of culture medium, and were incubated at 37°C under 5% CO_2 till 80–90% confluence.

2.4   Standard growth curve

To establish a standard growth curve in the experiment, an initial 0.4×10^5 cells were seeded in a 25-cm^2 flask. For each cell line, 8 flasks were seeded and 1 flask was harvested at intervals of 24 hr over 8 days. The cells were counted and the data was plotted using MS Excel.

2.5   Chromosomal study

For chromosomal study, cells were seeded at 4×10^3 cells/cm^2 in a 25-cm^2 TC flask. Colchicine (2.5 μg/ml) was added to the cultures at around 50–60% confluency, and allowed to stand for 2 hr at 37°C. The colchicine treatment resulted in rounding of cells, observed using an inverted microscope. Cells were harvested by the same trypsin-EDTA protocol described above. Chromosomal slides were prepared using hypotonic (0.075 M) KCl, fixative (acetic acid : methanol, 1:3) and air drying protocol described elsewhere (Gupta *et al.*, 2002) and stained in 2% Giemsa (pH 6.8) for 30 min.

2.6   Freezing

For long-term cryopreservation, cells were taken from healthy cultures. Cryovial (Nunc) of 1.8 ml volume contained 1×10^6 cells/ml in freezing media. The freezing media consisted of DMEM+Ham's F12 with 10% FBS and antibiotics. DMSO (10% V/V) was used as cryoprotectant. The cells were frozen at a controlled rate (approximately 1°C/min) using Mr. Frosty (Nunc). For short-term freezing, the cells were stored at -80°C in an ultra low temperature freezer (Forma), and for long-term preservation the frozen cryovials were transferred to a liquid N_2 container at -196°C.

2.7   Thawing and re-culturing

The cells were thawed at 39°C in a water bath. To establish the conditions for obtaining optimal post-thaw viability and re-growth, three trials were conducted.
1. A thawed cell suspension was poured into a centrifuge tube containing 10 ml culture media, and centrifuged at 1000 rpm for 10 min. After removing supernatant, cells were plated in a 60-mm Petri dish containing 5 ml of medium.
2. Cells after thawing were given three washings in 5 ml of medium by repeated centrifugation, before re-seeding them in a 60-mm Petri dish containing 5 ml of medium.
3. Cells immediately after thawing were plated directly onto a 60-mm Petri dish containing 5 ml of medium. The medium was replaced after 24 hr.

3.   RESULTS

In this study, skin samples were collected from a minimum of 10 males and 10 females of unrelated ancestry in different sheep flocks. It was found that transporting samples in culture media at 4°C produced better cell growth than transporting at body temperature. The cells appeared around tissue pieces within 4–6 days, and attained 80% confluency after 12–15 days, when processed within 72 hr of collection. However, older samples responded poorly and the desired growth in cells was attained much later. Sample collection using Dulbecco's Phosphate buffered saline (PBS) with Ca^{++}, Mg^{++}, 10% FBS and antibiotics, and transfer into culture media within 2 hr under aseptic conditions, avoided fungal and bacterial contamination to a greater extent than did direct collection of samples into culture media.

3.1   Purification of skin fibroblasts

Occasionally there were colonies of epithelial cells around the tissues, but cultures from samples properly scraped on both sides showed no epithelial cell contamination (Figure 1). Care at the time of first harvest could also minimize the problem of epithelial cell contamination in fibroblast culture. It was, however, seen that, by the 5th passage, no traces of epithelial cells were observed (Figure 2).

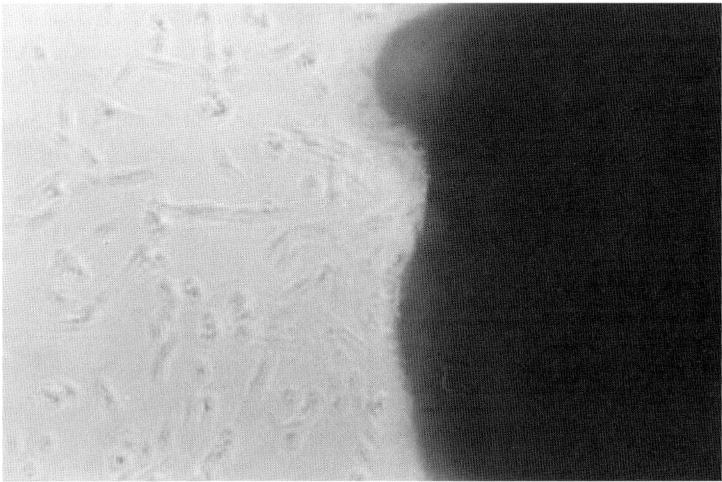

Figure 1. Primary skin fibroblast cells coming out of tissue.

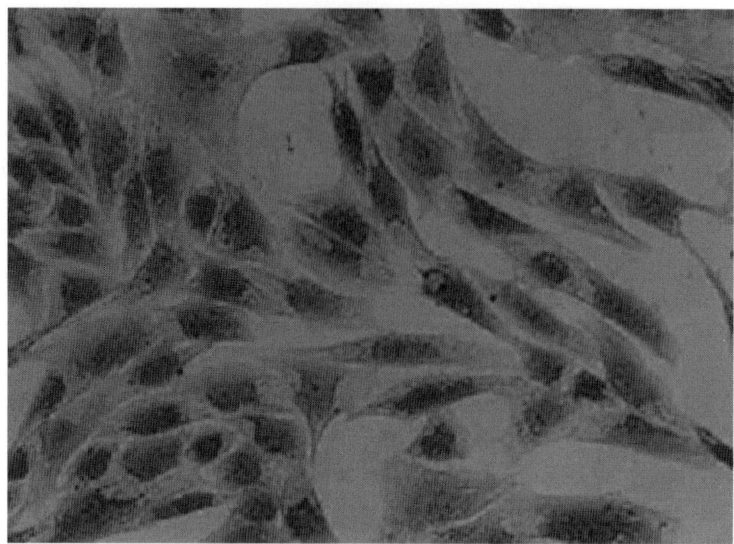

Figure 2. Crystal-violet stained secondary skin fibroblast cells.

3.2   Proliferation rate and cell viability

Cell viability ranged between 95 and 100% in different cultures, irrespective of breeds, animals and passages. The data on mean population doublings, proliferation rate and population doublings per day for Gaddi sheep samples is presented in Table 1. The mean population doublings per passage were 4 in male and 4.4 in female cell lines. The mean cell proliferation rates (r) were in the ranges of 0.94–0.67 and 1.15–0.56 for males and females, respectively. The differences between sexes were not significant. The cell proliferation rates were highest in first passage and then showed an age-related decline. Average cell doubling times were 30 hr in males and 29.63 hr in females in the first 5 passages.

3.3   Standard growth curve

The data for the standard growth curve in cell lines of sheep skin fibroblasts are presented in Table 2. Figure 3 shows that sheep skin fibroblasts cells had a sigmoid curve for Pugal sheep male cell line 1 (PSM1) and Pugal sheep male cell line 2 (PSM2) and an average of the their data. The skin fibroblast remained in lag phase for the first 2 days, when they settled on the solid surface of the culture vessel and entered into log phase (3rd to 7th days), when maximum growth took place. The growth curve reaches a plateau on the 8th day and showed decline subsequently.

3.4   Chromosomal profile

The chromosomal profile of each cell line was studied at first passage and then at every third passage. The diploid chromosomal count in proliferating cells up to the 5th passage was normal (2N=54), with no gross chromosomal aberrations recorded (Figure 4). Occasional metaphases showed a tetraploid (4N) genome (Figure 5), but their distribution was random.

3.5   Freezing and thawing

The cells frozen from cycling cells at 80–90% confluency showed superior post-thaw growth compared with cells from over-confluent cultures. From the latter type of culture, some 30–40% of cells remained floating in the media after thawing and re-seeding, even after short-duration freezing. The 10% DMSO in freezing media was found to be optimal for cell viability and growth potential after thawing. Controlled rate freezing at -1°C/min. showed better post-thaw cell viability and growth potential than did the rapid freezing protocol. Skin fibroblasts stored at -32°C in a deep freeze could be revived within 40 days of freezing, while those kept at -80°C in an ultra-low temperature deepfreeze could be revived even after 300 days. For longer storage periods, cells were kept in liquid nitrogen (-196°C).

Table 1. Cell proliferation data for Gaddi sheep skin fibroblast cells in culture.

Passage	Male			Female		
no.	[1]	[2]	[3]	[1]	[2]	[3]
1	4	0.94 ±0.08 (5)	25.52 ±0.51 (5)	4	1.15 ±0.06 (9)	20.87 ±1.65 (9)
2	8	0.89 ±0.11 (5)	26.70 ±0.43 (5)	9	0.92 ±0.05 (9)	26.08 ±1.50 (9)
3	12	0.79 ±0.13 (5)	30.31 ±1.31 (5)	13	0.78 ±0.08 (9)	30.77 ±2.76 (9)
4	16	0.72 ±0.21 (5)	33.23 ±2.76 (5)	18	0.66 ±0.09 (9)	36.36 ±3.83 (9)
5	20	0.67 ±0.13 (5)	36.47 ±1.34 (5)	22	0.56 ±0.07 (9)	41.85 ±4.76 (9)
Average	4.0 per passage	0.80 ±0.07 (5)	30.00 ±1.42 (5)	4.4 per passage	0.81 ±0.06 (9)	29.63 ±1.13 (9)

KEY: [1] Cumulative population doublings. [2] Proliferation rate, r. Figures in parenthesis indicate number of cell lines. [3] Population doubling per day. Figures in parenthesis indicate number of cell lines.

                                                    N. Gupta et al.

In three trials using different protocols for re-seeding after thawing, direct plating of thawed sample without removing DMSO and other contents of the freezing medium gave better post-thaw cell survival and proliferation (Table 3).

Table 2. Cell count ($\times 10^5$) for different skin fibroblast cell lines.

Day	Cell line PSM 1	Cell line PSM2	Average
0	0.400	0.400	0.40
1	0.194	0.79	0.186
2	0.654	0.763	0.707
3	2.770	2.600	2.68
4	5.160	6.112	5.637
5	12.24	11.75	11.98
6	13.410	13.716	13.563
7	12.681	13.905	13.292

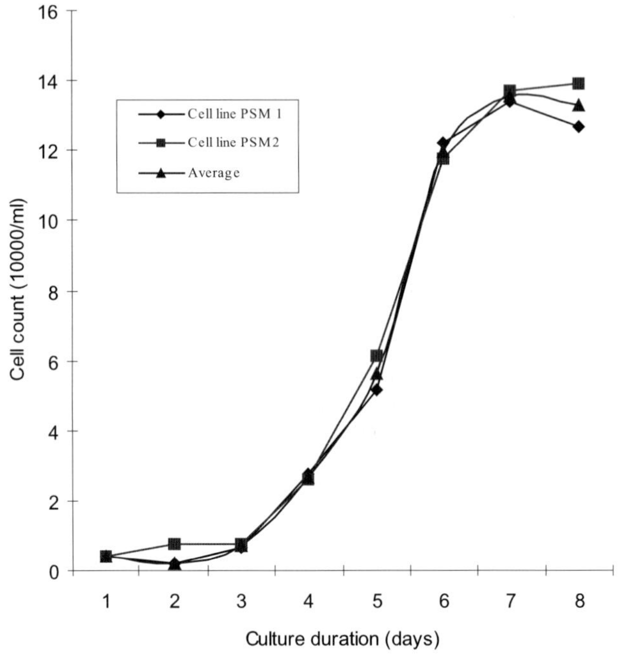

Figure 3. Standard growth curves of pugal sheep skin fibroblast cell lines.

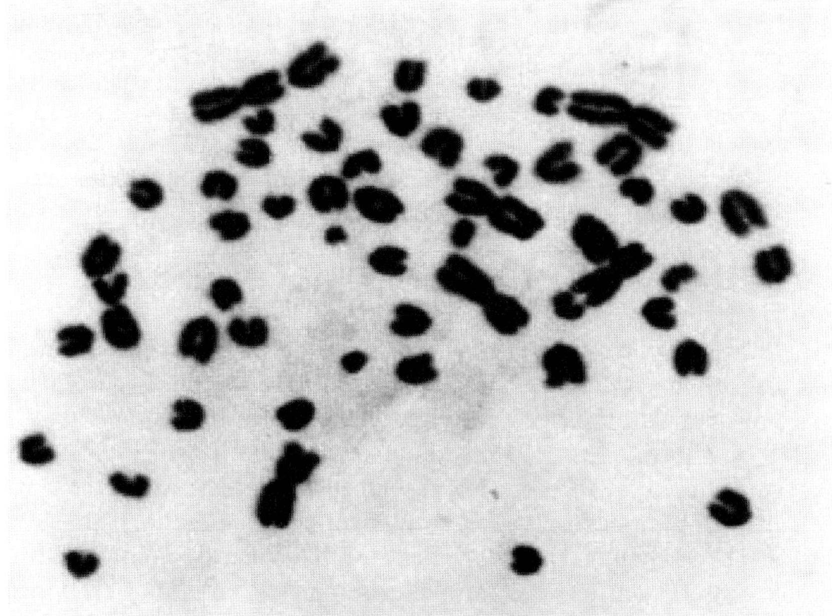

Figure 4. Metaphase chromosomes of Gaddi sheep showing normal 2N=54.

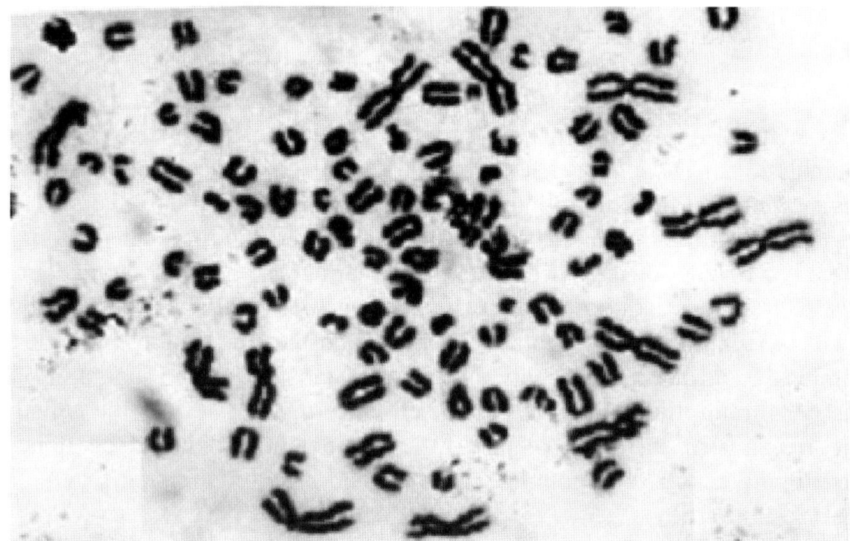

Figure 5. Metaphase chromosome of Gaddi sheep showing tetraploidy 4N=108.

Table 3. Effect of re-seeding method on post-thaw viability of skin fibroblast cells.

Treatment	No. of samples	Average culture duration (days)	Average initial population (cells/ml)	Average final population (cells/ml)
Single washing	6	2.8	1×10^6	1.11×10^6
Three washings	6	2.8	1×10^6	1.17×10^6
Direct plating	6	2.8	1×10^6	1.80×10^6

4.   DISCUSSION

The birth of Dolly, the cloned sheep, in 1997 was the first breakthrough in the use of this technology (Wilmut *et al.,* 1997). Since then, a large number of animals have been produced using NT technique (Wells *et al.,* 1997, 1998; Cibelli *et al.,* 1998; Kubota *et al.,* 2000; Baguisi *et al.,* 1999; Wakayama *et al.,* 1998). For genetic resources at high risk, where the current live animal and cryoconservation techniques are not feasible, the expert group in FAO (Hammond, 1997) considered that countries could begin now with the conservation of cells from skin samples. Conserving cells means conserving the genomes of a breed and providing insurance for the future. Cells from the endangered breeds can be grown in culture and stored in liquid nitrogen indefinitely (Kubota *et al.,* 2000).

It is often argued that cloning (copying) affects genetic diversity adversely, but the collection of samples from an unrelated 10 males and 10 females could avoid the risk of inbreeding (Brem, 1988). Our study was a first attempt to isolate the skin fibroblast cells from the desired number of animals of an endangered sheep breed, which should have the potential to revive a normal breeding population without losing biological diversity. The cryopreservation of 25–30 samples of each cell lines isolated from an effective population size where random mating takes place should ensure that sufficient variability exists in cloned progenies using these donor cells.

This technique has the practical advantage that the required number of samples could be taken in only a few days, transported to a laboratory and the cells multiplied and frozen. The cell culturing to multiply the cells and the freezing of the cells require little equipment and could be done in developing countries at nominal cost and with little requirement for training (Brem, 1988).

The efficiency of cloning is still very low, and there can be a number of internal as well as external factors affecting the re-programming potential of the cloned embryos generated by NT of somatic nuclei. Although somatic cell karyoplast and oocyte cytoplasm in NT embryos is very complex, the donor cell holds the key (Campbell, 1999). In culture media, various components and supplements affect the growth potential of cells. Cells can succumb to various biochemical, metabolic and genetic factors that retard their doubling potential in long-term culturing. It is very important to draw a standard growth curve of normal dividing cells for each cell line or cell type so that the effect of various intrinsic as well as external factors, including aging of cells in culture, can be monitored regularly to develop healthy and normal cell lines. Any significant deviation from the standard growth curve would indicate a toxic effect of the culture medium or its constituents, supply of CO_2 in incubator, infections or aging of cultures. In this study, the standard growth curve in sheep skin cell lines showed a similar pattern to that reported for established human and other animal cell lines (Dinnyes *et al.*, 2001; Kato *et al.*, 1998).

During long-term culture, the cells are likely to develop one or more chromosomal abnormalities. Use of a polyploid, aneuploid or translocation carrier nuclear donor cell would result in a high probability of premature termination of pregnancy in the case of NT reconstructed embryos. Therefore it must be ensured that the cells in different passages be checked for normal ploidy, so that they donate the normal karyoplast for producing viable clones. In this study, the chromosomal profile of each cell line was monitored at every third passage, and no gross chromosomal abnormalities were encountered in the first 5 passages. In comparison with these results, in earlier studies (Kubota *et al.*, 2000; White *et al.*, 1999), a higher frequency of aneuploidy was reported in cattle and sheep skin fibroblast cultures in 3–7 passages.

According to Hammond (Gupta *et al.*, 2002), maintenance of live populations of endangered breeds of livestock is the ideal, but there is reluctance to invest for breeds that are relatively less economic at present. In India, the deep freezing of ram semen is not available and collection and freezing of embryos is too costly to justify its utility in breed conservation. In such a situation, somatic cell banking technology is an alternative method for conservation of sheep genetic diversity in India.

5.   CONCLUSION

The basic concept in conservation using this technique is the isolation of suitable somatic cells in large numbers, growing them in a laboratory for a few cell generations and then preserving them for long periods in cryogenic storage so that the live population can be re-established by thawing these frozen cells and using them in NT reconstructed embryos and animal clones.

ACKNOWLEDGEMENTS

The authors are thankful to the Director, NBAGR, Karnal, for the facilities provided. Funding by World Bank under NATP is duly acknowledged

REFERENCES

Baguisi, A., Behboodi, E., Melican, D.T., Pollock, J.S., Destrempes, M.M., Cammuso, C., Williams, J.L., Scott, D.N., Porter, C.A., Midura, P., Palacios, M.J., Ayres, S.L., Denniston, R.S., Hayes, M.L., Ziomek, C.A., Meade, H.M., Godke, R.A., Gavin, W.G., Overström, E.W., Yann, E. 1999. Production of goats by somatic cell nuclear transfer. *Nature Biotechnology,* **17**: 456–461.

Brem, G. 1988. *Ex situ* cryoconservation of genome and genes of animals. pp. 73–84, *in:* A. Anderson (ed). *Conservation of Domestic Livestock.* Wallingford, UK: CAB International.

Campbell, K.H.S. 1999. Nuclear equivalence, nuclear transfer and the cell cycle. *Cloning,* **1**(1): 3–15.

Cibelli, J.B., Stice, S.L., Golucke, P.J., Kane, J.J., Jeny, J., Blackwell, C., Ponce de Leon, F.A. & Robl, J.M. 1998. Cloned transgenic calves produced from non-quiescent foetal fibroblasts. *Science,* **280**: 1256–1258.

Dinnyes, A., Dal, Y.P., Barber, M., Liu, L., Xu, J., Zhou, P.L. & Yang, X.Z. 2001. Development of cloned embryos from adult rabbit fibroblasts: effect of activation treatment and donor cell preparation. *Biology of Reproduction,* **64**: 257–263.

Gupta, N., Sharma, R., Taneja, R., Sisodia, B.S., Singla, S.K., Singh, G. & Gupta, S.C. 2002. Growth curve and ploidy levels in multiple passaging of buffalo skin fibroblast cells in culture. *Proceedings of the 4th Asian Congress on Buffaloes.* New Delhi, 25–28 February 2003.

Hammond, K. 1997. An interview: "Dolly the Sheep" cloning methods could hold key to breakthrough in conserving animal genetic diversity. Transcript of radio interview in conjunction with an FAO/Istituto Sperimentale per la Zootecnia-Monterotondo Workshop on *New developments in biotechnology and their implications for the conservation of farm animal genetic resources – Reversible DNA quiescence and somatic cloning.* Ministry of Agriculture, Rome, 26–28 November 1997. FAO, Rome.

Kato, Y., Tani, T., Solomaru, Y., Kurokawa, A., Kato, J., Doguchi, H., Yasue, H. & Tsunoda, Y. 1998. Eight calves cloned from somatic cells of a single adult. *Science,* **282**: 2095–2098.

Kubota, C., Yamakuchi, H., Todoroki, J., Mizoshita, K., Tabara, N., Barber, M. & Yang, X.Z. 2000. Six cloned calves produced from adult fibroblast cells after long-term culture. *Proceedings of the National Academy of Sciences, USA,* **97**: 990–995.

Lanza, R.P., Cibelli, J.B., Diaz, F., Moraes, C.T., Farin, P.W., Farin, C.E., Hammer, C.J., West, M.D. & Damiani, P. 2000. Cloning of an endangered species (*Bos gaurus*) using interspecies nuclear transfer. *Cloning,* **2**: 79–90.

Onishi, A., Iwamoto, M., Akita, T., Mikawa, S., Takeda, K., Awata, T., Hanada, H. & Perry, A.C. 2000. Pig cloning by microinjection of fetal fibroblast nuclei. *Science,* **289**: 1188–1190.

Polejaeva, I.A., Chen, S.H., Vaught, T.D., Page, R.L., Mullins, J., Ball, S., Dai, Y., Boone, J., Walker, S., Ayares, D.L., Colman, A. & Campbell, K.H.S. 2000. Cloned pigs produced by nuclear transfer from adult somatic cells. *Nature,* **407:** 86–90.

Wakayama, T., Perry, A.C.F., Zuccott, M., Johnson, K.R. & Yangimachi, R. 1998. Full term development of mice from enucleated oocytes injected with cumulous cell nuclei. *Nature,* **394**: 369–374.

Wells, D.N., Misica, P.M., Day, A.M. & Tervit, H.R. 1997. Production of cloned lambs from an established embryonic cell line: a comparison between *in vivo* and *in vitro* matured cytoplasts. *Biology of Reproduction,* **57**: 385–393.

Wells, D.N., Misica, P.M., Tervit, H.R. & Vivanco, W.H. 1998. Adult somatic cell nuclear transfer is used to preserve the last surviving cow of the Enderby Island cattle breed. *Reproduction Fertility and Development,* **10**: 364–378.

White, K.L., Bunch, T.O., Mitalipov, S. & Reed, W.A. 1999. Establishment of pregnancy after the transfer of nuclear embryos produced from the fusion of argali (*Ovis ammon*) nuclei in to domestic sheep (*Ovis aries*) enucleated oocytes. *Cloning,* **1**: 47–54.

Wilmut, I., Schnieke, A.E., McWhir, J., Kind, A.J. & Campbell, K.H.S. 1997. Viable offspring derived from fetal and adult mammalian cells. *Nature,* **385**: 810–813.

THE VACCINE PROPERTIES OF A BRAZILIAN BOVINE HERPESVIRUS 1 STRAIN WITH AN INDUCED DELETION OF THE gE GENE

Ana C. Franco,[1] Fernando R. Spilki,[1] Franciscus A.M. Rijsewijk[2] and Paulo M. Roehe.[1]

1. Centro de Pesquisa Veterinária Desidério Finamor-FEPAGRO, CP 2076, CEP 90001-970, Porto Alegre, Brazil.
2. Animal Sciences Group, Lelystad, The Netherlands.

Abstract: Aiming at the development of a differential vaccine (DIVA) against infectious bovine rhinotracheitis (IBR), a Brazilian strain of bovine herpesvirus type 1 (BHV1) with a deletion of the glycoprotein E (gE) gene was constructed (265gE⁻). Here we present the experiments performed with this strain in order to evaluate its safety and efficacy as a vaccine virus in cattle. In the first experiment, a group of calves was inoculated with 265gE⁻ and challenged with wild type virus 21 days post-inoculation. Calves immunized with 265gE⁻ virus and challenged with wild type virus developed very mild clinical disease with a significant reduction in the amount of virus excretion and duration. The safety of the 265gE⁻ during pregnancy was assessed using 22 pregnant cows, at different stages of gestation, that were inoculated with the 265gE⁻ virus intramuscularly, with 15 pregnant cows kept as non-vaccinated controls. No abortions, stillbirths or foetal abnormalities were seen after vaccination. The results show that the 265gE⁻ recombinant is attenuated and able to prevent clinical disease upon challenge. This recombinant will be further evaluated as a candidate virus for a BHV1 differential vaccine.

1.   INTRODUCTION

Differential vaccines (DIVA) have been used for the control of diseases caused by bovine herpesvirus 1 (BHV1) in developed countries (Van Oirschot *et al.*, 1996; Van Oirschot, Kaashoek and Rijsewijk, 1996; Van Oirschot, 1999). The use of DIVAs allows identification of infected animals

659

H. P. S. Makkar and G. J. Viljoen (eds.), Applications of Gene-Based Technologies for Improving Animal Production and Health in Developing Countries, 659-664.
© 2005 IAEA. Printed in the Netherlands.

in a vaccinated herd and may contribute significantly to the control of the infection in an affected region. Here we describe experiments carried out to assess the safety and efficacy of a Brazilian BHV1 from which the glycoprotein E was deleted (265gE⁻) (Franco *et al.*, 2002).

2.   MATERIAL AND METHODS

2.1   Safety and efficacy in calves

Four 3-month-old calves were intranasally inoculated with $10^{5.3}$ 50% tissue culture infective doses (TCID$_{50}$) of 265gE⁻ virus. Another three calves at the same age were kept as a non-vaccinated group. Two calves were kept as non-vaccinated, non-challenged controls. All animals except the controls were challenged with 10^7 50% TCID$_{50}$ of wild-type virus 21 days post-inoculation, and subsequently treated with dexamethasone (DX) 6 months post-inoculation in an attempt to reactivate latent virus. Observation of clinical signs, detection of antibodies and virus isolation were performed in all animals following vaccination, challenge and reactivation of latent infection.

2.2   Safety and efficacy during pregnancy

The 22 pregnant cows, of which 14 were seronegative and 8 seropositive for BHV1, at different stages of gestation, were inoculated with 10^7 50% TCID$_{50}$ of the 265gE⁻ virus by an intramuscular route, with 15 pregnant cows kept as non-vaccinated controls. The animals were observed for evidence of abortion or foetal abnormalities. Sera samples were taken to detect seroconversion in both experimental groups.

3.   RESULTS

3.1   Safety and efficacy of the 265 gE⁻ in calves

After challenge, mild clinical signs and lower virus titres were detected from vaccinated calves compared with non-vaccinated challenged animals (Figures 1a and 2a). Upon DX treatment, the mutant 265gE⁻ could not be reactivated from vaccinated calves (data not shown). Shedding of wild type virus and clinical signs were considerably reduced in vaccinated calves at reactivation (Figures 1b and 2b).

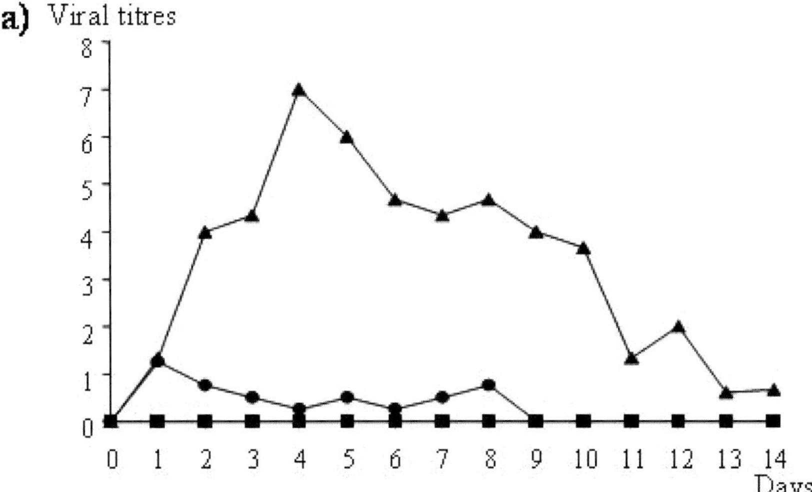

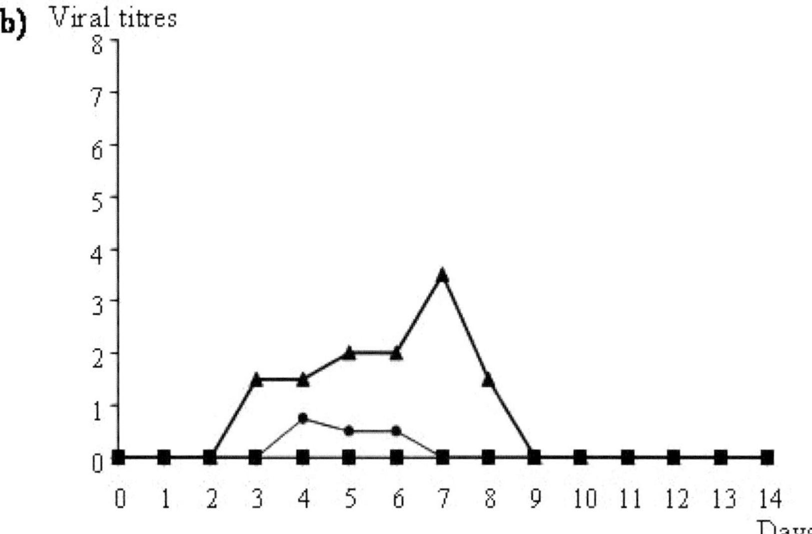

Figure 1. Nasal virus shedding (a) of vaccinated and unvaccinated calves after challenge with wild-type virus, and (b) upon re-activation. Infectious virus titres expressed in Log_{10} of 50% tissue culture infective doses per 50 µl ($TCID_{50}$).

KEY: ▲ = unvaccinated group; ● = vaccinated group; ■ = unvaccinated and non-challenged.

The neutralizing antibody profile of calves was similar in both vaccinated and non-vaccinated groups (Figure 3). Differences in antibody titres were never greater than fourfold between vaccinated and non-vaccinated animals, either after challenge or after reactivation.

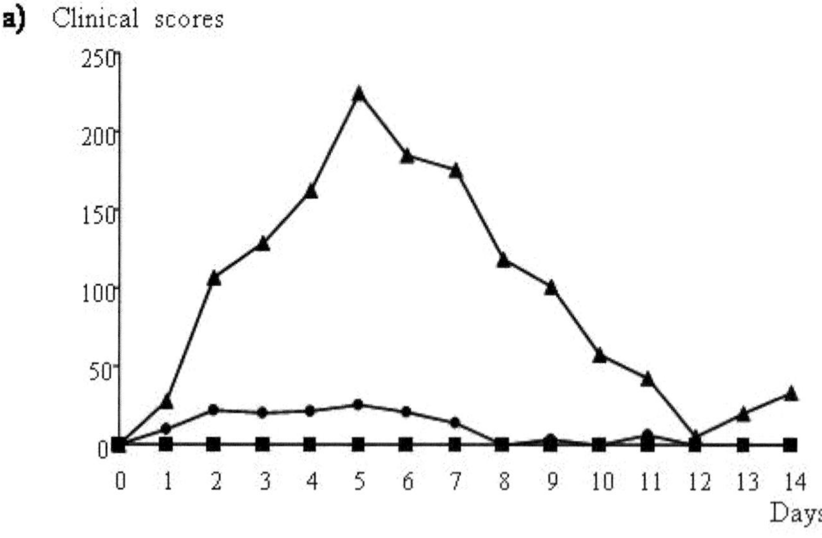

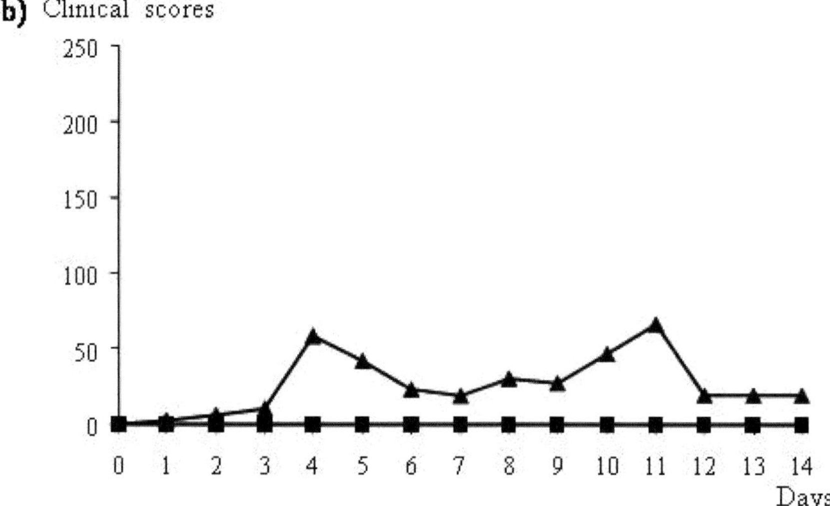

Figure 2. Mean clinical scores attributed to calves vaccinated and unvaccinated after challenge with wild-type virus (a); and after re-activation (b).
KEY: ▲ = unvaccinated group; ● = vaccinated group; ■ = unvaccinated and non-challenged.

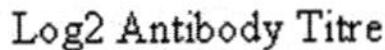

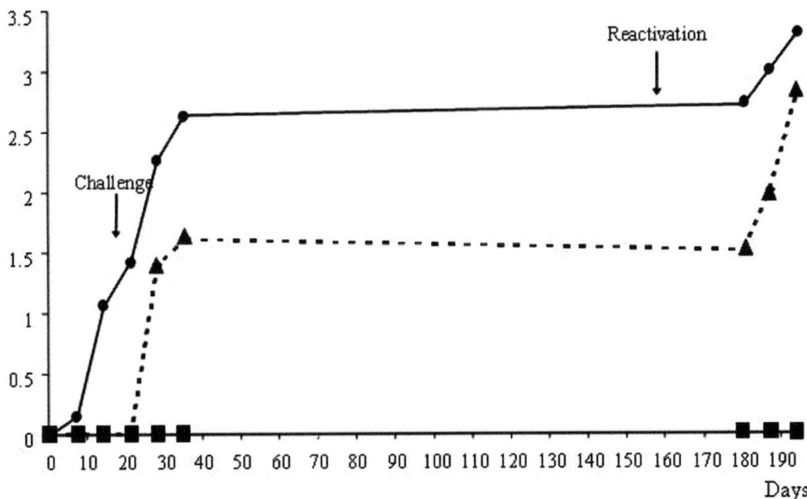

Figure 3. Neutralizing antibodies after inoculation of gE- virus in calves (day 0), challenge with wt virus (day 21) and after DX administration (180 days p.i.).
KEY: ▲ = unvaccinated group; ● = vaccinated group; ■ = unvaccinated and non-challenged. Titres expressed in $\log_{10}$ of the reciprocal of the neutralizing antibody titre.

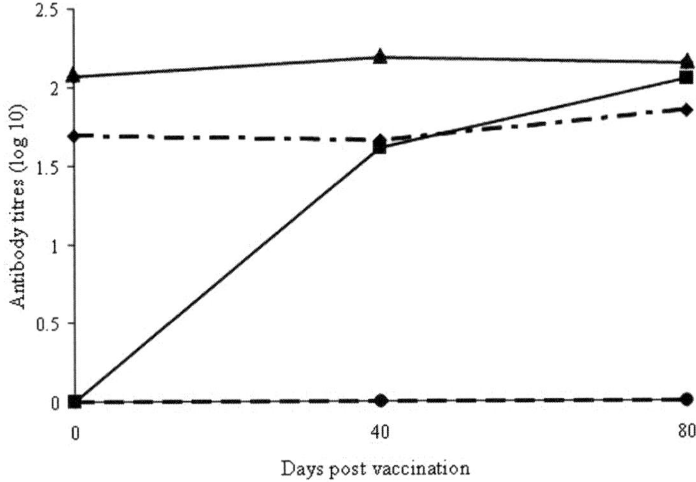

Figure 4. Titres of neutralizing antibodies (expressed in $\log_{10}$) in previously seronegative gE-negative vaccinated cows (■), previously seropositive vaccinated cows (♦), seropositive non-vaccinated cows (▲) and seronegative non-vaccinated cows (●).

3.2   Safety and efficacy during pregnancy

No abortions, stillbirths or foetal abnormalities were seen after vaccination of pregnant cows. Seroconversion was observed in all previously seronegative vaccinated animals. At the same time, the antibody titres in previously seropositive cows did not change significantly after vaccination (Figure 4).

4.   DISCUSSION AND CONCLUSION

The data presented here show that the 265gE⁻ virus was attenuated and induced a protective effect in calves challenged with wild-type virus. In addition, no embryonic or foetal losses were observed after application of the vaccine in pregnant cows. Seroconversion observed in pregnant cows indicates that the vaccinal virus induces the production of antibodies that might be possibly transferred to newborn calves.

It is concluded that the 265gE⁻ strain seems suitable for use as a vaccine virus, as it was shown to be attenuated and yet did not induce reproductive abnormalities in pregnant cows.

REFERENCES

Van Oirschot, J.T., Kaashoek, M.J., Rijsewijk, F.A. & Stegeman, J.A. 1996. The use of marker vaccines in eradication of herpesviruses. *Journal of Biotechnology*, **44:** 75–81.

Van Oirschot, J.T., Kaashoek, M.J. & Rijsewijk, F.A. 1996. Advances in the development and evaluation of bovine herpesvirus 1 vaccines. *Veterinary Microbiology*, **53:** 43–54.

Van Oirschot, J.T. 1999. Diva vaccines that reduce virus transmission. *Journal of Biotechnology*, **73:** 195–205.

Franco, A.C., Rijsewijk, F.A.M., Flores, E.F., Weiblen, R. & Roehe, P.M. 2002. Construction and characterization of a glycoprotein E deletion mutant of bovine herpesvirus type 1.2 strain isolated in Brazil. *Brazilian Journal of Microbiology*, **33:** 274–278.

OBTAINING CLASSICAL SWINE FEVER VIRUS E2 RECOMBINANT PROTEIN AND DNA-VACCINE ON THE BASIS OF ONE SUBUNIT

Oleg Deryabin,[1] Olena Deryabina,[1] Petro Verbitskiy[1] and Vitaliy Kordyum[2]
1. *Institute of Veterinary Medicine UAAS, 30, Donetskaya str., Kyiv, Ukraine.*
2. *Institute of Molecular Biology and Genetics NASU, 150, Zabolotnogo str., Kyiv, Ukraine.*

Abstract: Three forms of E2 recombinant protein were expressed in *E. coli*. Swine sera obtained against different forms of the recombinant protein were cross-studied with indirect ELISA. Using individual proteins as an antigen, only 15% of sera against other forms of protein reacted positively, while 100% of heterologous sera showed positive reaction with fused protein. Challenge experiments showed the existence of protective action only from the individual protein. Specificity and activity of sera obtained from the animals after control challenge was confirmed in a blocking variant of ELISA.

Genetic construction used a eukaryotic vector that contained the E2 protein gene. Immunization of mice with the resulting DNA induced synthesis of specific antibodies, the titre of which increased considerably after additional single immunization with the E2 recombinant protein, expressed in *E. coli*. This demonstrated the effectiveness of animal priming by DNA vaccine, and the possibility of using the E2 recombinant protein in *E. coli* for booster vaccination.

1.   INTRODUCTION

Classical Swine Fever (CSF) is a highly contagious disease causing considerable economic damage to countries with developed pig farming. It occurs periodically in European countries, and has not been eradicated from Asia and South America. The Office Internationale des Epizooties (OIE) classes this as a List A disease. Besides domestic pigs, wild boars are natural susceptible animals, and the CSF causative agent – a Pestivirus that belongs

H. P. S. Makkar and G. J. Viljoen (eds.), Applications of Gene-Based Technologies for Improving Animal Production and Health in Developing Countries, 665-671.

to the family of Flaviviridae – constantly circulates in nature and infection can re-appear in previously safe regions.

Two main approaches exist for fighting CSF: wide diagnostics and eradication of any seropositive animals found, or vaccination with live virus vaccines. In the latter case, the problem exists of differentiating post-infection and post-vaccine immunities. Hence the development of different types of marked vaccines is urgent.

The main contribution to the formation of protective immune response among the structural CSF virus proteins is made by a glycoprotein: gp 51–55 (E2) (Konig *et al.*, 1995). The aim of this work was to obtain in *E. coli* different variants of recombinant protein, reproducing highly conserved epitopes of CSF virus E2 protein.

2. INITIAL COMMENTS

The characteristics of the protein structural organization according to the model of van Rijn *et al.* (1994) are that the N-terminal end of the molecule forms 2 structurally independent subunits, where all the antigenic determinants are grouped in four domains – A (highly conserved) and D, B and C. Each subunit carries epitopes that react with virus-neutralizing antibodies. Independent one from another, they can also display synergism. We hypothesized that production of only one of the subunits in prokaryotes should make the problem of recombinant protein refolding easier.

2.1 Reproduction of the E2 protein subunit in *E. coli*

As the subunit with A and D domains is highly conserved, we reproduced it in a recombinant way with almost the whole C-terminal part. A primer pair was developed on the basis of the RNA sequence of the Brescia virulent strain (M31768, GeneBank) and used for the amplification of the E2 gene fragment (846 bp length) of the virulent ShiMen strain. This is one of the strains that are used in Ukraine for challenge of swine during the control of immunogenic properties of anti-CSF vaccine. PCR material was hydrolysed in turn with EcoRI and SacI restriction endonucleases, and a 775 bp fragment was cloned into the plasmid vector pUC18 (Kirilenko *et al.*, 1996). The cloned fragment in pUC18-775 was sequenced and phylogenetic analysis was performed with different CSFV strains (Kirilenko *et al.*, 1997).

For additional verification of the specificity of the fragment obtained, pathological material from piglets killed by CSF was studied using PCR with the primers selected. It was confirmed that only in cases of the acute form of the disease, the size of amplification products do not differ from the ShiMen fragment, and amplified fragments readily hybridized in severe conditions with EcoRI-SacI – fragment of pUC18-775 recombinant plasmid.

For the expression of the protein, plasmids pGEX-2T (Pharmacia), pET24a (+) (Novagen) and pBlu2SKM (InforMax) were used. As the result of expression, 3 variants of the E2 recombinant protein have been obtained, with molecular masses of 34 kDa, 59 kDa and 107 kDa (Figure 1). Proteins were synthesized as inclusion bodies.

For correct polypeptide folding *in vivo*, glycosylation is often necessary, but the contribution of oligosaccharides in the formation of antigenic specificity of CSF virus is not yet clear. It has been shown (Konig *et al.*, 1995) that amount and positions of cystein residues that participate in the formation of A and D domains are critical for preservation of antigenic and immunogenic properties of CSF virus. In all the variants of expressed recombinant protein, these amino acids positions were conserved.

2.2   Antigenic properties of E2 recombinant protein

In hyperimmune blood sera of mice immunized with different forms of the recombinant protein, virus-neutralizing antibodies (in titre 1/5) were detected only in the case of the 34 kDa protein. Swine hyperimmune sera obtained against different forms of the recombinant protein were cross-studied in indirect ELISA. When using 34 kDa as an antigen, only 15% of sera obtained for other forms of protein reacted positively, while with 107 kDa protein, 100% of sera for other proteins showed positive reaction. Sera of swine vaccinated with live-vaccine ASV and LC reacted positively in indirect ELISA with 34 kDa and 59 kDa, but not 107 kDa proteins.

2.3   Protective properties of recombinant proteins

Protective properties of recombinant proteins were assayed in experiments involving challenge of double vaccinated 3-month-old piglets with the CSF virus virulent strain "Washington". To increase the efficiency of swine immunization with the CSF virus recombinant protein, various adjuvant types have been tested. The best results were received with the use of mantonide ISA-25 developed by Seppic (France). The experiment with challenge showed the existence of protective action only from the 34 kDa protein. After the determination of recombinant protein dose used for the vaccination, infected animals had no clinical signs of the disease, including rectal temperature fluctuation (Figure 2).

Specificity and activity of sera obtained from the animals after control challenge was confirmed in a blocking variant of ELISA with Classical Swine Fever Virus Antibody Test Kit (IDEXX Laboratories) (Figure 3). The test sample is positive if it gives a blocking percent of 40% or more, and negative 30% or less.

3.   MODEL FOR DNA VACCINE

Proceeding from the results obtained, the fragment of CSF virus E2 gene was used for the creation a DNA-vaccine model (Deryabin *et al.*, 2002). The genetic construction was developed on the basis of eukaryotic vector pTR-UF, which apart from the CSF virus E2 protein gene contained CMV promoter and AAV inverted terminal repeats. The resulting plasmid DNA was used for intramuscular immunization of 2-month-old female mice, line Balb/C. Three times DNA administration in liposomes or in buffer induced synthesis of specific antibodies. Additional single immunization with the E2 recombinant protein, expressed in *E. coli*, considerably increased the titre of specific antibodies (Figure 4). This made evident the effectiveness of animal priming by DNA-vaccine, and the possibility of using the E2 recombinant protein in *E. coli* for booster vaccination.

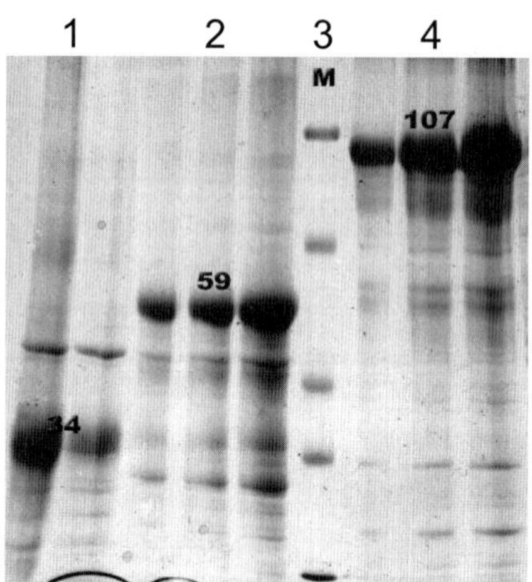

Figure 1. Three variants of the E2 recombinant protein with molecular masses 34 kDa (lane 1), 59 kDa (lane 2) and 107 kDa (lane 4). M = marker molecular masses (lane 3).

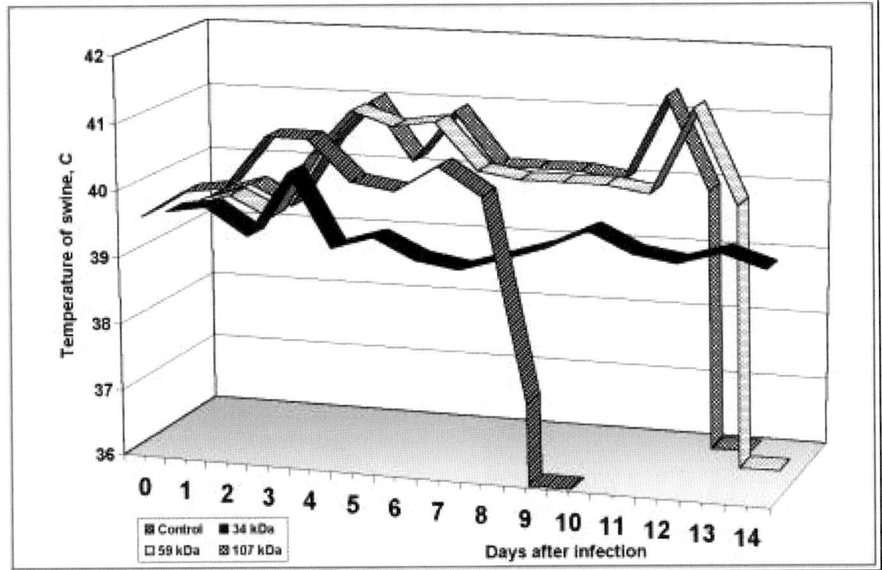

Figure 2. Dynamics of rectal temperature of 3-month-old pigs in challenge.

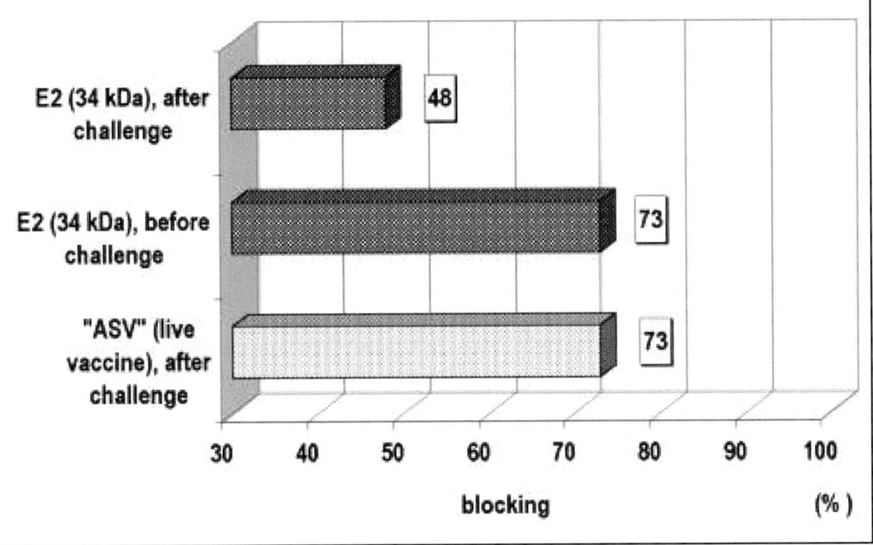

Figure 3. Sera activity obtained from the animals after control challenge in blocking variant of ELISA.

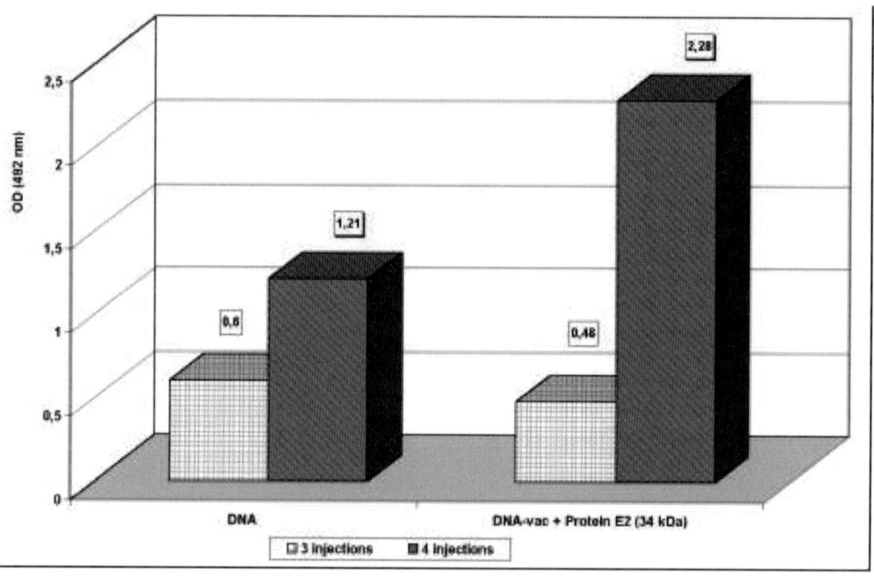

Figure 4. Results of mice immunization with DNA-vaccine against CSF virus.

4.   CONCLUSIONS

Among 3 forms of the CSF virus E2 recombinant protein expressed in *E. coli*, only one – an individual (non-fused) with molecular mass of 34 kDa – is able to protect 3-month-old piglets against CSF in control challenge with a virulent strain of the virus. The model of DNA-vaccine against CSF constructed provided, with intramuscular injection, synthesis of specific antibodies, the level of which increased significantly after the booster with E2 recombinant protein expressed in *E. coli*.

ACKNOWLEDGEMENTS

We would like to thank Dr Yu. S. Yakovlev for the opportunity to perform the studies on control challenge, and to Dr Z.R. Trozenko for help in analysis of the results.

REFERENCES

Deryabin, O.M.., Deryabina, O., Pokholenko, Y. & Tytok, T. 2002. Creation of the DNA-vaccine model against Classical Swine Fever. *Ukrainian Biochemical Journal,* **74**: 84.

Kirilenko, S.D., Deryabin, O.M., Kirilenko, O.L., Deryabina, O.G. & Busol, V.A. 1996. Cloning and super-expression of E1 structural protein gene fragment of Classical Swine Fever Virus ShiMen strain in *Escherichia coli. Biopolymers and Cell,* **12**: 93–99.

Kirilenko, S.D., Deryabin O.M., Deryabina O.G., Kirilenko O.L. & Troyanovsky B.M. 1997. Comparative phylogenetic analysis of coding part of the structural protein E1 genes of Shi-Min and other Classical Swine Fever Virus strains. *Journal of Cytology and Genetics,* **6**: 57–64.

Konig, M., Lengsfeld, T., Pauly, T., Stark, R. & Thiel, H.G. 1995. Classical Swine Fever Virus: Independent induction of protective immunity by two structural glycoproteins. *Journal of Virology,* **69**: 6479–6486.

van Rijn, P.A., Miedema, G.K.W., Wensvoort, G., van Gennip, H.G.P. & Moormann, R.J.M. 1994. Antigenic structure of envelope glycoprotein E1 of Hog Cholera Virus. *Journal of Virology,* **6**: 3934–3942.

DEVELOPMENT OF THERMOSTABLE PESTE DES PETITS RUMINANTS (PPR) VIRUS VACCINE AND ASSESSMENT OF MOLECULAR CHANGES IN THE F GENE

K.S. Palaniswami, A. Thangavelu and R. Velmurugan
Department of Veterinary Microbiology, Madras Veterinary College, Chennai – 600 007, India.

Abstract: Two Indian PPRV isolates were subjected to thermal hardening procedures to increase the proportion of temperature-resistant virions. Initial infectivity loss was compensated by titre increases on subsequent cell passages at 37°C. The immunogenicity of 'thermostable' viruses was assessed by virulent PPRV challenge and for safety by host animal inoculation and antibodies assessment. Vaccine viruses were not found using PCR on ocular and nasal swabs, although virus nucleic acid and antigens were demonstrated in spleen and lymph nodes by FAT and PCR. One vaccine strain (MIB187(T)) giving 100% protection (tested on only a few animals) was freeze dried and the minimum protective dose calculated. Changes in the virus genome after thermo-adaptation were examined using RT-PCR to amplify portions of the F gene, and three base changes were observed in the thermostable PPR strain (compared with the F gene sequence of the Nigerian PPRV strain). At room temperature, the titre and potency of the thermo-adapted vaccine remained constant up to one month at the $10^{5.5}$ TCID$_{50}$ level, and was $10^{4.5}$ TCID$_{50}$/100 µl after two months. Field trials with over 40 000 doses of the thermostable vaccine under various environmental conditions have given serum neutralization titres exceeding 2^3 and are assumed protective.

1.   INTRODUCTION

Peste des petits ruminants (PPR) is a highly contagious viral disease, causing high morbidity and mortality in small ruminants, equivalent to rinderpest in cattle. Hence small ruminants have to be protected against PPR

H. P. S. Makkar and G. J. Viljoen (eds.), Applications of Gene-Based Technologies for Improving Animal Production and Health in Developing Countries, 673-678.

to control losses, currently by using a homologous live attenuated PPR virus vaccine. However, as with many modified live lyophilized vaccines, the attenuated PPR vaccine must be stored under refrigeration to prevent thermal inactivation. The live PPR vaccine may not be stable enough for use in areas with inadequate cold storage facilities.

This cold chain requirement is the most costly component of PPR control and eradication programmes. Therefore a thermostable Vero-cell adapted PPR vaccine has to be developed to address several problems relating to the production and delivery of PPR vaccine.

2. MATERIALS AND METHODS

2.1 Heat selection of thermostable PPR virus

The PPR virus isolates were subjected to a process of temperature selection that was designed to increase the proportion of temperature-resistant virions, using the method described for Newcastle disease virus (V4 heat-resistant strain development by Ideris, Ibrahim and Spradbrow (1990); I_2 strain by Bensink and Spradbrow (1999); and the R2B strain by Dandawate *et al.* (2000)). The two vero-cell passaged (P_{30}) PPR isolates (MIB187, 197) were subjected to temperatures of 38°C, 39°C and 40°C for 15 min, 30 min and 1 hr, respectively, in succession by exposing the virus in a water bath. After each exposure a passage was done in vero cells. The temperature treatment was carried out for at least three cycles at every stage. The titre of thermostable virus population was assessed both before and after adaptation of PPR virus to higher temperature. The immunogenicity of these two thermostable PPR viruses was assessed by host animal inoculation along with virulent PPR virus challenge.

2.2 Molecular confirmation of F gene of thermostable PPR virus

Molecular changes in the virus genome of PPR virus after thermo-adaptation was studied by analysing the F gene by PCR and sequencing. Using reverse-transcription polymerase chain reaction (RT-PCR), a portion of the F gene of PPR viruses was amplified using the following primer sequences:

F_1 5′ ATCACAGTGTTAAAGCCTGTAGAGG 3′
F_2 5′ GAGACTGAGTTTGTGACCTACAAGC 3′
F_1A   5′ ATGCTCTGTCAGTGATAACC 3′

F$_2$A   5′ TTATGGACAGAAGGGACAAG 3′

A portion of F gene amplified PCR products were sequenced in an automatic sequencer. The nucleotides sequences of the F gene of thermostable PPR virus both before and after thermo-adoption were analysed.

2.3   Stability of thermostable PPR virus

The stability of thermostable PPR virus was compared with non-thermo-adapted PPR vaccine virus kept at both room temperature and incubated at 37°C for two months.

3.   RESULTS AND DISCUSSION

3.1   Selection of thermostable PPR virus

The vero-cell passaged (P$_{30}$) PPR isolates (MIB187 and MIB197) were subjected to high temperature (38°C, 39°C and 40°C for 15 min, 30 min and 1 hr respectively). The titres of thermostable virus population both before and after adaptation of PPR virus to higher temperatures are presented in Table 1.

The temperature treatment resulted in a massive reduction in virus titre, which indicated the elimination of the thermolabile population, leaving a thermostable virus population. The exposure of virus to elevated temperature to select thermostable population was done in sequential passages. The virus was adapted to 40°C, resulting in simultaneous attenuation as well as temperature adaptation. The resultant thermostable MIB 187T and MIB 197T were used for immunogenicity trials to choose the candidate strain for bulk production of the desired virus strain.

Table 1. Virus titres during thermo-adaptation (TCID$_{50}$/100 µl).

Temperature	38°C	39°C	40°C
MIB 187			
Before thermo-adaptation	$10^{5.8}$	$10^{2.9}$	$10^{1.28}$
After thermo-adaptation	$10^{1.23}$	$10^{1.8}$	$10^{3.2}$
MIB 197			
Before thermo-adaptation	$10^{7.3}$	$10^{2.5}$	$10^{1.23}$
After thermo-adaptation	$10^{1.8}$	$10^{2.3}$	$10^{2.58}$

The thermostable virus after the 70th passage (MIB 187T $10^{6.7}$ $TCID_{50}/100$ µl; MIB 197T $10^{5.7}$ $TCID_{50}/100$ µl) was tested for safety and immunogenicity. Both the candidate viruses were found to be safe. Virus excretion was studied by collecting ocular and nasal swabs and tested for virus presence by HA, PCR and neutralization, which were negative, thus confirming complete attenuation. The MIB187 T virus afforded 100% protection against challenge, while the MIB197 T virus gave only 70% protection. The MIB187 T virus was chosen as candidate virus for vaccine production.

3.2 Molecular confirmation of the F gene of thermostable PPR virus

The sequences of 242 nucleotides of the F gene of the thermostable PPR virus were analysed both before and after thermo-adaptation (Figure 1). Only three nucleotide base changes were noticed, at positions 1075, 1084 and 1085 in the thermostable PPR virus (MIB 187T strain) and one change was noticed at position 1068 with the other strain (MIB 197T). The nucleotide position numbers were identified by comparison with the F gene sequence of a Nigerian PPR virus strain available in GenBank. The result indicates that the stability of the F gene was not altered during thermo-adaptation in both the strains, though immunogenicity was maintained only in the MIB 187T strain. The nucleotide level changes ensured the genetic stability. Since there was no drastic change in the F gene there is no possibility of change in immunogenicity from that of the non-thermo-adapted vaccine strain. The thermostability might be due either to survival of a thermostable population alone, or to change in other proteins and in their regulatory sequences, such as promoters and non-coding areas. Hence the candidate strain can be used as a thermostable vaccine strain that maintains immunogenicity in vaccinated animals, confirming protection against challenge. Comparison of the deduced amino acids revealed 2 amino acid differences in MIB187 T, at positions 175 aspartic acid (hydrophilic to histidine (hydrophilic) and at 178 arginine (hydrophilic) to alanine (hydrophobic); and MIB197 T had one change at position 173 valine (hydrophobic) to isoleucine (hydrophobic) (Figure 2).

3.3 Stability of thermostable PPR virus

The tissue culture infectivity titre of thermostable PPR virus and non-thermo-adapted PPR vaccine virus kept at room temperature and at 37°C for two months are presented in Table 2. The results were promising for up to one month at 37°C. At room temperature, the aesthetic value and potency persists for up to two months. The titre of thermostable PPR vaccine virus kept at room temperature for one month was $10^{5.5}$ TCID$_{50}$, and after two months the titre was found to be $10^{4.5}$ TCID$_{50}$/100 μl. The stability of the freeze-dried thermostable PPR virus after 37°C incubation for 30 days has also proved the retention of the thermostable character.

The field trials with the experimental thermostable PPR vaccine have exceeded 40,000 vaccinations under variable environmental conditions and during disease outbreaks. The serum samples collected from the field trials have a serum neutralization titre exceeding 2^3 and were protective. The reconstituted vaccine could be used for 3 hours with 10^3 sheep infective dose.

916	MIB187	CAG	CTG	CTC	AGA	TAA	CTG	CAG	GAG	TCG	CCC	TTC	ATC	AAT	CAT	
		TGA	TGA	ACT	CCC	AAG	CAA	TTG	AGA	GTT	TAA	AAA	CCA	GTC	TTG	
		AGA	AGT	CGA	ATC	AGG	CAA	TAG	AAG	AAA	TCA	GAC	TTG	CAA	ATA	
		AGG	AGA	CCA	TAC	TGG	CAG	TAC	AGG	GCG	TCC	**AGG**	**ATT**	**ATA**	**TCAG**	1085
916	MIB187T	CAG	CTG	CTC	AGA	TAA	CTG	CAG	GAG	TCG	CCC	TTC	ATC	AAT	CAT	
		TGA	TGA	ACT	CCC	AAG	CAA	TTG	AGA	GTT	TAA	AAA	CCA	GTC	TTG	
		AGA	AGT	CGA	ATC	AGG	CAA	TAG	AAG	AAA	TCA	GAC	TTG	CAA	ATA	
		AGG	AGA	CCA	TAC	TGG	CAG	TAC	AGG	GCG	TCC	**AGC**	**ATT**	**ATA**	**TCGC**	1085
842	MIB197	ACT	CTG	ACA	CCT	GGG	CGT	AGA	ACT	CGC	CGT	TTT	GCT	GGA	GCT	GTT	CTG
		GCC	GGA	GTA	CAT	CAA	TCA	GCA	CTT	GGA	GTT	GCG	ACA	GCT	GCT	CAG	ATA
		ACT	GCA	GGA	GTC	GCC	CTT	TCG	AAT	CAG	TTG	ATG	AAC	TCC	CAA	GCA	ATT
		GAG	AGT	TTA	AAA	ACC	AGT	CTT	GAG	AAG	CAG	GGC	GCA	ATA	GAA	GAA	ATC
		AGA	CTT	GCA	AAT	AAG	GAG	ACC	ATA	CTG	GCA	GTA	**GTC**	CAG			1073
842	MIB197T	ACT	CTG	ACA	CCT	GGG	CGT	AGA	ACT	CGC	CGT	TTT	GCT	GGA	GCT	GTT	CTG
		GCC	GGA	GTA	CAT	CAA	TCA	GCA	CTT	GGA	GTT	GCG	ACA	GCT	GCT	CAG	ATA
		ACT	GCA	GGA	GTC	GCC	CTT	TCG	AAT	CAG	TTG	ATG	AAC	TCC	CAA	GCA	ATT
		GAG	AGT	TTA	AAA	ACC	AGT	CTT	GAG	AAG	CAG	GGC	GCA	ATA	GAA	GAA	ATC
		GCA	AAT	AAG	GAG	ACC	ATA	CTG	GCA	GTA	CAG	GGC	**ATC**	CAG			1073

Figure 1. Comparison of nucleotide sequences of PPR fusion gene (before and after thermo-adaptation)

1	M	N	S	Q	A	I	E	S	L	K	T	S	L	E	K	MIB 197
1	ATG	AAC	TCC	CAA	GCA	ATT	GAG	AGT	TTA	AAA	ACC	AGT	CTT	GAG	AAG	
1	M	N	S	Q	A	I	E	S	L	K	T	S	L	E	K	MIB 197T
1	ATG	AAC	TCC	CAA	GCA	ATT	GAG	AGT	TTA	AAA	ACC	AGT	CTT	GAG	AAG	
46	S	N	Q	A	I	E	E	I	R	L	A	N	K	E	T	MIB 197
46	TCG	AAT	CAG	GCA	ATA	GAA	GAA	ATC	AGA	CTT	GCA	AAT	AAG	GAG	ACC	
46	S	N	Q	A	I	E	E	I	R	L	A	N	K	E	T	MIB 197T
46	TCG	AAT	CAG	GCA	ATA	GAA	GAA	ATC	AGA	CTT	GCA	AAT	AAG	GAG	ACC	
91	I	L	A	V	Q	G	V	Q								MIB 197
91	ATA	CTG	GCA	GTA	CAG	GGC	GTC	CAG								
91	I	L	A	V	Q	G	I	Q								MIB 197T
91	ATA	CTG	GCA	GTA	CAG	GGC	ATC	CAG								
1	M	N	S	Q	A	I	E	S	L	K	T	S	L	E	K	MIB 187
1	ATG	AAC	TCC	CAA	GCA	ATT	GAG	AGT	TTA	AAA	ACC	AGT	CTT	GAG	AAG	
1	M	N	S	Q	A	I	E	S	L	K	T	S	L	E	K	MIB 187T
1	ATG	AAC	TCC	CAA	GCA	ATT	GAG	AGT	TTA	AAA	ACC	AGT	CTT	GAG	AAG	
46	S	N	Q	A	I	E	E	I	R	L	A	N	K	E	T	MIB 187
46	TCG	AAT	CAG	GCA	ATA	GAA	GAA	ATC	AGA	CTT	GCA	AAT	AAG	GAG	ACC	
46	S	N	Q	A	I	E	E	I	R	L	A	N	K	E	T	MIB 187T
46	TCG	AAT	CAG	GCA	ATA	GAA	GAA	ATC	AGA	CTT	GCA	AAT	AAG	GAG	ACC	
91	I	L	A	V	Q	G	V	Q	D	Y	I	R				MIB 187
91	ATA	CTG	GCA	GTA	CAG	GGC	GTC	CAG	GAT	TAT	ATC	AG				
91	I	L	A	V	Q	G	V	Q	H	Y	I	A				MIB 187T
91	ATA	CTG	GCA	GTA	CAG	GGC	GTC	CAG	CAT	TAT	ATC	GC				

Figure 2. Amino acid alignment.

Table 2. Comparison of stability of the thermostable and non-thermostable PPR virus at 37°C

	Weeks after exposure						
Virus titre (TCID$_{50}$)	1	2	3	4	5	6	7
Thermostable PPR vaccine	$10^{5.83}$	$10^{5.83}$	$10^{5.66}$	$10^{5.5}$	$10^{4.83}$	$10^{3.5}$	$10^{2.33}$
Non-thermostable PPR virus	$10^{4.83}$	$10^{4.66}$	$10^{2.33}$	$10^{1.5}$	$10^{0.83}$	–	–

4.　　CONCLUSION

Heat selection of PPR virus at 38°C, 39°C and 40°C and subsequent cultivation to proportionately increase the selected population at 37°C in vero cells was achieved. The resultant virus was thermostable at 37°C and at tropical room temperatures for 1 and 2 months, respectively. Such a thermostable vaccine virus could be transported and delivered without a cold chain. The costly cold chain requirement component of vaccination programmes would thus be avoided.

ACKNOWLEDGEMENTS

The work in our laboratory was supported by grants from the National Agricultural Technology Project, Indian Council of Agricultural Research (ICAR), New Delhi, India. The authors would also like to thank the authorities of Tamil Nadu Veterinary and Animal Sciences University, Chennai, India, for providing facilities.

REFERENCES

Ideris, A., Ibrahim, A.L. & Spradbrow, P.B. 1990. Vaccination of chickens against Newcastle disease with a food pellet vaccine. *Avian Pathology,* **19:** 371–384.

Bensink, Z. & Spradbrow, P. 1999. Newcastle disease virus strain I$_2$ – a prospective thermostable vaccine for use in developing countries. *Veterinary Microbiology,* **68**: 131–139.

Dandawate, P.D., Moghe, M.N., Tanwani, S.K. & Sharda, R. 2000. Isolation and characterisation of a thermostable derivative of R$_2$B strain of Newcastle disease virus. *Indian Veterinary Journal,* **77**: 649–651.

MOLECULAR CLONING OF A BANGLADESHI STRAIN OF VERY VIRULENT INFECTIOUS BURSAL DISEASE VIRUS OF CHICKENS AND ITS ADAPTATION IN TISSUE CULTURE BY SITE-DIRECTED MUTAGENESIS

M.R. Islam,[1] R. Raue[2] and H. Müller[2]

1. *Department of Pathology, Faculty of Veterinary Science, Bangladesh Agricultural University, Mymensingh 2202, Bangladesh.*
2. *Institute for Virology, Faculty of Veterinary Medicine, University of Leipzig, An den Tierkliniken 29, D-04103 Leipzig, Germany.*

Abstract: Full-length cDNA of both genome segments of a Bangladeshi strain of very virulent infectious bursal disease virus (BD 3/99) were cloned in plasmid vectors along with the T7 promoter tagged to the 5'-ends. Mutations were introduced in the cloned cDNA to bring about two amino acid exchanges (Q253H and A284T) in the capsid protein VP2. Transfection of primary chicken embryo fibroblast cells with RNA transcribed *in vitro* from the full-length cDNA resulted in the formation of mutant infectious virus particles that grow in tissue culture. The pathogenicity of this molecularly-cloned, tissue-culture-adapted virus (BD-3tc) was tested in commercial chickens. The parental wild-type strain, BD 3/99, was included for comparison. The subclinical course of the disease and delayed bursal atrophy in BD-3tc-inoculated birds suggested that these amino acid substitutions made BD-3tc partially attenuated.

1.   INTRODUCTION

Infectious bursal disease virus (IBDV) causes a highly contagious immunosuppressive and fatal disease known as infectious bursal disease (IBD), or Gumboro disease, in young chickens, usually at 3–6 weeks old

H. P. S. Makkar and G. J. Viljoen (eds.), Applications of Gene-Based Technologies for Improving Animal Production and Health in Developing Countries, 679-686.

(Lukert and Saif, 1997). IBDV is a dsRNA virus belonging to the family Birnaviridae, having a bisegmented genome (Müller, Scholtissek and Becht, 1979). There are two serotypes of IBDV (McFerran *et al.*, 1980). Serotype 2 is non-pathogenic. Virus strains belonging to serotype 1 are further categorized into classical virulent, antigenic variant and very virulent (vv) strains. Wild type IBDV, particularly vvIBDV, normally does not grow in cell culture. Yamaguchi and co-workers adapted two vvIBDV strains to replication in chicken embryo fibroblast (CEF) cell culture by repeated passages in chicken embryos and CEF cell culture, which resulted in marked attenuation (Yamaguchi *et al.*, 1996a). It was suggested that amino acid substitutions at positions 279 and 284 of the capsid protein VP2 could be responsible for tissue culture adaptation of vvIBDV (Yamaguchi *et al.*, 1996b; Lim *et al.*, 1999). It was, however, later demonstrated that amino acid exchanges at positions 253 and 284 resulted in tissue culture adaptation of a chimeric IBDV (Mundt, 1999) and a vvIBDV (UK661) (van Loon *et al.*, 2002). Here we report on the adaptation of another vvIBDV strain, BD 3/99 (Islam *et al.*, 2001), to CEF cell culture by site-directed mutagenesis resulting in amino acid substitutions at positions 253 and 284 of the VP2. We show that the rescued tissue culture adapted strain (BD-3tc) is partially attenuated for commercial chickens.

2.   MATERIALS AND METHODS

2.1   Construction of full-length clones

cDNAs of both segment A and segment B of a Bangladeshi strain of vvIBDV, BD 3/99 (Islam *et al.*, 2001), were synthesized in overlapping fragments by reverse transcription-polymerase chain reaction (RT-PCR) using appropriate primers (Table 1).

The overlapping regions of the fragments of segments A and B had unique *Hinc*II and *Bgl*II restriction sites, respectively. In the case of segment A, the *Eco*RI restriction site and T7 promoter sequence were tagged at the 5' end of the cDNA, while *Bsr*GI, the *Eco*RI and *Kpn*I restriction sites were added at the 3' end. In the case of segment B, the *Xba*I restriction site and T7 promoter sequence were tagged to the 5' end and *Sma*I, *Xba*I and *Xho*I restriction sites were added to the 3' end. A full-length cDNA clone of segment A was constructed in pBluescript vector utilizing *Eco*RI, *Hinc*II and *Kpn*I restriction sites. As pBluescript vector has its own T7 promoter sequence, the full-length cDNA of segment A was subsequently transferred from pBluescript to pUC 18 vector using the *Eco*RI restriction sites at both ends. At the same time a full-length cDNA clone of segment B was constructed in pQE 60 vector utilizing *Xba*I, *Bgl*II and *Xho*I restriction sites.

For sequencing segment A, six overlapping fragments covering the whole segment were amplified from the full-length cloned cDNA and from the original RNA by PCR and RT-PCR, respectively, using six pairs of primers (Islam, Zierenberg and Müller, 2001a). The PCR or RT-PCR products were sequenced from both ends. For sequencing segment B, the cloned full-length cDNA was sub-cloned in six fragments. Subclones obtained from two independently constructed full-length clones were sequenced (Islam, Zierenberg and Müller, 2001b).

2.2   Site-directed mutagenesis

A megaprimer PCR method (Barik, 1997) was used with primers (Table 2) for site-directed mutagenesis of nucleotides to introduce amino acid substitutions at position 253 and 284 of the VP2. The mutagenic primer and INCO-DC primer #3 were used for the synthesis of the megaprimer. Then the megaprimer and INCO-DC primer #4 were used for amplification of the BD 3/99 cDNA fragment containing the desired mutation. The fragment was digested with *Bsu*15 I and *Spe*I and inserted into the original full-length clone, replacing the corresponding region.

Table 1. Primers used in RT-PCR for amplification of cDNA to construct full-length clones of Segment A and Segment B of BD 3/99.

Primer	Primer sequence[1]	Characteristics
A1-s	5'-acc *gga att c*ta **ata cga ctc act ata GG**A TAC GAT CGG TCT GA-3'	*Eco*RI + T7 promoter + IBDV-A nt 1–17[2]
A1-as	5'-TGG GTG TCA TGG CGT CTT CCA CT-3'	IBDV-A nt 1847–1825
A2-s	5'-TGC AAT TGG GGA AGG TG-3'	IBDV-A nt 1543–1559
A2-as	5'-acg *cgg tac c*ga cag *gaa ttc* ggc tt*t gta ca*G GGG ACC CGC GAA CGG ATC CAA TT-3'	*Kpn*I + *Eco*RI + *Bsr*GI + IBDV-A nt 32613238
B1-s	5'-cta *gtc tag a*ta **ata cga ctc act ata GG**A TAC GAT GGG TCT GAC-3'	*Xba*I + T7 promoter + IBDV-B nt 1–18
B1-as	5'-GAT CCC GAG ATC TTT GCT GTA T-3'	IBDV-B nt 1860–1839
B2-s	5'-AGA CAG CGA GGA GTT CAA ATC AAT TGA GGA-3'	IBDV-B nt 1647–1677
B2-as	5'-acc *gct cga gtc tag acc c*GG GGG CCC CCG CAG GCG AAG-3'	*Xho*I + *Xba*I + *Sma*I + IBDV nt 2827–2808

NOTES: (1) IBDV-specific sequence is given in upper case; added restriction sites are in italics; and T7 promoter sequence is in bold face lower case. (2) The nt numbering is based on Bayliss *et al.* (1990) and modified according to Mundt and Müller (1995).

Table 2. Primers used for site-directed mutagenesis.

Primer	Sequence[1]	Nt position[2]
INCO-DC # 3-s	5'-AACAGCCAACATCAACG-3'	571–587
Mutagenic Q253H-as	5'-GTATAAGG<u>CCaTGG</u>ACGCTTG-3'	899–879
Mutagenic A284T-as	5'-TCAGT**GCCGG**tCGTTAGCCCA-3'	990–970
INCO-DC # 4-as	5'-GCTCGAAGTTGCTCACCC-3'	1247–1230

NOTES: (1) The mutated nucleotide is shown in lower case; the *Nco*I cleavage site is underlined; and the destroyed *Nae*I cleavage site is in bold. (2) The nt numbering is based on Bayliss *et al.* (1990) and modified according to Mundt and Müller (1995).

2.3 Rescue of infectious virus from cDNA

Plasmid vectors containing full-length cDNA of segments A and B of BD 3/99 were linearized with *Bsr*GI and *Sma*I enzymes, respectively. Capped cRNA was transcribed *in vitro* from the linearized plasmids under the control of T7 promoter and used for transfection of CEF cells, as described by Mundt and Vakharia (1996). Twenty-four hours later, a sample of transfected cells, grown on a coverslip, were examined by immunofluorescence for the expression of IBDV antigen. The remaining cells were examined regularly for the appearance of CPE. Seventy-two hours post-transfection the cells were frozen and thawed, and the culture supernatant was passaged in fresh CEF cells or inoculated in embryonated chicken eggs. RNA extracted from the culture supernatant or embryos was subjected to RT-PCR for VP2 hypervariable domain (Islam *et al.*, 2001), followed by appropriate restriction enzyme analysis.

2.4 Experimental infection

A total of 38 one-day-old Brown Nick chickens were commercially obtained and raised in relative isolation. At 5 weeks of age they were divided into three groups and placed in three separate houses. One group (n=20) was inoculated with the original wild-type parental strain BD 3/99 (BD-3wt) and another group (n=9) was inoculated with the molecularly cloned tissue culture adapted mutant (BD-3tc), while the third group (n=9) served as an uninfected control. Morbidity and mortality were recorded over a period of two weeks post-infection (p.i.). At days 3, 7 and 14 p.i., three birds from each group were selected randomly, killed and subjected to routine postmortem examination. The bursa/body weight ratio was determined on all three occasions.

3.   RESULTS

3.1   Rescue and molecular characterization of BD-3tc

The CEF cells co-transfected with capped RNA transcripts of cloned cDNA of both segments of BD-3/99 having two mutations in the VP2 gene to introduce amino acid substitutions at positions 253 (Q→H) and 284 (A→T) showed IBDV-specific immunofluorescence at 24 hr post-transfection. Extensive cytonecrotic changes were observed from 48 hr post-transfection, which probably represented the effects of transfection as well as virus-induced cytopathic effect (CPE). However, when the supernatant of the transfected cells was passaged in fresh CEF cells, a clear CPE was noticed by 48 hr p.i. RNA isolated from the CEF cells infected with the rescued virus (BD-3tc) tested positive in IBDV-specific RT-PCR. Cleavage of the PCR product by *Bsp*MI but not by *Sac*I indicated that the rescued virus was vvIBDV-like (Zierenberg, Raue and Müller, 2001). Furthermore, disappearance of the *Nae*I site and introduction of an *Nco*I site confirmed that the introduced mutations were retained by BD-3tc.

3.2   Pathogenicity of BD-3tc

In the BD-3wt infected group, 15 out of 20 birds (75%) developed clinical disease by day 3 or day 4 p.i., characterized by marked depression, ruffled feathers and diarrhoea. Two birds out of twenty (10%) died on day 5 or 6 p.i. BD-3tc inoculated birds showed no clinical signs. The birds of the uninfected control group remained healthy throughout the experiment.

The severity of gross lesions varied among the groups and among the birds within a group at a particular time of observation. Following inoculation with BD-3wt, marked oedematous swelling of the bursa of Fabricius, occasionally with congestion and haemorrhages, was seen at day 3 p.i., but the bursa appeared highly atrophied at day 7 and day 14 p.i. In BD-3tc groups, the bursa was slightly oedematous and swollen at day 3 p.i., almost normal at day 7 p.i., but atrophied at day 14 p.i. Petechial haemorrhages in the breast and thigh muscle were seen in some birds on all three occasions in the BD-3wt inoculated groups.

The bursa/body weight ratios, determined at days 3, 7 and 14 p.i., are presented in Figure 1. Bursa/body weight ratios of the BD-3wt and BD-3tc groups were slightly reduced at day 3 p.i., declining further at day 7 p.i., although the reduction was much greater in the BD-3wt group. However, at day 14 p.i., almost similar, markedly reduced bursa/body weight ratios were recorded in both BD-3wt and BD-3tc groups.

4.   DISCUSSION

BD-3wt was originally isolated from an outbreak of acute IBD in a poultry farm in Mymensingh, Bangladesh, in 1999 (BD 3/99), and the isolate was found to be molecularly and antigenically similar to other vvIBDV from Europe, Asia and Africa (Islam *et al.*, 2001). The original wild-type strain (BD-3wt) does not grow in tissue culture. However, BD-3tc, which was generated after site-directed mutagenesis of two amino acids (Q253H and A284T) in the capsid protein VP2, replicates in CEF cell culture. RT-PCR coupled with restriction enzyme analysis confirmed that the rescued virus was indeed derived from vvIBDV and that the desired mutations were present in the BD-3tc genome.

In the experimental infection trial, 75% of BD-3wt-inoculated birds developed clinical disease, but BD-3tc-inoculated birds remained apparently healthy. BD-3wt had been previously identified as vvIBDV on the basis of molecular and antigenic analyses (Islam *et al.*, 2001). Despite the development of typical clinical disease in the BD-3wt group, there was only 10% mortality. This was probably due to the presence of maternally derived antibodies at the time of inoculation.

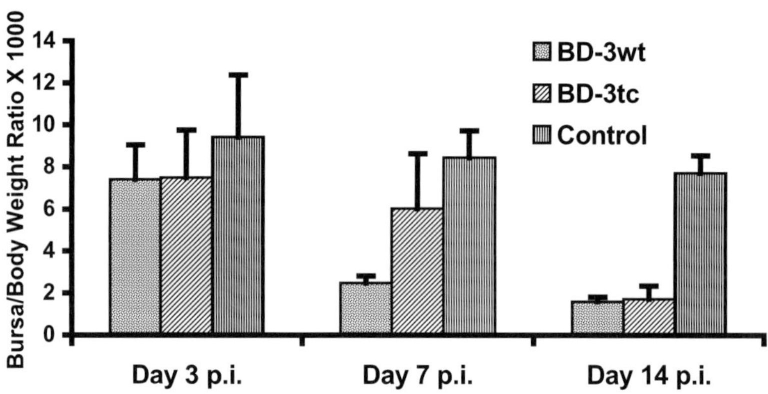

Figure 1. Bursa/body weight ratios in different groups determined at days 3, 7 and 14 p.i.

Infection with BD-3wt resulted in marked atrophy of the bursa of Fabricius at days 7 and 14 p.i. Although BD-3tc-infected birds remained clinically normal, they did develop bursal atrophy. However, the course of development of bursal atrophy was slower in the case of BD-3tc.

The results of the present study would indicate that tissue culture-adapted strain BD-3tc was partially attenuated for commercial chickens, as characterized by the subclinical course of the disease. Recently, van Loon *et al.* (2002) reported tissue culture adaptation of another vvIBDV strain (UK661) after amino acid exchanges at positions 253 and 284 of the VP2 and showed that this mutant (UK661tc) was also attenuated for specific-pathogen-free chickens. Delayed bursal atrophy in BD-3tc-infected birds, as observed in the present study, raised the question of whether the mutations of BD-3tc could revert during replication in the birds. This is now under investigation.

The findings of the present study show that a reverse genetics approach as applied in this study could be a useful tool for developing "tailor-made" molecularly or genetically engineered vaccine against IBDV.

ACKNOWLEDGEMENTS

M.R. Islam was supported by the Georg Forster Research Fellowship Programme of the Alexander von Humboldt Foundation, Germany. The investigations were further supported by the European Commission: INCO-DC Concerted Action Contract No. ERBIC18CT980330 – "Acute Infectious Bursal Disease in Poultry: Epidemiology of the Disease, Basis of Virulence and Improvement of Vaccination", and COST Action 839 – "Immunosuppressive Viral Diseases of Poultry.

REFERENCES

Barik, S. 1997. Mutagenesis and gene fusion by megaprimer PCR. pp. 173–182, *in:* B.A. White (ed). *Methods in Molecular Biology. PCR Cloning Protocols: From Molecular Cloning to Genetic Engineering*, Vol. 67. Totowa NJ: Humana Press Inc.,

Bayliss, C.D., Spies U., Shaw, K., Peters, R.W., Papageorgiou, A., Müller, H. & Boursnell, M.E. 1990. A comparison of the sequences of segment A of four infectious bursal disease virus strains and identification of a variable region in VP2. *Journal of General Virology,* **71**: 1303–1312.

Islam, M.R., Zierenberg, K., Eterradossi, N., Toquin, D., Rivallan, G. & Müller, H. 2001. Molecular and antigenic characterization of Bangladeshi isolates of infectious bursal disease virus demonstrates their similarities with recent European, Asian and African very virulent strains. *Journal of Veterinary Medicine, Series B,* **48**: 211–221.

Islam, M.R., Zierenberg, K. & Müller, H. 2001a. Regeneration and rescue of "very virulent" infectious bursal disease virus from cloned cDNA using cell culture and chicken embryo host systems in combination. *Bangladesh Journal of Microbiology,* **18**: 157–167.

Islam, M.R., Zierenberg, K. & Müller, H. 2001b. The genome segment B encoding the RNA-dependent RNA polymerase protein VP1 of very virulent infectious bursal disease virus (IBDV) is phylogenetically distinct from that of all other IBDV strains. *Archives of Virology,* **146**: 2481–2492.

Lim, B.L., Cao, Y., Yu, T. & Mo, C.W. 1999. Adaptation of very virulent infectious bursal disease virus to chicken embryonic fibroblasts by site-directed mutagenesis of residues 279 and 284 of viral coat protein VP2. *Journal of Virology,* **73**: 2854–2862.

Lukert, P.D. & Saif, Y.M. 1997. Infectious bursal disease. pp. 721–738, *in:* B.W. Calnek, H.J. Barnes, C.W. Beard, L.R. McDougald and Y.M. Saif (eds). *Diseases of Poultry.* 10th edition. Ames, Iowa: Iowa State University Press.

McFerran, J.B., McNulty, M.S., McKillop, E.R., Conner, T.J., McCracken, R.M., Collins, D.S. & Allan, G.M. 1980. Isolation and serological studies with infectious bursal disease viruses from fowl, turkey and duck: demonstration of a second serotype. *Avian Pathology,* **9**: 395–404.

Müller, H., Scholtissek, C. & Becht, H. 1979. The genome of infectious bursal disease virus consists of two segments of double-stranded RNA. *Journal of Virology,* **31**: 584–589.

Mundt, E. 1999. Tissue culture infectivity of different strains of infectious bursal disease virus is determined by distinct amino acids in VP2. *Journal of General Virology,* **80**: 2067–2076.

Mundt, E. & Müller, H. 1995. Complete nucleotide sequences of 5'- and 3'-noncoding regions of both genome segments of different strains of infectious bursal disease virus. *Virology,* **209**: 10–18.

Mundt, E. & Vakharia, V.N. 1996. Synthetic transcripts of double-stranded Birnavirus genome are infectious. *Proceedings of the National Academy of Sciences, USA,* **93**: 11131–11136.

van Loon, A.A.W.M., de Haas, N., Zeyda, I. & Mundt, E. 2002. Alteration of amino acids in VP2 of very virulent infectious bursal disease virus results in tissue culture adaptation and attenuation in chickens. *Journal of General Virology,* **83**: 121–129.

Yamaguchi, T., Kondo, T., Inoshima, Y., Ogawa, M., Miyoshi, M., Yanai, T., Masegi, T., Fukushi, H. & Hirai, K. 1996a. *In vitro* attenuation of highly virulent infectious bursal disease virus: some characteristics of attenuated strains. *Avian Diseases,* **40**: 501–509.

Yamaguchi, T., Ogawa, M., Inoshima, Y., Miyoshi, M., Fukushi H. & Hirai, K. 1996b. Identification of the sequence changes responsible for the attenuation of highly virulent infectious bursal disease virus. *Virology,* **223**: 219–223.

Zierenberg, K., Raue, R. & Müller, H. 2001. Rapid identification of "very virulent" strains of infectious bursal disease virus by reverse transcription-polymerase chain reaction combined with restriction enzyme analysis. *Avian Pathology,* **30**: 55–62.

TAPPING THE WORLD WIDE WEB FOR DESIGNING VACCINES FOR LIVESTOCK DISEASES

Custer C. Deocaris

Philippine Nuclear Research Institute, Commonwealth Avenue, Diliman, Quezon City, Philippines, but currently at Department of Chemistry and Biotechnology, University of Tokyo, 7-3-1, Hongo, Tokyo, Japan; and Gene Function Research Center, 1-1-1 Higashi, AIST Central 4, Tsukuba 305-8562, Japan.

Abstract: Post-genomic approaches in the development of new vaccines will fundamentally change how veterinarians prevent and treat diseases. One type of vaccine that has generated renewed interest is the subunit or synthetic vaccine, which has the advantage of rapid, safe and high-throughput production *via* chemical (as synthetic peptides) or recombinant approaches (as DNA, purified subunit or multigene vaccines). At the heart of such a vaccine are few but powerful epitopes that confer both the humoral and cell-mediated immune responses. Traditional biochemical assays have been used to map these epitopes; however, they are prohibitively labour and capital intensive. In contrast, *in silico* development of multivalent subunit vaccines is now possible through the availability of genomic information and the nascence of molecular immunoinformatics as a discipline. Algorithms are described in this paper to aid in identifying B and T cell epitopes for design of vaccines based on published available protein databases. From the mapped epitopes, synthetic mimotopes (or epitope-mimicking sequences) are concatenated using glycine bridges aimed at maintaining at least 90% of the secondary structures while minimizing steric hindrances between adjacent epitopes.

H. P. S. Makkar and G. J. Viljoen (eds.), Applications of Gene-Based Technologies for Improving Animal Production and Health in Developing Countries, 687-699.
© 2005 *IAEA. Printed in the Netherlands.*

1. INTRODUCTION

1.1 The field of molecular immunoinformatics and vaccine solutions

Molecular immunoinformatics is a pivotal post-genomic scientific discipline. Placed at the cutting edge of biomedical research, molecular immunoinformatics combines two disciplinary domains: bioinformatics, which is the application of informatics tools to biomolecules; and molecular immunology, which deals with understanding of the molecular pathways operative in an immune system.

Molecular immunoinformatics offers new opportunities and solutions to problems that beset traditional vaccine development. Ordinarily, immunodominant epitopes of vaccines are identified through techniques such as antigen mapping of overlapping peptide fragments, binding assays to the major histocompatibility complex (MHC), *in vitro* assays for evaluation of T cell response, cloning and expression of candidate fusion proteins, and *in vivo* testing of immunogenicity (Santonina *et al.*, 2002). To perform a preliminary antigenicity assay for a protein n amino acids long overlapped by decameric synthetic peptides, a laboratory has to synthesize $(n/10)$-1 peptides at a cost of US$ 250 per peptide. A regular epitope-driven vaccine development programme that uses a standard overlapping approach requires a budget of US$ 850,000 for peptide synthesis alone (De Groot *et al.*, 2001)! It would therefore be unrealistic for animal scientists from developing regions who need to develop their own repertoire of vaccines to engage in such R&D activity. However, the development of computational methods provides advantages in stringent and cost-effective screening for candidate peptides (epitopes) prior to assay development, engineering of the vaccine construct and testing in animal models. With the trend of transforming genomic databases into (non-proprietary) publicly accessible information, it is fortunate that many of these computational methods have also been made available for free on the Internet. In this sense, molecular immunoinformatics therefore *democratizes* the arena of vaccine development, giving more researchers a free hand to contribute to the growth of the technology. This paper describes a strategy in the rational molecular design of multivalent multigene vaccines, or synthetic mimotopes (epitope mimetics), using free Web-based tools – freewares, databases and servers that can be downloaded, mined and manipulated even by "non-experts" in the field.

1.2 The immune response to vaccination

Antigen recognition by T and B cells is central to the generation and regulation of an effective immune response against a pathogen *via* the cooperation of both humoral and cell-mediated immune pathways. Humoral response involves the production of antibodies in the blood and lymphatic fluids, where they bind specifically to foreign antigen. Cell-mediated immune response involves the activation and production of specialized cells – cytotoxic T lymphocytes (T_C), natural killer cells, etc. – that react with foreign antigens associated on the surface of infected cells. Both T and B cells are activated upon binding to specific antigens: T cells need to recognize antigens in the context of MHC molecules (or are MHC restricted), while B cells can bind to free antigens, but generally will also need helper T cell activation (T_H). Vaccination pre-empts pathogen invasion by inducing clonal selection of the immune cells and by transforming virgin cells into active and memory cells.

1.3 Strategy

To effectively induce both humoral and cell-mediated immune responses, our recombinant vaccine must carry immunodominant epitopes with the desired properties. First, it must carry candidate B cell epitopes capable of binding to the paratopes (antigen binding region of an immunoglobulin) rapidly (kinetic selection) and tightly (thermodynamic selection). Second, it must contain multiple epitopes presented by MHC I and II that will induce T_H1 and T_H2 pathways, respectively. Third, for the MHC I-peptide complex, the epitope must also bind to the T cell receptor (TCR) of T_H1 and T_C cells. Fourth, synthetic epitopes must not contain internal (immuno)proteosomal domains that would otherwise lead to vaccine deterioration once engulfed and processed by circulating antigen-presenting cells (APCs) or other cell types, such as myocytes during intramuscular injection. Fifth, the B cell epitopes must assume proper folding similar to its native state upon isolation and stringing with other peptides within the multigene construct. Distortions in the secondary structure (SS) may be prevented by prudent addition of tri-glycine bridges to minimize inter-epitope forces. A procedure for the proper determination of optimal combinations and (orientational) permutations of these epitopes is also employed to maximize secondary structural retention (SSR). Sixth, the final vaccine construct must not contain homology with the host animal's genome, to prevent autoimmunity. And lastly, a genetic adjuvant may optionally be attached to boost vaccine potency. The algorithm for engineering of multigene peptide vaccines is presented in Figure 1.

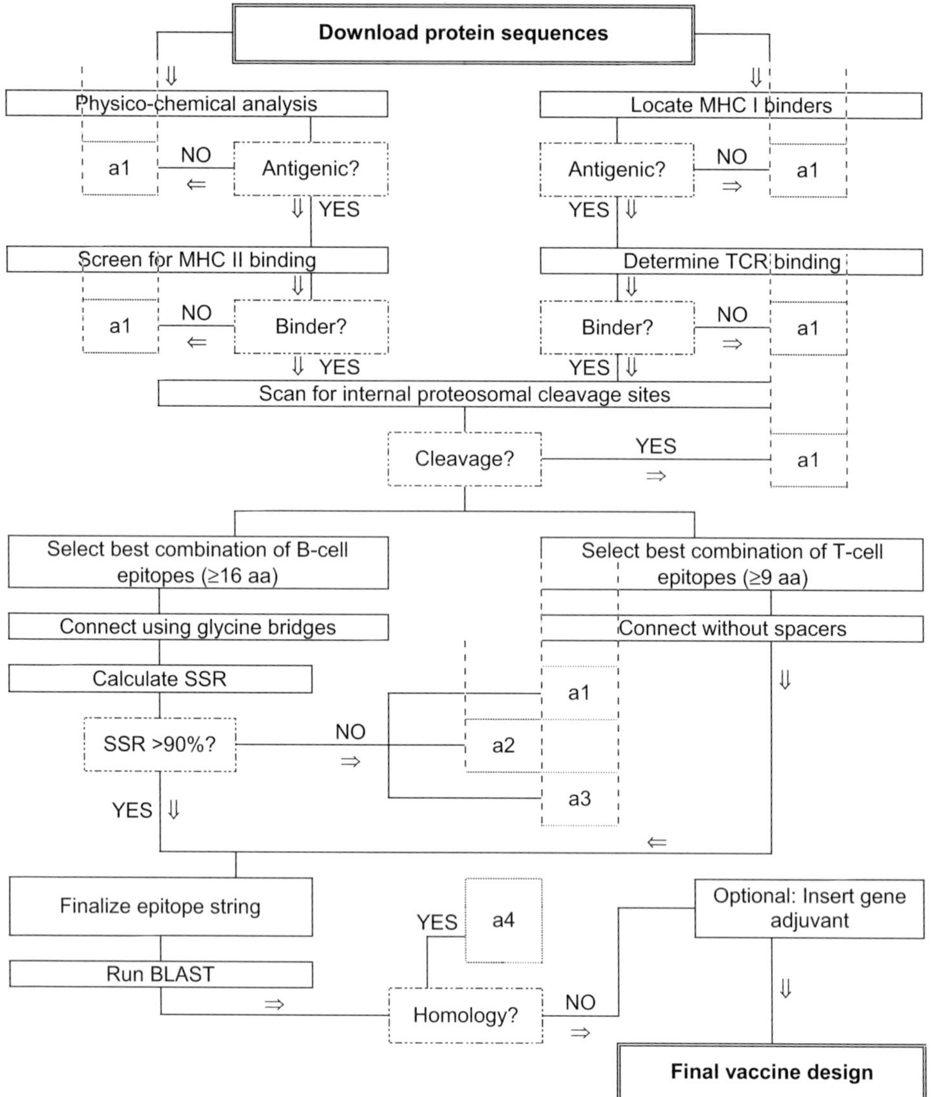

Figure 1. Algorithm of a rational epitope-driven vaccine design from the protein sequence of a pathogen.

KEY: a1 = discard; a2 = perform permutation of epitopes to improve SSR values; a3 = add amino acid sequence from N- or C-, or both, terminal of the epitope to improve SSR values and epitope stability; a4 = discard "self-reactive" epitopes and select other candidate epitopes.

2. COMPUTATIONAL METHODS

2.1 Protein databases

Bioinformatics tools designed to search large databases, such as the complete sequence file of a pathogen, are in the public domain, including for DNA (GenBank; the European Molecular Biology Laboratory; DDBJ), protein, (GenPept; Protein Information Resource; SWISS-PROT) and structure (Protein Data Bank) (Table 1). As of September 2003, GenBank, for example, contained more than 2,700 entries on immunogenic protein sequences. Most of these immunogenic protein data sources comprise data on origin, description, structural and functional properties, as well as links to other major public databases and literature.

2.2 Mapping regions that stimulate B cells

Predictive physico-chemical features of B cell epitopes have been derived from X-ray crystallographic studies on antigen-antibody complexes. Davies and Riechmann (1996) showed that interaction of protein antigens with antibodies contacts at least 16 residues of the antigen lysozyme.

Table 1. Web-based protein sequence databases.

Name	Brief Description and URL	Source
PIR	Classification-driven and rule-based approach in protein annotation with 283,000 entries. Integrates its three major databases, PSD, NREF and iPROclass. http://pir.georgetown.edu/pir/..	Wu *et al.*, 2003
SWISS-PROT	Standardized nomenclature, evidence-attributed and offers information on the life science aspects of 113,470 sequences. http://www.expasy.org/sprot/	Boeckmann *et al.*, 2003
ENTREZ	Comprehensive sequence database that contains publicly available DNA sequences for more than 119,000 different organisms. Integrates with the EMBL in the UK and the DDB in Japan. http://www3.ncbi.nlm.nih.gov/	Benson *et al.*, 2003
OWL	Composite and non-redundant protein database. Employs strict redundancy criteria and performs efficient similarity searches. http://www.bioinf.man.ac.uk/dbbrowser/OWL/	Attwood *et al.*, 2003
GenPept	Provided in a format similar to that of GenBank but is not an official release from the NCBI-GenBank. Current release was 74,752 entries as of Sept. 2002. ftp://ftp.ncifcrf.gov/pub/genpept/	Wu *et al.*, 2003

Because most, if not all, antigenic sites are located within surface-exposed regions of a protein, antigenic determinants can be localized using scales for hydropathy, inverted hydrophobicity scales, flexibility and local residue accessibility (Jameson and Wolf, 1988). The Web-based servers in Table 2 are useful, although downstream calculations and graphical analyses using an Excel worksheet are still required.

Apart from direct antibody-antigen interaction that mediates the clonal selection of a primed B cell that displays membrane-bound IgGs, the T_H2 pathway is also involved. Briefly, an antigen is engulfed by an antigen-presenting cell (APC) and processed in the lysosomal/endosomal compartments. Peptides of 8–10 residues are loaded in $\alpha\beta$-Ii components of MHC II and displayed on the surface of the cell (Rotzschke and Falk, 1994). Crystal structures of the MHC II with a bound peptide has shown tight interaction in the P6/P7 region of the peptide binding site and the conformation of peptides is somewhat similar to those bound with MHC II, and the possibility of predicting peptide alignment in the binding MHC II groove have been deduced (Zhu *et al.*, 2003) (Table 3).

Table 2. Web-based protein B cell antigenicity prediction servers.

Name	Brief description and URL	Source
EMBOSS	Open source software for protein motif identification. ftp://ftp.uk.embnet.org/pub/EMBOSS/	Rice, Longden and Bleasby, 2000
EpiPlot 1.0	Using joint-prediction, predicts B as well as T cell epitopes based on amphiphilcity, flexibility, hydrophilicity and antigenicity profiles. http://genamics.com/cgi-bin/genamics/software/	Menendez-Arias and Rodriguez, 1990
AntheProt	Fully interactive software package for protein sequence of secondary structure predictions, sites and function detection, physico-chemical profiles. http://antheprot-pbil.ibcp.fr/Documentation_antheprot.html	Geourjon, Deleage and Roux, 1991
Expasy	Uses 50 scales defining various physico-chemical and conformational parameters of amino acids. http://us.expasy.org/cgi-bin/protscale.pl	Gasteiger *et al.*, 2003
JaMBW 1.1	Java-based tool for antigenicity plotting. http://hometown.aol.com/lucatoldo/myhomepage/JaMBW/	Toldo, 1997
MTF	Prediction based on experimentally known B cell epitopes. Method has reported accuracy of 75%. http://mif.dfci.harvard.edu/Tools/antigenic.html	Kolaskar and Tongaonkar, 1990
Bcipep	Searches a database of more than 2,500 B cell epitopes compiled from literature. http://www.imtech.res.in/raghava/bcipep/	

Table 3. Web-based prediction and database servers for MHC I and II-binding.

Name	Brief description and URL	References
BIMAS	Matrix-based prediction of MHC I-binding. http://bimas.dcrt.nih.gov/molbio/hla_bind/	Parker, Bednarek and Coligan, 1994
SYFPEITHI	Prediction of MHC I and II binding based on 4000 sequence entries. http://syfpeithi.bmi-heidelberg.com/	Rammensee *et al.*, 1999
PREDEPP	Computational threading-based prediction of MHC I-binders. http://bioinfo.md.huji.ac.il/marg/Teppred/mhc-bind/	Schueler-Furnan *et al.*, 2000
EpiPredict	Prediction of MHC II binding from combinatorial peptide libraries. http://www.epipredict.de/	Jung *et al.*, 2001
Predict	Prediction of MHC I/II and TAP binding using 2-D databases. http://sdmc.lit.org.sg:8080/predict-demo/	Yu *et al.*, 2002
MHCPred	MHC I and II prediction tool from 3-D structure-activity experiments. http://www.jenner.ac.uk/MHCPred	Doytchinova and Flower, 2002
NetMHC	Prediction of HLA-A2 binding using artificial neural networks. http://www.cbs.dtu.dk/services/NetMHC/	Corbet *et al.*, 2003
MAPPP	Combined ORF, MHC I binding and proteosomal cleavage analysis. http://www.mpiib-berlin.mpg.de/MAPPP/	
RANKPEP	Position Specific Scoring Matrix analysis of MHC I and II binding. http://mif.dfci.harvard.edu/Tools/rankpep.html	Reche, Glutting and Reinherz, 2002
MHCPEP	Database for MHC binding peptides, last updated June 1998. http://wehih.wehi.edu.au/mhcpep/	Brusic *et al.*, 1998
ProPredI	On-line service for MHC I binding analysis using matrices for 47 MHC I and (immuno)proteosomes. http://www.imtech.res.in/raghava/propred1/	Singh and Raghava, 2003
ProPred	Matrix-based prediction of MHC II-binding. http://www.imtech.res.in/raghava/propred/	Singh and Raghava, 2001
Peptide Select	Downloadable program for physico-chemical analysis of peptides. http://cbi.swmed.edu/computation/cbu/PepSel/	
Vaccine Screening	Downloadable program that predicts MHC binding using matrices for 48 MHC I, 51 human MHC II, 7 mouse MHC I and 6 mouse MHC II. http://cbi.swmed.edu/computation/cbu/VacScr/	
IPD - MHC	Centralized MHC sequence repository from different species. http://www.ebi.ac.uk/ipd/mhc/	
TEPITOPE	Downloadable software for prediction of MHC II ligands of 25 alleles. http://www.vaccinome.com/pages/600800/	Bian, Reidhaar-Olson and Hammer, 2003

SVMHC	Uses Support Vector Machine for prediction of MHC I binding. http://www.sbc.su.se/svmhc/new.cgi	Dönnes and Elofsson, 2002
MMBPred	Predicts mutated MHC I-binding based on matrices for 47 MHC I alleles. http://www.imtech.res.in/raghava/mmbpred/	
NHLAPred	Predicts MHC ligands and CTL epitopes of 67 MHC alleles using neural networks and quantitative matrices. http://www.imtech.res.in/raghava/nhlapred/	

Table 4. Web-based prediction of proteosome and immunoproteosome cleavage sites.

Name	Brief description and URL	References
PAProC	Uses matrix-based analysis of cleavage data from yeast and human 20S proteosomes. http://paproc.de	Nussbaum *et al.*, 2001
FRAGPREDICT	Statistical inferencing on amino acid motifs and a kinetic model of the 20S proteasome. http://www.mpiib-berlin.mpg.de/MAPPP/expertquery.html	Holzhütter and Kloetzel, 2000
NetChop	Employs artificial neural networks for 1 110 MHC I ligands. http://www.cbs.dtu.dk/services/NetChop	Kesmir *et al.*, 2002

2.3 Mapping peptides involved in the T_C pathway

There are three main facets in mapping T_C stimulators: proteosomal cleavage, MHC I peptide anchorage and TCR recognition. First, precursor proteins from intracellular pathogens are delivered to the protein degrading machinery (proteasome) and are fragmented at the N-terminal regions by cytosolic peptidases. Some of these peptides are translocated to the endoplasmic reticulum *via* the TAP protein (transporter-associated antigen presentation). The TAP-peptide complex is then trimmed by the aminopeptidase associated with antigen processing (ERAAP) for binding to MHC I. Only 0.05% of peptides transported to ER bind to MHC I (Matsumura *et al.*, 1992). Unbound (empty) MHCs are highly unstable and are eventually degraded and recycled. MHC I contains a groove that binds strongly to 8–10 amino acid-long peptides for interaction with the TCRs of CD8+ T_C (Yang, 2003). The data about all of these crucial steps is now present in such amount that a computer can make generalized rules. Tables 3 and 4 show the several Web-based tools for predicting MHC-binding and proteosomal cleavage sites.

2.4 Vaccine engineering: epitope stringing, conformational retention analysis and super-antigen motifs

From this point, a vaccine can be designed by stringing identified immunostimulatory proteins like a string of beads. This method of basing vaccine design on terse protein regions of pathogens was recently termed as "reverse immunogenetics" or "epitope-driven vaccine design" (Hill and Davenport, 1996). Other complex vaccines, such as the multivalent minigene vaccine, containing T_H, T_C and B cell epitopes adjacent to each other have also been constructed and tested (Ling-Ling and Whitton, 1997).

An epitope-based vaccine construct contains epitopes (for either or both T and B cells) inserted consecutively within the construct. For T cell epitopes, the proper arrangement of epitopes is not very important and this may be done without intervening spacer amino acids. However, adjacent B cell epitopes would interfere with one another and affect conformation-specific binding with antibodies. In this case, triglycine bridges, for relieving steric hindrances, and the placement of individual epitopes, permutations (arrangement of epitopes) and combinations (choice of candidate B cell epitopes) are essential in improving the design. Calculation of secondary structure retentions (SSRs) [Eq. 1] using available structure prediction tools (Table 5) will show the behaviour of these epitopes when strung together.

$$\% \text{ SSR} = \frac{\text{No. of correct SS of native protein}}{\text{No. of residues in the peptide}} \times 100 \qquad \text{Eq. 1}$$

Various genetic adjuvants, e.g. streptococcal pyrogenic superantigen C (Sigmundsdottir, Gudjonsson and Valdimarsson, 2003), *Peptostreptococcus magnus* protein L (Genovese *et al.*, 2003), A and B subunits of cholera toxin and *E. coli* heat-labile enterotoxin (LT) (Arrington *et al.*, 2002), interleukin-12 (IL-12) (Lynch, Briles and Metzger, 2003), or interleukin-2 (Oh *et al.*, 2003), can be added to the vaccine construct to bolster its efficacy. These molecular designs may be delivered as synthetic peptides, recombinant proteins or DNA vaccines. Such new forms of recombinant vaccines, in contrast to live- and attenuated-types combine safety, purity and calculated efficacy. Veterinarians can also avoid exposing the vaccinate to the pathogen. Difficulty in long-term storage and transport of these biochemicals is not a problem as synthetic vaccines are stable at ambient temperatures and can make do without a stringent cold-chain requirement, a major problem of vaccine delivery in agricultural regions of developing countries.

Table 4. Web-based secondary structure prediction.

Name	Brief Description	Source
PPS	Structural predictions with >70% accuracy using artificial neural networks, hidden-Markov and prediction-based threading. http://www.embl-heidelberg.de/predictprotein/	Wu *et al.*, 2003
NPS	Joint-prediction analysis through a consensus of eight algorithms with final predicted accuracy of >70%. http://npsa-pbil.ibcp.fr/	Combet *et al.*, 2000
PROF	Cascades different structural classifiers using neural networks and linear discrimination and achieves an accuracy of 76.7%. http://www.aber.ac.uk/compsci/Research/bio/dss/prof/	Ouali and King, 2000
PHDsec	Uses profile network method, called PHDsec, rated at 72.1% average accuracy for globular proteins. http://www.public.iastate.edu/~pedro/pprotein_query.html	Rost, Sander and Schneider, 1994
PSA	Predicts protein 2- and 3-D structures based on probabilistic discrete state-space models and optimal filtering and smoothing. http://bmerc-www.bu.edu/psa/	Stultz, White and Smith, 1993

3. CONCLUDING REMARKS

Recombinant technology is relatively new to veterinary medicine. These recombinants can be tailored to contain multiple genetic inserts for use in animals. Epitope-driven multigene synthetic vaccines are still under development and the ability to apply these rapid advancements in the field through effective utilization of molecular immunoinformatic tools will be advantageous for scientists in developing countries. At present, there are but a few licensed recombinant vaccines in veterinary medicine, such as those against Lyme disease, pseudorabies, rabies, canine distemper, Newcastle disease, and a strain of avian influenza (Van Kampen, 2001). At the helm of these advances in vaccine development, the field is open for new players!

REFERENCES

Arrington, J., Braun, R.P., Dong, L., Fuller, D.H., Macklin, M.D., Umlauf, S.W., Wagner, S.J., Wu, M.S., Payne, L.G. & Haynes, J.R. 2002. Plasmid vectors encoding cholera toxin or the heat-labile enterotoxin from *Escherichia coli* are strong adjuvants for DNA vaccines. *Journal of Virology,* **76**: 4536–4546.

Attwood, T.K., Bradley, P., Flower, D.R., Gaulton, A., Maudling, N., Mitchell, A.L., Moulton, G., Nordle, A., Paine, K., Taylor, P., Uddin, A. & Zygouri, C. 2003. PRINTS and its automatic supplement, prePRINTS. *Nucleic Acids Research,* **31**: 400–402.

Benson, D.A., Karsch-Mizrachi, I., Lipman, D.J., Ostell, J. & Wheeler, D.L. 2003. Genbank. *Nucleic Acids Research,* **31**: 23–27.

Bian, H.J., Reidhaar-Olson, J.F. & Hammer, J. 2003. The use of bioinformatics for identifying class II-restricted T-cell epitopes. *Methods,* **29**: 299–309.

Boeckmann, B., Bairoch, A., Apweiler, R., Blatter, M.C., Estreicher, A., Gasteiger, E., Martin, M.J., Michoud, K., O'Donovan, C., Phan, I., Pilbout, S. & Schneider, M. 2003. The SWISS-PROT protein knowledgebase and its supplement TrEMBL in 2003. *Nucleic Acids Research,* **31**: 365–370.

Brusic, V., Rudy, G., Kyne, A.P. & Harrison, L.C. 1998. MHCPEP, a database of MHC-binding peptides: update 1997. *Nucleic Acids Research,* **26**: 368–371.

Combet, C., Blanchet, C., Geourjon, C. & Deléage, G. 2000. NPS@: Network Protein Sequence Analysis. *Trends in Biochemical Sciences,* **25**: 147–150.

Corbet, S., Nielsen, H.V., Vinner, L., Lauemoller, S., Therrien, D., Tang, S., Kronborg, G., Mathiesen, L., Chaplin, P., Brunak, S., Buus, S. & Fomsgaard, A. 2003. Optimization and immune recognition of multiple novel conserved HLA-A2, human immunodeficiency virus type 1-specific CTL epitopes. *Journal of General Virology,* **84**: 2409–2421.

Davies, J. & Riechmann, L. 1996. Affinity improvement of single antibody VH domains: residues in all three hypervariable regions affect antigen binding. *Immunotechnology,* **2**: 169–179.

De Groot, A.S., Sainth-Aubin, C., Bosma, A., Sbai, H., Rayner, J. & Martin, W. 2001. Rapid determination of HLA B*07 ligands from the West Nile Virus NY99 genome. *Emerging Infectious Diseases,* **7**: 706–713.

Dönnes, P. & Elofsson, A. 2002. Prediction of MHC class I binding peptides, using SVMHC. *BMC Bioinformatics,* **3**: Article no. 25.

Doytchinova, I.A. & Flower, D.R. 2002. Physicochemical explanation of peptide binding to HLA-A*0201 MHC: a 3-dimensional quantitative structure-activity relationship study. *Proteins,* **48**: 505–518.

Gasteiger, E., Gattiker, A., Hoogland, C., Ivanyi, I., Appel, R.D. & Bairoch, A. 2003. ExPASy: The proteomics server for in-depth protein knowledge and analysis. *Nucleic Acids Research,* **31**: 3784–3788.

Geourjon, C., Deleage, G. & Roux, B. 1991. ANTHEPROT: An interactive graphic software for analyzing protein structures from sequences. *Journal of Molecular Graphics,* **9**: 188–190.

Genovese, A., Borgia, G., Bjorck, L., Petraroli, A., de Paulis, A., Piazza, M. & Marone, G. 2003. Immunoglobulin superantigen protein L induces IL-4 and IL-13 secretion from human Fc epsilon RI+ cells through interaction with the kappa light chains of IgE. *Journal of Immunology,* **170**: 1854–1861.

Hill, A.V. & Davenport, M.P. 1996. Reverse immunogenetics: from HLA-disease associations to vaccine candidates. *Molecular Medicine Today,* **2**: 38–45.

Holzhütter, H.G. & Kloetzel, P.M. 2000. A kinetic model of vertebrate 20S proteasome accounting for the generation of major proteolytic fragments from oligomeric peptide substrates. *Biophysical Journal,* **79**: 1196–1205.

Kesmir, C., Nussbaum, A., Schild, H., Detours, V. & Brunak, S. 2002. Prediction of proteasome cleavage motifs by neural networks. *Protein Engineering,* **15**: 287–296.

Kolaskar, A.S. & Tongaonkar, P.C. 1990. A semi-empirical method for prediction of antigenic determinants on protein antigens. *FEBS Letters,* **276**: 172–174.

Jameson, B.A. & Wolf, H. 1988. The antigenic index: a novel algorithm for predicting antigenic determinants. *Computer Applications in the Biosciences,* **4**: 181–186.

Jung, G., Fleckenstein, B., von der Mulbe, F., Wessels, J., Neithammer, D. & Wiesmuller, K.H. 2001. From combinatorial libraries to MHC ligand motifs, T cell superagonists and antagonists. *Biologicals,* **19**: 179–181.

Lavoie, T.B., Drohan, W.N. & Smith-Gill, S.J. 1992. Experimental analysis by site-directed mutagenesis of somatic mutation effects on affinity and fine specificity in antibodies specific for lysozyme. *Journal of Immunology,* **148**: 503–513.

Ling-Ling, A. & Whitton, J.L. 1997. A multivalent minigene vaccine, containing B cell, CTL and Th from several microbes, induces appropriate responses *in vivo* and confers protection against more than one pathogen. *Journal of Virology,* **71**: 2292–2302.

Lynch, J.M., Briles, D.E. & Metzger, D.W. 2003. Increased protection against pneumococcal disease by mucosal administration of conjugate vaccine plus interleukin-12. *Infection and Immunity,* **71**: 4780–4788.

Matsumura, M., Fremont, D.H., Peterson, P.A. & Wilson, I.A. 1992. Emerging principles for the recognition of peptide antigens by MHC class I molecules. *Science,* **257**: 927–934.

McSparron, H., Blythe, M.J., Zygouri, C., Doytchinova, I.A. & Flower, D.R. 2003. JenPep: a novel computational information resource for immunobiology and vaccinology. *Journal of Chemical Information and Computer Sciences,* **43**: 1276–1287.

Menendez-Arias, L. & Rodriguez, R. 1990. A BASIC microcomputer program for prediction of B and T cell epitopes in proteins. *Computer Applications in the Biosciences,* **6**: 101–105.

Nussbaum, A.K., Kuttler, C., Hadeler, K.P., Rammensee, H.G. & Schild, H. 2001. PAProC: A prediction algorithm for proteasomal cleavages available on the WWW. *Immunogenetics,* **53**: 87–94.

Oh, Y.K., Park, J.S., Kang, M.J., Ko, J.J., Kim, J.M. & Kim, C.K. 2003. Enhanced adjuvanticity of interleukin-2 plasmid DNA administered in polyethylenimine complexes. *Vaccine,* **21**: 2837–2843.

Ouali, M. & King, R.D. 2000. Cascaded multiple classifiers for secondary structure prediction. *Protein Science,* **9**: 1162–1176.

Parker, K.C., Bednarek, M.A. & Coligan, J.E. 1994. Scheme for ranking HLA-A2 binding peptides based on independent binding of individual peptide side chains. *Journal of Immunology,* **152**: 163–175.

Rammensee, H., Bachmann, J., Emmerich, N.P., Bachor, O.A. & Stevanovix, S. 1999. SYFPEITHI: database for MHC ligands and peptide motifs. *Immunogenetics,* **50**: 213–219.

Reche, P.A., Glutting, J.P. & Reinherz, E.L. 2002. Prediction of MHC I binding peptides using Profile Motifs. *Human Immunology,* **63**: 701–709.

Rice, P., Longden, I. & Bleasby, A. 2000. EMBOSS: The European Molecular Biology Open Software Suite. *Trends in Genetics,* **16**: 276–277.

Rost, B., Sander, C. & Schneider, R. 1994. PHD – an automatic mail server for protein secondary structure prediction. *Computer Applications in the Biosciences,* **10**: 53–60.

Rotzschke, O. & Falk, K. 1994. Origin, structure and motifs of naturally processed MHC class II ligands. *Current Opinions in Immunology,* **6**: 45–51.

Santona, A., Carta, F., Fraghi, P. & Turrini, F. 2002. Mapping antigenic sites of an immunodominant surface lipoprotein of *Mycoplasma agalactiae,* AvgC, with the use of synthetic peptides. *Infection and Immunity,* **70**: 171–176.

Schueler-Furman, O., Altuvia, Y., Sette, A. & Margalit, H. 2000. Structure-based prediction of binding peptides to MHC class I molecules: application to a broad range of MHC alleles. *Protein Science,* **9**: 1838–1846.

Sigmundsdottir, H., Gudjonsson, J.E. & Valdimarsson, H. 2003. Interleukin-12 alone can not enhance the expression of the cutaneous lymphocyte associated antigen (CLA) by superantigen-stimulated T lymphocytes. *Clinical and Experimental Immunology,* **132**: 430–435.

Singh, H. & Raghava, G.P. 2003. ProPred1: prediction of promiscuous MHC Class-I binding sites. *Bioinformatics,* **19**: 1009–1014.

Singh, H. & Raghava, G.P. 2001. ProPred: prediction of HLA-DR binding sites. *Bioinformatics.* **17**: 1236–1237.

Stultz, C.M., White, J.V. & Smith, T.F. 1993. Structural analysis based on state-space modeling. *Protein Science,* **2**: 305–314.

Toldo, L.I. 1997. JaMBW 1.1: Java-based molecular biologists' workbench. *Computer Applications in the Biosciences,* **13**: 475–476.

Van Kampen, K.R. 2001. Recombinant vaccine technology in veterinary medicine. *Veterinary Clinics of North America Small Animal Practice,* **31**: 535–538.

Wu, C.H., Yeh, L.S.L., Huang, H., Arminski, L., Castro-Alvear, J., Chen, Y., Hu, Z.Z., Ledley, R.S., Kourtesis, P., Suzek, B.E., Vinayaka, C.R., Zhang, J. & Barker, W.C. 2003. The Protein Information Resource. *Nucleic Acids Research*, 31: 345–347.

Yang, Y. 2003. Generation of major histocompatibility complex class I antigens. *Microbes and Infection,* 5: 39–47.

Yu, K., Pterovsky, N., Schonbach, C., Koh, J.Y. & Brusic, V. 2002. Methods for prediction of peptide binding to MHC molecules: a comparative study. *Molecular Medicine,* **8**: 137–148.

Zhu, Y., Rudensky, A.Y., Corper, A.L., Teyton, L. & Wilson, I.A. 2003. Crystal structure of MHC class II I-Ab in complex with a human CLIP peptide: prediction of an I-Ab peptide-binding motif. *Journal of Molecular Biology,* **326**: 1157–1174.

COMPLEMENTING NUCLEAR TECHNIQUES WITH DNA VACCINE TECHNOLOGIES FOR IMPROVING ANIMAL HEALTH

Jenne Liza V. Relucio,[1] Maria Elena K. Dacanay,[1] Anna Cristina S. Maligalig,[1] Rizalina G. Osorio,[2] Erwin A. Ramos,[1] Ameurfina D. Santos,[1] Celia Aurora T. Torres-Villanueva[1]* and Custer C. Deocaris[2, 3]

1. National Institute of Molecular Biology and Biotechnology, University of the Philippines, Diliman, QuezonCity, Philippines.
2. Philippine Nuclear Research Institute, Commonwealth Avenue, Diliman, Quezon City, Philippines.
3. Currently with Department of Chemistry and Biotechnology, University of Tokyo, 7-3-1, Hongo, Tokyo, Japan; and Gene Function Research Center, 1-1-1 Higashi, AIST Central 4, Tsukuba 305-8562, Japan.
* Corresponding author.

Abstract   The use of nuclear methods can enhance several features of DNA vaccines in protecting livestock against pathogens. While DNA vaccines already have several advantages over their traditional predecessors (e.g. cheap production, stability over a wide range of temperature, amenability to genetic manipulation, and no risk of reversion to pathogenicity), conventional gene delivery systems make immunization of livestock and aquaculture populations tedious. For this reason, we are developing radiation-synthesized intelligent delivery systems for DNA vaccines. We encapsulated a reporter construct pCMV•SPORT-β-gal in radiation-synthesized κ-carrageenan-polyvinylpyr-rolidone microspheres IP20 (for stomach release) and IP18 (for intestinal release). The DNA-loaded polymers were orally administered to *Oreochromis niloticus* (black Nile tilapia), and whole organs were stained with X-gal to observe β-galactosidase activity. Intense staining was observed in the stomach regions with IP20, while minimal staining was observed with IP18. The gills, in contrast, did not express β-galactosidase activity. Our results show evidence of the successful gene delivery capabilities of radiation-synthesized microspheres.

701

H. P. S. Makkar and G. J. Viljoen (eds.), Applications of Gene-Based Technologies for Improving Animal Production and Health in Developing Countries, 701-708.
© 2005 *IAEA. Printed in the Netherlands.*

When monitoring the progress of an animal's immune response after DNA immunization, non-invasive and sensitive methods are preferred. We also evaluated chicken egg-yolk polyclonal antibody response (chIgY) after direct intramuscular inoculation of the Hepatitis B Surface antigen expression vector pRc/CMV-HBs(S). Radioimmunoassay (RIA) was done to maximize sensitivity for determining antibody levels. Polyclonal antibody titres were observed to have increased after six weeks. Results of the RIA using the chIgY were comparable to that of immunized sera. Our findings indicate that chIgY could offer a cheaper and more animal-friendly antibody source and could be derived with the advantage of epitope specificity through DNA vaccination.

1.   INTRODUCTION

In the Philippines, applications of nuclear technologies revolve mostly around medicine and agriculture. Radiation studies for the identification of biological markers and disease diagnosis are viable research options that local clinicians undertake, while agriculturists focus on inducing beneficial mutations in crops and monitoring the fitness of livestock. One novel use of nuclear techniques that bridge both medical and agricultural realms involves enhancing current immunization technologies, particularly DNA vaccines.

1.1   Immunization with DNA vaccines

DNA vaccines consist of immunogen-encoding sequences that are cloned into plasmid expression vectors, which are then amplified in cultures of transformed bacteria (Robinson and Torres, 1997). After purification, the vaccine is administered to the host, where it is able to produce immunizing proteins. The nature of these vaccines gives them advantages over their traditional predecessors (i.e. killed or live attenuated micro-organisms and their subunits). Their desirable features include strong B-cell and T-cell responses, better thermal stability than proteins, cheaper production, and minimal or non-existent risk of accidental release and infection (Kanellos *et al.*, 1999). Orally delivered DNA vaccines, in particular, confer additional advantages such as ease of administration, applicability for mass vaccination, and initiation of mucosal immunity (Roy *et al.*, 1999).

1.2   Evaluating DNA vaccine efficacy

The progression of an animal model immune response is observed to establish the efficacy of a DNA vaccine and determine whether effective protection from disease has been successfully achieved. Measuring antibody titres requires the use of a precise and accurate quantitative technique. Radio-immunoassay (RIA) is considered to be the best method for such purpose, as it exhibits the highest level of sensitivity compared with other immunoassays. Ordinarily, invasive serological techniques, such as blood extraction, are used to collect samples for analysis of antibody levels in the serum. In the medical and agricultural setting, however, non-invasive methods of sampling should be prioritized to avoid inflicting unnecessary stress on patients and animals.

2. INTELLIGENT POLYMERS

Previous studies have shown the feasibility of oral delivery of DNA vaccines using microparticles, chitosan and gelatin nanospheres (Eldridge *et al.*, 1990; Leong *et al.*, 1998). While the principle behind DNA vaccines relies on their successful delivery and eventual expression in the cell, gene delivery systems must not only deliver DNA efficiently – they should also be applicable to both basic research and clinical environments (Luo and Saltzman, 2000).

One DNA delivery technique that is gaining popularity is the use of intelligent polymers. These hydrogel-based substances have been tailored to respond (i.e. swell, shrink, take up or release DNA) to changes in environmental conditions (e.g. pH, ionic strength, temperature) (Savas and Güven, 2001). Examples of such are carrageenan-polyvinylpyrrolidone (PVP)-based polymers, which respond to varying degrees based on the concentration of the polymer components and environmental variables.

A common additive in food, cosmetics and pharmaceutical formulations, carrageenan has three different forms. One particular type – kappa carrageenan (κC) – produces strong hydrogels with PVP when subjected to gamma irradiation (van de Velde *et al.*, 2002). This property allows κC polymers to be utilized as delivery vehicles for DNA vaccines.

2.1 Suitability of κC-PVP for gene delivery

We used two types of κC-PVP intelligent polymers – IP 20 and IP18. IP20 swells at pH 8.0 and shrinks at pH 2.0, effectively releasing DNA in the stomach. IP18, in contrast, swells at pH 2.0 and shrinks at pH 8.0, thus mediating intestinal release of DNA. After graft co-polymerization of PVP (Kishida Co., Ltd.) and κ-carrageenan (Shemberg, Philippines) upon exposure to ^{60}Co γ-irradiation, the resultant polymers were then made to encapsulate the β-galactosidase expression construct pCMV•SPORT-β-gal (GIBCO-BRL, USA). The DNA-polymer complexes were buccally

 J.L.V. Relucio et al.

administered to *Oreochromis niloticus* (black tilapia). After five days, whole organs (gills, stomach, spleen, and small intestine) were stained with X-Gal (GIBCO-BRL, USA) to verify β-galactosidase expression, with positive expression indicated by a blue reaction. Results are summarized in Table 1.

For the IP20 treatment group (stomach release), all organs except the gills were heavily stained (Figures 1 to 3). Note that although the IP20 polymer is designed for gastric release, the small intestine from the IP20 treatment group was also heavily stained. Plasmid DNA may have easily reached the small intestine considering that the stomachs of generally herbivorous fish, such as *O. niloticus*, exhibit short-term gastric storages and evacuate their contents at a relatively fast rate (Jobling, 1995).

For the IP18 treatment group (intestinal release), the small intestine was heavily stained while the stomach was only faintly stained. The gills also remained negative for β-galactosidase (Figures 1 to 3).

Similar results between the IP18 and the negative control groups indicate that the staining of organs from the IP18 treatment group is possibly due to endogenous enzyme activity in the gut epithelial cells, rather than an effect of the delivered DNA.

Table 1. Summary of results of X-Gal staining of whole organs.

	Gills	Stomach	Small intestine
IP20 stomach release.	–	***	***
IP18 intestinal release.	–	*	**
(+) control, DNA in chitosan.	–	**	***
(-) control, polymer with Tris chloride-EDTA buffer.	–	*	**

KEY: – = no stain; * = faintly stained; ** = moderately stained; *** = heavily stained.

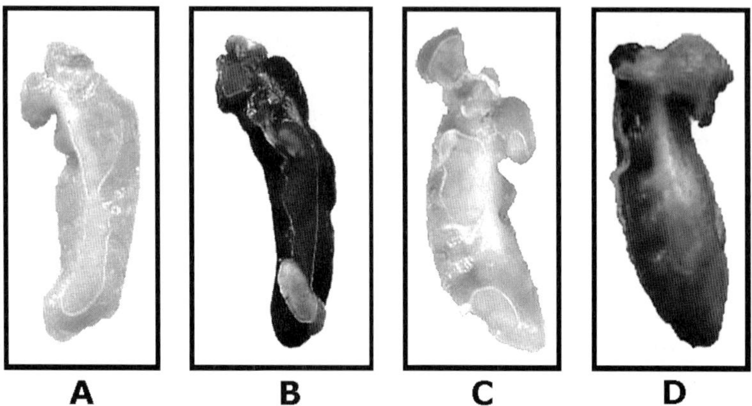

Figure 1. Stomachs of *O. niloticus* treated with DNA in κC-PVP polymers. (A) Negative control; (B) IP20; (C) IP18; and (D) positive control (DNA in chitosan nanospheres).

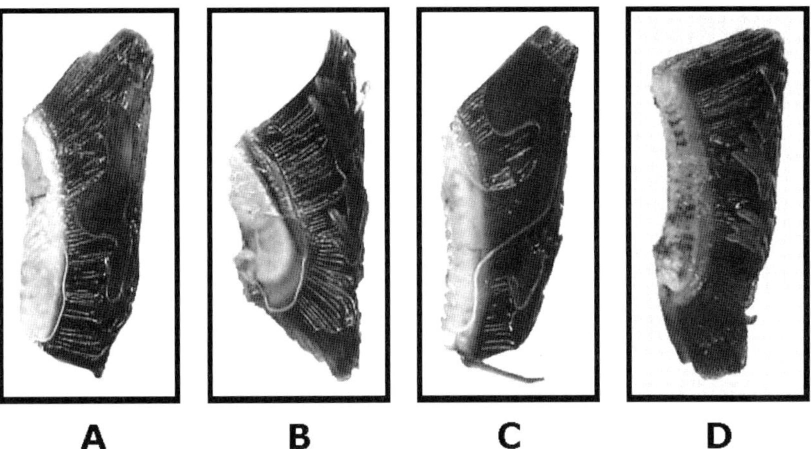

Figure 2. Gills of *O. niloticus* treated with DNA in κC-PVP polymers. (A) Negative control; (B) IP20; (C) IP18; and (D) positive control (DNA in chitosan nanospheres). Note absence of staining in all samples.

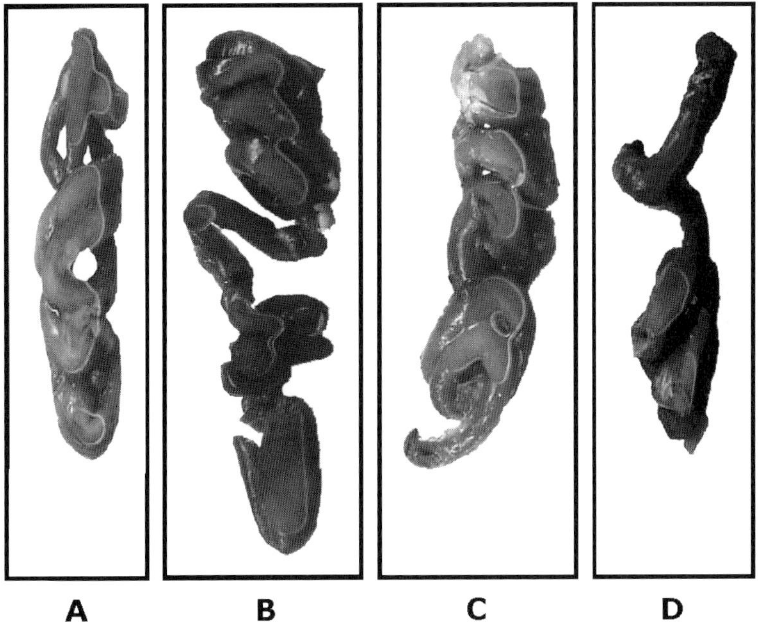

Figure 3. Small intestines of *O. niloticus* treated with DNA in κC-PVP polymers. (A) Negative control; (B) IP20; (C) IP18; and (D) positive control (DNA in chitosan nanospheres).

3. RADIO-IMMUNOASSAY WITH CHICKEN EGG YOLK ANTIBODY FROM A DNA VACCINE

Among agricultural livestock, the chicken has been able to produce antibodies when other animals give poor or no response. Physiologically, the egg yolks contain the same concentration of antibodies in the chicken sera and can be used as a convenient source of polyclonal antibodies (pAbs) (Schade, Halatsch and Henklein, 1991). Chickens produce immune responses to a vast variety of antigens, developing monoclonal (mAbs) and pAbs against numerous proteins (Michael *et al.*, 1998). To verify the efficacy of DNA vaccines, non-invasive sampling, followed by a radioimmunoassay (RIA), may be done to check antibody levels.

3.1 Radio-immunoassay for detecting maternal (yolk) response

The Hepatitis B Surface Antigen (HbsAg) expression vector pRc/CMV-HBs(S) (generously provided by Dr Ann Bakken, of Aldevron, Inc. [North Dakota, USA]) was directly inoculated into the thigh muscle of a laying Kabir hen, For eight weeks (two weeks before vaccine injection and six weeks after), eggs were collected and the chicken egg-yolk polyclonal antibody response (chIgY) was monitored for increases in antibody titres comparable to that of immunized sera. RIA using the isotope iodine-125 (^{125}I) was performed on the egg yolk samples to determine the amount of anti-HbsAg antibodies formed.

From the graphical representation of the RIA results (Figure 4), it could be seen that the yolk samples' counts per minute (cpm) increased over time as the hen developed an immune response against the antigen.

However, only the samples collected after the sixth week of immunization tested positive for true HbsAg antibodies. This lag time may be explained by the fact that it takes four to six weeks to develop a specific response against an antigen.

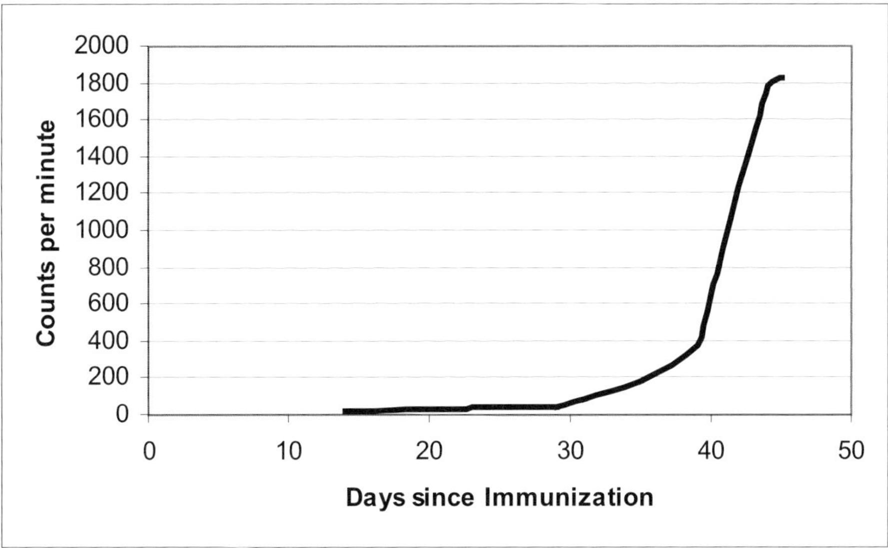

Figure 4. Graphical representation of counts per minute versus days post-immunization. The counts per minute of the samples consistently increased with the number of days after immunization with pRc/CMV-HBs(S). The cut-off value is 1525.44, wherein samples with higher counts per minute values are considered positive for anti-HbsAg antibodies.

4.   CONCLUSIONS

From these two studies, it has been demonstrated that nuclear techniques could indeed complement the emerging technology of genetic immunization. Intelligent polymers synthesized via ^{60}Co γ-irradiation can evidently be used as delivery systems for DNA vaccines in cultured fish populations. These polymers can apparently be modified to take up and release their DNA load under certain environmental conditions, thus having favourable implications for targeted treatment. For post-immunization analyses, RIA can provide high sensitivity and accuracy in monitoring the progress of the desired immune response. Coupled with non-invasive sampling methods, the egg yolk antibody based-RIA may soon be the method of choice for clinicians and agriculturists alike for monitoring their subjects' antibody titres.

REFERENCES

Eldridge, J.H., Hammond, C.J., Meulbroek, J.A., Staas, J.K., Gilley, R.M. & Tice, T.C. 1990. Controlled vaccine release in the gut-associated lymphoid tissues: orally administered biodegradable microspheres target the Peyer's patches. *Journal of Controlled Release,* **11**: 205–214.

Jobling, M. 1995. *Environmental Biology of Fishes.* London: Chapman & Hall.

Kanellos, T., Sylvester, I.D., Ambali, A.G., Howard, C.R. & Russell, P.H. 1999. The safety and longevity of DNA vaccines for fish. *Immunology,* **96**: 307–313.

Leong, K.W., Mao, H.Q., Truong-Lee, V.L., Roy, K., Walsh, S.M. & August, J.T. 1998. DNA-polycation nanospheres as gene delivery vehicles. *Journal of Controlled Release,* **53**: 183–193.

Luo, D. & Saltzman, W.M. 2000. Synthetic DNA delivery systems. *Nature Biotechnology,* **18**: 33–37.

Michael, N., Accavitti, M.A., Masteller, E. & Thompson, C.B. 1998. The antigen-binding characteristics of mAbs derived from *in vivo* priming of avian B cells. *Proceedings of the National Academy of Sciences, USA,* **95**: 1166–1171.

Robinson, H.L. & Torres, C.A.T. 1997. DNA vaccines. *Seminars in Immunology,* **9**: 271–283.

Roy, K., Mao, H.Q., Huang, S.K. & Leong, K.W. 1999. Oral gene delivery with chitosan-DNA nanoparticles generates immunologic protection in a murine model of peanut allergy. *Nature Medicine,* **5**: 387–391.

Savas, H. & Güven, O. 2001. Investigation of active substance release from poly(ethylene oxide) hydrogels. *International Journal of Pharmacology,* **224**: 151–158.

Schade, R., Halatsch, R. & Henklein, P. 1991. Polyclonal IgY antibodies from chicken egg yolk – an alternative to the production of mammalian IgG type antibodies in rabbits. *ATLA,* **19**: 403–419.

van de Velde, F., Knutsen, S.H., Usov, A.I., Rollema, H.S. & Cerezo, A.S. 2002. ^{1}H and ^{13}C high resolution NMR spectroscopy of carageenans: applications in research and industry. *Trends in Food Science and Technology,* **13**: 73–92.

USE OF DNA FROM MILK TANK FOR DIAGNOSIS AND TYPING OF BOVINE LEUKAEMIA VIRUS

R. Felmer, J. Zuñiga, M. Recabal and H. Floody
Instituto de Investigaciones Agropecuarias INIA-Carillanca. Unidad de Biotecnología, Casilla 58-D, Temuco, Chile

Abstract: With the aim of achieving a better understanding of the epidemiology of Bovine leukaemia virus (BLV) infection, we investigated the suitability of milk tank samples for effecting molecular epidemiology studies of BLV in a southern area of Chile. As part of a serological survey for BLV antibodies carried out in 280 herds, we selected 33 strong positive samples, from which DNA was isolated to perform a BLV-specific nested PCR. Using RFLP analysis, all 33 PCR products could be assigned to the known Australian or the Belgium subgroups. A phylogenetic tree resulting from the comparison of these sequences demonstrates the relations and differences among and within the subgroups.

1.   INTRODUCTION

Bovine leukaemia virus (BLV) is an exogenous retrovirus distributed worldwide. Most BLV-infected cattle remain clinically normal during their lifetime, with only 1–5% eventually developing lymphosarcoma. However, up to one-third of BLV-infected cattle may develop persistent lymphocytosis (PL), a polyclonal expansion of infected B-lymphocytes (Radostits, Blood and Gay, 1995). Eradication and control of the disease is based on early diagnostic and segregation of carriers. Serological tests, such as agar gel immunodifusion (AGID) and enzyme linked immunosorbent assay (ELISA), are the tests of choice for international trade because they are easy to perform, economically viable and capable of testing a large number of animals (Evermann and Jackson, 1997). However, serological tests do not

H. P. S. Makkar and G. J. Viljoen (eds.), Applications of Gene-Based Technologies for Improving Animal Production and Health in Developing Countries, 709-714.
© 2005 *IAEA. Printed in the Netherlands.*

discriminate maternal passive antibodies from an active immune response, and do not provide evidence of the infection in its early stages. Direct methods, in contrast, allow a reliable diagnosis in the initial stages of the disease or in newborn calves, avoiding the false-positive reactions caused by the presence of colostral antibodies and false-negative reactions caused by the immaturity of the immune system (Fechner *et al.*, 1996). Within the direct methods, polymerase chain reaction (PCR) is being increasingly used for the diagnosis of infection diseases. Because of its high sensitivity, PCR can be used in eradication campaigns where prevalence is low and it allows differentiation of seropositive calves that have acquired immunity by passive transfer of immunoglobulins through the colostrum. Lastly, PCR may detect variants of BLV provirus, which may not react with standard antibodies in routine serological tests (Beier, Blankenstein and Fechner, 1998).

2.   MATERIAL AND METHODS

ELISA. A commercial ELISA test (SVANOVIR, Sweden) was used in the serological survey, according to the manufacturer's instructions.

DNA isolation. DNA samples were isolated from the leukocyte fraction of 60 ml of milk by using a protocol developed in-house, consisting of lysis of leucocytes followed by a salt precipitation of the DNA.

Nested PCR for BLV. The PCR was performed in a total volume of 30 µl according to Beier *et al.* (2001). The first-round conditions were: denaturation at 94°C for 5 min, followed by 40 amplification cycles at 94°C for 30 sec; 57°C annealing for 30 sec; and 72°C for 60 sec; followed by a final extension step at 72°C for 10 min (using as 5'-tctgtgccaagtctcccagata-3' and 5'-aacaacaacctctgggaagggt-3' as external primers). For the second round of PCR, 3 µl of product was taken from the first amplification and re-amplified under the same conditions as before, except that the annealing temperature was increased to 68°C (using 5'-cccacaagggcggcgccggtt-3' and 5'-gcgaggccgggtccagagctgg-3' as internal primers). In order to visualize the PCR products, 10 µl of each sample was run on 1.5% agarose gel and stained with ethidium bromide.

Restriction fragment length polymorphism (RFLP) analysis. 10 µl of PCR products were digested with 5 U of *Bcl*I, *Pvu*II or *Bam*HI (Boehriner Mannheim, Germany) to assign the samples to the known BLV subgroups.

DNA sequencing and phylogenetic analysis. Purified PCR products, corresponding to dairy herds distributed in the Region de la Araucanía (southern Chile), were cloned into pGEM T-Easy vector (Promega) for sequencing analysis. Sequences of the clones were determined using the fluorescent dye deoxy-terminator cycle sequencing kit (Perkin Elmer Cetus,

Inc) and the ABI Prism 377 DNA Sequencer (Applied Biosystem). Eight of the sequences obtained were aligned to 21 published sequences (GenBank accession numbers K02120; K02251; D00647; S83530; M35238; M35242; M35239; M35240; AF399703; AF399704; AF067081; AY078387; AF547184; AF503581; AY151262; U87872; AF399702; NC_001414; AF033818; AY185360; and AF257515). Alignment of the sequences was established using ClustalX (version 1.8) software, and phylogenetic and molecular evolutionary analyses were conducted using MEGA version 2.1 (Sudhir *et al.*, 2001). The length of the compared sequences was 444 bp.

## 3.	RESULTS AND DISCUSSION

We assessed the suitability of samples from milk tanks as a source of DNA for quick diagnosis and typing of BLV in dairy herds. As part of a serological survey for BLV antibodies carried out in 280 herds, we selected 33 strong positive samples, from which DNA was isolated to perform a BLV-specific nested PCR. All 33 samples gave a positive PCR reaction, and further RFLP analysis enabled the PCR products to be assigned to 2 of the 3 known subgroups. The restriction pattern corresponding to the Belgium subgroup (Rice *et al.*, 1984) was observed in 17 samples, while the pattern of the Australian subgroup (Coulston *et al.*, 1990) was observed in 16 samples. No restriction pattern typical of the Japanese subgroup (Sagata *et al.*, 1985) was found (Figure 1 and Table 1).

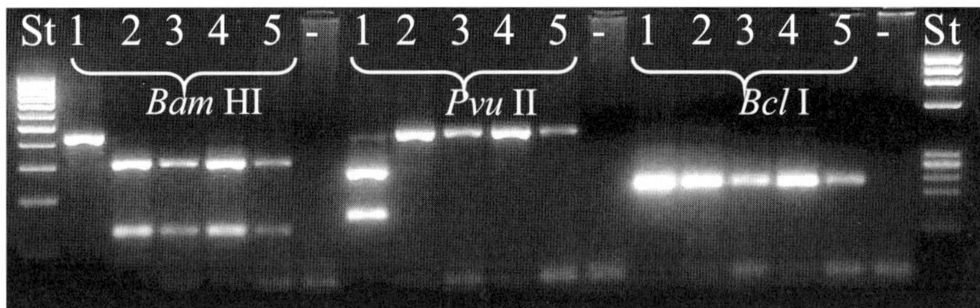

Figure 1. Restriction fragment length polymorphism (RFLP) analysis of PCR amplicons for BLV. Lanes 1–5 are PCR amplicons for BLV, corresponding to 5 different herd samples, digested with *Bam*HI, *Pvu*II and *Bcl*I, respectively. First and last lanes are 1 kb and Phi X-174 *Hae*III molecular markers, respectively. – = a negative herd sample.

Table 1. Assignment of the Chilean isolates to the known subgroups of BLV proviruses based on the restriction fragment length polymorphism (RFLP) analysis of a 444 bp env PCR fragment. Subgroups Belgian, Australian and Japanese were as described by Rice *et al.* (1985), Coulston *et al.* (1990) and Sagata *et al.* (1985), respectively.

Country	Isolates	Subgroup			
		Belgian	*Australian*	*Japanese*	*Other*
Chile	33	17	16	0	0

The phylogenetic analysis (Figure 2) clustered the sequences according to the known subgroups described earlier and confirmed the previous RFLP analysis and assignment of the eight Chilean isolates analysed to the Australian (3) and Belgian (5) subgroups. Interestingly, the Chilean clones formed distinct clusters within their respective subgroups, indicating a certain degree of sequence divergence. The five Chilean clones representing the Belgian subgroup had some very well conserved and unique nucleotide differences. These nucleotide differences occurred at position 70 of the 444 bp amplified product, where the nucleotide pair TG was replaced by the CA pair, and at position 174, where the nucleotide A was replaced by G. These unique changes affected the deduced amino acid sequence, replacing the conserved residues valine and glutamine by a methionine and an arginine, respectively.

These changes might have important implications for the infectivity of the virus, and further analysis is required on this point. From a practical standpoint, they could be used to monitor external introduction of the virus.

It is assumed that the genomic variability of BLV is very low and that variations are only tolerated at some nucleotide positions without loss of infectivity and ability to integrate in the host's genome (Sagata *et al.*, 1985), and that most of the differences in the nucleotide sequences are not followed by changes in amino acid sequences (Elwert, 1997). Nevertheless, our results showed that these changes are possible. Whether they have an effect on the infectivity of the virus remains to be determined.

In conclusion, the PCR technique used in this study might be useful not only as an aid in the diagnostic of BLV from milk but also in the typing of the virus, which provides a convenient way to better understand the epidemiology and distribution of BLV infection.

ACKNOWLEDGEMENTS

We thank Dr Gastón Muñoz for help with the phylogenetic analysis. This work was supported by a grant from *Fundación para la Innovación Agraria* (FIA), Ministerio de Agricultura, Chile.

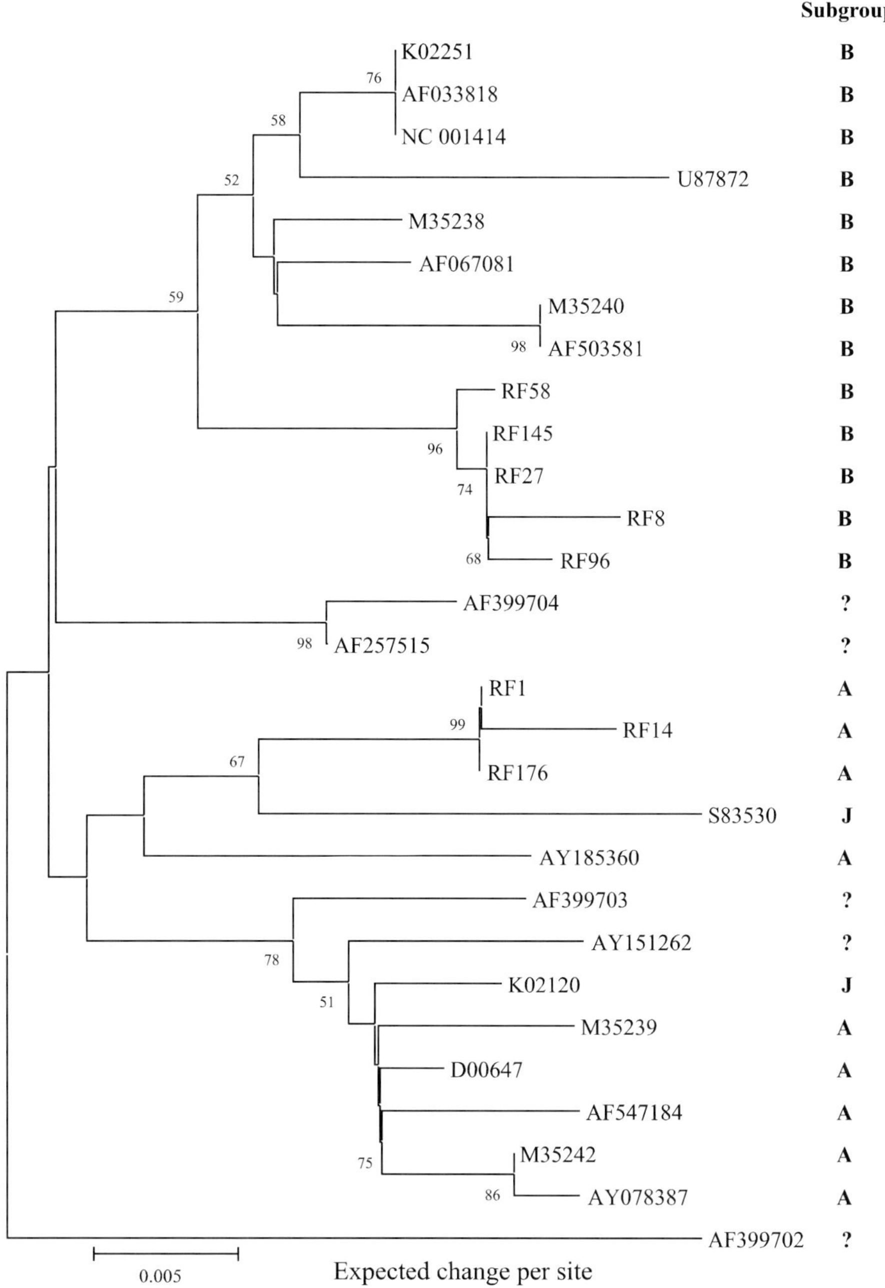

Figure 2. Neighbour Joining analysis of a 444 bp alignment of the BLV *env* gene. The Kimura-Two-Parameter model was used as substitution model. RF corresponds to the Chilean isolates, while the others correspond to all *env* BLV sequences deposited in GenBank (September 2003). B = Belgium; A = Australian; J = Japanese; ? = other.

REFERENCES

Beier, D., Blankenstein, P. & Fechner, H. 1998. Possibilities and limitations for use of the polymerase chain reaction (PCR) in the diagnosis of bovine leukemia virus (BLV) infection in cattle. *Deutsche Tierarztliche Wochenschrift,* **105**: 408–412.

Beier, D., Blankenstein, P., Marquardt, O. & Kuzmak, J. 2001. Identification of different BLV proviruses isolates by PCR, RFLP and DNA sequencing. *Berliner und Munchener Tierarztliche Wochenschrift,* **114**: 252–256.

Coulston, J., Naif, H., Brandon, R., Kumar, S., Khan, S., Daniel, R.C. & Lavin, M.F. 1990. Molecular cloning and sequencing of an Australian isolate of proviral bovine leukaemia virus DNA: comparison with other isolates. *Journal of General Virology,* **71**: 1737–1746.

Elwert, J. 1997. Genomanalyse verschiedener Isolate des Bovinen Leukosevirus unter besonderer Berücksichtigung des serologischen Status und der geographischen Herkunft des Rindes. Veterinary Dissertation. Berlin.

Evermann, J.F. & Jackson, M.K. 1997. Laboratory diagnostic tests for retroviral infections in dairy and beef cattle. *Veterinary Clinics of North America Food Animal Practice,* **13**: 87–106.

Fechner, H., Kurg, A., Geue, L., Blankenstein, P., Mewes, G., Ebner, D. & Beier, D. 1996. Related evaluation of polymerase chain reaction (PCR) application in diagnosis of bovine leukaemia virus (BLV) infection in naturally infected cattle. *Journal of Veterinary Medicine Series B – Zentralblatt Fur Veterinarmedizin Reihe B – Infectious Diseases and Veterinary Public Health,* **43**: 621–630.

Radostits, O., Blood, D., & Gay, C. 1995. Enzootic-bovine leukosis (bovine lymphosarcoma). pp. 954–965, *in:* O. Radostits, D. Blood and C. Gay (eds). *Veterinary Medicine.* 8th edition. London: Bailliere Tindall.

Rice, N., Stephens, R., Couez, D., Deschamps, J., Kettmann, R., Burny, A. & Gilden, R. 1984. The nucleotide sequence of the *env* gene and the post *env* region of the bovine leukaemia virus. *Virology,* **138**: 82–93.

Rice, N., Stephens, R., Burny, A. & Gilden, R. 1985. The *gag* and *pol* genes of bovine leukemia virus. Nucleotide sequence and analysis. *Virology,* **142**: 357–377.

Sagata, N., Yasunaga, T., Tsuzuku-Kawamura, J., Ohishi, K., Ogawa, Y. & Ikawa, Y. 1985. Complete nucleotide sequence of the genome of bovine leukemia virus: its evolutionary relationship to other retroviruses. *Proceedings of the National Academy of Sciences, USA,* **82**: 677–681.

Sudhir, K., Koichiro, T., Jakobsen, I.B. & Nei, M. 2001. MEGA2: Molecular Evolutionary Genetics Analysis software. Arizona State University, Tempe AZ, USA.

MOLECULAR MARKER STUDIES IN RIVERINE BUFFALOES, FOR CHARACTERIZATION AND DIAGNOSIS OF GENETIC DEFECTS

B.R. Yadav
Livestock Genome Analysis Laboratory, National Dairy Research Institute, Karnal – 132 001, India.

Abstract: The buffalo is probably the last livestock species to have been domesticated, with many genetic, physiological and behavioural traits not yet well understood. Molecular markers have been used for characterizing animals and breeds, diagnosing diseases and identifying anatomical and physiological anomalies. RFLP studies showed low heterozygosity, but genomic and oligonucleotide probes showed species-specific bands useful for identification of carcass or other unknown samples. Use of RAPD revealed band frequencies, band sharing frequencies, genetic distances, and genetic and identity indexes in different breeds. Bovine microsatellite primers indicate that 70.9% of bovine loci were conserved in buffalo. Allele numbers, sizes, frequencies, heterozygosity and polymorphism information content showed breed-specific patterns. Different marker types – genomic and oligonucleotide probes, RAPD and microsatellites – are useful in parent identification. Individual specific DNA fingerprinting techniques were applied with twin-born animal (XX/XY) chimerism, sex identification, anatomically defective and XO individuals. Molecular markers are a potential tool for geneticists and breeders to evaluate existing germplasm and to manipulate it to develop character-specific strains and to provide the basis for effective genetic conservation.

1.   INTRODUCTION

Developments in DNA technologies have made it possible to uncover a large number of genetic polymorphisms at the DNA sequence level and to use them as markers for evaluation of the genetic basis for observed

H. P. S. Makkar and G. J. Viljoen (eds.), Applications of Gene-Based Technologies for Improving Animal Production and Health in Developing Countries, 715-726.
© 2005 *IAEA. Printed in the Netherlands.*

phenotypic variability. The sequence markers possess unique genetic properties and methodological advantages that make them more useful and amenable for genetic analysis than other genetic markers. The possible applications of molecular markers in livestock improvement have shown promising scope in conjunction with conventional and transgenic breeding strategies (Mitra *et al.*, 1999). In conventional breeding strategies, molecular markers have several short-term or immediate applications (including parentage determination; genetic distance estimation; determination of twin zygosity and freemartinism; sexing of pre-implantation embryos; and identification of disease carriers) and longer-range applications (such as gene mapping and marker assisted selection). In transgenic breeding, molecular markers can be used as reference points for identification, isolation and manipulation of the relevant genes and for identification of the animals carrying the transgenes. Progress in development of molecular markers reinforces their potential for genetic improvement in livestock species. Consequently, enormous interest has been generated in determining genetic variability at the DNA sequence level in different livestock species, including buffaloes, and in assessing whether these variations can be exploited efficiently in conventional as well as in transgenic breeding strategies. The present discourse is a brief account of molecular markers and their various applications in livestock improvement.

The buffalo is probably the last livestock species domesticated. It is an important animal in tropical countries, particularly the Indian subcontinent, where a large number of breeds with many unique characteristics are available. However, many of its genetic, physiological and behavioural traits are not yet well understood. There is a need for precise markers for the characterization of animals and breeds, and for diagnosis of diseases and causes of anatomical and physiological defects. Developments in DNA technologies have made it possible to identify a large number of genetic polymorphisms at the DNA sequence level for evaluation of the genetic basis for observed phenotypic variability. The present discourse deals with investigations on different molecular markers in buffaloes for identification of individuals, correct parentage and genetic defects.

2.   MATERIAL AND METHODS

Genome analysis was carried out in seven important breeds of buffaloes, using different molecular markers. The breeds studied were Murrah, Nili-Ravi, Surti, Mehsana, Jaffrabadi, Nagpuri and Bhadhawari. The number of animals studied differed for different markers, ranging from 10 to 45 per breed. The markers included genomic probes, oligonucleotide probes, randomly amplified polymorphic DNA (RAPD) primers and bovine microsatellite markers. Methodology used was standard for DNA isolation, digestion, *in vitro* amplification, hybridization and autoradiography (Sambrook, Fritsch and Maniatis, 1989). Other protocols reflected the primers used, including genomic probes (Singh and Jones, 1986), restriction fragment length polymorphisms (RFLPs) (Mitra, 1994), RAPD (Shende and Yadav, 2004), oligonucleotides (Shashikanth, 1999) and microsatellite markers (Ganai and Yadav, 2001). The band patterns obtained were analysed to compare individuals, breeds and other distinguishing characteristics within and among breeds and species.

3.   RESULTS

The observations made using the various types of marker employed are presented according to study and technique, and breeds examined.

3.1   Restriction fragment length polymorphism

RFLPs were studied in Murrah and Nili-Ravi buffaloes using conventional hybridization and PCR technique for polymorphism. Low heterozygosity was found – mostly monomorphic bands – and might be due to the closed breeding policy and small population size of the animals examined.

3.2   DNA fingerprints revealed with genomic probes

The genomic probe Bkm and its derivative 2(8) carrying "GATA" repeats were found to produce DNA fingerprints (DFPs) (Figure 1). The probes also showed species-specific bands useful for identification of carcass or other unknown samples.

3.3   DNA fingerprints revealed with oligonucleotide synthetic probes

DFPs of Nili-Ravi and Murrah buffaloes used five different oligonucleotide probes – $(GT)_8$, $(GT)_{12}$, $(GTG)_5$, $(TCC)_5$ and $(GACA)_9$ – with five enzymes – *Alu*I, *Hinf*I, *Hae*III, *Mbo*I and *Eco*RI. All the probes gave multilocus hybridization patterns. Probes $(GT)_8$, $(TCC)_5$ and $(GTG)_5$ gave polymorphic DFPs. $(GTG)_5$ was the most polymorphic probe among the five probes studied (Figure 2). The band patterns showed allelic frequency between 0.22 and 0.29, band sharing of 0.45 and heterozygosity between 0.81 and 0.85 in Nili-Ravi and Murrah buffaloes. Randomly amplified polymorphic DNA (RAPD) techniques were used to study the Nagpuri and Murrah buffalo breeds (Figure 3). The polymorphic patterns revealed were used to establish band frequency, band sharing frequency, genetic distance, genetic identity index and mean average percentage (MAPD) in both the breeds (within and between). The average within-breed band sharing frequency was 0.739 ±0.032 in Nagpuri and 0.669 ±0.035 in Murrah. The between-breed band sharing was lower (0.490 ±0.062) than the within-breed band sharing. The overall average genetic distance was 0.464 ±0.15 between these two breeds. The genetic identity index was 0.632 ±0.076 between Nagpuri and Murrah buffaloes. The RAPD fingerprint analysis showed that the average percentage difference value varied for each primer and MAPD for these two breeds was found to be 50.97 ±6.15.

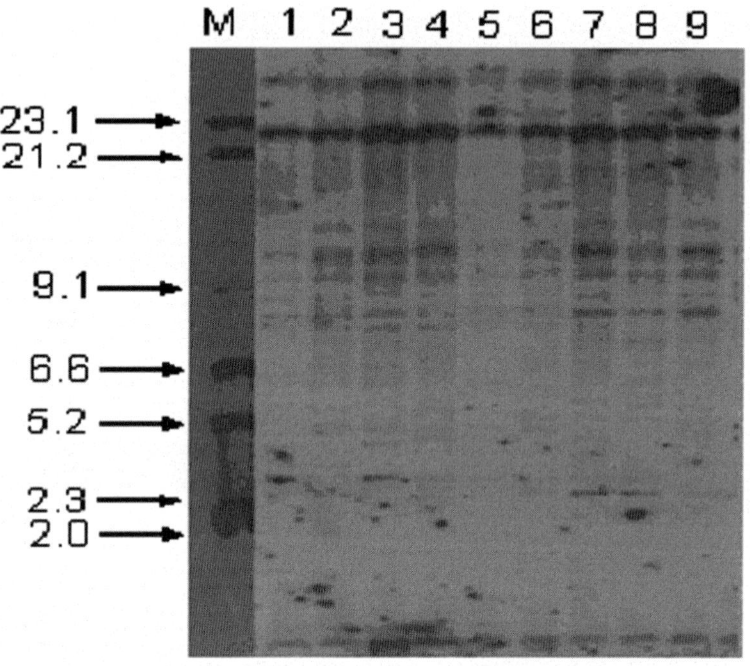

Figure 1. GATA repeats revealed by *Bkm* probes in buffalo genome.

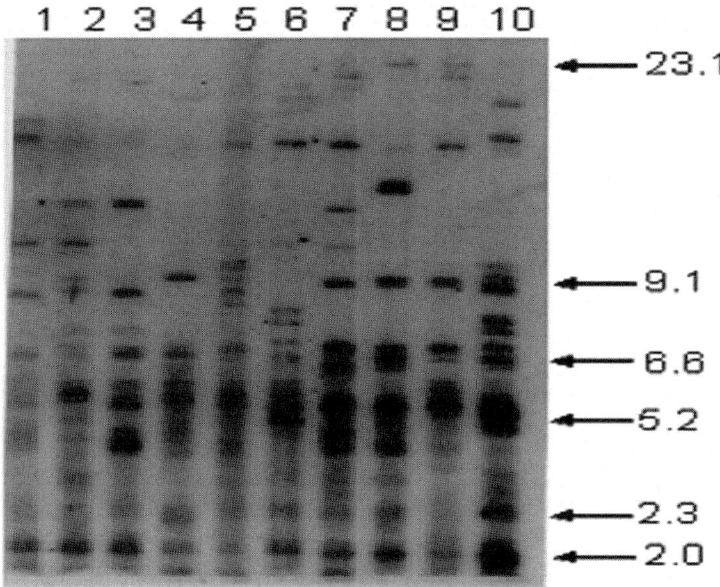

Figure 2. DNA fingerprinting using (GTG)₅ in combination with *Hinf*I in Murrah (Lanes 1–5) and Nili-Ravi (Lanes 6–10)

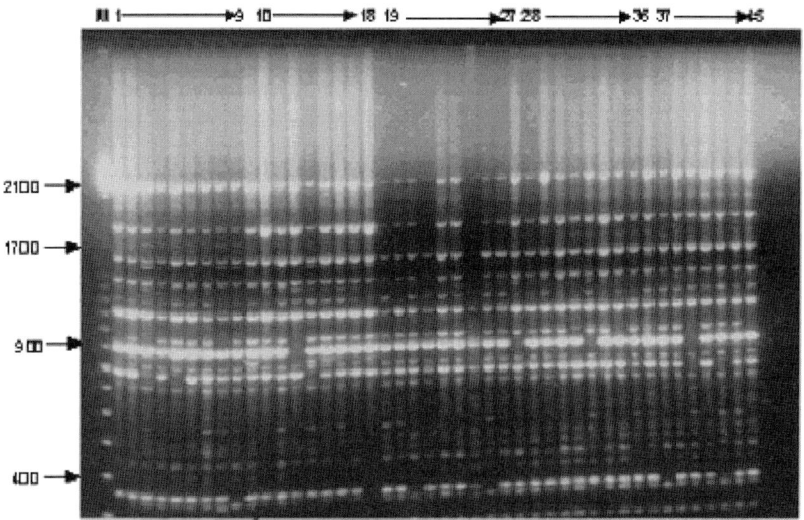

Figure 3. Polymorphism observed using RAPD primer in buffalo breeds.
KEY TO LANES: M = weight marker; 1–9 = Murrah; 10–18 = Nagpuri; 19–27 = Nili-Ravi; 28–36 = Bhadawari; and 37–45 = Surti.

3.4   Microsatellite (heterologous) primers

Bovine microsatellite primers were studied to differentiate breeds of water buffaloes (Figure 4). A detailed study was conducted on DNA samples from 40 individuals each of Murrah and Nili-Ravi. Among a total of 79 primers screened, 56 gave amplification products in buffalo, while in controls all the primers gave amplification. The observation showed that 70.89% of bovine loci were conserved in the case of buffalo. Out of 56 conserved microsatellite loci, 36 were polymorphic, i.e. informative (36/56 = 64.29%) and the rest were monomorphic. The numbers of alleles, their sizes, frequencies, heterozygosity and polymorphism information content (PIC) were calculated. The numbers of allele ranged from 2 (BM-044) to 6 (BMS-651) in Murrah, and from 2 (MB-077) to 5 (BMS-585) in Nili-Ravi. The size range of alleles at different loci varied from 104 (BMS-820) to 242 (MB-077) in Murrah, and from 100 (BMS-820) to 242 (MB-077) in Nili-Ravi. The heterozygosity of different microsatellite loci calculated from allele frequency ranged from 0.47 to 0.77 in Murrah and from 0.49 to 0.79 in Nili-Ravi. The average heterozygosity over different loci in Murrah and Nili-Ravi were 0.67 and 0.69, respectively. The PIC range varied from 0.375 (MB-077) to 0.734 (BMS-651) for Murrah, and from 0.358 (MB-077) to 0.754 (BMS-585) for Nili-Ravi. The pooled PIC for Murrah and Nili-Ravi ranged from 0.371 to 0.746

3.5   Parentage determination

The knowledge of correct parentage is a prerequisite in breeding programmes. Highly polymorphic DFP markers are quite useful for this purpose. Different types of makers, viz. genomic and oligonucleotide probes, RAPD and microsatellite markers, were used in identification of parents in two sire families in each of the Nili-Ravi and Murrah buffaloes. Microsatellite primers were found more useful, and particularly for verification of the semen used in artificial insemination.

3.6   Determination of freemartinism and other genetic defects

Individual specific DFP technique was applied to twin-born animal (XX/XY) chimerism, anatomically defective and XO individuals. PCR-RFLP assay using sex chromosome (Y)-specific primers enabled the identification of freemartin animals and other sex-specific defects.

The genetic improvement of animals needs marker-based information, which depends on the choice of an appropriate marker system for a given application. Molecular markers serve as a potential tool for geneticists and breeders to evaluate the existing germplasm and to manipulate it to create animals as desired and needed by society and for conservation for posterity.

## 4.	DISCUSSION

## 4.1	Restriction fragment length polymorphism

RFLPs were studied in Murrah and Nili-Ravi buffaloes (Mitra, 1994), using conventional hybridization and PCR techniques for polymorphism at growth hormone, k-casein and prolactin loci. However, low heterozygosity was found and mostly monomorphic bands, and this was attributed to the closed breeding policy and small population size.

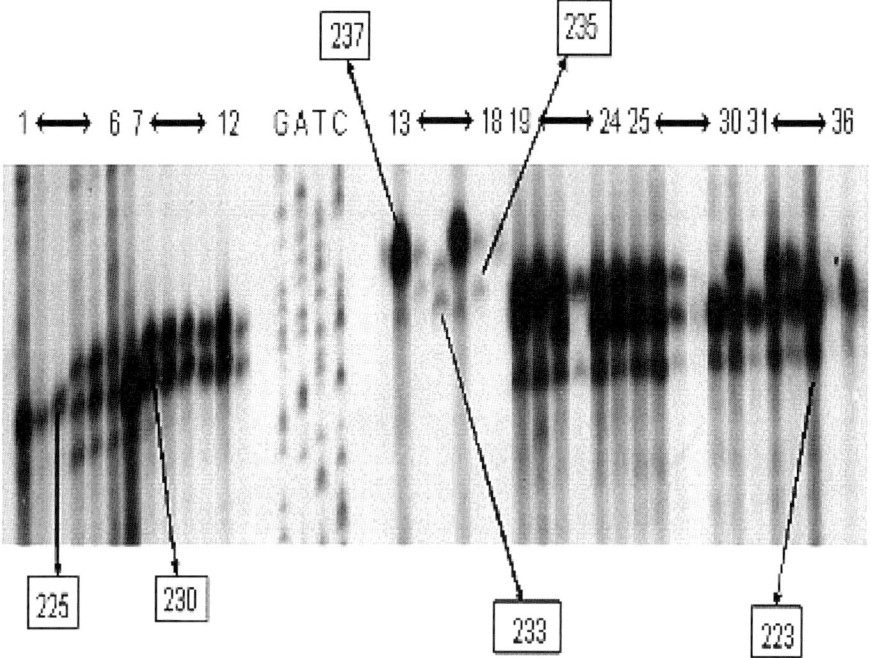

Figure 4. Polymorphic banding pattern in six breeds of buffalo revealed with a microsatellite marker (ILSTS001). The alleles observed were 223, 225, 230, 235, and 237.
KEY TO LANES: 1–6 = Murrah; 7–12 = Mehsana; 13–18 = Jafrabadi; 19–24 = Nagpuri; 25–30 = Nili-Ravi; 31–36 = Bhadawari.

4.2   Genomic probes

The genomic probe Bkm (Sambrook, Fritsch and Maniatis, 1989) and its derivative 2(8) carry long arrays of tetranucleotide repeats "GATA" and have been found to produce DFPs in buffaloes (Yadav and Balakrishnan, 1994). The probes also show species-specific bands useful for identification of carcasses or other unknown samples.

4.3   Oligonucleotide synthetic probes

John and Ali (1997) carried out DFP in buffaloes using $(CA)_n$, $(TGG)_6$, $(GGAT)_4$, $(GACA)_n$ and $(GATA)_n$ multilocus probes. They obtained no hybridization signal with $(GATA)_n$, while $(GACA)_n$ gave a monomorphic multilocus banding pattern. All the remaining three probes produced multilocus fingerprints, with $(TGG)_6$ the most polymorphic among the three probes.

DNA polymorphism has been reported (Shashikanth, 1999) in two breeds of buffaloes (Nili-Ravi and Murrah) with five different oligonucleotide probes – $(GT)_8$, $(GT)_{12}$, $(GTG)_5$, $(TCC)_5$ and $(GACA)_9$ – using five enzymes – *Alu*I, *Hinf*I, *Hae*III, *Mbo*I and *Eco*RI. All the probes gave multilocus hybridization patterns. Probes $(GT)_8$, $(TCC)_5$ and $(GTG)_5$ gave polymorphic DFPs, and $(GTG)_5$ was the most polymorphic among all five probes studied. The band patterns showed allelic frequency between 0.22 and 0.29, band sharing 0.45, and heterozygosity between 0.81 and 0.85 in Nili-Ravi and Murrah buffaloes.

GTG repeats selected and cloned as probes have been shown to generate unique fingerprint patterns (De, Yadav and Singh, 2001) when used for hybridization to *Hae*III digested buffalo genomic DNA (Shashikanth, 1999). $(GTG)_5$ repeat sequences have been identified in the buffalo genome, which can provide a homologous marker for genome characterization of different breeds of buffalo, genetic diversity and conservation studies. However, these markers are needed to saturate their linkage maps and genetic dissection of unique quantitative characters of buffalo species.

4.4   Random amplified polymorphic DNA

Using 14 arbitrary primers, earlier workers have shown a unique RAPD profile and revealed many common bands between buffalo and cattle (Appa Rao, Bhat and Totey, 1996). Genetic variation has been reported in cattle and buffalo breeds by employing RAPD techniques with 8 decamer primers (Aravindakshan and Nainar, 1998), who calculated band sharing (within and between breeds) and MAPD. They observed considerable homogeneity in individuals of a breed and lower inter-breed band sharing than in intra-breed. The MAPD analysis revealed lower values (24.16 ±3.55) between Murrah and Surti buffalo breeds than Jersey crossbred and Ongole breeds (28.10 ±10.53) of cattle.

The findings of a pilot study (Yadav *et al.*, 2002) revealed that the RAPD pattern of Murrah and Nili-Ravi breeds is quite similar to that of Jaffrabadi. Among 25 primers used, 11 primers were found to give fingerprints corresponding to PCR products ranging from 0.23 kb to 1.5 kb. Three of the primers gave banding patterns that clearly distinguished the three breeds used in this study. One primer revealed a band at 0.27 kb in Nili-Ravi pool DNA while another primer amplified a Jaffrabadi pool-specific RAPD fragment of 0.31 kb and a third amplified a Murrah pool-specific RAPD fragment of 0.60 kb

Studies in two breeds of buffalo, Nagpuri and Murrah, have revealed polymorphism by RAPD (Shende and Yadav, 2004). The information generated from the polymorphic patterns revealed by all the primers was used to determine band frequency, band sharing frequency, genetic distance, genetic identity index and MAPD in both breeds (within and between). The average within-breed band sharing frequency was 0.739 ±0.032 in Nagpuri and 0.669 ±0.035 in Murrah. The between-breed band sharing was lower (0.490 ±0.062) than within-breed. The overall average genetic distance was 0.464 ±0.15 between these two breeds. The genetic identity index was 0.632 ±0.076 between Nagpuri and Murrah buffaloes. RAPD fingerprint analysis showed that the average percentage difference value varied for each primer, and MAPD for these two breeds was found to be 50.97 ±6.15. On the basis of all these observation on differences in Nagpuri and Murrah buffalo breeds, the utility is confirmed of RAPD markers in differentiation of these breeds at molecular level.

4.5   Microsatellite (heterologous) primers

In the recent past, many studies have been reported in sheep, goat and other species where bovine microsatellites have been used and have revealed polymorphism. However, very few studies have been conducted in buffalo. It has been reported (Moore *et al.*, 1995) that a set of 80 bovine DNA-derived microsatellite primers amplified loci in both swamp and riverine types of water buffaloes (*Bubalus bubalis*). The primers were first chosen on the basis of their robustness (yield and reproducibility) in amplification, and tested on a large number of swamp- and riverine-type water buffaloes to determine allele numbers and sizes, and heterozygosity percentage. They found that the number of alleles generally correlated to levels of heterozygosity, with the exception of locus CSS MO 45. Heterozygosity levels were higher in buffalo in which the largest number of alleles was, observed, except at few loci. Yadav, De and Mitra (1998) reported the use of 22 bovine primers with the genomic DNA of 36 animals of Murrah buffalo. All the primers amplified the products, but 14 primers revealed polymorphism. The number of alleles varied from 2 to 5 for different microsatellite markers.

4.6   Determination of twin zygosity and freemartinism

Correct knowledge of zygosity of twins, particularly in monotocous animals, is very important. Monozygotic twins provide opportunities for epidemiological as well as genetic studies, and also help in transplant matching. Individual specific DFP techniques have potential application in determination of twin zygosity and demonstration of spontaneous XX/XY chimerism. Demonstration of XX/XY chimerism in heterosexual bovine twins by PCR-RFLP assay using sex chromosome-specific primers has enabled the identification of freemartin animals. This approach has several advantages over earlier reports.

Current livestock breeding strategies largely rely on the principle of selective breeding. In this method, genetic improvement can be brought about by increasing the frequency of advantageous alleles at many loci, though the actual loci are quite difficult to identify. However, the method does not allow genes to be assimilated or moved from distant sources, such as from different species or genera, due to reproductive barriers. Recent developments in molecular biology have given rise to the new technology of transgenesis, which eroded if not eliminated the breeding barrier between different species or genera. Transgenesis has opened many vistas in understanding behaviour and expression of genes. It has also made it possible to alter gene structure and modify function. Of many applications of transgenesis, the most outstanding one is the development of transgenic dairy animals for the production of pharmaceutical proteins in milk, and animals with altered milk composition.

The starting point for this technology is the identification of the genes of interest. In this, molecular markers can serve as reference points for mapping the relevant genes, which would be the first step towards their identification, isolation, cloning (positional cloning) and manipulation. After successful production of transgenic animals, appropriate breeding methods could be followed for multiplication of transgenic herds or flocks. Molecular markers can also be used for identification of the animals carrying the transgenes. Although most of the quantitative trait loci (QTLs) are polygenic in nature and transgenesis currently manipulates only single gene traits, the technology is highly promising, particularly in moving genes across breeding barriers.

5. CONCLUSIONS

The genetic improvement of animals is a continuous and complex process. Genetic polymorphism at the DNA sequence level has provided a large number of marker techniques with a wide variety of applications. This has in turn prompted further consideration of the potential utility of these markers in animal breeding. However, utilization of marker-based information for genetic improvement depends on the choice of an appropriate marker system for a given application. Selection of markers for different applications are influenced by several factors, including the cost involved, skill or expertise available, possibility of automation, radioisotopes used, reproducibility of the technique, and, finally, the degree of polymorphism. Currently the pace of development in applications for molecular markers is tremendous and suggests that the vibrant expansion in marker development will continue, at least for the foreseeable future. The newer techniques, such as non-gel-based screening systems using oligonucleotide arrays or "DNA chips", will have a great impact. It is expected that molecular markers will serve as a potential tool for geneticists and breeders to evaluate existing germplasm and to manipulate it to create animals that satisfy the requirements of society.

ACKNOWLEDGEMENTS

The author is grateful to the Indian Council of Agricultural Research for providing a National Fellow position, and to his various team workers, viz. Dr Shashikanth, Dr N.A. Ganai, Dr P.K. Shende, Mrs S.R. Zenu, Dr A. Pandey, Dr Soumi Sukla, Mr R.K. Tonk, Mr Parameet Kumar, Mr Nanku Singh and others in the Genome Analysis Laboratory for help and contributions in different ways to carry out the Genome Analysis work at NDRI.

REFERENCES

Appa Rao, K.B.C., Bhat, K.V. & Totey, S.M. 1996. Detection of species-specific genetic markers in farm animals through random amplified polymorphic DNA markers. *Biomolecular Engineering,* **13**: 135–138.

Aravindakshan, T.V. & Nainar, A.M. 1998. Genetic variation in cattle and buffalo breeds analysed by RAPD markers. *Indian Journal of Dairy Science,* **51**: 368–374.

De, S., Yadav, B.R. & Singh, R.K. 2001. Identification of GTG microsatellite DNA in riverine buffalo (*Bubalus bubalis*) genome. *Indian Journal of Animal Science,* **71**: 385–387.

Ganai, N.A. & Yadav, B.R. 2001. Genetic variation within and among three breeds of goats using heterologous microsatellite markers. *Animal Biotechnology,* **12**: 121–136.

John, M.V. & Ali, S. 1997. Synthetic DNA-based genetic markers reveal intra-species DNA sequence variability in *Bubalus bubalis* and related genomes. *DNA and Cell Biology,* **16**: 369–378.

Mitra, A. 1994. Studies on restriction fragment length polymorphisms in cattle and buffalo. PhD Thesis. NDRI-Deemed University, Karnal.

Mitra, A., Yadav, B.R., Ganai, N.A. & Balakrishnan, C.R. 1999. Molecular markers and their applications in livestock improvement. *Current Science,* **77**(8): 1045–1053.

Moore, S.S., Evans, D., Byrne, K., Barker, J.S.F., Tan, S.G., Vankan, D. & Hetzel, D.J.S. 1995. A set of polymorphic DNA microsatellites useful in swamp and river buffalo (*Bubalus bubalis*). *Animal Genetics,* **26**: 355–359.

Sambrook, J., Fritsch, E.F. & Maniatis, T. 1989. *Molecular Cloning: A Laboratory Manual.* 2nd Edition. Cold Spring Harbor Laboratory Press.

Shashikanth. 1999. Study on DNA polymorphism in cattle and buffalo. PhD. Thesis. NDRI - Deemed University, Karnal (India).

Shende, P.K. & Yadav, B.R. 2004. Genetic diversity analysis in Nagpuri vis-à-vis Murrah buffalo using randomly amplified polymorphic DNA markers. *Buffalo Journal,* **20**: 00–00 (In press).

Singh, L. & Jones, K.W. 1986. Bkm sequences are polymorphic in humans and are clustered in pericentric regions of various acrocentric chromosomes including Y. *Human Genetics,* **73**: 304-308.

Yadav, B.R. & Balakrishnan, C.R. 1994. DNA fingerprints of buffaloes as revealed by [GATA]$_n$ repeat specific probe. Abstract No. 105 (page 114) *in:* [Proceedings of] the 3rd International Conference on DNA Fingerprinting. Hyderabad, India, 13–16 December 1994.

Yadav, B.R., De, S. & Mitra, A. 1998. Bovine microsatellite primers reveal DNA polymorphism in Riverine buffalo. p. 26, *in: Proceedings of the Indo-Brazilian Workshop on Agricultural Biotechnology.* NDRI, Karnal, India, 27–29 April 1998.

Yadav, B.R., Thakur, M.K., Singhal, A. & Singh, R.P. 2001. Genetic variation in three breeds of water buffalo analyzed by RAPD primers. Abstract P.40. Presented in: *A Two-day Symposium on Perspectives in Genome Analysis.* CCM, Hyderabad, 23–24 February 2001.

HOST-RANGE PHYLOGENETIC GROUPING OF CAPRIPOXVIRUSES

Genetic typing of CaPVs

Christian Le Goff,[1] Emna Fakhfakh,[2] Amelie Chadeyras,[1] Elexpeter Aba Adulugba,[3] Geneviève Libeau,[1][*] Salah Hammami,[2] Adama Diallo[4] and Emmanuel Albina[1].

1. *CIRAD-EMVT – TA 30/G, 34398–Montpellier Cedex 5, France.*
2. *IRVT, La Rabta, 1006 Tunis, Tunisia.*
3. *National Veterinary Research Institute, Vom, Plateau State, Nigeria.*
4. *Animal Production Unit, FAO/IAEA Laboratories, A-2444 Seibersdorf, Austria.*
* *Corresponding author.*

Abstract: Because of their close relationship, specific identification of the CaPVs genus inside the Poxviridae family relies mainly on molecular tools rather than on classical serology. We describe the suitability of the G protein-coupled chemokine receptor (GPCR), for host range phylogenetic grouping. The analysis of 26 CaPVs shows 3 tight genetic clusters consisting of goatpox virus (GPV), lumpy skin disease virus (LSDV), and sheeppox virus (SPV).

1. INTRODUCTION

The *Capripoxvirus* genus comprises sheeppox virus (SPV), goatpox virus (GPV) and lumpy skin disease virus (LSDV). The three separate diseases caused by capripoxviruses (CaPVs) in sheep, goat and cattle have significant economic importance in large areas in Africa and Asia. Both sheeppox and goatpox are endemic in Africa, the Near East, South and East Asia, while lumpy skin disease is spread over most Africa but absent from Asia (Davies, 1981, 1991).

CaPVs are generally considered to be host specific, leading to outbreaks in the preferential host, even if experimental infections have shown that most strains can cause disease in more than one species. Sheeppox and goatpox

H. P. S. Makkar and G. J. Viljoen (eds.), Applications of Gene-Based Technologies for Improving Animal Production and Health in Developing Countries, 727-733.

exhibit similar clinical signs in sheep and goats, mainly characterized by generalized pox lesions throughout the skin, that can be confused with other exanthemas, e.g. contagious ecthyma. Lumpy skin disease is a subacute to acute cattle disease with appearance of skin nodules. During epizootics of CaPVs in domestic animals, comparable disease among wild ungulates has not been reported (Davies, 1991).

CaPVs cannot be distinguished serologically and can induce heterologous cross-protection. The common immunogenic properties of the capripoxvirus have been turned to account by the use of specific attenuated or low-virulence capripoxvirus isolates to protect all three host species from capripox infection (Kitching, Hammond and Black, 1986). It has been shown that they are ideal vectors to express recombinant antigens. The large size of the virus facilitates insertion of foreign genes. An LSDV vaccine strain has already been shown to be an effective dual vaccine that protects cattle against both lumpy skin disease and rinderpest (Romero *et al.*, 1993, 1994a, b) or small ruminants against sheep and goatpox and peste des petits ruminants (Berhe *et al.*, 2003; Diallo *et al.*, 2002).

Although closely related, restriction fragment pattern analysis, cross-hybridization studies, and, more recently, the complete genome sequencing of the three viruses, have shown that groupings of isolates correlated with animal species from which the viruses were isolated (Black, Hammond and Kitching, 1986; Gershon and Black, 1988; Kitching, Bhat and Black, 1989; Cao, Gershon and Black, 1995; Tulman *et al.*, 2002). These data indicate the close genetic relationship of SPV, GPV and LSDV, although they are phylogenetically distinguishable through genes involved in virulence and host-range functions. We describe here the suitability of the G protein-coupled chemokine receptor (GPCR), described by Cao, Gershon and Black (1995), for host-range phylogenetic grouping of CaPVs.

2.   METHOD

2.1   Virus strains

Original or first passage material of virulent or vaccine strains were selected. Eighteen CaPV strains from different African countries were compared. They included 11 strains isolated from sheep, 4 from goat and 3 from cattle. Among them, 4 vaccine strains were also studied: the Kenya sheep isolate (KS-1; Kitching, Hammond and Taylor, 1987), the Neethling strain of LSDV (Fick and Viljoen, 1994), and a sheep strain and a goat strain used for routine vaccination in Nigeria (sent by Dr Majiyagbe, National Veterinary Research Institute, Vom, Plateau State, Nigeria). They were compared with the corresponding sequences of 8 additional CaPV sequences and one swine pox virus (SWPV) registered in GenBank.

2.2   Viral DNA isolation

CaPV genomic DNA was extracted and purified from infected lamb testis cells (Qiagen DNeasy Tissue System) and amplified using primers derived from the KS-1 vaccine strain (Cao, Gershon and Black, 1995). Two primers, Chem 09 and Chem 2bis, were designed to delimit the genome at position 6976–8118 (Tulman *et al.*, 2001), while two others, Chem 06 and Chem 2bis, were positioned internal to this sequence (Table 1). Purified PCR products were amplified using a Blunt End Cloning kit (Roche), then used in dideoxy sequencing reaction. Reaction products were run on an automated DNA sequencer (Applied Biosystem, Prism 377).

2.3   Sequence alignment and phylogenetic analysis

The alignment of nucleotide sequences was created using the Clustal W program (Vector NTI, Informax Inc.). Phylogenetic trees were generated according to the neighbour joining method (Saitou and Nei, 1987) using the Phylip package. Bootstrap confidence values were calculated by Seqboot and Consense on 1000 replicates. SWPV was selected as an outgroup.

Table 1. Primer sequences.

Primer	Direction[1]	Start[2]	KS-1 homolog sequence 5′→ 3′
Chem 2bis	(a)	6961	ttt ttt tat ttt tta tcc aat gct aat act
Chem 10	(a)	7468	tga gac aat cca aac cac cat
Chem 06	(s)	7544	gat gag tat tga tag ata ccta gct gta gtt
Chem 09	(s)	8119	tta agt aaa gca taa ctc caa caa aaa tg

NOTES: (1) Forward (s) or reverse (a). (2) Nucleotide.

3. RESULTS

For taxonomy and evolutionary studies we have worked on a non-essential gene for virus growth encoding a homologue of a G protein-coupled chemokine receptor (GPCR), which sequence belongs to the Kenya sheep (KS-1) capripoxvirus isolate. The Q2/3L gene, known to be located in the terminal genomic region, is likely to affect viral virulence (Tulman *et al.*, 2002). This poxvirus-encoded gene affects the host immune response to viral infection because of its homology to mammalian chemokine receptors. From the published sequence of the Q2/3L gene of KS-1 vaccine strain, PCR primers were designed to amplify all the CaPV strains. To study the relationship among the 26 CaPVs, a representative phylogenetic analysis using the neighbour-joining method showed 3 tight genetic clusters consisting of GPV, LSDV and SPV. The different lineages were supported by the bootstrap values, suggesting a co-adaptation of the strains and their original host. The SWPV was considered an outgroup (Figure 1).

Unexpectedly, the KS-1 strain was closer to the LSDV strain cluster and the goat vaccine was closer to the sheep group. These vaccine strains may in fact originate from bovines and sheep, respectively. In the case of KS-1, the similarity with the LSDV genome was recently demonstrated (Tulman *et al.*, 2002). Alignment of the deduced amino acid sequence showed that one deletion of 7 amino acids was present in the sheep strains, while absent in the cattle and goat strains (position 10–16), and may have an interesting application in differential diagnosis. Another deletion is present at position 30–34, but only in two cattle strains (Figure 2).

4. CONCLUSION

These results suggest that the genomic mutations that occurred in the GPCR gene account for viral adaptation to interfere with the immune system in a host-specific manner. This gene provides one starting point for the understanding of the genetic basis of the CaPVs host range specificity. A particularly valuable application of the delineated primers would be their direct use for disease epidemio-surveillance and differential diagnosis. This could allow re-assessing the host-range specificity of capripoxviruses.

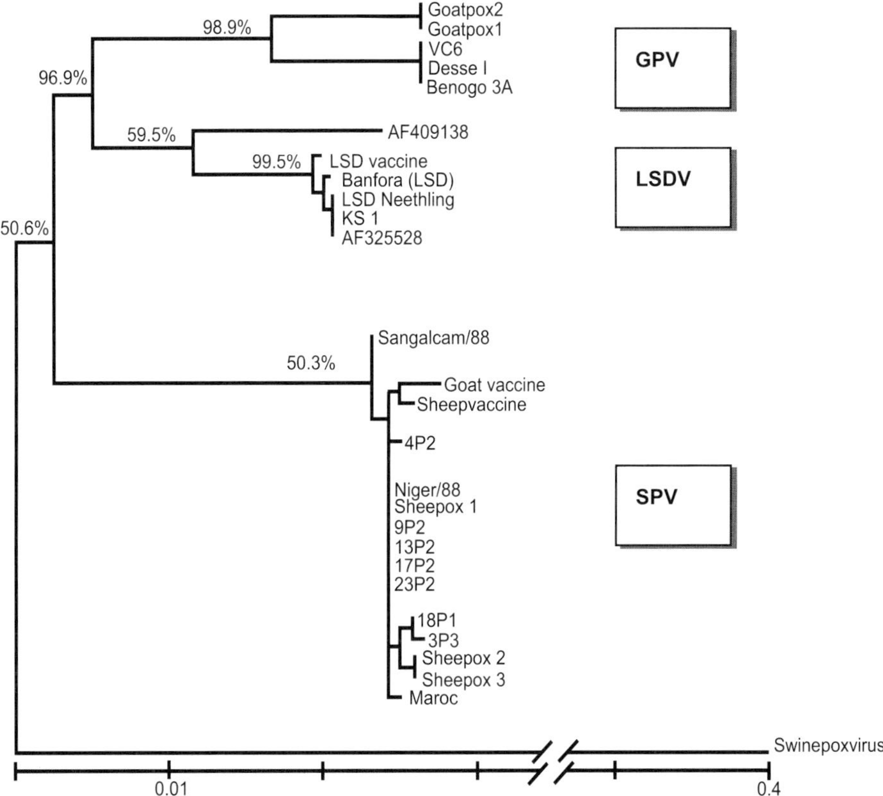

Figure 1. Phylogenetic tree of CaPVs based on the alignment of the nucleotide sequences (6976–8118) of the GPCR gene (NEIGHBOUR programme of the PHYLIP package). Percentages of 1000 bootstrap replicates supporting each group are indicated. The scale indicates the number of mutations out of around 1100 base pairs and is broken up to show the distant relation of the swinepox virus. Key: GPV = Goatpox virus; SPV = Sheeppox virus; LSDV = Lumpy skin disease virus.

Figure 2 — Alignment of amino acid sequences (positions marked 1, 10, 20, 30, 40, 50, 60, 70, 80, 90, 100, 112). Residues shown per strain (blank = identical/conservative; dashes = deletions):

Strain	Source	Residues (in order of position)
Goatpox 2	(g)	S YA V Q K T TS S T F
goatpox 1	(g)	S YA V Q K T TS S T F
VC6	(g)	N GYA VL Q K T T S F
Desse I	(g)	N GYA VL Q K T T S F
Benogo 3A	(g)	N GYA VL Q K T T S F
AF409138	(c)	S AI T Q T T T D S
LSD vaccine	(c)	S AI ——T Q T T T S D S
Banfora (LSD)	(c)	S AI ——T Q T T T S D S
LSD Neethling	(c)	S AI T Q T T T S D S
KS 1	(c)	S AI T Q T T T S D S
AF325528	(c)	S AI T Q T T T S D S
Goat vaccine	(g)	N GYA
Sheep vaccine	(s)	———————
Sangalcam/88	(s)	———————
4P2	(s)	———————
Niger/88	(s)	———————
Sheeppox 1	(s)	———————
9P2	(s)	———————
13P2	(s)	———————
17P2	(s)	———————
23P2	(s)	———————
18P1	(s)	———————
3P3	(s)	——————— S
Sheeppox 2	(s)	———————
Sheeppox 3	(s)	———————
Maroc	(s)	——————— S

Consensus (1): MNYILRIVSSYIMMNSSNITTIATTISIILSRISINKNWTIPSIYRNITAISNYKIAYNITYYSDMDDYEVNIVDIPHCDDGWYTISPGLITILYITIFFLGLPGNILV

Figure 2. Alignment of the deduced amino acid sequence of 26 CaPVs (partial representation of the 381 aa of the GPCR protein). The strains were isolated from: (g) goat, (c) cattle and (s) sheep. Blank areas represent identical and conservative residues. Dashes indicate the deletions.

REFERENCES

Berhe, G., Minet, C., Le Goff, C., Barrett, T., Ngangnou, A., Grillet, C., Libeau, G., Fleming, M., Black, D.N. & Diallo, A. 2003. Development of a dual recombinant vaccine to protect small ruminants against Peste des petits ruminants virus and capripoxvirus infections. *Journal of Virology,* **77**: 1571–1577.

Black, D.N., Hammond, J.M. & Kitching, R.P. 1986. Genomic relationship between capripoxviruses. *Virus Research,* **5**: 277–292.

Cao, J.X., Gershon, P.D. & Black, D.N. 1995. Sequence analysis of HindIII Q2 fragment of capripoxvirus reveals a putative gene encoding a G-protein-coupled chemokine receptor homologue. *Virology,* **209**: 207–212.

Davies, F.G. 1981. Sheep and Goatpox. pp. 733–748, *in:* E.P.J Gibbs (ed). *Virus Diseases of Food Animals.* Volume 2. London: Academic Press.

Davies, F.G. 1991. Lumpy Skin disease, an African capripox disease of cattle. *British Veterinary Journal,* **147**: 489–502.

Diallo, A., Minet, C., Berhe, G., Le Goff, C., Black, D.N., Fleming, M., Barrett, T., Grillet, C. & Libeau, G. 2002. Goat immune response to capripox vaccine expressing the hemagglutinin protein of peste des petits ruminants. *Annals of the New York Academy of Sciences,* **969**: 88–91.

Fick, W.C. & Viljoen, G.J. 1994. Early and late transcriptional phases in the replication of lumpy skin disease virus. *Onderstepoort Journal of Veterinary Research,* **61**: 255–261.

Gershon, P.D. & Black, D.N. 1988. A comparison of the genomes of capripoxvirus isolates of sheep, goats and cattle. *Virology,* **164**: 341–349.

Kitching, R.P., Bhat, P.P. & Black, D.N. 1989. The characterization of African strains of capripoxvirus. *Epidemiology and Infection,* **102**: 335–343.

Kitching, R.P., Hammond, J.M. & Black, D.N. 1986. Studies on the major common precipitating antigen of capripox virus. *Journal of General Virology,* **67**: 139–148.

Kitching, R.P., Hammond, J.M. & Taylor, W.P. 1987. A single vaccine for the control of capripox infection in sheep and goats. *Research in Veterinary Science,* **42**: 53–60.

Romero, C.H., Barrett, T., Chamberlain, R.W., Kitching, R.P., Fleming, M. & Black, D.N. 1994a. Recombinant capripoxvirus expressing the hemagglutinin protein gene of rinderpest virus: protection of cattle against rinderpest and lumpy skin disease viruses. *Virology,* **204**: 425–429.

Romero, C.H., Barrett, T., Evans, S.A., Kitching, R.P., Gershon, P.D., Bostock, C. & Black, D.N. 1993. Single capripoxvirus recombinant vaccine for the protection of cattle against rinderpest and lumpy skin disease. *Vaccine,* **11**: 737–742.

Romero, C.H., Barrett, T., Kitching, R.P., Carn, V.M. & Black, D.N. 1994b. Protection of cattle against rinderpest and lumpy skin disease with a recombinant capripoxvirus expressing the fusion protein gene of rinderpest virus. *Veterinary Record,* **135**: 152–154.

Saitou, N. & Nei, M. 1987. The neighbor-joining method: a new method for reconstructing phylogenetic trees. *Molecular Biology and Evolution,* **4**: 406–425.

Tulman, E.R., Alfonso, C.L., Lu, Z., Zsak, L., Kutish, G.F. & Rock, D.L. 2001. Genome of lumpy skin disease virus. *Journal of Virology,* **75**: 7122–7130.

Tulman, E.R., Alfonso, C.L., Lu, Z., Zsak, L., Sur, J.H., Sandybaev, N.T., Kerembekova, U.Z., Zaitsev, V.L., Kutish, G.F. & Rock, D.L. 2002. The genomes of sheeppox and goatpox viruses. *Journal of Virology,* **76**: 6054–6061.

POLYMERASE CHAIN REACTION (PCR) FOR RAPID DIAGNOSIS AND DIFFERENTIATION OF PARAPOXVIRUS AND ORTHOPOXVIRUS INFECTIONS IN CAMELS

A.I. Khalafalla,[1] M. Büttner[2] & H.-J. Rziha[2]
1. Department of Microbiology, Faculty of Veterinary Medicine, University of Khartoum, Shambat 13314 Khartoum North, Sudan.
2. Department of Immunology, Federal Research Centre for Virus Diseases of Animals, Paul-Ehrlich Str.28, D 72076 Tübingen, Germany.

Abstract: Rapid identification and differentiation of camel pox (CMP) and camel contagious ecthyma (CCE) were achieved by polymerase chain reaction (PCR) with primers that distinguish Orthopoxvirus (OPV) and Parapovirus (PPV). Forty scab specimens collected from sick camels and sheep were treated by 3 different DNA extraction procedures and examined by PCR. The sensitivity of the PCR was compared with that of electron microscopy and virus isolation in cell culture. Procedure 1, in which viral DNA was extracted directly from scab specimens followed by PCR, proved to be superior and more sensitive. Procedure 2 enables a fast specific diagnosis of PPV and OPV infections directly from scab materials without the need for DNA extraction. These assays provide a rapid and feasible alternative to electron microscopy and virus isolation.

1.   INTRODUCTION

Camels are an important livestock resource well adapted to hot and arid environments. Two poxvirus diseases are of economic importance: camel pox (CMP) caused by an Orthopoxvirus (OPV) and camel contagious ecthyma (CCE) caused by a Parapoxvirus (PPV). Clinically, CCE is indistinguishable from CMP, especially when both diseases co-exist in the

H. P. S. Makkar and G. J. Viljoen (eds.), Applications of Gene-Based Technologies for Improving Animal Production and Health in Developing Countries, 735-742.
© 2005 *IAEA. Printed in the Netherlands.*

same locality and when CCE undergoes a generalized course of disease. The risk caused by these diseases is not only due to mortality, which can reach 28% in CMP (Jezek, Kriz and Rothbauer, 1983) and 9% in young camel calves in CCE (Khalafalla and Mohamed, 1997), but also a loss in milk and meat production, labour, and quality of skin. The current classical methods for laboratory diagnosis of both diseases are unreliable and time consuming (virus isolation in cell culture or embryonated chicken eggs) or not available (electron microscopy) in countries where these diseases are endemic.

Nucleic acid hybridization techniques based on the polymerase chain reaction (PCR) are now widely used for detection and characterization of many viruses, including Poxvirus. A PCR assay has been reported that can identify OPVs, including CMP virus (Meyer, Pfeffer and Rziha, 1994), and two PCR assays have been developed for the detection of PPV infections (Inoshima, Morooka and Sentsui, 2000; Torfason and Gudnadottir, 2002). However, applicability of these PCR assays in PPV infection in camels has not yet been investigated. The aim of this investigation was to develop a rapid and reliable PCR technique for direct detection and differentiation of CCE and CMP in camels. Such a technique would overcome problems of classical methods.

2. MATERIAL AND METHODS

2.1 Viruses and field specimens

Skin scabs were collected from 33 CCE-suspected and 4 CMP-suspected camels in the Sudan (Khalafalla and Mohamed, 1997), and 3 skin scabs from sheep with Contagious Pustular Dermatitis (Orf) were used. Orf strain MRI (PPV), which has not been adapted to grow in cell culture, was kindly provided by David McHaig and Colin McInnes (The Moredun Research Institute, Pentland Science Park, Edinburgh, Scotland) and was used as a reference strain for PPV. Vaccinia virus Elstree strain was obtained from the virus stock of the Federal Research Centre for Virus Diseases of Animals, Tübingen, and used as a reference strain for OPV. Skin scabs were cut into small pieces, homogenized in mortar and pestle, and stored at +4°C until used. A specimen for Fowlpox virus (FPV) strain HR 1 identified by electron microscopy was included as negative control.

2.2 Electron microscopy and virus isolation

In order to compare the PCR protocols with classical techniques for diagnosis, a small volume of each sample was examined after negative staining in electron microscopy. Portions of the same samples were also inoculated in Bovine Embryonic Oesophageal Cells (KOP; from the cell collection of the Federal Research Centre for Virus Diseases of Animals, Greifswald/Riems, Germany) or Vero (African green monkey kidney) cells grown in Dulbecco`s modification of minimum essential medium (D-MEM) with the addition of 5% foetal bovine serum (FBS) and antibiotics. Virus identification was done by electron microscopy and an indirect immunoperoxidase (IP) test.

2.3 PCR procedure

2.3.1 DNA extraction

Three procedures for DNA extraction were used. In Procedure 1, viral DNA was extracted from skin scabs by a commercial DNA Isolation Kit (Puregene®, Gentra Systems, USA). Concentrations of 0.5–1.0 µg DNA per reaction were used in the PCR. In Procedure 2, 10 µl of scab homogenates were added directly to the PCR mixture minus the polymerase, as described by Ireland and Binepal (1998) for Capripoxvirus. In Procedure 3, the poxvirus particles in scab homogenate were purified through a 36% sucrose cushion and 10 µl of virus pellet was added to the PCR reaction mixture minus the polymerase. In both Procedures 2 and 3, a pre-amplification treatment by heating at 99°C for 15 minutes was used to release the DNA.

2.3.2 Oligonucleotide primers and PCR condition

Nucleotide sequences of primers used in the present study are listed in Table 1. Based on the sequences of Orf virus NZ 2 of the major envelope gene, described by Sullivan *et al.* (1994), primers 42KB5, 42KE3 and 42N ref were chosen and used in the first round of PCR (42KB5 and 42KE3) or a semi-nested PCR (42KB5 and 42N ref). Amplification was performed in 50 µl, containing 50 mM KCL, 20 mM Tris-HCL (pH 8.4), 1.5 mM $MgCl_2$, 0.02 mM dNTPs, and 45 nM of each primer. The Taq polymerase (2.5 units) (Invitrogen Life Technologies, The Netherlands) was added at 80°C, followed by 35 cycles of 1 minute at 95°C, 10 seconds at 60°C and 90 seconds at 72°C, with a final extension step for 8 minutes at 72°C.

Primers for the amplification of OPV-specific DNA (ATI up 1 and ATI low 1) (Table 1) and the PCR conditions were as described by Meyer,

Pfeffer and Rziha (1994). The reaction was carried out in a Biometra T3 Thermocycler and the PCR products were resolved by electrophoresis in 1% agarose gel. The described procedures were optimized using reference strains for OPV and PPV viruses (Vaccinia and MR I, respectively) and then performed for detection of viral DNA in scab specimens obtained from field outbreaks and in supernatants of cell cultures infected with PPV and OPV strains. A semi-nested PCR was performed on PPV-suspected scab specimens to enhance sensitivity of the first PCR and to confirm specificity.

Table 1. Name, sequence and size of primers.

Primer name	Sense	Sequence (5′→3′)	Product size (bp)
42KB5	+	TGATCAGGATCCTTAATTTATTGGCTTGCAGAACT	B5+E3: 1150
42KE3	–	GTACTTGAATTCGTTCTCCTCCATCCCCCTGGGCG	
42N ref	–	GTAGAAGGTGTTGTAGCGGTT	B5+N ref.2: 660
ATI up	+	AATACAAGGAGGATCT	ATI up + ATI low: 1596
ATI low	–	CTTAACTTTTTCTTTCTC	

3. RESULTS

No amplification product was detected when Fowl pox virus DNA or water controls were used as templates. Scab specimens collected from camels or sheep that tested positive for PPV by all techniques gave negative results when tested for OPV and *vice versa*. The size of the amplification products exactly corresponded to that obtained with the reference PPV strain NZ-2 (1150 bp in the first round PCR and 660 bp in the semi-nested PCR) and the reference OPV (1596 bp for Vaccinia and 881 for CMP) (Figures 1 and 2).

A total of 8 skin scab samples were negative by all diagnostic techniques used. Overall the results of the 5 techniques agreed on 21 specimens (53%). Comparison of the PCR assay using 3 DNA extraction procedures with classical methods of electron microscopy and virus isolation is shown in Figure 3. Additionally 10 cell culture supernatants collected from KOP and Vero cells infected with various PPV and OPV strains gave positive results when tested by Procedure 2.

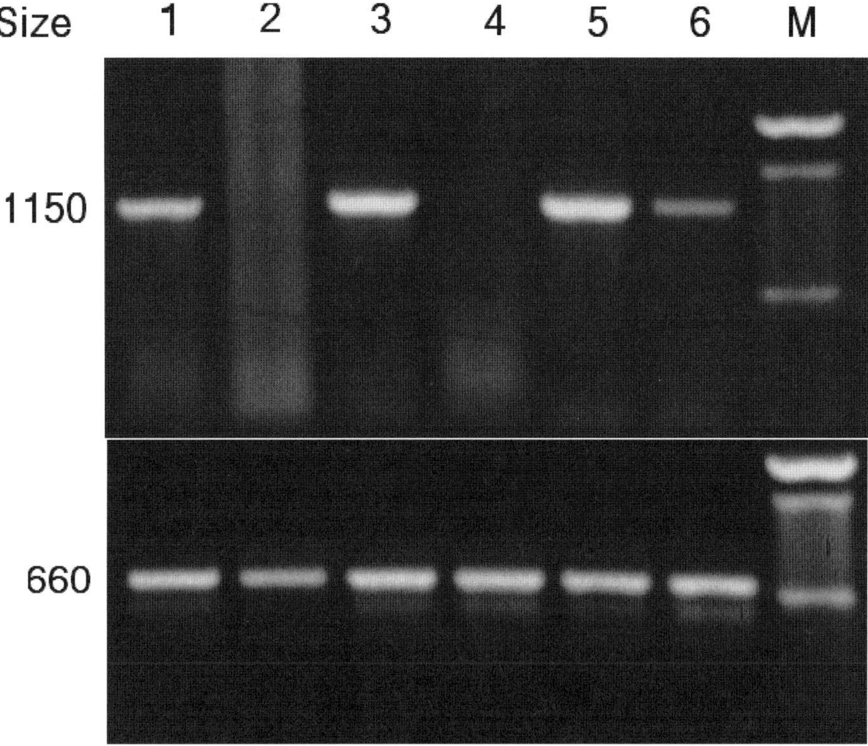

Figure 1. Visualization of Parapoxvirus PCR products by agarose gel electrophoresis. Above: First round PCR. Below: Semi-nested PCR. Lanes 1-6: Scab homogenates. Lane M: 100-bp molecular weight marker (Invitrogen). Product size in bp.

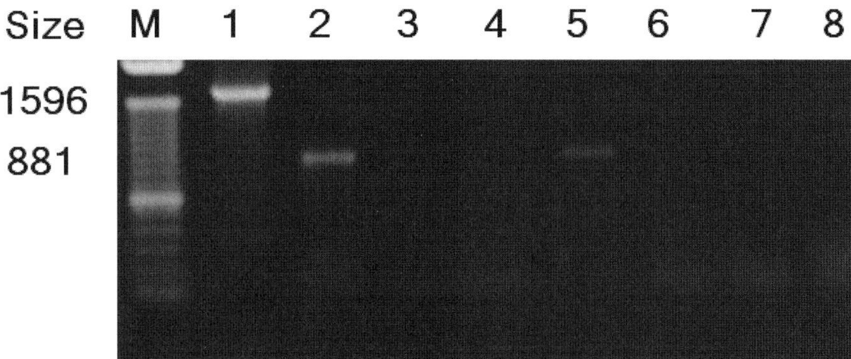

Figure 2. Visualization of Orthopoxvirus PCR products by agarose gel electrophoresis. Lane 1: Vaccinia virus, Elstree strain; lanes 2 and 5: CMP-infected scab specimens; lanes 3, 4 and 6: Parapoxvirus scab specimens (negative); lane 7: Fowl pox virus (negative control); lane 8: water (negative control); and lane M: 100-bp molecular weight marker (Invitrogen).

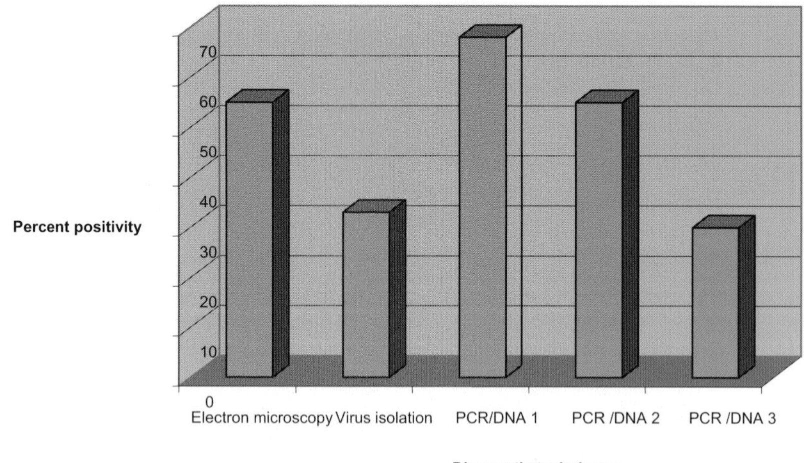

Figure 3. Comparison of electron microscopy, virus isolation and PCR using 3 DNA preparation procedures for detection of pox virus infections in camels

4. DISCUSSION

Viral diseases are difficult to control by chemotherapy. It is therefore necessary to apply rapid and reliable methods for diagnosis at an early stage, so that immediate control measures can be introduced in a timely manner. A number of diagnostic techniques for detection of the agents of CMP and CCE in skin specimens of camels have been reported. These include virus isolation in cell culture or embryonated chicken eggs, electron microscopy, immunocytochemistry, and ELISA (Munz *et al.*, 1986; Nothelfer, Wernery and Czerny, 1995; Azwai, Carter and Woldehiwet, 1995; Khalafalla, Mohamed and Ali, 1998). Most of these methods have disadvantages, such as being time consuming or expensive. Electron microscopy is the fastest technique for Poxvirus diagnosis, but has poor sensitivity, and the equipment is not always available in countries where these infections occur. Virus isolation and identification is a sensitive method for diagnosis, but it may take weeks to obtain results, by which time the results may no longer be useful.

In the present study, the sensitivity of a PCR assay using 3 different DNA extraction procedures for the detection of PPV and OPV in camels was compared with that of electron microscopy or virus isolation in cell culture. Procedure 1, in which viral DNA was extracted from scab specimens followed by PCR, proved to be superior and more sensitive than electron microscopy, virus isolation and the other two DNA extraction methods (Figure 3). In Procedure 2, the scab homogenates were used directly in the PCR reaction without a DNA extraction step. This procedure gave less sensitivity than Procedure 1, but comparable results to electron microscopy. In addition, this procedure is much faster and reduces the risk of DNA contamination that might occur during extraction and transfer of samples from one tube to another. Moreover, this assay is a useful and rapid confirmatory test, since it recognizes Orthopox and Parapox viruses in cell culture supernatants.

Virus isolation in cell culture detects only live virus particles, and therefore particles that have been inactivated during processing of scabs or transportation to the laboratory will not be detected. This explains why many scab specimens are PCR positive and virus isolation negative. Procedure 3 gave less sensitivity and detected only 44% of specimens positive by Procedure 1. Loss of virus particles during purification through the 36% sucrose cushion seems to be responsible for this result.

The PCR assay described in this report is a valuable addition to the current armoury of methods for diagnosis and differentiation of PPV and OPV infections in camels. Well-equipped laboratories with cell culture and electron microscopy facilities are not needed, and results can be obtained in a small laboratory within 24 hours of sample receipt. This is expected to help the implementing of prompt measures to control these important diseases, resulting in improved health and productivity of national camel herds. A multiplex format of this PCR could be envisaged in order to simplify the procedure.

ACKNOWLEDGEMENTS

This work was supported by the Alexander von Humboldt Foundation, Germany, through a George Forster Post-doctoral Fellowship to the first author. We also thank K. Mildner for her technical assistance in electron microscopy.

REFERENCES

Azwai, S.M., Carter, S.D. & Woldehiwet, Z. 1995. Immune response of the camel (*Camelus dromedarius*) to contagious ecthyma (Orf) virus infection. *Veterinary Microbiology,* **47**: 119–131.

Khalafalla, A.I. & Mohamed, M.E.H. 1997. Epizootiology of Camel Contagious Ecthyma in eastern Sudan. *Revue d'Elevage et de Medecine Veterinaire des Pays Tropicaux,* **50:** 99–103.

Khalafalla, A.I., Mohamed, M.E.M. & Ali, B.H. 1998. Camel pox in the Sudan: I – Isolation and identification of the causative virus. *Journal of Camel Practice and Research,* **5**: 229–233.

Jezek, Z., Kriz, B. & Rothbauer, V. 1983. Camel pox and its risk to the human population. *Journal of Hygiene Epidemiology Microbiology and Immunology,* **27: 29–42.**

Meyer, H., Pfeffer, M. & Rziha, H.J. 1994. Sequence alterations within and downstream of the A-type inclusion protein genes allow differentiation of orthopox virus species by polymerase chain reaction. *Journal of General Virology,* **75**: 1975–1981.

Munz, E., Schillinger, D., Reimann, M. & Mahnel, H. 1986. Electron microscopical diagnosis of ecthyma contagiosum in camels (*Camelus dromedarius*). First report of the disease in Kenya. *Journal of Veterinary Medicine, Series B,* **33:** 73–77.

Nothelfer, H.B., Wernery, U. & Czerny, C.P. 1995. Camel Pox: Antigen detection within skin lesions – immunohistochemistry as a simple method of etiological diagnosis. *Journal of Camel Practice and Research,* **2:** : 119–121.

Inoshima, Y., Morooka, A. & Sentsui, H. 2000. Detection and diagnosis of Parapoxvirus by the polymerase chain reaction. *Journal of Virological Methods,* **84:** 201–208.

Ireland, D.C. & Binepal, Y.S. 1998. Improved detection of capripoxvirus in biopsy samples by PCR. *Journal of Virological Methods,* **74:** 1–7.

Sullivan, J.T., Mercer, A.A., Fleming, S.B. & Robinson, A.J. 1994. Identification and characterisation of an orf virus homologue of the vaccinia virus gene encoding the major envelope protein p37K. *Virology,* **202**: 471–478.

Torfason, E.G. & Gudnadottir, S. 2002. Polymerase chain reaction for laboratory diagnosis of Orf virus infections. *Journal of Clinical Virology,* **24:** 79–84.

DEVELOPMENT OF A NEW LIVE ROUGH VACCINE AGAINST BOVINE BRUCELLOSIS

Diego J. Comerci,[1] Juan E. Ugalde[2] and Rodolfo A. Ugalde[2]

1. Grupo Pecuario, Unidad de Aplicaciones Tecnológicas y Agropecuarias, Centro Atómico Ezeiza, Comisión Nacional de Energía Atómica, Buenos Aires, Argentina.
2.Instituto de Investigaciónes Biotecnológicas, Universidad Nacional de Gral. San Martín, Parque Tecnológico Miguelete, INTI Ed. 24, San Martín, Buenos Aires, Argentina.

Abstract: *Brucella abortus* S19 is the most commonly used attenuated live vaccine to prevent bovine brucellosis. In spite of its advantages, S19 has several drawbacks: it is abortive for pregnant cattle, is virulent for humans, and re-vaccination is not advised due to the persistence of anti-lipopolysaccharide (LPS) antibodies that hamper the immunoscreening procedures. For these reasons, there is a continuous search for new bovine vaccine candidates. We have previously characterized the phenotype of the phosphoglucomutase (*pgm*) gene disruption in *Brucella abortus* S2308, as well as the possible role for the smooth LPS in virulence and intracellular multiplication. Here we evaluate the vaccine properties of an unmarked deletion mutant of *pgm*. Western blot analysis of purified lipopolysaccharide and whole-cell extract from Δ*pgm* indicate that it synthesizes O-antigen but is incapable of assembling a complete LPS. In consequence Δ*pgm* has a rough phenotype. Experimental infections of mice indicate that Δ*pgm* is avirulent. Vaccination with Δ*pgm* induces protection levels comparable to those induced by S19, and generates a splenocyte proliferative response and cytokines profile typical of a Th-1 response. The ability of the mutant to generate a strong cellular Th-1 response without eliciting specific O-antigen antibodies highlights the potential use of this mutant as a new live vaccine for cattle.

1.   INTRODUCTION

Brucella spp. are Gram-negative, facultative intracellular bacteria that cause a chronic zoonotic disease worldwide, known as brucellosis. *Brucella*

H. P. S. Makkar and G. J. Viljoen (eds.), Applications of Gene-Based Technologies for Improving Animal Production and Health in Developing Countries, 743-749.

abortus, the etiological agent of bovine brucellosis, causes abortion and infertility in cattle and undulant fever in humans (Corbel, 1997). Due to its intracellular localization, control of the infection requires a cell-mediated immune response, in which the Th-1 arm is relevant for protection (Jiang and Baldwin, 1993).

Brucella abortus S19 is the most commonly used attenuated live vaccine to prevent bovine brucellosis. The vaccine induces good levels of protection in cattle, preventing premature abortion. Although *B. abortus* S19 is by far the most used vaccine in eradication campaigns worldwide, it has two major problems: it produces abortion when administered to pregnant cattle and is fully virulent for humans; and it generates persistent anti-smooth lipopolysaccharide (LPS) antibodies that interfere with the discrimination between infected and vaccinated animals during immune-screening procedures (Sutherland and Searson, 1990). In order to avoid these problems, several strategies for the development of alternative vaccines have been described. One of them is the development of avirulent or attenuated strains lacking the O-antigen (i.e. rough strains) unable to induce antibodies that interferes with the diagnosis. The spontaneously rough strain 45/20 has the disadvantage of reverting to a smooth virulent phenotype. In consequence, this strain is no longer used as live vaccine. Another rough mutant strain, RB51, has been isolated after a series of passages in selective media (Schurig *et al.*, 1991). This strain has good vaccine properties, without interfering with diagnosis; however, it is rifampicin-resistant and the cause of its avirulence and rough phenotype is still not completely understood (Schurig, Sriranganathan and Corbel, 2002). Thus, several efforts to improve this vaccine have been made (Vemulapalli *et al.*, 2000).

In our laboratory we have previously cloned, sequenced and disrupted the gene coding for the enzyme phosphoglucomutase (*pgm*), responsible for the interconversion of glucose-6-phosphate to glucose-1-phosphate. The mutant does not synthesize the sugar nucleotide UDP-glucose and thus is unable to form any polysaccharide containing glucose, galactose or any other sugars whose synthesis proceeds through a glucose-nucleotide intermediate (Ugalde *et al.*, 2000). The mutant has a rough phenotype, and is avirulent in mice. These characteristics prompted us to evaluate the potential use of this strain as a live rough vaccine.

2.   GENERATION OF AN UNMARKED DELETION OF *pgm*

A problem in developing new live vaccines with mutant strains obtained by genetic engineering is the introduction of antibiotic resistant markers. To generate an unmarked deletion of *pgm* we constructed a dicistronic cassette with a promoterless *sacB* gene from *Bacillus subtilis* and the *accI* gene coding for a resistance to gentamicin. The cassette was introduced in a unique *Eco*RV site of the *pgm* cloned in pUC19 and the construct was electroporated in *B. abortus* S2308. The mutants were screened for gentamicin resistance and lack of growth in 10% sucrose. In the next step, a suicide plasmid containing a deleted copy of *pgm* was conjugated into a selected Gm^r-Sac^s intermediate strain and exconjugants were selected in a medium containing 10% sucrose and nalidixic acid. Selected colonies were tested for sensitivity to gentamicin and the genetic replacement events were confirmed by PCR and Southern blot. The resulting strain, named Δ*pgm,* was used for further studies.

3.   THE ROUGH PHENOTYPE OF Δ*pgm*

As in many other Gram-negative bacteria, LPS is an important component of the outer membrane. It has three domains: the lipid A, the oligosaccharide core and the O-side chain or O-antigen. The complete structure of the *Brucella* LPS has not been completely elucidated. The O-side chain is a linear homopolymer of α-1,2-linked 4,6-dideoxy-4-formamido-α-D-mannopyranosyl (perosamine) subunits with a degree of polymerization between 96 to 100 subunits. Sodium dodecyl sulphate (SDS)-PAGE and Western blot analysis of whole-cell extract and purified LPS from Δ*pgm* indicated that this strain is able to synthesize O-antigen of 45 kDa but is incapable of assembling a complete LPS, probably due to the presence of an altered core structure. In consequence, Δ*pgm* displays a rough phenotype.

4.   VIRULENCE OF Δ*pgm*

In order to analyse the virulence of Δ*pgm*, a high dose (10^7 CFU) of *B. abortus* S2308 and Δ*pgm* strains were inoculated intraperitoneally in mice. At 1, 3, 5 and 8 weeks post-inoculation (p.i.), groups of five mice were sacrificed, spleens removed and the number of viable *Brucella* examined. The number of viable bacteria recovered from spleens of mice inoculated with Δ*pgm* were, at all times tested, significantly lower than those inoculated with strain S2308, and was completely cleared at 8 weeks p.i., whereas the number of S2308 remained high (average of $5.5{\times}10^5$ CFU per spleen). These results indicate that even at high doses (vaccine dose), strain Δ*pgm* is avirulent for mice and is cleared from the animal in a short time.

5.   ANTIBODY RESPONSE AGAINST O-ANTIGEN

The presence in the mutant whole cell extracts of O-antigen molecules not attached to the LPS prompted us to study the antibody response against O-antigen of mice inoculated with 5×10^5 CFU of Δpgm or *B. abortus* S2308 strains. Specific anti-O antibodies were detected by Fluorescence Polarization Assay (FPA) (Nielsen *et al.*, 1998). Mice receiving *B. abortus* S2308 developed antibodies against the O-antigen that reached its maximal value at 49 days p.i (152.71 ±27.65 mP). In contrast Δpgm-vaccinated mice failed, as the saline control mice, to elicit antibodies against O-antigen at any time tested (92.33 ±3.66 mP and 92.2 ±7.10 mP at 49 p.i., respectively) (Figure 1). Moreover, the sera from Δpgm-inoculated animals failed to agglutinate at 1:25 dilution in a buffered antigen plate agglutination test (BPAT). In contrast, a fast and strong agglutination was observed with sera from the infected animals at 1:25 and 1:250 dilutions. These results indicate that the O-antigen present in Δpgm is incapable of eliciting a detectable specific antibody response.

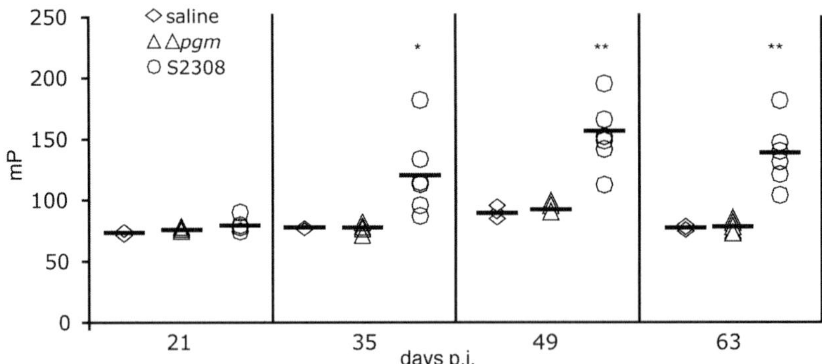

Figure 1. Elicitation of anti O-antigen antibodies. Groups of six mice were inoculated with 5×10^5 CFU of S2308, Δpgm or saline, and sera were collected at different times post-infection (p.i.). Titres of specific anti-O-antigen antibodies, expressed as millipolarization units (mP), were determined by fluorescence polarization assay. Horizontal bars represent the average values of the group. Probability: * *P*< 0.01, ** *P*< 0.001, compared with control S2308-inoculated mice.

6.   GENERATION OF CELLULAR IMMUNE RESPONSE IN MICE

To investigate the cellular immune response induced by *Δpgm* we analysed the proliferative response and the cytokine profile from splenocytes of vaccinated and non-vaccinated mice upon stimulation with heat-inactivated *B. abortus* S2308 whole cells. Eight weeks p.i., splenocytes recovered from mice vaccinated with *Δpgm* proliferated in a specific manner upon stimulation, in contrast with the non-vaccinated control group (60,142 ±7,443 cpm, *versus* 20,855 ±2,541 cpm; *P*<0.001). The lymphoproliferative responses of the vaccinated animals were concomitant with the secretion of high levels of IFN-γ (112.0 ng/ml *versus* 16.6 ng/ml in the non-vaccinated control group). IL-4 was not detected in the supernatants of splenocytes obtained from immunized or non-immunized animals. Taken together, these results indicate that immunization with strain *Δpgm* elicits a classical cellular Th-1 response.

7.   PROTECTION OF MICE WITH A *ΔPGM* LIVE VACCINE

To examine protection induced by *Δpgm,* a vaccine-challenge experiment was performed. Mice were vaccinated intraperitoneally with 10^7 CFU of *Δpgm,* phosphate buffer-saline (PBS) or 10^5 CFU of *B. abortus* S19. Eight weeks p.i., animals were challenged with 5×10^5 CFU of virulent strain *B. abortus* S2308. Protection was defined as the difference between the number of viable bacteria recovered from spleens of immunized mice compared with those receiving saline; results are summarized in Table 1. Vaccine efficacy was expressed as $\log_{10}$ units of protection. *Δpgm* generated significant protection 2 and 4 weeks post-challenge, with 2.25 and 1.93 protection units, respectively. As expected, *B. abortus* strain S19 also induced significant protection at 4 weeks (1.78 protection units). These results, alongside the inability of the mutant to elicit specific O-antigen antibodies, confirm the potential use of this mutant as a live vaccine for cattle.

Table 1. Protection against *B. abortus* S2308.

Treatment group (n = 5)	Mean $\log_{10}$ of brucellae ± SD in spleen at time (days) post-challenge		$\log_{10}$ of protection at time (days)	
	14	28	14	28
saline	5.35 ±0.14	4.73 ±0.41	---	---
Δpgm	3.10 ±0.37	2.80 ±0.87	2.25 [a]	1.93 [a]
S19	ND	2.95 ±0.48	ND	1.78 [a]

NOTE: [a] = P< 0.05 (significant) compared with value for control mice. ND = not determined.

8.   CONCLUSION

The absence of phosphoglucomutase impairs the biosynthesis of glucose-1-phosphate, which precludes the generation of the activated sugar-nucleotide needed in all biosynthetic reactions requiring activated glucose (mainly synthesis of polysaccharides, glycoproteins and glycolipids). In consequence, Δpgm has a rough phenotype, but is able to synthesize the O-antigen, a polymer of N-formylperosamine. Although Δpgm has a detectable level of O-antigen, it is not capable of inducing detectable specific antibodies. At the same time, the specific lymphocyte proliferative response upon stimulation with heat-inactivated *Brucella*, the high level induction of IFN-γ and the absence of IL-4 secretion strongly suggest that this strain induces a strong cellular Th-1-response in mice. Our vaccine-challenge experiments indicate that this strain induces protection levels equivalent to those induced by S19 (approximately 2 log units) and has a severe reduction in virulence at all times tested.

Preliminary results showed that once Δpgm was inoculated into previously *B. abortus* S19 vaccinated calves, it failed to induce antibody response against smooth LPS, as detected by routine agglutination assays, thus suggesting that O-antigen from Δpgm failed to elicit a humoral memory response in cattle. These characteristics suggest that Δpgm could be a useful vaccine strain to improve livestock immunological status by re-vaccination in specially designed eradication campaigns in countries with large cattle populations.

ACKNOWLEDGEMENTS

This work was supported by Grants from the Agencia Nacional de Promoción Científica y Técnica, SECyT, Argentina, PICT99 N° 01-06565 and PICT2000 N° 01-09194. D.J.C. is a researcher from the National Atomic Energy Commission, CNEA, Argentina. J.E.U. is a fellow of the Consejo Nacional de Investigaciones Científicas y Técnicas, CONICET, Argentina. R.A.U. is a member of the research arm of CONICET, Argentina. D.J.C. and J.E.U. contributed equally to this work.

REFERENCES

Corbel, M.J. 1997. Brucellosis: an overview. *Emerging Infectious Diseases,* **3**: 213–221.
Jiang, X. & Baldwin, C.L. 1993. Effects of cytokines on intracellular growth of *Brucella abortus*. *Infection and Immunity,* **61**: 124–134.

Sutherland, S.S. & Searson, J.   1990.   The immune response to *Brucella abortus*: the humoral immune response. pp. 65–81, *in:* J. R. Duncan and K. Nielsen (eds). *Animal Brucellosis*. Boca Raton, FL: CRC Press.

Schurig, G.G., Roop, R.M., Bagchi, T., Boyle, S., Buhrman, D. & Sriranganathan, N. 1991. Biological properties of RB51; a stable rough strain of *Brucella abortus*. *Veterinary Microbiology,* **28**: 171–188.

Schurig, G.G., Sriranganathan, N. & Corbel, M.J.   2002.   Brucellosis vaccines: past, present and future. *Veterinary Microbiology,* **90**: 479–496.

Vemulapalli, R., He, Y., Buccolo, L.S., Boyle, S.M., Sriranganathan, N. & Schurig, G.G. 2000. Complementation of *Brucella abortus* RB51 with a functional wboA gene results in O-antigen synthesis and enhanced vaccine efficacy but no change in rough phenotype and attenuation. *Infection and Immunity,* **68**: 3927–3932.

Ugalde, J.E., Czibener, C., Feldman, M.F. & Ugalde, R.A.   2000.   Identification and characterization of the *Brucella abortus* phosphoglucomutase gene: role of lipopolysaccharide in virulence and intracellular multiplication. *Infection and Immunity,* **68**: 5716–5723.

Nielsen, K., Gall, D., Lin, M., Massangill, C., Samartino, L., Perez, B., Coats, M., Hennager, S., Dajer, A., Nicoletti, P. & Thomas, F.   1998.   Diagnosis of bovine brucellosis using a homogeneous fluorescence polarization assay. *Veterinary Immunology and Immunopathology,* **66**: 321–329.

THE APPLICATION OF GENE-BASED TECHNOLOGIES IN THE STUDY OF NEWCASTLE DISEASE VIRUS ISOLATES FROM UGANDA

Maxwell. O. Otim,[1,2] Magne Bisgaard,[2] Henrick Christensen,[2] Poul Jorgensen[3] and Kurt Handberg.[3]

1. *Livestock Health Research Institute, LIRI, P. O. Box 96, TORORO, Uganda.*
2. *Department of Veterinary Microbiology, The Royal Veterinary and Agricultural University, Stigbøjlen 4, DK-1870 Frederiksberg C, Denmark.*
3. *Danish Veterinary Institute (DVI), Hangøvej 2, DK-8200 Århus N, Denmark.*

Abstract:

Molecular techniques were used to characterize 16 Newcastle disease (ND) Virus (NDV) isolates from ND outbreaks in chickens in eastern Uganda in 2001, and evaluate ND epidemiology, with emphasis on molecular aspects. F and HN genes, which are the major determinants of virulence, were studied. Strain pathogenicity was derived from genetic analysis of the F gene sequence and intracerebral pathogenicity index (ICPI).

Comparative genetic and phylogenetic tree analyses were performed on the HN genes of the isolates and some strains selected from GenBank. ClustalX 1.81 and Phylip were used for gene alignment analysis and the final phylogeny was produced by the neighbour-joining method. F gene cleavage site sequence analysis, phylogenetic analysis and biological characterization showed that the strains were very virulent and closely related, being of common ancestry.

All the Ugandan NDV isolates formed separate clades from the currently known genotypes, suggesting that they are a novel genotype, unrelated to those that have caused previous pandemics.

751

H. P. S. Makkar and G. J. Viljoen (eds.), Applications of Gene-Based Technologies for Improving Animal Production and Health in Developing Countries, 751-771.
© 2005 *IAEA. Printed in the Netherlands.*

1.   INTRODUCTION

Newcastle disease (ND) is a serious illness worldwide, primarily affecting birds, particularly chickens, and has been one of the major causes of economic loss in the poultry industry (Alexander, 1988). The aetiological agent of the disease, Newcastle disease virus (NDV), is a single-stranded, negative-sense, enveloped RNA virus of the Paramyxoviridae family in the order of Mononegavirales (Murphy *et al.,* 1995). The RNA is 15,186 nucleotides (nt) in size (de Leeuw and Peters, 1999) and contains six genes – 3′-NP-P-M-F-HN-L-5′ – encoding for six major polypeptides, namely: nucleocapsid protein (NP), phosphoprotein (P), matrix protein (MP), fusion protein (F), haemagglutinin-neuraminidase (HN), and large RNA-dependent polymerase protein (L) (Miller and Emmerson, 1988).

F protein is an important determinant for NDV pathogenicity (Russel *et al.,* 1990; Rott, 1992). It is synthesized as a nonfunctional precursor F0, which is proteolytically cleaved to yield F1 and F2 polypeptides by host proteases (Morrison, 1993). This is the basis for determining pathogenicity by molecular typing of different pathotypes (Garten *et al.,* 1980; Nagai *et al.,* 1976; Ogasawara *et al.,* 1992). The fusion protein cleavage site is related to NDV pathogenicity, and efficiency of proteolytic cleavage is dependent on the host cell and the virus strain (Nagai *et al.,* 1976, 1979).

Different pathotypes are characterized by differences in the amino acid sequences surrounding the fusion protein (F0) cleavage site, which hosts the molecular marker for virulence. The F0 of lentogenic strains possess two single basic amino acids at the cleavage site, 113 and 116, with leucine at 117, while sequences of mesogenic strains of intermediate virulence for chicken contain two pairs of basic amino acid residues or a single arginine and a lysine-arginine pair. The F0 of virulent NDV has two pairs of basic amino acids at the cleavage site, 112–113 and 115–116, with phenylalanine at 117, making their fusion protein susceptible to cleavage by an omnipotent protease, helping them to fuse with a wide range of cells, resulting in a fatal systemic infection (Alexander, 1991; Nanthakumar *et al.,* 2000).

The classification of NDV strains can be done by pathotyping in chickens and molecular typing. NDV strains can thus be classified into highly virulent (velogenic) strains, intermediate (mesogenic) strains, or non-virulent (lentogenic) strains on the basis of pathogenicity for chicken (Beard and Hanson, 1984).

Recently developed molecular techniques that identify pathotypes on the basis of the deduced amino acid sequences of the F protein cleavage site or restriction sites of the F gene (Hodder *et al.,* 1993; Ballagi-Pordany *et al.,* 1996) have been used to characterize NDV. The differences in F gene sequences that correlate with different virulent phenotypes have been prime targets for molecular approaches to identify and characterize NDV isolates (Oberdorfer and Werner, 1998). Along with biological virulence determination, OIE accepts reporting of the F protein cleavage site sequence of NDV isolates as a virulence criterion (Berinstein *et al.*, 2001).

Lomniczi *et al.*, (1999) used restriction enzyme site mapping of fusion (F) protein gene and sequence analysis to classify 45 NDV isolates into seven genotypes. Isolates from outbreaks in Western Europe between 1992 and 1996 belonged to genotypes VI and VII. Herczeg *et al.* (1999) identified two novel genetic groups, VIIb and VIII, from ND outbreaks in southern Africa.

Phylogenetic studies based on the sequences of both the F and HN NDV genes have also been used extensively in molecular epidemiology and characterization of NDV (Glickman *et al.*, 1988; Ballagi-Pordany *et al.*, 1996; Ke *et al.*, 2001) and to group NDV into specific lineages or clades (Wesbury, 2001).

Partial sequence data from the F0 cleavage site and HN proteins of NDV isolates have been used to reliably predict pathotypes and garner epidemiological information on NDV field isolates (Collin *et al.*, 1993; Sakaguchi *et al.*, 1989; Seal *et al.*, 1995, 2000). Gould *et al.* (2001) used the NDV HN gene to carry out a molecular epidemiological analysis of NDV from Australian ND outbreaks of 1998–2000.

The first ND outbreak in Uganda was documented in 1955, but it was not until 1986 that an NDV isolate was characterized as a virulent strain, using monoclonal antibodies (Mukiibi, 1992). The disease, however, remains endemic in Uganda, with wild birds and movements by poultry vendors suspected as possible sources of ND infection.

The aim of the present study was to use molecular techniques to genetically characterize (predict pathotype) and phylogenetically group NDV isolates collected from ND outbreaks in Uganda in 2001. The study analyses the haemagglutinin-neuraminidase (HN) and fusion (F) proteins of the virus and the genes that code them. Pathogenicity testing was also carried out in specific-pathogen-free (SPF) chicken.

2.   MATERIALS AND METHODS

2.1   Virus isolates

A total of 16 NDV field isolates from ND outbreaks in Uganda were analysed in the study. A panel of 4 antisera – NDV Ulster (NDV1 polyclonal antibody), 7D4 (LaSota monoclonal antibody), PMV3 (PMV3 monoclonal antibody) and U85 (PMV1 monoclonal antibody) – were used in the haemagglutination inhibition (HI) test. The LaSota strain was used as the positive control in HI and for reverse transcription-polymerase chain reaction (RT-PCR).

2.2   Field samples

2.2.1   General background information

Samples were collected from eight suspected ND outbreaks in three districts (Pallisa, Tororo and Soroti; Figure 1) in Uganda, between September and November 2001. Two criteria, based on spatial and temporal distribution of outbreaks, were used to define an outbreak's incidence (i.e. new outbreak or not). If a disease occurred in a given location with a radius of about 10 km, it was considered an outbreak. If there was another disease outbreak at the same time but in a different location outside the area where an outbreak had already occurred, the two were considered to be separate outbreaks. If disease outbreaks were reported at different times, they were considered as new outbreaks, even when they occurred in the same location, as long as the second outbreak occurred after the first one was considered finished.

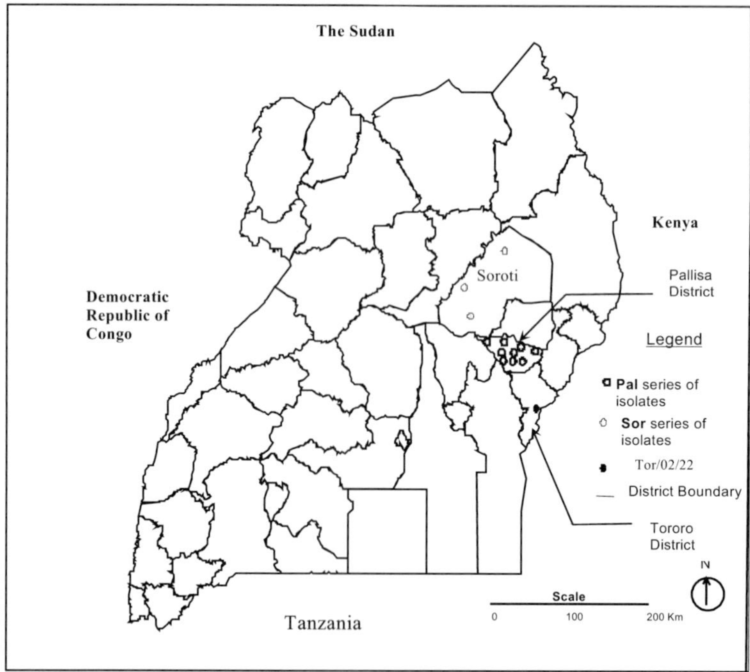

Figure 1. Epidemiological map showing areas from which NDV isolates were collected in Uganda, 2001.

2.2.2   Sample handling

Samples were collected from the lungs, trachea, heart, liver, spleen, kidney and intestines of sacrificed chickens suspected to have ND. Most had diarrhoea, and also showed nervous symptoms and respiratory difficulties. The tissue samples were kept in phosphate-buffered saline (PBS) containing antibiotics, penicillin (2000 units/ml); streptomycin (2 mg/ml); gentamycin (50 µg/ml); and mycostatin (1000 units/ml), and transported on ice to the laboratory. The samples were then stored at -20°C until processed for inoculation into chicken eggs.

To prepare the samples for inoculation, about 5 g of each tissue sample was pooled and finely ground using lake-sand in a mortar; 10 ml of cold PBS (pH 7.4) was then added to the suspension; and clarified by centrifuging at 1000 rpm for 10 minutes at room temperature. The supernatants obtained were filtered through a sterile 0.25 µm-pore filter, and 0.2 ml of the filtrate was inoculated into 10-day-old embryonated chicken eggs. The eggs were then incubated for 4–7 days and candled daily. Eggs containing a dead embryo were chilled overnight at 4°C and allantoic fluid harvested from them. The allantoic fluid was then tested for haemagglutination (HA) activity. Samples that were positive were then transferred to vials and sent to Aarhus Veterinary Institute, Denmark, for further analysis (Table 1).

2.3   Egg passage

In Aarhus, the isolates were further replicated in SPF eggs, as described elsewhere (Palmeri, 1989; Palmeri and Perdue, 1989). Briefly, 9-to-10-day-old SPF white leghorn embryonated eggs were inoculated with 0.2 ml of the allantoic fluid sample and incubated at 37°C. The eggs were candled daily; those with dead embryos or those 4 days post-inoculation were removed and refrigerated overnight. Allantoic fluid was harvested, and tested for any bacterial contamination by plating on blood agar. Allantoic fluids contaminated with bacteria were treated with antibiotic solution, allowed to stand for one hour and centrifuged at 13,000 rpm for 1 minute. The allantoic fluid was then transferred to vials and stored at -80°C till use.

Table 1. Isolates analysed in the study and their sources

Outbreak	Isolate	Outbreak location
1	Pal/01/03, Pal/01/02, Sor/01/04	Pallisa and Soroti
2	Pal/02/06, Pal/02/08, Tor/02/22	Pallisa and Tororo
3	Pal/03/05, Pal/03/21	Pallisa
4	Pal/04/05, Pal/04/06, Pal/04/07, Pal/04/08	Pallisa
5	Pal/05/15, Sor/05/09	Pallisa and Soroti
6	Sor/06/09	Soroti
7	Sor/08/14	Soroti

2.4   Monoclonal antibodies and antiserum

A panel of 4 antisera was used in the HI test: NDV Ulster (NDV1 polyclonal antibody), 7D4 (LaSota monoclonal antibody), PMV3 (PMV3 monoclonal antibody) and U85 (PMV1 monoclonal antibody).

2.5   HA and HI assays (Serological analysis)

HA and HI tests were performed on aseptically harvested allantoic fluids by the conventional microwell method using the 8 haemagglutination units (HAU) according to European Community Directive 92/66/EC (CEC, 1992). In back titrations, 8 HAU of the isolates were serially diluted in the V-bottom well microwell plastic plates from column 2 to 7. Column 8 was left as control. An equal volume of 1% erythrocytes from SPF chicken was added and the plates incubated at 4°C for 30 minutes, whereupon the virus titre was read off.

For the HI test, the available antisera were diluted 1:16 and twofold serial dilutions were made in PBS, pH 7.2, and 8 HAU of test antigen was added to each dilution and incubated at 4°C for 1 hour. An equal volume of 1% chicken erythrocytes in PBS was added as a test indicator, and the plates incubated for 30–40 minutes. The endpoint was determined as the last dilution with complete inhibition of HA activity.

2.6   Viral RNA extraction

RNA extraction from the 16 isolates was carried out using QIAGEN kit (The Rneasy principle and procedure). The method used was modified from that previously described by Yokozaki and Tahara (1995). Briefly, 300 µl of RTL buffer, containing 6 µl of mercaptoethanol, from the QIAgen RNeasy Kit were added to a 1.5 ml Eppendorf tube with 400 µl allantoic fluid. These were then mixed and left on the bench for at least 15 minutes. If the samples were cloudy, they were centrifuged (1 minute at 13,000 rpm) and transferred to new tubes. Then 700 µl 70% ethanol (room temperature) was added and mixed.

From the mixture from the foregoing step, 700 µl was added to an RNeasy spin-column. The samples were centrifuged for 30 seconds at 13,000 rpm. The run through was discarded, and the collection tube was put back. This step was repeated for the rest of the sample. The spin-columns were washed with 700 µl of RW1, centrifuged as above, and the collection tube was replaced with a new one. The samples were then washed with 500 µl RPE buffer and centrifuged again as above. This washing was repeated once. After the last step, the collection tube was emptied, put back on the spin-column and given an additional centrifugation at 13,000 rpm for 2 minutes. The spin-columns were put onto new 1.5 ml Eppendorff tubes and 50 µl of RNase-free water was added. The samples were centrifuged for 1 minute at 13,000 rpm. The spin-columns were discarded and the elute saved at -80°C until use.

2.7   Oligonucleotide primers used in reverse transcription-polymerase chain reaction

Oligonucleotide RT-PCR primers were designed according to published sequences of HN and F genes (Chambers *et al.*, 1996) to amplify regions of the fusion protein gene, including the fusion protein cleavage site, and a portion of the HN protein gene region. Sense and antisense primer pairs were generated by alignment, using ClustalX (1.81), from published NDV nucleotide sequences from GenBank. The primers (Table 2) were then ordered from DNA Technology (Aarhus, Denmark).

The NDV RNA extracted from the allantoic fluid was used to perform a single-tube RT-PCR and generate cDNA.

2.8   Reverse-transcription polymerase chain reaction

The reverse-transcription polymerase chain reaction (RT-PCR) reactions were carried out according to procedures provided by Titan® (Perkin Elmer Branchburg, NJ, USA). Briefly, the RT-PCR reactions were performed in a 50 µl tube containing 29.5 µl of RNase-free water, 10 µl 1X EZ buffer (2.5 mM manganese acetate solution), 1 µl dNTP, 1 µl NA polymerase/enzyme mix (5 units), 1 µl (5 µM) of primer and 1 µg of RNA. Reverse transcription was carried out at 50°C for 30 minutes. PCR reactions were subjected to 35 cycles consisting of denaturation for 1 minute at 94°C, annealing for 1 minute at 55°C, and extension for 1 minute at 72°C, with a final extension cycle at 72°C for 7 minutes. An additional 7 minutes at 60°C was included at the end of the programme to ensure complete extension.

After the completion of the PCR, 10 µl of the reaction mixture was loaded onto a 1.5% agarose gel, containing 10 mg/ml ethidium bromide, for electrophoresis and subsequent visualization by UV transillumination to confirm the presence of PCR products.

Table 2. Primers used in the study.

Label	Sequence	Target gene	Target site	Expected size
NDV2 (Forward)	5′-CTG CCA CTG CTA GTT GBG ATA ATC C-3′	F	Cleavage site	400
NDV3 (Reverse)	5′-GTY AAY ATA TAC ACC TCA TCY CAG ACW GG-3′	F	Cleavage site	400
NDV-HN-F2 (Forward)	5′-AGC ARG YYA TCT TAT CYA TCA ARG TGT CAA C-3′	HN		400
NDV-HN-R1 (Reverse)	5′-TYC TRA ATT CYC CRA AKA GRG TRT TRG ATA TTT-3′	HN		400

PCR amplification products were extracted and purified using the QIAquick® Gel Extraction Kit, according to the protocol described in the Spin Handbook. Some PCR products were also purified by Microspin-columns (Pharmacia). Briefly, for QIAquick® Gel Extraction, the cDNA fragments were excised from the agarose gel with a clean sharp scalpel and 3 volumes of buffer QG was added to each gel slice, mixed properly and incubated at 50°C in a water bath until the gel slices were completely dissolved, whereupon 1 gel volume of isopropanol was added to the sample. The sample was then transferred into QIAquick spin columns and centrifuged for 1 minute at 13,000 rpm, the flow-through discarded and the process repeated for the rest of the sample. Buffer QG (0.5 ml) was added to the spin column, and centrifuged for 1 minute at 13,000 rpm. The flow-through was discarded and the spin column centrifuged for another 1 minute. Buffer PE (0.75 ml) was added to the spin column and centrifuged as above to wash it. The QIAquick column was then transferred into a clean 1.5-ml microcentrifuge tube, 50 µl of Buffer EB was added to the centre of the QIAquick membrane and centrifuged at 13,000 rpm for 2 minutes. The spin column was discarded and 0.1 volume of sodium acetate (3M Sodium and 5M acetic acid) and 3 volumes of absolute ethanol were added to the elute, which was then saved for cycle sequencing.

Microspin column purification was done by pipetting the sample into the middle of a spin column placed in a 1.5-ml Eppendorf tube and centrifuging it at 3000 rpm for 2 minutes. This was repeated, and the purified sample was used for PCR cycle sequencing.

2.9 Cycle-sequencing of the amplification products

Nucleotide sequences were obtained by cycle sequencing of the purified PCR product using PCR primers and BigDye Terminator cycle sequencing. The cycle sequencing PCR was done in 20 µl of PCR reaction mix, containing 10 µl (200 ng) of ds cDNA, 1 µl of milli Q water, 1 µl (3.2 pmol) of primer, and 8 µl of BigDye Terminator. Twenty PCR cycles consisting of denaturation for 10 seconds at 96°C, annealing for 5 seconds at 50°C, extension for 4 minutes at 60°C, and finishing at 4°C were carried out.

The cycle sequencing product was precipitated by the ethanol/Na-acetate method. Briefly, 20 µl of the product was pipetted into a 1.5-ml Eppendorf tube, and 3 µl of 3 M Na-acetate (pH 4.6), 62.5 µl of 95% ethanol and 14.5 µl of milli Q water added. The tube was then vortexed, left for 15 minutes and centrifuged at maximum speed in an Ole Dich centrifuge with cooling. The supernatant was removed, and the pellet re-suspended in 70% ethanol and centrifuged at maximum speed for 10 minutes. The supernatant was removed and the pellet dried for between 5 and 15 minutes in a speed-vac. The final pellet was then re-suspended in 5 µl of loading buffer and loaded into acrylamide gel (18 g urea, 5 ml Long Ranger 40% acrylamide, 5 ml 10× TBE in 25 ml of milli Q water) for analysis in an ABI 377 automated DNA sequencer (Perkin Elmer Applied Biosystems).

2.10   Biological characterization

Because of the similarity at the F0 cleavage sites of the Ugandan NDV isolates, only one representative isolate (Pal/01/03) was used in the intra-cerebral pathogenicity index (ICPI) test. For the purpose of comparison, a Danish isolate (73.72070-1) of low ICPI was included in the pathogenicity test. Characterization was carried out according to the protocol of Alexander (1989). In brief, 0.05 ml of 10-fold dilutions of infective allantoic fluid of the isolates were injected intracerebrally into two groups each of 10 one-day-old SPF chicks, and observed for 8 days following inoculation. During the observation period, healthy chicks were scored as 0, apparently sick ones as 1, and dead chicks as 2. The quotient derived from the sum of scores and the numbers of observations represent the ICPI. An ICPI above 1.2 characterizes a velogenic strain; between 0.7 and 1.2 indicates a mesogenic strain; while an ICPI below 0.7 indicates a lentogenic strain.

2.11   Analysis of nucleotide and deduced amino acid sequences

Nucleotide sequence editing, analysis and prediction of amino acid sequences for the F and HN genes were conducted with Bioedit (sequence alignment editor) software, while the alignments were worked out using ClustalX 1.81 software. The final HN gene sequences were then analysed (Phylogeny; Phylip programs) on-line using DNAdist (distances from DNA sequences), and neighbour-joining methods. The final phylogenetic tree was then drawn using DRAWTREE.

3. RESULTS

3.1 Virus isolation and identification

All the samples isolated in SPF eggs tested positive with NDV Ulster (NDV1 polyclonal antibody) and U85 (PMV1 monoclonal antibody), but negative with 7D4 (LaSota monoclonal antibody) and PMV3 (PMV3 monoclonal antibody). The isolates were therefore identified as APMV-1 by HA and HI tests.

3.2 Sequencing of the cleavage site (F gene)

The primer pairs designed for the RT-PCR amplification of selected regions of the F and HN genes produced expected fragments. Visualization of the products and approximate product lengths can be seen in Figures 2a and 2b.

3.3 ICPI test

Isolate Pal/01/03 gave an ICPI value of 1.8 out of a possible maximum of 2, indicating that it is a highly virulent strain, while the ICPI for the Danish isolate 73.72070-1 was 0.25, showing it to be a lentogenic strain.

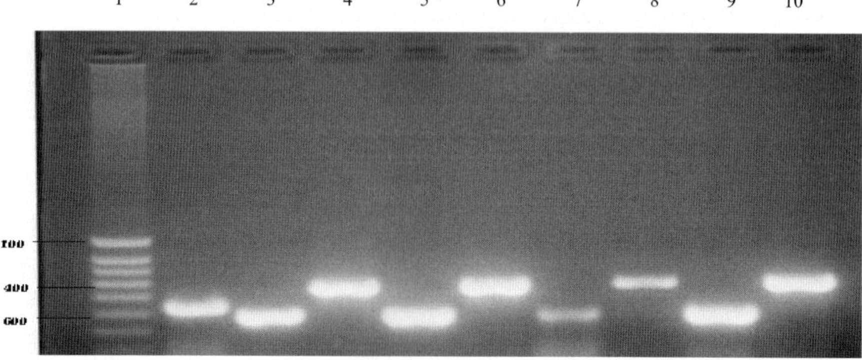

Figure 2a. Agarose gel electrophoresis of PCR products.
KEY: Lane 1: marker. Lanes with isolates amplified by F gene primers: 2 = Sor/05/09; 3 = pal/04/08; 5 = pal/04/05; 7 = pal/04/05; 9 = pal/01/03. Lanes with isolates amplified by HN primer: 4 = pal/04/08; 6 = pal/04/05; 8 = pal/04/05; 10 = pal/01/03.

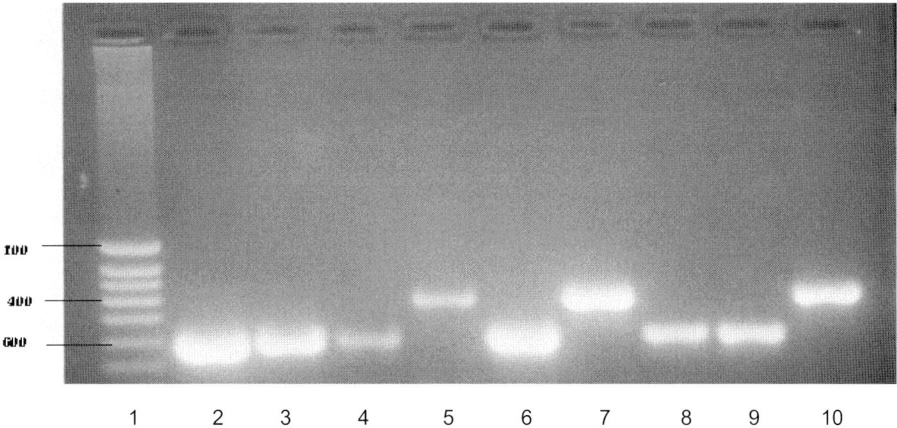

Figure 2b. Agarose gel electrophoresis of PCR products.
KEY: Lane 1: marker. Lanes with isolates amplified by F gene primers: 2 = pal/04/07; 3 = Pal/02/06; 4 = Sor/08/14; 6 = Sor/09/06; 8 = Sor/02/08; 9 = Sor/02/08. Lanes with isolates amplified by HN primer: 5 = Sor/08/14; 7 = Sor/09/06; 10 = Sor/02/08.

3.4   Nucleotide sequence and similarity analysis

F gene fragments of size of 394 bp and 389 bp HN gene fragments were amplified. The similarities of the predicted amino acid sequences of the amplified F gene protein of the isolates are indicated in Figure 3. The amino acid sequence deduced at the cleavage site of F0 protein was identical in all 16 isolates (Figure 3).

Only the HN nucleotide sequences of the Ugandan isolates and isolate NC_002617 from GenBank were first aligned and compared (Figure 4). Later, for phylogenetic analysis, other published HN NDV sequences (Table 3) obtained from GenBank were analysed together with the HN sequences of the 16 Ugandan NDV isolates.

Sequencing of the F gene was performed to identify nucleotides and deduce amino acids adjacent to the proteolytic cleavage site within the F protein. All the 16 Ugandan NDV isolates sequenced had the amino acid sequence ^{112}RRQKRF^{117} at the C-terminus of the F protein and phenylalanine at residue 117, the N-terminus of the F protein (Figure 3). There is therefore the existence of a pair of basic amino acids – R, arginine; and K, lysine – at residues 116 and 115, respectively, and a phenylalanine, F, at residue 117, as well as a basic amino acid, R, arginine, at residue 113, indicating a high virulence (Collins, Strong and Alexander, 1994; Aldous and Alexander, 2001) for the NDV isolates from Uganda for chickens.

Figure 3. Alignment of predicted amino acid sequences surrounding the fusion protein cleavage site. All the 16 Ugandan NDV isolates have the same amino acid residues RRQKR↓FVG. The basic amino acids around the cleavage site (↓) are shown in bold. The [112]RRQKRF[117] motif is consensus sequence around the F2/F1 cleavage site.

A phylogenetic tree based on the HN genes of 16 isolates in a 377-bp region and published NDV HN genes from GenBank was constructed (Figure 5). Phylogenetic analysis shows that the Uganda NDV isolates formed a monophyletic unit closest to genotype VIa (Figure 5).

Table 3. GenBank data on NDV isolates used in comparative analysis with the Ugandan NDV isolates.

NDV Strains	Host	Genotype	Accession no.
D26/76 (Japan)	Duck	I	M24692
V4 (Queensland)/66	Fowl	I	AF217084
Ulster 2C/76 (N. Ireland, UK)	Fowl	I	D00243
NI	NI	II	AF309036
NI	NI	II	M21409
Australia-Victoria (Australia)	Fowl	II	M22110
Beaudette C/45 (USA)	Fowl	II	X04719
Texas GB/48 (USA)	Fowl	II	M23407
LaSota/46 (USA)	Fowl	II	AF077761
Clone 30 (LaSota)	NI	II	Y18898
Russia	NI	II	Y19020
NI	NI	II	AF309418
Miyadera/51 (Japan)	Fowl	III	M18456
AUS Victoria/32 (Australia)	Fowl	III	M21881
ZA-5/68 (South Africa)	NI	III	AF136762
TW/59 (Taiwan, Prov. of China)	Fowl	III	AF083959
SIMF/64 (Russia)	Fowl	IV	Y19019
Italien/45 (Italy)	Fowl	IV	M18640
Texas (USA)	Fowl	IV	M33855
Herts 33 (United Kingdom)	Fowl	IV	M24702
NI	Fowl	IV	AJ243381
NI	Fowl	IV	AJ243385
CA/1085/71 (USA)	Fowl	V	AF001106
H-10/72 (Hungary)	Fowl	V	AF001107
GD/2/98/Go (PR China)	Geese	VI	AF456430
GD/3/98/Go (PR China)	Geese	VI	AF456431
GD/1/98/Go (PR China)	Geese	VI	AF456433
GD/5/98/Go (PR China)	Geese	VI	AF456434
ASTR/74 (Russia)	NI	VIa	Y18725
ASTR/74 (Russia)	Fowl	VIa	Y18728
ASTR/74 (Russia)	NI	VIa	Y19016
MZ-48/95 (Mozambique)	NI	VII	AF136779
Taiwan 95 (Taiwan, Prov. of China)	Fowl	VIIa	U62620
MZ-48 (Mozambique)	NI	VIIb	AF139150
ZA 360/95 (South Africa)	Ostrich	VIIb	AF109876
ZW 3422/95 (Zimbabwe)	Ostrich	VIIb	AF109877
AF 2240 (Malaysia)	NI	VIII	AF048763
AF 2240 (Malaysia)	NI	VIII	X79092
ZA-5/68 (South Africa)	Fowl	VIII	AF139133

NOTE: NI = no information available.

```
                  7560        7570        7580        7590        7600
             ....|....|  ....|....|  ....|....|  ....|....|  ....|....|
NC_002617    ATCCTTAGGC  GAAGACCCGG  TACTG-ACTG  TACCGCCCAA  CACAGTCACA
Pal/02/08    ----------  ..G.......  .G...-....  .......T..  T.....T...
Sor/01/04    ----------  ..G.......  .G...G....  .......T..  T.....T...
Pal/04/06    ----------  ..G.......  .G...-....  .......T..  T.....T...
Pal/03/21    ----------  ..G.......  .G...-....  ......TT..  T.....T...
Pal/03/05    ----------  ..G.......  .G...-....  ......TT..  T.....T...
Sor/05/09    ----------  A.G.......  .G...-....  ......TT..  T.....T...
Pal/01/03    ----------  CCCA......  .G...-....  .......T..  T.....T..G
Tor/02/22    ----------  ..G.......  .G...-....  .......T..  T.....T...
Pal/04/05    ----------  ..G.......  .G...-....  .......T..  T.....T...
Pal/04/08    ----------  ..G.......  .G...-....  .......T..  T.....T...
Pal/05/15    ----------  ..G.......  .G...-....  ....T..T..  T.....T...
Sor/08/14    ----------  ..G.......  .G...-....  .......T..  T.....T...
Pal/06/09    ----------  ..G.......  .G...-....  .......T..  T.....T...
Pal/02/06    ----------  ..G.......  .G...-....  .......T..  T.....T...
Pal/01/02    ----------  ..G.......  .G...-....  .......T..  T.....T...
Pal/04/07    ----------  TTGA......  .G...-....  .......T..  T.....T...

                  7610        7620        7630        7640        7650
             ....|....|  ....|....|  ....|....|  ....|....|  ....|....|
NC_002617    CTCATGGGGG  CCGAAGGCAG  AATTCTCACA  GTAGGGACAT  CCCATTTCTT
Pal/02/08    ..........  ..........  .G.C......  .....A....  .T........
Sor/01/04    ..........  ..........  .G.C......  .....A.T..  .T........
Pal/04/06    ..........  ..........  .G.C......  .....A....  .T........
Pal/03/21    ..........  ..........  .G.C......  .....A....  .T........
Pal/03/05    ..........  ..........  .G.C......  .....A....  .T........
Sor/05/09    ..........  ..........  .G.C......  .....A....  .T........
Pal/01/03    ..........  ..........  .G.C......  .....A....  .T........
Tor/02/22    ..........  ..........  .G.C......  .....A....  .T........
Pal/04/05    ..........  ..........  .G.C......  .....A....  .T........
Pal/04/08    ..........  ..........  .G.C......  .....A....  .T........
Pal/05/15    ..........  ..........  .G.C......  .....A....  .T........
Sor/08/14    ..........  ..........  .G.C......  .....A....  .T........
Pal/06/09    ..........  ..........  .G.C......  .....A....  .T........
Pal/02/06    ..........  ..........  .G.C......  .....A....  .T........
Pal/01/02    ..........  ..........  .G.C......  .....A....  .T........
Pal/04/07    ..........  ..........  .G.C......  .....A....  .T........

                  7660        7670        7680        7690        7700
             ....|....|  ....|....|  ....|....|  ....|....|  ....|....|
NC_002617    GTATCAGCGA  GGGTCATCAT  ACTTCTCTCC  CGCGTTATTA  TATCCTATGA
Pal/02/08    ...C..A..G  ..T..T....  ..........  ...C...C..  ..C.......
Sor/01/04    ...C..A..G  ..T..T....  ..........  ...C...C..  ..C.......
Pal/04/06    ...C..A..G  ..T..T....  ..........  ...C...C..  ..C.......
Pal/03/21    ...C..A..G  ..T..T....  ..........  ...C...C..  ..C.......
Pal/03/05    ...C..A..G  ..T..T....  ..........  ...C...C..  ..C.......
Sor/05/09    ...C..A..G  ..T..T....  ..........  ...C...C..  ..C.......
Pal/01/03    ...C..A..G  ..T..T....  ..........  ...C...C..  ..C.......
Tor/02/22    ...C..A..G  ..T..T....  ..........  ...C...C..  ..C.......
Pal/04/05    ...C..A..G  ..C..T....  ..........  ...C...C..  ..C.......
Pal/04/08    ...C..A..G  ..C..T....  ..........  ...C...C..  ..C.......
Pal/05/15    ...C..A..G  ..C..T....  ..........  ...C...C..  ..C......C
Sor/08/14    ...C..A..G  ..C..T....  ..........  ...C...C..  ..C.......
Pal/06/09    ...C..A..G  ..C..T....  ..........  ...C...C..  ..C.......
Pal/02/06    ...C..A..G  ..C..T....  ..........  ...C...C..  ..C.......
Pal/01/02    ...C..A..G  ..C..T....  ..........  ...C...C..  ..C.......
Pal/04/07    ...C..A..G  ..C..T....  ..........  ...C...C..  ..C.......

                  7710        7720        7730        7740        7750
             ....|....|  ....|....|  ....|....|  ....|....|  ....|....|
NC_002617    CAGTCAGCAA  CAAAACAGCC  ACTCTTCATA  GTCCTTATAC  ATTCAATGCC
Pal/02/08    .G..ACAT..  T.........  ..........  ..........  ...T.....T
Sor/01/04    .G..ACA...  ........T.  ..........  ..........  ...T.....T
Pal/04/06    .G..ACA...  ........T.  ..........  ..........  ...T.....T
Pal/03/21    .G..ACA...  ........T.  ..........  ..........  ...T.....T
Pal/03/05    .G..ACA...  ........T.  ..........  ..........  ...T.....T
Sor/05/09    .G..ACA...  ........T.  ..........  ..........  ...T.....T
Pal/01/03    .G..ACA...  ........T.  ..........  ..........  ...T.....T
Tor/02/22    .G..ACA...  ....T....T  ..........  ..........  ...T.....T
Pal/04/05    .G..ACA...  ....T....T  ..........  ..........  ...T.....T
Pal/04/08    .G..ACA...  ....T....T  ..........  ..........  ...T.....T
Pal/05/15    .G..ACA...  ....T....T  ..........  ..........  ...T.....T
```

```
                  ....|....|  ....|....|  ....|....|  ....|....|  ....|....|
                      7710        7720        7730        7740        7750
Sor/08/14         .G..ACA...  ....T....T  ..........  ..........  ...T.....T
Pal/06/09         .G..ACA...  ....T....T  ..........  ..........  ...T.....T
Pal/02/06         .G..ACA...  ....T....T  ..........  ..........  ...T.....T
Pal/01/02         .G..ACA...  ....T....T  ..........  ..........  ...T.....T
Pal/04/07         .G..ACA...  ....T....T  ..........  ..........  ...T.....T
                  ....|....|  ....|....|  ....|....|  ....|....|  ....|....|
                      7760        7770        7780        7790        7800
NC_002617         TTCACTCGGC  CAGGTAGTAT  CCCTTGCCAG  GCTTCAGCAA  GATGCCCCAA
Pal/02/08         ..........  ......G...  ....T..A.  ..A.......  .G.....A..
Sor/01/04         .........G  ......G...  ....T..A.  ..A.......  .G.....A..
Pal/04/06         ..........  ......G...  ....T..A.  ..A.......  .G.....A..
Pal/03/21         ..........  ......G...  ....T..A.  ..A.......  .G.....A..
Pal/03/05         ..........  ......G...  ....T..A.  ..A.......  .G.....A..
Sor/05/09         ..........  ......G...  ....T..A.  ..A.......  .G.....A..
Pal/01/03         ..........  ......G...  ....T..A.  ..A.......  .G.....A..
Tor/02/22         ..........  ......G...  ....T..A.  ..A.......  .G.....A..
Pal/04/05         ..........  ......G...  ....T..A.  ..A..G....  .G.....A..
Pal/04/08         ..........  ......G...  ....T..A.  ..A..G....  .G.....A..
Pal/05/15         ......G...  ......G...  ....T..A.  ..A..G....  .G.....A..
Sor/08/14         ..........  ......G...  ....T..A.  ..A..G....  .G.....A..
Pal/06/09         ..........  ......G...  ....T..A.  ..A..G....  .G.....A..
Pal/02/06         ..........  ......G...  ....T..A.  ..A..G....  .G.....A..
Pal/01/02         ..........  ......G...  ....T..A.  ..A..G....  .G.....A..
Pal/04/07         ..........  ......G...  ....T..A.  ..A..G....  .G.....A..
                  ....|....|  ....|....|  ....|....|  ....|....|  ....|....|
                      7810        7820        7830        7840        7850
NC_002617         CTCGTGTGTT  ACTGGAGTCT  ATACAGATCC  ATATCCCCTA  ATCTTCTATA
Pal/02/08         ...A...A.C  ..........  ....T.....  G.....TA..  G.....C.C.
Sor/01/04         ...A...A.C  ..........  ....T.....  G.....TA..  G.....C.C.
Pal/04/06         ...A...A.C  ..........  ....T.....  G.....TA..  G.....C.C.
Pal/03/21         ...A...A.C  ..........  ....T.....  G.....TA..  G.....C.C.
Pal/03/05         ...A...A.C  ..........  ....T.....  G.....TA..  G.....C.C.
Sor/05/09         ...A...A.C  ..........  ....T.....  G.....TA..  G.....C.C.
Pal/01/03         ...A...A.C  ..........  ....T.....  G.....TA..  G.....C.C.
Tor/02/22         ...A...A.C  ..........  ....T.....  G.....TA..  G.....C.C.
Pal/04/05         ...A...A.C  ..........  ....T.....  ......TA..  G.....C.C.
Pal/04/08         ...A...A.C  ..........  ....T.....  ......TA..  G.....C.C.
Pal/05/15         ...A...A.C  ..........  ....T.....  ......TA..  G.....C.C.
Sor/08/14         ...A...A.C  ..........  ....T.....  ......TA..  G.....C.C.
Pal/06/09         ...A...A.C  ..........  ....T.....  ......TA..  G.....C.C.
Pal/02/06         ...A...A.C  ..........  ....T.....  ......TA..  G.....C.C.
Pal/01/02         ...A...A.C  ..........  ....T.....  ......TA..  G.....C.C.
Pal/04/07         ...A...A.C  ..........  ....T.....  ......TA..  G.....C.C.
                  ....|....|  ....|....|  ....|....|  ....|....|  ....|....|
                      7860        7870        7880        7890        7900
NC_002617         GAAACCACAC  CTTGCGAGGG  GTATTCGGGA  CAATGCTTGA  TGGTGAACAA
Pal/02/08         .G..T.....  TG........  ..G.......  ..........  .AA.......
Sor/01/04         .G..T.....  TG........  ..G.......  ..........  .AA.......
Pal/04/06         .G..T.....  TG........  ..G.......  ..........  .AA.......
Pal/03/21         .G..T.....  TG........  ..G.......  ..........  .AA.......
Pal/03/05         .G..T.....  TG........  ..G.......  ..........  .AA.......
Sor/05/09         .G..T.....  TG........  ..G.......  ..........  .AA.......
Pal/01/03         .G..T.....  TG........  ..G.......  ..........  .AA.......
Tor/02/22         .G..T.....  TG........  ..G.......  ..........  .AA.......
Pal/04/05         .G..T.....  TG........  ..G.......  ..........  .AA.......
Pal/04/08         .G..T.....  TG........  ..G.......  ..........  .AA.......
Pal/05/15         .G..T.....  TG........  ..G.......  ..........  .AA.......
Sor/08/14         .G..T.....  TG........  ..G.......  ..........  .AA.......
Pal/06/09         .G..T.....  TG........  ..G.......  ..........  .AA.......
Pal/02/06         .G..T.....  TG........  ..G.......  ..........  .AA.......
Pal/01/02         .G..T.....  TG........  ..G.......  ..........  .AA.......
Pal/04/07         .G..T.....  TG........  ..G.......  ..........  .AA.......
                  ....|....|  ....|....|  ....|....|  ....|....|  ....|...
                      7910        7920        7930
NC_002617         GCAAGACTTA  ACCCTGCGTC  TGCAGTATTC  GATAGCAC
Pal/02/08         .....G..C.  .......T...  .........T  ...TA...
Sor/01/04         .....G..C.  .......T...  .........T  ...TA...
Pal/04/06         .....G..C.  .......T...  .........T  ...TA...
Pal/03/21         .....G..C.  .......T...  .........T  ...TA...
Pal/03/05         .....G..C.  .......T...  .........T  ...TA...
Sor/05/09         .....G..C.  .......T...  .........T  ...TA...
Pal/01/03         .....G..C.  .......T...  .........T  ...TAT..
Tor/02/22         .....G..C.  .......T...  .........T  ...TA...
Pal/04/05         .....G..C.  .......T...  .........T  ...TA...
Pal/04/08         .....G..C.  .......T...  .........T  ...TA...
Pal/05/15         .....G..C.  .......T...  .........T  ...TA...
Sor/08/14         .....G..C.  .......T...  .........T  ...TA...
Pal/06/09         .....G..C.  .......T...  .........T  ...TA...
Pal/02/06         .....G..C.  .......T...  .........T  ...TA...
Pal/01/02         .....G..C.  .......T...  .........T  ...TA...
Pal/04/07         .....G..C.  .......T...  .........T  ...TA...
```

Figure 4. Alignment of derived nucleotide sequences of the selected HN region. Only nucleotides that differ from NC_002167 are shown.

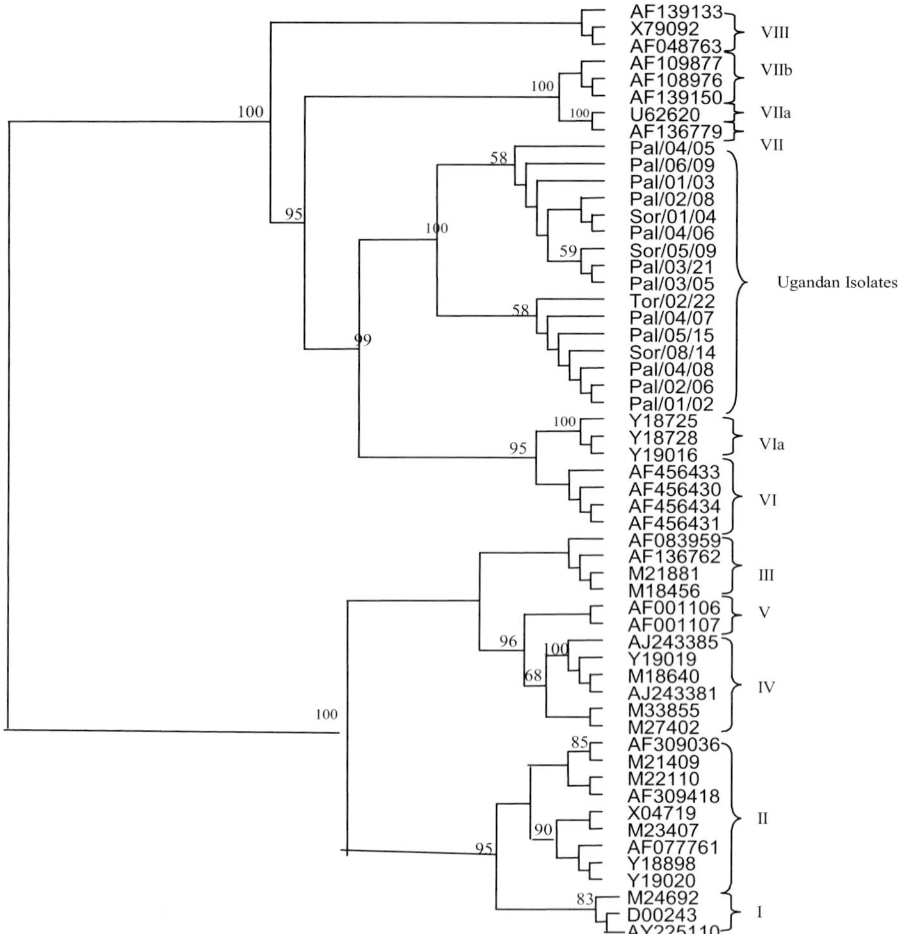

Figure 5. Phylogenetic tree of the nucleotide sequences of the Ugandan NDV isolates based on a 377-bp region of the HN gene. Genotypes and subtypes are indicated on the right with roman numerals. The tree was drawn using the computer program DRAWTREE. The distance matrix was created using the Juke-Canter model and constructed by neighbour-joining alignment of the predicted amino acid sequences

4.   DISCUSSION

In Uganda, the first documented evidence of ND occurred in 1955, although there had been earlier reports of a similar disease in the eastern region bordering Kenya (Mukiibi, 1992). A strain isolated in 1986, which caused 100% morbidity and 98% mortality, was characterized at Weybridge, UK, as virulent. Although ND is controlled in commercial poultry production by vaccination, the disease remains endemic and is the major cause of mortality in backyard poultry.

HA and HI tests were carried out to determine whether the isolates were related to NDV vaccine strains administered in the region (V4 and LaSota) or other PMV-1 strains. Studies by Wehmann *et al.,* (1999) showed that NDV isolated in Canada and Hungary were identical with the vaccine type used in the region. They noted that the isolates were recovered from non-vaccinated flocks, suggesting some degree of area spread within the region.

Studies comparing the deduced amino acid sequences of the F0 precursor of the ND viruses varying in virulence for chickens showed that the viruses that are virulent for chickens had the amino acid sequence ^{112}R/K-R-Q-K/R-R^{116} at the C-terminus of the F2 protein and F (phenylalanine) at residue 117, the N-terminus of the F1 protein, whereas the viruses of low virulence had sequences in the same region of ^{112}G/E-K/R-Q-G/E-R^{116} with L (leucine) at residue 117 (Collins *et al.*, 1994). The fact that all the ND virus isolates in this study have the amino acid sequence ^{112}RRQKRF^{117} at the C-terminus of the F2 protein and F (phenylalanine) at residue 117 qualifies them as virulent.

Pairwise alignment showed the lowest similarity of 97% among the Ugandan isolates. Although phylogenetic analysis has indicated that the Ugandan NDV isolates belong to the same genotype, sequencing has identified minor genetic heterogeneity in the HN gene. This could be attributed to variant genomes that are known to be produced during RNA virus infections (Domingo, Menendez-Arias and Holland, 1999), a phenomenon now embraced within the concept of quasispecies (Eigen, 1993).

The fact that all the 16 Ugandan NDV isolates formed a single clade different from the currently known genotypes suggests that they belong to a novel genotype. As noted by several authors (Seal *et al.,* 1995; King and Seal, 1997, 1998; Lomniczi *et al.,* 1997; Yang *et al.*, 1997; Herczeg *et al.*, 1999), viruses sharing temporal, geographic, antigenic or epidemiological parameters tend to fall into specific lineages or clades, a fact that has proved valuable in assessing the global and local spread of ND.

The very close similarity among the Ugandan NDV isolates therefore suggests a common source of infection. Wild bird movement is a suspected mode of spread of ND infection in Uganda. Flock owners also associate ND outbreaks with movements of poultry vendors, who travel with their poultry purchases from home to home. Preliminary field investigations have also indicated infected poultry purchased from markets or given as gifts to be possible sources of ND infection.

5. CONCLUSIONS

Serological analysis has shown that the virus strains involved in the ND epidemics in Uganda are the wild type and not of vaccine origin.

The intracerebral pathogenicity index (ICPI) of 1.8 and amino acid sequence ^{112}RRQKRF^{117} around the F2/F1 cleavage site indicate that the Ugandan NDV isolates are highly velogenic strains.

Phylogenetic analysis shows that the NDV isolates belong to a novel genotype previously not characterized.

The similarity in the F gene amino acid sequence, pathogenicity and clustering of the isolates in one clade suggests a common source of infection and ancestry.

ACKNOWLEDGEMENTS

This work was supported by the Livestock Systems Research Program funded by Danida. We thank American Society for Microbiology for permission to publish parts of this article. We are also grateful for contributions by Lisbeth Nielsen and Søs Hinge Siem of the Danish Veterinary Institute and Mukiibi Muka of Livestock Health Research Institute.

REFERENCES

Alexander, D.J. 1988. Newcastle disease diagnosis. pp. 147–160. *In:* D.J. Alexander (ed). *Newcastle Disease.* Boston, MA: Kluwer Academic Publishers.

Alexander, D.J. 1989. Newcastle disease. pp. 114–120, *in:* H.P. Purchase, L.H. Arp, C.H. Domermuth and J.E. Pearson (eds). *A Laboratory Manual for Isolation and Identification of Avian Pathogens.* 3rd edition. Kennett Square, PA: American Assoc. of Avian Pathologists.

Altschul, S.F., Madden, T.L., Schäffer, A.A., Zhang, J.H., Zhang, Z., Miller, W. & Lipman, D.J. 1997. Gapped BLAST and PSI-BLAST: a new generation of protein database search programs. *Nucleic Acids Research,* **25**: 3389–3402.

Aldous, E.W. & Alexander, D.J. 2001. Detection and differentiation of Newcastle disease virus (Avian Paramyxovirus Type 1). *Avian Pathology,* **30**: 117–128.

Ballagi-Pordany, A., Wehmann, E., Herczeg, J., Belak, S. & Lomniczi, B. 1996. Identification and grouping of Newcastle disease virus strains by restriction site analysis of a region from the F gene. *Archives of Virology,* **141**: 243–261.

Beard, C.W. & Hanson, R.P. 1984. Newcastle disease. pp. 425–470, *in:* M.S. Hofstad, H.J. Barnes, B.W. Calnek, W.M. Reid and H.W. Yonder (eds). *Diseases of Poultry.* 8th edition. Ames IA: Iowa State University Press.

Berinstein, A., Sellers, H.S., King, D.J. & Seal, B.S. 2001. Use of heteroduplex mobility assay to detect differences in the fusion protein cleavage site coding sequence among Newcastle Disease Virus. *Journal of Clinical Microbiology,* **39**: 3171–3178.

CEC [Commission of the European Communities]. 1992. Council Directive 92/66/EEC of 14 July 1992 introducing Community measures for the control of Newcastle disease. *Official Journal of the European Commission,* L260: 1–20.

Chambers, P. & Samson, A.C.R. 1980. A new structural protein for Newcastle disease virus. *Journal of General Virology,* **50**: 155–166.

Domingo, E.E., Menendez-Arias, C.L. & Holland, J.J. 1999. Viral quasispecies and fitness variations. pp. 141–161, *in:* E. Domingo, R. Webster and J. Holland (eds). *Origin and Evolution of Viruses.* New York NY: Academic Press.

Chambers, P., Millar, N.S., Platt, S.G. & Emmerson, P.T. 1986. Nucleotide sequence of the gene encoding the fusion glycoprotein of Newcastle disease virus. *Journal of General Virology,* **67**: 2685–2694.

Collins, M.S., Strong, I. & Alexander, D.J. 1994. Evaluation of the molecular basis of pathogenicity of the variant Newcastle disease viruses termed "Pigeon PMV-1 viruses". *Archives of Virology,* **134**: 403–411.

Eigen, M. 1993. Viral quasispecies. *Scientific American,* **269**: 32–39.

de Leeuw O. & Peeters B. 1999. Complete nucleotide sequence of Newcastle disease virus: evidence for the existence of a new genus within the subfamily Paramyxovirinae. *Journal of General Virology,* **80**: 131–136.

Felsentein, J. 1995. PHYLIP (Phylogeny Inference Package) Version 3.57c. Seattle: University of Washington.

Herczeg, J., Wehmann, E., Bragg, R.R., Travassos Dias, P.M., Hadjiev, G., Werner, O. & Lomniczi, B. 1999. Two novel genetic groups (VIIb and VIII) responsible for recent Newcastle disease outbreaks in Southern Africa, one (VIIb) of which reached southern Europe. *Archives of Virology,* **144**: 2087–2099.

Hodder, A.N., Selleck, P.W., White, J.R. & Gorman, J.J. 1993. Analysis of pathotype-specific structural features and cleavage activation of Newcastle disease virus membrane glycoproteins using antipeptide antibodies. *Journal of General Virology,* **74**: 1081–1091.

Ke, M.G., Liu, J.H., Lin, Y.M., Chen, H.J., Tsai, S.S. & Chang, C.P. 2001. Molecular characterization of Newcastle disease viruses isolated from recent outbreaks in Taiwan. *Journal of Virological Methods,* **97**: 1–11.

Lomniczi, B., Wehmann, E., Herczeg, J., Ballagi-Pordany, A., Kaleta, E.F., Werner, O., Meulemans, G., Jorgensen, P.H., Manté, A.P., Gielkens, A.L.J., Capua, I. & Damoser, J. 1998. Newcastle disease outbreaks in Western Europe were caused by an old (VI) and a novel genotype (VII). *Archives of Virology,* **143**: 49–64.

Millar, N.S., Chambers, P. & Emmerson, P.T. 1988. Nucleotide sequence of the fusion and haemagglutination neuraminidase glycoprotein genes of Newcastle disease virus, strain Ulster: molecular basis of variations of pathogenicity between strains. *Journal of General Virology,* **69**: 613–620.

Millar, N.S. & Emmerson, P.T. 1988. Molecular cloning and nucleotide sequencing of Newcastle disease virus. pp. 79–97, *in:* D.J. Alexander (ed). *Newcastle Disease.* Boston, MA: Kluwer.

Morrison, T., McQuain, C., Sergel, T., McGinnes, L. & Reitter, J. 1993. The role of the amino terminus F1 of the Newcastle disease virus fusion protein in cleavage and fusion. *Virology,* **193**: 997–1000.

Mukiibi, M.G. 1992. Epidemiology of Newcastle disease and the need to vaccinate local chickens in Uganda. pp. 155–163, *in:* P.B Spradbrow (ed). *Newcastle Disease in Village*

Chickens. Control with thermostable oral vaccines. Proceeding of the 39th International Workshop. Australian Centre for International Agricultural Research (ACIAR), Canberra, 6–10 October 1991.

Murphy, F.A., Fauquet, C.M., Bishop, D.H.L., Ghabrial, S.A., Jarvis, A.W., Martelli, G.P., Mayo, M.A. & Summers, M.D. (eds). 1995. Virus Taxonomy, Classification and Nomenclature of Viruses. Sixth report of the International Committee on Taxonomy of Viruses. *Archives of Virology* **10** [Suppl.]: 268–274.

Nagai, Y., Hamaguchi, M. & Toyoda, T. 1989. Molecular biology of Newcastle Disease Virus. *Progress in Veterinary Microbiology and Immunology,* **5**: 16 –64.

Nagai, Y. & Klenk, H.D. 1977. Activation precursors to both glycoproteins of Newcastle disease virus by proteolytic cleavage. *Virology,* **77**: 124–134.

Nagai, Y., Klenk, H.D. & Rott, R. 1976. Proteolytic cleavage of the viral glycoproteins and its significance for the virulence of Newcastle disease virus. *Virology,* **72**: 494–508.

Nagai, Y., Ogura, H. & Klenk, H.D. 1976. Studies on the assemblies of the envelope of Newcastle disease virus. *Virology,* **69**: 253–538.

Nanthakumar, T., Tiwari, A.K., Kataria, R.S., Butchaiah, G., Kataria, J.M. & Goswami, P.P. 2000. Sequence analysis of the cleavage site-encoding region of the fusion protein gene of Newcastle disease viruses from India and Nepal. *Avian Pathology,* **29**: 603–607.

Oberdörfer, A. & Werner, O. 1998. Newcastle disease virus: detection and characterization by PCR of recent German isolates differing in pathogenicity. *Avian Pathology,* **27**: 237–243.

Palmeri, S. & Perdue, M.L. 1989. An alternative method of oligonucleotide fingerprinting for resolving Newcastle Disease virus-specific RNA fragments. *Avian Disease,* **33**: 345–350.

Palmeri, S. 1989. Genetic relationships among lentogenic strains of Newcastle Disease Virus. *Avian Disease,* **33**: 351–356.

Rott, R. 1979. Molecular basis of infectivity and pathogenicity of myxoviruses. *Archives of Virology,* **59**: 285–298.

Rott, R. & Klenk, H.D. 1988. Molecular basis of infectivity and pathogenicity of Newcastle disease virus. pp. 99–112, *in:* D.J. Alexander (ed). *Newcastle Disease.* Lancaster, UK, & Norwell, MA: Kluwer Academic Publishers.

Rott, R. 1992. Molecular aspects of Newcastle disease virus pathogenicity. pp. 139–144, *in: Proceedings of a Workshop on Avian Paramuxoviruses.* Rauischolzhausen, Germany, 27–29 July 1992.

Russel, P.H., Samson, A.C.R. & Alexander, D.J. 1990. Newcastle disease virus variations. *Applied Virology Research,* **2**: 177–195.

Sakaguchi, T., Toyoda, T., Gotoh, B., Inocencio, N.M., Kuma, K., Miyata, T. & Nagai, Y. 1989. Newcastle disease virus evolution 1. Multiple lineages defined by sequence variability of the haemagglutinin-neuraminidase gene. *Virology,* **169**: 260–272.

Sanger, F., Nicklen, S. & Coulson, A.R. 1977. DNA sequencing with chain-terminating inhibitors. *Proceedings of the National Academy of Sciences, USA,* **74**: 5463–5467.

Seal, B.S., King, D.J. & Bennett, J.D. 1995. Characterization of Newcastle Disease Virus isolates by reverse transcription PCR coupled to direct nucleotide sequencing and development of sequence database for pathotype prediction and molecular epidemiological analysis. *Journal of Clinical Microbiology,* **33**: 2624–2630.

Seal, B.S., King, D.J. & Bennett, J.D. 1996. Characterization of Newcastle disease virus vaccines by biological properties and sequence analysis of the haemagglutinin-neuraminidase protein gene. *Vaccine,* **14**: 761–766.

Seal, B.S., King, D.J., Locke, D.P., Senne, D.A. & Jackwood, M.W. 1998. Phylogenetic relationships among highly virulent Newcastle Disease Virus isolates obtained from exotic birds and poultry from 1989 to 1986. *Journal of Clinical Microbiology,* **36**: 1141–1145.

Seal, B.S., King, D.J. & Meinersmann, R.J. 2000. Molecular evolution of the Newcastle disease virus matrix protein gene and phylogenetic relationships among the paramyxoviridae. *Virus Research,* **55**: 1–11.

Westbury, H. 2001. Newcastle disease virus: an evolving pathogen? *Avian Pathology,* **30**: 5–11.

Yang, C.Y., Chang, P.C., Hwang, J.M. & Sheih, H.K. 1997. Nucleotide sequence and phylogenetic analysis of Newcastle disease virus isolates from recent outbreaks in Taiwan. *Avian Disease,* **41**: 365–373.

Yokozaki, H. & Tahara, E. 1995. PCR analysis of RNA. pp. 202–212, *in:* E.R Levy and C.S. Herrington (ed). *Non-Isotopic Methods in Molecular Biology.* Oxford, UK: Oxford University Press.

MARKER DISCOVERY IN *TRYPANOSOMA VIVAX* THROUGH GSS AND COMPARATIVE ANALYSIS

Preliminary data and perspectives

Alberto M.R. Dávila, Luana T.A. Guerreiro and Silvana S. Souza
Departamento de Bioquímica e Biologia Molecular, Instituto Oswaldo Cruz, Fiocruz. Av. Brasil 4365, Rio de Janeiro, RJ, Brazil 21045-900.

Abstract: *Trypanosoma vivax* is a haemoparasite affecting the livestock industry in South America and Africa. Despite the high economic relevance of the disease caused by *T. vivax*, little work has been done on its molecular characterization, in contrast with human trypanosomes, such as *T. brucei* and *T. cruzi*. The present study reports the construction of a semi-normalized genomic library and the sequencing of 160 Genome Sequence Survey (GSS) ends of *T. vivax*. The analyses of this preliminary data show that this simple and rapid approach worked well to generate some potential new markers for this species.

1. INTRODUCTION

Trypanosoma vivax is a haemoparasite affecting the livestock industry in South America and Africa (Dávila and Silva, 2000; Jones and Dávila, 2001; Gardiner, 1989). According to Seidl, Dávila and Silva (1999), more than 11 million cattle, worth more than US$ 3 billion, are found in the Pantanal region of Brazil and in the lowlands of Bolivia. According to the same authors, if the recent outbreak in Poconé-MT (east-central Brazil) had gone untreated, the estimated losses would have exceeded US$ 140 000 on the seven ranches, US$ 200 million in the Pantanal and US$ 700 million regionwide. Despite the high economic relevance of the disease caused by *T. vivax*, few research reports have been found on its molecular characterization, in contrast with human trypanosomes, such as *T. brucei* and *T. cruzi*. The main reason is the difficulty of growing the parasite in

773

H. P. S. Makkar and G. J. Viljoen (eds.), Applications of Gene-Based Technologies for Improving Animal Production and Health in Developing Countries, 773-776.
© 2005 *IAEA. Printed in the Netherlands.*

laboratory rodents and *in vitro*. Only a very few strains (from West Africa) have been adapted to laboratory rodents. Furthermore, most field isolates cannot be characterized by tools such as RAPD, since parasitaemias are usually of very low titre, making it difficult to separate parasites from animal blood for subsequent extraction of parasite DNA. These characteristics have limited research on *T. vivax* in the past few decades, and consequently very few markers have been described for its molecular characterization. A search in GenBank in May 2003 showed only 22 entries for *T. vivax,* compared with 98,289, 38,577 and 23,507 available for *T. brucei, T. cruzi* and *Leishmania*, respectively. *T. vivax*'s (molecular) biology is also little understood, even considering major differences, such as the mechanical transmission in South America and both cyclical and mechanical transmission in Africa. The best strategies for gene discovery are Genome Sequence Survey (GSS) and Expressed Sequence Tags (ESTs) (El Sayed and Donelson, 1997). According to El Sayed *et al.* (2000), 10–15% of 4,500 ESTs of *T. brucei* presented similarity with some known gene of another organism deposited in GenBank. From those genes, ~50% (2,250) do not present significant similarity with the sequences deposited in GenBank; consequently, most of them probably represent trypanosomatid-specific genes, or potential targets for diagnostic assays, typing systems, rational drug design and, finally, potential targets for the future development of vaccines (Desquesnes and Davila, 2002; Dávila *et al.*, 2003).

In a recent consultation involving several experts on genomics, it was emphasized that *T. vivax* and *T. congolense* are underrepresented species in the molecular parasitology and genomics age, and that they should be considered priority candidates for genome sequencing (Dávila *et al.*, 2002, 2003). As a result, in order to discover new markers to be explored in the molecular characterization of *T. vivax*, it was decided to start a small GSS project.

2.   METHODS

For the discovery of new markers in *T. vivax*, it was decided to sequence 160 GSS ends. For that purpose, a semi-normalized genomic library was constructed. Basically, *T. vivax* genomic DNA from the ILDat2160 cloned stock (kindly provided by Dr Noel Murphy) was partially digested with Sau3A, cloned into the BamHI site of the pUC18 plasmid and transformed into *E. coli* DH5α. The library was then probed with the major repetitive regions described for *T. vivax* in the literature: mini-exon, 18S, 5.8S, satellite DNA and a gene coding for an antigen. Colonies that were negative with the probes were selected and checked for inserts. From those colonies presenting inserts sized 1.5–3 kb, 160 were randomly selected to have their ends sequenced in a Mega-Bace 1000 (Amersham-Biosciences) sequencer. GSSs were analysed with the Phred/Phrap/Consed package (http://www.phrap.org) for quality evaluation, vector removing and clustering. Sequence analyses by similarity were performed using several Blast algorithms with the standalone version of that package (ftp://ftp.ncbi.nlm.nih.gov/blast/executables/) using a Linux (RedHat 9.0) machine.

The NR, NT (ftp://ftp.ncbi.nlm.nih.gov/blast/db/) and "RepBase" (http://www.girinst.org/) databases were used for the Blast similarity search.

3.   RESULTS

The Blast survey shows that our data has 39.3%, 60% and 48.1% of "no hits" in Kineto, Repbase and NT databases using TblastX, respectively. Moreover, using BlastN, our data has 82.5% and 76.2% of "no hits" in the RepBase and NT databases. The BlastX search with the NR database showed 77.5% of "no hits" (Figure 1). The most abundant hits were those presenting high similarity (e-values better than 1e-8) to histone H4, dynein, protein kinase, actin-like protein, phosphoglycerate mutase, INGI retro-element, *T. brucei*–GRESAG, ubiquitin and P-glycoprotein. kDNA minicircle sequences were found among abundant hits showing e-values of more than 0.01.

4.   CONCLUSION

The preliminary data from this GSS approach combined with the semi-normalized library looks promising for the rapid discovery of new markers. We would expect a moderate decrease in the number of "no hits" if new Blast searches are performed when the data from the genomes of *T brucei*, *T cruzi* and *Leishmania* are complete and deposited in the public databases. While we have identified a number of new markers for *T. vivax*, in general most of our data represent new genomic regions that should be explored as species-specific markers, especially those with "no hits" in the databases. The minicircle sequences are being further explored for the design of a PCR-based diagnostic assay. More GSS are being sequenced, then ORF, conserved domains and motif analyses will be performed. Finally, all regions presenting high similarity to kinetoplastid databases can be further explored in a comparative genomics approach.

ACKNOWLEDGEMENTS

The team are grateful to IOC-Fiocruz, CNPq (Conselho Nacional de Pesquisa e Desenvolvimento), CIRAD (*Circulation of Trypanosomatidae ATP project*) and IFS for financial support.

We are also indebted to Rodolpho M. Albano and Daniela Cidade (UERJ) for their valuable help with sequencing.

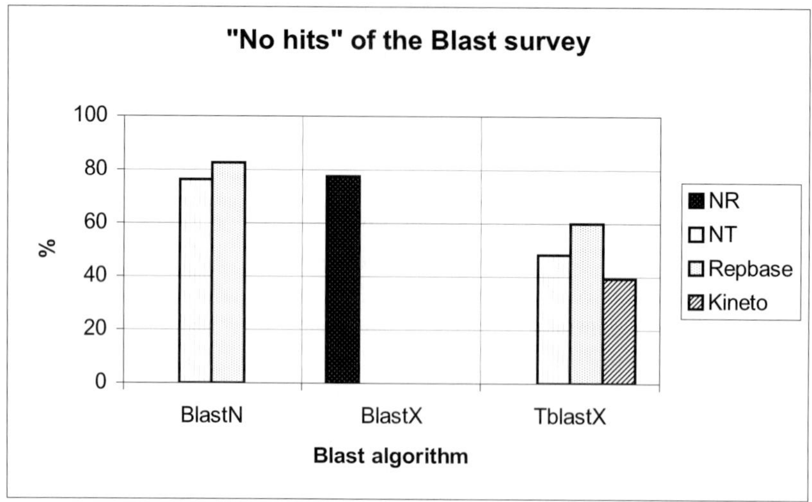

Figure 1. Percentage of "no hits" obtained from the Blast survey (n = 160).

REFERENCES

Dávila, A.M.R. & Silva, R.A.M.S. 2000. Animal trypanosomiasis in South America. Current status, partnership, and information technology. *Annals of the New York Academy of Sciences,* **916**: 199–212.

Dávila, A.M.R., Majiwa, P.A.O., Grisard, E.C., Aksoy, S. & Melville, S.E. 2002. Report of a consultation on the priorities and strategies for the analysis and exploitation of the genomes of *Trypanosoma vivax, T. congolense* and Tsetse. GFAR Tryps GPP Initiative. See: http://www.bioinformatica.ufsc.br/report-tvivax-tcongo-tsetse-genomics.pdf

Dávila, A.M.R., Majiwa, P.A.O., Grisard, E.C., Aksoy, S. & Melville, S.E. 2003. Comparative genomics to uncover the secrets of tsetse and livestock-infective trypanosomes. *Trends in Parasitology,* **19**: 436–439.

Desquesnes, M. & Davila, A.M. 2002. Applications of PCR-based tools for detection and identification of animal trypanosomes: a review and perspectives. *Veterinary Parasitology,* **109**: 213–231.

El Sayed, N.M. & Donelson, J.E. 1997. A survey of the *Trypanosoma brucei rhodesiense* genome using shotgun sequencing. *Molecular Biochemistry and Parasitology,* **84**: 167–178.

THE CENTAUR NETWORK'S CONTRIBUTION TO GENE-BASED TECHNOLOGY

Dissemination of information, international collaboration and training

Karel Hruska[1] and Kris. J. Wojciechowski[2]
1. CENTAUR Editor-in-Chief, Coordinator for Czech Republic and Visegrad Countries, Veterinary Research Institute, Brno, Czech Republic.
2. FAO-associated international consultant, CENTAUR Coordinator, Rome, Warsaw, Dublin.

Abstract: The CENTAUR Network (http://centaur.vri.cz) aims to increase awareness and knowledge among all those involved in veterinary-related activities in the public, private and academic sectors, with emphasis on harmonization of disease control standards and their application worldwide. The network encourages effective use of the Internet and to this end supports improvement in English ability, and promotes strengthening of ties among national and international entities (FAO; OIE; EU; etc.), international centres of excellence and programmes. Membership is free, as are extensive information resources.

1. BASIC CONCEPTS OF THE CENTAUR NETWORK

CENTAUR NETWORK was established by the Food and Agriculture Organization of the United Nations (FAO), but is now independent and managed by volunteers from various countries in Central and Eastern Europe. The worldwide impact of the CENTAUR NETWORK is expressed by the large number of network members from Western Europe, North and South America, Africa, Asia and Australia. Membership is free. Colleagues working in international organizations are among network members.

CENTAUR NETWORK aims at protecting the core network countries and Western European countries from transboundary animal diseases, and

H. P. S. Makkar and G. J. Viljoen (eds.), Applications of Gene-Based Technologies for
Improving Animal Production and Health in Developing Countries, 777-782.
© 2005 *IAEA. Printed in the Netherlands.*

addresses other major animal health matters, food safety and consumer protection correspondingly.

CENTAUR NETWORK coverage includes biotechnology in diagnostics and prevention of animal diseases; veterinary epidemiology, including molecular epidemiology and diagnostics; emergency diseases; zoonoses; food-borne diseases; food contamination and toxicology; genetically modified organisms; resistance of micro-organisms; veterinary administration; education; and research.

2.   CENTAUR NETWORK COMPONENTS

2.1   Biotechnology component

The biotechnology component was created through FAO consultancy missions in 1991–1995, with training meetings in Czech Republic, Hungary, Slovakia and Poland, and technical support from FAO through its Technical Cooperation Programme (TCP/RER/4551; 1995–1997). Later, CENTAUR was extended to cover Balkan, Baltic and other Eastern European Countries, including Ukraine, Belarus and Russia.

2.2   Epidemiological component

An FAO Project Formulation Mission to Croatia, Slovenia, Romania and Bulgaria took place in January 1999. As with the biotechnology component, it identified a lack of compatibility with the OIE and EU standards for infectious diseases control. A new Regional FAO TCP Project (TCP/RER/0066, 2001–2003) was implemented to cover Albania, Bosnia-Herzegovina, Bulgaria, Croatia, FYR Macedonia, Moldova, Romania and Turkey. Consultancy missions have been implemented in Moldova and Bosnia-Herzegovina, and workshops in Albania, Turkey and FYR Macedonia promoted computerized epidemiology, transboundary animal disease epidemiology software (TADinfo) and molecular epidemiology applications.

2.3   Food safety and consumer protection component

This component was initiated by the Veterinary Research Institute (VRI), Brno, Czech Republic, in 2002, and contributes to harmonization of legislation; improvement of technology for both production and distribution; control of food-borne diseases using molecular methods; control of pollution in the environment; prevention of contamination of feeds and food; and the improvement of animal health and welfare.

3.   CENTAUR NETWORK ACTIVITIES

CENTAUR NETWORK produces publications through VRI. The Centaur News Flash Information (CNFI) electronic bulletin provides immediate and daily information to registered network members covering 27 key subject areas.

The network assists its network members in contacting each other and exchanging knowledge using information and communication technology, and aims to improve Internet skills, e-mail connection and English-language proficiency.

The network assists network members in gathering updated and important scientific and technical information and materials from FAO, WHO, FAO/IAEA, OIE, EU, NATO (Scientific Affairs Division) and other international sources, and to establish contacts if required.

The network delivers important information related to bioterrorism hazards through CNFI (Topic 24: Emergency diseases and bioterrorism alerts).

The network works in cooperation with the Animal Genetic Resources Group (Animal Production Service, FAO) to ensure a biodiversity component within the organization.

Special links have been created with:

– The national veterinary authorities and departments of veterinary services of member countries, chambers of veterinary practitioners, leading research institutes, veterinary universities, practicing veterinarians, veterinary, medical and biological scientists, students, journalists and other related entities.

– European Union, Directorate General Science, Research and Development, and Directorate General Health and Consumer Protection, Brussels, Belgium.

– FAO Animal Health Service (Animal Production and Health Division), which is the largest technical agency of the United Nations, and facilitates application of the livestock component of the FAO global priority programme on Emergency Prevention System for Trans-boundary Animal and Plant Pests and Diseases (EMPRES). A special task of CENTAUR is to contribute to the transboundary animal disease (TAD) control in the region (based on the EMPRES principles of: Early Warning; Early Reaction; Enabling Disease Control and Coordination through improved compatibility of diagnostic methods). Vaccine production and quality control methods, as well as other methods related to animal health, are coordinated with the Office International des Epizooties – World Organization of Animal Health (OIE) and European Union standards. The FAO Reference Laboratories, Collaborating Centres and other linked structures may be approached through the network for assistance until the structure of CENTAUR laboratories and expertise becomes established.

- The Animal Health Service of the Joint FAO/IAEA (International Atomic Energy Agency) Division, in Vienna, Austria, under its Coordinated Research Programmes and Technical Cooperation Programmes, and for Enzyme-Linked Immunoassay (ELISA) and Molecular Techniques. The FAO/IAEA Seibersdorf Laboratory is also an OIE and WHO ELISA Collaborating Centre, and deals *inter alia* with standardization and validation of molecular diagnostic kits.
- The European Commission for the Control of Foot-and-Mouth Disease (EUFMD) Secretariat, located in AGAH, FAO, Rome.
- Commission on Genetic Resources for Food and Agriculture – Biotechnology Working Group, Agriculture Department, FAO, Rome.
- FAO Regional Animal Disease Surveillance and Control Network RADISCON, with 29 Near East and North African countries.
- The World Agricultural Information Centre (FAO WAINCENT), Rome.
- International Centre for Genetic Engineering and Biotechnology (ICGEB), Trieste, Italy.
- Office International des Epizooties – World Organization for Animal Health (OIE) Biotechnology Programmes, Paris.

Dissemination of scientific information, case reports, training, links with the international centres of excellence and cooperation between the network members are the permanent contributions of the CENTAUR network to strengthen and increase knowledge among its members. The major aim is also to present members with information on problems, personalities, institutions, and scientific achievement of the region. The network members themselves play an important role in contributing information. Efficient utilization of the Internet, e-mail and improvement in English-language proficiency is also strongly encouraged.

Most Central and Eastern European Countries have been undergoing intensive socio-economic change, and aim to join the European Union or to strengthen their trade and other relationships with the EU. In the last decade, the Balkan countries suffered devastating civil unrest, but are now recovering and rebuilding a normal economy. Initial links have been established with new countries willing to join the network, such as Estonia, Latvia and Lithuania. Also, in order to deal with the problems in the region more efficiently, as well as for the sake of geographic integrity, the

CENTAUR network encourages and welcomes the cooperation of Finland, Belarus, Russian Federation, Ukraine, Yugoslavia and Greece.

3.1   Catalytic role of CENTAUR

CENTAUR contributed to establishing the Interfaculty Studium of Biotechnology at Warsaw Agricultural University (WAU), and to exchange of scientists and cooperation among biotechnologists throughout the region, with annual International Biotechnology Seminars (since 1998) led by WAU, with upgrading of university curricula. It has been promoting biotechnology as a fundamental tool in reducing famine and malnutrition around the world, with regular exchange of scientific information, training of scientists, and links with international organizations and centres of excellence.

3.2   CENTAUR NETWORK PUBLICATIONS

In the CENTAUR network, CNFI is an international electronic bulletin published regularly, providing network members with timely information in the form of e-mail messages relating to their fields of interest. CNFI content is not limited to Central and Eastern Europe, but covers global animal disease- and food-safety-related events, and is distributed to registered users worldwide. The CNFI takes three forms:
— e-mail messages distributed to registered members according to their specific fields.
— CNFI bulletin summarizes general information for the CENTAUR NETWORK, and is available for download from the Centaur Web site (http://centaur.vri.cz).
— The Centaur Web site contains general information on the network and a number of topic-specific sections.

In 2002 and 2003 the CENTAUR network co-organized an international Internet Workshop on Good Research Practice. Similar workshops have been planned.

CENTAUR network members are welcome to contribute as authors of original papers or reviews submitted for publication in an international peer-reviewed journal for veterinary medicine and biomedical sciences, namely *Veterinarni medicina* (Vet Med-Czech), indexed in the Web of Science, Current Contents and other databases. Papers published in this journal are free in full pdf format at http://vetmed.vri.cz.

4.   CONCLUSIONS

The CENTAUR Network contributes to dissemination of information and knowledge and works towards promoting cooperation among network members and the world. The network members share an opinion that knowledge is based on information. It is believed that the Internet makes information available to the public, and Internet availability and English-language proficiency break down the borders between countries.

Registration of new CENTAUR network members is welcome. There are no geographical, educational, professional nor institutional affiliation limitations. There is no membership fee. We only request an e-mail address, Internet availability and basic knowledge of English. Visit http://centaur.vri.cz for a downloadable Word document form or use the on-line registration to join our growing network.

5.   FURTHER RESOURCES

Visit http://centaur.vri.cz to obtain detailed information on CENTAUR network countries, executive bodies, targets, and inputs; outputs supporting international collaboration; titles of articles, written by or for CENTAUR network members; reports and reviews; personalities; Supercourse-Vet PowerPoint presentations; and an index of the CENTAUR Newsletter Flash Information.